DATE DUE

Advances in
MARINE BIOLOGY

VOLUME 48

Series Editors

Alan J. Southward

Marine Biological Association, The Laboratory, Citadel Hill,
Plymouth, United Kingdom

Paul A. Tyler

School of Ocean and Earth Science,
University of Southampton,
Southampton Oceanography Centre,
Southampton, United Kingdom

Craig M. Young

Oregon Institute of Marine Biology,
University of Oregon, Charleston,
Oregon, USA

Lee A. Fuiman

Marine Science Institute,
University of Texas at Austin,
Port Aransas, Texas, USA

Aquatic Geomicrobiology

By

DONALD E. CANFIELD

Danish Center for Earth System Science and Institute of Biology, University of Southern Denmark, Odense, Denmark

ERIK KRISTENSEN

Institute of Biology, University of Southern Denmark, Odense, Denmark

BO THAMDRUP

Danish Center for Earth System Science and Institute of Biology, University of Southern Denmark, Odense, Denmark

WITHDRAWN
FAIRFIELD UNIVERSITY
LIBRARY

ELSEVIER
ACADEMIC
PRESS

AMSTERDAM • BOSTON • HEIDELBERG • LONDON
NEW YORK • OXFORD • PARIS • SAN DIEGO
SAN FRANCISCO • SINGAPORE • SYDNEY • TOKYO

Elsevier Academic Press
525 B Street, Suite 1900, San Diego, California 92101-4495, USA
84 Theobald's Road, London WC1X 8RR, UK

This book is printed on acid-free paper. 

Copyright © 2005, Elsevier Inc. All Rights Reserved.

No part of this publication may be reproduced or transmitted in any form or by any means, electronic or mechanical, including photocopy, recording, or any information storage and retrieval system, without permission in writing from the Publisher.

The appearance of the code at the bottom of the first page of a chapter in this book indicates the Publisher's consent that copies of the chapter may be made for personal or internal use of specific clients. This consent is given on the condition, however, that the copier pay the stated per copy fee through the Copyright Clearance Center, Inc. (www.copyright.com), for copying beyond that permitted by Sections 107 or 108 of the U.S. Copyright Law. This consent does not extend to other kinds of copying, such as copying for general distribution, for advertising or promotional purposes, for creating new collective works, or for resale. Copy fees for pre-2005 chapters are as shown on the title pages. If no fee code appears on the title page, the copy fee is the same as for current chapters. 0065-2881/2005, $35.00

Permissions may be sought directly from Elsevier's Science & Technology Rights Department in Oxford, UK: phone: (+44) 1865 843830, fax: (+44) 1865 853333, E-mail: permissions@elsevier.com. You may also complete your request on-line via the Elsevier homepage (http://elsevier.com), by selecting "Customer Support" and then "Obtaining Permissions."

For all information on all Academic Press publications visit our Web site at www.academicpress.com

ISBN: 0-12-026147-2 Case Bound
ISBN: 0-12-158340-6 Paperback

PRINTED IN THE UNITED STATES OF AMERICA
05 06 07 08 9 8 7 6 5 4 3 2 1

Working together to grow
libraries in developing countries

www.elsevier.com | www.bookaid.org | www.sabre.org

ELSEVIER BOOK AID International Sabre Foundation

This book is dedicated to our families:

Andreas, Ellen, Marianne, Daniela,
Mie, Karen, Lea, and Kirsten

CONTENTS

4. Carbon Fixation and Phototrophy

5. Heterotrophic Carbon Metabolism

6. The Oxygen Cycle

7. The Nitrogen Cycle

8. The Iron and Manganese Cycles

9. The Sulfur Cycle

10. The Methane Cycle

11. The Phosphorus Cycle

12. The Silicon Cycle

13. Microbial Ecosystems

PREFACE

This volume offers a comprehensive overview of the biogeochemistry of element cycling from the perspective of the organisms involved. We interface the disciplines of microbial ecology, in which biogeochemistry is often given only cursory attention, with biogeochemistry, in which the organisms are frequently viewed as mere catalysts. We report that microbes (meaning prokaryotes here) are exquisitely adapted to the environment in which they live, with often unexpected and complex behavior and physiologies allowing for extremely efficient energy utilization. Thus, individually and collectively, microbes conduct a bewildering array of different metabolisms aimed at efficient resource management, including the utilization of each other's waste products. Microbes have individual physiological adaptations allowing life in almost incomprehensible extremes of hot, cold, acid, base and salt. We learn that microbes have evolved intricate biochemical machineries capable of conducting efficient and complex chemistries. We also learn, through various pages of this volume, that evolution has been a major player in shaping the structure and function of the microbial world, and new tools in molecular biology are allowing us to explore the details of microbial evolution for the first time. Most importantly, we learn that microbes are the drivers of elemental cycling.

This volume is cross-disciplinary, so our first five Chapters include largely introductory material designed to set a common ground for the following Chapters on elemental cycling. Thus, we start in Chapter 1 by looking at microbial systematics and phylogeny from a molecular perspective. This is a rapidly evolving field, and the reader is forewarned that progress in this field is likely to continue at a rapid pace. This is followed by Chapter 2 outlining the general principles of microbial ecology, including aspects of microbial growth and population structure. Microbes operate in a chemical environment, and we devote Chapter 3 to outlining the basic chemical principles needed to understand microbial energetics and aspects of microbial competition in nature. This Chapter is followed by two Chapters (4 and 5) overviewing first phototrophy and carbon fixation and next heterotrophic carbon metabolism. The next Chapters (6 through 12) outline the interface between the biogeochemistry of elemental cycling and the organisms involved, focusing in turn on the cycling of oxygen, nitrogen, Fe and Mn, sulfur, methane, phosphorus and silicon. There is a great deal of interrelatedness between these Chapters, emphasizing the intricate and often extraordinary coupling of element cycles. The cycling of phosphorus (Chapter 11) and silicon (Chapter 12) involves microbial interaction, but not as much so as the other elements discussed. In these Chapters, the role of phosphorus and silicon as nutrients, as well as aspects of their cycling by eukaryotic organisms, is also emphasized. We finish with Chapter 13 exploring several specific microbial ecosystems.

This volume has been written both for students and for our colleagues. We hope to offer a source of reference and inspiration for those wishing to mix disciplines traditionally rooted in the biological or the geological sciences. There has been a heightened awareness recently of the role of microbes in controlling Earth-surface processes. This awareness is not new, but it is now taken seriously. Out of this, the field of geobiology has emerged. Again, this field is not new, but its prominence is clear. Although ours is not a volume specifically focusing on geobiology, we believe that the principles and core material outlined here have broad relevance to geobiological endeavors.

We have benefited from the expert advice of a number of good colleagues who have taken the time to read and comment on one or more of the Chapters. Thus, we are indebted to Daniel Alongi, Rudy Amann, Gary Banta, David Burdige, Dan Conley, Ralf Conrad, Raymond Cox, Kai Finster, Ronnie Glud, Ellery Ingall, Andy Knoll, Joel Kostka, Bente Lomstein, Claus Olesen, Jesper Pedersen, Eric Roden, Alan Southward and an anonymous reviewer for comments and advice. We are especially thankful to Raymond Cox and Ralf Conrad, who each endured multiple Chapters of the volume. We also thank Raymond Cox for suggesting our remarkably straightforward title. Our submission would have been considerably delayed beyond its already embarrassingly tardy date without the expert technical skills of Mette Andersen, who helped in so many ways. We also thank Yvonne Mukherjee for earlier help with the volume. Jacob Zopfi kindly allowed us to reproduce one of his photographs, and Peter Søholt patiently sorted through countless computer and networking problems. The extreme patience of our editor, suffering seemingly endless delays, cannot be overlooked. Finally, we thank our funding agencies, the Danish National Research Foundation (Dansk Grundforskningsfond) and the Danish National Research Council (SNF), for generous support of our work.

Donald E. Canfield
Erik Kristensen
Bo Thamdrup
Odense, Denmark

Chapter 1

Systematics and Phylogeny

1. INTRODUCTION

The classification of microorganisms[1] is a problem that has plagued microbiologists for more than a century. The problem is introduced by considering the macroscopic world of plants, animals, and fungi. An astounding array of morphological differences can be naturally grouped into two simple physiological types. Thus, all plants share photosynthesis as their basic means of energy metabolism but display an enormous range of obvious

[1]Microorganisms include a broad range of different types of organisms, including prokaryotes, and many single-celled eukaryotes such as protists and some algae. However, unless otherwise stated, we will use the term microorganisms synonymously with prokaryotes. We will try to avoid using bacteria as general designator for prokaryotes, as reclassification of the Tree of Life (Woese, 1990) has provided a very precise definition for *Bacteria* (see below) that does not encompass the whole of the prokaryote world. We frequently use the term bacteria to designate organisms within the bacterial domain.

0-12-026147-2

© 2005 Elsevier Inc.
All rights reserved

differences in morphology, size, reproductive strategy, and habitat. These differences can be used as indices of classification. Likewise, essentially all animals are aerobic heterotrophs, with classification based on similar considerations. Furthermore, through comparative biology, classification schemes for plants and animals can be constructed to reflect the evolutionary relationships (phylogeny) among organisms. For the microscopic world of prokaryotic organisms, that is, organisms lacking a membrane-bound nucleus and containing no organelles (with a few exceptions), the situation is quite the opposite. Rather limited differences in size, shape, and mode of locomotion are combined with an enormous range of physiological diversity. Furthermore, while some groups of microorganisms, such as cyanobacteria, for example, may be identified in ancient rocks, microbial fossils are of little use in constructing phylogenetic relationships. Fossil microbes are generally difficult to distinguish based on preserved morphology alone, and furthermore, the fossil record of microbes is spotty and poorly preserved in early Earth history, when the pace of new metabolic developments might have been most rapid.

Early microbial classification schemes sought a natural order from combined considerations of morphological and physiological characteristics (for example, earlier editions of *Bergey's Manual of Determinative Bacteriology*; Breed *et al.*, 1948). These schemes provided exact prokaryote identification by using a checklist of properties including cell shape, cell organization, basic physiology and details of cell motility, somewhat mimicking the strategy used to classify macroscopic organisms. In general, the properties of cell morphology and motility provided a higher order classification than did considerations of cell physiology. While these schemes produced an organized catalog of the prokaryote world, they provided no indication of evolutionary relationships, unlike the schemes used to classify plants and animals. Also, incorrect associations between organisms were sometimes generated. In one example, older classification schemes (Reichenbach and Dworkin, 1981) grouped filamentous cyanobacteria and filamentous colorless-sulfur bacteria together, with the filamentous nature of the organisms a principal consideration in forming the group. This grouping, like many others, has not survived in modern phylogenetic classification schemes. It is apparent that the metrics of comparison used to classify and organize members of the macroscopic world are insufficient to provide a phylogeny of organisms in the microscopic world.

A phylogenetic grouping of prokaryotic organisms became possible after the structure of DNA was discovered and its role in inheritance was made clear. It was a logical step to speculate, as Zuckerkandl and Pauling did in 1965, that "the comparison of the structure of homologous informational macromolecules allows the establishment of phylogenetic relationships" (Zuckerkandl and Pauling, 1965). Informational macromolecules include proteins, DNA and RNA, while homologous macromolecules are genes or proteins of common function inherited from the same ancestor. The cataloging of prokaryotic organisms (and indeed all organisms) based on the molecular sequences of genetic material (DNA and RNA) and proteins has advanced with the advent of sequencing techniques and the ability to sequence more complicated, and larger, molecules. The earliest sequences were dominantly of proteins, whereas now sequences for whole prokaryote genomes are becoming routine. In order to appreciate the construction and interpretation of phylogenies based on molecular sequence comparisons, we first briefly review some key aspects of basic genetics.

2. REVIEW OF BASIC GENETICS

All information related to the biochemical function of an organism is carried by the DNA molecule. DNA is double-stranded, with each strand consisting of four different nucleotides linked in long succession. For prokaryotes, DNA consists of, typically, 1 to 6 million nucleotides per strand. A nucleotide contained in the DNA molecule consists of one of four nitrogen bases: adenine (A), guanine (G), thymine (T), and cytosine (C), bonded to the sugar deoxyribose and a phosphate molecule (Figure 1.1). The two strands of DNA are connected by hydrogen bonds where A is linked to T and G is linked to C, forming base pairs (Figure 1.2).

The DNA molecule is organized into a series of genes encoding for the production of various proteins and RNA molecules as well as some non-coding regions sometimes referred to as "junk DNA." Proteins control

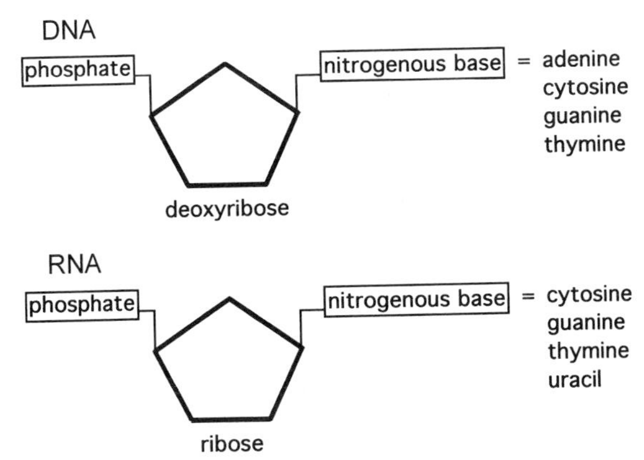

Figure 1.1 The basic structure of the nucleotides making up RNA and DNA. In RNA, the deoxyribose of DNA is replaced by ribose, and adenine is replaced by uracil.

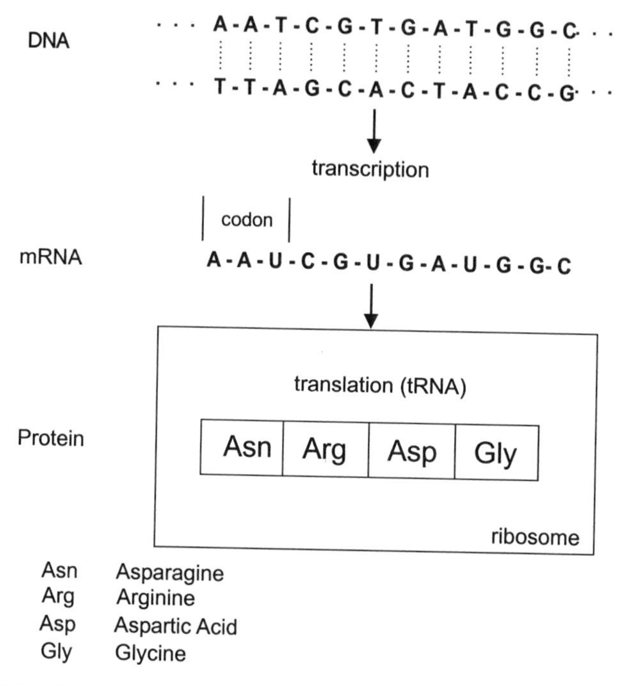

Figure 1.2 Cartoon displaying key aspects of the protein synthesis process.

the biochemical machinery of a cell, catalyzing countless chemical reactions. They may also serve a structural function in cell membranes and in various cytoplasmic components. RNA acts as an information shuttle, transferring the information from the DNA molecule to the site of protein synthesis, and it is also involved directly in protein synthesis itself (Figure 1.2). The molecular structure of RNA is quite similar to DNA, except that thymine is replaced by uracil (U), and the sugar ribose replaces deoxyribose (Figure 1.1). Furthermore, RNA is single stranded. Protein synthesis takes place in two principal steps (Figure 1.2). In the first step, genetic information from the DNA molecule is "transcribed" onto a messenger RNA molecule (mRNA). After formation, mRNA moves to the ribosome, consisting of various ribosomal RNAs (rRNA) intimately associated with a suite of ribosomal proteins. The ribosome is the place of protein synthesis within the cell. In the ribosome, information from the mRNA is "translated" into the amino acid sequence of a protein (Figure 1.2). Each three consecutive nucleotides from mRNA represent a codon that specifies for one of 20 possible amino acids that are moved into place by transfer RNA (tRNA). This process is continued until a stop codon is reached, signifying the end of the translation process. The basic process of transcription, translation and protein synthesis is common to all living organisms on Earth.

3. MOLECULAR BASIS FOR EVOLUTION

Darwin proposed that during evolution new traits are randomly acquired by organisms during the passing of generations (descent with modification), and if advantageous to survival, traits will be selected for and passed to subsequent generations (natural selection). In fact, there is a molecular basis for this proposition. Random changes (mutations) occasionally occur in the nucleotide sequence of the DNA molecule during the replication process. This changes the nucleotide sequence of the genes affected by mutation, the sequence of the RNA molecules transcribed from the genes and, in principle, the amino acid sequence of proteins encoded for by a gene.

In some cases, mutations of the DNA nucleotide sequence are "silent," as more than one three-nucleotide sequence (codon) may code for the same amino acid. Also, mutations can occur in the non-coding portion of the DNA, the "junk DNA," with no apparent consequence to the organism. However, mutations can lead to changes in the amino acid sequence of proteins, and this can affect the organism in various ways. In some cases, a change is deleterious to the function of the protein, and hence to the organism, and will not be accepted. In other cases, a change may not alter the function of the protein (or the RNA molecule) but will be transferred to subsequent generations and will provide a reminder of the mutation event. These changes in the sequences of homologous genes (and in the sequence of the protein or RNA encoded for by the gene) are the basis of the phylogenetic reconstructions imagined by Zuckerkandl and Pauling (see below). Gene mutations can also prove advantageous to the organism and can even promote new gene function or spawn the emergence of new species. In addition, the chance duplication of genes, followed by mutation and modification of the new gene, could also produce new gene function or could provide superior mediation of the original gene function (Maynard Smith, 1998). Gene deletions may also occur, altering the biochemical machinery of the organism.

Other forces acting at the molecular level may also promote evolution. Thus, the lateral transfer of genes between organisms may provide new opportunities for the host that could not be derived from a linear sequence of gene mutation and change (Maynard Smith, 1998). Lateral gene transfer can occur through a variety of paths; viral infection, in which the virus may swap genes with the host as a result of infection (see Chapter 3), is one example. Some organisms may also directly incorporate DNA from the environment into their genome.

At different levels, gene sequences, proteins and RNA molecules provide a comparative evolutionary history of organisms. The most information, which is also the most difficult to interpret, comes from whole-genome sequences. These provide an accumulated history of molecular-level changes that have shaped the evolution of an organism (Fitz-Gibbon and House, 1999; Lin and Gerstein, 2000; Clarke *et al.*, 2002; House *et al.*, 2003). The interpretation of whole-genome sequences is a rapidly evolving field. At another level, the comparison of sequences from homologous genes, RNA molecules and proteins provides the basis for constructing evolutionary relationships that have been extensively used to formulate organismal phylogenies among prokaryotes (e.g. Woese, 1987; Doolittle *et al.*, 1996; Brown and Doolittle, 1997).

4. CONSTRUCTING PHYLOGENIES

A great deal of thought and consideration has gone into the construction of organismal phylogenies from molecular sequence data. This matter is not trivial. Certain criteria must be met before a genetic element (gene, protein or RNA molecule) may be considered as appropriate for constructing phylogenies. These criteria have been carefully elucidated by Carl Woese and his associates (Woese, 1987; Olsen and Woese, 1993), and some of the important considerations are outlined below.

1. Phylogenies are generally constructed from genetic elements with the same function (homologous function). Furthermore, phylogenies constructed from genetic elements are only as inclusive as the distribution of the given genetic element among organisms. The more widely distributed the element, the more inclusive the phylogeny, and the less widely distributed, the more specialized the phylogeny. If a phylogeny is constructed, for example, from the nitrogenase enzyme, then the phylogeny will include only those organisms capable of fixing nitrogen.
2. Larger genetic elements provide more sequence differences among organisms. More differences provide better resolution and more statistically valid phylogenies.
3. The genetic element should accumulate mutations at a rate commensurate with the resolution required. Organisms separated by large evolutionary distances are best compared using sequences in which mutations accumulate only slowly. By contrast, a more rapid accumulation of mutations would benefit the comparison of organisms recently diverged from one another. In

this case, distant evolutionary relationships might become obscured by too many mutational changes.

4. Do the phylogenies constructed from the given genetic element reflect the phylogeny of the organism? They do if, as elegantly stated by Olsen and Woese (1993), "the clear majority of the essential genes in a genome share a common heritage." This should be true if, in turn, the lateral transfer of genetic material from one organism to another has not been the predominant means of genome construction. This issue will be considered in some detail below.

5. Related to the previous consideration, the construction of phylogenies implicitly assumes that modern organisms have been ultimately derived from one common ancestor (or a suite of ancestors freely exchanging a common genetic pool) (Woese, 2000).

Independent of the type of genetic element used to construct the phylogeny, the basis for phylogenetic reconstructions is comparative sequence analysis. As an example, the phylogeny of seven hypothetical organisms based on a 10-base sequence of either "X's" or "O's" is illustrated in Figure 1.3. It is assumed that organism "a" is the last common ancestor to all of the others. Also demonstrated in Figure 1.3 is how quickly phylogenetic resolution can be lost if the information content of the genetic element is too small.

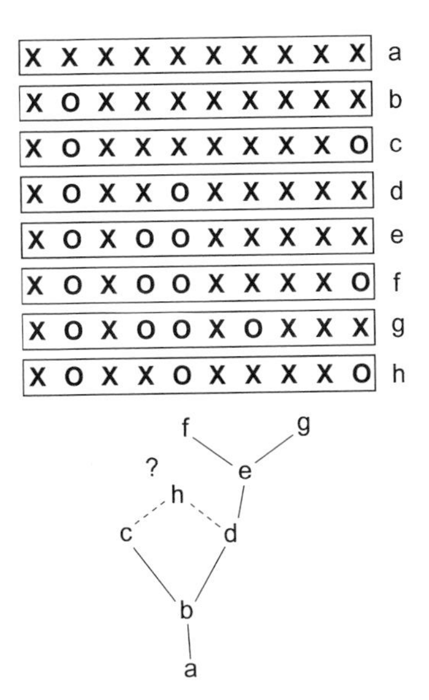

Figure 1.3 Phylogeny drawn from a hypothetical group of organisms with characteristic protein sequences designated either by an O or an X. The placement of organism h is uncertain with the limited sequence information available. It could have evolved from either c or d. Back mutations are not considered.

In practice, the analysis of molecular sequences to obtain phylogenetic information is quite complex and requires sophisticated computer algorithms. Corrections must be made for the possibility of back mutations and for homoplasy, the independent mutation of the same site in two lineages. Furthermore, analogous regions of the genetic element must be correctly identified and aligned for all of the organisms considered, and sequences must be statistically treated to provide the most probable order of emergence and its statistical validity (Hillis *et al.*, 1993; Huelsenbeck and Rannala, 1997).

5. PHYLOGENIES FROM RIBOSOMAL RNA MOLECULES

Phylogenies have been constructed from a range of different DNA, RNA, and protein sequences. By far, the most widely used and comprehensive phylogenies have been constructed from the small subunits of the rRNA molecule. As mentioned previously, the ribosome is the location of protein synthesis in all organisms. The ribosome of prokaryotic organisms is constructed of 30S and 50S subunits, yielding together 70S ribosomes. For eukaryotic organisms, 40S and 60S subunits make up 80S ribosomes. The "S" refers to Svedberg units and is the sedimentation coefficient of ribosomes or ribosomal subunits subjected to ultracentrifugation.

Each of the ribosome subunits is a complex of proteins and various rRNA molecules. For prokaryotes, the 30S subunit contains 16S rRNA and the 50S subunit contains both 5S and 23S rRNA, while in eukaryotes, the 40S subunit contains 18S rRNA and the 60S subunit contains 5.8S and 28S rRNA (Table 1.1). Each of the comparable rRNA subunits in prokaryotes and eukaryotes (e.g., 16S rRNA in prokaryotes and 18S rRNA in eukaryotes) is functionally equivalent and structurally similar (Figure 1.4). The

Table 1.1 Comparison of ribosomal RNA (rRNA) composition of prokaryotes and eukaryotes

	Prokaryotes	Eukaryotes
Overall size	70S	80S
Small subunit	30S subunit 16S rRNA (1500)	40S subunit 18S rRNA (2300)
Large subunit	50S subunit 5S rRNA (120) 23S rRNA (2900)	60XS subunit 5S rRNA (120) 5.8S rRNA (160) 28S rRNA (4200)

Number of bases in parenthesis. Modified after Madigan *et al.* (2003).

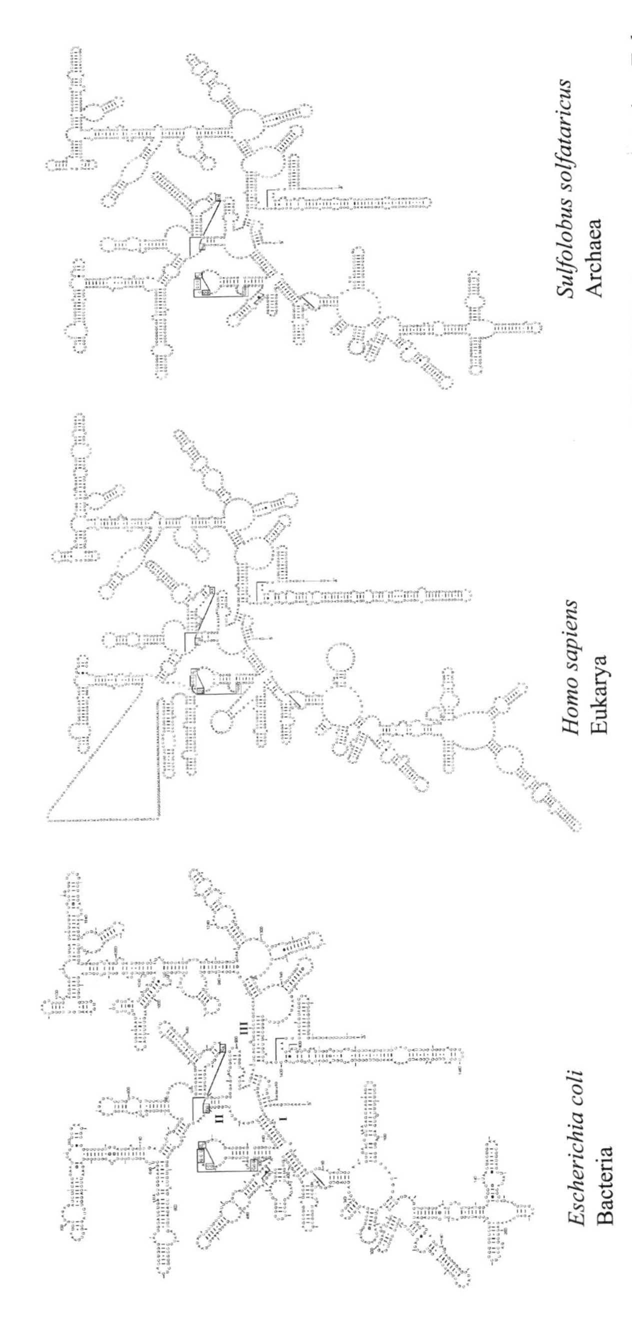

Figure 1.4 Sequences of 16S rRNA from members of the *Bacteria* and *Archaea* and 18S rRNA for a representative *Eukarya*. Emphasized is the secondary structure of the molecule, showing the broad similarity in structure for organisms throughout the Tree of Life. Structures from Maidak *et al.* (1999).

reader is referred to standard textbooks (e.g. Madigan *et al.*, 2003) for further discussion of ribosome function and structure.

Early phylogenies from rRNA molecules were constructed from sequence comparisons of the 5S rRNA subunit (Woese and Fox, 1977). These sequence comparisons demonstrated the potential for rRNAs to produce a template for organizing prokaryotic taxonomy and for generating a phylogeny for prokaryotic organisms as was never before possible. Unfortunately, the number of nucleotides in the 5S rRNA molecule is small (approximately 120), severely limiting the phylogenetic resolution of the molecule. As sequencing techniques developed, it became possible to sequence the larger 16S and 18S rRNAs, collectively known as the small subunit (SSU) rRNAs (Table 1.1). This development led to a widely accepted molecular-based "Tree of Life" and has forever changed our views of the taxonomic organization of the prokaryotic world, as well as our understanding of organismal evolution on Earth.

Molecular sequences from the SSU rRNA molecule have proven ideal for taxonomic organization and phylogenetic reconstructions for several reasons (Pace *et al.*, 1986; Woese, 1987; Olsen and Woese, 1993). First, the SSU rRNA molecule is distributed among all organisms on Earth. Second, the function of the ribosome, protein synthesis, is tightly constrained so the molecular sequences of the SSU rRNA are highly conserved, accumulating mutations only slowly; this allows comparison between organisms of distant evolutionary association. However, different regions of the SSU rRNA molecule accumulate mutations at different rates, providing resolution also for fairly closely associated organisms. Next, the SSU rRNA is relatively large (approximately 1500 nucleotides for prokaryotes and 2300 nucleotides for eukaryotes) (Table 1.1), allowing good phylogenetic resolution. Finally, the transcription–translation process of protein synthesis is so fundamental to the livelihood of an organism that the lateral transfer of rRNA between cells is unlikely to be a common mode of rRNA acquisition by the cell.

6. TREE OF LIFE FROM SSU rRNA COMPARISONS

At present (June, 2004) 97,000 prokaryote 16S rRNA sequences have been determined, aligned and archived for use by the scientific community (see http://rdp.cme.msu.edu). In fact, the number of different prokaryotic "species" known from molecular sequences is far greater than the number of "species" (more on the species concept in Chapter 2) actually cultured in the lab. This is because while it is relatively easy to extract DNA for molecular sequencing, it is difficult to find the appropriate culture conditions by which to enrich and isolate microorganisms. A general consensus guess is that only about 1% of the prokaryotic organisms present in nature have

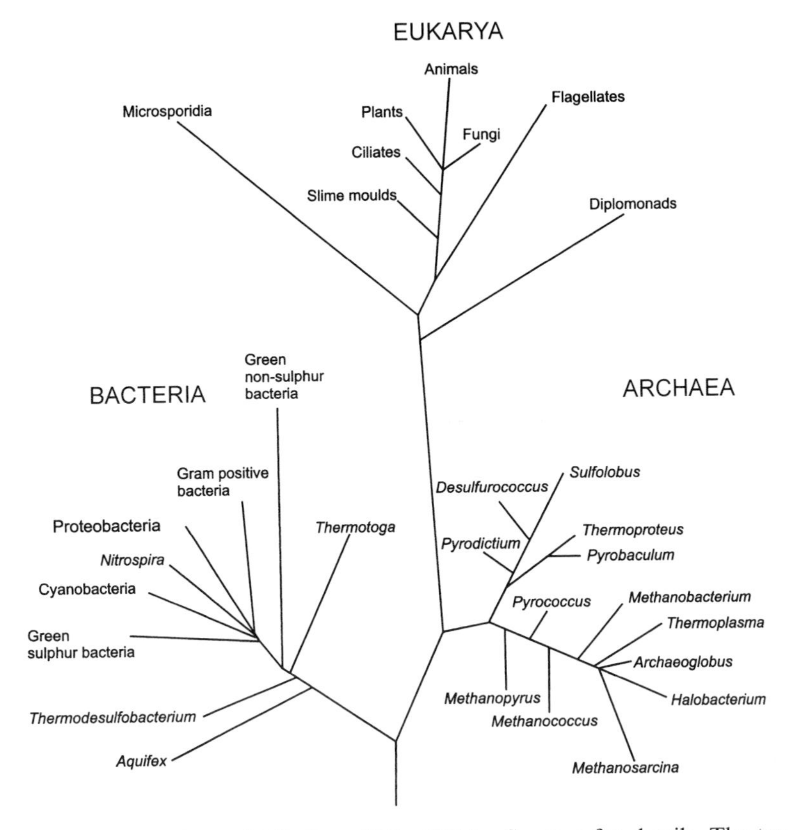

Figure 1.5 The Tree of Life from SSU rRNAs. See text for details. The tree is adapted from Olsen and Woese (1993), with additional lineages from Pace (1997).

been brought into culture so far (Pace, 1997). Nevertheless, from the available prokaryote and eukaryote sequences a Tree of Life emerges (Figure 1.5).

Molecular sequences from the SSU rRNAs naturally divide all life into three principal domains: *Bacteria*, *Archaea*, and *Eukarya*. Two of the domains, *Archaea* and *Bacteria*, are composed of prokaryotic organisms, while the *Eukarya* domain consists of the eukaryotic organisms. This tree has been rooted based on the presence or absence of a duplicated gene coding for the enzyme ATPase. This comparison suggests that the *Bacteria* are the out group (they do not contain the duplicated gene) (Gogarten *et al.*, 1989; Iwabe *et al.*, 1989), placing the root between the *Bacteria* and the domains *Eukarya* and *Archaea*. However, there is still much discussion about the placement of the root, and this issue is not yet fully resolved.

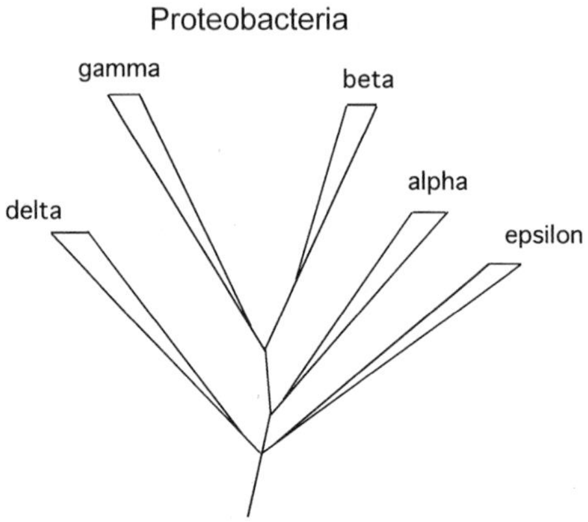

Proteobacteria

Figure 1.6 The branching order of the principal divisions within the proteobacteria (purple bacteria). See Figure 1.5 for the placement of the proteobacteria in the Tree of Life and the text for additional details.

Taken at face value, deeper branching lineages within the Tree of Life represent organisms evolved from more ancient ancestors than lineages branching further up on the tree. Therefore, the *Aquifex* lineage, for example, was presumably established before the earliest evolving cyanobacteria, and plants were established well after the emergence of diplomonads and slime molds. Likewise, the evolutionary history of a given lineage is represented by the branching order of major divisions, or groups of organisms, in the lineage. Thus, there are five major divisions within the proteobacteria (Figure 1.6): α, β, δ, ε, and γ, and of these, the γ and β subdivisions separated most recently, while the δ and the ε subdivisions are the most ancient.

The similarity or difference in the SSU rRNA sequences of two organisms is a measure of the evolutionary distance separating the two organisms. Trees are constructed so that the branch lengths represent the differences in SSU rRNA sequence relative to nearest neighbors, as well as to all other organisms on the Tree of Life. Evolutionary distance between two organisms is measured as the fractional difference between two sequences and is determined as the sum of the branch lengths connecting organisms through the shortest path. Thus, for sulfate-reducing organisms of the δ subdivision of the proteobacteria (Devereux and Stahl, 1993) (Figure 1.7), representing most known sulfate reducers, the evolutionary distance between species of the genus *Desulfobacter* is much less than the evolutionary distance between species of the genus *Desulfovibrio*. Furthermore, the evolutionary distance between *Desulfobacter* sp. and *Desulfobacterium* sp. is less than the distance between *Desulfobacter* sp. and *Desulfovibrio* sp. These evolutionary distances are all much shorter than the distance between members of the proteobacteria and members of the domains *Archaea* or *Eukarya* (see Figure 1.5).

Extraction, amplification and sequencing of DNA for phylogenetic analysis

The construction of phylogenies from molecular data begins, most frequently, with the extraction of DNA from organisms. In both lab cultures and environmental samples DNA is extracted after cell disruption, with various strategies, including beating the sample with small beads, sonication, freeze/thaw cycles and chemical lysis, used either singly or in combination. The DNA is subsequently purified by a variety of means, including phenol-chloroform extraction, dialysis and ethanol precipitation. After this, segments of DNA are amplified by polymerase chain reaction (PCR). The segment of DNA chosen (typically 100 to over 1000 base pairs in length) will be part or all of a gene sequence coding for specific enzymes or the RNA molecules used in constructing phylogenics. To begin a PCR cycle, DNA is melted, separating the strands, to which primers are attached. Primers consist of short stretches of nucleotides complementary to the regions of the DNA sequence flanking the target gene. Thus, the sequence of the desired gene(s) must be known so that primers can be constructed with good gene specificity yet wide applicability to a variety of organisms with the same gene.

The primers are added in great excess, and as the primer–DNA mixture cools, the primers fix to the DNA. After this, a thermostable DNA polymerase (the DNA polymerase from Thermus aquaticus is often used) extends the primer along the target gene of interest. This mixture is reheated, separating the DNA again into single strands and doubling the target sites for a new cycle of PCR. Typically, 30–40 cycles are run, yielding an increase in the number of sequences of 10^9 times or more. In environmental samples, a large yield of gene sequence is possible from organisms present in only small numbers. Obviously, contamination is also of great concern.

Since mutations in the SSU rRNA accumulate over time, in principle, the Tree of Life might be used to tell when in Earth history major episodes of biological evolution and innovation occurred. Important biological milestones include the evolution of individual metabolic capabilities such as methanogenesis, sulfate reduction, nitrogen fixation, denitrification and, of particular interest, oxygenic photosynthesis. Furthermore, the Tree of Life might help resolve when plants and animals first evolved.

Unfortunately, the use of SSU rRNA sequences to constrain the timing of major biological innovations has, thus far, proven to be more of a hope than a reality. It is clear that mutation rates may vary among different major lineages

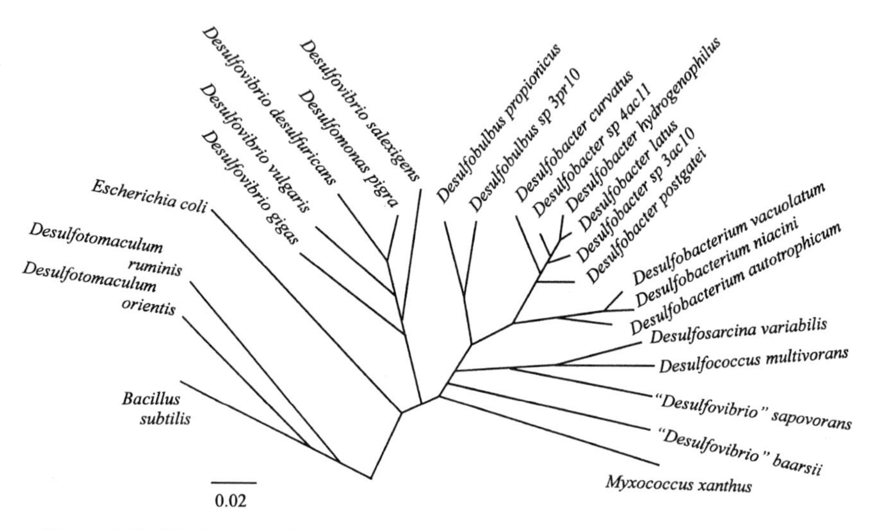

Figure 1.7 Phylogeny of members of selected genera of sulfate reducers from the proteobacteria and the gram-positive bacteria from 16S rRNA sequences. Redrawn from Devereux and Stahl (1993).

within the Tree of Life, and furthermore, mutation rates may also change over time. Early attempts to "calibrate" the Tree of Life utilized a general SSU rRNA mutation rate determined from SSU rRNA sequence differences among organisms calibrated with evolutionary events constrained by the fossil record (Ochman and Wilson, 1987). The calibration points were mostly from fossils of the Phanerozoic (geologic eon comprising the last 544 million years), and hence were young compared to the full time over which biological evolution has occurred on Earth. This approach, when extrapolated back in time, revealed unrealistically young ages for important evolutionary events such as the emergence of oxygenic photosynthesis (cyanobacteria). The value of SSU rRNA molecular clocks could improve with a better understanding of mutation rate variability within the tree, constrained, possibly, by calibration points further back in time. Also, alternative molecular clocks, based on comparisons of gene sequences coding for proteins, show promise in unraveling the timing of major biological innovations in Earth history (e.g. Feng *et al.*, 1997).

7. COMPARISONS WITH PREVIOUS TREES OF LIFE

Previous divisions of the natural world have been heavily influenced by what we can see. Whittaker and Margulis (1978) proposed that the diversity of life on Earth could be represented by five kingdoms: *Animalia, Plantae, Fungi,*

Protista, and *Monera* (prokaryotes), of which the first four are eukaryotic organisms, and the first three comprise most of the macroscopic living world. In an updated version, Margulis and Schwartz (1998) incorporated aspects of the SSU rRNA phylogeny but retained the five-kingdom division of life.

The SSU rRNA-based Tree of Life, in contrast with previous trees, emphasizes the diversity of prokaryotic organisms. Indeed, three of the prominent kingdoms in the tree of Whittaker and Margulis (1978), *Animalia, Plantae,* and *Fungi*, represent late-evolving lineages in the SSU rRNA Tree of Life, with little sequence variation separating them. As prokaryotes encompass such an enormous range of metabolic diversity and have been so prominent in the history of life on Earth, we feel that the three-domain Tree of Life, as derived from molecular sequences, provides a better phylogeny of the living world. Therefore, in the remainder of this work, we adopt the organization of the living world as provided by the SSU rRNA Tree of Life.

8. A BRIEF TOUR THROUGH THE TREE

8.1. Domain *Bacteria*

The *Bacteria*, one of the two prokaryotic domains, is divided into 30 currently recognized phyla. For example, the cyanobacteria, the proteobacteria (also known as the purple bacteria) and the green sulfur bacteria are all recognized as phyla. Some of these phyla, however, contain relatively few species, and taxonomic designations within the bacterial domain are at present rather imprecise. In many cases international standards have not been established for classification much beyond the genus level. This situation is changing, with new editions of *Bergey's Manual of Determinative Microbiology* and *The Prokaryotes* now being assembled and disseminated.

The bacterial domain is genotypically (based on genetic information) quite distinct from the *Archaea*, the other prokaryotic domain (see below), but there are many phenotypic characteristics (observable properties) that also distinguish the *Bacteria* from the *Archaea*. First, great metabolic diversity is housed in the *Bacteria*, including many physiologically unique groups of organisms. For example, among the prokaryotes, oxygenic photosynthesis (the cyanobacteria) is unique to the *Bacteria*. The *Bacteria* is also the principal domain for anoxygenic photosynthetic organisms. Other metabolisms such as methanotrophy[2] (methane oxidation), methylotrophy (oxidation of

[2]Methanotrophy may also occur among some methanogens during the anaerobic oxidation of methane, although the details of this process are still unclear (see Chapter 10).

one-carbon compounds, not including methane) and nitrification are also, as far as we currently know, restricted to the *Bacteria*. All known endospore-forming prokaryotes are found within the bacterial domain (gram-positive bacteria). Organisms from the bacterial domain also contain unique cell wall features. For example, with only a couple of exceptions, members of the *Bacteria* have the cell rigidifier peptidoglycan.

Other distinguishing features of the bacterial domain include the deep branching of high temperature-adapted organisms. These include the chemolithoautotrophic *Aquifex-Hydrogenobacter* group, sulfate reducers of the genera *Thermodesulfobacterium* and *Thermodesulfatator* (Moussard *et al.*, 2003) and the dominantly fermentative organisms of the Thermotogales group. All of these organisms have temperature optima in the 80–90 °C range (Stetter, 1996). Also deep branching are anoxygenic photosynthetic organisms of the family *Chloroflexiaceae*. It would appear that great metabolic diversity in hot environments accompanied early bacterial evolution (Stetter, 1996). An inspection of tree topology shows that the majority of the domain, including most of the bacteria with which we are most familiar, including members of the cyanobacteria, the proteobacteria, and the nitrospira group, emerged from approximately the same point in the tree. This could represent a massive diversification of bacterial life, possibly in association with cyanobacterial evolution (Knoll and Bauld, 1989; Canfield and Raiswell, 1999).

8.2. Domain *Archaea*

The *Archaea* is the second of the two prokaryotic domains and is organized into three principal kingdoms: *Crenarchaeota, Euryarchaeota*, and *Korarchaeota*. Of the three archaeal kingdoms, only the *Crenarchaeota* and *Euryarchaeota* have members with well-studied physiologies. The *Korarchaeota* are known only from molecular sequences (e.g., Barns *et al.*, 1996), with no cultured representatives so far. Molecular sequences have revealed, therefore, a fundamental new branch of life with possibly significant implications for the path of early evolution and physiological diversity within the *Archaea*. We wait eagerly for the isolation and physiological study of organisms from this kingdom.

Overall, physiological diversity within the *Archaea* is not as extensive as in the *Bacteria*. Nevertheless, there are some unique aspects. For example, methanogenesis is solely housed within the *Archaea*, and organisms with the highest temperature tolerance known, up to 121 °C, are also found in the *Archaea* (Kashefi and Lovley, 2003). Furthermore, within the *Archaea* are found the halophiles, organisms commonly growing at salt concentrations up to halite saturation (highly salt-tolerant organisms are, however, also found among the *Bacteria*). Some members of the halophiles provide the reddish pigmentation typical of salt-concentrating ponds and conduct a unique type of light-dependent energy conservation, a kind of photosynthesis, relying on the protein bacteriorhodopsin (see Chapter 4). This protein is functionally and structurally similar to the eye pigment rhodopsin. This type of photosynthesis is used to generate ATP and does not involve oxidation–reduction reactions. As with the *Bacteria*, certain cell wall features also characterize the *Archaea*. Thus, most *Archaea* have a cell wall surface layer consisting of protein or glycoprotein. Some methanogens contain a cell wall quite similar to the peptidoglycan cell walls of *Bacteria*. However, this pseudopeptidoglycan lacks muramic acid, a key component in bacterial peptidoglycan. Cell membranes in *Archaea* are also distinctive, as they contain ether-linked lipids as opposed to the ester-linked lipids found in organisms from the domains *Bacteria* and *Eukarya*.

Archaea have been isolated from environments considered to be the most extreme habitats for life on Earth. These environments include those with high temperature (hyperthermophiles), high salinity (halophiles), and extremes of pH. Recently, a much greater environmental range for *Archaea* has been established by the identification of *Archaea* as an important constituent of marine picoplankton (Delong, 1992; Karner *et al.*, 2001) and the discovery of symbiotic associations between *Archaea* and some marine animals (Preston *et al.*, 1996). Unfortunately, these *Archaea* have yet to be cultured.

Like the *Bacteria*, deep-branching organisms in the *Archaea* tend to be hyperthermophilic (thriving at high temperatures above 80 °C). The hyperthermophilic nature of deep-branching organisms has led to speculation that the last common ancestor to the three domains of life may also have been hyperthermophilic (Woese, 1987; Stetter, 1996) and, furthermore, that life itself may have originated in a high-temperature environment (Shock, 1996; Russell and Hall, 1997). This hypothesis has not been universally accepted (Galtier *et al.*, 1999) and, recently, low-temperature environments have also yielded molecular sequences from deep-branching *Archaea* (Delong, 1992; Delong *et al.*, 1994). Culturing of these organisms may generate further insights into the path of early biological evolution.

8.3. Domain *Eukarya*

Eukaryotic organisms are easily distinguished by their membrane-bound nucleus and by the presence of organelles.[3] Organelles are membrane-bound bodies, for example, mitochondria and chloroplasts, which specialize in carrying out specific functions within the cell. Compared to the prokaryotic domains, the metabolic diversity (as opposed to biochemical and morphological diversity) within the *Eukarya* is quite limited. Thus, eukaryotic organisms are oxygen-producing phototrophs, are aerobic heterotrophs, or in some cases, survive by fermentation (yeast, for example). Although eukaryotes dominate the macroscopic world, most of the SSU rRNA diversity within the domain is accounted for by microscopic organisms including protozoans (single-celled eukaryotes without a cell wall, including many human pathogens), unicellular algae, and some fungal types including yeast. Indeed, as noted previously, higher plants and animals occupy only late-evolving lineages within the domain.

Whereas most metabolic diversity has evolved within the prokaryotic domains, cellular complexity is a hallmark of eukaryotes and is a product of eukaryote evolutionary history. The earliest stages of eukaryote evolution are unclear, but in one prominent model (Sogin, 1991) an early proto-eukaryote, lacking a nucleus and with a fragmented, possibly RNA-based genome, engulfed an early archaeon. A fusion between the DNA-based genome of the archaeon and the fragmented genome of the proto-eukaryote formed the basis of the early eukaryote genome. The engulfed archaeon developed into the membrane-bound nucleus that distinguishes eukaryotes. This scheme is consistent with, and derived partly from, the relatedness between *Archaea* and *Eukarya* in the Tree of Life (Figure 1.5). In a somewhat similar proposition, Zillig *et al.* (1992) proposed that eukaryotes originated from a fusion between an early bacterium and an archaeon, with, as before, the archaeon becoming the nucleus.

More certain is the origin of some of the organelles, in particular, the mitochondria and chloroplasts. Both of these were derived from endosymbiotic associations between eukaryote cells and members of the *Bacteria* (Margulis, 1970). Thus, mitochondria arose from an endosymbiotic proteobacterium, whereas chloroplasts developed from endosymbiotic cyanobacteria. Each of these organelles contains their own rRNA whose sequences clearly support their endosymbiotic origins in the *Bacteria*.

The deepest-branching members of the *Eukarya* (Diplomonads, Microsporidia, and Trichomonads) are anaerobic protozoans lacking mitochondria (Sogin, 1989). It has therefore been argued that these lineages originated from primitive eukaryotes that evolved before the endosymbiotic event forming mitochondria. However, the nuclear genomes of many of these deep-branching eukaryotes contain what are apparently mitochondrial-derived genes

---

[3]What might be considered organelles are also found among certain prokaryote groups, including the chlorosomes of the green sulfur bacteria (see Chapter 9) and the anammoxozomes of some members of the genus *Planctomycetes* involved in anaerobic ammonia oxidation (see Chapter 7) (Sinninghe Damsté *et al.*, 2002). Other members of the *Planctomycetes* also contain membrane-bound bodies, including a membrane-bound nuclear structure.

(Sogin, 1997). Therefore, it seems more likely that these deep-branching eukaryotes once had mitochondria but later lost them. The loss of mitochondria may have resulted from a dominantly parasitic lifestyle and a retreat to anoxic habitats (Sogin, 1997). Even though eukaryotes stand alone as a principal domain in the Tree of Life, their evolution and defining features (nucleus and organelles) are ultimately tied to organisms of the prokaryotic domains.

9. HOW TRUE ARE SSU rRNA-BASED PHYLOGENIES?

Phylogenies drawn from the SSU of the rRNA molecule provide an unparalleled view into the evolution of prokaryotic life. Furthermore, these phylogenies have given us the three-domain Tree of Life, have been instrumental in defining *Archaea* as a separate domain, and have provided solid support for the endosymbiotic origins of mitochondria and chloroplasts in eukaryotic cells. With all this success, there is still debate as to whether the Tree of Life, as constructed from SSU rRNA, is correct (Doolittle, 1999). There are two major concerns. The first is statistical: as phylogenies are drawn from statistical analysis of molecular sequences, there is some concern whether the branching patterns depicted on SSU rRNA-based trees are statistically valid (van de Peer *et al.*, 1994). Statistical treatments are basically exercises in hypothesis testing, and the validity of the given phylogeny depends on the validity of the hypothesis and the extent to which the statistical test shows valid differences among various molecular sequences (Hillis *et al.*, 1993; Huelsenbeck and Rannala, 1997). Statistical problems in determining phylogenetic relationships should become reduced as more sequences become available, particularly in the deep branches, and as statistical treatments improve.

A potentially far more serious concern is whether it is appropriate, or even valid, to think about organisms as having a unique phylogeny (Doolittle, 1999). If the lateral transfer of genetic material from one organism to another has been a major mode of genome construction, then different genes would be expected to yield different, equally valid, phylogenies. Yet, none of these phylogenies could be taken to represent the evolutionary history of the organism; in fact, the organism would have a complex history defined by several different episodes of gene acquisition (Hilario and Gogarten, 1993). The evidence for lateral gene transfer among prokaryotic organisms is now indisputable, including divergent phylogenies obtained using different genes and the apparent acquisition of archaeal genes by bacterial genomes (Brown and Doolittle, 1997; Jain *et al.*, 1999; Nelson *et al.*, 1999). An emerging concept is that informational genes, those genes involved in the basics of cell function such as the transcription–translation process of protein synthesis (including ribosomal RNA genes), are relatively unsusceptible to gene transfer. By contrast, operational genes, those involved in cellular metabolism and "housekeeping," are more readily transferred (Jain *et al.*, 1999). The reality of lateral gene transfer has understandably given rise to a certain skepticism as to the validity of SSU rRNA-based phylogenies (Philippe and Laurent, 1998).

In principle, comparison of whole genome sequences should provide the best resolution of lateral gene transfer events between organisms and the most accurate organismal phylogenies. Thus far, whole genome sequences have provided equivocal answers to the question of organismal evolution and the phylogenetic validity of the SSU rRNA tree. Whereas lateral gene transfer is clearly indicated from whole genome sequences (e.g., Nelson *et al.*, 1999), phylogenetic analysis of whole genome gene families yields trees quite similar to the SSU rRNA tree (Fitz-Gibbon and House, 1999; Clarke *et al.*, 2002; House *et al.*, 2003). The principal domains are preserved, as is the differentiation of most phylogenetic groups as identified by SSU rRNA. Nevertheless, there are some important differences. For example, in the whole genome tree of House *et al.* (2003), an early evolution of the methanogens is not supported and some non-gram-positive bacteria cluster with the gram-positive group.

Whole genome phylogenetic analysis is in its infancy, and so far only a relatively small number of whole genomes have been sequenced and annotated (about 100, but the number is growing rapidly). Hopefully the pathways of genome acquisition by organisms will soon become clearer.

10. SSU rRNA SEQUENCES AS A TAXONOMIC TOOL

Despite current debate regarding the interpretation of organismal phylogenies, especially those obtained from deep branches in the SSU rRNA Tree of Life, organisms are not simply a homogeneous mix of shared genetic information. The Tree of Life provides countless examples of sensible groupings based on similar phenotypic expression or other distinct cell properties. As a conspicuous example, cyanobacteria represent a phylogenetically distinct monophyletic group with the common physiology of oxygenic photosynthesis.

If we look closely, related organisms based on SSU rRNA sequence comparisons (<5 to 10% difference) typically share common physiologies. For example, among the sulfate reducers of the Delta subdivision of the proteobacteria (Figure 1.7), species of the genus *Desulfobacter* group together and are related in their ability to oxidize acetate (see Chapter 9). Furthermore, all species of the genus *Desulfovibrio* are lactate utilizers, and these are all vibrio shaped. Within the archaeal domain, sulfate reducers of the genus *Archaeoglobus* all group together. These similarities, and many others, have encouraged the use of the SSU rRNA tree as a fundamental tool in taxonomic classification. Thus, a complete description of a new organism now requires a SSU rRNA sequence, and genus-level classification is typically based on where in the tree the organism resides.

This SSU rRNA-based taxonomy has also required some changes in previous taxonomies. For example, the organism *Desulfobotulus sapovorans* (Figure 1.7) was previously classified as a *Desulfovibrio* species based on similarity in phenotypic characters, and shape, to the organisms of the genus *Desulfovibrio*. However, SSU rRNA sequences clearly indicated a different evolutionary history for this organism, and hence a new genus name was needed. Other instances are outlined in the following Chapters. The physiological diversity of organisms known only from their SSU rRNA sequences (molecular isolates) is sometimes inferred from placement within the tree, but one should be wary of assigning physiologies to organisms before they have been isolated and studied in the laboratory.

Chapter 2

Structure and Growth of Microbial Populations

1. INTRODUCTION

This work, in part, documents how prokaryotes play an environmentally significant to dominant role in the cycling of numerous important redox-sensitive elements such as carbon, iron, manganese, oxygen, nitrogen, and sulfur. This Chapter explores some of the principles outlining the structure and growth of the populations conducting these transformations. We consider the factors regulating population growth and population size. We explore how populations with overall similar physiology can occupy an enormous range of environmental conditions, including extremes of temperature, salinity, and pH. We also look at factors influencing the ecology of microbial populations. Indeed, we know relatively little about the details of microbial interactions in nature, but we will overview what we do understand. In

0-12-026147-2
© 2005 Elsevier Inc.
All rights reserved

addition, we explore how cells communicate with each other and how, at least in some circumstances, they cooperate to effectively utilize available resources. The recognition that prokaryotes can display community behavior is an important revelation in microbial ecology (e.g., Shapiro, 1998; Miller and Bassler, 2001).

2. CONSIDERATIONS OF CELL SIZE

It is no secret that prokaryotes are small, with typical nominal cell diameters of between 0.5 μm and 2 μm. In fact, smallness has decided advantages, as prokaryotes obtain their nutritional requirements from the environment by diffusion. Other modes of transport such as advection or dispersion operate on spatial scales too large to influence transport of nutrients to microbes (Fenchel *et al.*, 1998). Indeed, over small-distance scales, diffusion is a remarkably effective transport process, whose timescale can be estimated from the Stokes-Einstein relationship. Here, the timescale (t) associated with transport over a characteristic length scale (L) depends on the length scale and the diffusion coefficient (D) (Equation 2.1):

$$t = \frac{L^2}{2D} \tag{2.1}$$

If we take $D = 1 \times 10^{-5}$ cm^2 s^{-1}, which is not uncommon for gases in solution, then values for t may be calculated for a range of characteristic-length scales, L, as shown in Table 2.1. It is apparent that diffusion is extremely rapid, with timescales of approximately 1 ms, over the length scales appropriate for transport to small prokaryotic organisms. Thus, small size allows rapid transport of nutrients to an actively growing cell.

Still, we may assume that even small cells can experience diffusion-limited growth in nature, particularly if the concentration of the limiting substrate is

Table 2.1 Time associated with diffusion over various characteristic-length scales

Length	Time
1 μm	0.5 ms
10 μm	50 ms
100 μm	5 s
1 mm	8 min
1 cm	14 hr
10 cm	58 d

low. The total diffusional flux of substrate to the surface of a spherical cell is given by

$$J = 4\pi Dr(C_\infty - C_o) \tag{2.2}$$

where r is the cell radius, D is the diffusion coefficient, C_∞ is the concentration of the substrate in the bulk reservoir and C_o is the concentration of the substrate at the cell surface (Fenchel *et al.*, 1998; Schulz and Jørgensen, 2001). The maximum diffusional flux occurs when the concentration of the limiting substrate is zero at the cell surface ($C_o = 0$):

$$J_{\max} = 4\pi DrC_\infty \tag{2.3}$$

We can now calculate the volume-specific rates of metabolism under diffusion limitation by dividing Equation 2.3 with the volume of the cell ($V = 4/3\pi r^3$), to yield the following:

$$\text{Specific rate} = 3DC_\infty/r^2 \tag{2.4}$$

Therefore, under diffusion limitation, the specific rate of cell metabolism should increase linearly with substrate concentration (C_∞) and decrease with the square of the cell radius (r^2). This is a strong indication that small cell size is a decided advantage in maintaining high specific rates of metabolism.

Yet, there is a limit to how small cells can be. This issue was brought into sharp focus by the report of McKay *et al.* (1996) of 20 nm to 100 nm objects from the Martian meteorite AHL84001, which they believed to be fossil cells. Theoretical considerations suggest that a minimum cell size should fall in the range of 170 nm. This value is obtained assuming a cell genome housing 250 genes (~300 kilobases), which is considered the minimal number needed to support heterotrophic cell growing in a nutrient-rich environment, in other words, a cell with minimal metabolic complexity (Adams, 1998). Furthermore, it is assumed that the cell contains 10% ribosomes, 20% proteins, and 50% water, by analogy with typical prokaryote cells (Adams, 1998). In this case, the cell contains 65 ribosomes and 65 proteins per gene. Further calculations reveal that a cell size of just 50 nm, with a 5 nm cell wall, can accommodate only two ribosomes, 520 protein molecules, and a DNA strand containing only about eight genes. This is a biosynthetic factory too small to function in a manner we understand (Adams, 1998).

We can compare these calculations with the smallest known prokaryotes in nature, often referred to as nanobacteria. The recently discovered archaeon *Nanoarchaem equitans* is coccoidal with an extremely small cell diameter of approximately 400 nm (Huber *et al.*, 2002). This organism has the smallest known genome among prokaryotes at 500 kilobases, and it lives in intimate association with an autotrophic sulfur-reducing archaeon of the genus *Ignicoccus*. *N. equitans* apparently represents the first member of a whole new kingdom of organisms within the *Archaea*. Other small *Archaea* are found among the genera *Thermofilum*, with rod diameters of 170 nm, *Thermoproteus,* with 300-nm-diameter elongated cells, and *Pyrobaculum,* with disk diameters of 200 to 300 nm and widths of 80 to 100 nm (Stetter, 1998). The pathogenic bacteria *Mycoplasma pneumoniae* has a cell diameter of only 200 nm, and very small bacteria have been discovered in human and cow blood, with typical diameters of 200 nm, or even smaller (Kajander and Çifçioglu, 1998). Indeed, some of these blood-associated nanobacteria pass through 0.1 μm filters (Kajander *et al.*, 1997), and transmission electron microscopy (TEM) observations show tiny cell forms as small as 50 nm in diameter (Kajander *et al.*, 1997). Therefore, numerous different small cells in nature approach the theoretical lower limit of 170 nm diameter, and some even seem to fall below this limit. However, it is an open question as to how cells function at diameters below 100 nm.

Despite the advantages of small size in allowing efficient nourishment of the cell, most prokaryotes grow larger than our theoretical lower cell-size limit. With a typical cell diameter of 1 to 2 μm, prokaryotes can accommodate relatively large genomes (2 to 4 megabases is common), allowing metabolic plasticity, which can be an advantage in chemically dynamic environments, as often found in nature. Therefore, metabolic flexibility is balanced against the efficiency of nutrient uptake. But what about prokaryotes growing to an extremely large size? The current world record holder for large size is the sulfide oxidizer *Thiomargarita namibiensis,* with cell diameters up to 750 μm, large enough to be seen by the naked eye (Schulz *et al.*, 1999). Most of the cell volume accommodates a large central vacuole (Figure 2.1), and active cell cytoplasm, concentrated at the external margins of the cell, comprises only 2% of the cell volume (Schulz *et al.*, 1999). The largest known heterotroph is *Epulopiscium fishelsoni*, found in the guts of tropical fish, with a maximum diameter of about 80 μm and a maximum length of 600 μm (Angert *et al.*, 1993).

Other large bacteria include the sulfide oxidizer *Thioploca araucae*, consisting of numerous cells of 30 to 40 μm in diameter making up filaments (also called trichomes) of up to 7 cm or more in length (Schulz *et al.*, 1996; see Chapter 9). Up to 100 individual filaments occupy a common sheath vertically oriented in the sediment. Some unsheathed filaments of *Beggiatoa*, also a sulfide oxidizer, are found in hydrothermal areas with diameters of up to 122 μm (Jannasch *et al.*, 1989). Similar to *Thiomargarita*, most of the volume of these large *Thioploca* (Figure 2.1) and *Beggiatoa* cells accommodates a central vacuole. Large size is also relatively common among filamentous cyanobacteria, with individual trichomes reaching diameters of 40 μm or more, as for example in some species of *Oscillatoria* (Schulz and Jørgensen, 2001).

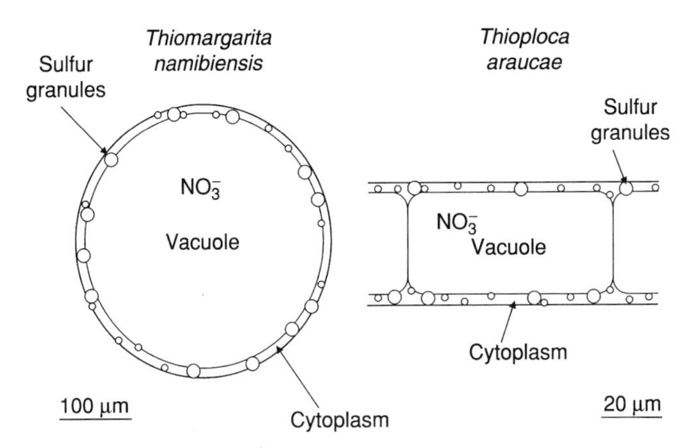

Figure 2.1 Relationships between the internal NO_3^--containing vacuoles and cell cytoplasm for *Thiomargarita namibiensis* and *Thioploca araucae*. Inspired from figures in Schulz *et al.* (1999) and Fossing *et al.* (1995).

Obviously, large size among these various organisms incurs some ecological advantage. Large vacuolated sulfide oxidizers concentrate nitrate into the vacuole for use as an electron acceptor in sulfide oxidation. With the cell cytoplasm concentrated in a thin layer near the cell membrane, the ratio between the volume of cytoplasm to the volume of vacuole will change with cell size in a manner similar to the surface area to volume ratio (surface area/volume $= 3/r$). Therefore, as cell size increases, there is a greater volume of vacuole compared to the volume of cytoplasm and a greater supply of nitrate per volume of cytoplasm. This could be particularly important for an organism such as *T. namibiensis,* which is not motile and probably fills its "nitrate tank" during periodic resuspension events (Schulz *et al.*, 1999). It then relies on this nitrate supply to oxidize sulfide in sulfide-rich sediments that are normally devoid of nitrate (Schulz *et al.*, 1999).

For *Thioploca*, the nitrate tank is filled as *Thioploca* filaments extend out of their sheath into nitrate-rich water overlying the sediment surface. Filaments then migrate down the sheath into the sulfide-containing sediment below (Jørgensen and Gallardo, 1999). Therefore, the large cell size of *Thioploca* allows the accumulation of a large storage reservoir of nitrate, while the long filaments and common sheath allow a unique strategy for shuttling the electron acceptor (nitrate) to the electron donor (sulfide). For cyanobacteria, large size and filamentous structure allow the construction of resilient microbial mats that are not easily broken apart in turbulent environments.

3. POPULATION GROWTH

3.1. Substrate uptake and growth

Like all life on Earth, prokaryote populations must grow to be successful. The initial stages of growth in the laboratory are often very slow, and in some cases, no growth is apparent. This is known as the "lag phase." During this phase, the organisms are adapting to the culture conditions. This could mean repair of damaged cells if the culture is old or the synthesis of new enzymes if the culture conditions are different from what the organisms have previously seen. A lag phase, however, is not always observed. As an example, a freshly grown culture of *Desulfovibrio desulfuricans* begins growth immediately after inoculation into fresh media (Figure 2.2). After the lag phase, populations normally experience a period of exponential growth. The basis behind exponential growth is a doubling of population size during every cycle of cell division. If the frequency of cell division remains constant, as occurs with a rich supply of nutrients, and in the absence of inhibiting substances, then population growth is exponential with

$$N = N_o e^{\mu t} \tag{2.5}$$

where N (cells ml^{-1}) is population size at time t, N_o is the starting population size, and μ (h^{-1}) is the specific growth rate, which is related to the population doubling time t_D, defined as $t_D = \ln 2/\mu$. A stationary phase follows (Figure 2.2), where batch cultures typically reach cell densities of 10^8 to 10^{10} cells ml^{-1}, after which growth apparently stops. In this phase of growth,

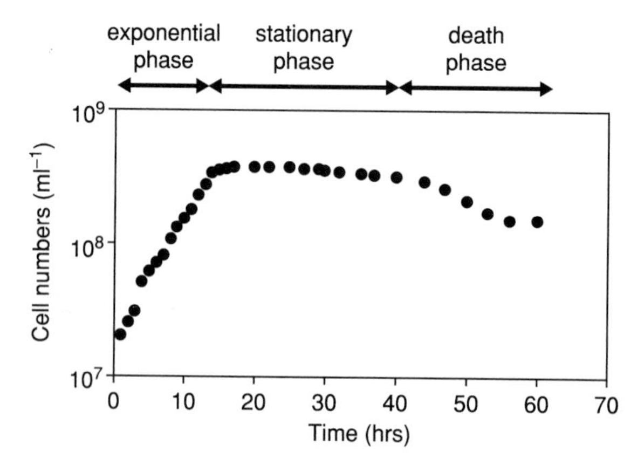

Figure 2.2 The stages of growth for a batch culture of the sulfate reducer *Desulfovibrio desulfuricans*. Results from Olesen and Canfield (unpublished).

cultures continue to metabolize substrate, but cell growth is nearly balanced by cell death. In this state a larger proportion of the energy gained during cellular metabolism is shuttled into cell function, with less used to support cell division. The stationary phase of growth begins after a vital nutrient or substrate becomes limiting in the culture media or when a harmful waste product accumulates to inhibitory levels. When cell growth cannot match cell death, the culture enters the "death phase" (Figure 2.2). Some cells may still multiply in this phase, but the medium may be too nutritionally poor, or too toxic, to support net growth.

Under all circumstances, there is a certain amount of energy used to support maintenance functions of the organism, including ongoing repair of cellular constituents, and the maintenance of chemical and electrical gradients needed to support cellular function (Pirt, 1975). Maintenance energy is a necessary expenditure by the organism, but it does not support growth. Thus, the energy derived from the metabolism of organisms under nutrient-limiting conditions is largely going toward maintenance functions, whereas when nutrients are abundant, energy is largely used for growth, although maintenance is still required. Maintenance energy, m_E, can be defined as the amount of energy dissipated per mol-equivalent of carbon substrate per time (kJ $Cmol^{-1}$ h^{-1}), and Tijhuis *et al.* (1993) have suggested, from a theoretical analysis, that m_E depends only on temperature. However, in the experiments of Scholten and Conrad (2000), maintenance energies were found to differ considerably from the theoretical values, suggesting that growth conditions other than temperature also influenced m_E.

The growth rates of prokaryote populations should logically depend, in some way, on the concentrations of critical substrates. Low concentrations should impose diffusion limitations on the cellular uptake of substrate, which limits growth rate. Also, if substrate concentration is too high, substrate transport sites at the cell surface become saturated, and increasing substrate concentration has no effect on substrate uptake rates. As part of his Ph.D. thesis, Jacques Lucien Monod (Monod, 1942) studied the relationship between growth rate and substrate concentration for strains of *Escherichia coli.* He found that the relationship could be described empirically through the now famous "Monod" equation:

$$\mu = \mu_{max} \frac{S}{K_S + S} \tag{2.6}$$

where μ (h^{-1}) is the specific growth rate, μ_{max} (h^{-1}) is the maximum specific growth rate, S (μM) is substrate concentration, and K_s (μM) is the half saturation constant, or the substrate concentration at half maximum growth rate. The magnitude of K_s indicates the affinity of the organism for a given substrate, with a lower K_s value meaning a higher substrate affinity. The

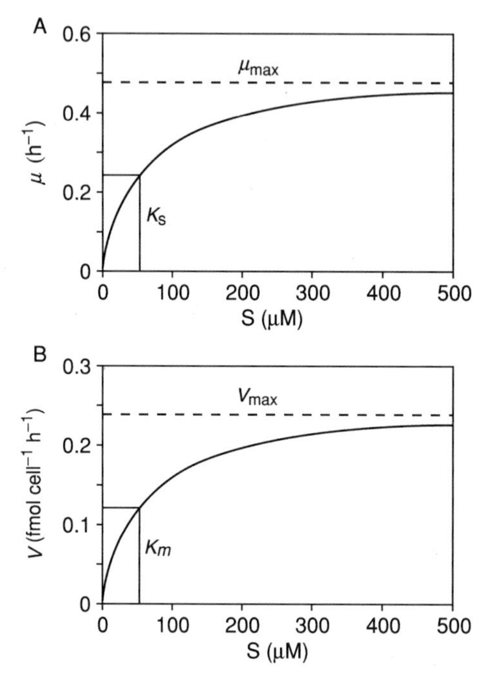

Figure 2.3 The relationship between μ_{max}, S, K_s and μ for microbial growth as given by the Monod equation (Equation 2.6) and between V_{max}, S, K_m and V for substrate utilization rate as given by the Michaelis-Menten equation (Equation 2.7).

relationship between these parameters is shown in Figure 2.3A. While the Monod equation was empirically derived, several theoretically based growth models produce a similar relationship (Panikov, 1995).

Microorganisms frequently also show similar relationships between substrate concentration and metabolic rate:

$$V = V_{max}\frac{S}{K_m + S} \tag{2.7}$$

where V is specific metabolic rate (μmol $g_{organisms}^{-1}$ h^{-1}, or fmol cell^{-1} h^{-1}), V_{max} is maximum metabolic rate, S is substrate concentration (μM), and K_m is the half saturation constant, or the substrate concentration at half maximum rate. The kinetic form shown in Equation 2.7 (Figure 2.3B) is often referred to as "Michaelis-Menten" kinetics, as Michaelis and Menten first derived a similar relationship from theoretical grounds for pure enzyme reactions (Michaelis and Menten, 1913). If two substrates are critical for metabolism, such as oxygen (electron acceptor) and ammonium (electron donor) for nitrifiers, the dual influence of substrate limitation on metabolic rate can often be approximated by the following modification of the Michaelis-Menten expression:

$$V = V_{max}\left[\frac{S_D}{K_{mD} + S_D}\right]\left[\frac{S_A}{K_{mA} + S_A}\right] \tag{2.8}$$

where the subscripts D and A refer to electron donor and electron acceptor, respectively.

The kinetic parameters describing the growth and metabolism of individual organisms are somewhat variable depending on the previous history of the organism, the extent of substrate limitation, and the timescale of the observation (Button, 1985). For example, organisms can adapt to substrate limitation by increasing the density of transporter enzymes, which would increase V_{max}. Generally, but not always, high values of K_m are correlated with high V_{max}, while low values of K_m are correlated with low V_{max}. These different adaptations to substrate availability allow organisms to exploit different ecological niches in a dynamic environment. Thus, organisms with high K_m and V_{max} can metabolize (and grow) rapidly with a sudden input of fresh substrate, while organisms with low K_m and V_{max} are adapted to situations of substrate limitation.

This general relationship between K_m and V_{max} can be rationalized as follows. Organisms with low K_m have transporter or metabolic enzymes with high substrate affinity, which also tend to bind the substrate tightly with a relatively long lifetime for the substrate–enzyme complex. This relatively long lifetime for the intermediate complex decreases V_{max} (Button, 1985). Conversely, organisms with a lower affinity for substrate have transporter or metabolic enzymes binding substrate less tightly. This leads to a higher K_m, but also to a higher V_{max} due to a shorter lifetime for the substrate–enzyme complex. The relationship between K_m and V_{max} for sulfate uptake by sulfate reducers is shown in Figure 2.4 (data from Widdel, 1988).

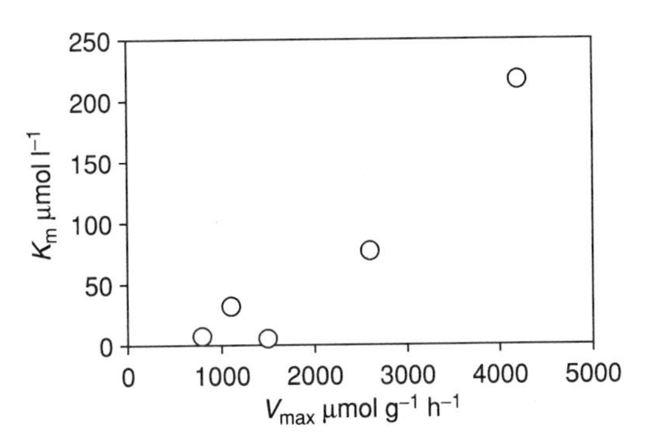

Figure 2.4 The relationship between K_m and V_{max} for acetate-utilizing sulfate reducers. Data from Widdel (1988).

The specific affinity, a_A (Equation 2.9), has been advocated as a fundamental parameter describing the affinity of prokaryotes for substrate (e.g., Button, 1986):

$$a_A = \frac{\mu_{max}}{K_s} \tag{2.9}$$

Specific affinity approximates the initial slope of a growth (μ) vs. substrate concentration (K) plot and provides an indication of how an organism's growth rate will respond to low substrate concentrations as generally found in nature. Hence, this parameter has ecological significance. Obviously, high specific affinities are advantageous for organisms growing under low substrate conditions. Nedwell (1999) found that within populations of bacteria and algae, specific affinities were heavily influenced by temperature. Thus, as temperature was reduced, specific affinity decreased, reflecting a stiffening of the cytoplasmic membrane, reducing the efficiency of transporter enzymes and proteins. We will discuss more of the relationship between temperature and metabolism below.

3.2. Growth yield

Growth yield, Y, expresses the amount of cell material formed during the utilization of a certain amount of substrate. This relationship is usually expressed in terms of mass, where

$$Y = \frac{\text{grams of dry cells formed}}{\text{grams of substrate consumed}} \tag{2.10}$$

A wide variety of different substrates can be chosen, including electron donor, electron acceptor or even a limiting nutrient. The definition of growth yield can also be modified to ratio cell formation against typical metabolic products, including, for example, ATP, O_2 or H_2. Growth yields are usually constant when organisms are in exponential growth (Fuchs and Kröger, 1999). Growth yield also tends to correlate with specific growth rate (μ), such that higher specific growth rates generate higher growth yields (up to a theoretical maximum, of course). In batch culture or chemostat experiments, high specific growth rates occur at relatively low population density where conditions are most favorable for growth (see above). At high population densities in batch cultures substrates become limiting and toxic waste products accumulate, slowing the rates of specific growth (see above) and reducing the growth yield. In this situation, a higher proportion of the energy gained during substrate utilization is channeled into maintenance rather than growth. A positive relationship between growth yield and specific growth rate probably also holds in nature.

Continuous culture

A continuous culture, or chemostat, provides a valuable means for exploring the growth physiology of microorganisms. A chemostat is a flow-through system in which growth substrates are introduced at a constant rate, allowing growth of a microbial culture. The volume is constant, so solution exits at the same rate as it is introduced, and in the output, microorganisms are removed at the same density as found within the chemostat. Therefore, in a chemostat, growth rates and population density are in a dynamic balance between rates of substrate input and rates of microorganism output. Microorganisms experience different growth conditions as input substrate concentrations and their rates of addition to the chemostat are manipulated. As substrates are usually consumed to low concentrations, input substrate concentration controls population density and substrate input rate controls growth rate. In a chemostat there is usually control over temperature, pH and oxygen levels when appropriate. In principle, chemostats develop to a steady state, with constant solution chemistry, microbial growth rates, and population size. This allows the exploration of growth physiology for organisms under constant conditions.

Population growth in a chemostat is assumed to follow Monod kinetics (Equation 2.6). The dilution rate, or residence time, of medium in the chemostat, D (time^{-1}), is equivalent to F/V, where F is the flow rate of medium into the chemostat (volume time^{-1}) and V is the volume. The growth rate of organisms is given by μX, where μ is the specific growth rate and X is population density (usually, grams dry weight liter^{-1}), and the loss or output rate is given by DX. The change in population density is given by

$$\frac{dX}{dt} = growth - output = \mu X - DX \tag{2.B1}$$

At steady state, $\mu = D$; this equality shows how the input rate of substrate controls population growth rate. In principle, chemostats will run until steady state is reached, and this will occur as long as the dilution rate does not become less than the critical dilution rate, D_c, which is the maximum dilution rate at which output balances growth (Equation 2.B2). This becomes an issue when substrate input concentration, S_R, becomes low relative to the specific growth rate of the organism (Harder *et al.*, 1977).

$$D_c = \mu_{max}\left(\frac{S_R}{K_S + S_R}\right) \tag{2.B2}$$

If $S_R \gg K_S$, then $D_c \approx \mu_{max}$, and D_c decreases as S_R decreases relative to K_S.

The substrate concentration in a chemostat depends on the balance between the input rate, the output rate, and how much is converted into biomass:

$$\frac{dS}{dt} = S_R D - SD - \frac{\mu X}{Y} \tag{2.B3}$$

where, in addition to the terms already described, S is the substrate concentration in the chemostat and Y is the growth yield (see Equation 2.10). Substituting the Monod equation (Equation 2.6) into Equation 2.B3 yields, at steady state ($dS/dt = 0$), an expression for S:

$$S = \frac{DK_S}{\mu_{max} - D} \tag{2.B4}$$

The steady state concentration of biomass is given as

$$X = Y(S_R - S) \tag{2.B5}$$

which yields the following expression after substituting from Equation 2.B4:

$$X = Y\left(S_R - K_S\frac{D}{u_{max} - D}\right) \tag{2.B6}$$

Thus, the biomass production and substrate concentrations in a chemostat can be predicted from a basic knowledge of grow parameters (μ_{max} and K_S), dilution rate (D), and input substrate concentration (S_R).

3.3. Growth in nature

Microorganisms are typically isolated from nature in nutrient-rich media and are studied under their optimal growth conditions. Under these conditions, there is a concern that microbial "weeds" are often isolated and that these organisms do not necessarily represent the major populations of microbes in nature. Even if environmentally relevant microbial strains have been isolated, aspects of their physiology and growth characteristics in nutrient-rich laboratory cultures probably differ from their situation in nature. This is because nutrients are much more limiting in the environment, and organisms may adapt physiologically to cope (e.g., Morita, 1997). We can further understand the ecology of prokaryotes in nature if we understand how fast they grow, what factors control their growth and how they adapt physiologically to nutrient stress.

Various methods have been employed to measure microbial growth rates in nature. The most widely used are the incorporation rates of radiolabeled nucleotides or amino acids into the production of DNA, RNA or protein. For example, ^{3}H-labeled thymidine incorporates into DNA, and the rate of incorporation should provide a measure of DNA production rate, which is linked to cell growth (e.g., Findlay *et al.*, 1984; Bell and Riemann, 1989).* Tritium-labeled adenine is incorporated into both DNA and RNA, providing a measure of total nucleic acid production, which is also linked to cell growth (e.g., Karl and Winn, 1984; Karl, 1993). Tritium-labeled leucine is incorporated into proteins, and since proteins make up a relatively constant fraction of bacterial biomass (about 60%), cell growth should be related to leucine uptake rates (e.g., Kirchman, 1992, 1993). All of these methods require assumptions about the ability of microbial populations to incorporate the labeled compound, as well as the relative rates of labeled compound incorporation versus incorporation rates of compound formed within the cell. Careful documentation of compound-specific activity is required, and balanced growth (all cellular constituents are produced at constant relative rates) is usually also assumed. The reader should consult the original literature for more detailed discussion of methodology.

Studies of microbial growth rates for water-column bacteria yield doubling times ranging from days to several months (Moriarty, 1986; Ducklow and Carlson, 1992). In surface marine sediments, doubling times of one to several days have been measured (Karl and Novitsky, 1988), and much longer doubling times are likely deeper in sediments where available substrates become quite limiting. Indeed, by a depth of 10 meters in deep-sea sediments, thymidine incorporation rates of around 100 fmol cm^{-3} d^{-1}

*It has been noted, however, that some important groups of anaerobic prokaryotes do not actively incorporate ^{3}H-labeled thymidine during growth (Wellsbury *et al.*, 1994).

(Parkes *et al.*, 2000) indicate cell doubling times of 1000 days, with a population density of about 10^8 cells cm^{-3} and a nominal thymidine incorporation ratio of 10^9 cells $nmol^{-1}$ thymidine incorporated (Findlay, 1993). At depths greater than 100 meters at the Blake Ridge, and at other deep-sea sites, thymidine incorporation rates of approximately 0.2 fmol cm^{-3} d^{-1} (Parkes *et al.*, 2000; Wellsbury *et al.*, 2000) indicate cell doubling times of over 100 years, with ambient population sizes of around 10^7 cells cm^{-3}, and the same incorporation ratio applied above.

Obviously, microbes in nature are impacted by moderate to severe nutrient limitation. However, substrate limitation occurs as demand outstrips supply, and demand is driven by population size as well as activity. Therefore, nutrient limitation also arises because of overpopulation, when microbes maintain relatively high populations in the face of restricted substrate supply and adapt physiologically to cope with this circumstance. Specific physiological adaptations to low substrate supply are numerous. Commonly, microbes reduce cell size in the face of starvation to increase their surface area to volume ratio, better allowing the efficient fueling of their active cytoplasm (e.g., Novitsky and Morita, 1976). The ribosome and RNA contents of cells also decrease, consistent with a slowing of protein synthesis and overall metabolic function (e.g., Kemp *et al.*, 1993; Morita, 1997). If extreme nutrient limitation is transient, cells may metabolize internal carbon reserves. In some instances, nutrient-limited cells may induce chemotaxis and flagellation (Beveridge, 1989) to help locate scarce substrates if they are unevenly distributed, and high-affinity uptake systems might also be induced to cope with low nutrient supplies (e.g., Jannasch, 1979). Frequently, nutrient-limited cells also become more robust and resistant to environmental extremes such as high pressure, large temperature variations, UV light, and oxidative agents (Morita, 1997). It has also been noted that viable cell numbers in a severely nutrient-limited population might decrease, with the remaining cells still metabolically active but unable to grow (Postgate and Hunter, 1962).

Observations on microbial adaptation to nutrient limitation are conducted on laboratory timescales, which cannot reproduce the slow growth conditions and extreme substrate limitation as is found, for example, in deeply buried sediments. Therefore, the metabolic response of these organisms to nutrient limitation is somewhat uncertain. However, observations demonstrate that approximately 1 to 2% of the total cell numbers as identified through DNA stains such as DAPI (4′,6-diamidino-2-phenylindole) or acridine orange are dividing in marine sediments buried in depths of over 100 m (Parkes *et al.*, 1990, 1994). Furthermore, in these same sediments, ^3H-thymidine incorporation can be measured, and radiotracer studies demonstrate biologically mediated sulfate reduction and methanogenesis (e.g., Parkes *et al.*, 2000). With these observations, especially the relatively high proportion of dividing cells, it seems likely that a high percentage of the total cells are viable and active. Nevertheless, attempts to enumerate viable cells with standard most probable number (MPN) techniques frequently show viable cell numbers several orders of magnitude lower than total cell numbers (e.g., Parkes *et al.*, 1994). This discrepancy likely demonstrates either our inability to find the proper growth media to induce a high proportion of cell growth, or our inability to wait long enough for slow-growing cells to respond physiologically to sudden nutrient-rich conditions.

3.4. Cell numbers and substrate levels in nature

We have seen above that microbes in nature often live under circumstances of nutrient limitation. This limitation certainly arises in part from limited substrate availability, but it also arises from high population numbers sharing the limited resource. In lake and marine water columns cell numbers typically vary between 10^4 and 10^6 cells cm^{-3}, and in surface sediments cell numbers are usually in the range of 10^8 to 10^{10} cells cm^{-3}. Cell numbers decrease with depth in sediments, but not strongly. Even at depths of hundreds of meters in deep-sea sediments, representing millions of years of deposition, cell numbers may still be in the range of 10^6 to 10^7 cells cm^{-3} (Parkes *et al.*, 2000). Microbial population size must represent the balance between growth and death, where growth is controlled by the physiological response of organisms to the available substrate concentrations, and death is controlled by processes such as starvation, viral infection, and grazing.

In what follows, we develop a simple model to explore the major processes controlling microbial population sizes in nature, focusing on acetate as the electron donor. A more complicated carbon flow would necessitate a more complex model, which would probably not alter the main conclusions offered here. We begin by assuming that microbial growth can be expressed by Michaelis-Menten-like kinetics (Equation 2.11):

$$R_G = YV_{max}\frac{SN}{(K_m + S)} \tag{2.11}$$

where population growth rate (R_G; cells cm^{-3} d^{-1}) is linked to the maximum specific metabolic rate (V_{max}; nmol C cell^{-1} d^{-1}) through the growth yield (Y; cells nmol C^{-1}), substrate concentration (S; nmol C cm^{-3}), half saturation constant (K_m; d^{-1}), and the population size (N; cells cm^{-3}). We assume cell death rate (R_D; cells cm^{-3} d^{-1}) is a simple first-order function of population size (N), with k_D (d^{-1}) the coefficient describing the death rate:

$$R_D = k_D N \tag{2.12}$$

Overall, the change in population size is the difference between growth and death rates (see also Lovley and Klug, 1986):

$$dP/dt = R_G - R_D = YV_{max}\frac{SN}{(K_m + S)} - k_D N \tag{2.13}$$

Thus far, we have only one equation (Equation 2.13) but two independent variables, population size, N, and substrate concentration, S. We therefore seek a further expression for substrate concentration. The concentration of substrate available to the microbial population will reflect the balance between substrate production rate and substrate consumption rate by the microbes:

$$dS/dt = k_C C - V_{max}\frac{SN}{(K_m + S)} \tag{2.14}$$

The first term expresses the rate of substrate availability, which is first-order with respect to the concentration of organic matter in the environment, C (nmol cm^{-3}), with k_C (d^{-1}) as the carbon oxidation coefficient. This is a normal representation of organic carbon oxidation rate (Berner, 1980). The second term expresses the rate of substrate oxidation by the microbial population, assuming that oxidation rate follows Michaelis-Menten kinetics.

If we assume steady-state ($dN/dt = 0$, and $dS/dt = 0$), Equation 2.13 simplifies to an expression for substrate concentration (Equation 2.15), which, surprisingly, does not depend directly on population size or organic matter reactivity (see also Lovley *et al.*, 1982):

$$S = \frac{k_D K_m}{YV_{max} - k_D} \tag{2.15}$$

To solve for population density, Equation 2.14 is rearranged to yield

$$V_{max}\frac{SN}{(K_m + S)} = k_c C \tag{2.16}$$

This is then substituted into Equation 2.13 to yield

$$N = \frac{YCK_C}{k_D} \tag{2.17}$$

Therefore, the size of a microbial population depends only on the growth yield, Y, the availability of organic carbon, C, carbon reactivity, k_C, and the coefficient describing death rate, k_D.

In what follows, we use Equation 2.17 to rationalize population sizes and substrate levels in natural settings. We assume a spherical cell diameter of

1 μm, a cell density of 1 g cm^{-3}, a cell dry weight of 30% and that 40% of the dry weight is organic carbon. These values yield a cell dry weight of 1.56×10^{-13} g cell^{-1} and a carbon content of 5.2×10^{-6} nmol C cell^{-1}. We further assume that all organic carbon mineralization proceeds through acetate, and with acetate as a carbon substrate, V_{max} values for sulfate reducers range from 830 to 9600 μmol g^{-1} h^{-1}, or, with the cell carbon content above, 3.1×10^{-6} to 36×10^{-6} nmol C cell d^{-1} (Widdel, 1988). Typical growth yields, Y, for acetate-utilizing sulfate reducers range from 4 to 10 g cell dry mass per mol acetate dissimilated (Widdel, 1988), or 0.13 to 0.33 moles of cell carbon per mole acetate. Utilizing the cell carbon content above, 25,000 to 63,500 cells are formed per nmol of acetate used.

We furthermore assume an active organic carbon content in the sediment of 0.5 wt %, which, with a porosity of 0.8 and a dry sediment density of 2.5, yields 2×10^5 nmol C cm^{-3}. Finally, typical K_m values for acetate-utilizing pure cultures of sulfate reducers lie in the range of 64 to 240 μM, while for mixed populations in anoxic sediments, K_m values for acetate utilization are much lower, around 3 to 5 μM (Lovley and Klug, 1986; Widdel, 1988). We will take these mixed population results as most representative, and with a sediment porosity of 0.8, they translate into K_m values of around 2.5 to 4 nmol cm^{-3}. We can now begin to use the equations. We set V_{max}, K_m, Y and C to the values shown in Table 2.2 (see also above). In our first example, we look at marine surface sediments of the continental margin, where values of k_C range from 3×10^{-3} to 3×10^{-4} d^{-1} (Boudreau, 1997), while values of k_D, though largely unexplored, are probably in the range of 0.1 to 0.01 d^{-1}. Taken together, acetate concentrations of 0.06 to 0.6 nmol cm^{-3} are indicated, with population densities of 2×10^8 to 2×10^9 cells cm^{-3} (Table 2.2).

Table 2.2 Values used for modeling substrate concentrations and population densities

V_{max}	2.5×10^{-5} nmol C cell^{-1} d^{-1}
K_m	4 nmol$_{acetate}$ cm^{-3}
Y	30,000 cells nmol$^{-1}_{acetate}$
k_D	Variable
k_C	Variable
C	Variable

Some results for continental margin scenario:

k_D (yr^{-1})	k_C (yr^{-1})	S (μM)	N (cells cm^{-3})
0.1	0.003	0.06	1.8×10^8
0.01	0.003	0.054	1.8×10^9
0.01	0.0003	0.054	1.8×10^8

Both of these predictions are within the range of observations (see also Chapter 3 for a discussion of substrate concentrations).

With increasing sediment depth, one might anticipate a decrease in the concentration of both organic carbon, C, and growth yield, Y, each of which could lower population numbers. Furthermore, population numbers would also decrease if the death rate constant, k_D, decreased more slowly than the reactivity of organic carbon, k_C. We take deep sediments from the Japan Sea as an example (Parkes *et al.*, 2000). As mentioned previously, thymidine incorporation rates deep in these sediments suggest cell-doubling times of about 100 years. If the population is at steady state, then a death rate constant of about $2 \times 10^{-5}\ d^{-1}$ is calculated. Sulfate reduction rates measured with radiotracer are very low, approximately 0.002 nmol $cm^{-3}\ d^{-1}$ (Parkes *et al.*, 2000), which translates into a k_C value of about $1 \times 10^{-7}\ d^{-1}$, assuming a reactive C content of about 20,000 nmol cm^{-3} (around 0.05 wt % with a porosity of 0.6). If we retain the value of Y from Table 2.2, then from Equation 2.17 a population density of 2×10^6 cells cm^{-3} is calculated, which is very close to the measured cell numbers (Parkes *et al.*, 2000). We also calculate an unreasonably low acetate concentration of around 1.4×10^{-4} nmol cm^{-3}. In deeply buried sediments, and in other natural environments with severe substrate limitation, substrates other than organic carbon may limit growth and activity, and the growth parameters for organisms may differ from our assumed values. Furthermore, in deeply buried deep-sea sediments thermogenic processes increase acetate concentration (Wellsbury *et al.*, 2000). All of these factors could elevate acetate concentrations over our predicted values.

4. ENVIRONMENTAL EXTREMES

Microorganisms in nature are found in an amazing spectrum of environments representing extreme departures from average Earth-surface conditions. However, as organisms are usually well adapted to their environment, what seems extreme to us may be quite comfortable for the organisms living there. Therefore, what is extreme is a matter of perspective. In the following, we focus on the most common departures from average Earth-surface conditions and consider some of the special biochemical and physiological adaptations microorganisms use to adapt to these environments. We do not consider all of the various extreme circumstances under which microorganisms can live. The discussion that follows is rather general, focusing on adaptation strategies; examples of specific microorganisms living under extreme conditions can be found in the individual Chapters on elemental cycling.

4.1. Temperature

The growth rates and metabolic rates of microorganisms respond to temperature in a profound way. The response of an individual organism to temperature changes is usually defined with three cardinal temperatures: a minimum growth temperature, an optimal growth temperature, and a maximum growth temperature (Figure 2.5). Typically, the optimal growth temperature is relatively close to the maximum growth temperature, and the steep drop between the two represents the influence of high temperatures on protein and membrane stability. Ultimately, as the optimal temperature is exceeded, repair mechanisms cannot keep up with the damage imposed by the high temperatures. At low temperatures, enzyme systems are slow, and membrane stiffening reduces the activity of membrane-bound transporter enzymes. Growth rates and metabolic rates follow similar temperature responses, although microorganisms can often metabolize at temperatures somewhat outside of their growth temperature range (Figure 2.5) (Isaksen and Jørgensen, 1996).

Increasing growth rates and metabolic rates, as observed in moving from the minimum to the optimal growth temperature, result from increasing

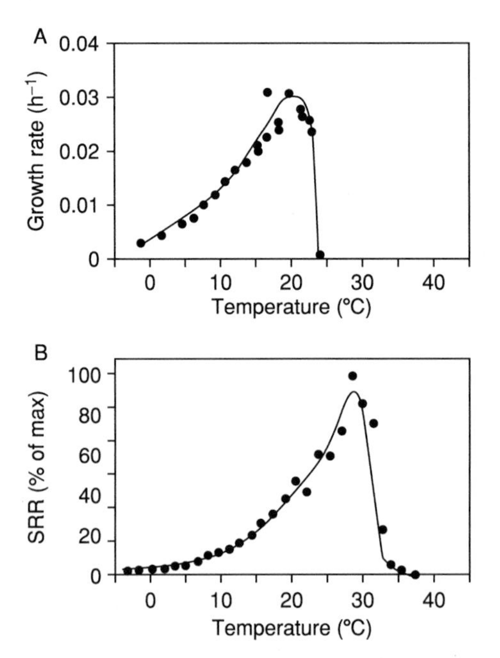

Figure 2.5 Growth rates (A) and sulfate reduction rates (B) as a function of temperature for a psychrophilic sulfate reducer isolated from Norsminde Fjord, Denmark. Redrawn from Isaksen and Jørgensen (1996).

enzyme activity. Within this temperature range both growth rates and metabolic rates are often modeled according to the Arrhenius equation:

$$v = Ae^{-\frac{E_a}{RT}} \tag{2.18}$$

where v is rate (growth rate or metabolic rate), A is a constant, E_a is an apparent activation energy (kJ mol^{-1}), R is the gas constant (8.31 J K^{-1} mol^{-1}), and T is temperature (K). Taking the natural logarithm of Equation 2.18, we find that:

$$\ln v = \ln A - \frac{E_a}{RT} \tag{2.19}$$

Therefore, a plot of ln v vs. $1/T$ will yield a slope of E_a/R, from which E_a can be calculated (Figure 2.6). Note that the Arrhenius equation was originally formulated to represent the kinetic response of pure chemical reactions to temperature, and in this case, E_a has a chemical meaning. When used in microbial systems, calculated E_a values are empirical parameters representing the total metabolic response of an organism to temperature with no grounding in chemical first principles.

A frequently quoted parameter is the Q_{10} response of an organism, which represents the proportional increase in metabolic rate or growth rate with a 10 °C increase in temperature. The Q_{10} response is related to the activation energy (E_a) by the following expression, where T_1 is the reference temperature (°K) and $T_2 = T_1 + 10$:

$$Q_{10} = e^{E_a(T_2 - T_1)/RT_1T_2} \tag{2.20}$$

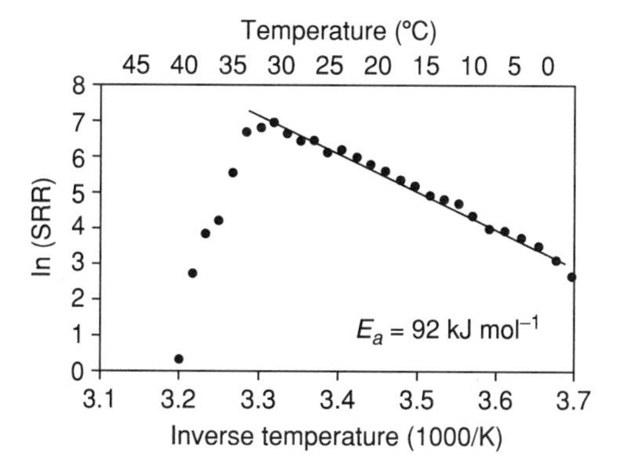

Figure 2.6 1/T vs. ln(sulfate reduction rate) for the same sulfate-reducing culture as in Figure 2.5. From the slope of this relationship, the activation energy, E_a, can be calculated (see Equation 2.18). Redrawn from Isaksen and Jørgensen (1996).

Activation energies for microbial populations, expressing either growth or metabolic rate, generally fall in the range of 50 to 110 kJ mol^{-1}, corresponding to Q_{10} values ranging from around 2 to 5. Note that with a constant E_a, Q_{10} values change slightly through the temperature range of growth.

Taken together, microbes in nature are known to function from subfreezing temperatures to 121 °C (see also Chapter 1). Organisms with maximum growth rates at <15 °C are known as psychrophiles, organisms with maximum growth rates from 15 °C to 45 °C are known as mesophiles, and organisms with maximum growth rates from 45 °C to 80 °C are thermophiles. Organisms with growth optima above 80 °C are known as hyperthermophiles (Table 2.3). Organisms living at the high and low temperature extremes require special biochemical adaptations in order to survive. For example, psychrophiles must maintain cell membrane fluidity in the face of low temperatures; they do this by modifying the membrane lipid composition

Table 2.3 Nomenclature describing different environmental adaptations

Environmental circumstance	Different adaptations	Notes
Temperature		
	Psychrophile	Max growth <15 °C
	Mesophile	Max growth 15 to 45 °C
	Thermophile	Max growth 45 to 80 °C
	Hyperthermophile	Max growth >80 °C
pH		
	Acidophile	Max growth pH < 5
	Neutrophile	Max growth pH 6 to 8
	Moderate alkaliphile	Max growth pH 8 to 9.5
	Obligate alkaliphile	Max growth pH >9.5
Salt		
	Mild halophile	Max growth 1 to 6% NaCl
	Moderate halophile	Max growth 6 to 15% NaCl
	Extreme halophile	Max growth >15% NaCl
Oxygen		
	Aerobe	
	Obligate	O_2 required
	Facultative	O_2 not required but preferred
	Anaerobe	
	Obligate	Cannot grow with O_2 present
	Facultative	Can tolerate O_2, but grows best without

(e.g., Scherer and Neuhaus, 2002). Specific adaptations include, but are not restricted to, synthesis of a higher proportion of unsaturated fatty acids in the membrane lipids and carbon-chain shortening. Enzymes tend to be more polar, with fewer hydrogen bonds and fewer ion pairs, which reduce hydrophobic interactions between enzyme subunits. Together, these adaptations allow greater enzyme flexibility in the cold. In addition, ribosome structure in psychrophiles is modified compared to mesophiles to aid protein synthesis at low temperatures (Scherer and Neuhaus, 2002).

At the other end of the temperature spectrum, thermophiles, and especially hyperthermophiles, have special adaptations to high temperatures. Compositional and structural changes in enzymes lead to higher thermal stability (e.g., Jaenicke and Sterner, 2002). Proteins have increased numbers of ion pairs and more hydrophobic interiors, and these adaptations, either singly or in combination, help resist unfolding. Nucleic acids tend to denature at high temperatures. For RNA, organisms may respond by increasing the G + C content, which imparts more stability. For DNA, special proteins may be produced that help to stabilize the DNA structure. Fast-acting repair systems are also utilized. In addition, the lipid composition of the cytoplasmic membrane may be heat stabilized by incorporating a high proportion of saturated fatty acids. This is by contrast with the psychrophiles, whose membrane lipids, as discussed above, incorporate short-chained unsaturated fatty acids and chain shortening to increase membrane fluidity.

4.2. pH

Individual organisms generally have a pH tolerance of 2–3 units, and most prokaryotes live with growth optima in the pH range of around 6 to 8. These are known as neutrophiles (Table 2.3). However, acidic environments, including sulfidic hot spring and acidic mine waters, may also house active microbial populations with growth optima at pHs of <6; organisms living under such conditions are known as acidophiles. Indeed, growth is known among iron-oxidizing and heterotrophic *Archaea* at pHs down to zero (e.g., Schleper *et al.*, 1995; Edwards *et al.*, 2000), which would seem to be the record low pH for microbial growth. Obligate alkaliphiles, living for example in soda lakes and alkaline soils, have optimal growth at pH values above around 9.5 (Krulwich, 2000).

Organisms growing under extremes of pH, both high and low, maintain cytoplasmic pH values within the neutrophilic range, and they face special problems in doing this in the face of strong pH gradients across the cell membrane. Several factors contribute to cytoplasmic pH regulation. The pH-buffering capacity of the cytoplasm itself helps to regulate pH, with buffering coming from the phosphate groups associated with RNA and

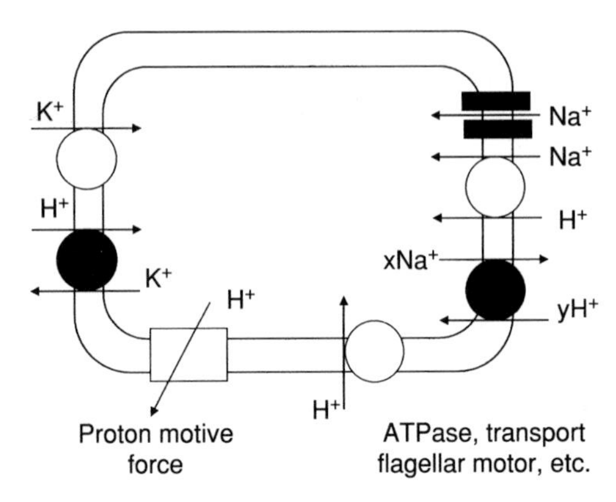

Figure 2.7 The main factors regulating the pH of a cell. The pH-sensitive processes are given by the darkened circles. Protons are translocated across the cell membrane generating a proton motive force used, for example, to generate ATP and to drive flagellar motion. An imbalance in H^+ concentration within the cell cytoplasm can be regulated by the antiport input of H^+ into the cell balanced by K^+ export, where K^+ is regenerated through K^+ import. The K^+/H^+ antiport is pH sensitive, and it is used to regulate cytoplasmic pH in acidophilic organisms. In alkaliphiles pH regulation is controlled by a Na^+ circuit. A pH-sensitive antiport exports Na^+ from the cell, replacing it with H^+. The amount of H^+ imported exceeds the Na^+ exported, to yield acidification of the cell cytoplasm. Na^+ replacement occurs through either a separate symport or a pH-sensitive Na^+ channel. Inspired from Booth (1999).

DNA and protein-associated side-chain amino acids. In addition, for acidophiles, a K^+ ion circuit regulates H^+ entry into the cell, and this circuit is pH sensitive (Figure 2.7). For alkaliphiles the internal pH is lower than the external pH, and these organisms couple the import of H^+ into the cell with the export of Na^+ from the cell (Booth, 1999) (Figure 2.7). For both acidophiles and alkaliphiles, surface-bound proteins are well adapted to the pH of the external environment, and cell membranes are also adapted to the external pH. Thus, when strongly acidophilic prokaryotes are exposed to neutral pH their membranes typically dissolve, demonstrating a strong requirement for low pH in these organisms.

4.3. Salt

All microorganisms maintain lower activities of water (a_w) within their cells compared to the external environment. Thus, water will tend to diffuse into the cell, establishing an osmotic pressure (turgor) that is necessary for cell growth and that must be maintained. Lower cytoplasmic water activity is established with a higher solute concentration in the cell compared to the external environment, and the solutes used are known as compatible solutes. Compatible solutes are either organic compounds or inorganic ions like K^+, which have several special properties. They must be highly soluble, and they must not interfere with enzyme activity in the cell. When organic compounds are used, they are frequently synthesized within the cell but may also be available from the environment. Obviously, the challenge to produce low water activity within the cell, and to produce high concentrations of compatible solutes, becomes more acute as the salinity of the external environment increases. Despite these hardships, diverse populations of halophilic organisms (see Table 2.3) are found at salinities up to halite (NaCl) saturation, corresponding to about 35% salt by weight.

We consider now how compatible solute choice influences the energetics and microbial ecology of halophilic organisms. When K^+ is used as a compatible solute (members of the aerobic archaeal family *Halocteriaceae*, and *Bacteria* from the order *Halanaerobiales*) (Oren, 1999), only a modest amount of energy is used to accumulate K^+ into the cell, around 1 ATP per 1.5 to 2 moles of KCl. The internal cellular biochemistry must, however, adapt to function in a high salt environment (Oren, 1999). By contrast, when organic compounds are used as compatible solutes, the energetic costs to the organism are huge, as one molecule of compatible solute requires between 30 to 109 ATPs (Oren, 1999). However, with compatible organic solutes, the cell cytoplasm is a relatively low-salt environment, and special biochemical adaptations to high salt are not required. Despite this modest advantage, due to the high-energy cost, organisms utilizing organic compounds as compatible solutes may face problems in obtaining enough energy to grow if they conduct low energy-yielding metabolisms. Microorganisms fitting into this category are methanogens forming methane from acetate or H_2 plus CO_2 (see Chapter 10), nitrifiers, sulfate reducers using acetate, and homoacetogens reducing CO_2 with H_2 to produce acetate (Oren, 1999). Indeed, no organisms conducting these metabolisms have been isolated, nor have their metabolisms been detected, at the high salt concentrations at which more energetic metabolisms are active (Oren, 1999).

5. POPULATION ECOLOGY

We wish to understand the population ecology of microbial ecosystems in the same way we do for macroscopic ecosystems. Thus, we wish to know what populations are present in an ecosystem, the population size, and how populations interact with each other. Basically, we would like to answer the simple question of who is doing what. Unfortunately, we have only a rudimentary understanding of the species diversity of microbial populations in nature. Part of the problem is that microscopic observations and culturing techniques give only a small glimpse of the diversity of the ecosystem. We know this because DNA extraction and amplification records a far greater diversity, but, unfortunately, we also obtain mostly unknown molecular isolates whose functional role in an ecosystem is unknown. Another part of the problem, as we shall see below, is that even an understanding of diversity does not tell us the activity level of individual population members.

However, the situation is not hopeless. Even though we cannot yet adequately describe the cast of players in a microbial ecosystem or their individual level of activity, we have a reasonably good appreciation for the main processes present, or in other words, the phenotypes represented in the population. Our phenotypic understanding of microbial ecosystems is based on process rate measurements such as carbon fixation rates, sulfate reduction rates, and nitrogen turnover rates, and on chemical profiles that respond to the activity of various metabolic pathways. Furthermore, we can frequently observe the presence of conspicuous population members such as cyanobacteria, sulfide oxidizers, and anoxygenic phototrophs. Our understanding of phenotypic diversity is also based on the exploration of model systems, like industrial fermenters, where many of the main pathways of carbon mineralization have been elucidated (see Chapter 3). Indeed, the phenotypes present in similar microbial ecosystems from around the world are likely quite similar, even though their species composition is unknown and probably quite variable.

In what follows, we summarize important aspects of the population ecology of microbial ecosystems. We discuss microbial diversity, and we describe some of the principal ways in which microbial populations interact. Finally, we explore microbial behavior, and we describe how microbial populations act in coordinated efforts to benefit the individual population, or the entire ecosystem.

5.1. Aspects of microbial diversity

5.1.1. General considerations

We begin with a few comments on microbial diversity. As for populations of macroscopic organisms, microbial populations under stress tend to show less diversity than unstressed populations (Atlas and Bartha, 1998; McCaig *et al.*, 1999). Furthermore, microbial populations living in extreme environments tend to display lower functional diversity, which probably also translates into lower species diversity. For example, photosynthesis is not sustained at temperatures above around 70 °C (Brock, 1994), and as we saw above, numerous metabolisms such as methanogenesis from CO_2 reduction with H_2 and acetate fermentation, nitrification, and sulfate reduction with acetate appear to be absent at high salt concentrations. Also, recent studies have demonstrated very limited species diversity in low pH acid mine systems (Tyson *et al.*, 2004). Outside of these generalizations, our appreciation for the true diversity of microbial populations in nature is limited by our ability to adequately quantify diversity.

5.1.2. Species diversity from molecular studies

Molecular techniques have provided evidence for far greater microbial diversity than previously imagined (e.g., Ward *et al.*, 1990; Pace, 1997) (see Chapter 1). This is satisfying as we come further in understanding the real diversity of microbial populations in nature, but is also frustrating, as most of the diversity represents unknown organisms. Therefore, we do not know the function of most of these unknown organisms in the ecosystem.

Nevertheless, a number of studies have attempted to evaluate species diversity from a growing body of molecular data, most typically 16S rRNA sequences (e.g., McCaig *et al.*, 1999; Nübel *et al.*, 1999; Hughes *et al.*, 2001). In evaluating diversity, one usually compares 16S rRNA sequences where molecular "species" are defined as different if the sequences are more than 97% different. This is an operational definition of a species, and "species" classified this way are often given the name of operation taxonomic units (OTU). From species abundance and frequency data, diversity indices can be calculated. One such index is the Shannon-Weaver index (H'), which provides an indication of the uncertainty in predicting the identity of a population member if one is chosen at random. The more diverse the population, the more uncertain is the identity of a randomly chosen individual (e.g., Pielou, 1969). The Shannon-Weaver index is expressed as

$$H' = -\sum_{i=1}^{s} p_i \ln p_i \tag{2.21}$$

where s is species and p_i is the proportion of the sample belonging to the ith species. The higher the value for H', the more diverse the sample. Species richness (d) is a further diversity index of interest:

$$d = \frac{s - 1}{\log N} \tag{2.22}$$

where s is the number of species and N is the number of individuals. With this index, a higher diversity occurs when there are a large number of species relative to individuals, particularly when N is large.

Of the microbial diversity studies utilizing molecular data, the study of Nübel *et al.* (1999) is noteworthy, as diversity was evaluated for a specific class of organisms, the cyanobacteria. Furthermore, diversity was estimated using not only 16S rRNA sequence data, but also microscopic identification of distinct cyanobacterial morphotypes and the extraction, separation, and identification of individual carotenoids. Eight different microbial mats were explored, and each of the three approaches used to identify cyanobacterial diversity gave a similar picture of diversity when comparing between sites. Other diversity studies have been more general, usually targeting a broad spectrum of the prokaryote world (e.g., McCaig *et al.*, 1999), but nevertheless also showing interesting differences in diversity when comparing between sites.

Even molecular studies, however, do not provide a true picture of microbial diversity, as there are often difficulties in extracting and amplifying DNA fragments from nature. This leads to an underestimate of true diversity. Furthermore, as quantitative amplification procedures are not yet fully developed, minor or even dormant members of the population will appear as

What is a prokaryote species?

According to the concept of species, individuals of a common species can reproduce with each other, but not with individuals of another species. In this way the gene pool of a given species is constantly mixed and shared, and any developing evolutionary innovations are rapidly dispersed. What then about prokaryotes who do not reproduce sexually and whose individuals do not swap genes? It would seem that under these circumstances localized populations could evolve independently of other populations, even those populations originating from the same immediate ancestor, if they are exposed to different environmental circumstances. Indeed, gene acquisition by lateral gene transfer operates locally, which should ensure that individual populations of similar heritage, but spatially displaced, will potentially evolve unique genomes with unpredictable trajectories. In an interesting example, whole genome analysis of two strains of *E. coli* (the pathogenic 0157:H7 strain and the non-pathogen K-12) revealed that about one-third of the genes are different and that many of the gene differences have probably arisen from rather recent gene transfer (Perna *et al.*, 2001).

It would seem that an infinite number of trajectories might be possible from any individual prokaryote, encompassing all possible metabolic capabilities. However, we know from the Tree of Life that a great deal of metabolic relatedness often accompanies phylogenetic relatedness based on SSU rRNA comparisons (see also Chapter 1). We also know that an individual "species" of sulfate reducer (as defined by SSU rRNA sequence), for example, will normally share a remarkably similar physiology (phenotype) to the same "species" from another part of the globe, although the two are countless generations removed. Why is this so? It seems that the answer might be at least partly ecological. We can conjecture that an individual prokaryote "species" is keenly adapted to a particular niche in the environment. If the population representing the species were to acquire too many new traits it might face stiff competition from organisms already better adapted to these particular traits. So, the population, or at least most of it, continues doing what it is best at. In this view, a "species" is stabilized genetically by particulars of environmental adaptation. Nevertheless, as we have seen with the example of *E. coli*, "species" can and do evolve quite different genomes.

important as major population members. Thus, it is difficult to quantify the involvement of individual community members within the ecosystem. Furthermore, most molecular isolates from nature are from organisms not yet cultured and whose physiology is therefore uncertain. This means our understanding of functional diversity lags behind even our understanding of species diversity.

5.1.3. Fluorescent in situ *hybridization (FISH)*

The exploration of community structure with fluorescent *in situ* hybridization (FISH) offers a different type of window into microbial diversity. FISH probes consist of a small stretch of nucleic acid attached to a compound that fluoresces when excited with UV light (Amann *et al.*, 1995). The nucleotide sequence of the probe is chosen to complement a stretch of SSU rRNA to which the FISH probe is hybridized. Under a fluorescence microscope, one sees the ribosomes in individual cells to which the FISH probe has bound. FISH probes are designed from known SSU rRNA sequences, and after examining sequences of related and distant organisms, small stretches that are unique, or nearly unique, to the target organisms of interest can often be identified. Thus, FISH probing provides a straightforward way to differentiate between members of the principal domains in environmental samples. Sequence stretches that are specific to closely related groups of organisms, or even individual species, can also often be identified.

The FISH technique can, in principle, provide a quantitative understanding of population structure, and FISH has yielded some spectacular results. For example, FISH probes have beautifully illustrated the three-dimensional structure of the sulfate-reducing and methanogenic populations believed responsible for anaerobic methane oxidation in marine sediments (Boetius *et al.*, 2000) (see Chapter 10). FISH has also revealed the close physical relationship between ammonia-oxidizing and nitrite-oxidizing chemolithotrophs (Schramm *et al.*, 1996) (see Chapter 7). These two populations are responsible for ammonia oxidation to nitrate. FISH has also revealed the close association between sulfide oxidizers and sulfate reducers as endosymbionts in oligochaete worms (Dubilier *et al.*, 2001).

Despite these successes, and many more, there is still a limit to what FISH probes can provide in defining microbial community structure. For example, many FISH probes are rather unspecific, and they hybridize with organisms outside of the intended group. This, however, is more a problem of probe design than an inherent problem with FISH. Other difficulties with FISH arise from a poor understanding of the diversity of microbes in nature. If probes are used to specify for related groups of organisms they will frequently hybridize with some organisms that have yet to be isolated (most organisms have not). We cannot be sure whether these uncultured organisms behave similarly to their cultured relatives. Furthermore, since only a small percentage of natural microbial diversity has been identified, we cannot anticipate the appropriate nucleotide sequence, much less the physiology, of organisms that are currently unknown.

5.2. Microbial interactions

From years of careful observation, a wide variety of different types of microbial interactions can be described in nature. The types of interactions are numerous, often complex, and highly interesting. According to the list of interactions presented in Table 2.4, they can be separated into seven general types depending on how the two interacting populations are affected. Furthermore, when the interactions become very intimate, and are obligate for one or both of the populations, another type of interaction, symbiosis, is also observed. We here explore the different types of interactions with examples from the microbial world.

5.2.1. Competition

Competition for available resources, inducing negative effects on the participating populations, is intense in the microbial world. For example, organisms with similar overall physiologies such as sulfate reducers compete with

Table 2.4 Microbial interactions in nature

	Effect of interaction	
Type of interaction	Population 1	Population 2
General		
Competition	−	−
Synergism, syntrophy	+	+
Predation	+	−
Parasitism	+	−
Commensalism	0	+
Amensalism	0	−
Neutralism	0	0
Obligate		
Symbiosis		
Commensalism		
Mutualism		
Parasitism		

After Atlas and Bartha (1998), with modifications.

each other. They can do so by fine-tuning their growth to the particulars of substrate availability and thereby exclude other population members with similar substrate requirements. In chemostat experiments with two populations sharing the same substrate, the population with the highest specific growth rate (a function of both K_s and μ_{max}) eventually will outgrow the other population (Jannasch, 1967). The less successful population will wash out of the chemostat, leaving the successful population behind. The successful population under substrate limitation could be the one with highest substrate affinity (lowest K_s), or when substrate is non-limiting, the population with the highest μ_{max} (Figure 2.8).

As a strategy to increase competitive fitness, prokaryotes in nature tend to be generalists, relying not on one, but on a variety of different substrates (or even types of metabolisms) for survival. In this way, they can compete on several fronts. Population members might also adapt to very specific substrates available only in limited abundance, and for which competition is not keen. In another strategy, members of the population might stay dormant until appropriate conditions materialize. For example, heterotrophs with low substrate affinity, but also high maximum growth rates (μ_{max}), might wait for a sudden input of organic carbon before actively metabolizing. In this way they can, at least temporarily, out-compete organisms with high substrate affinity but lower maximum growth rates (see above).

Organisms with very different physiologies might also compete for substrate. As is shown in Chapter 3, certain anaerobic populations tend to

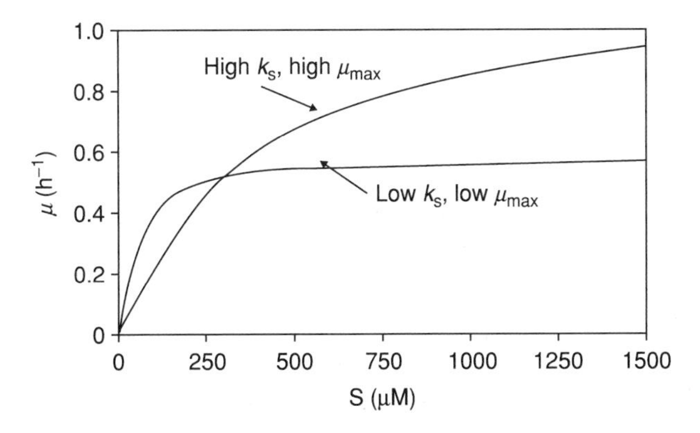

Figure 2.8 Demonstration of how various combinations of maximum growth rate (μ_{max}) and substrate half-saturation constants (K_s) can prove advantageous to growth for different organisms depending on substrate availability. Thus, at low substrate concentrations, the organism with a low K_s (50 μM) and low μ_{max} (0.6 h^{-1}) is favored, while at high substrate concentrations, the organism with a high K_s (400 μM) and high μ_{max} (1.2 h^{-1}) is favored.

survive with the minimal amount of energy available for the maintenance of metabolic functions. Although this may seem like a poorly chosen lifestyle, it serves to remove substrates (such as acetate and H_2) to very low concentration and therefore to exclude other populations with metabolisms gaining less or no energy at these low substrate levels. Thus, iron reducers, for example, can maintain the available substrates in the environment at their survival level, providing insufficient energy for sulfate reducers and methanogens to survive.

Competition might not always be based on energy gain. Some organisms with different metabolisms may compete for a common substrate, with the outcome depending on the physical characteristics of the environment. For example, anoxygenic phototrophic purple sulfur bacteria oxidize sulfide phototrophically (see Chapter 9), and the so-called colorless-sulfur bacteria oxidize sulfide in a chemoautotrophic process by the reaction of sulfide with oxygen or nitrate. Both of these types of organisms are found, for example, closely associated within the sulfide–oxygen interface in microbial mats (Jørgensen and Des Marais, 1986). Whether sulfide oxidation is dominated by phototrophic organisms or non-phototrophic sulfide-oxidizing organisms depends on the availability of light in the near-IR range at the sulfide–oxygen interface. In the microbial mats studied by Jørgensen and Des Marais (1986), IR light penetrated deeper, favoring phototrophic sulfide oxidation, in the mat with the most loosely packed overlying oxygenic phototrophic population.

In another example, chemoautotrophic anammox bacteria, oxidizing ammonia with nitrate, compete with heterotrophic denitrifiers for nitrate in anoxic settings (see Chapter 7). The criteria controlling this competitive interaction in sediments, with steep opposing gradients of critical chemical constituents such as oxygen, nitrate, and ammonia, are not well understood. However, there is a general tendency for anammox to be relatively more important when oxygen penetration is deeper (Thamdrup and Dalsgaard, 2002). In this case, a thicker anoxic nitrate-containing zone is probable, which could benefit anammox bacteria. This is because anammox bacteria are strict anaerobes, whereas many denitrifying strains can withstand micro-oxic conditions. With active carbon mineralization and shallow oxygen penetration, nitrate does not significantly penetrate into the anoxic zone, restricting the activity of anammox organisms. Recent work has shown that the relationship between anammox and denitrification turns from competition to commensalism (see below) in thick anoxic nitrate-containing water columns, such as what might be found in oceanic oxygen minimum zones. Here, the ammonia liberated from organic nitrogen during denitrification is supplied to anammox bacteria, producing a tight coupling between these two processes (Dalsgaard *et al.*, 2003).

5.2.2. Synergism and syntrophy

A synergistic relationship is one in which two members of a population benefit from each other's existence, but the relationship is not obligatory (Table 2.4). When the relationship is one of nutritional interdependence, it is referred to as syntrophy. A good example of syntrophy in microbial populations is the coupling between organisms producing H_2 and those consuming it. This relationship is known as interspecies H_2 transfer (see Chapters 3 and 5). Hydrogen is produced during the fermentation of organic compounds in anoxic environments. As H_2 is a reaction product, its accumulation in the environment renders the fermentation reactions less thermodynamically favorable, until finally, with high enough H_2 partial pressures, fermentation stops altogether (see Chapter 3). However, H_2 is also an excellent electron donor for a variety of anaerobic respiration processes, including metal oxide reduction (Fe and Mn), sulfate reduction, and methanogenesis. Thus, fermentation produces a substrate beneficial to a variety of respiring anaerobes, and consumption of H_2 by these organisms reduces the H_2 partial pressure, allowing the fermentation to continue.

In another example of probable syntrophy, sheathed *Thioploca* spp. filaments contain sulfate reducers of the genus *Desulfonema* (Fossing *et al.*, 1995). *Thioploca* is a sulfide oxidizer utilizing nitrate as the electron acceptor (see Chapter 9). Therefore, *Thioploca* (probably) supplies organic matter to benefit *Desulfonema,* while *Desulfonema* supplies sulfide to benefit the *Thioploca.*

5.2.3. Predation and parasitism

A predator is an organism that feeds on other organisms, and many types of protozoans are the principal predators of prokaryotes in nature. Protists actively engulf prokaryotes (or other food particles) in a process known as phagocytosis, in which a food particle (e.g., prokaryote) is "consumed" in special feeding organelles located at the cell surface. Prokaryotes are delivered to the protist by filter feeding, direct interception, or passive diffusion (Fenchel, 1987). Filter feeding is accomplished by the active transport of water through a filter of cilia, or rigid tentacles on the surface of the protist, which strain small cells (and other food particles) from the environment. In direct interception, fluid flow within the medium carries particles to the surface of the feeding protist. When feeding by passive diffusion, food particles migrate to the protist, either by Brownian motion or through the prokaryote's own motility. Once ingested, food particles form a vacuole, which fuses to membrane-bound enzyme sacs called lysosomes, accomplishing the digestion of the particle.

The impact of protozoans on prokaryote populations can be enormous. In coastal waters, the abundance of flagellated protozoa is sufficient to filter from 10 to 100% of the water column prokaryote population every day (Fenchel, 1987). Such efficient removal keeps prokaryote populations relatively low, allowing the persistence of rapid growth rates. Protozoans are also important feeders of prokaryotes in sediments consisting of well-sorted sand with minimal clay and silt. Here, the protists have ample room to move, and they feed in the interstitial space. However, in fine-grained silts and clays there is insufficient room for protozoan motility, and as a result they are concentrated in the flocculent sediment surface layer, while generally absent in the deeper sediment layers (Fenchel, 1987).

Some prokaryote populations exhibit what may also be considered predatory behavior. For example, various myxobacteria (members of the δ-subdivision of the proteobacteria) thrive in nature by lysing living cells with a variety of hydrolytic exoenzymes, and taking up the cell constituents released. However, these organisms do not depend on living tissues, and they are therefore generally referred to as scavengers rather than predators (Reichenbach and Dworkin, 1992).

Numerous different prokaryotic parasites are disease-causing agents in plants and animals, and some prokaryotes can parasitize other prokaryotic organisms. For example, the gram-negative bacteria *Bdellovibrio* actively hunts and kills its prey to accomplish its parasitic lifestyle. In the attack phase, it is a flagellated non-reproductive cell that enters the periplasmic space of other gram-negative host cells (Figure 2.9). There it loses its flagellum and grows, feeding off the host, into a reproductive septate filament. After growth ceases, the filament separates, lysing the host and releasing a number of individual attack cells, ready to repeat the cycle (Dworkin, 1992).

Viral infection is another sort of parasitism, and it is a major cause of prokaryote mortality in nature. When a virus infects a prokaryote cell, it is known as bacteriophage. Viruses are basically very small (20 to 300 nm diameter) sacs of double- or single-stranded DNA and/or RNA, bound in a protein membrane. As they have no metabolism and cannot replicate on their own, they are not considered life as it is normally defined. In simple terms, viruses use the host's metabolic machinery, in combination with their own genomes, to replicate. To begin a viral attack, the virus needs an appropriate receptor site on which to attach, and after attachment, the viral genome is injected into the host, leaving the protein coat outside of the cell. With a virulent virus, one that destroys the host cell, the host's metabolic machinery is redirected into replicating the viral genome and into the assembly of the structural proteins forming the body of the mature phage particle. Nucleotides forming the host genome may also be harvested into the assembly of new viral genomic material. After the assembly of the phage

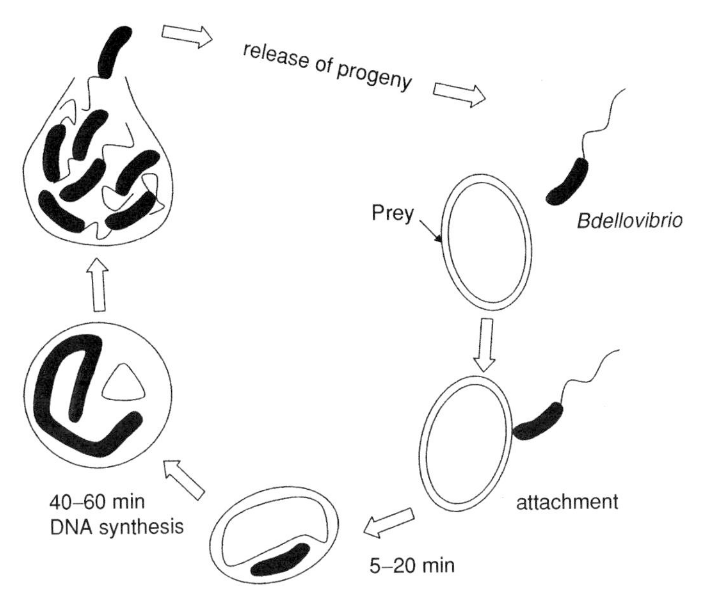

Figure 2.9 Predatory and parasitic lifestyle of *Bdellovibrio bacteriovorus*. *Bdellovibrio* can only reproduce after invasion into a host cell. Here it feeds off the host, grows, and divides into a number of new attack cells, which leave the now decimated host seeking new prey. Inspired from Madigan *et al.* (2003).

particle is complete, the host cell is lysed, releasing the many-formed new phage particles back into the environment.

Temperate viruses form another type of virus–host interaction. After entering into the cell, the virus may integrate into the host chromosome and replicate along with the host chromosome. In this way, the virus is dormant and does not influence the function of the host cell, except perhaps to inhibit the virulent expression of new incoming phage of the same type by expressing special repressor proteins. However, the same virus can enter a virulent stage, at which time the phage genome becomes expressed. New phage particles are formed, resulting in the lysis and death of the host cell. An excellent discussion of viruses and viral infection can be found in Madigan *et al.* (2003).

Active viral infection acts to control microbial population size in nature, and also contributes to the carbon cycle. Thus, after cell lysis following virus attack, cytoplasmic material is released as DOM (dissolved organic matter), which is quickly utilized by other bacteria, stimulating cell growth. This cycle is known as the "viral loop" (Riemann and Middelboe, 2002), and the DOM released as a result of virus infection can constitute a major source of substrate for bacterial growth in planktonic bacterial communities (Middelboe *et al.*, 2002; Riemann and Middelboe, 2002).

5.2.4. Commensalism

Commensalism is the interaction between populations in which one gains from the interaction and the other is unaffected (Table 2.4). In the microbial world, commensalism is mostly related to nutrition, that is, when metabolic products of one microbial population can be used by other microbial populations with no specific gain to the first population. A specific example of commensalism between denitrifiers and anammox bacteria has been presented (see above). Other examples include the production of reduced redox-sensitive species, which, when oxidized, can fuel the growth of other microbial populations. Thus, methanogens produce methane, which can be oxidized by methanotrophs (see Chapter 10), and sulfate reducers produce sulfide, which can be oxidized by a variety of sulfide-oxidizing organisms (see Chapter 9). The microbial loop represents commensalism between eukaryotic algae and aerobic prokaryotic heterotrophs. In this case the algae produce excretion products that are used by the prokaryote population (see Chapter 9).

5.2.5. Amensalism

Amensalism defines a relationship in which the activity of one population is harmful to another. For microbes, this typically results when the products of one type of metabolism are detrimental to another. Examples include the production of oxygen by cyanobacteria, inhibiting anaerobic organisms, or the production of inhibitory organic compounds as metabolic byproducts. For example, ethanol, a fermentation product, is inhibitory to many microorganisms, particularly at higher concentrations. The production of acid during sulfide oxidation, particularly in surface sediments where the pH can be driven very low, creates an extreme environment inhibitory to a wide variety of microorganisms. Some organisms also produce antibiotics that exclude other organisms. We stress, however, that while some microbial populations may be excluded in relationships of amensalism, other populations will thrive. Thus, anaerobes may be excluded in the presence of oxygen, but aerobes will thrive, and while ethanol may be inhibitory to some microbial populations, others can actively use it as a substrate. Also, while low pH might inhibit a great number of different microbial populations, a large number of populations are well adapted to exploit this circumstance.

5.2.6. Neutralism

Neutralism is a lack of interaction between microbial populations. This could occur if populations are spatially separated or if they promote different types of metabolisms that are not interrelated. For example, in oxic water columns aerobic heterotrophs probably exist without significant interaction with nitrifying bacteria (oxidizing ammonia liberated during organic matter mineralization) or methanotrophs (oxidizing methane from sediment sources, for example). These relationships, however, could become competitive in sediment environments in which oxygen becomes limiting. In general, it is difficult to identify examples of neutralism in nature due to the extensive interrelationships between microbial populations. Thus, at an oxygen–sulfide interface neutralism might be expected, for example, between chemolithoautotrophic organisms such as nitrifiers or sulfide oxidizers and heterotrophs such as sulfate reducers. However, on closer inspection, chemolithoautotrophic organisms produce organic matter that can fuel heterotrophic metabolism. Therefore, even in this situation, a relationship between the populations exists, and strict neutralism probably does not occur.

5.2.7. Symbiosis

Various definitions are used to describe symbiosis, and the borders between syntrophism, parasitism, and commensalism on one hand, and symbiosis on the other hand, are rather blurred. For our purposes, symbiosis is a sustained and intimate physical association between individual species. The interaction need not benefit both partners, as is commonly assumed, but when it does, the association is usually referred to as mutualism. Symbiotic partners can also engage in commensalism, in which one of the partners benefits and the other is unaffected by the relationship. Finally, the symbiosis may be parasitic, with one partner benefiting at the expense of the other. However, cases of parasitic symbiosis are rare in the prokaryote world but are common, for example, among plants. Thus, members of the non-photosynthetic plant family *Orobanchaceae* bury their roots into the tissues of their host plants to obtain nutrition for growth. The parasitic *Orobanchaceae* obviously benefits in this association, while the host is negatively impacted by yielding nutrition to the parasite.

Symbiotic relationships between prokaryote partners are apparently rather rare in nature (Overmann and Schubert, 2002), and while cases of sustained physical associations between prokaryotes can be found (Figure 2.10) (Overmann and Schubert, 2002), the nature of the association is often rather uncertain. A possible example of prokaryote symbiosis is presented above, where sulfate-reducing *Desulfonema* filaments are found within the sheaths of the sulfide oxidizer *Thioploca* sp., although the interdependence and specificity of this relationship have yet to be elucidated. Interdependency also is probable between numerous different consortia involving members of the anoxygenic phototrophic green sulfur bacteria. In this symbiosis, the

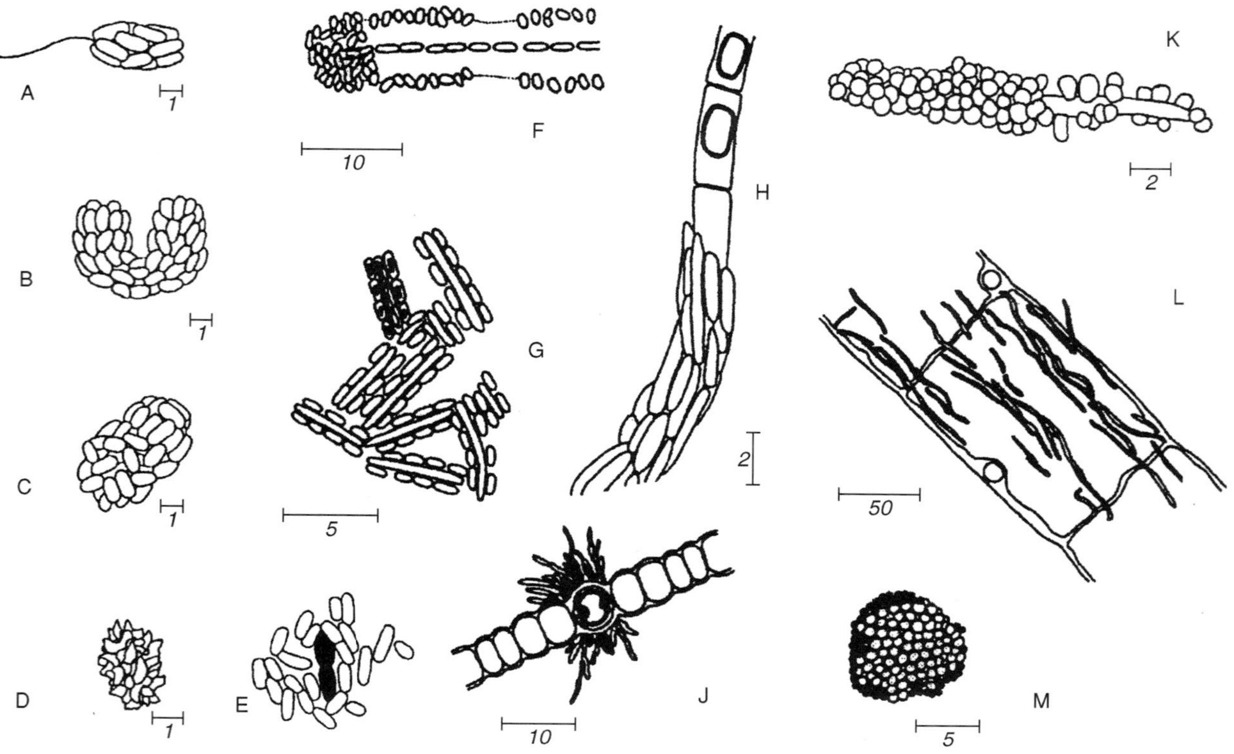

Figure 2.10 Examples of various consortia formed between prokaryotes in nature: (A) morphotype of *Chlorochromatium aggregatium* and *Pelochromatium roseum*; (B) *Chlorochromatium glebulum*; (C) morphology of *Chlorochromatium magnum* and *Pelochromatium roseo-viride*; (D) morphotype of *Chlorochromatium lunatum* and *Pelochromatium selenoides*;

phototroph forms the epibiont, completely covering a central bacterium, which at least in some instances comes from the β-subdivision of the proteobacteria (Overmann and Schubert, 2002) (see also Chapter 9). The exact relationship between these two consortia partners is not clear, so it is not known whether they engage in commensalism or mutualism. However, there is obvious signaling between the two partners, as the consortia exhibit a tactic response to the wavelength optima of the phototrophic partner, but motility is provided only by the flagellum of the central bacterium (see Chapter 9) (Overmann and Schubert, 2002).

Mutualistic symbiotic relationships between prokaryotes and eukaryotes are quite common. Cyanobacteria (as well as algae) join forces with fungi to form lichens. In this association, the phototroph is the primary producer, supplying organic matter and fixed nitrogen (when the phototroph is a cyanobacterium), and the fungus is the consumer, supplying nutrients back to the phototroph for further utilization. The fungus also provides protection for the phototroph. The interactions between the phototroph and fungus are quite specific and distinctive, and the lichens have their own genus and species names. Other mutualistic relationships include the partnerships formed from endosymbiotic sulfide-oxidizing bacteria and various animals including those from hydrothermal vents (see Chapter 4), marine oligochaetes (Dubilier *et al.*, 2001), and bivalves (Gros *et al.*, 2000). These relationships may involve very specific physiological adaptations on the part of the animal to house and nurture the prokaryote hosts. For example, the deep-sea hydrothermal vent tube worm *Riftia pachytila* has evolved without a mouth or intestinal tract, and it obtains its nutrition from sulfide-oxidizing endosymbionts (see Chapter 4), with active transport systems for O_2 and sulfide to fuel the sulfide-oxidizing population. Mutualistic symbiosis is also found between prokaryotes and protozoans, where protozoan populations may house, for example, methanogenic partners as endosymbionts (Fenchel and Findlay, 1995). Hydrogen is produced by fermentation within the hydrogenosomes of the protist, and it is consumed by the methanogen,

---

(E) *Chlorochromatium aggregatium* after disaggregation revealing the central bacterium; (F) *Cylindrogloea bacterifera*; (G) *Chloroplana vacuolata*. Gas vacuolation of both the green sulfur bacteria and the central colorless bacteria is shown for only a few of the cells; (H) consortium from the hindgut of the termite *Reticulitermes flavipes*. Upper portion reveals the chain of central bacteria containing endospores; (J) *Anabena* sp. filament with chemotrophic bacteria covering a heterocyst; (K) dental plaque; (L) *Thioploca* sp. filament covered with sulfate-reducers of the genus *Desulfonema*; (M) archaeal-bacterial consortia where the central cluster of Archaea is covered by a layer of sulfate reducers. This consortia is believed to regulate the anaerobic oxidation of methane. Scale bars are in μM. From Overmann and Schubert (2002). Reproduced with permission.

providing an obvious food source for the methanogen. The methanogen removes H_2, thus favoring the fermentation of the hydrogenosome. Consistent with this interdependence, if methanogenesis is inhibited, the growth rate of the protist may suffer, particularly if it is large (Fenchel and Findlay, 1995).

Strombidium purpureum is a protozoan housing a purple non-sulfur phototrophic bacterium (Fenchel and Bernard, 1993). The phototroph utilizes H_2 produced by the protist, obviously benefiting the phototroph, but also the protist by lowering the H_2 partial pressure, favoring continued fermentation. The phototroph may also provide a food source to the protist. In the dark, the protist seeks low oxygen levels of 1 to 4% air saturation, where the phototroph oxidizes H_2 and fatty acids by oxidative phosphorylation (see Chapter 4). Therefore, the phototroph can remain metabolically active day and night, utilizing two different metabolisms. This unusual behavioral adaptation greatly expands the environmental range of the protist (Fenchel and Findlay, 1995).

Some ciliates may also house sulfate reducers as ectosymbionts, attached to external cell surfaces (Fenchel and Findlay, 1995). The sulfate reducers apparently use substrates (e.g., H_2, acetate) coming from the ciliate, and the relationship between the two organisms would appear to be one of commensalism. The sulfate reducers gains from the association with the ciliate, and the cost or gain to the ciliate is less obvious.

5.3. Horizontal gene transfer

The transfer of genetic material between microbes in nature alters the genome of host populations and promotes evolutionary change. As explored in Chapter 1, horizontal gene transfer has been a prominent mode of genome modification through the history of life. Horizontal gene transfer can provide new metabolic possibilities to recipient organisms, but at the same time, it complicates our efforts to reconstruct the history of life from single gene trees such as those constructed from SSU rRNA sequences. Given the potential significance of horizontal gene transfer in building microbial genomes, here we briefly overview some of the pathways by which genetic material may be transferred between prokaryotes in nature.

Some prokaryotes are able to uptake DNA from the environment in a process known as transformation. Cells can only uptake DNA when they are competent, and in some cases competence is induced by a quorum-sensing circuit (see below) when cell numbers become high enough. The DNA incorporated may come from the lysis of other cells after viral attack or starvation. After the DNA is taken up by the competent cell it may be incorporated into the host genome.

Viral infection is another vector for DNA transfer between microorganisms, in a process known as transduction. In this case, the genome of a virus particle incorporates some DNA from a host. The virus particle may incorporate either specific or nonspecific stretches of host DNA, replacing some of the viral DNA. There is the possibility that this DNA can be incorporated into a new host during viral infection, particularly with a temperate virus. There is a low probability of DNA transfer by transduction, as the probability of incorporating part of the host genome into a virus particle is low, and the probability of permanently incorporating some of this prokaryote DNA into another host is also low. Nevertheless, this transfer mechanism can occur, and given the astronomical number of viral infections in prokaryotes in nature, it is probably not uncommon.

Another pathway of gene transfer is a process called conjugation, in which plasmids from one cell are transferred to another, and the genetic material of the plasmid becomes incorporated into the host genome. Plasmids are double stranded, normally circular stretches of DNA existing in the cell cytoplasm independent of the cell chromosome. Plasmids accomplish numerous important functions for the cell, including the production of antibiotics as well as antibiotic resistance, and they can also code for important parts of carbon metabolism. Plasmids are transferred between cells during conjugation by cell-to-cell contact, and the process of conjugation is encoded by the plasmid itself. Conjugation can be a very efficient pathway for genetic material exchange. For example, some plasmids transfer antibiotic resistance between cells during cell-to-cell contact, and others transfer virulence. In principle, any process encoded by a plasmid can be transferred to other members of the same population, or even different distantly related populations (see Madigan *et al.*, 2003, for a full discussion).

6. SOCIAL BEHAVIOR AND CELL DIFFERENTIATION

We can view social behavior as interactions between population members that benefit the population (Crespi, 2001). For example, sulfide oxidizers of the genus *Thiovulum* can attach to solid surfaces with a slime thread (see Chapter 9) and spin rapidly around this tether, enhancing the transport of oxygen and sulfide to the whole *Thiovulum* population. This is a specific behavior enhancing the metabolic activity of the population. Many individual populations can also differentiate both functionally and morphologically in ways to benefit the whole population (e.g., Shapiro, 1998). For example, a number of filamentous cyanobacteria differentiate a portion of their cells into special heterocysts, where nitrogen fixation occurs. With this physiological (and morphological) adaptation, the cyanobacterial filaments can meet their own fixed nitrogen needs.

Another example of morphological differentiation is found among members of the myxobacteria (see also above), which are aerobic heterotrophs commonly found in soils, living among decaying vegetation and animal dung (e.g., Reichenbach and Dworkin, 1992). In their vegetative state myxobacteria are motile rod-shaped cells with gliding motility. Under nutrient stress, they undergo an impressive and complex process of cell differentiation forming fruiting bodies, which are frequently stalked (Figure 2.11). Complex signaling and cell communication coordinate the formation of fruiting bodies. The process begins as cells aggregate together, forming mounds, after which morphogenesis occurs. This includes the secretion of a slime stock (in species that produce such a stock) and the migration of cells to the top of the stock where the head of the fruiting body is formed. Here, cells differentiate into myxospores, which are sometimes encased in walled structures known

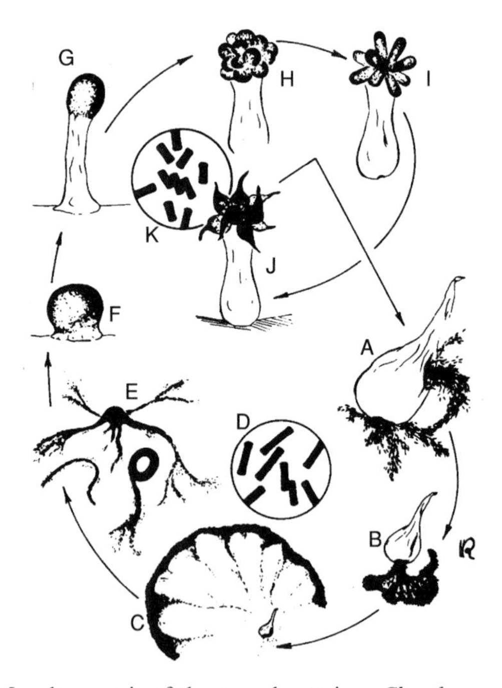

Figure 2.11 Morphogenesis of the myxobacterium *Chondromyces apiculatus*. (A) germinating sporangiole; (B) and (C) swarm colony development; (D) vegetative cells; (E) aggregation of vegetative cells within the swarm; (F) knob formation stage; (G) excretion of slime stalk, with cells concentrating in the terminal knob; (H) terminal mass begins to differentiate; (I) club-like structures form; (J) turnip-shaped sporangioles form; (K) myxospores. From Reichenbach and Dworkin (1992), with permission.

as cysts. Fruiting body cells exhibit enhanced resistance to environmental extremes such as drying and UV radiation. When conditions again become favorable for growth, the sporangiole holding the myxospores ruptures, releasing the myxospores, and the vegetative cycle begins. Yet another example of cell differentiation and community behavior among single populations is colony development by *E. coli*, in which different zones within the colony display cells of unique shape, size, and patterns of arrangement (Shapiro, 1998).

Some individual populations also release and detect chemical signaling molecules, known as autoinducers, in a process called quorum sensing (e.g., Miller and Bassler, 2001; Schauder and Bassler, 2001). As populations grow past a minimum size, autoinducers reach a threshold concentration at which genes are expressed controlling many types of microbial behavior, including luminescence, virulence, antibiotic production, and biofilm formation (Schauder and Bassler, 2001). In general, the quorum-sensing circuit involves the production of a specific protein that in turn produces an autoinducer compound that diffuses freely across the cell membrane. The autoinducer compound accumulates in solution at a concentration proportional to microbial numbers and density. The autoinducer combines with specific receptor proteins within the cell when a critical high concentration of autoinducer is reached. At this point the receptor protein triggers the induction of gene expression (Figure 2.12). In this way, populations can coordinate

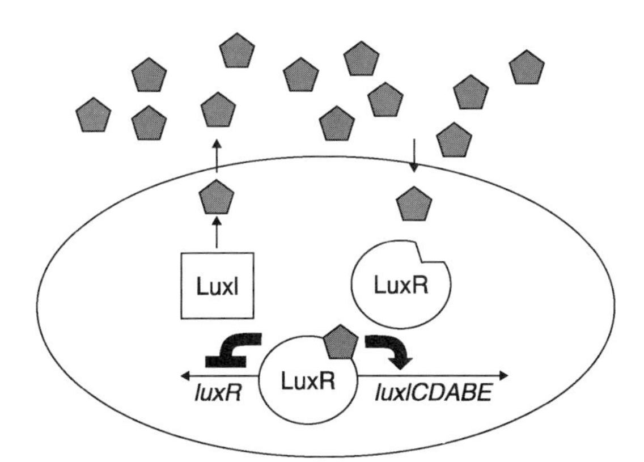

Figure 2.12 Quorum-sensing circuit for *Vibrio fischeri*. Autoinducer (pentagons) accumulates in solution after production by *LuxI*. The concentration of autoinducer is sensed by *LuxR*, which, at a threshold level, activates the *LuxICDABE* operon, which initiates bioluminescence. There is a positive feedback through *LuxICDABE,* causing an increase in autoinducer production (through increasing *LuxI* expression) and increasing, ultimately, bioluminescence. To control light production, the *LuxR*-autoinducer complex also inhibits the expression of *LuxR*, providing a necessary negative feedback. Inspired from Bassler and Miller (2001).

gene expression at cell densities that are advantageous to the population (Miller and Bassler, 2001; Schauder and Bassler, 2001).

The first demonstration of quorum sensing was with the bioluminescent marine bacterium *Vibrio fischeri*, which lives in symbiotic association with various marine animals (Nealson and Hastings, 1979). An example is the association between *V. fischeri* and the nocturnal squid *Euprymna scolopes* (Ruby, 1996; Bassler and Miller, 2001). *V. fischeri* lives within a special light organ on the underside of the squid, and the squid balances the illumination from *V. fischeri* to match the illumination from the moon so that the squid casts no shadow on the sea bottom. This helps protect the squid from predators. The squid gains by having a beneficial source of light, and *V. fischeri* gains from a ready source of nutrients within the light organ of the squid. Quorum sensing controls the bioluminescence from *V. fischeri*. Within the light organ of the squid, the autoinducer compounds produced by *V. fischeri* accumulate to a concentration high enough to induce bioluminescence. By contrast, when *V. fischeri* is present in the environment outside of the squid light organ (the squid sheds the bacteria at sunrise every morning), it produces no bioluminescence because the population density is too small. As a result, autoinducer concentrations are too low to activate bioluminescence gene expression.

In another example of quorum sensing, the human pathogen *Pseudomonas aeruginosa* uses quorum sensing to regulate the expression of a variety of virulence factors that interfere with protein synthesis and promote host tissue destruction (Parsek and Greenberg, 1999; Bassler and Miller, 2001). Presumably, high population densities give the best opportunity for *P. aeruginosa* to infect its host successfully. Numerous other examples of quorum sensing may also be found in nature, and this may indeed be a fundamental aspect of microbial communication (Bassler and Miller, 2001).

Chapter 3

Thermodynamics and Microbial Metabolism

ADVANCES IN MARINE BIOLOGY VOL 48
0-12-026147-2
© 2005 Elsevier Inc.
All rights reserved

1. INTRODUCTION

Life, like all chemical processes, obeys the laws of thermodynamics. Indeed, life has evolved to exploit them. For example, all life on Earth has associated reaction pathways, such as the utilization of adenosine triphosphate (ATP) (see Section 8.4), where seemingly thermodynamically unfavorable reactions can be promoted. Furthermore, life has evolved to survive by using countless energetically favorable oxidation-reduction reactions provided by the natural environment. Also, the competition between microbial populations in nature, as well as their mutualistic cohabitation, is often governed by thermodynamic considerations. Thermodynamic considerations explain the mechanisms of chemical transport of materials into and out of microbial cells, and the cell membrane itself typically houses a symphony of chemical processes engaged in energy conservation. An appreciation of chemical thermodynamics, therefore, provides us with a window into basic aspects of microbial metabolism. Furthermore, many fundamental principles of microbial ecology are explained by thermodynamics.

In this Chapter we consider basic aspects of chemical thermodynamics as relevant for understanding microbial metabolisms in nature and for defining the chemical environments of the microbial world. In addition, we consider cellular architecture and its relationship to how organisms gain energy for their growth and metabolism. Finally, we discuss some of the basic aspects of cellular metabolism and explain how different metabolisms are named.

2. STATE FUNCTIONS

2.1. Enthalpy

The First Law of Thermodynamics states that energy cannot be created or destroyed. As a natural outcome of this law (see Stumm and Morgan, 1996, for formal derivations), a function called enthalpy, ΔH, is defined. Enthalpy is equivalent to the heat added to, or subtracted from, a system as a result of a chemical process at constant pressure; it is also known as the heat of reaction. By convention, when a system evolves heat, ΔH is negative, and the reaction is exothermic. By contrast, when heat is absorbed, ΔH is positive, and the reaction is endothermic. Enthalpy is one of several functions defining the state of a system regardless of the system's prior history. Therefore, enthalpy is called a state function. We define ΔH^o relative to a standard state (STP), taken usually as one atmosphere (1 ATM) total pressure and 25 °C (298.15 K). Other standard states may be defined, but this one is of particular relevance for many (but not all) biological systems at the Earth's surface, and it is the standard state for which most thermodynamic data are available. Enthalpies for individual compounds or elements are given, by convention, as the heat of reaction necessary to form, isothermally, one mole of a substance from its elementary components at standard state. Enthalpies of formation at standard state are designated as ΔH_f^o, and the ΔH_f^o of the most stable phase of the element is taken as 0 kJ/mole. For example, the ΔH_f^o for elemental sulfur (S^o), oxygen gas (O_2), and graphite ($C_{graphite}$) are all 0 $kJ\,mol^{-1}$, whereas the ΔH_f^o for diamond ($C_{diamond}$) is not, at 1.88 $kJ\,mol^{-1}$. Diamond is the high-pressure phase of elemental carbon.

If we write a general reaction (Equation 3.1) in which n_l is the number of moles of compound of element X_1,

$$n_1X_1 + n_2X_2 \rightarrow n_3X_3 + n_4X_4 \tag{3.1}$$

the enthalpy, ΔH^o, for this reaction is calculated from Hess's Law of Summation as

$$(n_3\Delta H^o_{f,X_3} + n_4\Delta H^o_{f,X_4}) - (n_1\Delta H^o_{f,X_1} + n_2\Delta H^o_{f,X_2}) \tag{3.2}$$

or, more generally,

$$\Delta H^o = \Sigma(n_i\Delta H^o_{fi})_{products} - \Sigma(n_i\Delta H^o_{fi})_{reactants} \tag{3.3}$$

As a specific example we consider the dissolution of NaCl, written as $NaCl_{(s)} \rightarrow Na^+_{(aq)} + Cl^-_{(aq)}$. The ΔH^o for this reaction is $\Delta H^o = (1 * \Delta H^o_{f,Na^+} + 1 * \Delta H^o_{f,Cl^-}) - (1 * \Delta H^o_{f,NaCl}) = 3.63\ kJ\,mol^{-1}$. This is an endothermic reaction. Values of ΔH_f^o for compounds and elements of biological and geochemical interest are compiled in Appendix 2 and Appendix 3.

2.2. Entropy and Gibbs free energy

The endothermic dissolution of NaCl demonstrates that chemical reactions need not be exothermic to be spontaneous. This is because the spontaneity of a reaction is governed also by another attribute of the reacting system; this is an outcome of the Second Law of Thermodynamics. The second law may be stated in various ways, but for our purposes a good definition is "for any spontaneous process there is an increase in the entropy of the universe." We thus define entropy, S, as a new state function. "The universe" consists of the reacting system and its surroundings; therefore, the change in the entropy of the universe, or ΔS_{total}, may be broken into its component parts as follows:

$$\Delta S_{\text{total}} = \Delta S_{\text{surroundings}} + \Delta S_{\text{system}} \tag{3.4}$$

At constant temperature and pressure, the entropy change in the surroundings is equivalent to the heat added to the surroundings by the system divided by the temperature (K) at which the heat is added:

$$\Delta S_{\text{surroundings}} = \Delta H_{\text{system}}/T \tag{3.5}$$

Equations 3.4 and 3.5 are combined and rearranged to yield

$$T\Delta S_{\text{total}} = -(\Delta H_{\text{system}} - T\Delta S_{\text{system}}) \tag{3.6}$$

Since ΔS_{total} must be positive for a spontaneous change, so must $T\Delta S_{\text{total}}$. In turn, the term ($\Delta H_{\text{system}} - T\Delta S_{\text{system}}$) must be negative for spontaneous change. From here we define a new state function, called Gibbs free energy, G, such that

$$\Delta G = \Delta H - T\Delta S, \text{or at standard state, } \Delta G^{\text{o}} = \Delta H^{\text{o}} - T\Delta S^{\text{o}} \tag{3.7}$$

If ΔG for a reaction is negative, then the reaction is spontaneous, and it is known as an exergonic reaction. If ΔG is positive, then the reaction is spontaneous written in the other direction, or requires energy to proceed in the direction written, and the reaction is known as endergonic. If ΔG is equal to 0, then the system is at equilibrium. The driving force for a reaction depends on its ΔG, so the more negative ΔG, the more favorable the reaction. The Gibbs free energy for reaction at standard state, ΔG^{o}, is calculated in a way comparable to how enthalpy, ΔH^{o} is calculated (Equation 3.3):

$$\Delta G^{\text{o}} = \Sigma(n_i \Delta G^{\text{o}}_{\text{f}i})_{\text{products}} - \Sigma(n_i \Delta G^{\text{o}}_{\text{f}i})_{\text{reactants}} \tag{3.8}$$

and through a calculation of Gibbs free energy, we can evaluate whether a chemical reaction should be thermodynamically favorable. We therefore obtain a powerful tool for understanding the chemical circumstances favoring specific microbial metabolisms. We can also predict the possibility of new microbial metabolisms not previously described. However, standard state (STP) Gibbs free energy, ΔG^{o}, does not represent the normal chemical circumstances experienced by microorganisms in nature. Most significantly, concentrations of chemical species are generally much less than the unit molar concentrations used to calculate ΔG^{o}.

Thus, for any component of a system whose concentration deviates from standard conditions, the Gibbs free energy of that component can be determined from the following:

$$\Delta G_i = \Delta G^{\text{o}}_i + RT\ln a_i \tag{3.9}$$

where R is the gas constant (=8.314 J mol^{-1} K^{-1}), T is temperature in Kelvin (K), and a_i is a quantity known as the activity. Activity is related to concentration, c (molar), through the activity coefficient (dimensionless), γ.

$$a_i = \gamma_i c_i \tag{3.10}$$

For a gas,

$$f_i = \gamma_{fi} P_i \tag{3.11}$$

where f_i is fugacity, γ_{fi} is the fugacity coefficient and P_i is pressure. Activity coefficients and fugacity coefficients represent the deviation of the component of interest from ideal behavior in solutions and gases. For ions in solution, for example, the electrical interactions with other ions and the crowding of ions in concentrated solutions cause the ions to interact at effective concentrations different from the actual concentrations in the solution. These issues will be considered in more detail in Section 3.5.

2.2.1. Gibbs free energy under environmental conditions

Equation 3.9 considers how the Gibbs free energy of individual components of a system is influenced by deviations from unit molar concentration, but what about the reacting system as a whole? Consider the following general reaction, which could represent any biologically important chemical process:

$$aA + bB \leftrightarrow cC + dD \tag{3.12}$$

The components are indicated by the capital letters, and the stoichiometric amounts of each component are indicated by the lowercase letters. The Gibbs free energy of this system, ΔG, is related to the Gibbs free energy at standard state, ΔG^o by the following expression:

$$\Delta G = \Delta G^o + RT\ln(a_C^c a_D^d)/(a_A^a a_B^b) \tag{3.13}$$

Let us consider the thermodynamics of the rather newly discovered microbial disproportionation of elemental sulfur (see Chapter 9). Sulfur-disproportionating organisms are anaerobic and have been found in abundance in anoxic surface sediments (Thamdrup *et al.*, 1993). The general reaction is written as

$$4H_2O_{(l)} + 4S^o_{(s)} \rightarrow 3H_2S_{(aq)} + SO^{2-}_{4(aq)} + 2H^+_{(aq)}$$

The ΔG^o for this reaction is, from Equation 3.8,

$$\Delta G^o = \underset{H^+_{(aq)}}{2[0]} + \underset{SO^{2-}_{4(aq)}}{[-744.6]} + \underset{H_2S_{(aq)}}{3[-27.87]} - [\underset{S^o_{(s)}}{4(0)} + \underset{H_2O_{(l)}}{4(-237.18)}] = 120.5 \text{ kJ mol}^{-1}$$

This reaction is not thermodynamically favorable under standard conditions, yet we know the organisms promoting S^o disproportionation thrive in nature. Let us explore the free energy change actually experienced by S^o disproportionating organisms in nature. We assume a temperature of 25 °C. Furthermore, the concentration of $H_2S_{(aq)}$ is taken as 100 μM, a value not atypical in near-surface marine sediments, and a seawater $SO^{2-}_{4(aq)}$ concentration of 28 mM is used. We assume a sediment pH value of 7.5, and furthermore, we assume that activity equals concentration. The activities of liquid water and solid phases are set as 1 by definition, and from Equation 3.13 we calculate the ΔG for S^o disproportionation under environmental conditions as

$$\Delta G = 120.5 + (8.314 \times 10^{-3})\ (298)\ \ln(a^3_{H_2S} a_{SO_4} a^2_{H^+}) = -42.4 \text{ kJ mol}^{-1}$$

This is a thermodynamically favorable reaction and explains how S^o disproportionation can occur in sediments where the concentrations of reacting species differ markedly from standard state.

3. EQUILIBRIUM

All thermodynamically favorable chemical reactions will proceed, barring kinetic barriers, until the distribution of reacting components in the system reaches equilibrium. The distribution of chemical species at equilibrium is defined by the equilibrium constant, K_{eq}, which, for the general reaction presented in Equation 3.12, is given by

$$K_{eq} = (a^c_C a^d_D)/(a^a_A a^b_B) \tag{3.14}$$

Furthermore, at equilibrium, the ΔG for a reaction is equal to 0. Thus, after setting ΔG to 0 and substituting Equation 3.14 into Equation 3.13, we obtain the following:

$$\Delta G^o = -RT \ln K_{eq} \tag{3.15}$$

This expression may be rearranged and used to calculate equilibrium constants from ΔG^o.

$$\ln K_{eq} = -\Delta G^o/RT$$

or,

$$K_{eq} = e^{-\Delta G^o/RT} \tag{3.16}$$

4. INFLUENCE OF TEMPERATURE ON THERMODYNAMIC PROPERTIES

Microorganisms, of course, generally live at temperatures different from the standard state temperature (25 °C) at which thermodynamic properties are typically defined. Therefore, to fully appreciate the thermodynamics surrounding microbial metabolism in a given environment, thermodynamic calculations should be corrected for temperature deviations from the standard state. As long as deviations are within approximately 20 °C of the standard state temperature, the influence of temperature on ΔG^o can be evaluated directly from Equation 3.7, whereas the influence of temperature on K_{eq} can be evaluated from Equation 3.16. For temperature differences greater than approximately 20° from the standard state, ΔH and ΔS must be recalculated from heat capacity data, and these new values then can be used to recalculate ΔG and K_{eq}. This topic falls beyond the subjects we wish to emphasize in the present Chapter.

5. ACTIVITY COEFFICIENT CALCULATIONS

Ionic species interact with each other in solution relative to their respective activities, not concentrations (Equation 3.10). Therefore, in order to evaluate the state of chemical equilibrium of ionic components of a biological system, we must calculate the activities of the components, which requires a calculation of activity coefficients. Derived from first principles, the Debye-Hückel equation is frequently used to calculate activity coefficients. The activity coefficient for any component "i" is calculated as follows:

$$-\log\gamma_i = AZ_i^2(I)^{\frac{1}{2}} \tag{3.17}$$

where A is a constant depending only on temperature and pressure (Table 3.1), Z_i is the charge of the ion, and I is the ionic strength of the solution. This is calculated as

$$I = \frac{1}{2}\Sigma m_i Z_i^2 \tag{3.18}$$

Table 3.1 Parameters used for the Debye-Hückel equation at 1 atm

Temp (°C)	A
0	0.4883
5	0.4921
10	0.4960
15	0.5000
20	0.5042
25	0.5085
30	0.5130
35	0.5175
40	0.5221
45	0.5271
50	0.5319
55	0.5371
60	0.5425

From Garrels and Christ (1965).

where m_i is the molar concentration of the species of interest. The Debye-Hückel equation is, however, only valid for dilute solutions with ionic strengths (I) of $<5 \times 10^{-3}$ M. Empirically based extensions of the Debye-Hückel expression have been proposed, and a common alternative, used for ionic strengths up to 0.5 M, is a simplified version of the Davies equation:

$$-\log\gamma_i = [AZ_i^2(I)^{\frac{1}{2}}/(1 + (I)^{\frac{1}{2}})] + 0.2I \tag{3.19}$$

As an example, let us calculate the activity coefficient for Ca^{2+} in lake water with the following major ion composition:

Ionic species	mM
Na^+	5
K^+	1
Ca^{2+}	1
Mg^{2+}	1
HCO_3^-	1
Cl^-	5
SO_4^{2-}	2

The ionic strength for the solution is $I = \frac{1}{2}(5{*}1^2 + 1{*}1^2 + 1{*}2^2 + 1{*}2^2 + 1{*}1^2 + 5{*}1^2 + 2{*}2^2) \times 10^{-3} = 14 \times 10^{-3}$ M, and using the Davies equation, the activity coefficient for Ca^{2+} at 20 °C is 0.58. This is quite a deviation from ideal behavior.

6. GAS SOLUBILITY AND HENRY'S LAW

We know from common experience that gases dissolve into water. We also know that the amount of gas held by water is highly sensitive to salt and temperature. Thus, concentrated brine holds very little gas, and we know that as we bring a kettle of water to boil massive bubble formation occurs, representing solution degassing, well before the boiling actually begins. The amount of gas dissolved into water is expressed by equilibrium thermodynamics similar to what we have already seen. Thus, at equilibrium, and for dilute solutions and low gas pressures, the relationship between the partial pressure of gas in the gaseous phase and the amount of gas dissolved in water is given by Henry's Law and is expressed as

$$C_{(aq)i} = P_i K_{Hi} \tag{3.20}$$

where $C_{(aq)i}$ (M) is the concentration of the gas, in aqueous solution, P_i (atm) is the partial pressure of the gas and K_{Hi} is the Henry's Law constant (M atm^{-1}). A list of the Henry's Law constants for gases of biological interest at 25 °C is presented in Table 3.2. In concentrated solutions, and at high pressures, the concentration, $C_{(aq)i}$, should be replaced with activity, a_i (Equation 3.10), and partial pressure, P_i, should be replaced with fugacity, f_i (Equation 3.11). We will assume that fugacity equals partial pressure for atmospheric constituents.

Table 3.2 Henry's law constants (K_H, M atm^{-1}), salting out coefficient (k_i) and heats of solution (ΔH_{sol}) for some gases of biological interest at 25 °C

Gas	K_H	k_i	(kJ mol^{-1}) ΔH_{sol}
CO_2[a]	3.4×10^{-2}	0.095	19.6
NH_3	5.7×10^{1}		
H_2S	1.0×10^{-1}	0.02	
N_2	6.5×10^{-4}	0.131	8.8
O_2	1.3×10^{-3}	0.122	12.5
CO	9.6×10^{-4}	0.134	15.9
CH_4	1.3×10^{-3}	0.092	13.4
NO_2	1.0×10^{-2}		
NO	1.9×10^{-3}		11.3
N_2O	2.5×10^{-2}		20.1
H_2O_2	1.0×10^{5}		
O_3	9.4×10^{-3}		
H_2	7.8×10^{-4}		1.3

[a] $CO_{2(g)} + H_2O_{(g)} \leftrightarrow H_2CO_{3(aq)}$.

Data from Stumm and Morgen (1996) and Millero (1996).

6.1. Influence of salt on gas solubility

The "salting out" of gases occurs in solution as the ionic strength, I, of the solution increases. The activity coefficient, γ_{gl}, for a gas in solution is approximated as

$$\gamma_{gl} = 10^{k_i I} \tag{3.21}$$

and the concentration of a gas in solution $C_{(aq)i}$ is related to the concentration in an infinitely dilute solution, $C^{o}_{(aq)i}$, through the activity coefficient as follows:

$$C_{(aq)i} = C^{o}_{(aq)i} / \gamma_{gl} \tag{3.22}$$

Values for k_i vary from gas to gas, but 0.1 is typical (Table 3.2). So, let us compare the solubility of O_2 in both dilute solution and sea water. The concentration of $O_{2(aq)}$ in dilute solution in equilibrium with atmospheric oxygen at 25 °C ($p_{O_2} = 0.21$ atm) is, from Equation 3.20 (and the K_H value from Table 3.2), 2.64×10^{-4} (M), or 264 μM. The k_i (Equation 3.21) value for O_2 is 0.122 (Table 3.2), and the ionic strength of sea water, I, is 0.7. Thus, an activity coefficient, γ_{gl}, for $O_{2(aq)}$ in seawater of 1.22 is calculated, yielding an air equilibrium concentration for $O_{2(aq)}$ of 217 μM, a significant reduction in solubility over the dilute solution.

6.2. Influence of temperature on gas solubility

All gases decrease their solubility as water temperature increases, by amounts varying from gas to gas. The influence of temperature on gas solubility may be approximated from the Clausius-Clapeyron equation in the form

$$\ln(C_0/C_1) = \Delta H_{sol}/R[1/T_1 - 1/T_0] \tag{3.23}$$

which rearranges to the following:

$$\ln(C_1) = \ln(C_0) - \Delta H_{sol}/R[1/T_1 - 1/T_0] \tag{3.24}$$

where ΔH_{sol} is the heat of solution for the gas in water at 25 °C (Table 3.2), R is the gas constant, C_0 is the saturation concentration of the gas at 25 °C, C_1 is the concentration at the desired temperature, T_1 is the desired temperature (K), and T_0 is the standard state temperature of 298.15 K. This equation is useful for relatively small deviations around the standard state temperature.

7. OXIDATION-REDUCTION

The life process is intimately coupled to oxidation-reduction reactions, including electron transfers associated with carbon fixation, fermentation, proton potential generation leading to ATP formation, and countless biochemical processes within the cell. Also, many important microbial respiration reactions are coupled to the oxidation of organic carbon with electron acceptors such as oxygen, nitrate, and sulfate. The localization of organisms promoting these different respiration reactions in the environment depends on the availability of the electron acceptor and also very much on the thermodynamics of the respiration reactions. Therefore, understanding oxidation-reduction allows us to appreciate aspects of cellular metabolism, as well as the ecology of microbes in nature.

7.1. Half reactions and electrode potential

Oxidation-reduction reactions, also known as redox reactions, involve electron transfer. If a chemical species gains electrons it is said to be reduced, whereas if it loses electrons it is said to be oxidized. Electrons do not accumulate in solution, so all oxidation reactions must be coupled to reduction reactions, and vice versa. It is easiest to consider oxidation and reduction reactions separately. For example, the photosynthetic production of oxygen gas can be broken into its component oxidation and reduction reactions. Each of these is known as a half reaction and represents a redox pair or redox couple:

$$4H^+ + CO_2 + 4e^- \rightarrow CH_2O + H_2O \text{ (reduction, gain of } e^-)$$

$$2H_2O \rightarrow O_2 + 4H^+ + 4e^- \text{ (oxidation, loss of } e^-)$$

$$CO_2 + H_2O \rightarrow CH_2O + O_2 \text{ (overall, reduction + oxidation)}$$

The ease with which chemical species gain or lose electrons varies greatly, and the ability of a chemical species to gain or liberate electrons is referred to as the electrode potential, E, often called the redox potential.

By convention, electrode potentials are compared for reactions written as reduction reactions. Furthermore, electrode potentials are generally reported relative to a standard redox reaction known as the standard hydrogen electrode, SHE. The SHE represents the following half reaction, written as a reduction reaction:

$$2H^{+}_{(aq)} + 2e^{-} \rightarrow H_{2(g)} (\text{SHE}, a_{H^{+}} = 1, p_{H_2} = 1\,\text{atm})$$

under the standard conditions of 1 atm $H_{2(g)}$ and 1 M $H^{+}_{(aq)}$ activity. The SHE is given a standard electrode potential, E^{o}, of 0.000 volts. The standard electrode potential of other redox pairs at unit activity may be determined directly by coupling to a SHE, or alternatively, relative to other reference electrodes such as the calomel electrode ($Hg_2Cl_{2(s)} + 2e^{-} \rightarrow 2Hg_{(l)} + 2Cl^{-}_{(aq)}$, $E^{o} = 0.241$ V), with appropriate corrections back to the SHE scale. When written relative to the SHE, electrode potential is designated as E^{o}_{H}.

The oxidized form of the redox pair with the most positive E^{o}_{H} value is the strongest oxidant. By contrast, the reduced form of the redox pair with the most negative E^{o}_{H} value is the strongest reductant. A list of E^{o}_{H} values for important half reactions of biological interest is shown in Table 3.3, arranged in order of decreasing electrode potential. An ordered arrangement like this is known as an electrochemical series. Of particular importance is the fact that the oxidized form of a redox pair will oxidize the reduced form of a redox pair lower in the electrochemical series. Therefore, sulfate can act as an oxidant for methane, but not for Fe^{2+}. Of course, the conditions expressed in Table 3.3 are standard conditions that are not normally met in the environment. We will see below how to make calculations under natural conditions.

Table 3.3 E^{o}_{H} values for some redox pairs of biological interest at 25 °C, calculated from Equation 3.25 and at standard state

Note	Oxidized form		Reduced form	E^{o}_{H} (v)
	$NO_2^{-} + 4H^{+} + 4e^{-}$	→	$1/2N_{2(g)} + 2H_2O$	1.51
	$O_{2(aq)} + 4H^{+} + 4e^{-}$	→	$2H_2O$	1.27
	$NO_3^{-} + 6H^{+} + 5e^{-}$	→	$1/2N_{2(g)} + 3H_2O$	1.24
Pyrolusite	$MnO_2 + 4H^{+} + 2e^{-}$	→	$Mn^{2+} + 2H_2O$	1.23
	$NO_3^{-} + 10H^{+} + 8e^{-}$	→	$NH_4^{+} + 3H_2O$	0.88
Amorphous oxide	$Fe(OH)_3 + 3H^{+} + e^{-}$	→	$Fe^{2+} + 3H_2O$	0.88
	$NO_3^{-} + 2H^{+} + 2e^{-}$	→	$NO_2^{-} + H_2O$	0.85
	$Fe^{3+} + e^{-}$	→	Fe^{2+}	0.77
	α-$FeOOH + 3H^{+} + e^{-}$	→	$Fe^{2+} + 2H_2O$	0.67
	$SO_4^{2-} + 8H^{+} + 6e^{-}$	→	$S^{0}_{(s)} + 4H_2O$	0.35
	$SO_3^{2-} + 2H^{+} + 2e^{-}$	→	$H_2S_{(aq)} + 3H_2O$	0.34
	$S^{0}_{(s)} + 2H^{+} + 2e^{-}$	→	$H_2S_{(aq)}$	0.29
	$SO_4^{2-} + 10H^{+} + 8e^{-}$	→	$H_2S_{(aq)} + 4H_2O$	0.23
	$HCO_3^{-} + 9H^{+} + 8e^{-}$	→	$CH_{4(aq)} + 3H_2O$	0.21
Acetate	$HCO_3^{-} + 9/2H^{+} + 4e^{-}$	→	$1/2C_2H_3OO^{-} + 2H_2O$	0.19
Ethanol	$HCO_3^{-} + 7H^{+} + 6e^{-}$	→	$1/2C_2H_5OH + 5/2H_2O$	0.17
Lactate	$HCO_3^{-} + 14/3H^{+} + 4e^{-}$	→	$1/3C_3H_5O_3^{-} + 2H_2O$	0.16
Glucose	$HCO_3^{-} + 5H^{+} + 4e^{-}$	→	$1/6C_6H_{12}O_6 + 2H_2O$	0.10
	$2H^{+} + 2e^{-}$	→	$H_{2(g)}$	0.00

Table 3.4 Thermodynamics of glucose oxidation with oxygen

Reaction	E^o_H(v)	ΔG^o (kJ)	$\ln k_{eq}$
$O_2 + 4H^+ + 4e^- \rightarrow 2H_2O$	1.27	−490.68	198
$1/6C_6H_{12}O_6 + 2H_2O \rightarrow 5H^+ + HCO_3^- + 4e^-$	−0.10	40.4	−15.6
$1/6C_6H_{12}O_6 + O_2 \rightarrow HCO_3^- + H^+$	1.17	−450.28	182.4

Values for the electrode potential are independent of the number of electrons transferred in the balanced equation. They also are additive. As an example, we can consider the electrode potential associated with the oxidation of glucose with oxygen (Table 3.4). We compile first the electrode potentials for the individual half reactions and then sum these to obtain the electrode potential for the overall reaction. Note that the oxidation of glucose to carbon dioxide is an oxidation reaction (Table 3.4), and the sign of the electrode potential associated with this reaction is reversed from the value presented in Table 3.3.

7.2. Gibbs free energy and electrode potential

Electrode potentials are related to the Gibbs free energy, ΔG, through the following relationship:

$$\Delta G = -nFE, \text{or relative to SHE, } \Delta G^o = -nFE^o_H \tag{3.25}$$

where n is the number of electrons transferred in the reaction, either half reaction or coupled oxidation-reduction reaction, and F is Faraday's constant, with a value of 96.53 kJ volt^{-1}. Thus, as opposed to the electrode potential, the free energy change of a reaction depends on reaction stoichiometry and the number of electrons transferred. As discussed earlier, thermodynamically favorable reactions are given by negative values for ΔG. Therefore, positive values for electrode potential, E or E^o_H, represent favorable reactions, and negative values for E or E^o_H indicate that the reaction is favorable in the opposite direction.

An example is given in Table 3.4, where the standard state free energy change, ΔG^o, associated with the oxidation of glucose with oxygen is calculated from the electrode potential, E^o_H, for the individual redox couples using Equation 3.25. The ΔG^o is −450.3 kJ mol^{-1} per mole of O_2, and the reaction is clearly favorable. The ΔG^o of a coupled oxidation-reduction reaction may also be calculated from the ΔG^o of the individual reactants and products as shown in Equation 3.8, or from entropy and enthalpy data as shown in Equation 3.7.

7.3. Equilibrium constant and electrode potential

The electrode potential is also related to the equilibrium constant, K_{eq}, for the reaction. Thus, we can combine Equation 3.15 with Equation 3.25 to yield the following relationship:

$$\ln K_{eq} = nFE_H^o/RT \tag{3.26}$$

Using this equation, the equilibrium constants for the individual half reactions, as well as for the overall reaction expressing the oxidation of glucose with oxygen, are shown in Table 3.4. For balanced oxidation-reduction reactions, K_{eq} may also be computed from free energy data, as shown in Equation 3.15.

7.4. Electrode potential in non-standard conditions

The standard conditions represented by E_H^o values are rarely found in nature, except perhaps in some extreme examples of acid production in abandoned metal sulfide mines. In order to represent realistic natural conditions, and to accommodate the variability of chemical environments found in nature, we must calculate electrode potentials for situations far removed from the standard state. Consider a reduction half reaction of the following general form:

$$a\mathrm{A}_{oxid} + be^- + c\mathrm{H}^+ \rightarrow d\mathrm{A}_{red} - g\mathrm{G} \tag{3.27}$$

Here, A refers to a redox-active species undergoing reduction, and G is a possible non-redox active reaction product, while *a*, *c*, *d*, and *g* are stoichiometric coefficients. The electrode potential for this reaction under non-standard conditions is determined from the Nernst equation:

$$E = E_H^o + (RT/nF)\ln(a_{A_{oxid}}^{a} a_{H^+}^{c})/(a_{A_{red}}^{d} a_{G}^{g}) \tag{3.28}$$

Note that the oxidized form of the redox pair is in the numerator while the reduced form is in the denominator, and the electron does not enter into the equation. To see how this equation is used, consider the reduction of nitrate to nitrogen gas at 25 °C, a pH of 7, a nitrate concentration of 30 μM and a partial pressure of nitrogen gas of 0.78 atm. We assume concentration equals activity and partial pressure equals fugacity, and with E_H^o values from Table 3.3:

$$\mathrm{NO_3^-} + 6\mathrm{H}^+ + 5e^- \rightarrow \frac{1}{2}\mathrm{N}_{2(g)} + 3\mathrm{H_2O} \tag{3.29}$$

Thus,

$$E = 1.24 + [(8.314 * 10^{-3} * 298.15)/(5 * 96.53)] \ln[(30 * 10^{-6})(10^{-7})^6/(0.78)^{\frac{1}{2}}] = 0.69 \text{ V} \quad (3.30)$$

The electrode potential associated with nitrate reduction to nitrogen gas under typical environmental conditions is very different from the electrode potential relative to the SHE, as shown in Table 3.3.

Commonly, electrode potentials for biological and environmental systems are calculated relative to pH = 7. This is done to more faithfully represent the chemistry of the environment or of a cell, as opposed to the 1 M H^+ activity used for the SHE. Electrode potentials calculated in such a fashion are designated variably as $E^{o}(w)$, E_{m7} or E'_0, and the calculation is easily accomplished with Equation 3.28. Frequently, E'_0 values are arranged in an "electron tower" such as the one shown in Figure 3.1, and as for the electrochemical series presented in a tabular form (Table 3.3), the oxidized form of a redox pair can oxidize the reduced form of a redox pair lower on the tower. Still, electrode potentials calculated relative to a neutral pH are

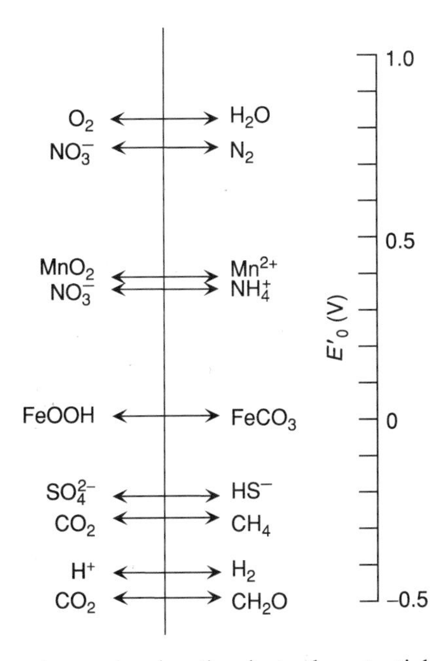

Figure 3.1 Electron tower showing the electrode potential of various oxidation-reduction pairs of environmental interest at a pH of 7, but otherwise at standard state. Concept after Fenchel *et al.* (1998).

approximations of the natural environment. Significantly, unit activity is assumed for reactants and products other than H^+, and furthermore, excursions from neutral pH are normal. Electrode potentials should be calculated for the chemistry of the specific environment of interest.

7.5. *pε*

Geochemists traditionally express redox intensity relative to the dimensionless parameter *pε*, which gives the potential activity of electrons in solution and is defined as

$$p\varepsilon = -\log(a_{e^-}) \quad (3.31)$$

The activity of electrons, a_{e^-}, is only hypothetical, as already discussed; electrons do not accumulate free into solution. Rather, *pε* expresses the tendency of a redox pair to either liberate or accept electrons. The derivation of *pε* and its practical use is beyond the scope of the current Chapter; however, a straightforward relationship exists between *pε* and electrode potential:

$$p\varepsilon = [F/(2.303RT)]E, \text{ and } p\varepsilon^o = [F/(2.303RT)]E^o_H \quad (3.32)$$

8. BASIC ASPECTS OF CELL BIOCHEMISTRY

8.1. Energy gain, catabolism, and anabolism

Prokaryotes are clever little chemists. They exploit, with complex biochemical machinery, numerous energy-yielding chemical interfaces met within the environment. Indeed, microbial enzymes such as nitrogenase, promoting nitrogen fixation, and Rubisco, promoting carbon fixation in the Calvin cycle, easily perform chemical reactions that frustrate the bench chemist. Ultimately, usable energy within a cell is derived from electrons transferred in oxidation-reduction reactions. Light drives energy-gaining oxidation-reduction reactions in photosynthesis (see Chapter 4), while in the absence of light, energy may be gained from electron transfer between primary electron donors such as organic carbon and primary electron acceptors such as oxygen. This is known as respiration. Energy can also be gained from the fermentation of organic compounds, where the same organic molecule acts as both the electron donor and the electron acceptor. The breakdown of organic and inorganic compounds by an organism, whether by respiration or by fermentation, is known as catabolism. Much of the energy gained during cellular catabolism, or from light (photosynthesis), is used for the biosynthesis of cell constituents from simple molecules. The process of biosynthesis, therefore, needs energy, and it is known as anabolism.

8.2. Mobile electron carriers

Regardless of the process from which the energy is derived or how it is used, the transfer of electrons in a cell relies on numerous different electron carriers. Mobile, freely diffusible electron carriers, of which coenzymes NAD^+/NADH and $NADP^+$/NADPH are the most common, are involved in oxidation-reduction reactions within the cell necessitating the transfer of hydride ($H^- = H^+ + 2e^-$). These co-enzymes are similar, with a low electrode potential, E'_0, of −0.32 V (Figure 3.2); however, NAD^+/NADH is used principally in catabolic pathways while $NADP^+$/NADPH is used in anabolic pathways. The oxidized forms of these electron carriers gain electrons, and become reduced, from redox pairs with lower electrode potential.

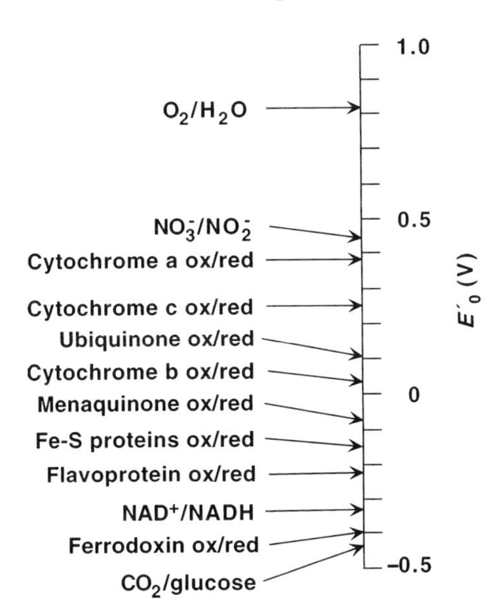

Figure 3.2 Electron tower showing the electrode potentials of various redox couples involved in electron transport chains leading to ATP formation by oxidative phosphorylation. Electrons may be transferred up the tower from redox couples with lower electrode potential to those with progressively higher electrode potentials. Electrode potentials are calculated at a pH of 7, but otherwise at standard state. Data from Thauer *et al.* (1977) and Madigan *et al.* (2003).

Once reduced, they can donate electrons to redox couples with a higher electrode potential. For example,

Substrate oxidation:

$$NAD^+ + H^+ + 2e^- \rightarrow NADH$$
$$Sub_{(red)} \rightarrow Sub_{(ox)} + 2e^-$$
$$\overline{Sub_{(red)} + NAD^+ + H^+ \rightarrow Sub_{(ox)} + NADH}$$

Substrate reduction:

$$NADH \rightarrow NAD^+ + H^+ + 2e^-$$
$$Sub_{(ox)} + 2e^- \rightarrow Sub_{(red)}$$
$$\overline{Sub_{(ox)} + NADH \rightarrow Sub_{(red)} + NAD^+ + H^+}$$

A real-world example is the reduction of pyruvate to lactate, coupled to the oxidation of NADH to NAD^+:

$$\text{Pyruvate} + NADH + H^+ \rightarrow \text{lactate} + NAD^+$$

8.3. Membrane-bound electron carriers and oxidative phosphorylation

Electron carriers are also bound in the cell membrane, which is a semi-permeable barrier separating the inside of the cell from the environment (Figure 3.3). In most prokaryotic cells, a rigid protective layer, the cell wall, is found just outside of the cell membrane. Membrane-bound electron carriers are arranged in a series, comprising an electron transport system, and they promote the transfer of electrons between an electron donor and an electron acceptor. The transfer of electrons, however, is not direct, and numerous small steps are used to ensure that energy is conserved in a form that can be used by the cell. Several of the enzymes in an electron transport chain direct positively charged protons to the outer surface of the cell membrane (Figure 3.4), generating a voltage gradient across the membrane. A pH gradient is also established, and the combined electrical and proton gradients are known as a proton motive force. The relaxation of these gradients is carefully controlled through an enzyme known as ATPase, which couples the energy gained from the import of protons across the cell membrane, to the synthesis of ATP. The import of three to four protons is coupled to the production of one ATP. This process of ATP generation is known as oxidative phosphorylation and is the principal means of ATP formation during respiration and photosynthesis (see Chapter 4).

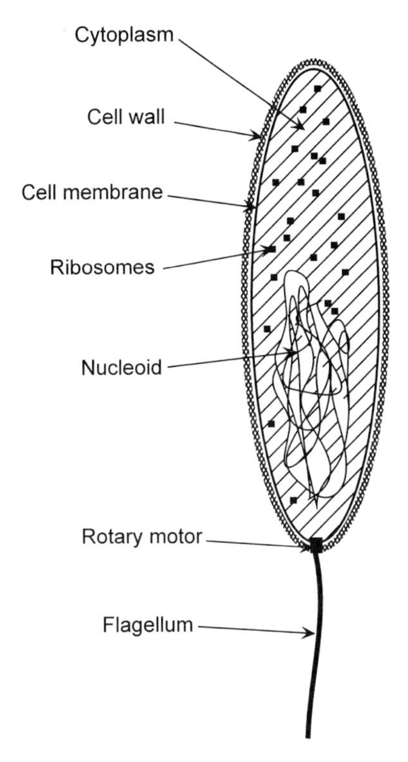

Figure 3.3 Key components of a gram-negative prokaryotic cell. Figure inspired by Margulis and Schwartz (1998).

8.4. ATP

As mentioned above, the energy gained from cellular catabolism and photosynthesis is derived ultimately from coupled oxidation-reduction reactions. These reactions are carefully controlled within the cell to maximize the transfer of chemical energy to the formation of ATP. ATP is constructed from the nucleoside adenosine (a ribose sugar combined with the nitrogen base adenine; see Chapter 1) connected to a triphosphate group through a phosphate ester linkage (Figure 3.5). The hydrolysis of the terminal phosphate on ATP, forming ADP (Figure 3.5), has a high-energy yield with a ΔG° of approximately -32 kJ mol^{-1} of ATP (Thauer *et al.*, 1977). ATP drives to completion, in cooperation with the appropriate enzymes, otherwise thermodynamically unfavorable reactions. Consider the following generic example of an unfavorable biosynthetic reaction:

$$A + B \rightarrow C + D \text{ (unfavorable)} \tag{3.33}$$

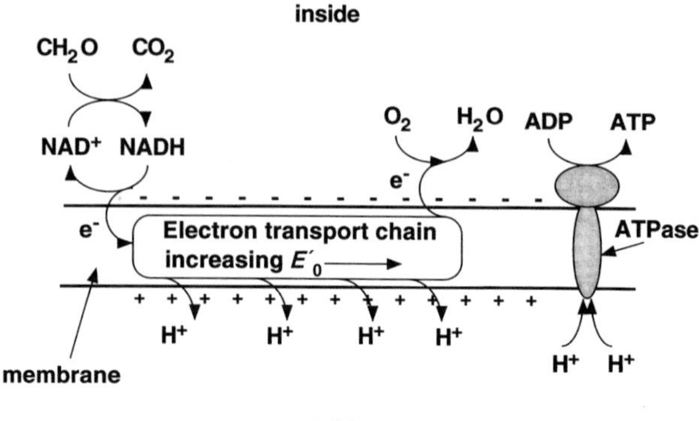

Figure 3.4 The principal features of ATP generation by oxidative phosphorylation. Electrons are derived at a low redox potential from the oxidation of a reduced electron donor. These electrons are passed through a series of membrane-bound redox couples, known as an electron transport chain, in which energy is conserved by translocating protons to the outer surface of the membrane. ATP is generated through the energy produced by the controlled mobilization of protons back into the cell through the enzyme ATPase. Electrons are finally consumed through the reduction of an electron acceptor, in this case oxygen, at a high redox potential.

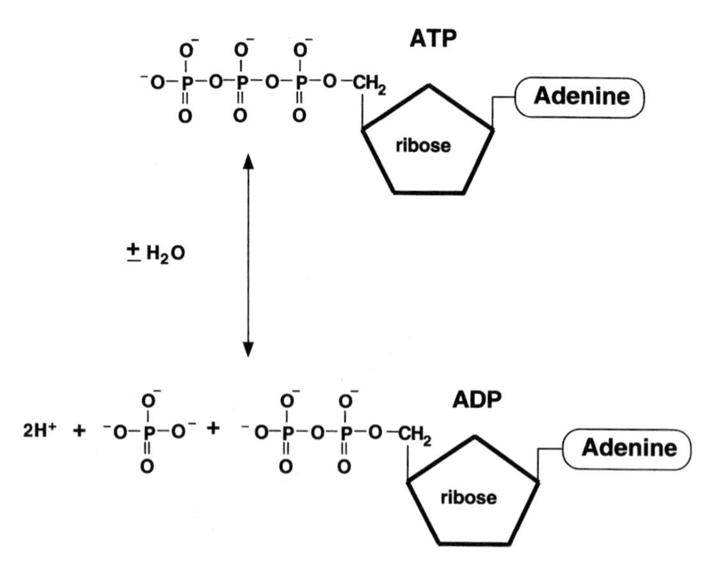

Figure 3.5 Schematic drawings of ATP and ADP and the relationship between the two.

This reaction sequence can be broken down into two favorable reactions with the release of energy during the hydrolysis of ATP and the formation of the high-energy intermediate compound, A-P. This is illustrated in Equations 3.34 and 3.35, which, upon addition, give the reaction in Equation 3.36, made favorable due to the hydrolysis of ATP to ADP.

$$\text{A} + \text{ATP} \rightarrow \text{A-P} + \text{ADP (favorable)} \tag{3.34}$$

$$\text{A-P} + \text{B} \rightarrow \text{C} + \text{D} + \text{P (favorable)} \tag{3.35}$$

$$\text{A} + \text{ATP} + \text{B} \rightarrow \text{C} + \text{D} + \text{ADP} + \text{P (favorable)} \tag{3.36}$$

An example is the reaction of glucose plus fructose to yield sucrose and water with an unfavorable ΔG^{o} of 23 kJ mol^{-1}:

$$\text{glucose} + \text{fructose} \rightarrow \text{sucrose} + H_2O,\ \Delta G^{o} = 23\ \text{kJ mol}^{-1}$$

However, when this reaction is coupled to the energy released during the hydrolysis of 2ATP to 2ADP, the formation of sucrose becomes favorable:

$$\begin{aligned}\text{glucose} + \text{fructose} + 2\text{ATP} &\rightarrow \text{sucrose} + 2\text{ADP}\\ &+H_2O + 2\text{P}, \Delta G^{o} = -41\ \text{kJ mol}^{-1}\end{aligned}$$

8.5. Fermentation and ATP generation

As mentioned previously, during fermentation organic compounds undergo coupled oxidation and reduction reactions, with no utilization of external electron acceptors such as oxygen or nitrate. Numerous different types of fermentation reactions are accomplished by microorganisms, and a few common fermentation pathways are presented below, including the fermentation of ethanol to acetate and H_2 gas (Equation 3.37), the fermentation of glucose to ethanol and CO_2 (Equation 3.38), the fermentation of glucose to lactate (Equation 3.39), and the fermentation of acetate to CO_2 and methane (Equation 3.40):

$$CH_3CH_2OH + H_2O \rightarrow CH_3COO^- + 2H_2 + H^+ \tag{3.37}$$

$$C_6H_{12}O_6 \rightarrow 2C_2H_6O + 2CO_2 \tag{3.38}$$

$$C_6H_{12}O_6 \rightarrow 2C_3H_4O_3^- + 2H^+ \tag{3.39}$$

$$H^+ + CH_3COO^- \rightarrow CO_2 + CH_4 \quad (3.40)$$

The oxidation-reduction reactions involved in these fermentation reactions are obvious, except perhaps for the fermentation of glucose to lactate (Equation 3.39), in which the oxidation and reduction occurs between the carbon atoms in glucose and in lactate. Thus, if glucose is written as $HCO(HCOH)_4H_2COH$ and lactate as $CH_3(HCOH)COO^-$, we see that the methyl carbon in lactate is more reduced (charge of -3), and the carboxyl carbon is more oxidized (charge of $+3$), than any of the carbon atoms in glucose (range of -1 to $+1$).

Generally, oxidative phosphorylation is not used to generate ATP during fermentation. A notable exception is the fermentation of acetate to methane and CO_2 (acetoclastic methanogenesis), in which a unique biochemistry generates a proton potential that is used to form ATP through ATPase (see Chapter 10). In most cases, however, ATP is formed during fermentation through the formation of phosphorylated intermediates in a process known as substrate level phosphorylation. The ATP yield during fermentation is not high. For example, the fermentation of glucose generally yields 2–4 ATPs per molecule of glucose fermented, whereas the oxidation of glucose with oxygen produces 32 ATPs (Fenchel *et al.*, 1998). However, the main advantage to fermentation is that no external electron acceptor is required. As we shall see in Chapter 5, fermentation plays a critical role in the anaerobic degradation of organic material.

When H_2 is produced during fermentation, the energetics of the process depend critically on the ambient concentration of H_2. Thus, under standard conditions, the fermentation of ethanol to acetate and H_2, as shown in Equation 3.37, has a positive Gibbs free energy change, ΔG^o, of 49.52 $kJ\,mol^{-1}$ ethanol. This reaction is clearly not favorable, and even at a pH of 7, with equal concentrations of ethanol and acetate, ΔG is still unfavorable (from Equation 3.13) at 9.55 $kJ\,mol^{-1}$ ethanol. With an H_2 partial pressure of 0.1 atm, the reaction becomes barely favorable with a ΔG of $-1.87\ kJ\,mol^{-1}$ ethanol, and it becomes increasingly more favorable as pH_2 decreases (Figure 3.6).

Due to the low solubility of H_2 gas in water (Table 3.2), an $H_{2(g)}$ partial pressure of 0.1 atm is equivalent to only 80 μM $H_{2(aq)}$ at 25 °C. It is obvious that to maintain active fermentation in natural environments, some mechanism must be in place to limit the accumulation of H_2 in solution. Therefore, active H_2 production also requires active H_2 removal, and this is accomplished with microbial metabolisms coupling H_2 as an electron donor with a variety of different electron acceptors.

Indeed, H_2 provides an excellent electron donor to numerous types of microbial metabolisms, including methanogenesis by CO_2 reduction,

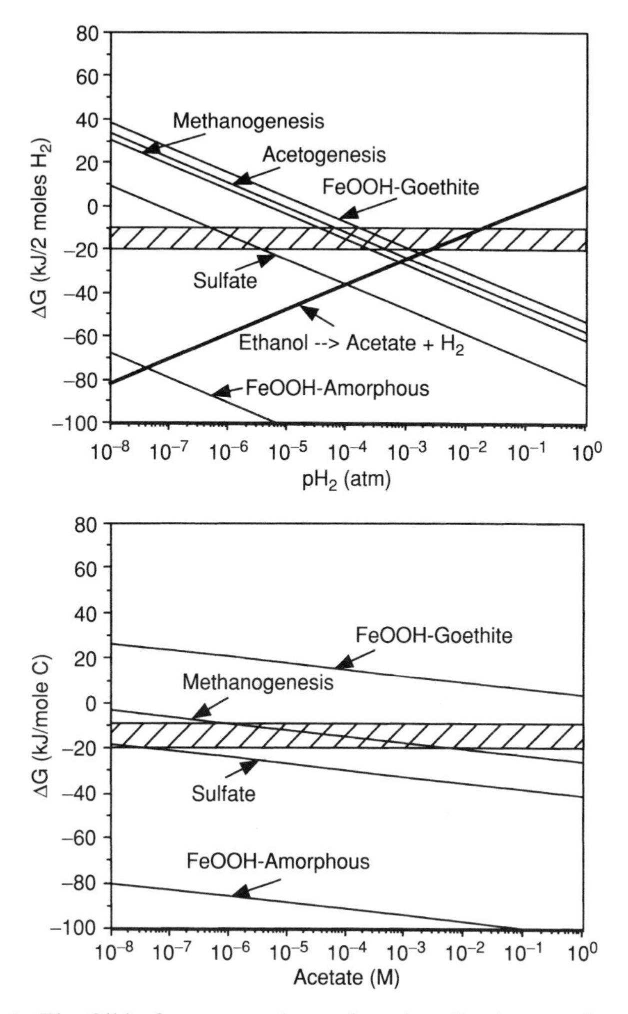

Figure 3.6 The Gibbs free energy change for mineralization reactions with vàrious electron acceptors using H_2 and acetate as electron donors. Free energy has been calculated for reactions yielding four electrons transferred, at a pH of 7, and for reasonable environmental concentrations of reactants and products. Also shown is the free energy change associated with the fermentation of ethanol to acetate and H_2.

acetogenesis, sulfate reduction, iron reduction, manganese reduction, and others (Table 3.5). The transfer of H_2 between fermenting organisms and organisms utilizing H_2 is known as interspecies H_2 transfer. This is a syntrophic relationship (see Chapter 2) and is just one of many types of mutually beneficial metabolic associations found in nature. Similar to H_2, the

Table 3.5 Examples of H_2 consuming respiratory reactions in nature

Reaction	Organisms
$O_2 + 2H_2 \rightarrow 2H_2O$	Hydrogen bacteria
$2\ Fe(OH)_3 + H_2 + 4H^+ \rightarrow 2Fe^{2+} + 6H_2O$	Fe reducers
$MnO_2 + H_2 + 2H^+ \rightarrow Mn^{2+} + 2H_2O$	Mn reducers
$CO_2 + 4H_2O \rightarrow CH_4 + 2H_2O$	Methanogens
$2CO_2 + 4H_2 \rightarrow CH_3COO^- + H^+ + 2H_2O$	Acetogens
$H_2 + S^o \rightarrow H_2S$	Sulfur reducers
$4H_2 + SO_4^{2-} + 2H^+ \rightarrow H_2S + 4H_2O$	Sulfate reducers

accumulation of other fermentation products, such as acetate, may also affect the thermodynamics of the fermentation process. Thus, active fermentation also requires active removal of fermentation products other than H_2 (Lovley and Phillips, 1987).

8.6. Minimum energy for growth

To sustain growth, organisms need to utilize a reaction with a ΔG considerably lower than zero (Thauer *et al.*, 1977). The threshold for microbial growth is usually considered as the energy needed to produce ATP, and as mentioned previously, under standard state conditions the production of ATP from ADP has a free energy of -32 kJ mol^{-1}. However, a larger ΔG of about -50 kJ mol^{-1} is required to produce ATP under the chemical conditions of a growing cell, where the concentrations of ATP, ADP, and phosphate deviate considerably from standard state (Schink, 1997). In addition, accounting for the energy lost as heat, ATP formation requires a ΔG of approximately -70 kJ mol^{-1} of ATP synthesized (Schink, 1997). This, however, is not the minimal energy needed for microbial growth. Recall that the formation of one ATP during oxidative phosphorylation is coupled to the mobilization of three to four protons across a semipermeable membrane. Therefore, the minimal metabolically convertible energy is considered to be the energy needed to translocate one proton, or to form 1/3 to 1/4 ATP. This is therefore around -20 kJ per 1/3 to 1/4 mole of ATP.

Anaerobic systems in nature are often poised at what appears to be a threshold near the minimal energy necessary to sustain microbial growth (Conrad *et al.*, 1986; Conrad, 1999). For example, when respiration reactions are written as four electron transfers (equivalent to the oxidation of one organic carbon; see below), the free energy gain associated with anaerobic metabolism during sulfate reduction, methanogenesis, and acetogenesis is consistently around -20 kJ mol^{-1} of organic carbon oxidized (Table 3.6).

Table 3.6 ΔG values for anaerobic mineralization processes *in situ* and in laboratory experiments with sediment slurries. Values are calculated from the chemistry of the environment or the slurry experiments

Process	ΔG (kJ per mole org C)[a]		Reference
Sulfate reduction	−23	±1.2	Hoehler *et al.* (1998)
Methanogenesis	−20	±0.6	Hoehler *et al.* (1998)
Methanogenesis	−15	±4	Lovley and Goodwin (1998)
Acetogenesis	−18	±1.1	Hoehler *et al.* (1998)

[a]Or equivalent, two moles of H_2 are equivalent to one mole of organic carbon.

As we shall see, this has implications for the competition between different anaerobic microbial populations for electron donors in the environment. Note that nitrate reduction and Mn reduction (and possibly also Fe reduction in some cases) conduct their metabolisms at energy yields considerably more negative than the minimal threshold discussed here (see below and also Hoehler *et al.*, 1998).

9. ENERGETICS OF ORGANIC MATTER MINERALIZATION DURING RESPIRATION

9.1. Free energy gain

The free energy gain associated with the oxidation of electron donors such as H_2, acetate or other organic compounds varies with the different electron acceptors. A careful consideration of these differences helps us to understand the stratification of microbial communities in environments such as sediments and anoxic water columns. To illustrate this point, we calculate the free energy gain associated with the oxidation of H_2 and acetate under standard state conditions (Table 3.7). The order of the sequence varies somewhat depending on the electron donor used. Also, the specific electron donors used in the calculation, H_2 and acetate, are more appropriate for anaerobic metabolisms (without O_2) than for aerobic metabolisms (utilizing O_2) (see Chapter 5). Nevertheless, we see that, consistent with numerous previous discussions (e.g., Berner, 1980), the greatest free energy gain is associated with oxic respiration, whereas the lowest free energy gain is associated with methanogenesis. Therefore, based strictly on energetic considerations, oxic respiration is the most favorable process of organic carbon mineralization, whereas methanogenesis is the least favorable. This sequence in free energy gain is roughly coincident with the depth distribution

Table 3.7 Standard Gibbs free energy calculated for the principal respiratory pathways of organic matter mineralization in nature, with H_2 and acetate as electron donors

	kJ per reaction	
Reaction	ΔG^0 (H_2)	ΔG^0 (acetate)[a]
Oxic respiration		
$O_2 + 2H_2 \rightarrow 2H_2O$	−456	–
$O_2 + 1/2C_2H_3O_2^- \rightarrow HCO_3^- + 1/2H^+$	–	−402
Denitrification		
$4/5H^+ + 4/5NO_3^- + 2H_2 \rightarrow 2/5N_2 + 12/5H_2O$	−460	–
$4/5NO_3^- + 3/5H^+ + 1/2C_2H_3O_2^- \rightarrow 2/5N_2$ $+HCO_3^- + 1/5H_2O$	–	−359
Mn reduction (pyrolusite)		
$4H^+ + 2MnO_2 + 2H_2 \rightarrow 2Mn^{2+} + 4H_2O$	−440	–
$7/2H^+ + 2MnO_2 + 1/2C_2H_3O_2^- \rightarrow 2Mn^{2+}$ $+HCO_3^- + 2H_2O$	–	−385
Fe reduction (freshly precipitated amorphous FeOOH)		
$8H^+ + 4FeOOH + 2H_2 \rightarrow 4Fe^{2+} + 8H_2O$	−296	–
$15/2H^+ + 4FeOOH + 1/2C_2H_3O_2^- \rightarrow HCO_3^-$ $+4Fe^{2+} + 6H_2O$	–	−241
Sulfate reduction		
$H^+ + 1/2SO_4^{2-} + 2H_2 \rightarrow 2H_2O + 1/2H_2S$	−98.8	
$1/2H^+ + 1/2SO_4^{2-} + 1/2C_2H_3O_2^- \rightarrow$ $1/2H_2S + HCO_3^-$	–	−43.8
Methanogenesis		
$1/2H^+ + 1/2HCO_3^- + 2H_2 \rightarrow H_4 + 3/2H_2O$	−74.8	–
$1/2H_2O + 1/2C_2H_3O_2^- \rightarrow CH_4 + 1/2HCO_3^-$	–	−19.9

[a]Values are standardized to a four e^- transfer equivalent to the oxidation of one mole of organic carbon with a charge of 0, as in carbohydrates. Calculation conditions: 25 °C and unit activity for all reactants and products.

of electron acceptor utilization in sedimentary environments (Froelich *et al.*, 1979; Canfield *et al.*, 1993a,b). Thus, oxic respiration occurs highest in the sediment column, followed generally by denitrification, and so on, until finally methanogenesis occurs after the other electron acceptors are depleted.

9.2. Competition for electron donors

We can explore the underlying ecological reasons for this tendency of microbial communities to stratify in nature by considering the competition between microbial communities for substrate. This competition is dictated by the energetics of the respiratory processes and, therefore, has an underlying thermodynamic rationale. We begin by calculating the free energy gain associated with the various significant respiratory processes, with both H_2 and acetate as electron donors. These calculations are performed at a pH of 7, and with realistic concentrations for all of the dissolved and gaseous species used or produced during bacterial metabolism (Figure 3.6).

The energetics of oxic respiration, denitrification, and Mn reduction are all highly favorable for environmental concentrations of H_2 and acetate, and they are not included in Figure 3.6. Considerations other than competition for an electron donor probably determine the stratification of these microbial populations. By contrast, the thermodynamic favorableness of Fe reduction, sulfate reduction, methanogenesis, and acetogenesis is highly dependent on electron donor concentration (Figure 3.6). This dependency, and the relative differences in the energetics of the processes, forms the basis for the competitive exclusion of one respiratory process over another. Thus, with H_2 as an electron donor, an H_2 partial pressure of 10^{-7} atm (equivalent to around 0.08 nM $H_{2(aq)}$, a typical value for a sediment supporting active Fe reduction; see below) allows a highly favorable free energy gain for Fe reduction of approximately -75 kJ per 2 moles of H_2 oxidized (for reactions see Table 3.7). However, whereas Fe reduction may proceed at 10^{-7} atm H_2, the other processes are not energetically favorable. Therefore, if Fe reducers can metabolize at H_2 partial pressures below those at which the other processes are thermodynamically favorable, then Fe reduction will dominate, and the other anaerobic respiration pathways will be inhibited. The energetics of Fe reduction, however, depend critically on the nature of the solid iron phase being reduced. Thus, while the reduction of amorphous FeOOH is favorable at a pH_2 of 10^{-7} atm, the reduction of crystalline goethite is highly unfavorable.

After amorphous iron oxides become utilized, pH_2 should rise until the energetics of the next respiration process becomes favorable. Thus, sulfate reduction becomes favorable at a pH_2 of around 10^{-5}. This level of H_2 will produce a free energy gain of -20 kJ mol^{-1} per 2 moles of H_2 oxidized, which should be sufficient to fuel microbial growth. Importantly, the other respiration processes, methanogenesis and acetogenesis, are not thermodynamically possible. Thus, if sulfate reducers can maintain a pH_2 near their threshold for growth, then methanogens and acetogens are inhibited. These processes become favorable after sulfate is depleted and pH_2 rises further. In a similar way, the maintenance of low acetate concentrations by Fe reducers can inhibit sulfate reducers (again, depending on the nature of the iron oxide), and sulfate reducers can inhibit methanogens at somewhat higher acetate concentrations (Figure 3.6).

In nature, it appears that Fe reducers, sulfate reducers, methanogens, and acetogens metabolize at near the minimum energy needed for ATP

Table 3.8 Concentrations of H_2 and acetate in sediments supporting different respiratory processes

Process	H_2 (nM)	Acetate (μM)
NO_3^- respiration	0.05–0.04	
Fe reduction	0.05	±0.1
Sulfate reduction	0.5–3.0	±0.2
Methanogenesis	2–12	±0.8
Acetogenesis	150 ± 50	

Data from Hoehler *et al.* (1998), Lovley and Phillips (1987a,b), and Lovley and Goodwin (1988).

generation and growth (Lovley and Goodwin, 1988; Hoehler *et al.*, 1998; Conrad, 1999). By doing so, the concentrations of electron donors are maintained too low for other processes with a lower energy yield. Supporting this scenario is a strong relationship between H_2 concentration and the type of microbial metabolism occurring in sediments (Table 3.8).

This, however, is only part of the story. The calculations presented in Figure 3.6 have been made at a constant pH of 7. In the environment, pH may range greatly, but values between 6 and 9 are common. Furthermore, the thermodynamics of some of the respiratory processes are highly pH dependent, and of the processes considered here, Fe reduction is the most highly affected. Thus, Fe reduction with goethite is favorable compared to sulfate reduction only at pHs below around 6.3 (Figure 3.7), and Fe reduction with amorphous FeOOH is favorable at pHs below 9. These calculations assume an acetate concentration of 10^{-6} M and a pH_2 of 10^{-4} atm (realistic average values for sediments; see above). Environmental pH is therefore an important controlling factor on the significance of Fe reduction in nature. We underscore the necessity of carefully considering the thermodynamics of the microbial processes of interest in any given environment.

10. NAMING ENERGY METABOLISMS

We have already discussed how catabolic (also called dissimilatory) processes and light provide the energy for the anabolic (also called assimilatory) synthesis of cellular material. A vast array of different energy-providing metabolisms exist in nature, and a common nomenclature has been adopted whereby these metabolisms are named based on their (1) energy source, (2) electron sourcer, and (3) carbon source (Figure 3.8). Thus, energy may be provided either by light, whereby the organism is known as a phototroph, or from chemical energy in the absence of light, whereby the organism is known as

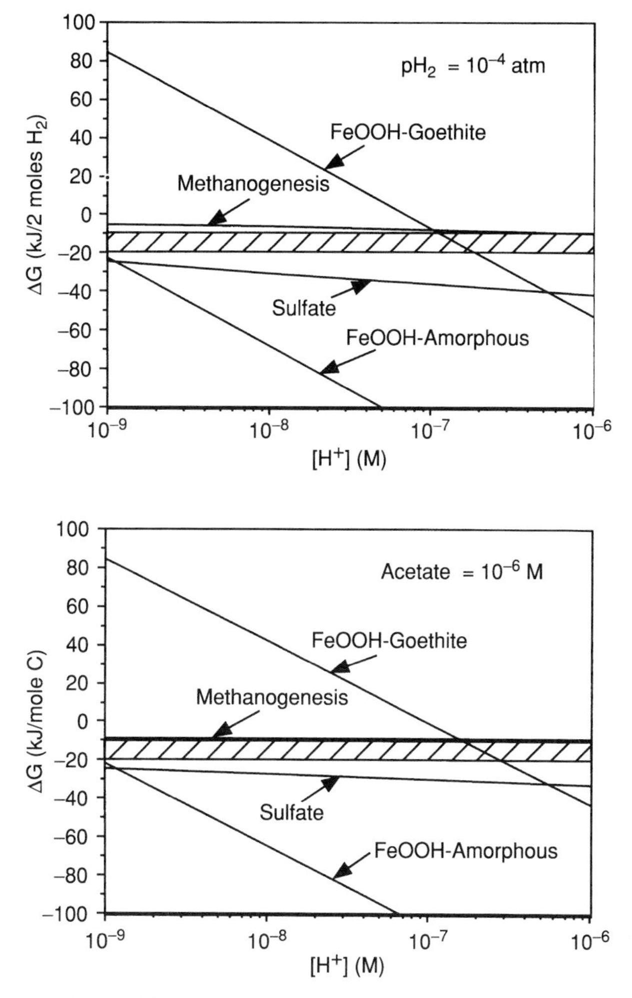

Figure 3.7 Relationship between the free energies associated with Fe reduction, sulfate reduction, and methanogenesis and pH, with both H_2 and acetate as electron donors.

a chemotroph. The electron source may be an inorganic compound, whereby the organism is a lithotroph, or an organic compound, whereby the organism is an organotroph. If the carbon source is CO_2, the organism is an autotroph, and if it uses organic compounds, it is a heterotroph.

In principle, all three descriptors should be used to name an organism's metabolism in the following order: energy source → electron source → carbon source. For example, an organism using chemical energy, an

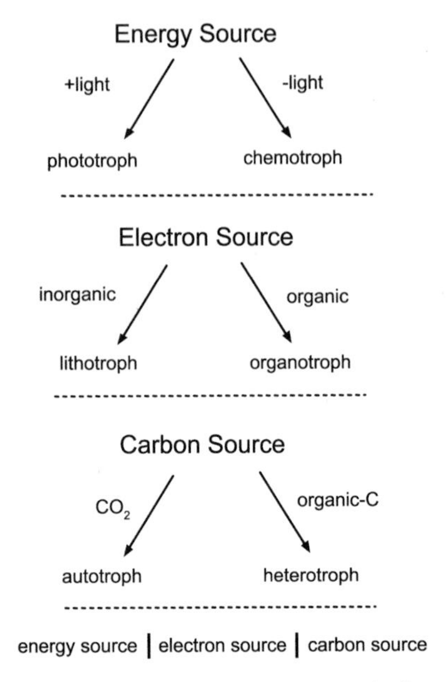

Figure 3.8 Naming energy metabolisms.

inorganic electron donor, and CO_2 for carbon is a chemolithoautotroph. This is a common type of metabolism at interfaces of electron donor and electron acceptor, such as, for example, the O_2–H_2S interface in sediments or the water column. An organism using light energy, an inorganic electron donor, and an organic source of carbon is known as a photolithoheterotroph. Many anoxygenic photosynthetic purple bacteria can be classified this way (see Chapter 9).

Chapter 4

Carbon Fixation and Phototrophy

1. INTRODUCTION

The interplay between autotrophic and heterotrophic processes provides the basis for the extremely efficient biological utilization of the available solar and chemical energy within the environment. Thus, oxygenic photosynthesis, limited only by the availability of water, light, and nutrients, fixes CO_2 into organic biomass and provides nearly all of the primary production of organic matter at the Earth's surface (Figure 4.1). Oxygenic photosynthesis is an example of photolithoautotrophy. Organic matter so produced is utilized and decomposed by heterotrophic organisms, and when the decomposition is by anaerobic processes, utilizing inorganic electron acceptors such as NO_3^-, SO_4^{2-}, Fe oxides, and Mn oxides, reduced species such as NH_4^+, H_2S, Fe^{2+}, and Mn^{2+} accumulate (see subsequent Chapters). These compounds typically establish reaction fronts with more oxidized species where organisms can, in many instances, promote thermodynamically favorable redox reactions and gain energy for their growth (Figure 4.1). Examples include, but are not restricted to,

ADVANCES IN MARINE BIOLOGY VOL 48
0-12-026147-2

© 2005 Elsevier Inc.
All rights reserved

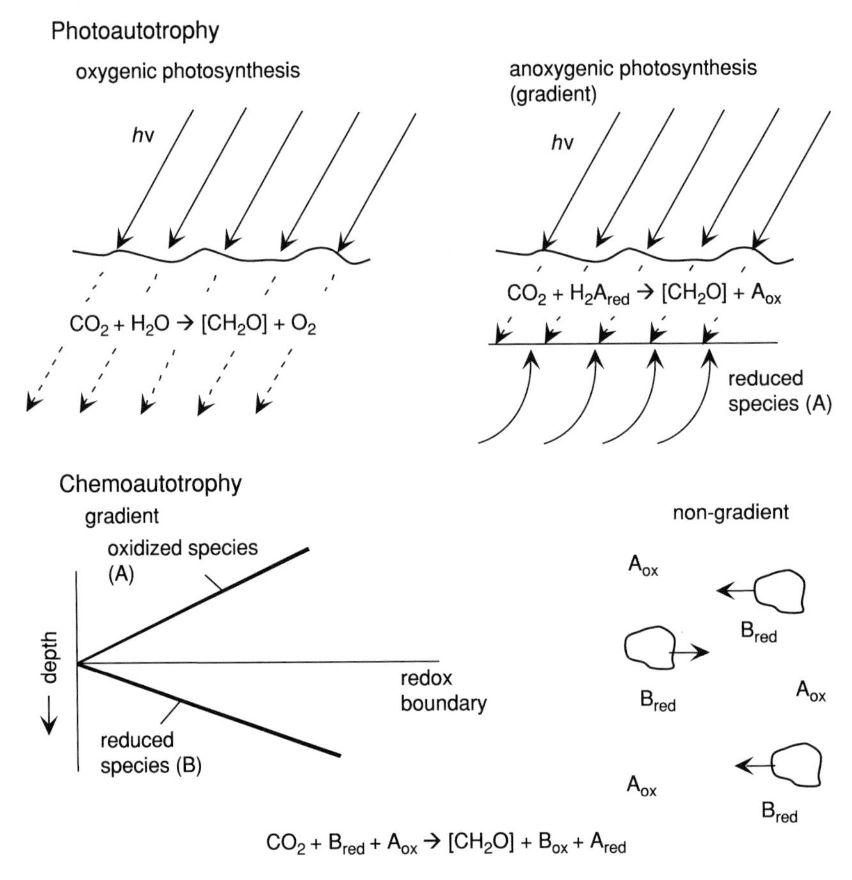

Figure 4.1 Examples of different environmental situations supporting phototrophic and chemotrophic CO_2 fixation.

interfaces between O_2 and H_2S, O_2 and NH_4^+, NO_3^- and H_2S and NO_3^- and Fe^{2+}. Indeed, large populations of organisms may be supported by the chemical energy generated at these redox gradients. Mats and layers of colorless sulfur bacteria, as sometimes found atop and within highly reducing sediments, are highly conspicuous examples (Jørgensen, 1977b; Jørgensen, 1982a; Fossing *et al.*, 1995) (see Chapter 9). Organisms promoting the reaction between inorganic chemical species are often able to fix inorganic carbon as chemolithoautotrophs, or if organic matter is also required, as mixotrophs.

Chemolithoautotrophy need not take place, however, in chemical gradients (Figure 4.1). Electron donors may also be provided from disbursed sources; H_2 produced during fermentation is a good example (see Chapters 3 and 5). Hydrogen gas is an extremely good electron donor, and rapid utilization by microorganisms leaves only very low concentrations in the environment (see Chapter 3). Therefore, large vertical gradients of H_2, as is often found with NH_4^+ and H_2S in anoxic environments, are difficult, if not impossible, to establish. Once produced, H_2 may fuel a variety of autotrophic metabolisms coupled to, for example, methanogenesis, acetogenesis, and sulfate reduction (see Chapters 3, 9, and 10).

When the reduced species H_2S, Fe^{2+}, and H_2 are found in the presence of light (Pfennig, 1977; Widdel *et al.*, 1993), phototrophic organisms may use them to fuel CO_2 fixation and to incorporate organic compounds into cellular biomass. These organisms are known as anoxygenic phototrophs, as no oxygen is produced during their metabolism; rather, oxidized counterparts to the reduced compounds are produced (see Chapters 8 and 9). Classical examples of habitats for anoxygenic phototrophs include cyanobacterial mats and the anoxic water columns of lakes and anoxic marine basins (Figure 4.1; see also Chapters 9 and 13). Anoxygenic phototrophs are not, however, bound to redox gradients or even in proximity to anoxic habitats. Aerobic anoxygenic photoheterotrophs (AAPs) are obligate aerobes, related to the photosynthetic purple bacteria (see Chapter 9), that use light and bacteriochlorophyll *a* (BChl *a*) to drive ATP production. This enhances the efficiency of dissolved organic carbon (DOC) utilization within the cell (Kolber *et al.*, 2001), as less of the DOC needs to be oxidized to fuel the production of ATP. When starved of organic carbon, CO_2 fixation can be induced, although the details of this process are presently limited. Somewhat similarly, although based on a completely different type of photosynthetic system, photosynthetic prokaryotes have been discovered utilizing bacteriorhodopsin as a light-harvesting pigment, which acts as a proton pump driving ATP synthesis. These anoxygenic phototrophs are also aerobes and heterotrophs and are widely distributed as members of the *Bacteria* through the surface ocean (Béjè *et al.*, 2001; Venter *et al.*, 2004). Previously, this type of photosynthesis was only known among the halobacteria within the *Archaea* (see Chapter 1). The ability of bacteriorhodopsin-utilizing phototrophs to fix CO_2 is unknown.

Thus, both chemotrophic and phototrophic processes are used by organisms to recycle inorganic compounds and organic matter efficiently in the environment, maximizing the use of available energy. The remainder of this Chapter focuses on the processes whereby organisms fix CO_2 into organic biomass, leaving the discussion of heterotrophic and mixotrophic processes to subsequent Chapters. This Chapter, therefore, explores how energy is gained through both chemotrophic and phototrophic metabolisms and how this energy is used to fix CO_2. Some attention is also given to the principal known CO_2 fixation pathways. We also speculate as to the evolutionary history of the CO_2 fixation processes. Aspects of isotope fractionation during CO_2 fixation are explored, and some specific examples of chemolithoautotrophic-driven systems are provided.

2. CHEMOAUTOTROPHIC ENERGY CONSERVATION

The energy used to drive carbon fixation by chemotrophs is conserved from thermodynamically favorable reactions between inorganic electron donors and electron acceptors, or from light reactions (Figure 4.2). The basis for energy conservation at the cellular level is quite similar to energy conservation during respiration, except that the electron donor (e.g., H_2S, Fe^{2+} or H_2^{+}) is inorganic. Ultimately, reducing power from the electron donor is used to generate NAD(P)H, some of which, through a series of electron transport reactions among membrane-bound electron carrier proteins (see Chapter 3), produces a proton motive force conducive for ATP formation (oxidative phosphorylation). Recall that ATP is formed by the controlled translocation of protons with the enzyme ATPase across a membrane over which a proton motive force is established (Chapter 3).

Most carbon-fixing chemotrophs have a special problem in producing a sufficiently reduced redox potential to produce NAD(P)H. This is because most of the electron donors driving chemolithotrophic reactions have too positive an electrode potential to reduce $NAD(P)^{+}$ to NAD(P)H (Figure 4.3). Only H_2 produces an electrode potential sufficiently negative to form NAD(P)H directly. For other chemolithoautotrophs, energy must be used to shuttle electrons backward through a reverse electron transport system to produce an electrode potential sufficiently negative to produce NAD(P)H (Figure 4.3). The utilization of a reverse electron transport system is energy expensive, and it severely limits the amount of energy that can be channeled by the organism into growth. Thus, organisms utilizing electron donors with particularly positive redox potentials, such as iron oxidizers, for example, must oxidize a large amount of electron donor to produce cell material.

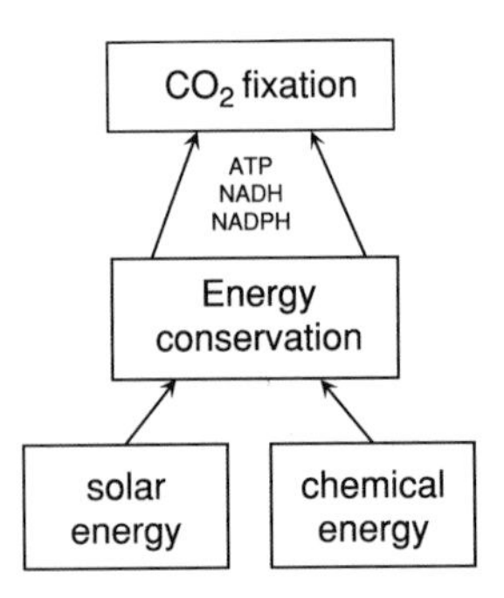

Figure 4.2 The principal steps in CO_2 fixation, emphasizing the similarity of the processes regardless of whether the energy is derived from light or from oxidation-reduction reactions.

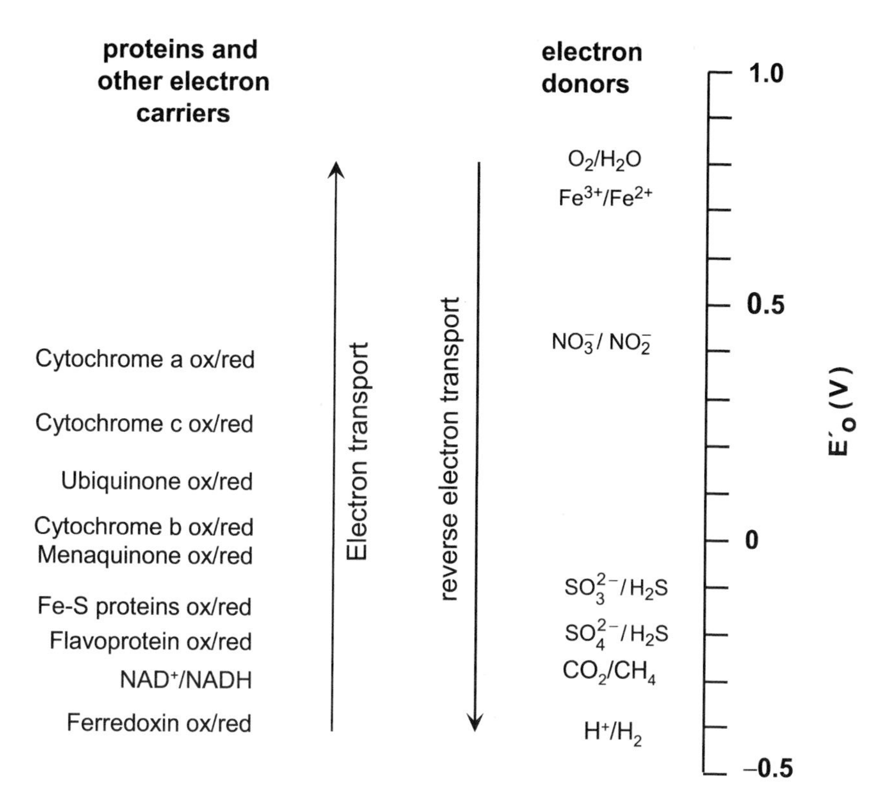

Figure 4.3 Electron tower showing some important electron carriers involved in electron transport and some important electron donor redox pairs. Electrode potentials are calculated at a pH of 7, but otherwise at standard state (see Chapter 3 for details). Electron donor redox couples are generally not reducing enough to form NADH directly, so reverse electron transport is frequently required.

3. PHOTOTROPHIC ENERGY CONSERVATION

Phototrophs conserve energy from light reactions to produce ATP and reducing power for carbon fixation or to supplement the energy needed for organic carbon uptake and utilization by the cell. In all photoautotrophs (excluding phototrophic *Archaea* and *Bacteria* using bacteriorhodopsin), light is harvested for eventual conservation as usable energy in an antenna complex consisting of a variety of light-sensitive pigments, including chlorophylls,

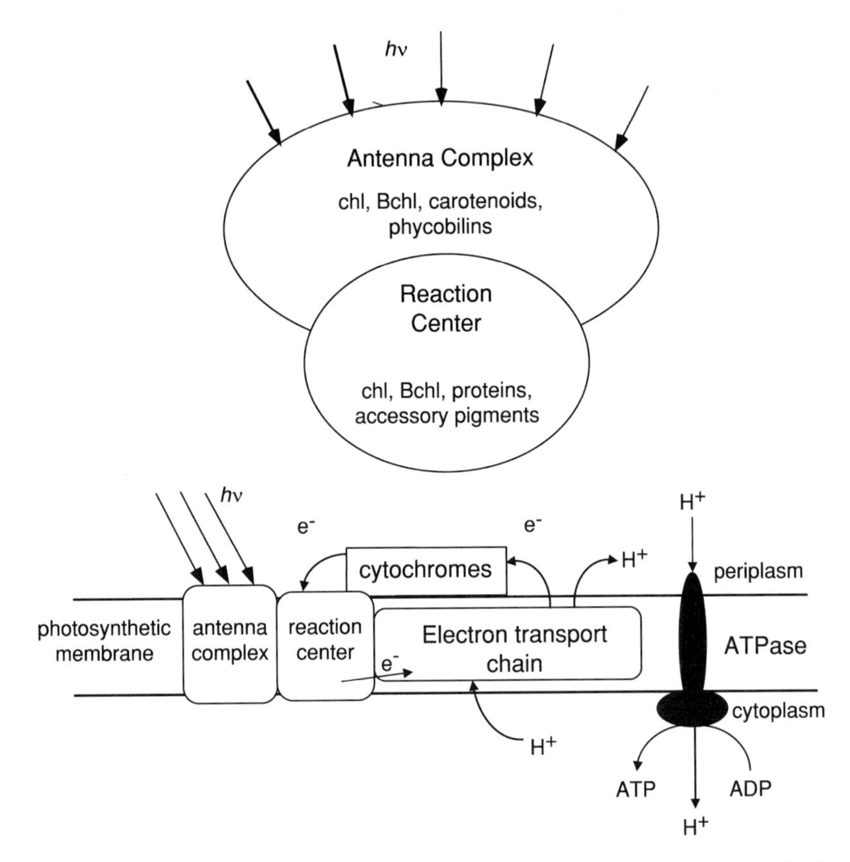

Figure 4.4 The photosynthetic unit in anoxygenic phototrophs, composed of an antenna complex and a reaction center. Also shown is the relationship of the photosynthetic unit to the electron transport chain promoting ATP formation in photosynthetic purple bacteria. Note that cyclic electron flow is indicated with electrons from the reaction center being returned through the electron transport chain. The processes forming reducing power (NADPH) for carbon fixation are not shown. Adapted from Madigan *et al.* (2003).

bacteriochlorophylls, carotenoids, and phycobilins (Figure 4.4). Chlorophyll molecules are unique to oxygen-producing phototrophs, whereas bacteriochlorophylls are utilized by anoxygenic phototrophs. The carotenoids and phycobilins are accessory pigments that contribute to light harvesting, and in the case of carotenoids, they also serve to protect the primary light-harvesting pigments from damage by excessive solar radiation. Each of the primary and secondary light-harvesting pigments has absorption properties allowing the capture of energy from only a portion of the visible and near-infrared range of the electromagnetic spectrum (see Chapter 9). However, when various pigments are used together, for example, bacteriochlorophyll molecules in association with carotenoids, a much greater portion of the available light energy may be harvested and transferred to the reaction center (Figure 4.5).

The energy captured by the antenna complex is transferred to the reaction center, where reaction center chlorophyll or bacteriochlorophyll, together with a variety of proteins and accessory pigments, promotes the final transfer of light energy to ATP formation and in some cases reducing power (Figure 4.4). Two types of reaction centers are found in anoxygenic photosynthetic organisms. In one type, characterized by anoxygenic photosynthetic purple bacteria and *Chloroflexus* spp., the principal electron carriers are pheophytin and quinone; these are therefore known as pheophytin-quinone type reaction centers (Figure 4.6). In another type of reaction center, characterized by green sulfur bacteria and heliobacteria, membrane-bound iron-sulfur proteins act as early electron acceptors; these are, therefore, Fe-S reaction centers (Figure 4.6). The photosynthetic process in oxygenic phototrophs utilizes two coupled reaction centers (Figure 4.7), termed photosystem I and photosystem II. Photosystem II has an electron transport chain

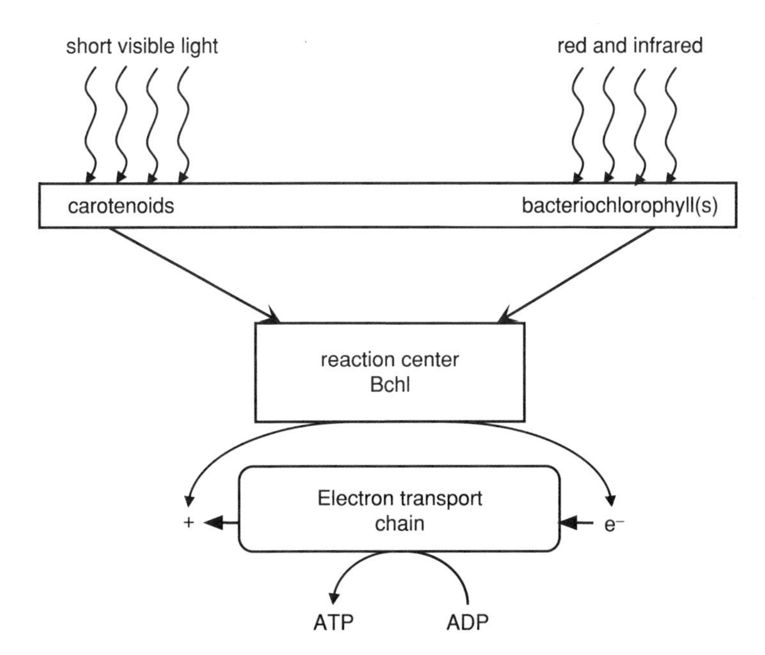

Figure 4.5 Accessory pigments (carotenoids), in concert with bacteriochlorophyll, can increase the range of light available to anoxygenic phototrophs. In oxygenic phototrophs, phycobilins increase the range of usable light filling in much of the green and yellow parts of the visible spectrum, compared to the blue and red absorption lines of Chl *a*. See text and Chapter 9 for details. Inspired by Kondratieva *et al.* (1992).

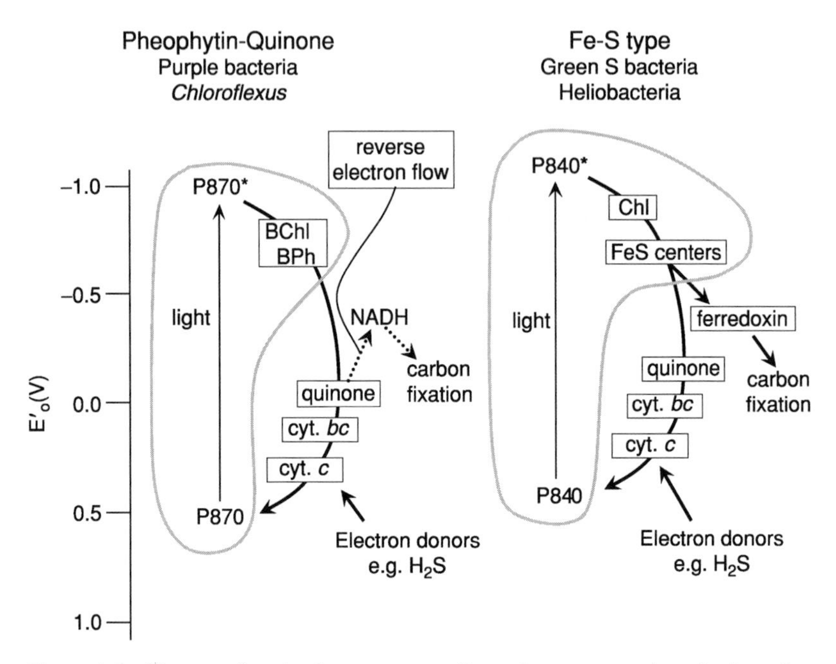

Figure 4.6 Electron flow in the two types of reaction centers, pheophytin-quinone and Fe-S, used by bacteriochlorophyll-containing anoxygenic phototrophs. P designates the different types of reaction centers, P870 for purple bacteria and *Chloroflexus*, P840 for green S bacteria, and P798 for heliobacteria (not shown). BChl refers to bacterio-chlorophyll and Chl to chlorophyll, while cyt refers to various cytochromes. The processes occurring in the light are enclosed within the heavy grey envelope. Note that for the pheophytin-quinone type reaction center the quinone is too oxidized to form NADH directly, so a reverse electron transport chain must be used. However, the ferrodoxin in the Fe-S type reaction center is reducing enough to form NADH directly. Electron flow may be both cyclic and non-cyclic. In non-cyclic electron flow with the pheophytin-quinone reaction center, quinone in the reaction chain donates an electron to NADH (dotted line). Electrons lost in NADH formation are replenished though the oxidation of a reduced electron donor such as H_2S. The electron flow in the Fe-S reaction center is mostly non-cyclic, and again, electrons lost are replenished from a reduced electron donor. Adapted from a figure made available by Raymond Cox.

that is quite similar to the pheophytin-quinone type reaction center in anoxygenic phototrophs, while electron transport in photosystem I bears a strong resemblance to that utilized by the Fe-S type reaction center (Figure 4.7). It is believed that the coupled photosystems of oxygenic phototrophs originated by the merging of the two preexisting anoxygenic photosynthetic reaction centers (Blankenship, 1992).

The steps in energy conservation during photosynthesis include the light-mediated excitation of the reaction center chlorophyll (or bacteriochlorophyll) producing a strong reductant (Figures 4.6 and 4.7). This excited molecule can

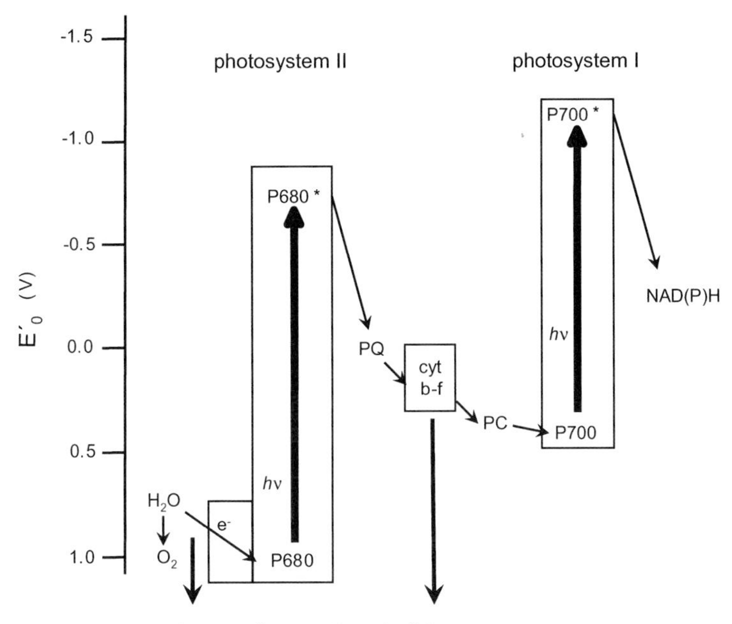

Figure 4.7 Highly simplified cartoon of the "Z" scheme of electron transport used in oxygenic photosynthesis. Two separate but connected photosystems are used. The oxidation of H_2O to O_2 (through a Mn cluster) donates electrons to photosystem I. These electrons replace those lost from the highly reducing light-excited P680*, which donates electrons to photosystem II through a series of electron carriers. The light-excited photosystem II chlorophyll P700* donates electrons to the reduction of $NAD(P)^+$ to NAD(P)H, which is used further in biosynthesis. In this scheme PQ refers to a combination of pheophytins and quinones, of which cyt b-f is the cytochrome b_6f complex. Many electron carriers in the path are not shown; the reader is referred to sources such as Blankenship (2002) and Falkowski and Raven (1997) for a more complete description. Note the similarity between the two photosystems in oxygenic photosynthesis and the two different photosystems in anoxygenic photosynthesis (Figure 4.6). Redrawn from a figure kindly made available by Raymond Cox.

transfer electrons to an electron acceptor with a low redox potential, which, through subsequent electron transfers between membrane-bound proteins, generates a gradient in charge and proton concentration across the photosynthetic membrane, promoting ATP formation through proton translocation by ATPase (Figure 4.4). This is analogous to ATP generation during respiration and by chemolithoautotrophic organisms. With purple bacteria and some green bacteria (*Chloroflexus* spp.), the electrons utilized to generate the proton motive force are used to reduce the oxidized reaction center pigments from which they were derived. There is no net loss or gain of electrons in this process, which is referred to as cyclic electron flow (Figures 4.4 and 4.6).

With cyclic electron flow only ATP is produced directly. The reducing power (NADPH) necessary to drive CO_2 reduction to organic matter is derived from an electron donor by a pathway similar to chemolithoautotrophic organisms (see Section 2 above). As discussed previously, except for H_2, most of the electron donors used by anoxygenic phototrophs (e.g., Fe^{2+}, H_2S, S°, $S_2O_3^{2-}$) have an electrode potential that is too positive to form NAD(P)H directly. Therefore, reverse electron flow must be utilized for NAD(P)H formation (Figure 4.3). As with energy conservation among chemolithoautotrophic organisms, only H_2 has a sufficiently reducing electrode potential for direct formation of NAD(P)H (Figure 4.3). The energy used to drive the reverse electron flow is the trans-membrane proton potential generated at the photosynthetic reaction center.

In some green sulfur bacteria and heliobacteria, NAD(P)H may be formed directly from electrons made available from reaction center bacteriochlorophyll. In this case, the electron flow is noncyclic, as NAD(P)H formation constitutes a loss of electrons from the reaction center (Figure 4.6). These electrons are replenished by the oxidation of reduced electron donors, not including H_2O (Table 4.1). For organisms with the Fe-S reaction center, NADH may be formed directly from reduced ferrodoxin (Figure 4.6).

In oxygenic phototrophs, a noncyclic electron flow is also associated with ATP and NAD(P)H formation (Figure 4.7). To trace the path of electron flow, we start with photosystem I, in which light converts P700 to a strongly reducing excited state. The excited chlorophyll loses an electron (oxidizing the P700), which is transferred through a chain of electron carriers to reduce $NAD(P)^+$ to NAD(P)H. A similar light-dependent process in photosystem II produces highly reducing excited P680. This excited chlorophyll loses an electron, which reduces the oxidized P700 in photosystem I via another chain of electron carriers. The oxidized P680 in photosystem II is in turn reduced by electrons from H_2O, forming O_2. This is an unusual reaction, and oxidized P680 is the strongest known biological oxidant. The photosynthetic apparatus is arranged in a membrane such that protons are translocated during light-induced electron transfer. The return of protons through the membrane is used to generate ATP. Excellent discussions of photosynthesis and reaction center biochemistry can be found in Falkowski and Raven (1997) and Blankenship (1992).

In some cyanobacteria under reducing conditions (e.g., *Oscillatoria limnetica*; Cohen *et al.*, 1975), photosystem II becomes inactivated, and anoxygenic photosynthesis occurs using only photosystem I. Reducing power is generated with H_2S (or in some cases H_2; Cohen *et al.*, 1986) as an electron donor. Anoxygenic photosynthesis by some cyanobacteria may be an adaptive strategy to clear H_2S from the environment so that oxygenic photosynthesis can be

Table 4.1 Electron donors during photosynthesis and their oxidized products

	e^- donor	Product
Oxygenic photosynthesis	H_2O	O_2
Anoxygenic photosynthesis	H_2	H_2O
	H_2S	S^0, $S_2O_3^{2-}$, SO_4^{2-}
	S^0	SO_4^{2-}
	$S_2O_3^{2-}$	SO_4^{2-}
	Fe^{2+}	Fe oxides

better established (Cohen *et al.*, 1986). In other cases, as for example with *Oscillatoria limnetica*, prolonged anoxygenic photosynthesis may be sustained. However, even after prolonged periods of anoxygenic photosynthesis, photosystem II may be activated when sulfide is absent, re-establishing oxygenic photosynthesis (Cohen *et al.*, 1986).

4. CARBON FIXATION PATHWAYS

We have just considered how energy and reducing power are made available to fuel carbon fixation. We now consider the principal pathways by which carbon is fixed into organic matter by microorganisms. The actual fixing of CO_2 into organic material constitutes the second step in the carbon fixation process (Figure 4.2). For excellent additional discussion, consult the Ph.D. dissertation of Christopher House (House, 1999).

4.1. Reductive pentose phosphate cycle (the Calvin-Benson-Bassham cycle)

The reductive pentose phosphate cycle is the predominant pathway of carbon fixation on Earth, used by all oxygenic photosynthetic organisms, including cyanobacteria, land plants, and algae (Table 4.2). It is also the pathway of carbon fixation for purple anoxygenic phototrophs as well as numerous chemolithoautotrophic and mixotrophic organisms of the bacterial domain, including sulfide oxidizers, nitrifiers, and various iron oxidizers (Fuchs, 1989; Buchanan, 1992; Tabita, 1995). The reductive pentose phosphate pathway has also been found active among members of the anoxygenic photosynthetic genus *Oscillochloris*, located within the deep-branching phylum *Chloroflexi* (Pierson, 2001). Rubisco (ribulose-1,5-bisphosphate

Table 4.2 Carbon fixation pathways and the metabolisms involved

Pathway	Metabolism (e-acceptor or donor)	Affiliation (domain)	Notes
Reductive pentose phosphate cycle (Calvin-Benson-Bassham-cycle)	Oxygenic photosynthesis	Cyanobacteria (B) Plants (E) Algae (E)	Strong association with oxygen—either oxygen production, or chemoautrophic metabolism under aerobic or microaerophilic conditions (also using NO_3^-). Many anoxygenic phototrophic purple bacteria are facultative aerobes.
	Anoxygenic photosynthesis	Proteobacteria (B)	
	Sulfide oxidation (O_2, NO_3^-)	Proteobacteria (B)	
	Nitrification (O_2)	Proteobacteria (B)	
	Fe oxidation (O_2, NO_3^-)	Proteobacteria (B)	
	Mn oxidation (O_2)	Proteobacteria (B)	
	H_2 oxidation (O_2)	Gram-positive bacteria (B) (mycobacteria) Proteobacteria (B) (*Azospirillium*)	

Reductive citric acid cycle (reverse TCA cycle)	Anoxygenic photosynthesis	Green S bacteria (B)	Near reverse of the citric acid cycle (Krebs cycle), which is used extensively in respiration (fermentation). Present in aerobic and anaerobic metabolisms. However, no obligate aerobes.
	Sulfate reduction (H_2)	Proteobacteria (B) (*Desulfobacter*)	
	S° S oxidation (O_2)	*Sulfolobales* (A)	
	S° reduction (H_2)	*Sulfolobales* (A)	
		Thermoproteales (A)	
	Oxygen reduction (H_2)	*Aquifex-Hydrogenobacter* (B)	
Reductive acetyl-CoA pathway	Sulfate reduction (H_2)	Proteobacteria (B)	Generally associated with H_2 metabolism. Oxygen-sensitive enzymes in pathway.
		Gram positive (B)	
		Archaeoglobus (A)	
	Methanogenesis	Euryarchaeta (A)	
	Acetate formation (H_2)	Proteobacteria (B)	
	(homoacetogenic bacteria)	Gram positive (B)	
3-Hydroxypropionate pathway	Anoxygenic photosynthesis	*Chloroflexaceae* (B)	Apparently limited phylogenic distribution.
	S° oxidation (O_2)	*Sulfolobales* (A)	
	S° reduction (H_2)	*Sulfolobales* (A)	

(A) *Archaea*, (B) *Bacteria*, (E) *Eurkarya*.

carboxylase/oxygenase) activity (Rubisco is the key enzyme in the reductive pentose phosphate cycle; see below) has also been identified among some aerobic halophilic *Archaea* (Altekar and Rajagopalan, 1990), although little is known about the role of Rubisco in these organisms and whether it is linked to active CO_2 fixation. A gene encoding for a primitive form of Rubisco also has been found among the genome of several strictly anaerobic members of the *Archaea* (see Tabita, 1999) and among the green sulfur phototrophs (*Chlorobium* sp.) (Hanson and Tabita, 2001). These Rubisco-like proteins, however, are apparently not involved in CO_2 fixation.

Thus, although Rubisco is widespread among both aerobes and anaerobes, the reductive pentose phosphate cycle is apparently restricted to chemolithoautotrophs utilizing oxygen or nitrate as an electron acceptor or to phototrophs (both oxygenic and anoxygenic). Obligate anaerobic chemolithoautotrophs are not known to use this carbon fixation pathway. Many of the anoxygenic purple phototrophs utilizing the reductive pentose phosphate cycle can also grow as microaerophilic to aerobic chemoorganoheterotrophs (Pfennig, 1977; Imhoff, 1995). There seems, therefore, to be a strong relationship between oxygen (or nitrate) metabolism and organisms utilizing the reductive pentose phosphate cycle to fix CO_2 (Fuchs, 1989).

The pathway of this cycle is outlined in Figure 4.8, and the key step in carbon fixation is the carboxylation of ribulose-1,5-bisphosphate, promoted by the enzyme Rubisco. Rubisco has both a carboxylase (CO_2 fixation) and oxygenase (oxidation with O_2) activity (this aspect of Rubisco function, and its kinetic properties, are considered in more detail later in the Chapter). Following carboxylation, forming 3-phosphoglycerate, the reductive part of the cycle begins, forming glyceraldehyde 3-phosphate. One-sixth of the glyceraldehyde 3-phosphate is utilized by the cell for biosynthesis, whereas the remaining is re-generated to ribulose-1,5-bisphosphate, completing the cycle.

4.2. Reductive citric acid cycle (reverse tricarboxylic-TCA cycle)

The reductive citric acid cycle is used in carbon fixation by the anoxygenic photosynthetic green sulfur bacteria (*Chlorobiaceae;* see Chapter 9). It is also utilized by some sulfate reducers from the δ subdivision of the proteobacteria and the deep-branching hydrogen-oxidizing bacteria of the *Aquifex-Hydrogenobacter* group (Table 4.2) (Fuchs, 1989; Kandler, 1994) in the bacterial domain. In addition, the reductive citric acid cycle has been found in the hyperthermophilic sulfur-metabolizing *Sulfolobus* and *Thermoproteus* in the crenarchaeotal kingdom of the archaeal domain.

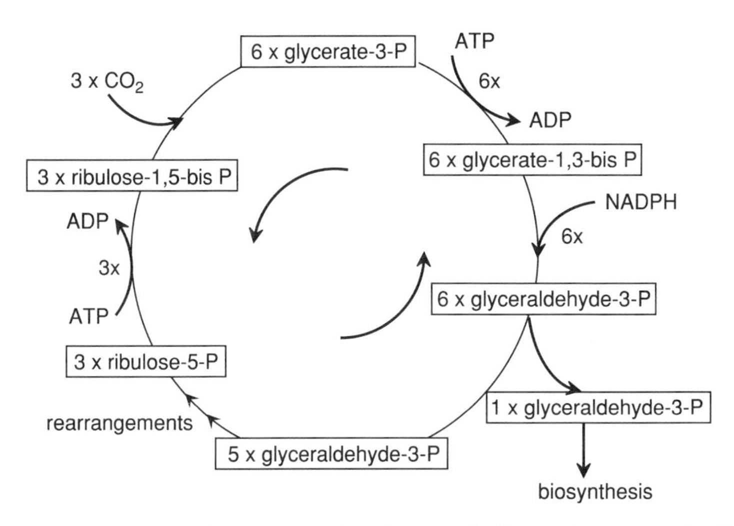

Figure 4.8 The reductive pentose phosphate cycle, better known as the Calvin-Benson-Bassham cycle. A key step is carboxylation of ribulose-1,5-bis phosphate with the enzyme Rubisco. Through the cycle nine ATPs and six NADPHs are used to form one 3-C compound used in biosynthesis.

Taken together, the pathway is distributed among both anaerobic (*Chlorobium*, sulfate reducers, *Thermoproteus*) and microaerophilic organisms (*Aquifex-Hydrogenobacter*).

The reductive citric acid cycle is, in most principal aspects, a reverse of the well-known Krebs cycle (TCA cycle), a primary respiratory pathway. In considering the reductive citric acid cycle, we begin with oxaloacetate (Figure 4.9). After two CO_2 additions and several reductive steps, citrate is produced, which is split by the enzyme citrate lyase, producing acetyl-CoA and regenerating oxaloacetate. The acetyl-CoA may be used in biomass synthesis. Key enzymes in the pathway are citrate lyase and α-ketoglutarate synthase.

4.3. Reductive acetyl-CoA pathway

The reductive acetyl-CoA pathway is found among many anaerobic organisms utilizing H_2 as an electron donor. Examples include sulfate reducers and homoacetogenic organisms (producing acetate from H_2 and CO_2 or formate) from the proteobacterial and the gram-positive groups in the bacterial domain. In addition, the archaeal sulfate reducers of the

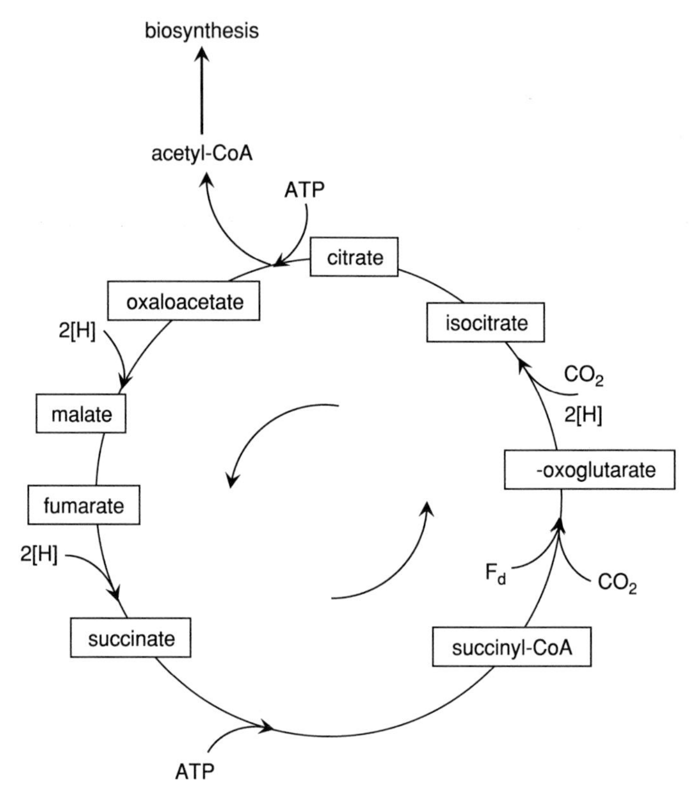

Figure 4.9 The reductive citric acid cycle for *Chlorobium* sp., after Evans *et al.* (1966). See text for details.

genus *Archaeoglobus* and the methanogens, also *Archaea*, use this pathway (Table 4.2).

A highly simplified representation of the reductive acetyl-CoA pathway, as used by methanogens, is presented in Figure 4.10. Note that this pathway is not a cycle. Key steps are the reduction of CO_2 to a methyl carbon through a series of steps involving an enzyme containing methanopterin (see Chapter 10) and the reduction of another CO_2 to a carbonyl carbon, through enzyme-bound CO. The enzyme promoting this reduction is carbon monoxide dehydrogenase. These steps combined provide both the oxidized and reduced carbon found in acetate. Also shown in Figure 4.10 are bifurcation points leading to the production of methane and acetate. Ultimately, biosynthesis occurs from the acetyl-CoA produced in the pathway (Fuchs, 1989).

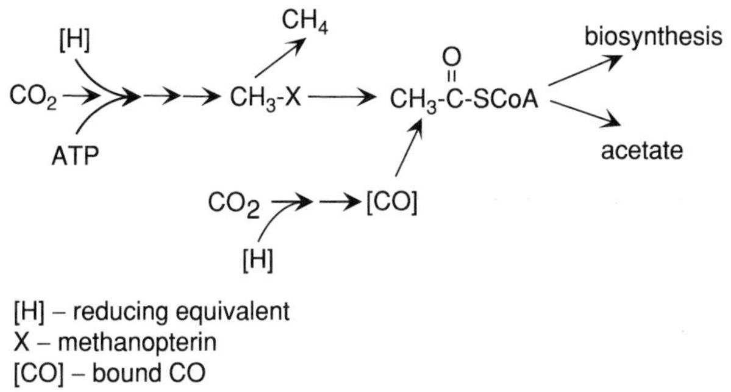

[H] CH_4 biosynthesis
CO_2 → → → CH_3-X → CH_3-C(=O)-SCoA
ATP acetate
CO_2 → → [CO]
[H]

[H] – reducing equivalent
X – methanopterin
[CO] – bound CO

Figure 4.10 A highly simplified cartoon of the reductive acetyl-CoA pathway for acetogenic and methanogenic bacteria. Adapted from Fuchs (1989).

4.4. 3-Hydroxypropionate cycle

The 3-hydroxypropionate cycle is operative in the anoxygenic phototrophic bacterium *Chloroflexus aurantiacus* (Holo, 1989; Strauss and Fuchs, 1993), and it may also operate in some members of the archaeal order *Sulfolobales* (Ishii *et al.*, 1996; Menedez *et al.*, 1999) (Table 4.2). The basis for this pathway is the conversion of acetyl-CoA to 3-hydroxypropionate (Figure 4.11), which is further carboxylated and converted ultimately to malyl-CoA. Malyl-CoA is cleaved into acetyl-CoA and glyoxylate with the enzyme malyl-CoA lyase (Strauss and Fuchs, 1993). The glyoxylate is used in cell synthesis through a yet-unspecified path. It will be interesting to see whether the 3-hydroxypropionate cycle is further distributed among prokaryotic organisms of the bacterial and archaeal domains.

5. ENERGETICS OF CARBON FIXATION

The ATP demand for each of the four carbon fixation pathways is expressed, per CO_2 fixed, in Table 4.3. The most energy expensive is the reductive pentose phosphate cycle, whereas the least energy is used to fix CO_2 in the reductive acetyl-CoA pathway. The energy demand in these different carbon fixation pathways is, to some extent, consistent with the habitat and physiology of the organisms utilizing the pathway. For example, the energy-demanding reductive pentose phosphate cycle is utilized by organisms metabolizing oxygen (or in some cases nitrate), all oxygenic phototrophs, and some anoxygenic phototrophs. Energy is generally abundant for photosynthetic organisms, and aerobic organisms (as well as nitrate-utilizing organisms) obtain a high energy yield for their metabolisms.

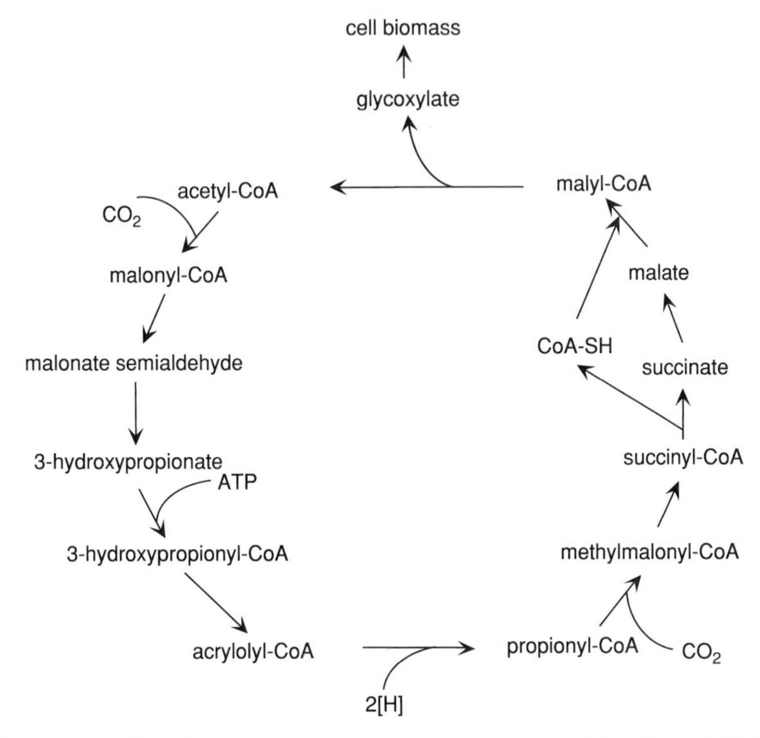

Figure 4.11 The 3-hydroxypropionate cycle as suggested by Holo (1989) for *Chloroflexus aurantiacus*.

Table 4.3 Energy demand for the different CO_2 fixation pathways

Pathway	ATP/CO_2
Reductive pentose phosphate	9/3
Reductive citric acid	5/3
Reductive acetyl-CoA	4/3
3-Hydroxypropionate	6/3

By contrast, the reductive acetyl-CoA pathway is known only among anaerobic organisms, including many methanogens, whose metabolic energy yield is low. It is therefore an advantage, and perhaps even a necessity, that the carbon fixation processes for these strict anaerobes have a low energy demand. The energy demand of the reductive citric acid cycle is also relatively low, reflecting, probably, the predominantly anaerobic nature of the organisms utilizing this pathway. By contrast, the relatively high energy demand of the 3-hydroxypropionate pathway can be rationalized by the photosynthetic nature of *Chloroflexus*. Furthermore, both *Chloroflexus* and members of the *Sulfolobales* utilizing the 3-hydroxypropionate pathway gain energy from metabolic processes utilizing oxygen (Segerer and Stetter, 1992; Pierson and Castenholz, 1995).

6. EVOLUTION OF CARBON FIXATION PATHWAYS

Kandler (1994) explored evolutionary relationships among the different carbon fixation pathways by identifying the lineages on the Tree of Life containing each of the different pathways. An update of this approach is presented in Figure 4.12. The occurrence of deep-branching lineages supporting carbon fixation in both prokaryote domains suggests that carbon

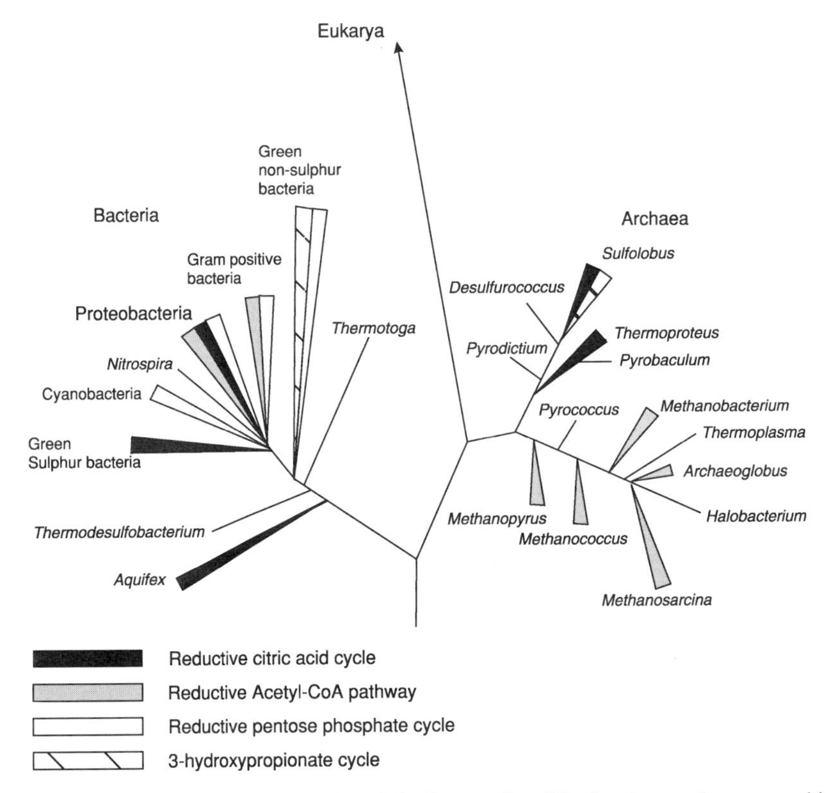

Figure 4.12 The branching order of the four major CO_2 fixation pathways used in autotrophic and mixotrophic organisms in the Tree of Life. Information used in drawing the figure comes from Table 4.2. See Chapter 1 and text for details.

fixation is an ancient process, possibly emerging before the divergence of the three domains of the Tree of Life (Kandler, 1994). Assuming for the moment that the depth of the emergence of a pathway in the Tree of Life speaks to the timing of the emergence the pathway, then all four pathways have deep-branching representatives, indicating a possible early evolution of each. Close inspection, however, renders this analysis less than compelling. First, the reductive pentose phosphate cycle is known so far only in a single deep-branching genus, *Oscillochloris*, within the family *Chloroflexiaceae*. It is possible that this pathway was acquired by *Oscillochloris* from lateral gene transfer. This issue could be resolved, however, from parallel phylogenies from the Rubisco gene, which to our knowledge have not yet been sequenced from members of *Oscillochloris*.

A similar situation exists for the 3-hydroxypropionate pathway. This pathway is known only from a few strains among the deep-branching genus *Chloroflexus*, also within the family *Chloroflexiaceae*, and among a few members of the *Sulfolobales*, which do not branch deep within the *Archaea* (see Chapter 1). The apparently limited distribution of the pathway suggests that it was not particularly successful in fueling carbon fixation. We must also entertain the possibility that it came long after the *Chloroflexus* lineage was established. Importantly, phylogenies constructed from reaction center protein sequences suggest a late development for the photosynthetic reaction center within *Chloroflexus* (Blankenship, 1992). The antiquity of the 3-hydroxypropionate pathway may become better established if it is discovered among other members of the *Chloroflexiaceae*, or if it is discovered in other prokaryote lineages.

By contrast, the reductive acetyl Co-A pathway is widely distributed among lineages in both prokaryote domains, and its deep branching among the methanogens might suggest that it was a bona fide early innovation in carbon fixation. Fuchs (1989) has argued that the reductive acetyl-CoA pathway contains many features expected in an ancestral carbon fixation process. Prominent among these is a low energy requirement, as would benefit an organism occupying a strictly anoxic environment utilizing H_2 (supporting methanogenesis and sulfate reduction) as an electron donor. Early anaerobic H_2-based autotrophy could have been supported in early Earth hydrothermal ecosystems, well before the evolution of oxygenic photosynthesis (Holland, 1984; Stetter, 1996).

The deep branching and wide distribution of the reductive citric acid cycle suggest that it may also have been an important early pathway of carbon fixation, for both phototrophic and chemotrophic organisms. As the reductive citric acid cycle is closely related to the widely distributed respiratory Krebs cycle, one can imagine how it might have developed from a pre-existing respiratory metabolism. The deep-branching lineage of *Aquifex-Hydrogenobacter*, with its microaerophilic habitat, might suggest an early emergence of oxygen-respiring organisms and the development of the reductive citric acid cycle among these organisms. The reductive citric acid cycle is, however, better known among anaerobic organisms (green sulfur bacteria, *Thermoproteus*, sulfate reducers; see Table 4.2). It is possible that organisms of the *Aquifex-Hydrogenobacter* lineage, while utilizing the reductive citric acid cycle early in the history of the lineage, were originally anaerobic and developed the ability to respire oxygen later in evolutionary history (Fenchel and Finlay, 1995). This would be consistent with the much later emergence of oxygen-producing cyanobacteria.

If we accept that current evidence is not particularly compelling for an early emergence of the reductive pentose phosphate cycle, then this C fixation pathway apparently emerged into an environment with at least two, and possibly three, preexisting pathways. At least one, and possibly two, of these (the reverse citric acid cycle and the 3-hydroxypropionate cycle) was already used in photosynthetic organisms. The high-energy demand of the reductive pentose phosphate cycle would have acted against its development. Furthermore, another disadvantage, as we explore in more detail below, is the oxygenase activity of the key enzyme Rubisco in the pathway, which can severely reduce the efficiency of carbon fixation.

Some argue that the reductive pentose phosphate cycle evolved among anaerobic chemolithoautotrophic organisms (Badger and Andrews, 1987). This would partly explain the oxygenase activity of Rubisco, with no selective pressure against it (Badger and Andrews, 1987; Tolbert, 1994). Apparent support for this hypothesis comes from the identification of Rubisco-like genes among the genomes of anaerobic members of the *Archaea* (Tabita, 1999; Watson *et al.*, 1999). These genes encode for primitive Rubisco-like proteins distantly related to the bona fide Rubisco used in the reductive pentose phosphate cycle (Tabita, 1999). Whether these genes actually express for proteins is unclear, and if they do, the function of the proteins is also unclear. If expressed, the proteins are apparently not involved in CO_2 fixation by the reductive pentose phosphate cycle, as other critical enzymes in this cycle have not been identified and carbon fixation has not been demonstrated (Tabita, 1999).

Thus, while Rubisco-like molecules may have originated early in the history of life among anaerobic organisms, there are no known modern obligate anaerobic chemolithoautotrophic organisms utilizing the reductive pentose phosphate cycle for fixing CO_2. Furthermore, organisms utilizing the reductive pentose phosphate cycle, including the anoxygenic photosynthetic purple bacteria, apparently all emerged either consecutively with the evolution of cyanobacteria or thereafter (Figure 4.12) (see Chapter 1; Canfield and Raiswell, 1999).

It is plausible that the reductive pentose phosphate cycle evolved as an innovation to support carbon fixation by oxygenic photosynthesis, incorporating a modified pre-existing Rubisco enzyme (see below). After its initial development, the reductive pentose phosphate cycle became widely distributed among aerobic and microaerophilic chemolithotrophic organisms and the anoxygenic phototrophic purple bacteria. As mentioned previously, many of the anoxygenic photosynthetic purple bacteria utilizing the reductive pentose phosphate pathway are facultative aerobes, and they can respire oxygen heterotrophically (Imhoff, 1995). This provides metabolic diversity and would seem to be a selective advantage.

7. FURTHER MUSINGS ON RUBISCO

The principal enzyme responsible for carbon fixation in the reductive pentose phosphate cycle, Rubisco, is a curious enzyme indeed. Not only does the enzyme catalyze the carboxylation of ribulose-1,5 bisphosphate, leading ultimately to the production of cellular biomass, but it also catalyzes the oxygenation of ribulose-1,5 bisphosphate, oxidizing this compound and negating the effects of the carboxylation reaction. Thus, the enzyme competes with itself, and as noted by Miziorko and Lorimer (1983),"That one enzyme should catalyze the primary steps of two major, but diametrically opposing, metabolic pathways is a curious situation indeed. We know of no precedent for this." The carboxylase and oxygenase activity of Rubisco operate simultaneously, at rates depending on the relative concentrations of CO_2 and O_2 available to the Rubisco and the nature of the Rubisco in the organism. At equal concentrations of CO_2 and O_2, the carboxylase activity is between 10 and 100 times more active than the oxygenase activity (Badger and Andrews, 1987) (Table 4.4). The lowest relative rates of carboxylase activity are found among Rubiscos (form II Rubiscos) comprised of only two large subunits, restricted to certain proteobacteria and dinoflagellates (Badger and Andrews, 1987; Badger *et al.*, 1998). The more common Rubiscos (form I), with eight large subunits and eight small subunits, also found among the proteobacteria and oxygenic phototrophs (Tabita, 1999), generally show higher relative carboxylase activity, although a wide range of relative activity is observed.

Another class of Rubiscos (form III and form IV) has been found among anaerobic *Archaea* and members of the genus *Chlorobium* (see above; Hanson and Tabita, 2001). Although these Rubiscos are probably not directly involved in CO_2 fixation, the enzyme obtained from the Rubisco gene in one of these organisms, the methanogen *Methanococcus jannaschii*, has a demonstrable carboxylase activity (Watson *et al.*, 1999), and its activity is severely suppressed in the presence of oxygen, unlike the form I and form II Rubiscos (Watson *et al.*, 1999). The *M. jannaschii* Rubisco

Table 4.4 Relative specificity of Rubisco for O_2 and CO_2

Organism groups	Rubisco type	τ^a
Rhodophyta	Form I (red type)[b]	129–238
Chrysophyta	Form I (red type)	77–114
Higher plants	Form I (green type)	78–90
Green algae	Form I (green type)	54–83
α/β proteobacteria	Form I (red type)	60–75
β/γ proteobacteria	Form I (green type)	37–56
Cyanobacteria	Form I (red type)	35–56
Pyrrhophyta	Form II	37
Proteobacteria	Form II	9–15
Methanogen (*M. jannaschii*)	Form III	~0.5

[a]τ, sometimes also called the relative specificity, expresses the relative rates of the carboxylase and oxygenase activity of Rubisco at equal aqueous concentrations of CO_2 and O_2. Formally, $\tau = (V_{CO_2} K_{O_2})/(V_{O_2} K_{CO_2})$, where V is the maximum velocity of the carboxylase or oxygenase reactions and K is the respective Michaelis-Menton constant.
[b]From molecular phylogenies of the Rubisco enzyme form I, Rubisco is divided into "red" and "green" types. "Green type" Rubisco includes higher plants, green algae, cyanobacteria, and some purple bacteria. "Red type" Rubisco includes α and β members of the proteobacteria and non-green algae.
References: Badger *et al.* (1998); Watson *et al.* (1999).

also has a very low specificity of CO_2 (Table 4.4) and a demonstrable oxygenase activity. The oxygenase activity, however, is very slow given the reduced activity of the enzyme in the presence of oxygen.

Other Rubiscos also have rather low relative specificities for the carboxylase reaction; these tend to be found among microaerophilic chemolithotrophic bacteria and anoxygenic phototrophs for which an active oxygenase has relatively little significance (Badger and Andrews, 1987) (Table 4.4). For oxygenic phototrophs, the oxygenase activity is a real problem. To combat this, some organisms, for example, the C3 plants, have evolved Rubiscos with relatively high CO_2 specificities (Table 4.4). Still, it is estimated that the oxygenase activity in C3 plants, in a process known as photorespiration, is 25 to 50% the rate of CO_2 fixation (Zelitch, 1975). This means that the CO_2 fixation is only 50 to 75% as efficient as it could be in the absence of the oxygenase. Despite a concerted effort at bioengineering, the rather poor performance of land-plant Rubisco has not been improved upon (Morell *et al.*, 1992).

The relatively poorer performance of cyanobacterial and algal Rubiscos has been compensated for by active carbon-concentrating mechanisms acting to increase the internal cellular levels of CO_2 (Badger and Andrews,

1987; Raven, 1991), thereby reducing the influence of the oxygenase activity. Other adaptations to combat poor Rubisco performance include the development of C4 metabolism in higher plants, in which CO_2 is initially fixed by the enzyme PEP carboxylase forming oxaloacetate (a C4 compound, giving the C4 pathway its name). The oxaloacetate can be reduced to malate, which diffuses into "bundle-sheath cells," containing Rubisco, and is consequently oxidized back to CO_2 and pyruvate, thus enhancing the CO_2 partial pressure in the vicinity of the Rubisco. This C4 pathway has also been discovered in a marine diatom (Reinfelder *et al.*, 2000).

It is apparent that the oxygenase activity of Rubisco is seemingly an innate property of the enzyme. Indeed, oxygenase activity is found even among Rubiscos in anoxygenic photosynthetic and chemoautotrophic prokaryotes living in low-oxygen environments, as well as in strictly anaerobic *Archaea*. Furthermore, through the course of evolution, oxygenic photosynthetic organisms have adopted numerous strategies to overcome the inherent imperfections in the Rubisco enzyme.

Could any early advantage have been gained from the Rubisco oxygenase activity? We have speculated that the reductive pentose phosphate cycle, to which the enzyme belongs, originated with oxygen-producing cyanobacteria. By comparison with modern benthic cyanobacterial communities, in which oxygen partial pressures of up to 1 atm can be found (see Chapter 13) (Revsbech *et al.*, 1983; Canfield and Des Marais, 1993), early comparable communities could have been sites of high oxygen concentrations even in a low-oxygen early Earth atmosphere. For modern oxygenic phototrophs, high oxygen levels pose potential problems, including the generation of highly reactive oxygen species such as hydrogen peroxide (H_2O_2), superoxide radical anions (O_2^-), hydroxyl radicals (•OH), and singlet oxygen ($^1O_2$). These highly toxic species may damage essential proteins, lipids, and the nucleic acids comprising DNA (more on this in Chapter 6). Modern organisms have developed numerous defense mechanisms against these reactive oxygen species, including detoxifying enzymes and oxygen radical scavengers (Raven *et al.*, 1994). However, unless elaborate oxygen defense mechanisms developed simultaneously with the ability to produce oxygen, high oxygen levels in early cyanobacterial mat communities could have posed a potential problem to the health of the community. Early Rubiscos incorporated into the reductive pentose phosphate cycle in newly evolved cyanobacteria may have expressed only a weak carboxylase activity (evidenced by the weak carboxylase activity for Rubisco expressed from in *M. jannaschii*; see above) combined with a low specificity for CO_2 as seen in form II Rubiscos used by many microaerophilic proteobacteria. We suggest that early cyanobacteria capitalized on these properties of Rubisco to reduce oxygen build-up in microbial mats and thereby limit the damaging effects of elevated oxygen concentrations. Once adequate defense mechanisms against oxygen toxicity evolved, oxygenic phototrophs began to exploit numerous strategies to restrict the activity of the oxygenase in Rubisco.

We leave the subject with a parting thought. Whether deliberate or not, the oxygenase activity of Rubisco has a potential regulating effect on atmospheric O_2 levels. Most C3 plants, numerous aquatic macroalgae, and some important oceanic phototrophs such as *Emiliania huxleyi* will reduce their carbon production with increasing concentrations of O_2 relative to CO_2 (Badger and Andrews, 1987; Badger *et al.*, 1998; Raven *et al.*, 1994) (Table 4.5). For example, for the soybean *Glycine max* a doubling of O_2 concentration at constant partial pressure of CO_2 decreases plant production by one-half, whereas halving levels of O_2 doubles plant production. A relationship between O_2 levels and plant production represents a potential negative feedback on increasing O_2 levels (Tolbert, 1994). The nature of the feedback is obviously complicated, as both CO_2 and O_2 levels act to regulate the relative activity of Rubisco's oxygenase. Nevertheless, for a broad range of primary producers, O_2 levels should exert an influence on rates of primary production, which could be an additional consideration in constructing models of atmospheric O_2 concentration into the geologic past. Berner *et al.* (2000) have already shown that the oxygenase activity of Rubisco influences the magnitude of carbon isotope fractionation, which, in turn, affects the modeling of atmospheric O_2 concentrations from the carbon isotope record.

Table 4.5 Effect of O_2 on biomass production rates for oxygen-sensitive oxygenic phototrophs. Biomass production is relative to production at normal atmospheric oxygen

Organism	CO_2	0.09	0.48	1.0	1.9	4.6	Oxygen relative to atmosphere
Glycine max (soy bean)	30 Pa		2.09	1.0	0.55		
Panicum bisulcatum (millet)	32 Pa		1.17	1.0	0.67		
Chlamydomonas reinardtii[a]	38 Pa			1.0		0.36	
	35 Pa			1.0		0.33	
Euglena gracilis	30 Pa	2		1.0			

[a]No effect of oxygen level on biomass production was observed at low CO_2 partial pressures where the CO_2-concentrating mechanism was active.

References: Raven *et al.* (1994); Merrett and Armitage (1982).

This model, however, does not take into account the influence of oxygenase activity on carbon production rates.

8. ISOTOPE FRACTIONATION DURING CARBON FIXATION

The fixation of CO_2 into organic compounds requires, as we have seen, the chemical reduction of the oxidized carbon in CO_2 to a reduced valence state. This requires the breaking of C-O bonds by enzymes. The two stable isotopes of carbon in nature are ^{12}C (98.89%) and ^{13}C (1.11%). For lighter masses, a higher proportion of total bond energy is partitioned into vibrational modes, making the ^{12}C-O bond slightly more susceptible to enzymatic cleavage than the ^{13}C-O bond, providing a higher proportion of ^{12}C for fixation into organic carbon during CO_2 reduction. This kinetic isotope effect results in ^{13}C-depleted organic matter compared to the original CO_2 source (Park and Epstein, 1960).

In discussing the isotopic composition of carbon we compare the ^{13}C:^{12}C ratio of the species of interest to the ^{13}C:^{12}C ratio of a standard, generally taken to be PDB belemnite (a fossil-containing limestone from the Pee Dee Belemnite Formation in South Carolina). Formally, the isotopic composition of a carbon-containing species is given as the parts per thousand (‰) deviation from a standard:

$$\delta^{13}C_{sam} = 1000\left[\frac{\left({}^{13}C/{}^{12}C\right)sam}{\left({}^{13}C/{}^{12}C\right)std} - 1\right] \tag{4.1}$$

The fractionation associated with a chemical reaction is expressed as ε_p, and is given formally as

$$\varepsilon_p = 1000\left[\frac{(\delta_{prod} + 1000)}{(\delta_{react} + 1000)} - 1\right] \approx \delta^{13}C_{react} - \delta^{13}C_{prod} \tag{4.2}$$

where "prod" (product) refers to the cell biomass formed from the reduction of the reactant CO_2, which is designated by "react" in the equation. Fractionation is also given, approximately, as the isotope difference between the reactant and the product in the reaction (Equation 4.2).

By contrast with the detailed studies of isotopic fractionation during carbon fixation by eukaryotic plants and algae (e.g., Farquhar *et al.*, 1989; Popp *et al.*, 1998), isotopic fractionations accompanying carbon fixation in prokaryotic organisms have been poorly studied, although work by House *et al.* (2003) has considerably expanded the database. A survey of available information is presented in Table 4.6, where only fractionations associated with the production of cellular material are shown. Additional fractionations of interest may be associated with carbon fixation processes such as the formation of methane or acetate by methanogens and homoacetogenic bacteria utilizing the reductive acetyl-CoA pathway. The fractionations associated with the production of these soluble compounds are large, and are considered in Chapter 10.

The largest fractionations during autotrophic cell biomass production are associated with the reductive acetyl-CoA pathway; up to 35.5‰ for *Desulfobacterium autotrophicum*, a proteobacterial sulfate reducer (Preuss *et al.*, 1989), and up to 34‰ for methanogens (Fuchs *et al.*, 1979). However, small and even reverse fractionations have also been reported during methanogenesis under carbon-limiting conditions (Fuchs *et al.*, 1979).

Large fractionations are also associated with the reductive pentose phosphate cycle, and the type I Rubisco as used in plants gives an experimentally determined maximum of approximately 25 to 27‰ (Farquhar *et al.*, 1989; Popp *et al.*, 1998). Recently, pure type II Rubisco has been purified from endosymbionts in the tubeworm *Riftia pachyptila*, yielding a lower fractionation of 19.5 during CO_2 fixation (Robinson *et al.*, 2003). These limits are nearly reached with both oxygenic and anoxygenic phototrophic prokaryotes (Table 4.6). However, much-reduced fractionations are also produced, which probably reflects a limited supply of CO_2 to fuel the carbon fixation reactions (Farquhar *et al.*, 1989; Des Marais and Canfield, 1994). In the face of CO_2 limitation, there are two main factors acting to reduce fractionation. One factor is mostly physical and may be visualized as follows. With low CO_2 supply, the metabolism of CO_2 becomes limited by the rate of exchange of CO_2 across the cell membrane. With extreme CO_2 limitation, only a small proportion of the CO_2 entering the cell will exchange back out. In the end-member case, all of the CO_2 entering the cell will be fixed, and even if Rubisco discriminates against ^{13}C to its maximum extent, the isotopic composition of the cellular material must be equal to the CO_2 that entered the cell. The internal cellular pool of CO_2, however, will be ^{13}C enriched compared to the fixed carbon. The full expression of Rubisco fractionation will only be realized when the CO_2 is not limiting and when the internal and external pools of CO_2 are in complete exchange equilibrium.

The second factor acting to suppress isotope fractionation under CO_2-limiting conditions is biological and quite deliberate on the part of the organism. In cyanobacteria and many species of aquatic algae, carbon limitation can result in active pumping of bicarbonate across the cell membrane (Badger and Price, 1992). If this is the case, the isotopic composition of the cell material more closely reflects the isotopic composition of the bicarbonate pumped into the cell (Des Marais and Canfield, 1994).

Table 4.6 Isotope fractionation associated with the four principal CO_2 fixation pathways

Pathway	ε_p(‰)[a]	Notes	Ref.
Reductive pentose phosphate	2 to 23.5	Cyanobacteria – *Anabaena variabilis* – range encomposes different extents of carbon limitation	1
	4 to 19	Cyanobacterial mats	1
	19.6 to 22.5	*Chromatium* sp.	2, 3, 4
	12.4 to 20.5	*Rodospirillum rubrum*	2, 3
	10.6	*Rhodopseudomonas capsulate*	2
	12.4	*Thiobacillus novellas*	5
Reductive citric acid	2.5 to 12.2	*Chlorobium* sp.	2, 3
	10	*Desulfobacter hydrogenophilus*	2, 8
	8	*Thermoproteus neutrophilus*	2, 8
	7.9	*Aquifex aeolius*	5
	4.4	*Pyrobaculum aerophilum*	5
	3.5	*Thermoproteus neutrophilus*	5
Reductive acetyl-CoA	18 to 24	*Methanobacterium thermoautotrophicum* • high CO_2 availability	5, 6
	13	• low CO_2 availability	6
	34	• low CO_2 availability	7
	35.5	*Desulfobacterium autotrophicum*	8
	4.7	*Archaeoglobus fulgidus*	5
	10.2	*Archaeoglobus lithotrophicus*	5
	6.1	*Ferroglobus placidus*	5
	22.5	*Methanococcus igneus*	5
	8.2 to 20.0	*Methanococcus jannaschii*	5
	9.2 to 30.4	*Methanococcus thermolithotrophicus*	5
	14.0 to 21.9	*Methanopyrus kandleri*	5
	13.5 to 25.3	*Methanosarcina barkari*	5
	15.4	*Methanothermus fervidus*	5
3-Hydroxypropionate	13.7	*Chloroflexus aurantiacus*	9
	2.6	*Acidianus brierleyi*	5
	2.1	*Metallosphaera sedula*	5
	−0.7	*Sulfolobus solfataricus*	5

[a] $\varepsilon_p = 10^3\left[\frac{(\delta^{13}C_{CO_2}+1000)}{(\delta^{13}C_{cell}+1000)}-1\right] \approx \delta^{13}C_{CO_2} - \delta^{13}C_{cell}$.

References: (1) Des Marais and Canfield, 1994; (2) Quandt *et al.*, 1977; (3) Sirevåg *et al.*, 1977; (4) Madigan *et al.*, 1989; (5) House, 1999; (6) Fuchs *et al.*, 1979; (7) Fuchs, 1989; (8) Preuss *et al.*, 1989; (9) Holo and Sirevåg, 1986.

Both phototrophic and non-phototrophic organisms utilizing the reductive citric acid cycle apparently fractionate much less in the formation of cellular biomass than either the reductive pentose phosphate cycle or the reductive acetyl-CoA pathway (Table 4.6). However, fractionations by organisms utilizing the reductive citric acid cycle have not been well studied, and the influence of carbon limitation on the fractionations has not been explored. The information available for the 3-hydroxypropionate pathway suggests similar fractionations to the reductive citric acid cycle, but again, the database is extremely limited.

In summary, it appears that the large fractionations possible with the reductive acetyl-CoA pathway are greater than any of the other carbon fixation pathways and might, therefore, be diagnostic. However, carbon limitation may reduce the extent of fractionation during carbon fixation. Thus, each of the carbon fixation pathways experiencing severe carbon limitation can probably produce cellular material depleted in ^{13}C by $< 10‰$ relative to the starting CO_2. Therefore, low extents of ^{13}C depletion are not particularly diagnostic of any of the fixation pathways. We stress, however, that very little is known about the systematics and extents of fractionation during carbon fixation by prokaryotic organisms. A careful exploration of isotope fractionation during carbon fixation by prokaryotes is a ripe field for future endeavor.

9. CASE STUDIES

The ecology of photosynthetic and chemoautotrophic organisms conducting CO_2 fixation will be woven into many of the future discussions of element cycling. We do, however, wish to highlight a few specific examples of active ecosystems fueled by chemoautotrophic organisms.

9.1. Hydrothermal vent systems

Deep-sea hydrothermal chemoautotrophic vent ecosystems are well known (Corliss *et al.*, 1979). At these sites, chemically reducing substances from the Earth's mantle are delivered to oxygen-containing ocean water, creating a vast opportunity for chemolithoautotrophic microbial production (van Dover, 2000). Table 4.7 lists some of the electron donors and electron acceptors available to fuel autotrophy in these systems (Jannasch and Mottl, 1985). Of the potential electron donors, many are primary mantle-derived substances (Von Damm, 1990; Canfield and Raiswell, 1999); others such as thiosulfate and nitrite are oxidized intermediates probably not transported as primary constituents of hydrothermal mantle-derived fluids.

It is obvious that many of the chemoautotrophic reactions are based on the utilization of oxygen or nitrate as an electron acceptor. Much of the biomass production in these ecosystems may not be considered as primary because the O_2 ultimately comes from oxygenic photosynthesis. Some of the

Table 4.7 Electron donors and electron acceptors for chemolithoautotrophs fueling metabolisms at hydrothermal vents

Electron donor	Electron acceptor	Organisms
S^{2-}, S^{o}, $S_2O_3^{2-}$	O_2	Sulfur-oxidizing prokaryotes
S^{2-}, S^{o}, $S_2O_3^{2-}$	NO_3^-	Denitrifying and sulfur-oxidizing prokaryotes
H_2	O_2	Hydrogen-oxidizing prokaryotes
H_2	NO_3^-	Denitrifying hydrogen prokaryotes
H_2	S^{o}, SO_4^{2-}	Sulfur- and sulfate-reducing prokaryotes
H_2	CO_2	Methanogenic and acetogenic prokaryotes
NH_4^+, NO_2^-	O_2	Nitrifying prokaryotes
Fe^{2+}, (Mn^{2+})	O_2	Iron- and manganese-oxidizing prokaryotes
CH_4, CO	O_2	Methylotrophic and carbon monoxide-oxidizing prokaryotes

Adapted from Jannasch and Mottl (1985).

metabolisms, however, such as those using H_2 as an electron donor and CO_2, S^{o} and SO_4^{2-} as electron acceptors, do not require oxygen and could be considered true primary producers of cellular biomass (Stetter, 1996; Canfield and Raiswell, 1999).

Primary or not, the chemoautotrophic production of organic matter at hydrothermal vents is impressive and creates an oasis of life in an otherwise deserted sea floor. Hydrothermal circulation may also fuel a vast subsurface ecosystem based, probably, on H_2 metabolism (Lilley *et al.*, 1993; Furnes and Staudigel, 1999). The evidence for the existence of these deep subsurface ecosystems includes anomalous concentrations of methane and ammonia in hydrothermal fluids, indicating organic matter decomposition (Lilley *et al.*, 1993). Ocean crust alteration features also indicate biological mediation of the alteration process (Furnes and Staudigel, 1999). Furthermore, low-temperature fluids effusing from mid-ocean ridge hydrothermal circulation systems contain hyperthermophilic organisms (Summit and Baross, 2001; Kelley *et al.*, 2002). These organisms testify to an active subsurface community fueled by the chemical interactions between water, basalt, reducing fluids from the mantle and oxygenated sea water.

Hydrothermal vent ecosystems are typified by dense populations of animals harboring sulfide-oxidizing endosymbiotic bacteria (Corliss *et al.*, 1979; Jones, 1981; van Dover, 2000). Also, dense mats of hydrogen sulfide-oxidizing bacteria, such as mats of *Beggiatoa* in the Guaymas Basin (Jørgensen *et al.*, 1992), are found in some hydrothermal vent areas. The animal populations have drawn special attention because, for the most conspicuous species, the main food source is sulfide-oxidizing prokaryotes. For example, the tube worm *Riftia pachyptila* has evolved without a mouth, stomach or intestinal tract and instead obtains its food by the metabolism of sulfide oxidizers growing within the body of the animal (Cavanaugh *et al.*, 1981). Blood vessels intertwine with the sulfide oxidizers, transporting both oxygen, in the form of hemoglobin, and protein-bound sulfide. The microbial biomass may constitute up to 50% of the wet weight of the worm. In another example, the giant clam *Calyptogena magnifica* houses large populations of sulfide oxidizers in its gill tissue. Harvesting of these microbes by the clam constitutes a significant source of nutrition for the animal (Grassle, 1986).

9.2. Chemoautotrophic cave ecosystem

A chemoautotrophic-based ecosystem has been discovered in limestone caves located in southeastern Romania (Sarbu *et al.*, 1996). The main electron donor fueling the chemoautotrophic microbial community is H_2S, with probable additional biomass production associated with aerobic methanotrophy. Both H_2S and CH_4 are derived from a deep subsurface aquifer. The nature of the microbial community has not been identified in detail, but it is composed of sulfide oxidizers and fungi, floating together as a dense mat between the interface of sulfidic cave waters and an oxygen-containing cave atmosphere. The mat is kept afloat by bubbles, presumably methane, trapped underneath it (Sarbu *et al.*, 1996). Enzymes of the reductive pentose phosphate cycle, in particular Rubisco, have been found active within the mat community. Therefore, a chemoautotrophic base for carbon production seems likely.

This dense floating mat, while interesting and unexpected in itself, provides the food source for more than 48 obligate cave-dwelling invertebrate species, of which 33 are endemic to the cave. The invertebrate community consists of both grazers and carnivorous species. The relationship between the cave-dwelling invertebrates and the mat material was convincingly established with carbon and nitrogen stable isotope analysis of organic material.

9.3. Subsurface biosphere

The rocks are alive! In recent years, there have been numerous dramatic demonstrations of active microbes in hard rock, deeply buried marine sediments, and ancient sedimentary environments, where no life was previously thought to exist. Thus, relatively high cell numbers of 10^4 to 10^7 cm^{-3} are routinely found at depths of hundreds of meters in deep-sea sediments (Parkes *et al.*, 1994) (see Chapter 2). Furthermore, this population of microbes is active, probably diverse, and supports sulfate reduction with low but measurable rates determined with the radiotracer $^{35}SO_4^{2-}$ technique (Parkes *et al.*, 1994) (see Chapters 2 and 9). Microbial sulfate reduction also has been found within Cretaceous-aged shale and sandstone sequences in New Mexico (Krumholz *et al.*, 1997). In these cases, the source of energy for sulfate reduction is the organic matter trapped within the sediments or the shales. The organic matter is probably indigenous and, therefore, quite old. Chemolithoautotrophic activity is likely both in deeply buried marine sediments and shales, but the main source of energy fueling microbial metabolisms is the organic matter originally deposited at the sediment surface and buried along with the sedimenting particles.

By contrast, numerous other subsurface environments support microbial activity without any obvious indigenous organic fuel. Thus, microorganisms may accelerate, or even promote, the alteration of basaltic glass extruded onto the sea floor (Furnes and Staudigel, 1999; Furnes *et al.*, 2001a) at subsurface depths up to 500 m. Evidence for a microbial role in the alteration process includes unique alteration textures indicative of microbial excavation of the glass. These features are different from those obtained when glass is altered in the absence of organisms. Other arguments supporting microbial activity include the presence of ^{13}C-depleted organic matter (Furnes *et al.*, 2001b) and the presence of microbial morphologies and DNA, as revealed by DAPI (4′,6-diamidino-2-phenylindole) staining of rock fragments. DAPI binds to DNA, giving a fluorescent complex.

The reasons for the microbial alteration of the glass are not clear. However, in a nutrient-depleted environment, organisms could be scavenging for P, or possibly limiting trace metals. Also unclear are the energy sources exploited by the organisms for growth. One possibility might be chemolithoautotrophy based on the reduction of CO_2 with H_2. Possible primary metabolisms include methanogenesis (Equation 4.3), acetogenesis (Equation 4.4), and sulfate reduction (Equation 4.5), with sulfate originating from circulating sea water.

$$CO_2 + 4H_2 \rightarrow CH_4 + 2H_2O \tag{4.3}$$

$$CO_2 + 2H_2 \rightarrow CH_3COOH \tag{4.4}$$

$$SO_4^{2-} + 4H_2 + 2H^+ \rightarrow H_2S + 4H_2O \tag{4.5}$$

Hydrogen gas might come directly from the mantle or, perhaps more likely, from the oxidation of ferrous iron silicates within the ocean's crust with the reduction of H_2O in a process known as serpentinization. Serpentinization occurs most readily in mafic and ultramafic rocks containing abundant ferromagnesian minerals (Stevens, 1997):

$$H_2O + (FeO)_x(SiO_2)_y \rightarrow H_2 + x(FeO_{3/2}) + y(SiO_2) \qquad (4.6)$$

Indeed, from sulfur stable isotope measurements, Alt and Shanks (1998) have convincingly demonstrated that sulfate reduction in ultramafic ocean crust is probably driven by serpentinization reactions.

Similar processes of H_2 generation likely drive numerous other deep subsurface microbial communities such as those found in the continental Columbia River Basalt Group (Stevens and McKinley, 1995) and deep granitic rocks from the Fennoscandian shield in southeastern Sweden (Pedersen, 1997). In both of these studies, diverse and rich microbial populations are indicated. Furthermore, microbial metabolisms involved in primary organic matter production through chemolithoautotrophic H_2 oxidation (Equations 4.3–4.5) produce biomass that can fuel numerous organotrophic processes. Overall, the subsurface biosphere is estimated to contain living biomass nearly equal in magnitude to all of the living plant life at the Earth's surface (Whitman *et al.*, 1998). As only a fraction of the subsurface biomass can be sustained from plant primary production at the Earth's surface, substantial primary production through chemolithoautotrophy is indicated and, indeed, required.

Chapter 5

Heterotrophic Carbon Metabolism

1. INTRODUCTION

Carbon is central to all life, and it is the most actively cycled element in the biosphere. Most carbon is tied up as inorganic carbon in limestone or in fossil organic pools such as bitumen and kerogen in shales (Table 5.1). The carbon in living biomass represents only a small fraction of the total carbon inventory on Earth (0.002%) and of the active carbon pool (2%) represented by unaged dead organic matter, dissolved organic carbon, inorganic carbon, and atmospheric CO_2 (Hedges and Keil, 1995). Numerous transformation modes and pathways deliver carbon to the different pools (Table 5.2). Inorganic carbon oxidized from organic matter or weathered from limestone enters the atmosphere as CO_2 or enters aquatic environments mostly as HCO_3^-. Atmospheric CO_2 tends to equilibrate with the carbonate system of the oceans (Takahashi *et al.*, 1997; Fasham *et al.*, 2001), and therefore, these two carbon pools are coupled. Inorganic carbon enters living biomass (LPOC [living particulate organic carbon]) via carbon fixation by autotrophic organisms (e.g., plants and algae; see Chapter 4).

0-12-026147-2
© 2005 Elsevier Inc.
All rights reserved

Table 5.1 Major carbon reservoirs on Earth

Type of pool	Location	Pool size (g C)
CO_2 gas	Atmosphere	6.6×10^{17}
Living biomass (LPOC)	Aquatic and terrestrial	9.5×10^{17}
Dissolved organics (DOC)	Aquatic	1.5×10^{18}
Dead organics (DPOC)	Sediments and soils	3.5×10^{18}
Dissolved CO_2	Aquatic	3.8×10^{19}
Carbonates ($CaCO_3$, $Mg(CO_3)_2$)	Carbonaceous rock	6.0×10^{22}
Fossil organics	Sedimentary rocks	1.5×10^{22}

From Bowen (1979); Hedges and Keil (1995).

Table 5.2 Major transformations of carbon in the lithosphere

Pathway	Transport mechanism
Atmospheric CO_2 → Oceanic HCO_3^-	Diffusive transport
Inorganic C (CO_2, HCO_3^-) → Organic C	Photo- and chemolithoautotrophy
Inorganic C (CO_2, HCO_3^-) → Carbonates	Precipitation, mostly biological
Organic C → Inorganic C (CO_2, HCO_3^-)	Decomposition, mineralization
Fresh organic C → Fossil C	Sedimentation and burial with time
Fossil C → Atmospheric CO_2	Human combustion of fossil fuels

Decomposing biomass is converted back to inorganic carbon through hydrolysis/fermentation (dissolved organic carbon [DOC] formation) and subsequent mineralization (CO_2 formation) by heterotrophic organisms (Canfield, 1993; Kristensen and Holmer, 2001).

Not all the dead particulate organic carbon (DPOC) in sediments and soils is available for biological decomposition, either because of a refractory chemical structure or because environmental conditions are unfavorable for microbial activity. The carbon escaping mineralization will be buried permanently in deep sediments and soils, and after enough time (millions of years), it will end up as fossil organics in sedimentary rock, fossil hydrocarbons, and natural gas. On long time scales, fossil carbon is returned to the Earth's surface for subsequent oxidation. The human circumvention of this natural carbon cycle through the burning of coal and other fossil fuels is well known; it currently increases the atmospheric concentration of CO_2 about 0.5% per year, bringing present CO_2 levels to 34% above their pre-industrial average value (Duffy *et al.*, 2001; Soon *et al.*, 2001).

Microorganisms grow and divide, and this requires energy, nutrients, and often organic building blocks from the environment. As outlined in Chapter 3, energy-gaining activities are referred to as catabolic (dissimilatory or energy) metabolism, whereas the synthesis of cellular material is referred to as anabolic (assimilatory) metabolism. These two types of metabolism are coupled in the sense that microorganisms use a large share of the energy gained during catabolism to fuel anabolic carbon assimilation (Table 5.3).

A variety of terms describe catabolic processes, including fermentation, respiration, decomposition, degradation, decay, and mineralization. Fermentation, as we saw in Chapter 3, occurs when organic matter acts as both the electron acceptor and the electron donor in organic matter breakdown. Decomposition, degradation, and decay are three similar terms describing the overall enzymatic processes mediating (1) the dissolution of particulate organic polymers into large macromolecules (hydrolysis), (2) the cleavage of these into smaller moieties (fermentation), and (3) the terminal oxidation (respiration) of the organic carbon by various electron acceptors, such as oxygen, manganese oxides, nitrate, iron oxides, and sulfate (Table 5.4). More generally, respiration refers to all catabolic reactions generating ATP involving both organic and inorganic primary electron donors and electron acceptors, including, therefore, also chemolithoautotrophic and chemolithoheterotrophic processes (see also Chapter 4). The term mineralization refers to the terminal oxidation of organic carbon into inorganic (mineral) form.

Aerobic and anaerobic heterotrophic microorganisms are responsible for a large proportion of the organic carbon oxidation occurring in aquatic environments. This Chapter focuses on these heterotrophic processes of carbon mineralization. We look at the relationship between microbial activity and carbon source and consider the role of microorganisms in the

Table 5.3 ATP use for a microbial cell grown on glucose

Process	Available ATP (%)
Synthesis	
Polysaccharides	6.5
Protein	61.1
Lipid	0.4
Nucleic acids	13.5
Transport mechanisms	18.3

From Fenchel *et al.* (1998).

Table 5.4 Schematic overview of chemotrophic catabolic (dissimilative) processes mediated by microorganisms

	e^- donor							
e^- acceptor	H_2	CH_2O	CH_4	HS	NH_4^+	N_2	NO_2	Fe^{2+}
CH_2O	1	1	–	–	–	–	–	–
CO_2	2	2	–	–	–	–	–	–
SO_4^{2-}	3	3	3	–	–	–	–	–
Fe^{3+}	4	4	4	9	–	–	–	–
NO_3^-	5	5	5	9	11	–	–	–
Mn^{4+}	6	6	6	9	12	–	–	–
O_2	7	7	7	9	10	–	10	8

1, fermentation; 2, methanogenesis; 3, sulfate reduction; 4, iron reduction; 5, denitrification; 6, manganese reduction; 7, aerobic respiration; 8, iron oxidation; 9, sulfide oxidation; 10, nitrification; 11, anammox; 12, anoxic nitrification.

detritus food chain. We also outline the various pathways of both aerobic and anaerobic organic carbon mineralization, and we consider the processes leading to the preservation of organic carbon in sediments. This Chapter is a springboard to subsequent Chapters, which discuss the specifics of individual elemental cycling by microbes.

2. THE DETRITUS FOOD CHAIN

2.1. The chemical nature of detritus

In many ecosystems a significant part of the organic production is not consumed or assimilated by animals but is instead converted to the pool of dead organic matter—the detritus pool (Figure 5.1) (see Wetzel *et al.*, 1972). The fraction of primary production entering the detritus pool as particulate and dissolved organic matter is highly dependent on the composition of the plant community. Phytoplankton is the predominant detritus source in the open ocean and deep lakes, whereas macroalgae and vascular plants are more important in coastal marine areas, shallow lakes, and rivers. The chemical composition of plants and algae is well defined, and 90–100% of the organic matter consists of carbohydrates (cellulose, alginate), proteins, lipids, and lignin (Table 5.5). Lignin is only present in significant amounts in terrestrial vascular plants (10–30%), whereas phytoplankton may contain more than 50% protein (Hodson *et al.*, 1984; Hedges *et al.*, 2002). Animal

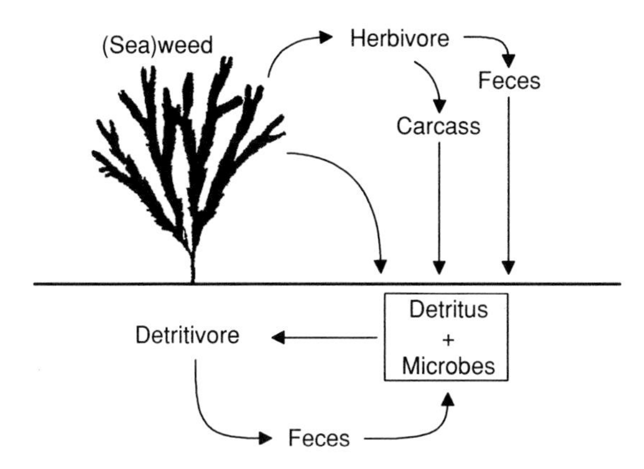

Figure 5.1 Fate of organic production by primary producers, in this case, seaweed. The plant biomass will either be grazed by herbivores or enter the food chain as dead organic matter. Plant remains enter the detritus pool together with herbivore feces and carcasses. Detritus is either decomposed by microorganisms or consumed by detritivores together with microorganisms. Finally, the detritivore feces (and carcasses) return to the detritus pool together with carnivore remains, if present.

tissues, on the other hand, are very rich in protein (up to 80%; Kristensen, 1990). Prokaryotes contain about 50% protein, and together with carbohydrates and lipids, about 80% of prokaryote biomass can be accounted for (Madigan *et al.*, 2003). The remainder is composed of nucleic acids and cell wall constituents (murein, teichoic acid, and lipopolysaccharide).

In total, detritus consists of ungrazed plant material, along with macrofaunal feces, carcasses, exuviae, and dead microbial cells. The fraction of recognizable biochemical constituents decreases as detritus passes through the food chain. In the eastern and equatorial Pacific, for example, the chemical composition of living phytoplankton and zooplankton is well characterized (>80%, Table 5.5). However, only about 30–60% of the organic matter can be characterized in the flocculent layer at the sediment surface at depths of 3500 to 4000 m, and this decreases to 20% or less at 10 cm depth in the sediment (Wakeham *et al.*, 1997; Wang *et al.*, 1998). The largely uncharacterized and relatively refractory organic carbon fraction was once thought to be formed by spontaneous polycondensation reactions among small reactive intermediates released during the enzymatic breakdown of biomolecules, forming, in some cases, highly refractory "geopolymers." Evidence suggests, however, that this uncharacterized fraction probably has more complex formation pathways than previously recognized (Hedges *et al.*, 2000).

Table 5.5 Carbohydrate, lignin, lipid, and protein content as percentage of total organic matter in (A) selected plants, animals, and microorganisms (adapted from Kristensen, 1990), and (B) phytoplankton, zooplankton, the top 2–4 mm loose flocculent layer of sediment (sediment floc), the upper cm of the sediment, and the 10–14 cm depth interval of sediment at 4100 m depth in the eastern Pacific (from Wang *et al.*, 1998)

Material	Carbohydrate	Lignin	Lipid	Protein	Σ
(A)					
Chlorella sp.	33.4	0	16.1	46.4	95.9
Skeletonema costatum	32.5	0	7.3	57.8	97.6
Ulva lactuca	77.0	0	0.1	22.7	99.8
Fucus vesiculosus	90.7	~0	2.6	6.6	99.9
Aquatic angiosperm	81.5	~0	2.1	16.1	99.7
Barley straw	82.6	13.6	1.0	1.0	98.2
Tree-leaves	77.1	12.5	2.9	6.7	99.2
Pine wood	71.7	26.7	1.5	~0	99.9
Polychaete worm	4.8	0	9.0	78.5	92.3
Bivalve	20.0	0	4.0	71.1	95.1
Prokaryote	16.6	0	9.4	52.4	78.7
(B)					
Phytoplankton	48	0	19	24	91
Zooplankton	14	0	39	46	99
Sediment floc	19	0	3	35	57
0–1 cm sediment	16	0	3	17	36
10–14 cm sediment	5	0	2	11	18

2.2. Organisms decomposing detritus

Microorganisms play a dominant role in the breakdown of detritus, for which there are several reasons. First, microbes can hydrolyze a diverse assemblage of organic compounds, some of which are totally indigestible by animals. Second, due to their small size and consequently large surface-to-volume ratio, they have the ability to take up and utilize carbon sources and mineral nutrients efficiently in dilute solutions (see Chapter 2). Furthermore, small cells can create intimate contact with solid surfaces and thus minimize loss of products from exoenzyme activity. Finally, microbes can maintain an efficient metabolism and mineralize organic substrates under anoxic conditions. Prokaryotes are important decomposers of detritus in most aquatic environments, but fungi may contribute significantly in certain environments, particularly in freshwater settings. Thus, fungi can account for up to 96% of the microbial biomass associated with vascular plant litter entering lakes and rivers (Baldy *et al.*, 1995). By contrast, in seagrass beds and salt marshes fungi usually account for no more than 20% of the total microbial biomass associated with litter, whereas values as high as 80% have been recorded from mangrove forests (Blum *et al.*, 1988). The role of fungi is reduced even further in anoxic sediments, particularly if the sediments are sulfidic (Mansfield and Bärlocher, 1993).

The amount of primary production consumed by herbivorous animals in aquatic environments is positively correlated with the nutrient (i.e., protein) content of the primary producer, and it is inversely correlated with the primary producer C:N ratio (Table 5.6). Animals, having a C:N ratio of 4–6, require a diet with a C:N ratio below about 17 (Russell-Hunter, 1970). Thus, the high C:N ratio and the high content of structural carbohydrates (cellulose and hemicellulose) associated with lignin in vascular plants, and for certain polysaccharides in macroalgae, restricts the herbivory of these plant types. It is not surprising, then, that most animals lack digestive enzymes for hydrolyzing structural plant components and that only a few aquatic animals are capable of utilizing the tissues of vascular plants. Phytoplankton and benthic microalgae, on the other hand, are readily grazed and digested by zooplankton and benthic animals.

Many herbivores feed on a variety of substrates, both living and dead, in order to balance their nutrition. For example, the grapsid crab *Sesarma messa* can process 30–70% of the total leaf production in certain mangrove forests by eating both attached leaves and fallen leaf litter (Robertson *et al.*, 1992). Because leaves from mangrove trees are poor in nutrients, these crabs must supplement their diet with nutrient-rich detritus (Skov and Hartnoll, 2002). One mechanism of enhancing the nutritional value of plant material is "microbial gardening." Thus, fallen leaf litter is pasted by *S. messa* along burrow walls to foster microbial colonization and to facilitate leaching of tannins. After several weeks, the microbially processed litter/detritus is consumed by the crabs.

Table 5.6 Fraction of primary production, which is consumed directly by herbivores. The C:N ratio of the various plants is give for comparison

Ecosystem	Plant species	Fraction consumed	C:N ratio
Mangrove forest	*Rhizophora* sp.	~2%	50–70
Salt marsh	*Spartina* sp.	~9%	40–50
Seagrass bed	*Zostera* sp.	~8%	20–30
	Thalassia sp.	5–10%	20–30
Macroalgae bed	*Fucus* sp.	10–20%	15–20
	Ulva sp.	30–100%	7–10
Intertidal flats	Diatoms	20–50%	6–8
Open ocean	Phytoplankton	50–100%	6–8

Compiled from Robertson (1991); Pfeiffer and Wiegert (1991); Alongi (1998); Nienhuis and Groenendijk (1986); Klumpp *et al.* (1992); Geertz-Hansen *et al.* (1993); Kristensen (1993).

Animals can also supplement their nutrition either by browsing fine detritus particles rich in prokaryotes and microalgae from the sediment surface (Micheli, 1993) or by ingesting large amounts of sediment and utilizing the associated microorganisms. As an example of the latter, the head-down conveyor belt-feeding detritivorous polychaete *Arenicola marina* feeds on sediment detritus and associated microorganisms from 10–30 cm in depth (Figure 5.2). The prokaryote density in the feeding pocket around the head of the worm is only two-thirds of the surface density, and the selective feeding on fine particles by the worm increases the prokaryote density slightly in its esophagus. About half of the detritus-associated prokaryotes present in the esophagus are then digested during passage through the

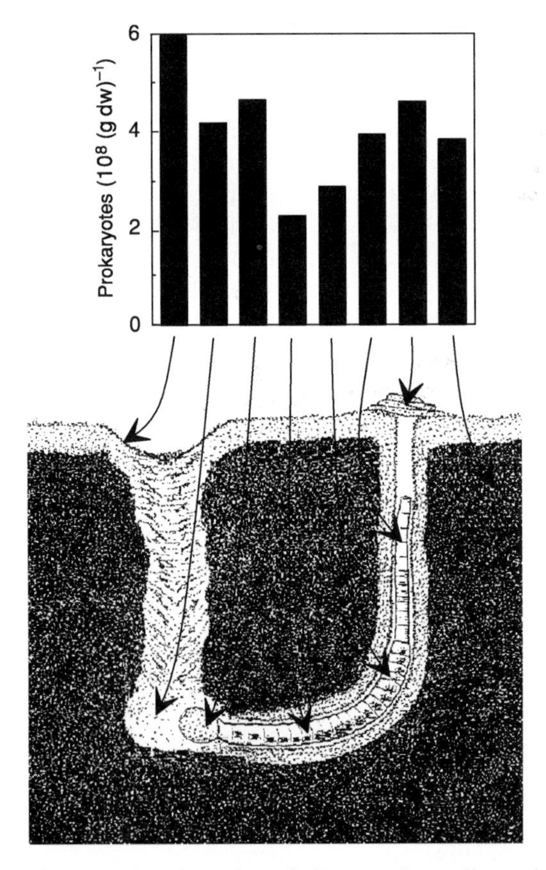

Figure 5.2 Prokaryote abundance in a shallow marine sediment inhabited by the lugworm *Arenicola marina*. Prokaryote numbers are given for surface sediment, feeding pocket, lugworm pharynx, lugworm gut (foregut, midgut and endgut), fecal cast, and anoxic surrounding sediment. Redrawn from Andresen and Kristensen (2002).

stomach. Subsequently, prokaryote regrowth on detrital particles occurs during passage through the intestine, and the egested feces contain as many prokaryotes as found in the esophagus. The digestion of prokaryotes in the stomach supports only 8% or less of the nutritional need of the worm (Andresen and Kristensen, 2002). The remaining nutrition originates from microphytobenthos or, alternatively, from digestion of detritus particles and meiofauna found in the sediment.

In general, the degradation of detritus is enhanced through mechanical disintegration of particles through wave action, or maceration by feeding animals, which increases the surface area of the particles. The rates of microbial decomposition of freshly fragmented leaf litter and macroalgae may increase several-fold, mainly due to leaching of labile DOC (Gunnarson *et al.*, 1988). Furthermore, maceration and passage of leaf litter through animal guts creates an excellent substrate for microbial colonization. For example, fecal material from leaf-eating mangrove crabs maintains a similar or higher microbial decomposition potential as compared to the original fragmented litter material (Kristensen and Pilgaard, 2001).

2.3. Detritus and dissolved organic matter (DOM)

Dissolved organic matter (DOM) is the quantitatively most important part of the detritus pool in aquatic environments. DOM accounts for >95% of total organic matter in the water column of oceans, lakes, and rivers (Table 5.1). It is involved in short- and long-term trophic interactions in plankton and acts as a dynamic part of the sediment organic carbon pool (Kirchman *et al.*, 1991). However, the distinction between particles and true solutes is difficult due to a continuum of particle sizes, and it is common to define DOM operationally as the fraction of organic material passing through glass fiber filters with nominal retention of 0.5–0.7 μm (Nagata and Kirchman, 1992). As the size of true DOM is less than 10 kDa, this definition includes macromolecules and microparticles (colloids) in the size range of 1 nm (~10 kDa) to 0.5–0.7 μm (Buffle, 1990) and even viruses and small prokaryotes.

DOM is a fundamental part of the carbon flow in aquatic systems, but its composition and geochemistry are poorly understood. One major reason is the difficulty in chemically characterizing this highly dilute and complex mixture of organic compounds. Less than 20% of DOM has been grouped into major biochemical classes such as carbohydrates, amino acids, and lipids (Amon *et al.*, 2001; Burdige, 2002). Combined carbohydrates constitute about 90% of the well-characterized fraction of DOM, while amino acids contribute roughly 10%, and lipids (fatty acids) are mostly found in trace amounts of <1% (Table 5.7). Humic compounds account for about

Table 5.7 Concentration ranges (μM) for dissolved organic carbon (DOC) in aquatic environments. In addition, the contribution of major well-characterized biopolymers is shown. Compiled from various sources

	Ocean		Lake		
	Coastal	Open	Eutrophic	Oligotrophic	River
DOC	100–500	40–80	500–2000	100–300	400–2000
Carbohydrate	20–80	5–15	70–400	20–60	40–200
Amino acids	1–12	~1	5–10	~1	10–80
Fatty acids	<1	<<1	~1	<<1	~1
Humic compounds	20–125	10–20	200–700	40–130	200–1000

50% of the total DOM pool in freshwater environments and only 15–25% in marine waters.

The sources of DOM in aquatic environments should provide clues as to its chemical composition, but there are several limitations to this approach (Lee and Henrichs, 1993, Bauer *et al.*, 2002). First, there are many potential sources of organic matter, and there are only semi-quantitative estimates of their relative importance. Second, the composition of organic matter from the various sources is often unknown. Third, biological and chemical processes substantially alter DOM after it is released to water and sediment. Nevertheless, a closer examination of DOM sources may provide evidence about its nature. The major identified sources of DOM in aquatic environments are phytoplankton exudation in open waters, macrophyte exudation in shallow waters, the damage or lysis of microbial, plant and animal cells, and microbial and animal excretion. DOC is also delivered to aquatic environments from land, largely via rivers, and from sediment porewaters (Burdige, 2002).

A substantial fraction of the organic carbon synthesized by phytoplankton is released as DOC. In some cases, over 50% of primary production is so released, but typical values are within the range of 3–30% (Strom *et al.*, 1997; Søndergaard *et al.*, 2000). Phytoplankton exudates probably constitute the single most important source of DOM in the photic zone of the water column. Aquatic macrophytes (vascular plants and seaweeds), on the other hand, are known to release only about 3–6% of their photosynthetically fixed carbon as DOC (Brylinsky, 1977). There are a number of possible mechanisms for DOM release by plants. Exudation may arise as photosynthetic intermediates diffuse through cell membranes. In this case, the exudate should reflect the composition of small molecules in the intracellular fluid, such as glycollate. Exudation of large molecules, on the other hand, can occur by plants producing extracellular polysaccharides. Loss of DOM as a result of cell lysis following virus attack, or damage and sloppy feeding by zooplankton and macrophyte grazers, yields a DOC composition similar to the intracellular fluid. The DOM released by phytoplankton is found in all molecular weight fractions, with, in the study of Chrost and Faust (1983), 19% ending up in the <0.5 kDa fraction, 30% in the fraction between 10 and 30 kDa, and 15% in the >300 kDa fraction.

Prokaryotes produce and excrete many high-molecular-weight compounds not found in other types of organisms (Mannino and Harvey, 2000). These include teichoic acids, polymers made up of sugars and phosphate that often include D-alanine and glycerol, peptidoglycans or mureins, highly branched complex molecules consisting of sugar and amino sugar chains cross-linked by peptide bridges, and capsular polysaccharides composed of neutral and acidic monosaccharides and amino sugars. In addition, hydrolytic exoenzymes excreted by prokaryotes may generate low-molecular-weight DOM by hydrolysis of larger molecules and particles in the water (Amon *et al.*, 2001). DOM originating from vascular plants, and the DOM entering from land via rivers or from wetlands, differs greatly in structure from marine and lacustrine DOM derived from phytoplankton (Klinkhamer *et al.*, 2000). In particular, DOM from vascular plants contains substantial quantities of phenolic polymers derived from lignin.

Despite its large quantitative importance in pelagic environments, DOM is usually found in relatively low concentrations (Table 5.7). For example, in offshore regions of the Gulf of Mexico, surface water DOC concentrations range from 60 to 80 μM, and they increase substantially to more than 500 μM in nearshore regions (Guo *et al.*, 1995) (Figure 5.3). The combined

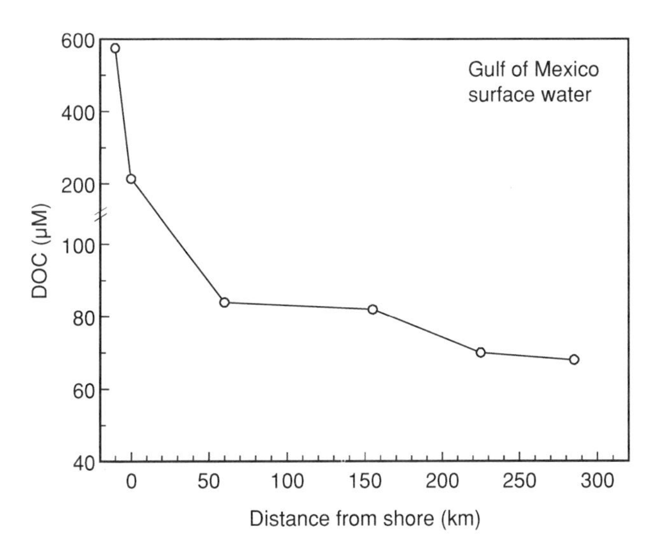

Figure 5.3 DOC concentrations in surface waters from the Gulf of Mexico as a function of distance from the shore. Redrawn from Guo *et al.* (1995).

influence of DOC input from terrestrial sources, high rates of phytoplankton production, and a high flux of DOC from shallow water sediments is believed to be responsible for the high DOC levels in coastal areas. If we consider the vertical distribution of DOC in the offshore regions of the Gulf of Mexico, surface water concentrations of 65 to 80 μM drop sharply below 150 m to less than 50 μM (Guo *et al.*, 1995) (Figure 5.4). In these profiles we can differentiate an upper layer of net DOC production, a zone of downward transport, and a zone of net DOC consumption in the deep layer.

Heterotrophic prokaryotes are the primary consumers of marine DOC in seawater, and they preferentially utilize specific components of the bulk DOC pool (Cherrier *et al.*, 1999; Bauer *et al.*, 2002). Thus, the average turnover time for oceanic DOC is in the range of 30–300 years, but turnover times of <10 years are noted for colloids in the euphotic zone of coastal shelf regions, while DOC persists for 6000–8000 years in the deep ocean. There is a general relationship between turnover time and particle or colloid size. Thus, large molecules (i.e., >10 kDa) tend to be the most labile, with

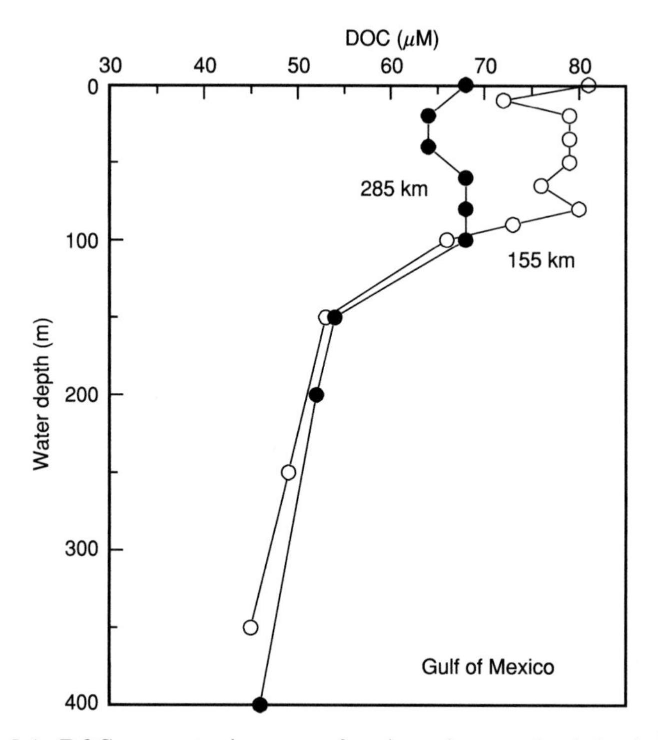

Figure 5.4 DOC concentrations as a function of water depth in the Gulf of Mexico at distances of 155 km and 285 km from the shore. Redrawn from Guo *et al.* (1995).

smaller molecules (<1–3 kDa) being less biodegradable (Santschi *et al.*, 1995; Burdige and Gardner, 1998; Mannino and Harvey, 2000).

"Classical" models of pelagic food chains described a unidirectional transfer of organic matter and energy, from phytoplankton to zooplankton and further to higher carnivores. The quantitative role of water column prokaryotes was not appreciated in these models. However, it has become apparent that prokaryotes constitute 20–30% of the total plankton biomass and that they process a substantial part of the primary production (Fenchel *et al.*, 1998). Thus, prokaryotes are actively engaged in the "microbial loop" (Azam *et al.*, 1983), in which a large fraction of primary production is released as DOM and is subsequently utilized by microbes (Figure 5.5). The microbial production is then consumed by microzooplankton (heterotrophic protists) and is thus channeled into the "classical" pelagic food chain. The importance of the microbial loop varies among aquatic environments, and it is generally considered most important under oligotrophic conditions. The microbial loop is least important in productive upwelling

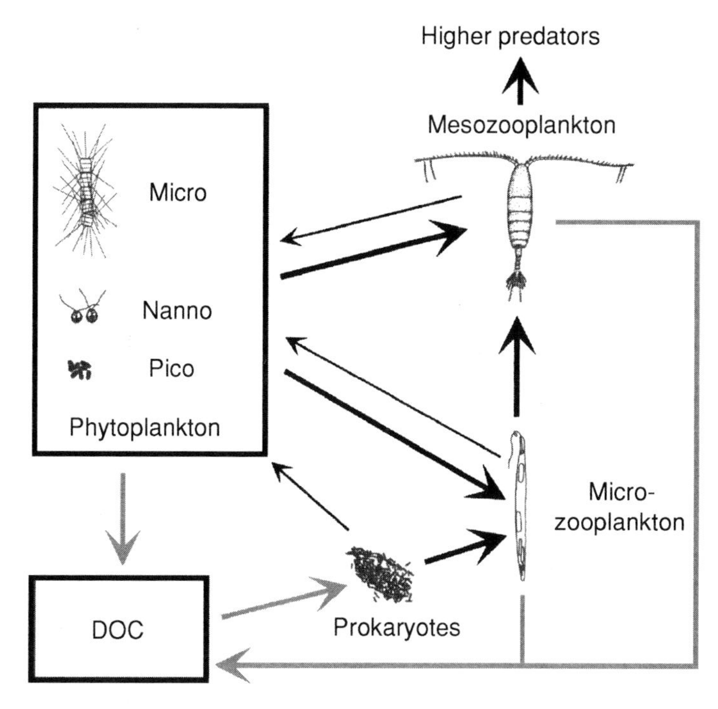

Figure 5.5 Microbial loop involving phytoplankton (micro: >5 μm, nano: 1–5 μm, pico <1 μm), zooplankton (meso: >200 μm, micro <200 μm), and prokaryotes. Grey arrows indicate net DOC pathways, thick black arrows indicate POC pathways, and thin black arrows indicate net nutrient pathways.

systems and during the initial stages of algal blooms (Fenchel *et al.*, 1998). Although the concept of microbial loop was developed within the context of plankton dynamics, it is also valid for photosynthetic sediment systems such as microbial mats.

Concentrations of DOC in sediment porewaters are typically 10 times or more higher than in the overlying water (Burdige, 2002), implying substantial DOC production during sedimentary organic carbon decomposition. Degradation of particulate organic matter in sediments occurs by a series of microbially mediated hydrolytic, fermentative, and respiratory processes, which produce and consume DOM intermediates (Figure 5.6). The concentration of DOC typically increases with sediment depth, and it is controlled by the balance between production, consumption, and upward

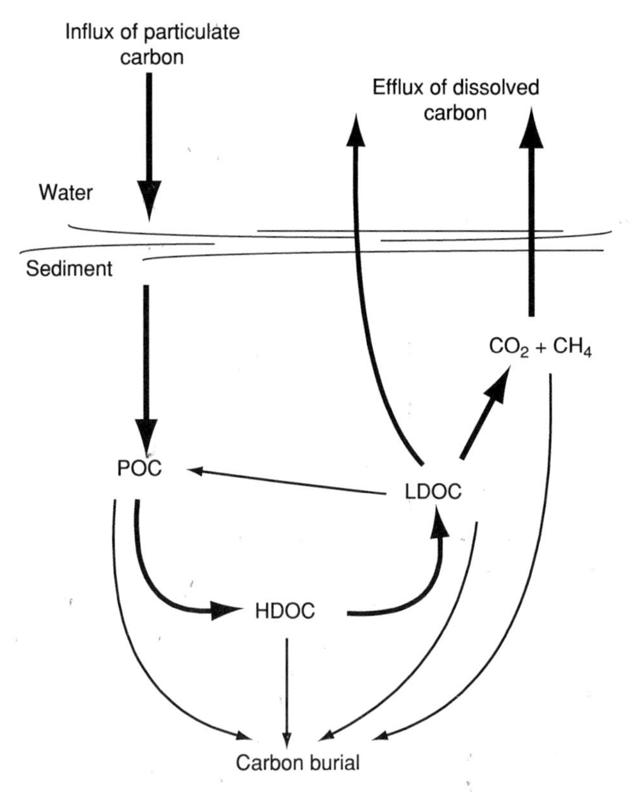

Figure 5.6 A schematic presentation of organic matter transformations in sediments. Only the major pathways are shown. Arrow thickness gives some sense for the quantitative role of the processes. POC, particulate organic carbon; HDOC, high-molecular-weight dissolved organic carbon; LDOC, low-molecular-weight dissolved organic carbon. Modified from Arnosti *et al.* (1994).

molecular diffusion. Generally, an asymptotic concentration is approached in deeper sediment layers. The maximum porewater concentration of DOC is positively correlated with carbon oxidation rate, and it is therefore related to the input rate of reactive organic matter (Burdige, 2002). In a specific example, higher porewater DOC concentrations are found in sediments of the Chesapeake Bay, with a high organic carbon input rate, compared to deeper sediments on the continental margin, where the organic carbon input rate is lower (Burdige and Gardner, 1998) (Figure 5.7).

Sediment DOC, which is only slightly better characterized than water-column DOC, is a heterogeneous mixture of labile and refractory compounds, ranging in size from relatively large macromolecules such as proteins and lipids to smaller molecules such as individual amino acids and short-chain fatty acids. DOC becomes more refractory with increasing sediment depth, possibly due to geopolymerization of more labile moieties through condensation reactions (the melanoidin or "browning" reaction;

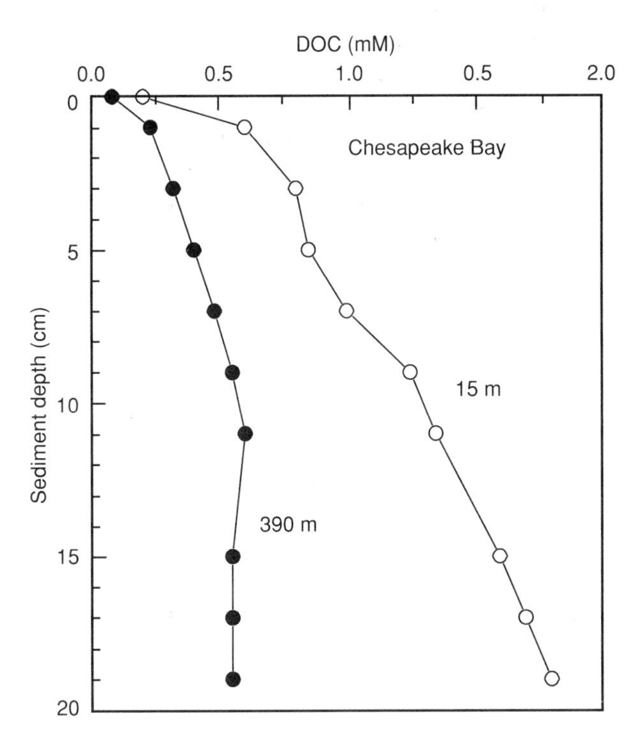

Figure 5.7 Porewater DOC concentrations as a function of sediment depth in shallow-water sediments from the Chesapeake Bay (15 m water depth) and in deep continental shelf regions offshore of the bay. Redrawn from Burdige and Gardner (1998).

Hedges *et al.*, 2000), or complexation reactions (Michelson *et al.*, 1989). High sediment DOC concentrations represent a DOC source to overlying waters. Accordingly, it has been suggested that the benthic flux of old and refractory DOC to the deep ocean can partly explain the discrepancy between the old age (~8000 yr) of deep water DOC and the average oceanic mixing time (~1000 yr) (Hedges, 1992).

3. AEROBIC CARBON OXIDATION

Organisms utilizing oxygen are aerobes, and they can be either obligate aerobes, with an absolute requirement for oxygen, or facultative aerobes, in which oxygen is not required for metabolism, but preferred. Thus, facultative organisms can adapt their catabolism to operate with oxygen, or they can utilize other metabolisms such as fermentation or denitrification when oxygen is absent. If organisms can grow under only low oxygen conditions, they are referred to as microaerophilic.

The general mechanisms by which microorganisms degrade particulate organic polymers in aquatic environments are reasonably well understood. When oxygen is present, all catabolic reactions occur by aerobic processes (i.e., aerobic respiration). Microorganisms first attach to particles in order to hydrolyze biopolymers via substrate-specific cell-bound ectoenzymes or exoenzymes released into restricted microniches afforded by the particles (Figure 5.8). The low-molecular-weight byproducts are then assimilated by the particle-bound organisms, or they are released into the surrounding water if rates of biopolymer hydrolysis are particularly high. The intracellular mineralization process involves the generation of reduced pyridine nucleotides through the tricarboxylic acid (TCA) cycle with a terminal acetate oxidation to carbon dioxide (Madigan *et al.*, 2003). The reducing power in the form of hydrogen atoms or electrons derived from the organic matter is conveyed to oxygen by an electron transport system (see Chapters 3 and 6) to form water. In the transfer of reducing power via the electron transport chain, some of the energy is liberated and conserved in ATP during oxidative phosphorylation (see Chapter 3). Most aerobic heterotrophs are capable of completely oxidizing particulate polymers to carbon dioxide, water, and inorganic nutrients, according to the following stoichiometry (Froelich *et al.*, 1979):

$$(CH_2O)_x(NH_3)_y(H_3PO_4)_z + xO_2 \rightarrow xCO_2 + yNH_3 + zH_3PO_4 + xH_2O,$$
$$\Delta G^0 = -3190 \text{ kJ mol}^{-1} \text{ for glucose} \qquad (5.1)$$

Although oxygen serves the role of terminal electron acceptor, a unique feature of aerobic decomposition is the formation and action of highly

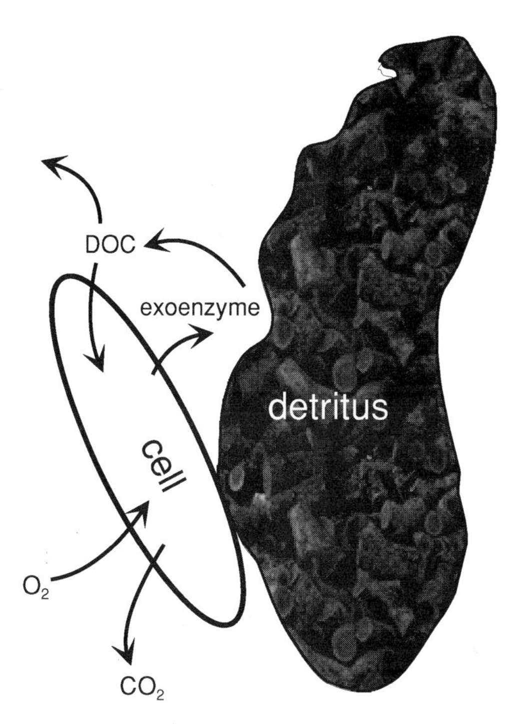

Figure 5.8 The action of aerobic prokaryotes that attach to organic particles and excrete ectoenzymes or exoenzymes. The dissolved hydrolysis products (DOC) are then assimilated by the microbes for use in catabolic and anabolic processes or are released into the surrounding water.

reactive oxygen-containing radicals such as superoxide anion ($\cdot O_2^-$), hydrogen peroxide (H_2O_2), and hydroxyl radicals ($\cdot OH$). These are capable of breaking bonds and depolymerizing relatively refractory organic compounds such as lignin and thus promote the aerobic decomposition process (Canfield, 1994). At the same time, these oxygen intermediates are extremely toxic because they are powerful oxidizing agents and destroy cellular constituents very rapidly. Many microorganisms possess enzymes that afford protection against toxic oxygen products (see Chapter 6).

The extent to which aerobic carbon oxidation occurs in aquatic sediments depends on a variety of factors, including organic carbon input rate, organic carbon composition, particle size, sedimentation rate, the depth of the water column through which the organic carbon descends, temperature, and the degree of stratification and oxygenation of the water column. Most of the highly reactive organic matter produced in surface waters of the ocean interior, and in deep oligotrophic lakes, is oxidized as it sinks through the oxic water column. Microbes rapidly colonize the sedimenting detritus, and more than 90% of the organic carbon is decomposed aerobically before reaching the sediment (Suess, 1980). Some large eutrophic water bodies, such as the Black Sea and many fjords, have permanent to intermittent oxygen depletion in the bottom layer. Anoxia arises, usually, due to some sort of stratification that restricts oxygen supply relative to demand by organic carbon mineralization processes. In these systems, a considerable amount of the water column carbon oxidation can occur by anaerobic metabolisms.

Sediments are reducing environments covered by only a thin oxic surface layer where aerobic microorganisms are active. Sediments in productive shallow waters are generally characterized by oxygen penetration depths of millimeters compared with centimeters or decimeters for sediments in the ocean interior (Figure 5.9) (Glud *et al.*, 1994). The penetration depth of oxygen is controlled by the balance between downward transport rates of oxygen from the overlying water and oxygen consumption rates within the sediment. Oxygen is transported into sediments by molecular diffusion, by the bioturbating and irrigating activities of benthic organisms, and by bottom current-induced flushing when the sediments are sandy (Huettel and Gust, 1992). Consumption of oxygen is accomplished by benthic fauna and inorganic reactions, but mainly by aerobic microbial metabolisms (Figure 5.10). While some of these microbial metabolisms are directed

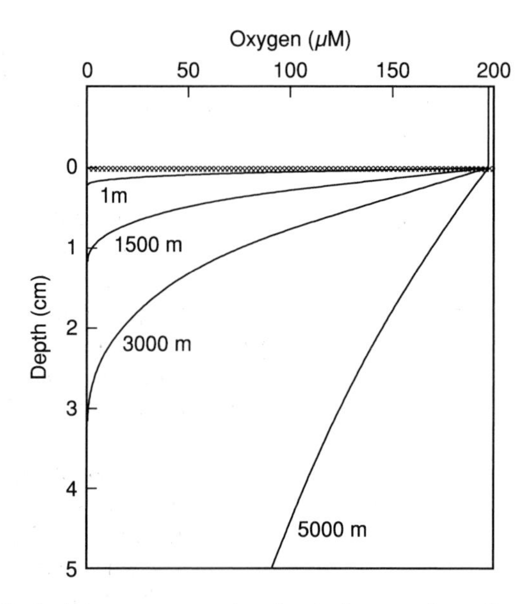

Figure 5.9 Typical oxygen penetration depths into marine sediments at water column depths from 1 to 5000 m. The horizontal line indicates the sediment–water interface. The oxygen level in the overlying water is, for simplicity, fixed to 197 μM in all environments. Modified from Jørgensen and Revsbech (1985) and Glud *et al.* (1994).

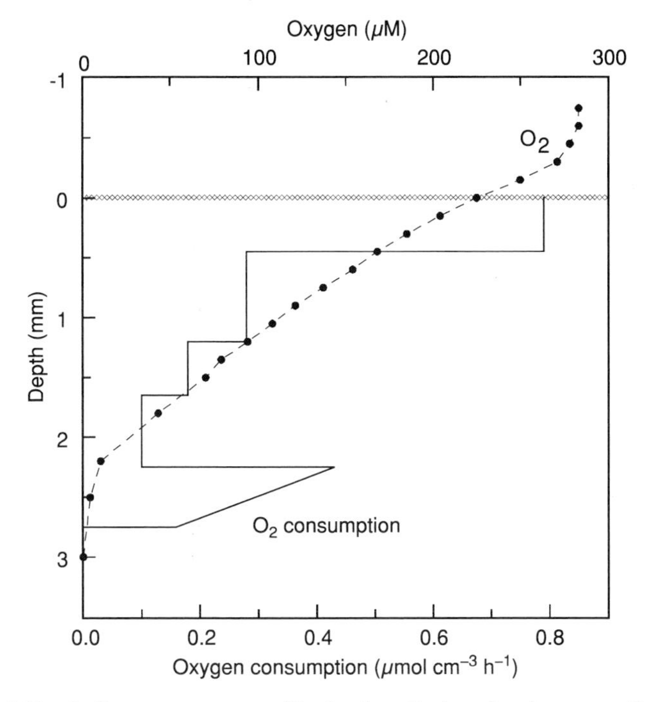

Figure 5.10 Sediment oxygen profile (broken line) and volume-specific rates of oxygen consumption (solid line) are shown. Oxygen consumption down to 2 mm is primarily due to aerobic respiration, whereas the peak between 2 and 3 mm in depth is caused by the reoxidation of reduced inorganic metabolites. The horizontal line indicates the sediment–water interface. Modified from Jensen *et al.* (1993).

toward heterotrophic carbon oxidation, a significant fraction is coupled to the oxidation of reduced metabolites from anaerobic processes occurring in the underlying sediment. These issues are explored in more detail in the following sections.

4. ANAEROBIC CARBON OXIDATION

Below the oxic zone in stratified water bodies such as the Black Sea and meromictic lakes, and in aquatic sediments, deposited detritus degrades through anaerobic microbial food chains (Figure 5.11). In such oxygen-deficient environments, catabolic reactions are coupled to terminal electron acceptors other than oxygen. Anaerobic decomposition is accomplished by mutualistic consortia of different types of heterotrophic microorganisms,

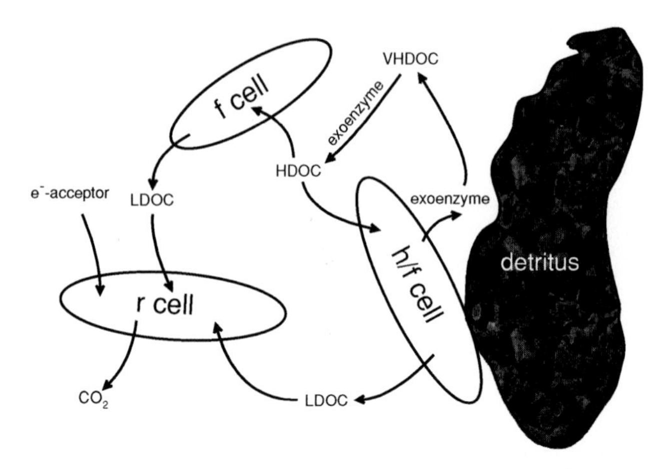

Figure 5.11 A consortium of organic matter-degrading microbes. Hydrolytic/fermenting organisms (h/f cell) attach to particles and excrete ectoenzymes or exoenzymes. The very-high-molecular-weight dissolved organic carbon (VHDOC) released by initial hydrolysis is further hydrolyzed to high-molecular-weight dissolved organic carbon (HDOC) by exoenzymes. The molecules of HDOC are small enough to be assimilated by the fermenting organisms (f cell) with the release of low-molecular-weight dissolved organic carbon (LDOC). Prokaryotes capable of anaerobic respiration (r cell) can then perform the final oxidation of LDOC to CO_2 with various electron acceptors.

as no single type of anaerobic organism is capable of complete oxidation of particulate organic polymers. In anaerobic microbial food chains, only a fraction of the energy in the substrate is utilized by each physiological type of organism during the decomposition process. The remaining chemical energy is left in excreted metabolites, which are assimilated and further metabolized by other organisms.

4.1. Hydrolysis/fermentation

The hydrolysis of particulate polymers such as polysaccharides, proteins, nucleic acids, and lipids is the first step in the anaerobic decomposition of organic carbon. In anoxic environments, the initial enzymatic attack on biopolymers is dependent on the activity of a relatively select number of microorganisms. The initial hydrolysis must take place outside the cell because prokaryotes can only transport substrates with a maximum molecular weight of about 600 Da across their cell membranes (Weiss *et al.*, 1991). For comparison, glucose molecules have a molecular weight of 180 Da. To utilize polymers larger than 600 Da, prokaryotes must secrete cell surface-attached (ectoenzymes) or free extracellular enzymes (exoenzymes) to hydrolyze macromolecules prior to transport into the cell. The large and normally complex polymeric organic molecules are first split into smaller, water-soluble moieties. For example, starch and glycogen are hydrolyzed by amylases to glucose, maltose, and other products. Cellulose is hydrolyzed by cellulase to cellobiose and glucose, and lipids such as triglycerides and esters of glycerol are hydrolyzed by lipase enzymes to glycerol and fatty acids. Furthermore, proteins are hydrolyzed by proteases to peptides and amino acids (Madigan *et al.*, 2003). These smaller molecules are then taken up and fermented by the hydrolyzing and/or fermenting microbes.

Fermentation reactions, and some aspects of the thermodynamics surrounding them, have been already discussed in Chapter 3, and the reader is referred to that Chapter for a complete discussion. We reemphasize here some of the important fermentation processes involved in anaerobic carbon mineralization. These include the following examples:

Glucose fermentation to ethanol:

$$C_6H_{12}O_6 \rightarrow 2CH_3CH_2OH + 2CO_2, \quad \Delta G^0 = -244.1 \text{ kJ mol}^{-1} \tag{5.2}$$

Ethanol fermentation to acetate:

$$CH_3CH_2OH + H_2O \rightarrow CH_3COOH + 2H_2, \quad \Delta G^0 = +34.8 \text{ kJ mol}^{-1} \tag{5.3}$$

Lactate fermentation to acetate:

$$CH_3CHOHCOOH + H_2O \rightarrow CH_3COOH + CO_2 + 2H_2, \quad \Delta G^0 = +23.9 \text{ kJ mol}^{-1} \tag{5.4}$$

Propionate fermentation to acetate:

$$CH_3CH_2COOH + 2H_2O \rightarrow CH_3COOH + CO_2 + 3H_2, \quad \Delta G^0 = +105.1 \text{ kJ mol}^{-1} \tag{5.5}$$

Amino acid degradation (Strickland reaction):

$$\underset{\text{alanine}}{CH_3CHNHCOOH} + \underset{\text{glycine}}{2CH_2NHCOOH} \rightarrow \underset{\text{acetate}}{3CH_3COOH} + 3NH_3 + CO_2, \quad \Delta G^0 = +619.3 \text{ kJ mol}^{-1} \tag{5.6}$$

Together, these fermentation reactions are performed by a number of fungi, yeast, and prokaryotes such as clostridia, enterobacteria, lactobacilli, streptococci, and propionibacteria.

Since organic compounds and hydrogen are typical products of fermentation, the complete mineralization of organic matter under anoxic conditions depends on other physiological types of microbes, namely, the anaerobic respirers. Nevertheless, the hydrolyzing and fermenting organisms are important in anoxic environments because they are the only organisms that can perform initial hydrolysis and thus utilize complex biopolymers. Anaerobic respiring microorganisms, on the other hand, are capable of using only a limited number of low-molecular-weight substrates.

4.2. Anaerobic respiration

The organic compounds and hydrogen released during hydrolysis and fermentation are further oxidized to H_2O and CO_2 by a number of respiring microorganisms using a variety of inorganic compounds as electron acceptors. The individual anaerobic respiration processes generally occur in sequence with depth in sediments according to the energetics associated with the process and the availability of electron acceptors. A typical order of electron acceptor use is Mn oxides, NO_3^-, Fe oxides, SO_4^{2-}, and CO_2 reduction (Figure 5.12). The actual sequence is determined partly by the thermodynamics of the process in the environment, and as outlined in Chapter 3, the canonical order as presented above may shift depending on pH and the actual concentrations of reactants and products present. As also discussed in Chapter 3, organisms utilizing a specific electron acceptor can frequently control the concentrations of key electron donors such as H_2 and acetate. This allows the organisms to maintain an energy gain sufficient to support their own growth but insufficient to fuel less energetically favorable metabolisms. Finally, many of these different respiration reactions are not mutually exclusive. Thus, Fe reducers can co-exist with sulfate reducers, and both of these can co-exist with methanogens (Canfield, 1993). The stoichiometries of denitrification, manganese respiration, iron respiration, sulfate reduction with acetate, and acetate fermentation to methane are shown below (see also Chapter 3 and the individual Chapters on element cycling):

Denitrification:

$$CH_3COOH + 1.6NO_3^- + 1.6H^+ \rightarrow 2CO_2 + 0.8N_2 + 2.8H_2O \tag{5.7}$$

Mn reduction:

$$CH_3COOH + 4MnO_2 + 8H^+ \rightarrow 4Mn^{2+} + 2CO_2 + 6H_2O \tag{5.8}$$

Fe reduction:

$$CH_3COOH + 8FeOOH + 16H^+ \rightarrow Fe^{2+} + 2CO_2 + 14H_2O \tag{5.9}$$

Sulfate reduction:

$$CH_3COOH + SO_4^{2} \rightarrow 2HCO_3^- + H_2S + 2H_2O \tag{5.10}$$

Acetoclastic methanogenesis:

$$CH_3COOH + 4H_2 \rightarrow 2CH_4 + 2H_2O \tag{5.11}$$

As mentioned above, the strict vertical distribution of electron acceptors as depicted in Figure 5.12 is an oversimplification of true spatial distributions in nature. In addition to thermodynamic considerations (see Chapter 3), the influence of sediment inhomogeneities, such as worm burrows and macrophyte roots, affects the vertical structure of electron acceptor distribution (Chambers, 1997; Aller, 2001). In many cases, a downward transport of electron acceptors such as oxygen, nitrate, and oxidized metals creates radial profiles of oxic respiration, denitrification, and metal oxide reduction in sediments otherwise dominated by sulfate reduction or methanogenesis (Gribsholt *et al.*, 2003; Nielsen *et al.*, 2003) (Figure 5.13).

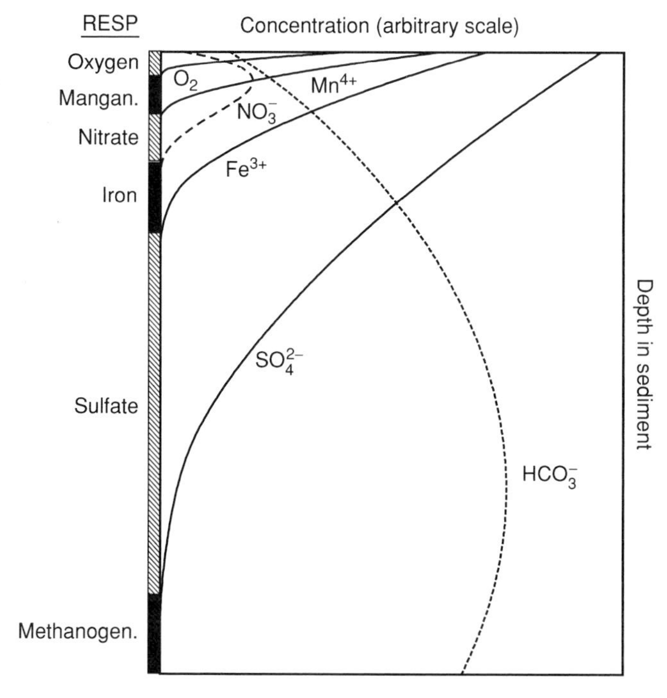

Figure 5.12 Idealized vertical distribution of electron acceptors in marine sediments. The bar at the left-hand side indicates the vertical distribution of respiration pathways.

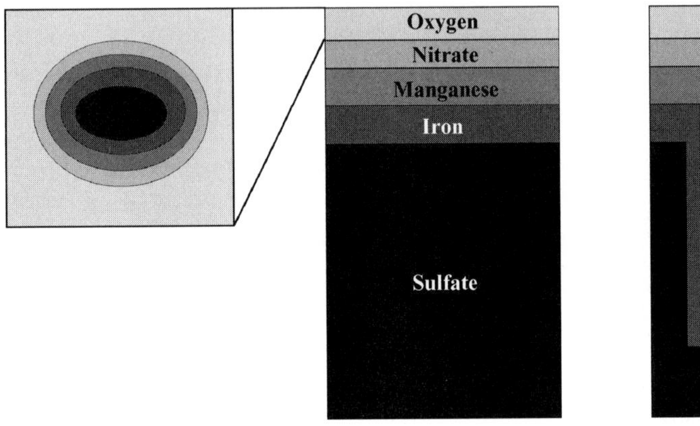

Figure 5.13 Schematic presentation of the vertical distribution of respiration pathways in marine sediment. Left panel: expanded oxic zone with anoxic microniche (e.g., fecal pellet); middle panel: idealized sediment without inhomogeneities; right panel: sediment with the presence of a biogenic structure like a polychaete burrow.

Organic "hot spots" associated with fecal pellets, for example, are known to create anoxic microniches where processes such as denitrification and sulfate reduction can occur in otherwise oxic surface sediments (Jørgensen, 1977a; Brandes and Devol, 1995) (Figure 5.14). As a consequence of these inhomogeneities, the distribution of electron acceptors can appear to overlap, and in some cases even be inverted, compared with the typical distribution shown in Figure 5.12. Bioturbated and rooted sediments are therefore mosaics of physicochemical and biological microenvironments. The area of the sediment–water interface is considerably increased by the presence of burrow-dwelling invertebrates and vascular plant roots, as is the volume of sediment undergoing oxic respiration, denitrification, and metal oxide reduction (see Figure 5.13).

4.3. Partitioning of total benthic metabolism into respiration pathways

There are inherent difficulties in partitioning total benthic metabolism into individual respiratory processes. The main problem is our inability to directly measure rates of many of the important oxidation pathways. We can measure rates of total metabolism (determined as total O_2 or CO_2 flux; Jørgensen, 1983; Kristensen and Hansen, 1999), sulfate reduction (determined by ^{35}S assay; Fossing and Jørgensen, 1989) nitrate respiration

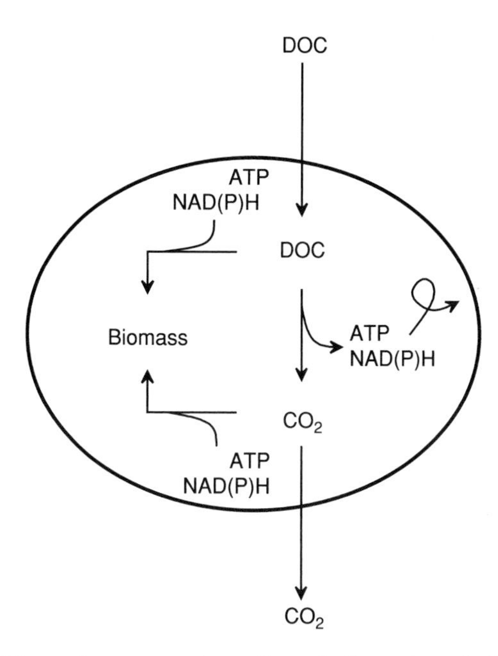

Figure 5.14 Schematic presentation of catabolic and anabolic processes in a microbial cell.

(determined by ^{15}N assay; Nielsen, 1992), and methane production from a variety of carbon sources without disturbing the sediment. The remaining pathways, O_2 respiration, Mn reduction, and Fe reduction, are inferred indirectly from porewater profiles, from oxidant consumption and metabolite production in sediment incubations, or by subtraction of the known processes from the total sediment metabolism (Canfield *et al.*, 1993a,b; Aller *et al.*, 1998).

Recognizing these limitations, a number of recent studies have quantified the partitioning of organic carbon mineralization by the different respiration pathways in a variety of marine sediments (Table 5.8; see also discussion in Chapters 8 and 9). As originally suggested by Jørgensen (1983), oxygen respiration can account for 50% or more of the total carbon oxidation in shallow coastal sediments. Many attributes of shallow environments, including fully oxic conditions in the bottom water, rapid water currents, and high rates of bioturbation and irrigation, help promote relatively extensive aerobic respiration in some sites despite the generally shallow penetration depth of oxygen (usually only a millimeter or two in shallow sediments; see Chapter 6). As we move into deeper waters O_2 becomes even relatively more important (see also Chapter 9). This reflects the deep oxygen penetration in deep-sea sediments, giving a long time for aerobic

Table 5.8 Partitioning of total sediment carbon oxidation into pathways using different electron acceptors

Location	Water depth (m)	Bottom water O_2 (% sat)	Total metabolism (mmol C m^{-2} d^{-1})	O_2 (%)	NO_3^- (%)	Mn^{4+} (%)	Fe^{3+} (%)	SO_4^{2-} (%)
Saltmarsh[1]	0	100	80	~0	~0	~0	95	5
Mangrove[2]	0	100	46	3	1	~0	78	18
Trop. Seagrass[2]	0	100	47	65	<1	?	~0	35
Norsminde Fjord[3]	1	100	40	55	23	?	?	23
Kattegat[4]	14	90	16	63	1	0	3	33
Aarhus Bay[5]	16	30–100	22	45	2	5	5	43
Northeast Greenland[6]	36	~100	6	38	4	0	25	33
Kattegat[4]	43	90	11	3	8	0	53	36
Svalbard[7]	115	~100	12	13	3	0	26	58
Chile margin[8]	122	0	60	0	0	0	0	100
Skagerrak[9]	190	80	16	14	3	0	32	51
Skagerrak[9]	380	80	10	17	4	0	51	28
Laurentian Trough[10]	525	75	5	17	4	18	2	59
Skagerrak[9]	695	80	12	3	6	90	0	1
Goban Spur[11]	1400	76	2	96	2	0	0	2
Chile margin[8]	2000	70	5	8	7	0	29	56
Panama Basin[12]	3890	30	1	0	0	100	0	0
Equatorial Pacific[13]	4000	50–100	0.4	95	3	<1	<1	1
Deep sea[14]	5000	50–100	<0.1	~80	~20	~0	~0	~0

[1]Kostka *et al.* (2002b); [2]Kristensen *et al.* (2000); [3]Jørgensen and Sørensen (1985); [4]Rysgaard *et al.* (2001); [5]Jørgensen (1996); [6]Rysgaard *et al.* (1998); [7]Kostka *et al.* (1999); [8]Thamdrup and Canfield (1996); [9]Canfield *et al.* (1993a); [10]Boudreau *et al.* (1998); [11]Lohse *et al.* (1998); [12]Aller (1990); [13]Bender and Heggie (1984); [14]Jørgensen (1983).

heterotrophs to oxidize sedimentary carbon and providing little reactive organic matter to the sediment below the zone of oxygen depletion.

The contribution of sulfate reduction, on the other hand, varies from less than 20% to greater than 50% of the total carbon oxidation in shallow sediments. The significance of sulfate reduction diminishes as we proceed into deep water (Table 5.8; see also Chapter 9), except where bottom water oxygen is low or absent. Nitrate respiration (denitrification) is found in a zone just below the depth of oxygen penetration and generally accounts for <10% of the total carbon mineralization (Table 5.8). In agricultural areas, however, with a strong anthropogenic NO_3^- source, high NO_3^- concentrations of up to 1 mM (Jørgensen and Sørensen, 1985) can support high rates of nitrate respiration. These rates can be comparable to rates of sulfate reduction, and in some cases nitrate reduction can dominate total sediment metabolism. Nitrate reduction is also important in oxygen minimum zones of the global ocean, where nitrate concentrations are elevated (around 30 μM) and oxygen concentrations are suppressed (Canfield, 1993).

Manganese reduction appears to be an insignificant carbon mineralization pathway except for special sites with unusually high Mn oxide concentrations, such as those found in the central Skagerrak between Denmark and Norway or in the Panama Basin (Table 5.8; see also Chapter 8). In these areas, Mn reduction dominates carbon oxidation. Iron can be a very important electron acceptor in carbon oxidation in coastal areas where active bioturbation replenishes iron oxides to sediment depth (see Chapters 8 and 9 for a further discussion) and in certain iron-rich intertidal sediments where rooted macrophytes and bioturbating infauna provide a substantial reoxidation potential (Kostka *et al.*, 2002b). The significance of Fe reduction in deeper sediments of the continental slope and rise is not well established.

5. CARBON ASSIMILATION

In the process of microbial growth, inorganic and (usually) organic constituents are converted into cell biomass at the expense of energy (Figure 5.14). Many of the basic constituents (Table 5.9) of prokaryote cells such as amino acids, nucleotides, monosaccharides, and fatty acids can often be found in the environment, and they can be directly assimilated by organisms. More commonly, the breakdown of organic compounds through catabolic processes generates intermediates such as pyruvate, phosphoenol pyruvate, acetyl coenzyme A, and oxaloacetate, which are then channeled into biomolecule production. Organic carbon taken up by prokaryotes will, therefore, be partly used in catabolic reactions to generate energy and reducing power and partly used in anabolic reactions for biomass synthesis. Some respired

Table 5.9 Elemental and chemical composition of prokaryotes

Element	Dry wt. (%)	Compound class	Dry wt. (%)
C	40–50	Protein	55
O	16–20	polysaccharide	5
N	12–15	Lipid	9
H	8–10	DNA	3
P	2–6	RNA	20
S	~1	Other	8

From Schlegel (1976); Madigan *et al.* (2003).

CO_2 may also be channeled into biomass production (Cajal-Medrano and Maske, 1999) (Figure 5.14). For example, Romanenko (1965) found that aerobic heterotrophic prokaryotes generally assimilate CO_2 at a rate of 19 mmol per mol O_2 consumed.

Because anabolic processes are reductive, they must rely on reducing equivalents and energy for the generation of biopolymers. The reducing equivalents are formed through catabolism and are stored in the form of NADH or NADPH. The oxidative reactions of catabolism are also associated with the release of free energy, some of which is stored in a biologically useful form as ATP (see also Chapter 3).

Prokaryote cell yield is often proportional to ATP production as seen by positive relationships between respiration rate and growth rate in many natural microbial populations (Figure 5.15). This occurs when the cells are well nourished. However, as discussed in the Chapter 2, a strict coupling between anabolism and catabolism is not always observed, and prokaryotes can metabolize and produce energy in the absence, or near absence, of growth (e.g., Tempest, 1978; Russell and Cook, 1995). This happens, for example, when nutrient availability limits the production of cellular biomass.

Thus, microbes channel variable amounts of metabolic energy into growth, and this balance is described by the microbial growth efficiency (MGE), which expresses the balance between microbial production (MP) and microbial respiration (MR) in carbon units:

$$MGE = MP/(MP + MR) \tag{5.12}$$

Published values of MGE in aquatic environments range from <0.1 to >0.7 (del Giorgio and Cole, 1998). Although there is a large variation in MGE values, there is a general increase along gradients of eutrophication (or primary production) (Figure 5.16). It seems that growth is energetically more costly in oligotrophic systems, probably because of substrate

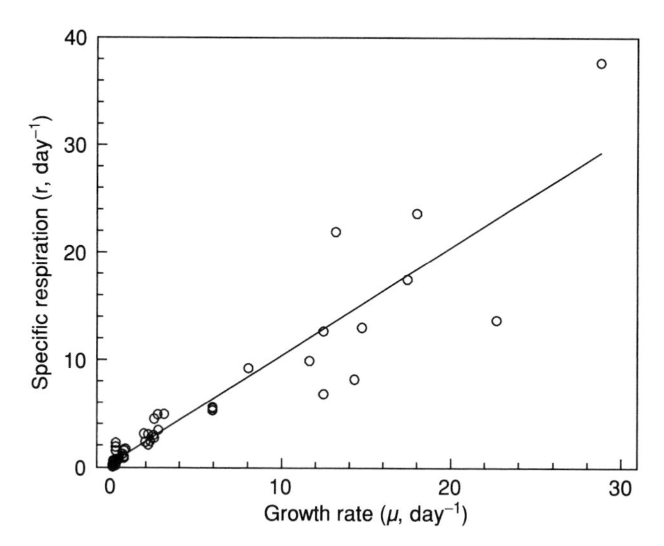

Figure 5.15 Relationship between growth rate and specific respiration of bacterioplankton. Redrawn from Cajal-Medrano and Maske (1999).

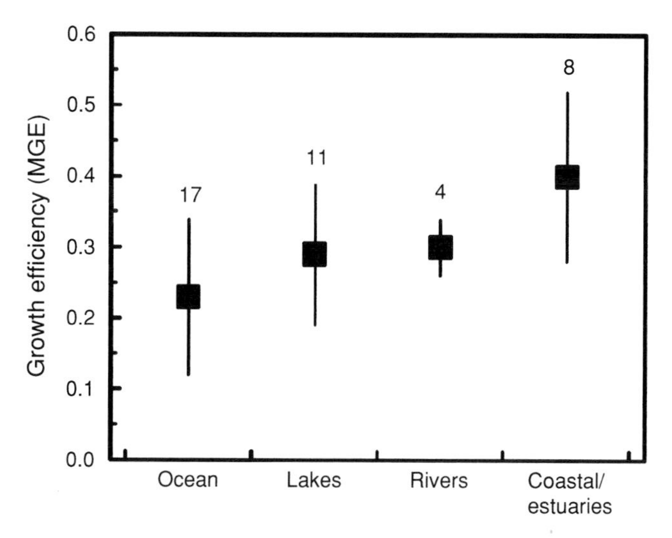

Figure 5.16 Microbial growth efficiency (MGE) of bacterioplankton from various aquatic environments. The data are shown as the mean ± S.D. of the number of data given. Based on data compiled in del Giorgio and Cole (1998).

limitation. Thus, key components from which to construct cell biomass are missing, and, therefore, proportionally more energy is used on cell maintenance rather than cell growth. This is consistent with the suggestion of del Giorgio and Cole (1998) that MGE in the environment is regulated by a combination of factors, including the quality of organic matter, nutrient availability, and the particular energetic demands of the organism. In their view, bacterioplankton in oligotrophic lakes and the oceans are exposed to generally low concentrations of dissolved substrates (organic carbon and nutrients), causing cell growth to be co-limited by energy, carbon, and nutrients. The relatively high-maintenance and other non-growth energy requirements are met by the catabolism of relatively oxidized low–molecular-weight organic compounds, which provide neither enough energy nor enough carbon to sustain growth. As systems become enriched in nutrients, and primary production increases, both the rate of supply and the quality of DOC increase, as does the availability of nutrients, with a general increase in microbial growth rates and MGE (Figure 5.17).

The energy released per unit organic carbon respired anaerobically is lower than during aerobic respiration (see Chapter 3). Consequently, much less energy is available for anabolic processes, resulting in a reduced MGE. Blackburn and Henriksen (1983) found values of MGE around, or below, 0.3 for a variety of marine sediments for which sulfate reduction was the dominant respiration process. Similarly, Pedersen *et al.* (1999) observed that

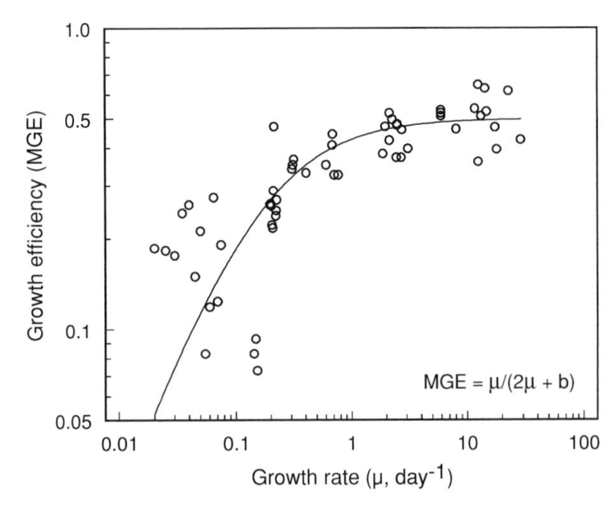

Figure 5.17 The relationship between growth rate and growth efficiency of bacterioplankton. The line is fit according to the equation shown. Redrawn from Cajal-Medrano and Maske (1999).

MGE gradually decreased from 0.55 to 0.19 over a 25-day period after addition of seagrass detritus to an experimental sediment system with an oxic surface layer and anoxic deeper layers. They argued that as time progressed a greater proportion of carbon oxidation in the sediment was mediated by anaerobes such as sulfate reducers. The low MGE found in anoxic sediments from coastal areas probably results from significant anaerobic metabolism rather than a low supply of DOC and nutrients, as these sediments are usually rich in organic matter and dissolved nutrients.

6. DEGRADABILITY AND DECOMPOSITION KINETICS

The microbial degradation of organic matter occurs over time scales ranging from hours, for the breakdown of simple biomolecules, to millions of years, for the mineralization of very refractory organics in deep-sea sediments. The chemical composition of organic matter is an essential rate-controlling parameter for microbial degradation (Figure 5.18). Natural organic polymers have different susceptibilities to prokaryote (enzymatic) attack, and this variation controls the degradation rate (Fenchel *et al.*, 1998). The most rapidly decomposed substances are amino acids, simple sugars, and short-chain carboxylic acids, which decompose on time scales of hours to

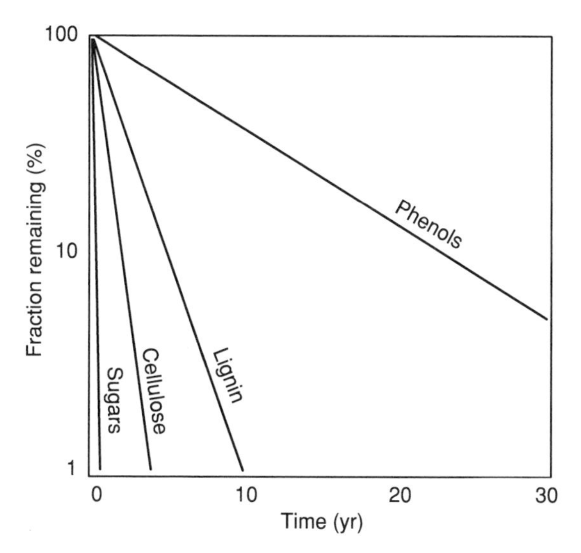

Figure 5.18 Idealized presentation of the decomposition rates of important biomolecules. Redrawn from Fenchel *et al.* (1998).

weeks. Polysaccharides such as cellulose and sugar–amino acid condensation products (melanoidins) decompose on time scales of years.

Certain structural characteristics of biomolecules are known to decrease biodegradability. Thus, extensive polymerization and branching can generate bonds not readily hydrolyzed, and heterocyclic, polycyclic, and aromatic compounds are inherently difficult to decompose, especially by anaerobic metabolisms (Hedges *et al.*, 2000). A typical compound containing many of these recalcitrant features is lignin (a structural compound intimately associated with cellulose in vascular plants), which is practically non-degradable under anoxic conditions and is degraded only very slowly under oxic conditions.

Organic matter in sediments and natural waters is a heterogeneous mixture ranging from fresh and reactive plant and animal remains to poorly reactive polymeric compounds. Some of the poorly reactive compounds represent the recalcitrant remains of aquatic organisms, and others are formed geochemically by condensation reactions (Hedges *et al.*, 2000). Natural microbial communities tend to first use the most easily degradable organic compounds. As labile fractions are consumed, less usable components accumulate, and the remaining organic matter's ability to support microbial activity is reduced.

It is very difficult to determine precisely the chemical makeup of natural organic matter assemblages, so there is no easy *a priori* way to assess the reactivity of these assemblages in nature. Instead, organic matter decomposition kinetics are addressed empirically through models. The first kinetic model describing the degradation of organic matter in sediments was proposed by Berner (1964). In this first-order (G) model, it was assumed that organic matter decomposes at a rate directly proportional to its concentration:

$$\mathrm{d}G/\mathrm{d}t = -kG \leftrightarrow G = G_0[\exp(-kt)] \tag{5.13}$$

where G is the concentration of the metabolizable pool of organic matter at time t, G_0 is the initial concentration of organic matter, and k is the first-order decay constant. According to this model, both the concentration of reactive organic matter and the rate of organic matter decomposition decrease exponentially with time.

While Equation 5.13 has enjoyed success in describing the decomposition of simple substrates, the simple first-order dependence of oxidation rate on organic carbon concentration is not consistent with the observations (Figure 5.19). Recall that in nature the reactivity (not just the concentration) of decomposing organic material decreases with time. To accommodate this fact, Jørgensen (1978b) proposed a modified version of the original G model in which organic matter is divided into various groups of compounds of different reactivity, each of which undergoes first-order decomposition. This

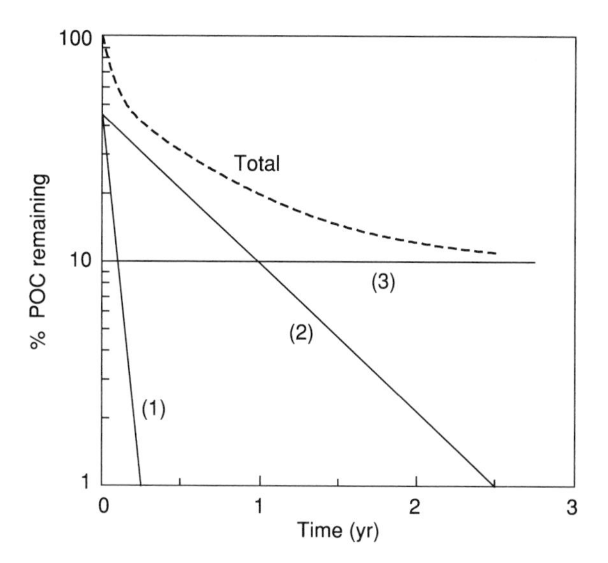

Figure 5.19 Idealized presentation of mixed detritus decomposition. The broken line represents the overall decomposition pattern (total) of a detritus pool consisting of three fractions: a rapidly degradable fraction (1), a slowly degradable fraction (2), and a non-degradable fraction (3). The overall (total) decomposition rate is not linear in a semilogarithmic plot when several fractions of different reactivity are decomposing.

multi-G model was expressed by Berner (1980) (see also Chapter 9) as the following:

$$dG/dt = -\Sigma k_i G_i \leftrightarrow G_T = \Sigma G_{0i}[\exp(-k_i t)] \qquad (5.14)$$

where G_T is the total pool of organic carbon at time t, G_i is the concentration of the ith reactive component, G_{0i} is the initial concentration of this component, and k_i is the corresponding decay constant. Selective removal according to the reactivity of each group accounts for the change in the reactivity and the amount of organic matter remaining with time.

To apply this model successfully, the number of reactive types and their associated k values must be determined empirically, typically by analyzing sediment oxidation rate data. However, usually no more than two or three different reactivity types can be identified from a given set of data, regardless of the actual number involved. To solve this problem, a number of alternative models based on the G model approach have been proposed. For example, in the power model of Middelburg (1989), the first-order decay constant in the simple one-G model is allowed to decrease as a power function of time. In the more complex reactive continuum model of Boudreau and Ruddick (1991), the decay of sedimentary organic matter is presented as a spectrum (a continuous distribution of an infinite number) of reactive types, which can be characterized by a variable function of the decay constant. A fundamental property of the continuum theory is that it can generate reaction order for the decay of the total mixture greater than one. This high order is related to the predominance of the more refractory components of the continuum relative to the more reactive components.

Both the multi-G model and the later developments can successfully describe decomposition patterns of sedimentary organic matter in many cases (Figure 5.20). The strongest argument favoring the multi-G model is its conceptual and mathematical simplicity.

7. CARBON PRESERVATION

Global inventories indicate that essentially all carbon in the Earth's crust is stored in sedimentary rocks (>99.9%), of which about 20% is organic (Hedges and Keil, 1995). Only about 0.05% of the carbon in the Earth's upper crust cycles in active surface pools (Table 5.1), the greatest fraction of which (~90%) is inorganic carbon dissolved in sea water. The other dynamic pools are one to two orders of magnitude smaller and include atmospheric carbon dioxide, soil carbonate, soil humus, plant biomass, dissolved organic matter in sea water, and carbon preserved in surface sediments (Table 5.1). Most organic carbon in the biosphere is continually recycled, but a small amount escapes microbial mineralization and becomes preserved in sediments. Long-term organic carbon burial removes from the aquatic system not only carbon, but also associated nutrients, which can conceivably influence rates of primary production. The long-term burial of organic carbon also acts as a net source of atmospheric oxygen (Berner and Canfield, 1989). For our purpose here, buried organic carbon must have escaped microbial oxidation processes, and we seek to understand how.

The preservation of organic carbon in sediments is defined as the fraction of the sedimentary flux of carbon that becomes buried and can formally be given in the following (Henrichs and Reeburgh, 1987):

$$\% \text{ carbon preserved} = (\text{burial flux}/\text{deposition flux}) \times 100 \qquad (5.15)$$

The burial flux of carbon is computed as the concentration of organic carbon remaining after early diagenesis is complete (i.e., residual carbon content below the diagenetically active zone), multiplied by the sedimentation rate. The deposition carbon flux is the concentration of organic carbon in particles reaching the sea floor, multiplied by the rate of particle deposition.

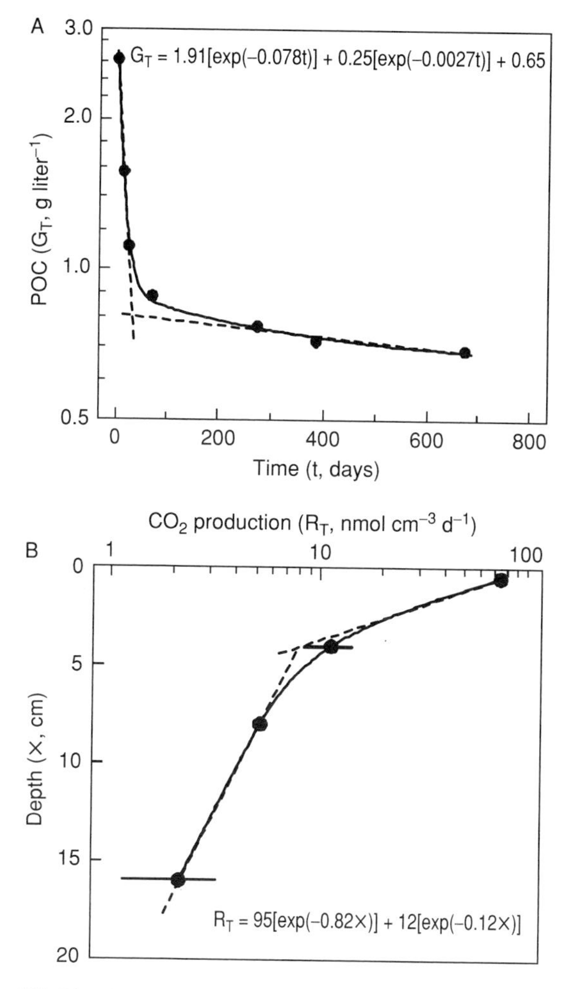

Figure 5.20 (A) The decomposition rates of plankton (primarily diatoms) in sea water with time are shown. The solid line is fit assuming two pools of degradable carbon and a residual pool according to the equation given. The broken lines represent the decay rate of the rapidly and slowly decomposing fractions. The non-reactive pool accounts for 0.65 g liter^{-1}. Modified from Westrich and Berner (1984). (B) The vertical pattern of carbon oxidation in sediment off the west coast of Mexico. The solid line is fit assuming two pools of degradable carbon according to the equation given. The broken lines represent the attenuation of carbon oxidation of the fast and slow fractions. Modified from Kristensen *et al.* (1999).

Organic matter preservation in marine sediments ranks as one of the most complex and controversial issues in contemporary sediment biogeochemistry. One major school of thought argues that the most important controlling factors include rates of primary production, water column depth, organic matter sources, reaction histories, and sediment accumulation rate (Suess, 1980; Westrich and Berner, 1984; Calvert and Pedersen, 1992; Henrichs, 1992). For example, Calvert and Pedersen (1992) argued that as sedimentation rate increases, the flux of organic carbon to sediments also increases. This, in turn, increases the demand for the most important electron acceptors used to oxidize organic carbon, including oxygen, nitrate, and sulfate. At high rates of deposition, electron acceptors become depleted at fairly shallow sediment depths, providing less time for mineralization and increasing preservation. In this view, the presence or absence of oxygen has relatively little bearing on preservation.

The other major school of thought accepts that sedimentation rate has an important role in regulating organic carbon preservation but argues that oxygen availability also has a strong influence (Hedges *et al.*, 1988, 1999; Canfield, 1989b, 1994; Hulthe *et al.*, 1998). The idea is that organic carbon will be preferentially oxidized in the presence of oxygen and preserved in the absence of oxygen, since decomposition rates are much more efficient with oxygen (Kristensen and Holmer, 2001). This is particularly true in slowly depositing sediments where oxygen may contact sediment for a long period of time. Therefore, the oxygen exposure time (OET) has become an important parameter related to the extent of organic carbon preservation. OET is defined as the time elapsed before deposited organic matter in sediments is buried below the oxic zone (Hedges *et al.*, 1999).

We evaluate carbon preservation here using data from sites spanning a wide range of sedimentation rates and oxygen levels (Figure 5.21). Consistent with Canfield (1994), it appears that burial rate has an important influence on the preservation of organic carbon in sediments. At high sedimentation rates ($>10^{-1}$ g cm^{-2} yr^{-1}) 50% or more of the deposited carbon escapes oxidation and is buried, whereas only 1% is preserved at low sedimentation rates ($<10^{-3}$ g cm^{-3} yr^{-1}). Euxinic sediments (sediments underlying basins of restricted circulation with permanently anoxic and sulfidic conditions near the bottom), however, tend toward greater preservation than well-oxygenated normal marine environments of similar sedimentation rate, provided that the rates of sedimentation are lower than about 0.1 g cm^{-2} yr^{-1}. This means that anaerobic processes are less efficient than oxic processes at decomposing organic carbon when the organic matter is not fresh, but reasonably well degraded (Kristensen and Holmer, 2001). Therefore, there are merits to each school of thought on carbon preservation. There are, however, factors unrelated to oxidation pathways but still controlled by the depositional environment that may also influence preservation (Hedges *et al.*, 1999).

Following the work of Mayer (1994), Hedges and Keil (1995) argued that the preservation of organic carbon in marine sediments is significantly controlled by the adsorption of dissolved organic compounds onto mineral surfaces. They noted that only a fraction (<10%) of the total organic matter

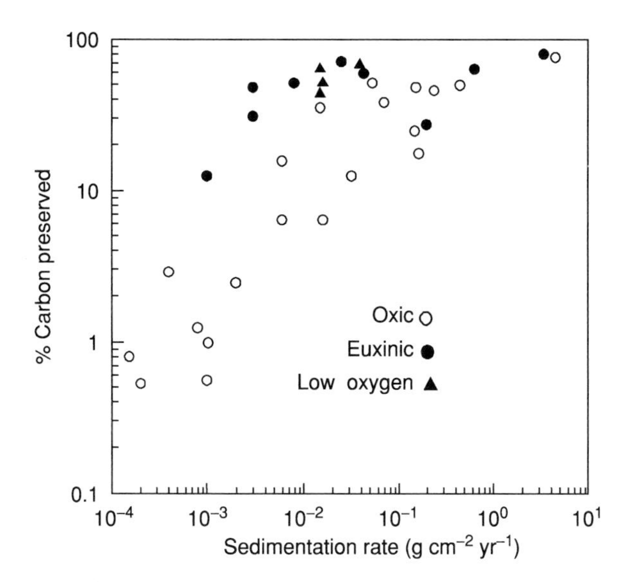

Figure 5.21 Carbon preservation as a function of sedimentation rate in normal marine sediments with oxic overlying water, euxinic sediments from the Black Sea, the Baltic Sea and the Cariaco Trench, several seasonally euxinic sediments from coastal areas, and low oxygen sites from the California borderland basins. Modified from Canfield (1993).

in marine sediments is present as discrete particles, implying that a major portion of the organic matter must be adsorbed to minerals. Additionally, in many instances, the organic carbon concentrations asymptote to values equivalent to a monolayer coverage of the sediment particles. Therefore, the adsorption of dissolved organics onto sediment particles appears to stabilize the organics with respect to decomposition. Mineral particles in sediments are in many cases highly irregular and appear to have a major fraction of their surface area in mesopores less than 10 nm in width. Although the reasons behind the stabilization of surface-adsorbed organics have not been established, a number of mechanisms have been proposed. Mayer (1994) argued that hydrolytic enzymes are excluded from or inactivated within confined mesopores. Another explanation is that the activity of enzymes is sterically limited by close association of substrate molecules to mineral surfaces via chemisorptive bonds (Hedges and Keil, 1995).

However, in the deep sea, where sediment layers may be exposed to oxygen for extended periods (>1000 years), the organic content is considerably less than expected for a monolayer coverage. This led Hedges and Keil (1995) to propose that all sediments receive three basic organic components: (1) hydrolyzable (fresh) organic matter that is completely mineralized regardless of redox conditions, (2) oxygen-sensitive organic matter (monolayer coatings) that degrades slowly in the presence of oxygen, but not under anoxic conditions, and (3) totally refractory organic matter. The slow degradation of adsorbed organics under oxic conditions could result from reaction with H_2O_2 or other oxygen radicals, which are sufficiently small and aggressive to alleviate the mesopore or steric protection of monolayer coatings. In slowly depositing oxygen-containing sediments there is ample time for the aerobic degradation of the surface-absorbed pool. By contrast, rapid sediment deposition results in shallow oxygen penetrations, which, combined with high sedimentation rates, yields shorter oxygen exposure times and preservation of the surface-adsorbed pool.

Chapter 6

The Oxygen Cycle

1. INTRODUCTION

Oxygen gas, O_2 ("oxygen" hereafter), is often treated as a prerequisite for life, and anoxic environments are often seen as "dead". While oxygen is essential to almost all eukaryotes, this volume describes a great diversity of metabolic strategies for life without oxygen, and for many organisms, the high oxygen concentrations at the Earth's surface are toxic and may in fact be deadly. Much biological evolution has taken place in the absence of oxygen, and seen in the perspective of Earth's history, we and other organisms thriving in the oxygenated world should be considered extremophiles.

The abundance of oxygen profoundly influences life. Thermodynamically, oxygen is the most favorable abundant electron acceptor, and the free energy

ADVANCES IN MARINE BIOLOGY VOL 48
0-12-026147-2
© 2005 Elsevier Inc.
All rights reserved

yield from the oxidation of organic matter is about four times higher with oxygen than with sulfate or ferric iron at pH 7. Because of their higher energy yield, aerobic food chains are longer (typically five to six trophic levels) than anaerobic food chains, which only extend to two trophic levels; high oxygen concentrations are also a prerequisite for the large body size of macroorganisms (Runnegar, 1986; Fenchel and Finlay, 1995; Knoll, 2003).

Oxygen in the atmosphere originates from oxygenic photosynthesis by cyanobacteria and eukaryotic algae and plants (Chapter 4). Over time, however, the production of oxygen and organic matter is closely matched by their consumption through mineralization processes (Chapter 5). Thus, although the residence time of oxygen in the atmosphere with respect to photosynthesis and mineralization is short, on the order of 1500 years, the oxygen concentration changes only on geological time scales (Table 6.1). A primary control of atmospheric oxygen is therefore the balance between the rate by which new organic matter escapes reoxidation through burial in sediments and the amount of oxygen consumed in oxidation of organic carbon in sedimentary deposits during weathering (Berner, 1989).

The large amount of oxygen in the atmosphere, and the oxygenation of the ocean, is a relatively new phenomenon, dating back to the end of the Proterozoic Eon, ~700 to 540 million years ago (Figure 6.1). Atmospheric oxygen is thought to have been a few percent of the present level through much of the Proterozoic (2500 to 540 million years ago), and only the mixed surface layer of the oceans was oxic, while the bottom water contained

Table 6.1 Global oxygen budget[a]

Oxygen production (mol O_2 yr^{-1})	
Terrestrial gross[b]	2.0×10^{16}
Terrestrial net[c]	4.7×10^{15}
Marine gross[d]	4.9×10^{15}
Marine net[e]	2.3×10^{15}
Total gross photosynthesis (A)	2.5×10^{16}
Burial of organic carbon and pyrite (mol O_2 equiv. yr^{-1})[f] (B)	1.3×10^{13}
Atmospheric O_2 inventory (mol O_2)[g] (C)	3.7×10^{19}
Residence time, w. r. t. photosynthesis/respiration (C/A)	1460 yr
Residence time, w. r. t. burial/weathering (C/B)	2.8×10^{6} yr

[a]Some values have been converted from carbon units assuming a respiratory quotient of 1.
[b]Including oxygen consumed by photorespiration, 0.6×10^{16} mol yr^{-1}. Bender *et al.* (1994).
[c]Field *et al.* (1998).
[d]del Giorgio and Duarte (2002).
[e]Calculated as oxygen production balancing the export of organic matter from the photic zone. After del Giorgio and Duarte (2002).
[f]Berner (1984).
[g]Schlesinger (1997).

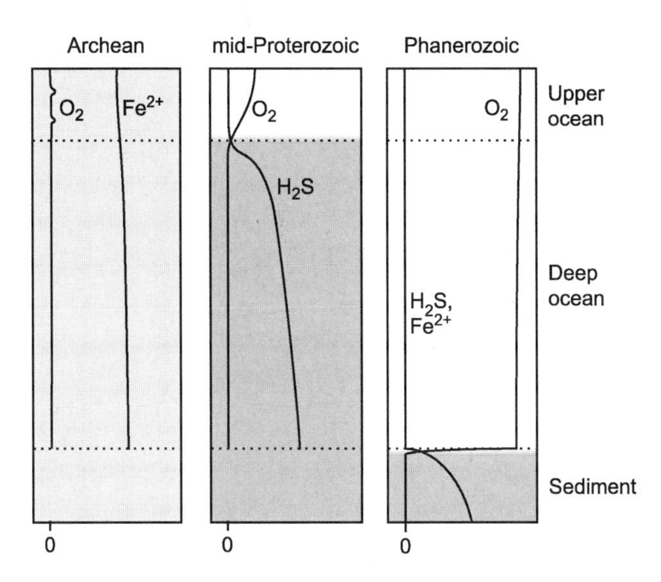

Figure 6.1 Schematic representation of vertical oxygen distributions in the ocean during major stages of Earth history. Anoxia and the presence of ferrous iron or hydrogen sulfide is indicated by grey shading. During the Archean, oxygen was restricted to oases of oxygenic photosynthesis possibly associated with cyanobacterial blooms and benthic microbial mats. Modified from Anbar and Knoll (2002).

hydrogen sulfide or ferrous iron (Canfield, 1998). During the Archean Eon (3.8 to 2.5 billion years ago), when much of the extant microbial diversity probably evolved, oxygen was virtually absent from the atmosphere and was mainly found in oxygen oases created by oxygenic photosynthesis in the surface oceans and in photosynthetic microbial mat communities. For reviews of the biological and geological considerations underlying this scenario, see Knoll (2003) and Anbar and Knoll (2002).

Throughout Earth's history, the anoxic part of the biosphere has contracted, but even today, oxygen is absent below a few millimeters depth in many aquatic sediments (Figure 6.2). In some lakes and protected marine basins, such as the Black Sea, anoxia is found in the water column. Low oxygen concentrations, or even anoxia, are also found at intermediate depths in some regions of the ocean, forming oxygen-depleted oxygen minimum zones as seen along the west coast of South America. Here, oxygen is depleted at depths of 100–200 m through the respiration of organic matter that rains as algal detritus and fecal pellets from the highly productive surface water. Oxygen reappears at greater depths, where the input from oxygen-rich deep water exceeds the demand from the attenuated organic flux.

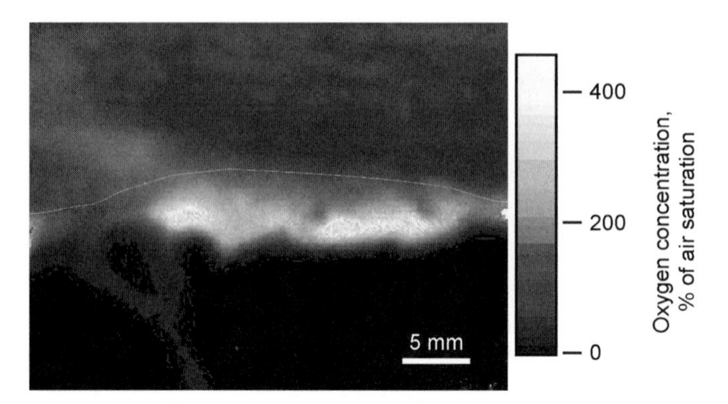

Figure 6.2 A two-dimensional map of oxygen concentrations determined *in situ* in a shallow sediment covered by photosynthesizing cyanobacteria and diatoms. The sediment–water interface is indicated by the curve. To the left, oxygen penetrates into the sediment through burrows of a polychaete worm. Oxygen concentrations in the photosynthetic layer from the center to the right reach more than four times air saturation. The image represents a vertical section obtained at noon with a planar oxygen optode. With planar optodes, oxygen is quantified photographically through its quenching effect on a fluorescent dye. The dye is contained in a thin film mounted on a glass plate through which it is illuminated and photographed. Reprocessed from Glud *et al.* (2001).

The interface separating oxic and anoxic environments in both sediments and anoxic water columns is the site of intense microbial activity. Thus, reduced metabolites produced during anaerobic mineralization, including reduced manganese and iron, sulfide, methane, and ammonium, are reoxidized at the oxic–anoxic interface through respiratory and abiotic processes (Figure 6.3). These processes efficiently capture most of the reduced compounds, so that oxygen is the ultimate acceptor of electrons released during mineralization in the system. In this Chapter, we discuss oxygenic photosynthesis by bacteria, the physiology of oxygen respiration, and the effects of oxygen on other microbial processes. Photosynthesis is further discussed in Chapter 4, and, in the context of microbial mats, in Chapter 13. The effects of oxygen on organic carbon oxidation are discussed in Chapter 5, and organisms and reactions involved in the reoxidation of secondary metabolites are covered in the following Chapters.

2. CHEMICAL CONSIDERATIONS

The element oxygen is mainly found in the oxidation state of −2, such as in water, oxide and hydroxide minerals, and in a wide variety of inorganic and organic compounds. Zero-valent oxygen is found in oxygen gas and in

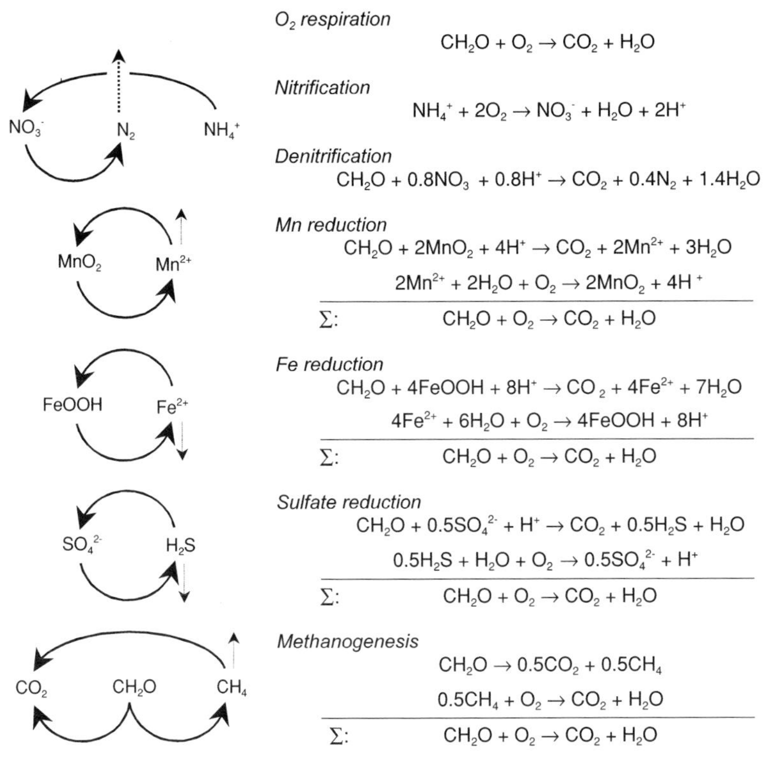

Figure 6.3 Organic carbon oxidation reactions and the reoxidation of inorganic metabolites with oxygen. As indicated by the summations, the stoichiometry of anaerobic mineralization coupled to reoxidation is the same as for direct oxidation of organic carbon with oxygen. Small vertical arrows indicate the inorganic metabolites that escape reoxidation. Modified from Canfield *et al.* (1993a).

atmospherically important ozone, O_3. The oxidation state -1 is represented by hydrogen peroxide, H_2O_2, which is typically found at submicromolar concentrations in aquatic systems. Other minor, yet important, species include the superoxide ($O_2^{\cdot-}$) and hydroxyl ($OH^{\cdot}$) radicals, where $^{\cdot}$ indicates an unpaired electron. Hydrogen peroxide and the oxygen radicals form as intermediates in the reduction of oxygen, and their high reactivity toward organic compounds makes them highly toxic to organisms.

Oxygen gas is moderately soluble in pure water with an air-equilibrium concentration of 284 μM at 20 °C. The solubility decreases with increasing temperature and salinity (see Chapter 2), and the equilibrium concentration in μmol kg^{-1} at 1 atm total pressure can be estimated as a function of these variables from the following equation (García and Gordon, 1992):

$$C_0^* = \exp[5.80818 + 3.20684T_s + 4.11890T_s^2 + 4.93845T_s^3 + 1.01567T_s^4 + 1.41575T_s^5 - (7.01211 + 7.25958T_s + 7.93334T_s^2 + 5.54491T_s^3)S_{1000} - 0.132412S_{1000}^2] \tag{6.1}$$

where T_s is a scaled temperature: $T_s = \ln[(298.15 - t)(273.15 + t)^{-1}]$, t is temperature in °C, and S_{1000} is salinity S times 0.001 ($S_{1000} = 10^{-3}\ S$). This equation is valid for the range ($t_{\text{freezing}} \leq t \leq 40\,°\text{C}, 0 \leq S \leq 42$). Photosynthesis frequently leads to oxygen supersaturation, which may reach several times air saturation in microbial mats (Figure 6.2; see also Chapter 13), leading sometimes to gas bubble formation.

The kinetics of the abiotic reaction of oxygen with organic matter is very slow, and such reactions are probably not significant for oxygen cycling. By contrast, oxygen reacts quite rapidly with reduced iron and sulfur species such as aqueous ferrous iron, hydrogen sulfide, and sulfite, whereas abiotic reactions with reduced manganese, ammonium, and methane are slow or absent. Because the abiotic reactions generally involve the transfer of fewer than four electrons, oxygen is not reduced all the way to water in one step, so these reactions are a source of reactive intermediates. For example, during the series of reactions involved in the oxidation of micromolar amounts of ferrous iron in air-saturated sea water, superoxide and hydrogen peroxide reach concentrations of the order of 1 nM and 0.1 μM, respectively (King *et al.*, 1995; Rose and Waite, 2002):

$$Fe^{2+} + O_2 \rightarrow Fe^{3+} + O_2^{\cdot-} \tag{6.2a}$$

$$Fe^{2+} + O_2^{\cdot-} + 2H^+ + \rightarrow Fe^{3+} + H_2O_2 \tag{6.2b}$$

$$Fe^{2+} + H_2O_2 \rightarrow Fe^{3+} + OH^{\cdot} + OH^- \tag{6.2c}$$

$$Fe^{2+} + OH^{\cdot} \rightarrow Fe^{3+} + OH^- \tag{6.2d}$$

A number of different terms are often used to classify oxygen levels and their relation to metabolism as well as general redox conditions. We give here an overview with definitions and discussion of the most important terminology:

1. Environments are classified as either oxic or anoxic, depending on the presence or absence of oxygen. Operationally, anoxia is usually assigned to environments in which oxygen is not detected by standard techniques, such as the Winkler titration, with a typical detection limit of ~1 μM. As we shall see below, however, half-saturation concentrations for

oxygen respiration by microorganisms may be lower than this, and depending on the context, it may be necessary to apply more sensitive techniques to assess anoxia. Suboxic environments are defined as those free of both oxygen and hydrogen sulfide. This is the zone dominated by manganese and iron cycling (Berner, 1981). The term is ambiguous, however, as it is often applied to depths in stratified water columns with oxygen concentrations slightly above the detection limit.

2. The terms oxidized and reduced refer to a positive or negative redox potential, respectively, as measured with a platinum electrode versus the standard hydrogen electrode. This classification is completely operational, and can be only very coarsely related to the water chemistry since only some solutes interact with the platinum electrode, and the signal may be affected by coatings and bacterial colonization. There is a rough correspondence between the transition from oxidized to reduced conditions and the appearance of free hydrogen sulfide. Thus, the oxidized zone includes both oxic and anoxic, nonsulfidic ("suboxic") environments.

3. Organisms are classified as aerobic or anaerobic depending on whether or not they require oxygen for growth. Microaerophilic organisms are those that thrive at low oxygen concentrations but not at air-saturated levels. Most microaerophiles grow at oxygen concentrations $<10\%$ of air saturation, while an extreme example is provided by *Bacteroides fragilis*, which benefits from oxygen respiration only at $<1\%$ of air saturation (Baughn and Malamy, 2004). Organisms may be either facultative or obligate members of these classes. Furthermore, anaerobes able to survive for extended periods, or even to grow through fermentative metabolism under oxic conditions, such as some lactic acid bacteria, are known as aerotolerant.

3. BACTERIAL OXYGENIC PHOTOSYNTHESIS

Among the prokaryotes, oxygenic photosynthesis is known only from the cyanobacteria. In these organisms, just as in eukaryotic algae and plants, water is oxidized to oxygen by chlorophyll *a* in the reaction center of photosystem II, and CO_2 is reduced to organic matter by Rubisco in the reductive pentose phosphate cycle (see Chapter 4). Indeed, eukaryotic chloroplasts have evolved from a cyanobacterial endosymbiont, and as a result of similar functionality, cyanobacteria occupy ecological niches very similar to those of the microalgae. In fact, cyanobacteria were previously known as blue-green algae. Two prominent metabolic traits of cyanobacteria are unique among oxygenic phototrophs: the ability of some species to fix N_2 (see Chapter 7) and that of some species to perform anoxygenic photosynthesis with hydrogen sulfide (see Chapter 4).

Cyanobacteria are ubiquitous in the photic part of the biosphere. They are of particular significance in aquatic systems, where they are estimated to account for 20–40% of the phototrophic biomass in the oceans and to account for a similar percentage of oxygen production (Partensky *et al.*, 1999a; Pearl, 2000). Furthermore, through their dominating role in nitrogen fixation, cyanobacteria are the main source of new nitrogen for primary production in many ecosystems (Pearl and Zehr, 2000; Karl *et al.*, 2002) (see Chapter 7). Cyanobacteria may have been of even greater importance early in Earth's history, as oxygenic photosynthesis likely first evolved in a cyanobacterial ancestor. Furthermore, stromatolite fossils indicate that benthic photosynthetic communities probably dominated by cyanobacteria were widespread during much of the Proterozoic Eon (Knoll, 2003).

3.1. Characteristics of cyanobacteria

The cyanobacteria form a monophyletic clade in the domain *Bacteria*, branching off the Tree of Life at approximately the same place as the proteobacteria and several other late-branching lineages (see Chapter 1). Based on morphotypes and biogeography, ~2000 species have been described, though this number may be revised after a modern, genetically based reclassification. They are characterized by oxygenic photosynthetic systems located in membranes within the cell, known as thylakoids. In addition to chlorophyll *a*, most cyanobacteria contain phycobilins as accessory pigments, including the blue phycocyanin and the red phycoerythrin (Madigan *et al.*, 2003). These are open-chain tetrapyrroles that are derived from porphyrin rings and are linked to protein moieties. Some cyanobacteria, sometimes referred to as prochlorophytes, contain only small amounts of phycobilins or lack them completely. Instead, they contain both chlorophyll *a* and *b*, like green algal chloroplasts. One of these genera, *Prochlorococcus*, has special divinyl modifications of the chlorophylls, known as chlorophyll a_2 and b_2 (La Roche *et al.*, 1996). The prochlorophytes are not a monophyletic group but are phylogenetically scattered among other cyanobacteria. They are also not closely related to the chloroplasts (Palenik and Haselkorn, 1992; Urbach *et al.*, 1992).

Carbon fixation through the Calvin-Benson-Bassham cycle in cyanobacteria involves a form I Rubisco with relatively low selectivity toward CO_2 (see Chapter 4). Rubisco is organized in semicrystalline arrays in a region of the cell bounded by a protein membrane and known as the carboxysome. Here, CO_2 is supplied through the activity of the enzyme carbonic anhydrase, which catalyzes the equilibration between carbonic acid and CO_2 (Kaplan and Reinhold, 1999). Inorganic carbon is taken up through active transport of both bicarbonate and CO_2 and involves both constitutive and inducible uptake systems (Kaplan and Reinhold, 1999; Benschop *et al.* 2003).

Photosynthetic production of organic carbon in excess of the immediate metabolic needs is stored as the polysaccharide glycogen, which serves as a substrate for oxygen respiration in the dark (Smith, 1982; Matthijs and Lubberding, 1988). A few cyanobacteria are able to grow in the dark through oxygen respiration using a small number of exogenous substrates, primarily mono- and disaccharides such as glucose and sucrose. Many can grow photoheterotrophically, assimilating a similar suite of substrates while using light as an energy source (Stal and Moezelaar, 1997). Some cyanobacteria living in symbiosis with plants also obtain most if not all of their organic carbon from their host (Rai *et al.*, 2000).

In environments such as microbial mats, cyanobacteria are exposed to anoxia at night. They survive through fermentation of their endogenous carbohydrates to acetate in some cases, or by a mixed acid fermentation to H_2, ethanol, and organic acids. Fermentation of exogenous substrates has been observed in a few cases, but growth only occurs at millimolar concentrations of substrate, and sustained anaerobic growth is therefore not likely in natural environments (Stal and Moezelaar, 1997). Some species reduce inorganic electron acceptors such as elemental sulfur, ferric iron or dimethylsulfoxide (DMSO) under anoxic conditions. However, these oxidants appear to serve as electron sinks for the reductive branch of the fermentation pathway, with no associated proton translocation, and are therefore not used in respiration (Stal and Moezelaar, 1997) (see also Section 4). As we saw in Chapter 4, some cyanobacteria can perform anoxygenic photosynthesis under anoxic conditions in the light using H_2S or H_2 as the electron donor (Cohen *et al.*, 1986).

Many cyanobacteria fix N_2, though others do not, instead relying on the uptake of combined nitrogen species such as nitrate and ammonium. The requirement for strict anoxia in nitrogen fixation poses special challenges to cyanobacteria, and as discussed Chapter 7, they show a number of different adaptations to solve this problem.

Due to large variations in morphology and size, visual identification of cyanobacteria is possible to a greater extent than for most other prokaryotes (e.g., Whitton and Potts, 2000). Unicellular forms are often coccoid or oval, and may be free or assembled in colonies with various degrees of ordering. Cell diameters range from 0.5–0.7 μm in *Prochlorococcus*, the smallest known oxygenic phototroph, to 40 μm in *Chroococcus*. Filaments with or without sheaths are another important morphotype, and these are also found individually or assembled in bundles, tufts or in globular masses such as the gelatinous balls of *Nostoc* spp., which may reach several centimeters in diameter. Filaments may contain specialized cells including heterocysts for nitrogen fixation (see Chapter 7) and akinetes, which are large, resistant, spore-like cells that form under adverse conditions. While flagella are absent, many species can move by gliding, and some can adjust their buoyancy by means of gas vesicles, which are air-filled, protein-lined compartments within the cell.

3.2. Cyanobacteria in the environment

Cyanobacteria can be found in abundance in both freshwater and marine photic habitats and are a major component of phototrophic microbial mats. The group includes extremophiles that grow under hypersaline conditions up to 300‰ parts per thousand salt by weight. (e.g., *Aphanothece halophytica*; Oren, 2000), at temperatures up to 74 °C (*Synechococcus* spp.; Ward and Castenholz, 2000) and at pH values from 4 to near 10 (Brock, 1994; Namsaraev *et al.*, 2003). The group also includes some of the most desiccation-tolerant bacteria (Potts, 1999). For example, the unicellular *Chroococcodiopsis* spp. is found in hot and cold deserts where liquid water is ephemeral; it has been recovered from samples that were dry for several years. Cyanobacteria in such habitats, and in shallow aquatic systems, further exhibit a high degree of tolerance to ultraviolet radiation (Castenholz and Garcia-Pichel, 2000). This tolerance is in part conferred through special UV-screening compounds such as scytonemin, which gives a dark color to the thin cyanobacterial crusts growing on rocks (Garcia-Pichel and Castenholz, 1991).

Common genera in fresh and coastal waters include filamentous *Anabaena* and *Nodularia*, as well as the unicellular, colony-forming *Microcystis*, all of which form summer blooms in lakes and coastal waters. These organisms under some, as yet not fully understood, circumstances produce toxins that pose a hazard to other organisms, including humans (Dow and Swoboda, 2000; Kaebernick and Neiland, 2001). In the open oceans, the filamentous *Trichodesmium* spp. and the picoplanktonic *Synechococcus* spp. and *Prochlorococcus* spp. are prominent cyanobacterial representatives. *Trichodesmium* is a cosmopolitan of the oligotrophic tropical and subtropical oceans, where it may form extensive surface blooms. The blooms are readily observed due to the aggregation of the filaments in characteristic tufts or puffs. Its ability to fix N_2, which is realized in the absence of heterocysts and during the day, is believed to contribute decisively to its success. *Trichodesmium* may be one of the most important sources of new nitrogen in the open ocean (Capone *et al.*, 1997; see Chapter 7), though recently discovered unicellular cyanobacteria may also be important (Zehr *et al.*, 2001).

With diameters of only 0.9 and 0.6 μm, respectively, the unicellular *Synechococcus* and *Prochlorococcus* escaped scientific notice until relatively recently (Johnson and Sieburth, 1979; Waterbury *et al.*, 1979; Chisholm *et al.*, 1988), yet these are the most numerous cyanobacteria in the oceans and probably contribute significantly to oceanic primary production (Partensky *et al.*, 1999a,b). Species of *Synechococcus* are found from polar to tropical waters, and they are typically concentrated within the upper part of the photic zone, while those of *Prochlorococcus*, being restricted to subtropical and tropical latitudes, are distributed to greater depths and may even be found below the euphotic zone (Figure 6.4).

Different strains of *Prochlorococcus* dominate different depths of the water column, and two phylogenetic groups have been discerned that show specific adaptations to life both in surface waters and in deeper waters (Figure 6.4) (Moore and Chisholm, 1999). Thus, strains adapted to deeper water grow optimally at lower irradiances and are photoinhibited by light

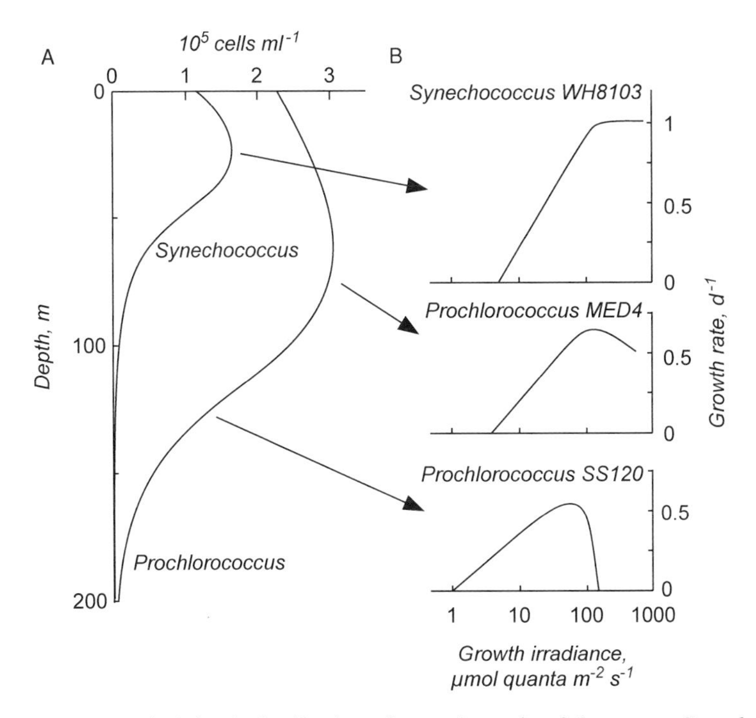

Figure 6.4 Typical depth distribution of cyanobacteria of the genera *Synechococcus* and *Prochlorococcus* in the ocean (A) related to their adaptation to light (B). Cell numbers represent typical maximum counts. As *Synechococcus* cells are typically larger than *Prochlorococcus*, the maximum biomasses of the two genera are similar. For *Prochlorococcus*, two subpopulations are often discerned with adaptations to life in shallower and deeper waters (see text). Growth irradiance curves are for individual but representative strains, including a surface-adapted and a subsurface-adapted *Prochlorococcus*. Panel A redrawn from Partensky *et al.* (1999b); panel B redrawn from Moore *et al.* (1995).

levels where the surface strains just reach their maximal growth rate. The light dependences are related to differences in the ratio of the characteristic chlorophylls with higher (Chl b + Chl b_2)/Chl a_2 in the subsurface strains. The strains also differ in their inorganic nitrogen utilization, with deep subsurface strains using both ammonium and nitrite, while surface strains use only ammonium (Moore *et al.*, 2002). This pattern matches depth differences in nitrogen availability. Thus, nitrite and ammonia are available deep in the photic zone, where ammonia is liberated during organic matter mineralization, and nitrite is formed as the ammonia is nitrified to nitrate. Ammonium is the only nitrogen source in the upper photic zone, where nitrification is insignificant owing to rapid assimilation of the ammonium (see also Chapter 7). The inability of *Prochlorococcus* to utilize nitrate may also be understood as adaptative since nitrate is not present in the oligotrophic surface waters. In deeper waters, where nitrate is present but little light energy is available, assimilatory nitrate reduction may require too much energy compared to nitrite or ammonium assimilation (Moore *et al.*, 2002). Thus, overall, *Prochlorococcus* appears to follow a minimalist strategy, in which elimination of unnecessary functions permits a reduction of genome size as well as a minimization of cell size (Fuhrman, 2003; Rocap *et al.*, 2003). Small size allows for more efficient scavenging of nutrients (e.g. Jørgensen, 2000) (see also Chapter 2), which is a critical factor in oligotrophic waters.

3.2.1. Environmental controls of cyanobacterial photosynthesis

With the exception of nitrogen fixation in diazotrophic species, cyanobacteria are qualitatively similar to eukaryote microalgae in their requirements for light and nutrients. We briefly discuss the influence of these factors, and refer the reader to more detailed discussions of the ecophysiology of aquatic photosynthesis in Falkowski and Raven (1997).

Photosynthesis rates are typically proportional to light intensity (irradiance) at low light levels. A minimum irradiance, known as the light compensation intensity, is required for photosynthesis to balance dark respiration and sets the minimum amount of constant light required for growth. As irradiance increases, the photosynthetic apparatus is eventually saturated, and rates may remain constant or decrease at supraoptimal irradiances. Compensation intensities range from $\sim 1\ \mu$mol quanta m^{-2} s^{-1} in subsurface *Prochlorococcus* strains (Figure 6.4) to $\sim 150\ \mu$mol quanta m^{-2} s^{-1} in *Trichodesmium* (Kana, 1993; Roenneberg and Carpenter, 1993). The response of the cyanobacteria to changes in light is dynamic and involves changes in pigment contents, which tend to moderate the immediate effects of light changes (photoacclimation; see Falkowski and Raven, 1997; MacIntyre *et al.*, 2002). The daily variation of photosynthesis in some cyanobacteria is apparently also affected by endogenous timing through a biological clock (circadian rhythm; Mori and Johnson, 2001). Thus, the light compensation intensity in *Trichodesmium thiebautii* varies in a daily manner even when the organism is kept in constant light or darkness (Roenneberg and Carpenter, 1993).

In fresh water with a low external supply of fixed nitrogen relative to phosphorus, the cyanobacteria are often able to meet the nitrogen requirements of photosynthesis through diazotrophy, so that phosphorus becomes the limiting nutrient (Howarth *et al.*, 1988b; Falkowski and Raven, 1997). The unique ability for diazotrophy hence contributes to the success of cyanobacteria (often heterocystous types) in freshwater habitats. In the oceans, by contrast, nitrogen is limiting, yet diazotrophic cyanobacteria are not major contributors to primary production. How can this be? One possible explanation is that nitrogen fixation itself is limited by other nutrients, most likely by iron (Howarth *et al.*, 1988a; Rueter, 1988). As discussed in Chapter 8, iron is a limiting nutrient in parts of the oceans, and cyanobacteria may be particularly affected by iron shortage. Owing to a high iron content of nitrogenase and other parts of the nitrogen-fixing machinery, the iron requirements of nitrogen-fixing phototrophs may be 100-fold higher than the requirements for those that utilize fixed nitrogen sources (Raven, 1988; Kustka *et al.*, 2002) (Chapter 7). Furthermore, cyanobacteria in general have relatively high iron requirements, which can be related to high photosystem I (PSI)/photosystem II (PSII) ratios, with much more iron associated with PSI. A comparison of the iron demands of *Trichodesmium* spp. in culture with iron fluxes to the oceans suggests that this organism would be iron limited in 75% of the global ocean (Berman-Frank *et al.*, 2001a,b).

3.2.2. Evolutionary and geobiological considerations

Cyanobacteria are unique among the prokaryotes in having a true fossil record, which provides important clues to the timing of their evolution and dominance in the environment. Microfossils with great similarity to modern cyanobacteria have been found in deposits dating back to the early Proterozoic Eon, 2.1 billion years ago (Hofmann, 1976; Knoll, 2003). The organisms lived in settings seemingly similar to modern shallow-water benthic cyanobacterial habitats, and their identification is supported by the presence in filamentous fossils of cells resembling the akinetes of extant species. From this time, and even from the late Archean 2.7 billion years ago, geologists have found stromatolites that are interpreted as fossilized phototrophic microbial, likely cyanobacterial, communities (Buick, 1992;

Knoll, 2003). The early occurrence of cyanobacteria is further supported by the finding of long-chain 2-methylhopanoids, biomarker molecules characteristic of cyanobacteria, in 2.7- and 2.5-billion-year-old marine deposits (Brocks *et al.*, 1999; Summons *et al.*, 1999). Filamentous structures in even older rocks have also been interpreted as cyanobacterial fossils (Schopf, 1992), but this view is controversial (Brasier *et al.*, 2002; Schopf *et al.*, 2002; Knoll, 2003).

The coupled photosystems of cyanobacteria are believed to originate from the two types of reaction centers found in different anoxygenic phototrophs, with lateral gene transfer possibly involved in bringing these types together (Xiong *et al.*, 2000; Raymond *et al.*, 2002) (see also Chapter 4). The evolutionary pathway of the unique water-splitting complex, however, remains obscure. Hydrogen peroxide or reduced manganese has been suggested as electron donors for evolutionary intermediates, but no extant organisms are known to utilize these reductants (Blankenship and Hartman, 1998; Kálmán *et al.*, 2003) (see also Chapter 8).

With cyanobacteria branching from the Tree of Life relatively late, it appears that much of the diversification of the prokaryotes had occurred by the time cyanobacteria had evolved, by the late Archean or even earlier. Based on the abundance of cyanobacterial remains and the relative scarcity of eukaryotic fossils through much of the Proterozoic (Knoll, 2003), it further appears that cyanobacteria were the main biological agents behind the oxygenation of the Earth's surface during this eon (Figure 6.1). Further evidence for cyanobacterial abundance, or even prominence, in the late Archean comes from biomarkers, where 2-methylhopanoids are found in far greater amounts than steranes of probable algal origin (Brocks *et al.*, 2003). Thus, modern cyanobacterial communities such as microbial mats are of particular interest as model systems for an analysis of the early biogeochemical evolution of Earth.

4. MICROBIAL OXYGEN CONSUMPTION

Oxygen is consumed in many different enzymatic processes, in which it may function primarily as an electron acceptor, with conversion to water or hydrogen peroxide, or as a co-substrate, which is incorporated into another molecule (Unden, 1999). Enzymes catalyzing the former type of reaction are called oxidases, while those catalyzing oxygen incorporation are called oxygenases. Dioxygenases catalyze the cleavage of molecular oxygen with the reductive incorporation of both oxygen atoms into the substrate, while monooxygenases add one oxygen atom to the substrate as a hydroxyl group and reduce the other to water.

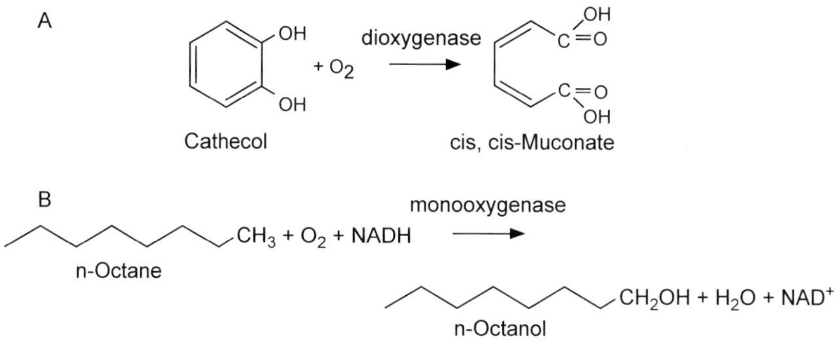

A

Cathecol + O_2 —dioxygenase→ cis, cis-Muconate

B

n-Octane $CH_3 + O_2 + NADH$ —monooxygenase→ n-Octanol $CH_2OH + H_2O + NAD^+$

Figure 6.5 Examples of reactions catalyzed by oxygenase enzymes during the degradation of aromatics (A; cathecol dioxygenase) and long-chain alkanes (B; *n*-alkane monooxygenase).

Dioxygenases are important in the aerobic degradation of aromatic compounds, including the central step of opening the aromatic ring (Figure 6.5). Likewise, monooxygenases serve important functions in specialized metabolisms, including the initial step in methane oxidation and nitrification catalyzed by methane monooxygenase and ammonia monooxygenase (see Chapters 7 and 10) and in the degradation of long-chain alkanes (Figure 6.5). Monooxygenases require a reductant such as NADH as a co-substrate. Oxygen also serves as a co-substrate for aerobic biosynthesis of compounds such as pyrimidines, ubiquinones, tetrapyrroles, and sterols.

Anaerobes must utilize alternative biochemical pathways for the metabolic functions performed by oxygenases in aerobes (or dispense with such reactions altogether, as is the case with sterol synthesis), and these alternatives are generally much less efficient. For example, the anaerobic degradation of aromatics is costly, requiring ATP for opening of the ring (Fuchs, 1999). Such factors contribute to differences in the rate and extent of carbon oxidation under aerobic and anaerobic conditions (see Chapter 5).

4.1. Oxidases

Oxidases include the membrane-bound components that terminate aerobic respiratory chains, as discussed in greater detail below, as well as soluble non-respiratory enzymes that catalyze reduction of oxygen with the oxidation of various substrates in the cytoplasm or outside of the cell. With the latter type, oxygen may be reduced to hydrogen peroxide or completely to water. In aerotolerant anaerobes, oxygen reduction by non-respiratory oxidases may serve as an electron sink for the consumption of reduced substances such as NADH or lactate:

$$\underset{\text{lactate}}{CH_3CHOHCOOH} + O_2 \xrightarrow{\text{lactate oxidase}} \underset{\text{pyruvate}}{CH_3COCOOH} + H_2O_2 \quad (6.3)$$

No proton translocation is associated with this oxygen consumption (which is therefore non-respiratory). The regenerated pyruvate and NAD^+ serves as intracellular electron acceptors for the oxidative branch of the fermentative pathway, which is associated with the synthesis of ATP through substrate-level phosphorylation (Buckel, 1999). By stimulating this oxidative branch, processes such as that in Equation 6.3 may increase the yield of ATP. An alternative benefit from non-respiratory oxygen consumption may be the removal of oxygen to relieve oxidative stress (see below).

Extracellular oxidases of the multi-copper type appear to be effective in bacterial manganese oxidation (see Chapter 8) and are used in the degradation of lignin by fungi. Multi-copper oxidases catalyze the four-electron reduction of oxygen to water coupled to successive one-electron substrate oxidations (Call and Mücke, 1997; Xu *et al.*, 2000). The fungal oxidases, called laccases, oxidize phenolic compounds to highly reactive phenoxy radicals that can subsequently convey the spontaneous breakdown of the lignin polymer (Xu *et al.*, 2000). The phenol may be either part of the lignin polymer or a low-molecular-weight phenolic compound, which, once activated, may attack parts of the lignin that are not accessible to the enzyme. Further oxygen-dependent reactions involved in lignolysis include the production of hydrogen peroxide and reactive Mn^{3+} complexes as powerful oxidants (Call and Mücke, 1997; Fuchs, 1999; Schlosser and Höfer, 2002). Lignin is virtually non-degradable under anoxic conditions, but even in the presence of oxygen, degradation is slow relative to other major classes of organics.

As we saw in Chapter 4, the CO_2-fixing enzyme Rubisco is also a (mono)-oxygenase and catalyzes photorespiration in oxygenic phototrophs. Oxygenic phototrophs harbor another, similarly curious, light-dependent oxygen-consuming process, the Mehler reaction, in which oxygen in photosystem I is reduced to superoxide, which dismutates to hydrogen peroxide and oxygen (see Equation 6.4). Hydrogen peroxide from this reaction is rapidly reduced to water by ascorbate. Because the electrons for oxygen reduction originate from the splitting of water by PSII, there is no net exchange of oxygen, but electron transport generates a proton motive force, which may be used for ATP synthesis (Asada, 1999). Physiological functions of this water–water cycle may include the dissipation and/or conservation of light energy absorbed by PSII, which prevents photochemical damage at times when this energy cannot be used for the reduction of CO_2 (Falkowski and Raven, 1997; Ort and Baker, 2002). Photorespiration and the Mehler reaction are important in terrestrial photosynthesis, where they are estimated to account for 30 and 10% of gross oxygen production, respectively, while they are thought to be of less significance in aquatic photosynthesis (Bender *et al.*, 1994; Falkowski and Raven, 1997).

4.2. Reactive oxygen species

Although the terminal oxidases of oxygen respiration produce few of the highly reactive intermediates, aerobic organisms are continuously exposed to such compounds. The intermediates originate from inorganic reactions (e.g., Equation 6.2) and from reactions of oxygen with cell constituents. In *Escherichia coli*, an intracellular superoxide production rate of $5\,\mu M\ s^{-1}$ is estimated, corresponding to ~0.2% of the oxygen respiration rate, mainly resulting from the adventitious reaction of oxygen with flavo- and quino-proteins of the electron transport chain (Imlay, 2002). For comparison, the organism cannot tolerate intracellular superoxide concentrations much greater than 0.1 nM (Imlay and Fridovich, 1991). Endogenous hydrogen peroxide production is $\sim 14\,\mu M\ s^{-1}$, but only impedes growth at $2\,\mu M$ (Seaver and Imlay, 2001). As an example of their detrimental effects, superoxide may destroy iron sulfur proteins, releasing iron, which may react with hydrogen peroxide to form a hydroxyl radical (Equation 6.2c) that damages DNA (Imlay, 2002).

Consequently, aerobic and aerotolerant organisms generally possess a variety of antioxidant systems that efficiently remove these toxic compounds (Valentine *et al.*, 1998). Superoxide and hydrogen peroxide are consumed through dismutation (disproportionation) by the enzymes superoxide dismutase and catalase, respectively:

$$2O_2^{\cdot-} + 2H^+ \rightarrow H_2O_2 + O_2 \quad (6.4)$$

$$2H_2O_2 \rightarrow 2H_2O + O_2 \quad (6.5)$$

Hydrogen peroxide reduction to water is further catalyzed by peroxidases, typically with NADH as the reductant:

$$H_2O_2 + NADH + H^+ \rightarrow 2H_2O + NAD^+ \quad (6.6)$$

Peroxidase-catalyzed reduction is the main sink for endogenous hydrogen peroxide in *E. coli*, while the primary function of catalase appears to be the removal of larger exogenous doses (Seaver and Imlay, 1999, 2001). The extremely reactive hydroxyl radical is not enzymatically consumed, but its formation is avoided by the removal of its precursors, superoxide and hydrogen peroxide. While a few aerobes lack catalase, all have superoxide dismutase. Superoxide dismutase is absent from some aerotolerant anaerobes, but in these, superoxide is dismutated by Mn^{2+} complexes. These may be seen as primitive precursors of superoxide dismutase, which often contains Mn^{2+} as a cofactor (Unden, 1999).

4.3. Oxygen metabolism in anaerobes

Anaerobic organisms in many natural habitats experience periodic exposure to oxygen, either regularly, such as during diel variations in phototrophic microbial mats, or more randomly. Random exposure to oxygen can occur, for example, as a result of bioturbation. Anaerobes may also actively position themselves close to oxic–anoxic interfaces in order to intercept favorable substrates and electron acceptors that are produced in the oxic environment (e.g., Sass *et al.*, 2002) (see also the discussion of microbial mats in Chapter 13). Anaerobic metabolism is generally inhibited by low levels of oxygen, often by direct interaction of oxygen with critical enzymes. Some organisms, such as the methanogens, may die after a relatively brief oxygen exposure, while others survive extended oxic periods (e.g., Cypionka, 2000). One prerequisite for such survival should be the presence of antioxidant systems. Superoxide dismutase and catalase activities are indeed found in several anaerobes (e.g., Brioukhanov *et al.*, 2002), but an alternative pathway for superoxide removal has been found in some anaerobes including the archaeon *Pyrococcus furiosus* and the sulfate-reducing bacterium *Desulfoarculus baarsii* (Jenney *et al.*, 1999; Lombard *et al.*, 2000). Instead of dismutation, superoxide is reduced to hydrogen peroxide by a superoxide reductase with NADH as the reductant. One advantage to this pathway could be that oxygen is not regenerated as in the dismutation (Equation 6.4). This would only be significant, however, if a major part of the oxygen consumption went via superoxide. Another advantage could be that the reductase maintains superoxide at lower concentrations, thus offering better protection for superoxide-sensitive enzymes (Imlay, 2002).

Some organisms generally considered to be strict anaerobes, including some sulfate reducers and acetogenic bacteria, may reduce oxygen at high rates (Cypionka, 2000; Karnholtz *et al.*, 2002; Boga and Brune, 2003) (see also Chapter 9). In these organisms, oxygen may be reduced by NADH oxidases and/or through an electron transport chain (Lemos *et al.*, 2001). While there is some evidence for proton translocation, and thereby energy conservation, in oxygen-consuming sulfate reducers (Dilling and Cypionka, 1990), growth from this process has not been demonstrated. The most likely advantages of oxygen consumption by anaerobes are protection against oxidative stress and the reestablishment of anoxia required for growth.

5. OXYGEN RESPIRATION

Oxygen respiration is by far the most important oxygen-consuming reaction in living organisms and in the biogeochemical oxygen cycle. In oxygen respiration, the reduction of oxygen to water is the terminal step in a series of coupled oxidation-reduction reactions involving cell membrane-bound redox couples that make up an electron transport chain. Part of the free energy available from the reactions is used to translocate protons across the membrane, and the proton-motive force thus established drives the production of ATP. The exact components of the electron transport chains, as well as their configuration, vary among different organisms, but some classes of enzymes and coenzymes are generally present and are reviewed in the following sections.

5.1. Electron transport chains

With organic matter as the substrate, electrons are mainly delivered to the electron transport chain by reduced intermediates such as NADH. Thus, the net reaction catalyzed by the electron transport chain may be written as follows:

$$2NADH + O_2 + 2H^+ \rightarrow 2NAD^+ + 2H_2O \tag{6.7}$$

NADH is derived from the reduction of NAD^+ coupled to the oxidation of the organic or inorganic substrate. The redox potential for $NADH/NAD^+$ is -0.32 V, while H_2O/O_2 is at 0.82 V; this allows the oxidation of NADH with oxygen to proceed through a long electron transport chain with the potential for translocation of several protons (Figure 6.6). To follow the electrons as they flow through the respiratory chain, we begin with the catalytic oxidation of NADH by a membrane-bound dehydrogenase. Electrons are passed through a flavoprotein and a Fe-S-protein to a quinone that diffuses freely within the membrane. The reduced quinone (a quinol) may deliver electrons to a quinol:cytochrome *c* oxidoreductase complex, itself consisting of a number of cytochromes, which in turn reduces a *c*-type cytochrome on the outside of the cytoplasmic membrane. Finally, the electrons are transferred across the membrane from cytochrome *c* to oxygen via a cytochrome *c* oxidase. A shorter alternative route involves the direct coupling of quinol oxidation to oxygen reduction by a quinol oxidase (Unden, 1999). In both

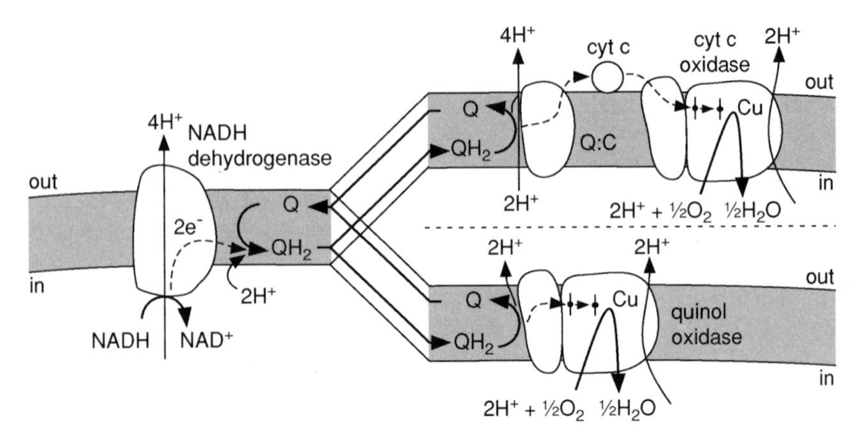

Figure 6.6 Main elements of the electron transport chain involved in oxygen respiration in *Paracoccus denitrificans*. Electrons are delivered through the quinone cycle (Q/QH_2) either to a quinol oxidase complex (bottom) or to a quinone:cytochrome *c* oxidoreductase (Q:C; top), which, via cytochrome *c*, passes electrons to a cytochrome *c* oxidase. After Unden (1999) and Madigan *et al.* (2003).

these oxidases, oxygen is reduced at a unique site composed of a copper ion and a high-spin heme group. The metal center is well embedded in the protein and located inside the cytoplasmic membrane, thus avoiding the release of reactive oxygen species (Pereira *et al.*, 2001). Another, structurally distinct type of quinol oxidase, known as type *bd* oxidase, carries only heme and no copper (Poole, 1994).

Branched electron transport chains are the rule in aerobic prokaryotes and may include up to at least five different terminal oxidases in one organism (García-Horsman *et al.*, 1994; Poole, 1994). *Paracoccus denitrificans,* for example, has two cytochrome *c* oxidases as well as a quinol oxidase, and *E. coli* has no cytochrome *c* oxidase but two quinol oxidases, one heme-copper and one *bd* type. Oxygen concentration is a main factor partitioning the electron flow between the alternative oxidases (see following section).

In facultative anaerobes such as *P. denitrificans* and *E. coli*, further branches of the electron transport chain lead to the terminal reductases for nitrate, nitrite and fumarate. These are active in the absence of oxygen (Figure 6.7). Likewise, there are several entry points other than NADH dehydrogenase for electrons to the respiratory chain, including dehydrogenases of formate, succinate and lactate, as well as hydrogenase. The respiratory quinones serve as a hub receiving electrons from the (de)hydrogenases and passing them on to the terminal reductases or oxidases.

Aerobic chemolithotrophs generally employ the same mechanisms for energy conservation as the organotrophs. Electrons from the inorganic substrates

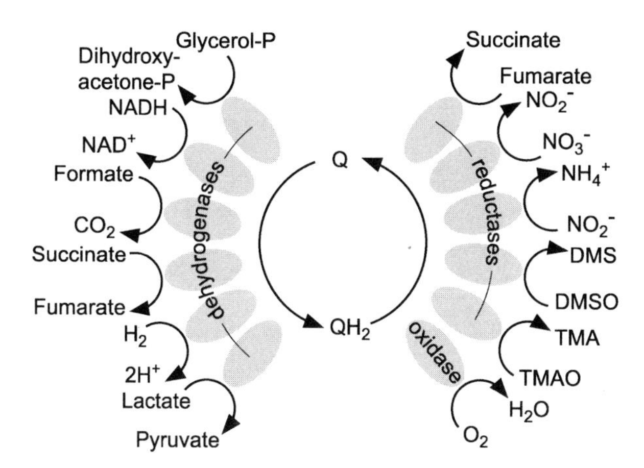

Figure 6.7 The quinone cycle (Q/QH_2) as a hub passing electrons from multiple dehydrogenases to multiple reductases and oxidases in *E. coli*. After Fuchs (1999).

are delivered to the electron transport chain at sites that are related to the redox potential of the donor (Kuenen, 1999; Madigan *et al.*, 2003). Because many inorganic redox couples have redox potentials less negative than $NADH/NAD^+$ (see Figure 3.1 in Chapter 3), electron transport chains are typically shorter, with fewer components involved. For example, the oxidations of nitrite and of Fe^{2+} at low pH involve only cytochrome *c* and a cytochrome *c* oxidase.

Protons are moved across the membrane at several sites along the electron transport chain. In the long chain of *P. denitrificans* (Figure 6.6), translocation occurs at the dehydrogenase complex, the quinone cycle and the quinol:cytochrome *c* oxidoreductase, and at the terminal cytochrome *c* oxidase. A total of 10 protons may be moved per two electrons transported from NADH (Equation 6.7). In *E. coli*, lacking quinol:cytochrome *c* oxidoreductase, the maximum number of protons is eight per two electrons for the same reaction. This yield, however, is only realized with one of two alternative NADH dehydrogenases and one of the two quinol oxidases, since only these alternatives are proton pumps. Thus, the amount of energy conserved depends on the configuration of the electron transport chain, which is regulated in response to the availability of substrates and oxygen (Unden and Bongaerts, 1997; Unden, 1999).

5.2. Regulation and kinetics

Oxygen is the preferred electron acceptor for facultative anaerobic bacteria, and the expression of the enzymes needed for anaerobic respiration is usually repressed in the presence of oxygen. Oxygen levels also regulate the expression of some enzymes involved in aerobic metabolism. For example, the type *bd* quinol oxidase of *E. coli* is preferentially expressed at low oxygen concentrations, whereas the heme copper-type oxidase dominates at higher oxygen levels. This regulation reflects the K_m values for oxygen of the two oxidases. Thus, the *bd* type dominating at low oxygen levels is optimized for such concentrations, with a K_m value for oxygen in the range of 0.003 μM to 0.008 μM, and it is inhibited at $\sim$1 μM of oxygen. The heme-copper quinol oxidase, with a K_m of $\sim$0.2 μM, conveys oxygen reduction at higher concentrations (Rice and Hempfling, 1978; D'mello *et al.*, 1996).

The minimum oxygen concentration required for oxygen respiration in microorganisms has not yet been determined. In addition to the high-affinity type *bd* oxidase of *E. coli*, K_m values for oxygen of 3 nM to 8 nM are also found for terminal heme-copper oxidases in *Bacillus cereus* and the nitrogen-fixing *Bradyrhizobium japonicum* (Preisig *et al.*, 1996; Contreras and Escamilla, 1999), suggesting that oxygen may be respired at levels that low, or even lower.

Many microorganisms living in oxygen gradients are able to position themselves at a preferred oxygen concentration through a variety of chemotactic responses to oxygen, known as aerotaxis (e.g., Fenchel, 2002; Thar and Kühl, 2003). In obligate anaerobes and microaerophiles, negative aerotaxis can be of vital importance, but full aerobes may find an advantage in seeking low oxygen concentrations (e.g., in the 1–10 μM range), since this reduces oxidative stress without limiting oxygen metabolism.

6. PHYLOGENY AND EVOLUTION OF OXYGEN RESPIRATION

Oxygen respiration is present in all three domains of the Tree of Life, with the same major classes of electron transport chain components (Castresana and Saraste, 1995). In the *Bacteria*, oxygen respiration is present in most major lineages, including all proteobacterial subdivisions and deep-branching thermophiles such as *Thermus* and *Aquifex*, but excluding the green sulfur bacteria (see Chapter 9). The process is less dominating in the *Archaea*, where it is present in the major orders of the *Crenarchaeota*, but has not been found in the methanogenic orders of the *Euryarchaeota*. In the *Eukarya*, anaerobiosis is restricted to some protists, but it now appears that all these anaerobes may be descended from aerobic ancestors through a secondary loss of mitochondria, which suggests that this entire domain is aerobic in its origin (Fenchel and Finlay, 1995; Silbermann *et al.*, 2002).

The wide phylogenetic distribution of oxygen respiration suggests an early evolutionary origin for this metabolism. This is further supported by sequence analysis for enzymes involved in oxygen respiration, which indicates a single origin for aerobic respiration (Castresana and Saraste, 1995). This analysis includes the heme-copper type oxidases but excludes oxidases of the *bd* type. Type *bd* oxidases show no homology to heme-copper oxidases, and sequence-based *bd*-oxidase phylogenies diverge substantially from the canonical Tree of Life, suggesting frequent horizontal transfer of *bd*-oxidase genes (Osborne and Gennis, 1999; Baughn and Malamy, 2004).

Because some types of heme-copper oxidases are found in both the *Bacteria* and the *Archaea*, diversification of these oxidases may have occurred before the separation of the domains. In an alternative scenario, based on both sequences and functional analysis, the heme-copper oxidases evolved early in the diversification of the *Bacteria* (Pereira *et al.*, 2001). Since the archaeal heme-copper oxidases show greatest similarity to those of some gram-positive bacteria, which in turn appear to be derived from other oxidases found only in the *Bacteria*, the aerobic *Archaea* in this scenario obtained their oxidases through lateral gene transfer (Figure 6.8). It has been hypothesized that the heme-copper oxidases evolved from a nitric oxide (NO) reductase (Saraste and Castresana, 1994; de Vries and Schröder, 2002) (see also Chapter 7). The heme-copper reaction center has similarities

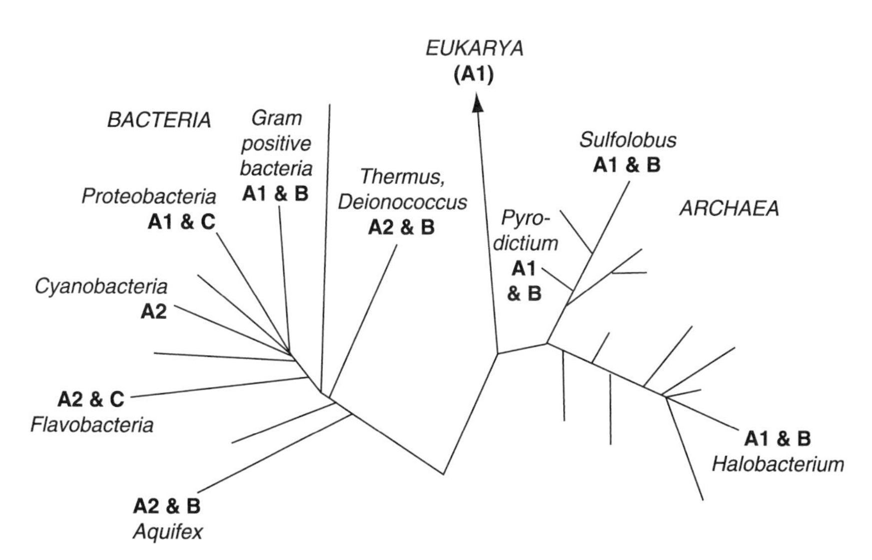

Figure 6.8 The distribution of the main heme-copper oxidase families *A1*, *A2*, *B*, and *C* in the Tree of Life (Figure 1.5). Type *A1* is proposed to have evolved from type *A2*. Archaeal oxidases *A1* and *B* have the greatest similarity to those of the gram-positive bacteria, suggesting that they were acquired by *Archaea* from the gram positives by lateral gene transfer (Perreira *et al.*, 2001). Modified after Pereira *et al.* (2001).

to the reaction center of NO reductase in denitrifiers, which has iron in place of the copper ion but does not function as a proton pump.

Nitric oxide reductase could have evolved to serve in the detoxification of NO formed abiotically in the atmosphere of the early anoxic Earth (Kasting and Walker, 1981; de Vries and Schröder, 2002), thus paving the road for oxidase evolution. Nonetheless, the early and apparently single origin of oxygen respiration suggested previously contrasts with the relatively late evolution of oxygenic photosynthesis, which is associated with the appearance of cyanobacteria (Section 3.2.2). With oxygenic photosynthesis as the only substantial source of oxygen in the biosphere, it is not clear what may have driven the evolution of oxygen respiration in a pre-cyanobacterial world. The atmospheric photochemical production of oxygen is very small, $\sim$0.5 μmol m^{-2} d^{-1} (globally 10^{10} mol yr^{-1}; compare to Table 6.1), much of which may not reach the Earth's surface, and bacteria would have to compete with abiotic reactions for the traces of oxygen produced this way (Kasting and Walker, 1981). An early origin is also indicated for Mn/Fe superoxide dismutase with a phylogeny that is largely, though not entirely, congruent with the small subunit (SSU) rRNA-based phylogenies (Brown and Doolittle, 1997). Because superoxide is generated from oxygen (e.g., Equation 6.2), this enzyme probably functions at similar oxygen levels as does oxygen respiration.

Could there have been other sources of oxygen? Though speculative, one possibility is hydrogen peroxide. Chemical models of the early Earth atmosphere suggest that hydrogen peroxide formed in the atmosphere may have rained to the surface at rates of 1 to 10 μmol m^{-2} d^{-1} (Kasting *et al.*, 1984). In environments rich in ferrous iron and other reductants, input of hydrogen peroxide would lead to the formation of the extremely reactive hydroxyl radicals (Equation 6.2). Thus, the hydrogen peroxide rain may have forced the evolution of catalases for oxidative stress protection (Blankenship and Hartman, 1998). Catalases catalyze the disproportionation of hydrogen peroxide to water and oxygen at extreme efficiency (Equation 6.5). In environments experiencing significant rainfall, sufficient oxygen may have formed through this reaction to sustain some oxygen respiration. As we have seen, oxygen respiration can utilize oxygen at extremely low levels.

7. MICROBIAL RESPIRATION IN NATURAL ENVIRONMENTS

7.1. Aerobes in natural aquatic environments

Aerobic organisms are found in the extremes of the chemical and physical limits of life as long as oxygen is present. Thus, although many known hyperthermophiles are anaerobes, some, including *Pyrolobus fumarii,* with a maximum temperature of 113 °C, grow as microaerophiles (Blöchl *et al.*, 1997). The aerobic iron oxidizer *Ferroplasma acidarmanus* lives at pH 0 (see Chapter 8), while alkaliphilic *Bacillus* spp. grow at pH values around 11 (Horikoshi, 1999). Aerobic respiration is the only chemotrophic metabolism found in the extreme salinity of halite saturation (Oren, 2001).

In oxic aquatic environments where more amenable conditions prevail, prokaryote numbers range from 10^4 ml^{-1} in the ocean interior, to over 10^6 ml^{-1} in surface waters, to 10^9 ml^{-1} in sediments, and up to 10^{11} ml^{-1} in photosynthetic microbial mats. Although representatives of many bacterial divisions and major archaeal groups are present, a few groups are usually of widespread importance and together often completely dominate the total microbial community (Giovannoni and Rappé, 2000). The most detailed information on community structure comes from planktonic communities from fresh waters and the oceans, where numerically important groups of *Bacteria* include the α-, β-, and γ-Proteobacteria and the *Cytophaga-Flavobacterium* cluster (Table 6.2).

Within the different lineages, abundant organisms, the organisms typically belong to one of a few clusters of closely related species. In the ocean, two such common clades are the α-proteobacterial SAR11 and *Roseobacter* clusters (Eilers *et al.*, 2000; Morris *et al.*, 2002). While species of *Roseobacter* are in pure culture, members of the SAR11 clade were originally identified in clone libraries from the Sargasso Sea and did not have close relatives in culture (Giovannoni *et al.*, 1990). Members of the SAR11 cluster are ubiquitous in the world's oceans and are frequently very abundant, contributing up to half of the prokaryote cells in surface sea water (Morris *et al.*, 2002). Members of the SAR11 cluster were later cultivated using media based on natural sea water with no or only small additions of organic substrates, and have been assigned to the new genus *Pelagibacter*

Table 6.2 Examples of numerically important phylogenetic groups of aerobic heterotrophs in aquatic environments

Clade	Typical habitat
α-proteobacteria	Sea water
Pelagibacter (SAR11) clade	Sea water
Roseobacter clade	Sea water
β-proteobacteria	Fresh water
γ-proteobacteria, SAR86 clade	Sea water
Cytophaga-Flavobacterium clade	Coastal waters, aggregates, sediments
Planctomycetes	Aggregates, sediments
Crenarchaeota	Sea water

Sources: Llobet-Brossa *et al.* (1998), Giovannoni and Rappé (2000), Cottrell and Kirchman (2000, 2003), Karner *et al.* (2001), Ploug *et al.* (2002).

(Rappé *et al.*, 2002). With a very small cell volume of ~0.01 μm^3 (length 0.4–0.9 μm, diameter 0.1–0.2 μm), these organisms are well adapted to aerobic heterotrophy in substrate-poor ocean waters. Important marine surface water *Roseobacter* affiliates synthesize bacteriochlorophyll *a* under oxic conditions and may conserve energy through cyclic photophosphorylation in addition to aerobic respiration (Giovannoni and Rappé, 2000), while other *Roseobacter*-affiliated phylotypes abundant in temperate and polar waters are not phototrophic (Selje *et al.*, 2004).

A large fraction of the organisms in natural microbial communities is not closely related to known isolates, and therefore little is known about the metabolic specifics of the different dominant organisms or clades, or about factors governing their abundance. Some general patterns are, however, emerging. Thus, among the proteobacteria, members of the α subdivision are particularly abundant in sea water, while β-proteobacteria often prevail in fresh water (Figure 6.9). Though γ-Proteobacteria from genera such as *Alteromonas*, *Vibrio*, and *Oceanospirillum* typically dominate viable counts of aerobic bacterioplankton, cultivation-independent quantifications by means of cloning and fluorescent *in situ* hybridization (FISH) (see Chapter 2 for a discussion of FISH) have shown that these organisms are numerically inferior but capable of rapid growth in the substrate-rich media that are used for most probable number counting (Giovannoni and Rappé, 2000).

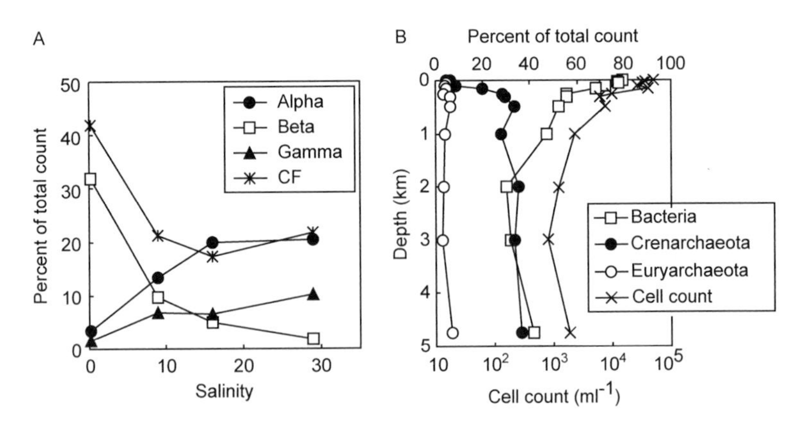

Figure 6.9 Distribution of major bacterial clades along environmental gradients as determined by fluorescent *in situ* hybridization (FISH). (A) Abundance of α-, β-, and γ-proteobacteria and members of the *Cytophaga-Flavobacterium* lineage (CF) relative to total cell counts in the waters of the Delaware Estuary, plotted as a function of salinity. Modified from Cottrell and Kirchman (2003). (B) Relative abundance of members of the *Bacteria*, *Crenarchaeota*, and *Euryarchaeota*, as well as total cell counts in the water column of the Pacific Ocean near Hawai'i. Modified from Karner *et al.* (2001).

Grazing may further contribute to their low abundance, since the cells are large relative to other bacterioplankton and therefore are more easily captured by nanoflagellates (Beardsley *et al.*, 2003).

Cultivated members of the *Cytophaga-Flavobacterium* cluster are specialists in the degradation of high-molecular-weight compounds such as cellulose and chitin; this may also be an important part of their niche in nature (Kirchman, 2002). Accordingly, these bacteria are particularly abundant in organic aggregates and in sediments, where macromolecules are the main substrate (Llobet-Brossa *et al.*, 1998; Nold and Zwart, 1998). Another common group in these environments is the planctomycetes, which may have a similar metabolic specialization. In contrast to these examples, virtually nothing is known about the physiology and ecology of the *Crenarchaeota*, which make up an estimated 20% of the prokaryotic cells in the world's oceans (Figure 6.9; Fuhrman *et al.*, 1992), except that they take up amino acids and fix CO_2 for lipid synthesis (Ouverney and Fuhrman, 2000; Wuchter *et al.*, 2003). Their abundance is very intriguing, since all cultivated non-thermophilic *Archaea* are strict anaerobes or extremophiles with respect to pH or salinity.

7.2. Controls of oxygen respiration in aquatic environments

7.2.1. Oxygen

As we saw previously, K_m values of terminal oxidases vary among species and among alternatives within each organism, so that the kinetics of oxygen respiration in natural systems depends both on the species composition and on which oxidases are expressed at a given time. The kinetics of oxygen consumption in natural communities is further affected by factors including diffusional limitations, which result in higher concentrations in the bulk water than those experienced by the microbes and, in some systems, by non-respiratory oxygen consumption (see below). Together with analytical difficulties in resolving the low oxygen concentrations, these factors will bias the apparent K_m toward higher values relative to the cellular K_m. Despite these limitations, natural microbial communities are typically only oxygen-limited at concentrations of a few micromolar (~1% of air saturation) or lower (Hao *et al.*, 1983; Ellis *et al.*, 1989). In sediments, this is evidenced by a linear decrease in oxygen concentration to below detection when oxygen supply is shut off (Figure 6.10). In oceanic waters, a diversion of the electron flow from oxygen respiration toward denitrification, which sets in when oxygen concentrations drop below ~3 μM, may indicate incipient oxygen limitation at this concentration (Codispoti *et al.*, 2001) (see also Chapter 7).

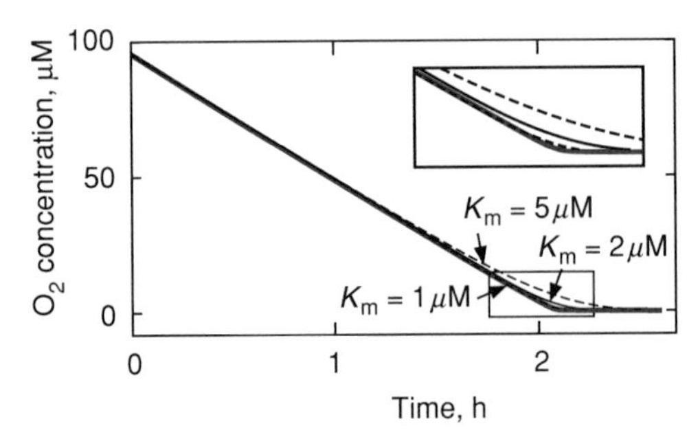

Figure 6.10 Oxygen consumption in a suspension of marine surface sediment from Aarhus Bay, Denmark (grey curve) compared to theoretical consumption curves (thin black curves) generated under the assumption of Michaelis-Menten kinetics with different K_m values for oxygen. For each curve, V_{max} was chosen to provide the best fit to the initial linear decrease in oxygen concentration. The measured values are almost hidden behind the theoretical curve for $K_m = 1\,\mu M$, indicating an apparent K_m close to this value. Organotrophic respiration was the main oxygen-consuming reaction. Oxygen was measured continuously. Oxygen measurements from Thamdrup *et al.* (1998).

7.2.2. Electron donors

Bulk rates of oxygen respiration in oxygen-replete aquatic environments range from $\sim 2\,\mu M\ y^{-1}$ in the ocean interior to $\sim 10\,\mu M\ s^{-1}$ in organic-rich sediments and photosynthetic microbial mats (Revsbech and Jørgensen, 1986; del Giorgio and Duarte, 2002). Microbes are generally the main contributors to this respiration, and the range reflects the difference in availability of energy substrate, which controls most of the variation in respiration and growth rates observed in aquatic systems. In pelagic waters, a main energy source for prokaryotes is dissolved organic carbon (DOC), which is released from phytoplankton and other members of the food web through excretion, grazing or lysis by virus (see Chapters 2 and 5). Particulate organic carbon (POC) in sinking aggregates appears to play a smaller role in open waters but may be the dominant substrate in aggregate-rich waters such as estuaries (Turley and Mackie, 1994; Lefèvre *et al.*, 1996; Ploug *et al.*, 2002). Next to organic carbon, the ammonium released during organic matter degradation is the most important electron donor for oxygen respiration in fully oxic systems. Nitrification is linked to organotrophy through the stoichiometric composition of organic matter. With phytoplankton-derived organics of the typical Redfield composition, nitrification accounts for 16% of the oxygen consumption (22 of 138 moles):

$$(CH_2O)_{106}(NH_3)_{16}H_3PO_4 + 138O_2 \rightarrow 106HCO_3^- + 16NO_3^- + H_2PO_4^- + 16H_2O + 123H^+ \quad (6.9)$$

In sediments, POC is also a major source of substrate for oxygen respiration, though a large fraction of the oxygen respiration in the vicinity of oxic–anoxic interfaces may be lithotrophic with sulfide, ferrous iron or ammonium as electron donors (see below).

7.2.3. Nutrients

In sediments and aphotic waters, nutrients are not expected to limit aerobic prokaryotes, but there are some examples of nutrient limitation of microbial growth in the photic zones of fresh waters and the oceans where nutrient levels are low due to photosynthesis (Thingstad, 2000). Phosphorus limitation of microbial growth occurs in lakes during summer stratification and may also occur in marine systems (Vadstein, 2000), and heterotrophic prokaryotes may be iron-limited in parts of the open ocean (Pakulski *et al.*, 1996). Competition for nutrients between heterotrophs and phytoplankton adds considerable complexity to the interpretation of food web structure and carbon cycling in low-nutrient environments (Kirchman, 2000; Thingstad, 2000).

7.2.4. Temperature

Oxygen respiration is affected by temperature in a manner similar to other microbial processes, with typical Q_{10} values for natural communities of 2 to 3 (Thamdrup and Fleischer, 1998; Thamdrup *et al.*, 1998). The influence of seasonal temperature variations on cellular respiration rates will be modulated, as yet quite unpredictably, by changes in community composition and size, as well as other mutually interacting factors (see Chapter 9 for a similar discussion concerning sulfate reduction). An interesting issue is whether heterotrophic bacterioplankton are more strongly inhibited than primary producers at temperatures near freezing, which would have implications for the effect of low temperature on carbon cycling (Pomeroy and Deibel, 1986). There are some examples of particularly slow microbial growth at very low temperatures, but this does not apparently much alter the biogeochemical cycling of elements in permanently cold environments relative to warmer locations (Nedwell, 1999; Pomeroy and Wiebe, 2001).

Oxygen microgradients and the laws of diffusion

When near particle surfaces, such as in sediment pore waters, oxygen—like other solutes—is transported by molecular diffusion. Diffusion results from the random movement of molecules and will therefore tend to even out concentration differences. Thus, the vertical flux $J(x)$ of oxygen across an interface at depth x in a sediment is proportional to the vertical concentration gradient as described by Fick's 1st law:

$$J(x) = \phi \cdot D_s \cdot \frac{\partial C}{\partial x}$$

where ϕ is the sediment porosity and D_s is the diffusion coefficient of oxygen in the sediment, which is related to the salinity- and temperature-dependent free solution diffusion coefficient D through the equation

$$D_s = D \cdot \theta^{-2}$$

Listings of free diffusion coefficients for many gases and solutes in water, as well as their temperature dependence, have been compiled by Boudreau (1997). The term θ is the tortuosity, which takes into account the fact that diffusion in sediment interstices cannot follow straight paths. The tortuosity can be determined from the porosity using the following empirical relationship (Boudreau, 1997):

$$\phi^2 = 1 - 2ln(\phi)$$

To derive net rates of oxygen consumption as a function of depth, $R(x)$, from the oxygen distribution we use Fick's 2nd law:

$$\frac{dC}{dt} = D_s \frac{\partial^2 C}{\partial x^2} + R(x)$$

At steady state, $dC/dt = 0$, and this law states that the net rate of consumption is proportional to the second partial derivative, the curvature, of the concentration profile. Hence, at steady state, a convex concentration profile indicates net production, a concave profile indicates net consumption, and a linear profile ($\partial^2 C/\partial x^2 = 0$) indicates that there is no net consumption or production. Detailed discussions of the mathematical description of diffusion and other transport processes in sediments are given by Berner (1980) and Boudreau (1997).

Solute transport in stagnant water columns is dominated by eddy diffusion, essentially the random movement of small water parcels. Net consumption rates of solutes can therefore be estimated from concentration profiles using the equations above with K_z, the vertical eddy diffusion coefficient, substituted for D_S. K_z is the same for all solutes, but depends on the stability of the water column, which in turn is influenced by density gradients and the kinetic energy of the water mass. Thus K_z varies between basins and with depth in each basin (e.g., Lewis and Landing, 1991), and rate estimates in systems with eddy diffusion will generally be less accurate than those from systems dominated by molecular diffusion.

7.3. Oxic–anoxic interfaces

The oxic world is limited by sharp boundaries where oxygen concentrations often decrease from over 100 μM to zero within a range of millimeters to centimeters, as in sediments, or decimeters to meters, as in water columns of stratified basins. Anoxic oases may also exist in sinking organic aggregates and in the cores of the pelagic oxygen minimum zones underlying productive regions of the oceans. The region surrounding the oxic–anoxic interface harbors a particularly large metabolic diversity, which, on the oxic side, includes lithotrophs that oxidize reduced inorganic compounds emanating from the anoxic side, such as ammonium, ferrous iron, sulfide or methane. Although some interfaces seem to be stable on longer time scales, the "gradient organisms" that thrive near the oxic–anoxic interface and try to position themselves for optimal concentrations of oxygen and substrates are often met by rapidly changing conditions. An obvious example is phototrophic microbial mats, where the interface migrates between the surface and some millimeters in depth in response to the light–dark cycle (see Chapter 13). In aphotic sediments, the location of the interface also fluctuates due to the burrowing and irrigating activity of the infauna (Figure 6.2). With typical oxygen consumption rates of $\sim$10 μM min^{-1} in coastal marine sediments (e.g., Glud *et al.*, 2003), changes in supply or consumption of oxygen will affect oxygen concentrations within minutes.

In extreme cases, the oxygen respiration near the oxic–anoxic interface can be coupled to a single lithotrophic reoxidation reaction, as in sediments covered by mats or veils of sulfide-oxidizing bacteria (see Chapter 9). In general, however, there are no direct ways to quantify the individual oxygen-consuming processes near oxic–anoxic interfaces. We can, however, quantify the importance of the different electron donors by comparing the total oxygen consumption rate to the oxidation rates of the individual inorganic donors. Organotrophic oxygen respiration is then determined by difference:

$$\text{organotrophic } O_2 \text{ respiration} = \text{total } O_2 \text{ consumption} - \sum_{\substack{NH_4^+, Mn^{2+},\\ Fe^{2+}, H_2S, CH_4, \ldots}} \text{inorganic reoxidation} \qquad (6.10)$$

where the summation designates the oxidation of inorganic compounds in oxygen equivalents (see Chapter 5).

One practical approach to separating the oxygen-consuming reactions at oxic–anoxic interfaces is quantification of the fluxes of oxygen and inorganic electron donors toward the interface. This approach is particularly useful in systems such as microbial mats and stagnant water columns, in which the electron donors are all soluble. The fluxes can then be inferred from concentration gradients, assuming that the solutes are transported by molecular diffusion, as in mats, or by the analogous eddy diffusion, as in water columns (see Box). An example of an oxygen budget from such an environment is presented in Figure 6.11. For further discussion of processes in stratified water columns, see the Chapter 13.

While the oxidation of ammonium and methane occurs exclusively through respiratory microbial processes, abiotic reactions potentially contribute to the oxidation of manganese, ferrous iron, hydrogen sulfide, and particulate sulfides (e.g., Equation 6.2). As discussed in the Chapters 8 and 9, the oxidation of manganese and sulfide with oxygen is mostly catalyzed by microbes in most aquatic environments, whereas there is no general understanding of the relative importance of abiotic and biological iron oxidation.

7.4. Oxygen consumption in sediments

Animal activity complicates the analysis of oxygen-consuming reactions in sediments. The bottom-dwellers affect the transport of oxygen and reduced solutes through burrowing and irrigation (Figure 6.2), and by particle

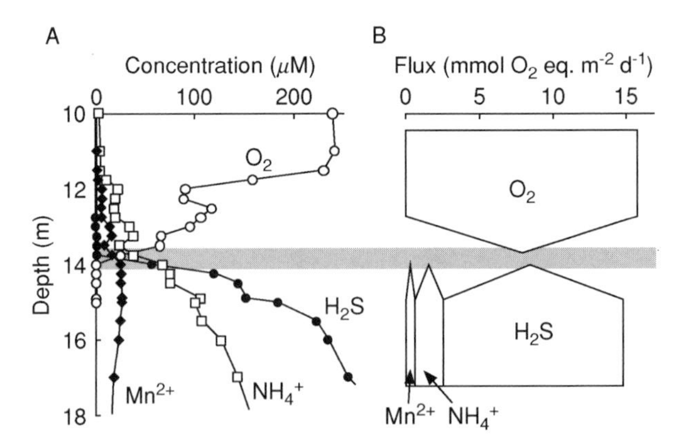

Figure 6.11 Oxygen consumption in the chemocline of Mariager Fjord, Denmark. (A) Depth distributions of oxygen and the major reduced inorganic species around the chemocline. (B) Rates, in oxygen equivalents, of oxygen consumption and the oxidation of soluble manganese(II), ammonium, and hydrogen sulfide as determined from the concentration gradients. Nitrate and manganese oxides were of minor importance as oxidants, and most of the oxygen consumption was coupled to hydrogen sulfide oxidation, predominantly by aerobic sulfide-oxidizing bacteria. Modified from Zopfi *et al.* (2001).

mixing they bring particulate compounds containing reduced iron or sulfur into contact with oxygen so that these substances may account for much of the lithotrophic or abiotic oxygen consumption (e.g., Thamdrup *et al.*, 1994b).

In most sediments underlying well-oxygenated bottom water, oxygen is the ultimate acceptor of almost all the electron equivalents released during the oxidation of organic matter. Thus, the benthic uptake of oxygen is equivalent to the total sediment metabolism. Even if most of the organic matter is oxidized anaerobically, almost all of the resulting reduced inorganic metabolites are reoxidized within the sediment, and only minor amounts escape through burial or to the bottom water (Figure 6.3) (Thamdrup and Canfield, 2000) (see also Chapter 5 and the following Chapters). We can therefore substitute the rate of anaerobic carbon mineralization minus the rate of reduced inorganic metabolites escaping reoxidation for the reoxidation term in Equation 10 with all terms expressed in oxygen equivalents:

$$\text{organotrophic } O_2 \text{ respiration} = \text{total } O_2 \text{ consumption} - \left(\sum_{\text{pathways}} \text{anaerobic C mineralization} - \sum_{\substack{N_2, Mn^{2+}, Fe^{2+}, \\ H_2S, CH_4, \ldots}} \text{efflux/burial} \right) \quad (6.11)$$

Because the summed loss term from efflux and burial is typically small, the partitioning of benthic oxygen consumption between organotrophy and reoxidation reactions is, to a first-order approximation, the same as the partitioning of carbon oxidation between aerobic and anaerobic processes (e.g., see Table 5.8 in Chapter 5), and the same factors govern these two ratios.

The mass balance in Equation 6.11 assumes a steady state between production of reduced inorganic metabolites and their reoxidation, which does not always apply. An extreme violation of this assumption is found in the mobile muds off the mouth of some tropical rivers, where reoxidation is largely restricted to discrete resuspension events that may affect the sediment to a depth of 1 m (Aller *et al.*, 1996). Also, sediments not affected by resuspension may build up an "oxygen debt" of reduced iron and sulfur species, which only later react with oxygen (e.g., Thamdrup *et al.*, 1994b). Unless reoxidation is focused to rare events that are not captured in the measurements of total oxygen consumption, such imbalances should average out in time series studies.

Except for the most rapidly accumulating sediments, most organic matter deposited to the sediment surface is mineralized during early diagenesis (see Chapter 5). Thus, sediment oxygen uptake rates largely reflect the organic depositional flux, ranging from ~200 mmol m^{-2} d^{-1} in shallow depositional areas to ~0.1 mmol m^{-2} d^{-1} in the marine abyss, with an exponential-like decrease with depth (see also Table 9.3 in Chapter 9; Figure 6.12). The decrease in oxygen flux is accompanied by an exponential increase in the depth of the oxic–anoxic interface from ~1 mm in shallow sediments to more than 1 m into abyssal sediments (see Figure 5.9; Figure 6.12). There is, therefore, a marked shift in the relative importance of organotrophic oxygen respiration with water depth in marine sediments.

At high oxygen penetration depths, essentially all oxygen consumption is by organotrophic respiration (and the associated nitrification), but as the oxic layer thins and anaerobic carbon mineralization increases in importance, more and more of the oxygen is diverted to the reoxidation of inorganic metabolites, and in many cases these processes totally dominate, as in some coastal sediments (Table 5.8). Between these extremes, there is not a good picture of the relative significance of organotrophic oxygen respiration. The largest uncertainty is associated with the contributions of dissimilatory manganese and iron reduction, which recent studies suggest to be of widespread importance (see Chapter 8). As discussed in Chapter 9, sulfate reduction contributes 55% of the global carbon mineralization of marine sediments, mainly through its importance in coastal and continental shelf sediments. The available data suggest that manganese and iron reduction are also of particular importance in continental shelf sediments, locations where most of the marine benthic carbon mineralization takes place. The average contribution of iron reduction to carbon mineralization of 17% (see Chapter 8) together with the 55% from sulfate reduction and approximately

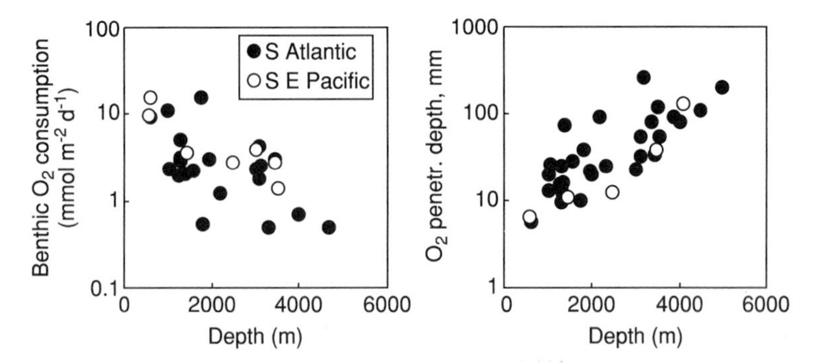

Figure 6.12 Compilation of benthic oxygen consumption rates and oxygen penetration depths as a function of water depth in the South Atlantic and Southeast Pacific Oceans. Note logarithmic scales. Redrawn from Wenzhöfer and Glud (2002).

5% from denitrification (Table 7.8 in Chapter 7) leaves only 23% of the benthic marine carbon mineralization to oxygen respiration. This number is a crude estimate that is likely to change as more information on the relative roles of the different anaerobic processes accumulates.

To determine the contribution from aerobic microbes, we need to further divide the organotrophic oxygen respiration (Equation 6.10) between micro- and macroorganisms. Again, it is difficult, if not impossible, to quantify faunal respiration rates directly within the sediment, but an estimate can be obtained from the biomass and the mass-specific respiration rates of the animals laboratory-determined (e.g., Piepenburg *et al.*, 1995). In such estimates for fine-grained fresh water and coastal marine sediments, the fauna typically account for 5 to 30% of the total oxygen consumption, indicating that a substantial fraction of the organotrophic oxygen respiration may be due to macroorganisms (Piepenburg *et al.*, 1995; Hansen and Kristensen, 1997; Hansen *et al.*, 1998; Glud *et al.*, 2000, 2003). This partitioning is, however, not a sign of direct competition for oxygen, because the animals can actively pump oxygenated water, while microbes rely on molecular diffusion from their surroundings. Indeed, the amount of oxygen that reaches the sediment through advection created by the fauna typically exceeds the requirements of the animals themselves (e.g., Glud *et al.*, 2003). In this respect, aerobic microbes may benefit from the irrigating activity. Conversely, the fauna may also stimulate anaerobic carbon mineralization by transporting labile organic matter to the anoxic parts of the sediment (e.g., Aller, 1990). It is difficult to generalize about the effect of fauna on specific microbial processes.

The partitioning of benthic oxygen consumption between microbial and macrofaunal organotrophic respiration and reoxidation of inorganic metabolites is shown in several examples in Figure 6.13 (the microbial contribution here also includes respiration by micro- and meiofauna). The large fraction of oxygen consumed by the reoxidation of inorganic metabolites raises the question of which processes are involved in these oxygen-consuming reactions. To put it shortly, we don't know! The relative importance of the anaerobic carbon oxidation pathways and the associated production of reduced inorganic metabolites can be determined (e.g., Table 5.8 in Chapter 5). However, an array of secondary redox processes occurring under anoxic conditions may transfer electron equivalents from the primary metabolites to other compounds before electrons are eventually delivered to oxygen. For example, hydrogen sulfide from sulfate reduction may be oxidized by iron oxides, generating ferrous iron, which in turn can be oxidized with manganese oxides, with reduced manganese finally reducing oxygen (see Chapter 8). We gain most insight to this web of processes through diagenetic transport/reaction models incorporating what we know about the kinetics of the many reactions. Application of such a model to

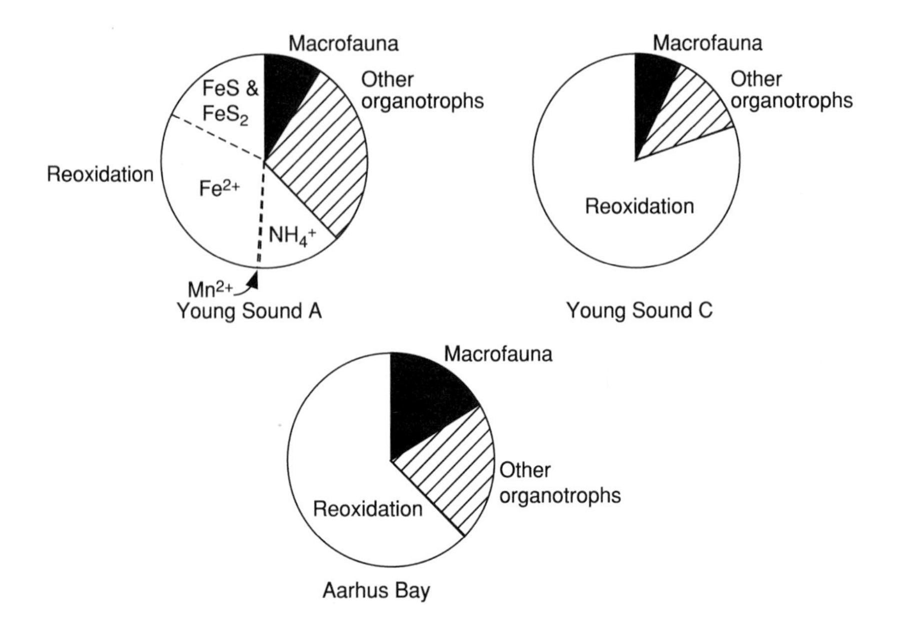

Figure 6.13 Partitioning of benthic oxygen consumption in three coastal marine sediments in Young Sound, Northeast Greenland, and Aarhus Bay, Denmark. Contributions from macrofaunal respiration are maximum estimates based on biomass. Reoxidation reactions include both microbially catalyzed oxygen consumption and abiotic oxygen consumption coupled to the reoxidation of inorganic metabolites. At Young Sound Station A the relative importance of various inorganic electron donors, ammonium, reduced manganese, and iron and iron sulfides was determined through numerical reaction-transport modeling. After Glud *et al.* (2000, 2003) and Berg *et al.* (2003).

sediment from Young Sound, Greenland, indicates that ferrous iron is the most important electron donor for oxygen, although at this site sulfate reduction is the most important anaerobic respiration process (Figure 6.13). In other sediments, reduced manganese may be most important in donating electrons to oxygen, while dissimilatory manganese reduction is unimportant in carbon oxidation (Aller, 1994b).

8. STABLE OXYGEN ISOTOPES AND MICROBIAL OXYGEN TRANSFORMATIONS

The stable isotopes of oxygen are ^{16}O, ^{17}O, and ^{18}O, with abundances of 99.63%, 0.038%, and 0.200%, respectively (Faure, 1986). Variations in the isotope abundances of atomic oxygen in water and minerals are mainly influenced by chemical and physical factors (Faure, 1986), but biological processes exert a major control on the isotopic composition of molecular oxygen. There is no isotope fractionation during oxygenic photosynthesis (Table 6.3). However, globally, the O_2 produced by land plants is enriched in ^{18}O by 4.4 ‰ relative to O_2 from marine photosynthesis (Farquhar *et al.*, 1993). This is a combined influence of the ^{18}O depletion in rain water, which is a strong function of latitude (Craig, 1961), and the $\delta^{18}O$ enrichment in chloroplast water relative to ground water. The major O_2-consuming processes, oxygen respiration, photorespiration, and the Mehler reaction (see above), all preferentially use the lighter $^{16}O^{16}O$ before $^{16}O^{18}O$, with fractionations of 15 to 25‰ (Table 6.3). As a result of these fractionations, atmospheric O_2 is isotopically heavier than sea water oxygen, with $\delta^{18}O_{O_2} - \delta^{18}O_{H_2O} = 23.5$‰, a fractionation known as the Dole effect (Dole *et al.*, 1954).

The $\delta^{18}O$ of O_2 dissolved in surface sea water varies annually from higher winter values in equilibrium with the atmosphere ($\delta^{18}O$ ~24‰) to lower summer values due to the addition of 0‰ oxygen from photosynthesis. These patterns hold potential information about marine primary production, which may be particularly useful in water with low rates. (Bender and Grande, 1987). Because O_2 respiration increases the $\delta^{18}O$ of O_2, counteracting the effect of photosynthesis, the approach requires very accurate estimates of the respiratory isotope effects. These isotope effects are difficult to establish, however, and as a result this approach has not been extensively applied (Kiddon *et al.*, 1993).

On the global scale, the differences in isotope fractionation patterns between marine and terrestrial systems potentially leave signatures in atmospheric O_2, reflecting the relative importance of primary production in the two realms. Somewhat surprisingly, the isotopic composition of O_2 in ice

Table 6.3 Isotope effects associated with oxygen metabolism[a]

Process	Isotope effect (‰)[b]
Photosynthesis	~0
Dark respiration	
Prokaryotes, microalgae, zooplankton	20 ± 3
Animals	5–10
Photorespiration	
Rubisco	20.8
Glycolate oxidase	22.2
Mehler reaction	15.1

[a]After Guy *et al.* (1993), Kiddon *et al.* (1993).
[b]$\varepsilon = 10^3(\alpha_{react\text{-}prod} - 1) \approx \delta^{18}O_{react} - \delta^{18}O_{prod}$, where $\alpha_{react\text{-}prod}$ is the fractionation factor.

cores shows an almost constant Dole effect over the past 130,000 years, indicating that the relative importance of marine and terrestrial production has not changed, although marine productivity has varied substantially during glacial–interglacial cycles (Bender *et al.*, 1994). There is, however, a large discrepancy between the present-day Dole effect (23.5‰), and the effect of 20.8‰ predicted from the quantification of the oxygen cycle and the associated isotope effects. This further substantiates that more knowledge is required for useful interpretations of oxygen isotope systematics.

Chapter 7

The Nitrogen Cycle

1. INTRODUCTION

Nitrogen is a key constituent of many important biomolecules, such as amino acids, nucleic acids, chlorophylls, amino sugars and their polymers, and it is essential to all living organisms. Nitrogen has the property of an

ADVANCES IN MARINE BIOLOGY VOL 48
0-12-026147-2
© 2005 Elsevier Inc.
All rights reserved

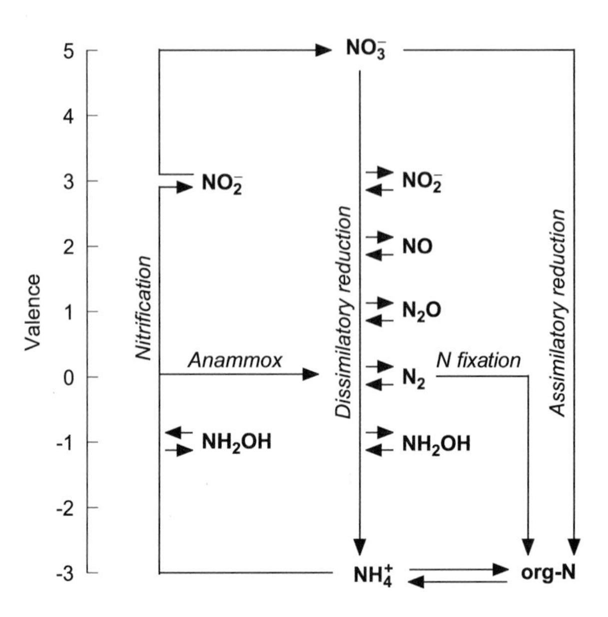

Figure 7.1 The microbial nitrogen cycle with an indication of the valence state of the various nitrogen-containing compounds involved.

eight-electron difference between its most oxidized and reduced compounds (Figure 7.1). Thus, the redox cycling between nitrogen compounds forms the basis for numerous microbial metabolisms. Many of these microbial processes, in turn, control the availability of nitrogen in the environment and hence are significant in regulating the activities of primary producing eukaryotes, which require a ready source of nitrogen for growth.

In a quick survey of the various microbial nitrogen-transforming processes, we start with N fixation. In this process, prokaryotes transform atmospheric dinitrogen (N_2) into ammonia (NH_3), a biologically available form that can be incorporated into biomolecules (Sprent and Sprent, 1990). The difficulty in N fixation is breaking the strong triple bond holding the two N atoms in N_2 together. Most organic nitrogen is recycled into inorganic form by a process known as ammonification in which nitrogen-containing biomolecules are degraded by microorganisms or digested by animals. The released ammonium (NH_4^+) can either be re-assimilated by microbes or plants and transformed into new biomolecules, or it can be oxidized by an assemblage of largely chemolithoautotrophic prokaryotes. The resulting oxidized nitrogen forms, nitrite (NO_2^-) and nitrate (NO_3^-), can either be assimilated by microorganisms and plants or be denitrified to N_2 by a number of heterotrophic prokaryotes using oxidized nitrogen as electron acceptor and organic carbon, for example, as electron donor. The resulting

N_2 is returned to the large atmospheric pool and thus, on short term, is lost for further biological transformations. Most bioavailable nitrogen is recycled several times between autotrophic and heterotrophic organisms, because the rates of nitrogen input to the biosphere by N fixation, and output by denitrification, are at least an order of magnitude slower than the internal cycling rates. This conventional view of the nitrogen cycle has recently been amended with the anammox process, in which ammonium oxidation is coupled to nitrite reduction, leading to the production of N_2. This process may be of significance in the nitrogen cycle of aquatic enviroments.

In this Chapter we explore how the interplay between microbial processes and the geochemical environment controls the cycling of nitrogen in aquatic environments. Thus, we look at how various microbial pathways promote the transformation of nitrogen compounds and how environmental factors regulate the operation and intensity of these pathways. In addition, we also explore how nitrogen isotopes are fractionated during microbial transformation. First, however, we consider aspects of the global nitrogen cycle and the influence of human activities.

2. THE GLOBAL NITROGEN CYCLE AND HUMAN PERTURBATIONS

The largest reservoir of nitrogen at the Earth's surface, including the crust and the atmosphere, is in igneous rocks, where nitrogen is primarily found as ammonium substituted within potassium-rich minerals (Table 7.1) (Krohn *et al.*, 1988). Rock weathering liberates this nitrogen, which then becomes available to living organisms. Nitrogen in sediments and sedimentary rocks is the next largest pool. Here, too, nitrogen is mostly fixed as ammonium in secondary silicate minerals (Blackburn, 1983), and this nitrogen source also becomes available during weathering. Of comparable size is the reservoir of atmospheric N_2, which accounts for 78% of the gas in the atmosphere. The

Table 7.1 Major nitrogen reservoirs on earth

Type of pool	Location	Pool size (g N)
N_2 gas	Atmosphere	3.8×10^{21}
Living biomass (LPON)	Aquatic and terrestrial	1.3×10^{16}
Dead organics (DPON + DON)	Aquatic and terrestrial	9.0×10^{17}
Inorganic (NH_4^+, NO_2^-, NO_3^-)	Aquatic and terrestrial	2.4×10^{17}
Inorganic (fixed NH_4^+)	Sediments and sedimentary rock	4.0×10^{21}
Inorganic (NH_4^+ within minerals)	Igneous rock	1.4×10^{22}

From Delwiche, 1970; Blackburn, 1983; Madigan *et al.*, 2002.

biologically available forms of fixed nitrogen mainly consist of dissolved NH_4^+, NO_2^-, and NO_3^- in aquatic and terrestrial environments, but this pool is small compared with the atmospheric reservoir (0.006%) and the reservoir comprising dead organic detritus (~25%) (Vitousek *et al.*, 1997). Living biomass is the smallest nitrogen reservoir, only about 1% of the size of the dead organic detrital pool.

Pool sizes tend to be inversely related to their biological importance. The large igneous and sedimentary pools, for example, are not actively cycled by organisms, although rock weathering can contribute a locally significant source of nitrate to surface and ground waters (Holloway *et al.*, 2001). The biological processes of N fixation and denitrification interact with the large atmospheric N_2 pool (Figure 7.2), but even here, the turnover time of the atmospheric N_2 pool is slow. The inorganic ions, NH_4^+, NO_2^-, and NO_3^-, are distributed in aqueous solution throughout the ecosphere, and they form small actively cycled reservoirs. Assimilation, mineralization,

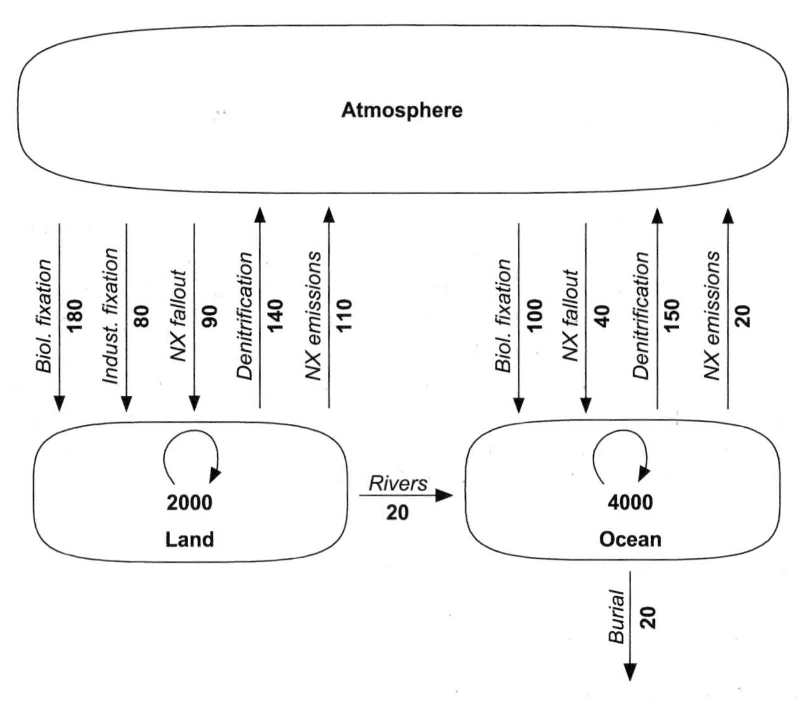

Figure 7.2 The global nitrogen cycle with an indication of the most important transfer processes. Rates are given as 10^{12} g N yr^{-1}. NX indicates inorganic combined nitrogen. Based on data from Roswall (1983); Codispoti and Christensen (1985); Seitzinger and Giblin (1996); Vitousek *et al.* (1997); Seitzinger and Kroeze (1998). Other recent estimates of marine dentrification are as high as 450 Tg Ny^{-1} indicating a large possible imbalance of the marine N budget (Codispoti *et al.*, 2001).

and nitrification (internal cycles on land and in the ocean as shown in Figure 7.2) are quantitatively the most important processes linking the inorganic reservoir with the likewise small and actively cycled reservoirs of living and dead organic nitrogen (Jaffe, 2000).

Biological nitrogen fixation on land amounts to about 180 $\times$ 10^{12} g yr^{-1}, which includes about 40 $\times$ 10^{12} g yr^{-1} from agriculturally managed legume crops (Figure 7.2). This rate exceeds rates of biological nitrogen fixation in the marine environment, with an estimate of about 100 $\times$ 10^{12} g yr^{-1} (Gruber and Sarmiento, 1997). Industrial nitrogen fixation into synthetic fertilizer from the Haber-Bosch process produces roughly 80 $\times$ 10^{12} g yr^{-1} of fixed nitrogen, similar in magnitude to marine biological N fixation. It is obvious that humans have heavily impacted the global nitrogen cycle. Another example of this impact comes from the cycle of nitric oxide (NO), which can be transformed in the atmosphere into nitric acid, a major component of acid rain (Vitousek *et al.*, 1997). Fossil fuel burning emits more than 20 $\times$ 10^{12} g N yr^{-1} as NO, while deforestation through burning contributes another 10 $\times$ 10^{12} g N yr^{-1} as NO. Furthermore, a substantial fraction of the total of 5 to 20 $\times$ 10^{12} g yr^{-1} of NO nitrogen emitted from soils is human related. Overall, 80% or more of NO emissions worldwide are generated by human activities.

To consider the ammonia cycle, nearly 70% of the global emissions to the atmosphere are human related. Ammonia volatilization from fertilized fields contributes an estimated 10 $\times$ 10^{12} g N yr^{-1}, release from domestic animal wastes liberates about 32 $\times$ 10^{12} g N yr^{-1}, and forest burning contributes a further 5 $\times$ 10^{12} g N yr^{-1}. Synthetic nitrogen fertilizer input has increased fourfold over the last four decades (about 20 $\times$ 10^{12} g N yr^{-1} in 1960) and is expected to rise from a current level of 80 $\times$ 10^{12} g N yr^{-1} to 134 $\times$ 10^{12} g N yr^{-1} in 2020 (Vitousek *et al.*, 1997). Additional non-biological sources of fixed nitrogen to the Earth's surface include volcanic outgasing and atmospheric fixation through ionizing radiation and electrical discharge.

3. BIOLOGICAL NITROGEN FIXATION

Only specialized prokaryotes contain the enzyme nitrogenase and the ability to fix N_2 into a biologically useful combined form (Sprent and Sprent, 1990). These organisms, known as diazotrophs, are found in both prokaryote domains. They may be found both free-living and in symbiotic association with plants. Nitrogen-fixing prokaryotes utilize a wide variety of different energy metabolisms, including oxygenic photosynthesis, sulfate reduction, methanogenesis, anoxygenic photosynthesis, and chemolithoautotrophy (Table 7.2).

Table 7.2 Selected list of free-living microbial genera, which contain N-fixing species or strains. The list is separated into different metabolic types of microorganisms with indication of their preferred habitat

Metabolism	Genus or type	Environment
Aerobes	*Azotobacter*	Sediment/water
Facultative anaerobes	*Klebsiella*	Sediment/water
(not fixing when aerobic)	*Paenibacillus*	Microbial mat/rhizosphere
	Enterobacter	Sediment/animal gut
	Escherichia	Animal gut
Microaerophiles	*Xanthobacter*	Microbial mat/sediment
(when fixing N_2)	*Thiobacillus*	Microbial mat/sediment
	Azospirillum	Rhizosphere
	Aquaspirillum	Water
Anaerobes	*Clostridium*	Sediment
	Desulfovibrio	Sediment
	Methanosarcina	Sediment
	Methanococcus	Sediment
Phototrophs	*Chromatium*	Microbial mat/sediment
	Chlorobium	Microbial mat/sediment
	Thiopedia	Microbial mat/sediment
	Rhodospirillum	Microbial mat/sediment
	Rhodopseudomonas	Sediment
	Oscillatoria	Microbial mat/water
	Nodularia	Water
	Anabaena	Water
	Nostoc	Microbial mat
	Calothrix	Microbial mat
	Gloeotheca	Microbial mat

Modified from Capone (1988) and Postgate (1998).

Combined, a major nutrient for plant primary production, is generally found in short supply in marine areas, where significant loss of occurs through denitrification (see Section 6). Nitrogen is also lost through burial as organic N and as adsorbed and mineral-bound ammonium in sediments (Figure 7.3). Nitrogen fixation is therefore needed to replace lost nitrogen. The overall process of N fixation is exergonic at standard conditions (Equation 7.1), but the great stability of the N≡N bond in N_2 makes it extremely unreactive at room temperature.

$$3H_2 + N_2 \rightarrow 2NH_3,\ \Delta G^0 = -33.4\ \text{kJ mol}^{-1} \qquad (7.1)$$

Indeed, in the industrial Haber-Bosch process this reaction will only occur when operated at temperatures between 300 and 400°C and at pressures

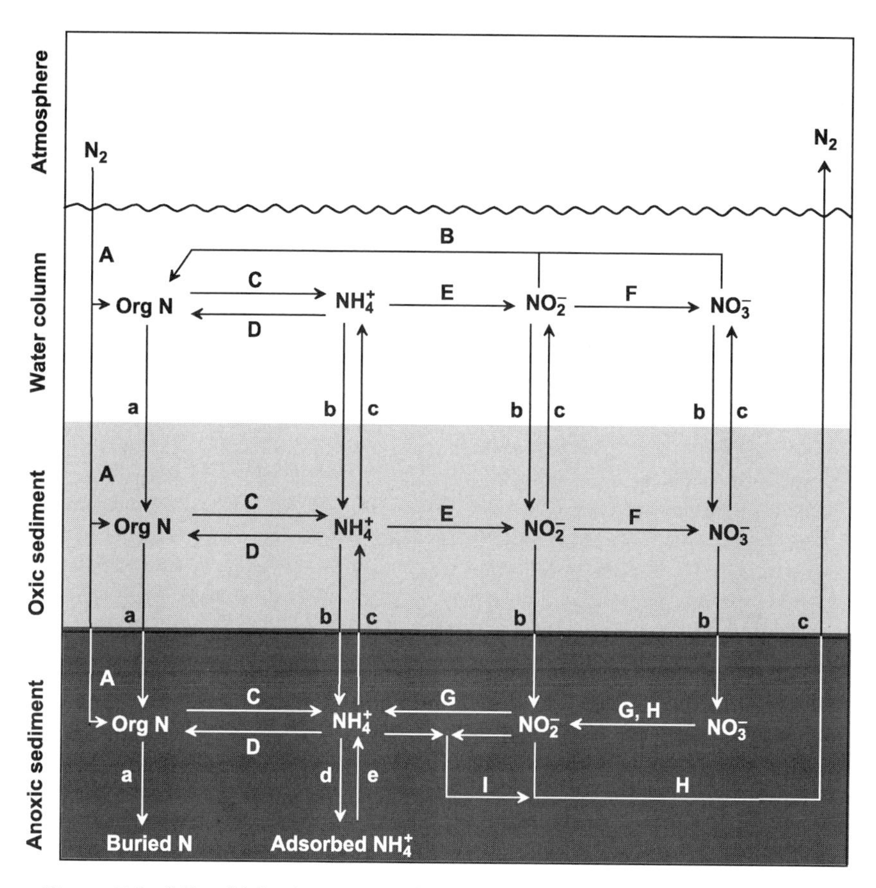

Figure 7.3 Microbial nitrogen cycling in aquatic environments showing the major transformations within (uppercase letters) and between (lowercase letters) anoxic sediment, oxic sediment, the water column, and the atmosphere. A, nitrogen fixation; B, NO_x assimilation; C, ammonification; D, NH_4^+ assimilation; E, NH_4^+ oxidation; F, NO_2^- oxidation; G, NO_3^- ammonification; H, denitrification; I, anammox; a, burial; b, downward diffusion; c, upward diffusion; d, NH_4^+ adsorption; e, NH_4^+ desorption.

between 35 and 100 MPa. Prokaryotes outperform the industrial process with the nitrogenase enzyme complex. Thus, nitrogenase reduces the triple-bonded N≡N molecule to NH_4^+ at normal environmental temperatures and pressures, but it is not completely N_2 specific, as other triple-bonded molecules such as acetylene (HC≡CH) and hydrogen cyanide (HC≡N) are also reduced.

Measuring N fixation

The development of the acetylene (C_2H_2) reduction assay (Stewart *et al.*, 1967; Hardy *et al.*, 1968) revolutionized the study of N fixation by providing a sensitive and simple procedure for determining nitrogenase activity. The procedure relies on the low specificity of nitrogenase for its natural substrate (N_2) and its capacity to reduce other triply bonded small molecules. Acetylene is reduced in preference to N_2 by nitrogenase, rapidly forming ethylene (C_2H_4):

$$HC \equiv CH + 2H^+ + 2e^- \rightarrow H_2C = CH_2$$

The C_2H_4 generated in the C_2H_2 reduction assay is quantified by gas chromatography.

There are, however, some potential concerns with this assay. Most importantly, the concentrations of C_2H_2 required in the assay may be inhibitory to a broad range of both diazotrophs and non-diazotrophs (Capone, 1988). Another concern relates to stoichiometry of the reaction. In principle, the two electrons transferred during the reduction of C_2H_2 to C_2H_4 are three times less than the six electrons required to reduce N_2 to NH_4^+. Thus, three moles of C_2H_2 reduced should be equivalent to one mole of N_2 fixed. Physiological, toxicological, and enzymological factors may result in deviations from this ideal stoichiometry. For N-fixation rates measured in aquatic sediments, the ratio of C_2H_2 reduction to N_2 reduction may range from 10:1 to 100:1 (Seitzinger and Garber, 1987). Ideally, the C_2H_2 reduction assay should be calibrated against $^{15}N_2$ uptake measurements before it can faithfully provide quantitative measurements of N fixation (Seitzinger and Garber, 1987).

3.1. The nitrogenase enzyme

Nitrogenase is large (up to 300 kDa) compared with many other enzymes, and relatively slowly reacting, with one enzyme taking 1.25 s to form two NH_4^+ molecules (Equation 7.2) (Postgate, 1998). The formation and maintenance of the nitrogenase enzyme complex require a major investment of protein (up to 30% of total cell protein; Haaker and Klugkist, 1987), energy (ATP), and trace metals (Mo and Fe). The trace metals incorporate into the active core of two different components associated with nitrogenase, a Mo-Fe protein known as dinitrogenase and an Fe protein known as dinitrogenase reductase. Additionally, some organisms have Mo-free dinitrogenases with cofactors containing V and Fe, or Fe only, which are structurally and functionally similar to the Mo-Fe dinitrogenase but appear to be less efficient (Eady, 1996). The Fe-only proteins may have been of particular significance through much of the Proterozoic Eon of Earth history (2.5 to 0.54 billion

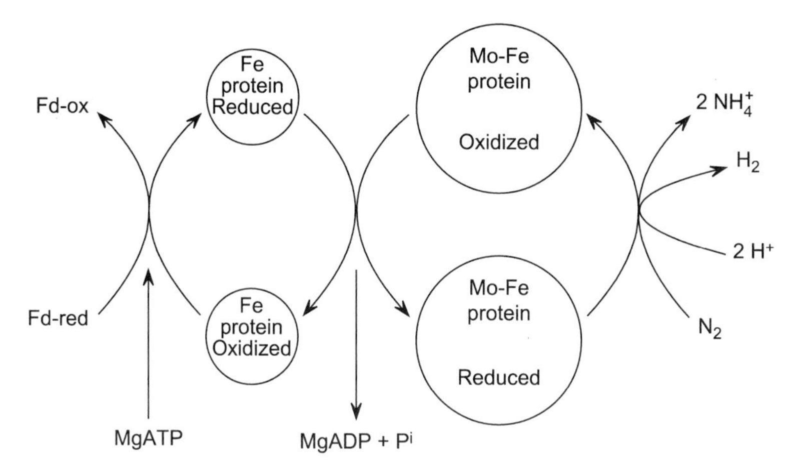

Figure 7.4 Schematic representation of the nitrogenase system, which catalyzes the reduction of molecular nitrogen to ammonia. The enzyme system contains dinitrogenase reductase (Fe protein) and dinitrogenase (Mo-Fe protein).

years ago), when seawater Mo concentrations were probably quite low (Anbar and Knoll, 2002).

The Fe protein consists of two identical subunits coded for by the *nifH* gene and contains approximately four Fe atoms and four labile S atoms. The Mo-Fe protein consists of four subunits, which are pairs of two different types: the α subunit coded for by the *nifD* gene and the β subunit coded for by the *nifK* gene. The Mo-Fe protein complex contains two Mo, 21–35 Fe and 18–24 labile S atoms per molecule (Postgate, 1998).

The reduction of N_2 to NH_4^+ proceeds as follows (Figure 7.4): a low redox potential molecule such as ferredoxin (at least -430 mV) donates an electron to the Fe protein, which enables it to react with MgATP. Meanwhile, the N_2 molecule to be reduced combines with the Mo-containing part of the Mo-Fe protein. The two proteins now join to form the active enzyme complex. Electrons flow singly from the Fe protein to the Mo-Fe protein. Thus, the two proteins have to meet and separate eight times to accommodate the six electrons transferred to reduce one N_2 molecule to two NH_4^+ molecules and the two electrons needed to reduce two protons to H_2 (see Figure 7.4). The formation of H_2 during the reduction of N_2 is an inherent property of the enzyme. Overall, the N fixation reaction, including its cost in ATP, may be written as the following:

$$N_2 + 9H^+ + 8e^- + 16ATP \rightarrow 2NH_4^+ + H_2 + 16ADP \qquad (7.2)$$

3.2. Ammonium assimilation

The ammonium formed during N fixation is incorporated into cellular material by one of two pathways (Gottschalk, 1986). Most nitrogen-fixing organisms produce glutamate as their initial product of NH_4^+ assimilation, and in one pathway glutamate is formed from the reductive amination of α-oxoglutarate by the enzyme glutamate dehydrogenase (GDH):

$$\mathrm{NAD(P)H} + \mathrm{NH_4^+} + \alpha\text{-oxoglutarate} \rightarrow \mathrm{NAD(P)^+} + \text{glutamate} + \mathrm{H_2O} \quad (7.3)$$

GDH has only a moderate affinity for NH_4^+, and since NH_4^+ generally represses nitrogenase activity, a rapid assimilation is imperative. Thus, an alternative two-step high-affinity assimilation pathway that maintains a low NH_4^+ concentration is sometimes used (Figure 7.5). Initially, NH_4^+ is added to glutamate by an ATP-requiring step with the enzyme glutamine synthetase (GS):

$$\text{glutamate} + \mathrm{NH_4^+} + \mathrm{ATP} \rightarrow \text{glutamine} + \mathrm{ADP} \quad (7.4)$$

Glutamine is transformed back into glutamate by a second reaction, catalyzed by the enzyme glutamate synthase or GOGAT (glutamine-α-oxoglutarate-amino-transferase):

$$\text{glutamine} + \alpha\text{-oxoglutarate} + \mathrm{NADPH} \rightarrow 2\ \text{glutamate} + \mathrm{NADP} \quad (7.5)$$

The cost of the GS-GOGAT pathway is one ATP per glutamate formed, but the benefit to the organism is rapid assimilation of NH_4^+ from low concentrations in the environment, preventing suppression of nitrogenase activity.

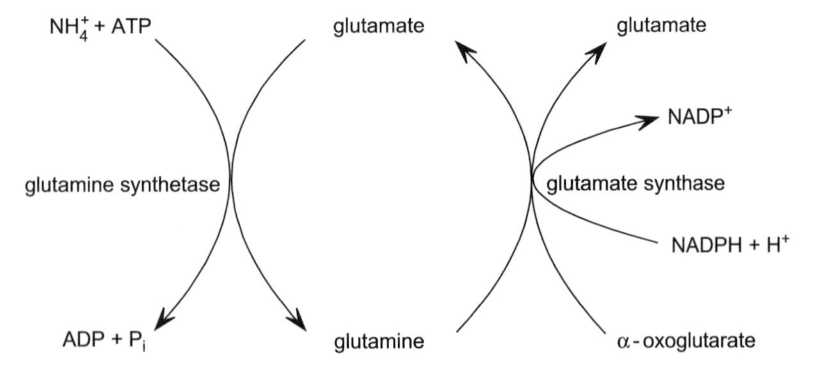

Figure 7.5 Assimilation of ammonia by the glutamine synthetase/glutamate synthase (GS/GOGAT) enzyme complex. Modified from Gottschalk (1986).

3.3. The oxygen problem

The biochemical properties of nitrogenase present the organism with a number of physiological problems. Most important is the O_2 sensitivity of the two main proteins. Thus, both the Mo-Fe protein and the Fe protein are irreversibly inactivated by O_2. All diazotrophs are therefore obliged to protect nitrogenase from exposure to O_2. Numerous strategies have been adopted to avoid O_2, including (1) life without O_2, (2) high rates of O_2 respiration, (3) conformational protection, and (4) heterocyst utilization (Postgate, 1998; Madigan *et al.*, 2003). Thus, obligate anaerobes such as *Clostridium pasteurianum* and *Desulfovibrio desulfuricans* can fix nitrogen whenever their normal metabolism is active. Facultative anaerobes such as *Klebsiella* spp. and *Enterobacter* spp. are only capable of diazotrophic growth under anoxic conditions and must rely on other nitrogen sources in the presence of O_2.

Aerobic (e.g., *Azotobacter* spp.) and microaerophilic (e.g., *Azospirillum* spp.) diazotrophs can fix nitrogen only when their respiration decreases the O_2 concentration near the cell to low enough levels to protect the functioning enzyme. Therefore, a characteristic feature of aerobic diazotrophs is exceptionally high O_2 respiration rates. As an example, *Azotobacter* spp. have cell-specific O_2 uptake rates several times higher than that needed for normal catabolism (Poole and Hill, 1997). These high respiration rates, however, reduce O_2 levels to the point where nitrogenase can fix N_2 (see also Chapter 6, Section 5.2). Some *Azotobacter* species are also able to maintain undamaged nitrogenase under oxic conditions by conformational protection. In this case, nitrogenase proteins undergo a reversible conformational change to a state that is unaffected by O_2. In this way, the enzyme becomes inoperative but remains undamaged, and can readily resume nitrogen fixation as soon as O_2 disappears.

Some phototrophic cyanobacteria, which live in oxic environments and generate O_2 in the light, have solved the problem of nitrogenase inactivation by separating photosynthesis and nitrogen fixation in space and time. Cyanobacteria such as *Anabaena cylindrica* have specialized thick-walled cells called heterocysts at regular intervals along their filaments. Nitrogenase activity is normally restricted to the heterocyst, which lacks an oxygen-evolving apparatus. The heterocyst walls allow N_2 to diffuse into the cell but limit O_2 diffusion to the point where any oxygen entering the heterocyst can be scavenged by respiration (Walsby, 1985). Some non-heterocystous cyanobacteria conduct nitrogen fixation at night, when O_2 generation cannot occur. The unicellular cyanobacteria *Gloeotheca* spp. can fix nitrogen in the light and dark, but when grown in alternating light–dark cycles it fixes 20 times more nitrogen in the dark than in light (Postgate, 1998). In this type of organism, and in other non-heterocystous cyanobacteria, O_2 is removed to low concentrations by rapid respiration in cell colonies.

Members of the cyanobacterial genus *Trichodesmium* are the most important nitrogen fixers in the ocean. *Trichodesmium* spp. are filamentous and non-heterocystous with individual cells organized along trichomes, which frequently bundle into large aggregates. Nitrogenase is found only in a fraction of the cells, yet the oxygen-evolving apparatus is found in all cells, and at any given moment during the day, when N fixation occurs, a significant fraction of the nitrogenase is inhibited by oxygen (Berman-Frank *et al.*, 2003). To accomplish nitrogen fixation, *Trichodesmium* spp. temporally separates nitrogen fixation and oxygen production (Berman-Frank *et al.*, 2001b). Nitrogen fixation is concentrated at midday, when there is a lull in oxygen production. Also at midday there is an increase in light-dependent oxygen consumption by the Mehler reaction (see Chapter 6). The combination of low oxygen production and high rates of oxygen consumption around noon reduce the oxygen concentrations in the cell and allow for nitrogenase activity. The rather contradictory occurrence of daytime N fixation by *Trichodesmium* spp. occurs because of the careful temporal regulation of N fixation and oxygen production and the high rates of oxygen consumption by the light-dependent Mehler reaction.

3.4. Phylogeny of nitrogen-fixing organisms

Nitrogen fixation is found among members of the *Bacteria* and the *Euryarchaeota* kingdom of the *Archaea* (Young, 1992). Within these groups, nitrogenase activity has been reported, or inferred, in more than 100 genera distributed over many of the major phylogenetic divisions. Such a wide distribution suggests that the process has a very ancient origin (see also - Berman-Frank *et al.*, 2003). Phylogenies have been constructed from 16S rRNA and from several of the different *nif* genes associated with nitrogenase. In this section we discuss phylogenies from the *nifH* gene, a principal gene in the N fixation pathway, and compare these with 16S rRNA phylogenies.

The *nifH* phylogenetic tree organizes into several clusters (Figure 7.6), including the cyanobacteria, the actinomycetes, the α-proteobacteria, and the β and γ-proteobacteria. All of these are aerobes. There are also two clusters of anaerobes, one including members of the *Bacteria* and the other including members of the *Archaea*. In addition, the aerobic gram-positive genus *Paenibacillus* clusters with the cyanobacteria and is well separated from the anaerobic members of gram positives represented by the genus *Clostridium*. In general, the *nifH* phylogeny is concordant with the 16S rRNA phylogeny. This supports the thesis that, for the most part, nitrogenase genes have descended vertically from an ancient progenitor (Chien and Zinder, 1994, 1996). Therefore *nifH* evolution, as well as the evolution of

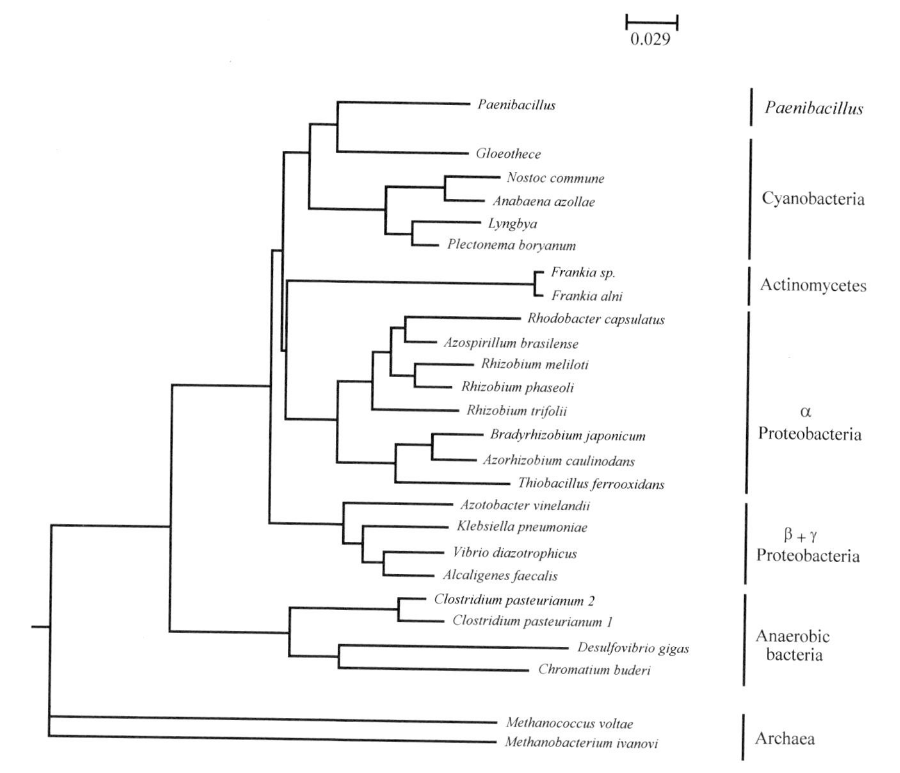

Figure 7.6 Phylogeny constructed for amino acid sequences from a fragment of the *nifH* gene, analyzed by the neighbor-joining method. Scale bar indicates 2.9% sequence divergence. Modified from Achouak *et al.* (1999).

other *nif* genes (Fani *et al.*, 2000), is due mostly to the constrained evolution of an ancient gene rather than lateral gene transfer. However, the placement of *Paenibacillus* in the *nifH* phylogeny is inconsistent with its gram-positive location in the 16S rRNA phylogeny and points to the possibility of lateral gene transfer of this gene.

There are some other unusual aspects of *nifH* gene phylogeny. Some *Bacteria*, such as *Rhodobacter capsulatus, Desulfobacter curvatus*, and *Clostridium pasteurianum*, contain multiple copies of the *nifH* gene. In each of these cases, one *nifH* copy clusters with its natural neighbors based on 16S rRNA, and the other clusters elsewhere. For example, one of the *nifH* genes for both *D. curvatus* and *C. pasteurianum* clusters with the methanogens (Braun *et al.*, 1999). Multiple *nifH* gene copies may have arisen during an ancient gene duplication event, as is apparently the case for other *nif* genes (Fani *et al.*, 2000). If so, it is unclear why multiple *nifH* genes from the same organism followed such different phylogenic trajectories. Also unclear is whether the multiple *nifH* genes actually accomplish the same function.

3.5. Nitrogen fixation in aquatic environments

Cyanobacteria are responsible for most planktonic nitrogen fixation in open marine and lake waters, but rates are high only when cyanobacteria dominate the planktonic biomass, which occurs frequently in eutrophic lakes and estuaries (Table 7.3). Nitrogen-fixing cyanobacteria need not be free living. In wetlands, for example, cyanobacteria such as *Anabaena* spp. live in a symbiotic relationship with the floating water fern *Azolla* spp., and in this relationship nitrogen fixation rates are greatly enhanced. In oligotrophic and mesotrophic lakes as well as the open ocean, rates of pelagic nitrogen fixation tend to be very low. Thus, the role of nitrogen fixation in supplying N to primary producers in open water ecosystems is quite variable and usually low (generally <1% of total nitrogen input). Nitrogen fixation, however, may account for as much as 80% of the N inputs to intertidal and eutrophic systems (Howarth *et al.*, 1988b).

In freshwater lakes of moderate productivity, nitrogen limitation rarely occurs since generally abundant heterocystous cyanobacteria fix nitrogen and

Table 7.3 Rates of nitrogen fixation by planktonic and benthic organisms in a variety of marine and freshwater environments (μmol N m^{-2} d^{-1}). The estimated contribution of nitrogen fixation (pelagic and benthic) to the total nitrogen demand by primary producers in the systems is presented

	Plankton	Benthos	Contribution
Open oceans[a]	0.4–18	~0	0–0.5%
North Atlantic/Pacific[b]	137–197	~0	20–75%
Estuaries	3–352	2–305	0.5–3%
Saltmarsh	—	50–1507	73–82%
Mangrove forest	—	96–400	3–13%
Seagrass bed	—	164–392	3–28%
Oligotrophic lake	~0		~0%
Mesotrophic lake	3–18	0.8–55	0–0.1%
Eutrophic lake	39–861		0.5–7%
Rice paddy	156–3000	117–431	10–80%

[a]Based on a global compilation of nonbloom *Trichodesmium* abundance and average per trichome nitrogen fixation rate (Capone and Carpenter, 1982).

[b]Based on linear combination of nitrate and phosphate distributions along isopycnal surfaces and water age in the North Atlantic (Gruber and Sarmiento, 1997), or on the nitrogen isotope ratio ($^{15}N/^{14}N$) expressed as $\delta^{15}N$(‰) of sinking particles in the North Pacific (Karl *et al.*, 1997).

From Capone (1988), Howarth *et al.* (1988), Roger and Ladha (1992), and Gruber and Sarmiento, 1997).

alleviate nitrogen limitation. By contrast, current N budgets for the marine realm suggest that N-fixing organisms are not active enough to balance N losses by denitrification, and hence the oceans may be losing fixed nitrogen (Figure 7.2). However, past estimates of oceanic nitrogen fixation rates may have been low by an order of magnitude or more (Gruber and Sarmiento, 1997; Karl *et al.*, 1997). At the same time, global estimates of denitrification are rising, so the state of balance of the present-day marine nitrogen budget is still in doubt (Codispoti *et al.*, 2001).

Rates of nitrogen fixation are low to moderate in the sediments of most lakes and estuaries, but may be high in particularly organic-rich estuarine sediments (Table 7.3). In mesotrophic and eutrophic lakes as well as estuarine sediments, most benthic nitrogen fixation is mediated by heterotrophic and chemolithoautotrophic prokaryotes in the absence of light. However, benthic nitrogen fixation in oligotrophic lakes and shallow coastal systems is often dominated by cyanobacteria. Benthic nitrogen fixers are generally unimportant for the nitrogen budget of mesotrophic and eutrophic lakes and estuaries, but may be an important nitrogen source in nutrient-poor tropical marine lagoons and oligotrophic lakes. Nitrogen fixation in wetlands (e.g., salt marshes and mangrove forests) and seagrass beds (e.g., *Zostera* sp. and *Thalassia* sp.) may be several-fold higher than in comparable unvegetated sediments (McGlathery *et al.*, 1998; Nielsen *et al.*, 2001). This is due to stimulated nitrogen fixation in the rhizosphere fueled by labile root exudates derived from plant photosynthesis. However, the contribution of newly fixed nitrogen in these very productive vegetated ecosystems is, in most cases, small relative to the total nitrogen input.

Nitrogen fixation, therefore, appears to be important in making up deficits in nitrogen availability, relative to phosphorus, in many lakes and possibly the open ocean, whereas many estuaries and coastal seas are nitrogen limited due to the relatively low rates of nitrogen fixation found in these productive systems.

4. MICROBIAL AMMONIFICATION AND NITROGEN ASSIMILATION

4.1. Ammonification

Nitrogen in living and dead organic matter occurs predominantly in the reduced amino form, principally in proteins and nucleotides (e.g., DNA, RNA, and ATP). On average, prokaryote cells contain about 55% proteins and 23% polynucleotides, on a dry weight basis (Madigan *et al.*, 2003). When these compounds are hydrolyzed and catabolized by heterotrophic organisms, nitrogen is ultimately liberated in the form of NH_4^+ (ammonification, also known as nitrogen mineralization), which can be assimilated and incorporated back into biomolecules by a variety of aerobic and anaerobic organisms. Alternatively, the ammonium can be oxidized by nitrifying microorganisms under oxic conditions (Figure 7.3) (Jaffe, 2000). As with carbon (see Chapter 5), the rate-limiting step in nitrogen mineralization is the extracellular hydrolysis of macromolecules into smaller units. The initial microbial hydrolysis of proteins and polynucleotides yields oligopeptides, amino acids, oligonucleotides, and nucleotides, all having C:N ratios below 6. Subsequently, intracellular fermentative and respiratory processes deaminate these small dissolved molecules, resulting in the release of NH_4^+.

Ammonification occurs in both oxic and anoxic aquatic environments (Herbert, 1999). Oxic environments include most lakes, swamps, rivers, and marine water bodies as well as their underlying near-surface sediments. Anoxic environments include euxinic and stratified water bodies and subsurface sediments. The partitioning between aerobic and anaerobic ammonification in aquatic environments depends, to a large measure, on the depth of the oxic water column. This controls the time elapsed before organic particles introduced into the photic zone reach anoxic environments, either by sinking into anoxic water masses or by burial into anoxic sediment. In shallow coastal environments overlying intensively bioturbated sediments, about 50% of the ammonification occurs in the oxic water column, and the remainder occurs by both aerobic and anaerobic processes within the sediment. The balance between aerobic and anaerobic ammonification varies, but 50% or more anaerobic ammonification is found in some cases. In the deep sea, more than 99% of the organic nitrogen in sinking particles usually is mineralized in the oxic water column (Suess, 1980).

Ammonification always occurs in conjunction with heterotrophic carbon mineralization. Because nitrogen-containing polymers are generally degraded more easily than carbon-containing structural cell components such as cellulose and lignin, nitrogen is mineralized preferentially to carbon. As a consequence, the average C:N ratio of decomposing organic matter increases gradually during microbial decomposition (Blackburn and Henriksen, 1983). Such preferential removal of nitrogen during the early aerobic stages of decomposition provides less degradable and nitrogen-poor organic matter for later anaerobic decomposition (Figure 7.7). The increase in C:N ratio may occur over short depth scales in sediments, compared to the water column, because sediment accretion rates are slow compared to the sinking rates of particles in water.

4.2. Deamination and ammonium incorporation

The first step in the ammonification of amino acids is enzymatic deamination, in which amino groups are released and transformed into ammonium. This is accomplished by a diverse range of aquatic microorganisms, including

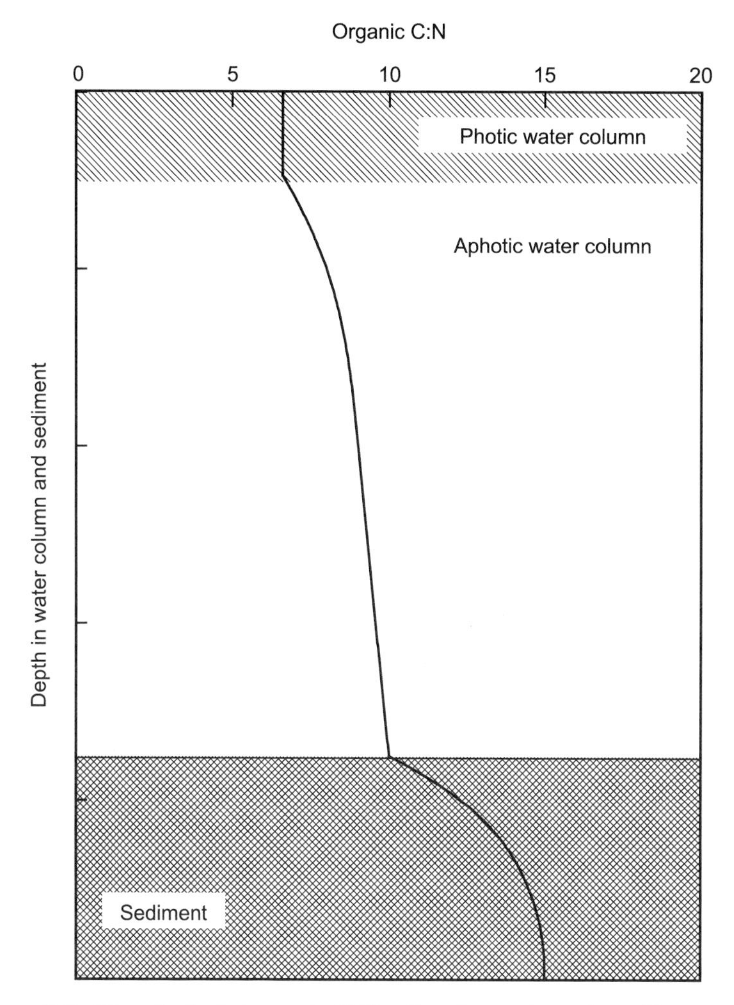

Figure 7.7 Hypothetical change in C:N ratio of organic matter with progressive organic matter mineralization (expressed here as depth) in aquatic environments. Organic matter with a C:N molar ratio of about 6.6 (Redfield, 1958) is produced by phytoplankton in the photic zone of the water column. Organic matter sinking below the photic zone becomes nitrogen deplete as nitrogen-rich compounds such as proteins and nucleotides are degraded preferentially to structural carbon-rich compounds. The rate of nitrogen stripping decreases gradually with depth, in concert with the decreasing content of labile and nitrogen-rich compounds. Particles reaching the sediment surface typically have a C:N ratio of 8–10 (Richardson, 1996). The preferential nitrogen mineralization continues within the sediment, but at a much lower rate. The C:N ratio of organic matter increases to a level of up to 15 within the upper tens of centimeters of sediment.

members of the genera *Pseudomonas, Vibrio, Proteus, Serratia, Bacillus*, and *Clostridium*, as well as many actinomycetes and fungi (Herbert, 1999). Amino acid deamination primarily occurs as an oxidative process catalyzed by NAD-linked dehydrogenases (Gottschalk, 1986):

$$\mathrm{R{-}CH(NH_2)COOH} + \mathrm{NAD^+} + \mathrm{H_2O} \longrightarrow \mathrm{R{-}C({=}O)COOH} + \mathrm{NH_4^+} + \mathrm{NADH} \tag{7.6}$$

Alternatively, oxidative deamination in respiring organisms can occur as a reaction with oxygen. Deamination of amino acids can also occur as a non-oxidative process catalyzed by dehydratases and is associated with the removal of water.

The hydrolysis and deamination of nucleotides are more complex due to the heterocyclic structure of the molecules. The degradation of purines and pyrimidines is not well studied, but the process is assumed to occur via either an oxidative pathway, where both urea and NH_4^+ are released, or via a reductive pathway, where only NH_4^+ (and probably amino acids) is produced as shown below for cytosine (Vogels and van der Drift, 1976; Gottschalk, 1986; Therkildsen *et al.*, 1996):

$$\mathrm{NH_3} + \mathrm{{}_2HN{-}C({=}O){-}NH_2} \longleftarrow \text{cytosine} \longrightarrow \mathrm{NH_3}\ (+\text{alanine}) \tag{7.7}$$

Urea formed by the oxidative pathway is further mineralized to CO_2 and NH_4^+ by the action of the enzyme urease according to the following:

$$\mathrm{{}_2HN{-}C({=}O){-}NH_2} + 2\mathrm{H_2O} + 2\mathrm{H^+} \longrightarrow \mathrm{CO_2} + 2\mathrm{NH_4^+} \tag{7.8}$$

The NH_4^+ ions released during mineralization can be assimilated into amino acids, nucleotides, and other nitrogen-containing biomolecules by numerous plants and microorganisms. The biochemistry of NH_4^+ incorporation into organic matter is similar to what was previously described for nitrogen fixers. In brief, when the concentration of NH_4^+ in the environment is high it is assimilated by reductive amination of α-oxoglutarate, forming glutamate by the action of the enzyme glutamate dehydrogenase (GDH) (Gottschalk, 1986). It is clear that GDH, with a K_m for NH_4^+ of ~0.1 M, cannot be involved in the assimilation of NH_4^+ in most aquatic environments, where the concentration of NH_4^+ usually is low ($\ll$1 mM). Under such conditions, a combination of glutamine synthetase and glutamate synthase is responsible for glutamate formation from α-oxoglutarate and

NH_4^+. In this pathway, glutamate synthase (GOGAT) catalyzes the reductive transfer of the amino group of glutamine to α-oxoglutarate. Glutamate is always the initial amino acid formed in the assimilation process. From this compound, the amino group can be transferred through transamination to other α-oxoacids, thus forming their corresponding amino acids.

Measuring ammonification and nitrogen incorporation

Temporal changes in the NH_4^+ pool reflect net ammonification or net nitrogen incorporation. However, when more precision is required, or when gross rates of NH_4^+ turnover are desired, $^{15}NH_4^+$ methods can be used (Blackburn, 1979a). Thus, by assaying the change in NH_4^+ pool sizes and ^{15}N dilution through time after adding $^{15}NH_4^+$ to the NH_4^+ pool, gross rates of ammonification (N_d) and nitrogen incorporation (N_i) can be quantified with model calculations. Dilution of $^{15}NH_4^+$ during incubation indicates that the label has been incorporated into new biomass or has been diluted by ammonification. Net ammonification ($N_o = N_d - N_i$) can be calculated from the temporal change in NH_4^+ concentration (C_t):

$$C_t = N_o t + C_0$$

Based on the initial NH_4^+ concentration (C_0) and labelling percentage ($\%^{15}N_0$), and the NH_4^+ concentration (C_t) and labelling ($\%^{15}N_t$) at time t, the following relationship can be established:

$$\ln(\%^{15}N_t) = \ln(\%^{15}N_0) - N_d/N_o \ln(C_t/C_0)$$

A plot of $\ln(C_t/C_0)$ against $\ln(\%^{15}N_t)$ provides a slope of $-N_d/N_o$, from which the rate, N_d, is easily calculated by the use of the measured N_o. N_i can then be calculated from ($N_i = N_d - N_o$; see above). When the method is applied to oxic environments the incorporation term, N_i, also includes $^{15}NH_4^+$, which is lost by oxidation.

One major problem arises when this method is used in sediments. A variable amount of $^{15}NH_4^+$ disappears immediately into the sediment matrix, and cannot be extracted with KCl or other salts, which are commonly used to liberate adsorbed NH_4^+ from sediment particles.

4.3. Nitrogen mobilization and immobilization

The free NH_4^+ pool is generally small, but dynamic, in aquatic environments with high production and consumption rates. Microbial mineralization of organic nitrogen occurs in conjunction with microbial incorporation (assimilation) of soluble inorganic nitrogen. The balance between these two processes (mineralization and incorporation) is determined by a variety of factors, including the ease of hydrolysis, the chemical composition of the organic compounds, the growth rates of the organisms, and environmental factors such as redox conditions. Fresh protein- and polynucleotide-rich organic materials derived from phytoplankton (bulk C:N $\sim$ 7) and animal tissues (C:N $\sim$ 5) are readily hydrolyzed and mineralized to CO_2 and NH_4^+, providing an excess of dissolved inorganic nitrogen. A large proportion of this liberated NH_4^+ is mobilized into the environment. By contrast, during the initial microbial hydrolysis and metabolism of structural carbohydrates such as cellulose-rich tissues from vascular plants poor in nitrogen (C:N > 20) there is a demand for incorporation of dissolved inorganic nitrogen into cell biomass. Little of the nitrogen may be mobilized. In later stages of decomposition and population growth, when most of the available carbon is oxidized and the microbial population declines, the mineralization of nitrogen-rich bacterial biomass can provide excess nitrogen to the environment (Figure 7.8).

The intimate association between organic carbon and nitrogen and the balance between immobilization and mobilization of nitrogen during microbial mineralization can be visualized with a C:N ratio model (modified from Blackburn, 1979b). This non-steady-state model considers the relationship between the C:N ratio of the organic substrate used by heterotrophic microbial populations ($C{:}N_S$), the C:N ratio of the growing microbial cells ($C{:}N_B$), and the efficiency of carbon incorporation (E_C) into new cells (Figure 7.9). Organic carbon is degraded at a rate of $C_d = C_o + C_i$, where C_o is the rate of oxidation to CO_2 and C_i is the rate of incorporation into microbial cells. Similarly, organic nitrogen is degraded at a rate of $N_d = N_o + N_i$, where N_o is the rate of net mineralization and N_i is the rate of incorporation. Since the nitrogen transformations are difficult to measure, they can be deduced from carbon transformations as follows: $E_C = C_i/C_d$, $N_d = C_d/C{:}N_S$ and $N_i = E_C C_i/C{:}N_B$. When all nitrogen assimilated by microbes is in the form of NH_4^+, then the C:N ratio of substrates ($C{:}N_S$) and the mobilization/immobilization of nitrogen, $MI_N = N_i/N_d$, are related according to the following equation:

$$MI_N = E_C C{:}N_S/C{:}N_B \qquad (7.9)$$

Mobilization occurs when $MI_N < 1$ and immobilization occurs when $MI_N > 1$. If it is assumed that E_C is 0.5 for aerobes and 0.2 for anaerobes (Pedersen *et al.*, 1999), and that $C{:}N_B$ is 5 (Goldman *et al.*, 1987), then nitrogen mobilization will occur only when $C{:}N_S$ is below 10 under oxic conditions and below 25 under anoxic conditions (Figure 7.10). These predictions are in close agreement with observations from laboratory

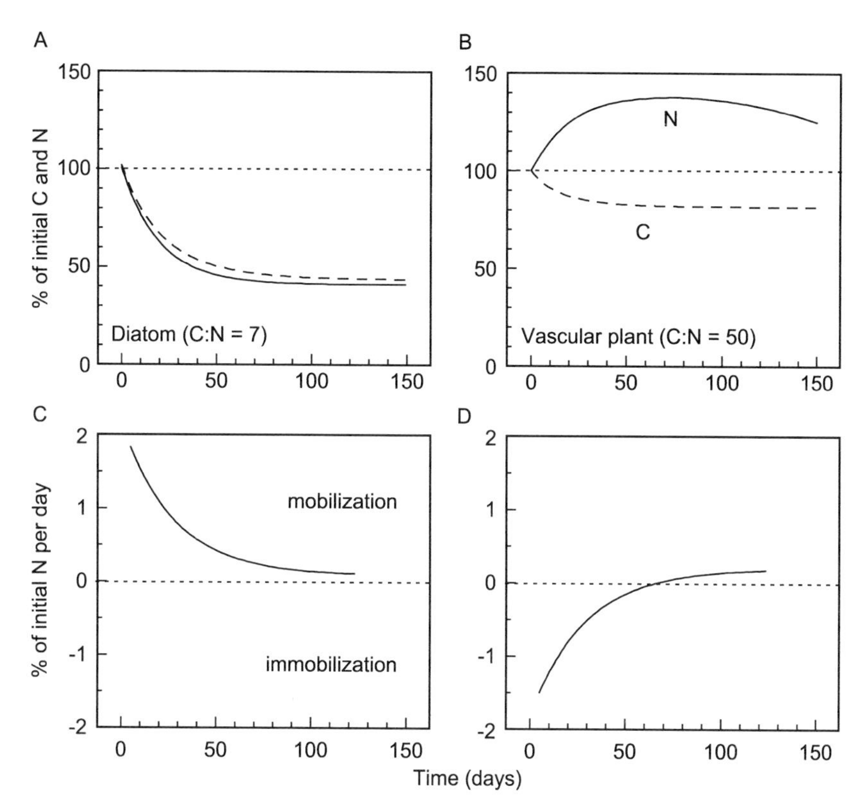

Figure 7.8 Hypothetical changes in carbon and nitrogen pools with time in degrading plant materials. (A and C) Decomposition of diatoms (initial C:N ratio of 7); (B and D) decomposition of vascular plant material (initial C:N ratio of 50). (A and B) Temporal patterns of carbon (broken lines) and nitrogen (full lines) in the degrading materials given as percent of the initial content; (C and D) percent of the initial nitrogen content in the degrading materials that have been mobilized or immobilized per day.

experiments on cultured aerobic marine prokaryotes (Goldman *et al.*, 1987; Tezuka, 1990) and on anaerobic sediment microbes (Blackburn, 1979b).

4.4. Anaerobic nitrogen mineralization and ammonium behavior in sediments

Nitrogen mineralization in sediments is divided between the upper oxic layer and the deeper anoxic layer, which in coastal and shelf sediments can be of particular significance due to the generally shallow depths of oxygen

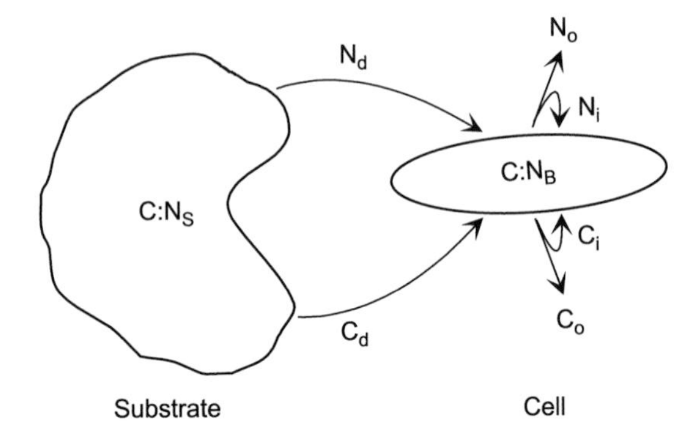

Figure 7.9 Schematic presentation of carbon and nitrogen mineralization with an indication of the terms used in the C:N ratio model of Blackburn (1979b). See text for more details.

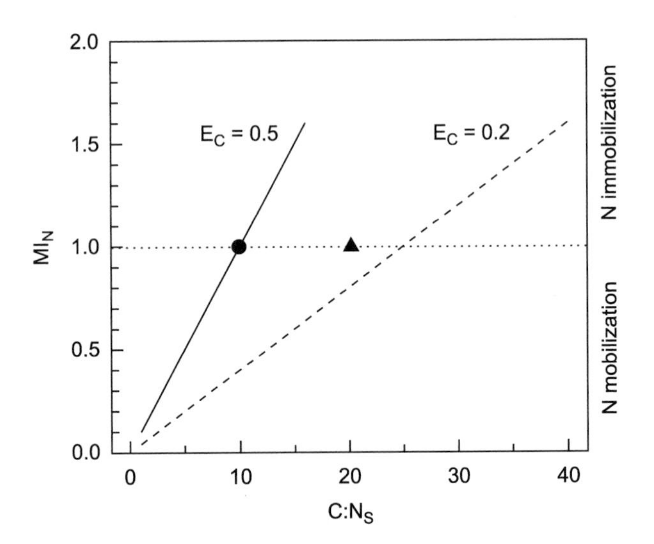

Figure 7.10 The relationship between C:N ratio of substrates (C:N$_S$) and MI_N (see text for explanation). Two examples are given: when carbon incorporation efficiency (E_C) is 0.5 (aerobic prokaryotes) and 0.2 (anaerobic prokaryotes). The C:N ratio of prokaryotes (C:N$_B$ in text) is fixed at 5. The intercept between $MI_N = 1$ (dotted line), and regression lines represent the conditions at which no net exchange of NH_4^+ occurs. The symbols represent measured values from laboratory experiments on cultured marine aerobes (filled circle; Goldman *et al.*, 1987) and on sediment anaerobes (filled triangle; Blackburn, 1979b).

penetration in these environments (see Chapter 6). Anoxic pathways of nitrogen mineralization depend little on the terminal carbon mineralization pathway. Thus, while we think of sulfate reduction and methanogenesis as important pathways of carbon mineralization in anoxic marine and freshwater sediments, nitrogen mineralization is largely uncoupled from the terminal carbon oxidation process. In anoxic sediments, nitrogen mineralization occurs during the initial hydrolysis and/or fermentation step of anaerobic decomposition (Kristensen and Hansen, 1995). Large biomolecules are fermented to small nitrogen-poor organic molecules (fatty acids such as acetate), while amino acids and nucleotides are deaminated to NH_4^+ and fatty acids (Figure 7.11). The overall coupling between carbon oxidation and nitrogen mineralization (i.e. CO_2 and NH_4^+ production), as typically observed in

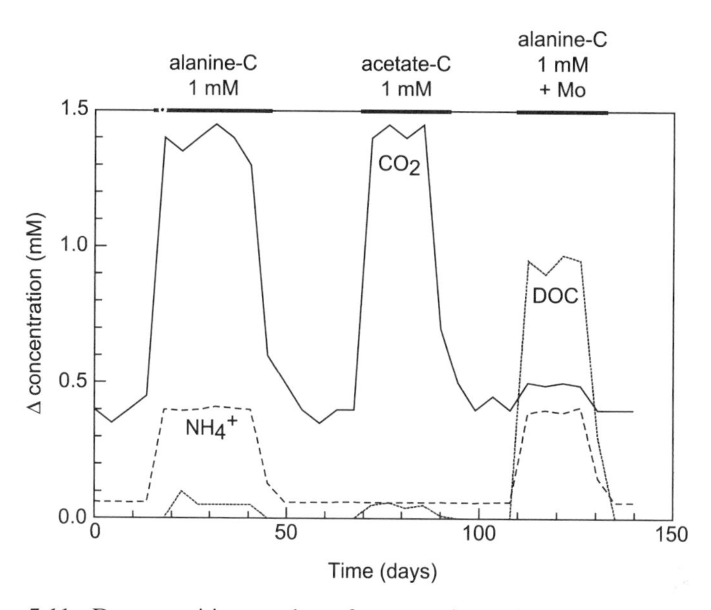

Figure 7.11 Decomposition products from anoxic marine sediment using a plug flow-through reactor system (as developed by Roychoudhury *et al.*, 1998). The concentration scale (Δ concentration) shows the difference between water exiting and entering the sediment and represents the net carbon and nitrogen transformations by the anaerobic community. When unamended, the sediment produced 0.4 mM CO_2, 0.06 mM NH_4^+, and no DOC. Alanine and acetate added at a concentration of 1 mM C in the inflowing water was mineralized almost completely by the microbial community with only traces of DOC. The mineralization products reflect the stoichiometry of the substrates (1 mM CO_2 and for alanine 0.33 mM NH_4^+). Sulfate reduction can be inhibited by addition of 10 mM molybdate. When molybdate (+Mo) was added together with alanine, almost all carbon was recovered in organic form as DOC with only traces of CO_2, whereas all nitrogen was mineralized completely to NH_4^+. Data from K. S. Hansen (unpublished).

sediment porewaters or incubation experiments (e.g., Canfield *et al.*, 1993a,b), is therefore rather indirect and a result of the rapid turnover of the fermentation products.

Detritus is rapidly decomposed in shallow sediment depths, and consequently more than 80% of the total nitrogen mineralization (N_d) generally occurs within the upper 0.1 m of the sediment (Figure 7.12). Whereas nitrogen incorporation rates typically decrease with depth, the percentage of the total mineralized nitrogen (N_d) incorporated into biomass (N_i) may increase, as shown in Figure 7.12. In this case, incorporation ranges from 10% at 0–2 cm to 40% at 12–14 cm. This increase with depth is due to decreasing nitrogen content (or increasing C:N ratio) of the organic detritus and lower degradability of the detritus. The microorganisms assimilate and incorporate more of the mineralized nitrogen to maintain growth in the deeper layers if carbon incorporation efficiency (E_C) remains constant.

The concentration of NH_4^+ in anoxic sediments is always higher than in the overlying water when there is net production. The produced NH_4^+ diffuses, or is advected by animals or bottom currents, to the sediment surface. Here it can be oxidized by nitrifying bacteria (see Section 7.5), assimilated by benthic microalgae (when there is light) or released to the overlying water. Owing to both nitrification and assimilation, NH_4^+ fluxes to the overlying water are generally lower than the total net production rates (N_o) in the sediment (Table 7.4).

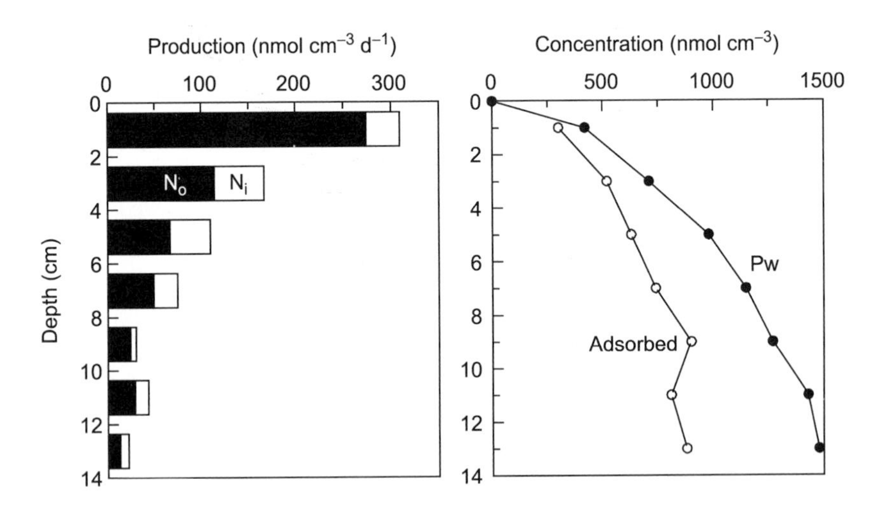

Figure 7.12 Depth dependence of NH_4^+ production in a coastal sediment from the Limfjord, Denmark. (A) Indication of how much of the mineralized N is liberated to the surroundings (N_o) and how much is incorporated into organisms (N_i). (B) Concentrations of NH_4^+ in the porewater (Pw) and adsorbed to particles (adsorbed). Modified from Blackburn (1979a).

Table 7.4 The relationship between water depth (m) and total sedimentary net NH_4^+ production and NH_4^+ flux across the sediment–water interface in various marine environments. Rates are given as mmol m^{-2} d^{-1}. Negative values indicate uptake

Sediment location	Depth	Net NH_4^+ production	NH_4^+ flux	Reference
Mangrove forest, Thailand	0	1.00	−0.40	1
Knebel Vig, Denmark	4	2.00	1.00	2
W. Kattegat, Denmark	20	3.60	0.90	3
Bering/Chukchi Shelf, USA	50	2.10	0.80	4
E. Kattegat, Denmark	70	2.80	1.37	3
Washington Shelf, USA	90	3.83	0.48	5
Washington Shelf, USA	180	0.49	0.10	5
Mexican Shelf, Mexico	240	0.35	0.00	6
Svalbard Shelf, Norway	300	0.25	0.14	7
South Island Shelf, New Zealand	500	0.62	0.10	8
Svalbard Shelf, Norway	530	0.06	0.00	7
Svalbard Shelf, Norway	1010	0.07	0.00	7
Mexican Shelf, Mexico	1020	0.08	0.00	6

References: 1, Kristensen *et al.* (2000); 2, Lomstein *et al.* (1998); 3, Blackburn and Henriksen (1983); 4, Henriksen *et al.* (1993); 5, Christensen *et al.* (1987); 6, Kristensen *et al.* (1999); 7, Blackburn *et al.* (1996); 8, Kaspar *et al.* (1985).

When the availability of NH_4^+ to microorganisms for assimilation is considered, we must also keep in mind that a fraction of the mineralized NH_4^+ becomes reversibly adsorbed onto sediment particles (Mackin and Aller, 1984) (Figure 7.12). The equilibrium distribution of NH_4^+ between porewater and particles has been successfully described with a linear adsorption coefficient, K_S (Rosenfeld, 1979; Mackin and Aller, 1984) (Figure 7.13):

$$K_S = \frac{(1-\phi)}{\phi} \rho_s \frac{C_{ads}}{C_{pw}} \tag{7.10}$$

where C_{ads} is the concentration of adsorbed NH_4^+ (μmol g^{-1} dry solids), C_{pw} is the concentration of porewater NH_4^+ (μmol cm_{pw}^{-3}), ϕ is porosity ($cm_{pw}^3/cm_{total\ sed}^3$), and ρ_s is sediment density (g cm_{solids}^{-3}). Mackin and Aller (1984) have reported a relatively consistent K_S value of 1.3 ± 0.1 for a wide range of marine environments, meaning that approximately 57% of the produced NH_4^+ is adsorbed to sediment particles while 43% remains in dissolved form. Values for K_S may be somewhat lower in biogenic and very porous sediments. There is the possibility, however, that in some cases K_S may decrease with increasing NH_4^+ concentrations as adsorption sites on the particles become highly loaded (van Raaphorst and Malschaert, 1996).

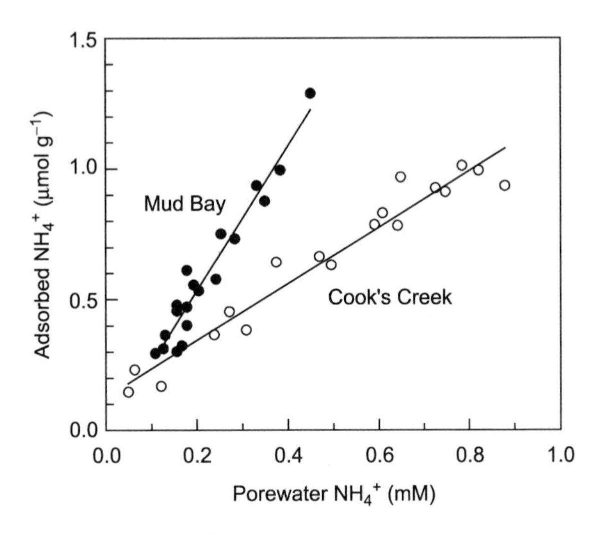

Figure 7.13 Examples of NH_4^+ adsorption from typical anoxic marine sediments. Modified from Mackin and Aller (1984).

4.5. "New" versus "regenerated" nitrogen in pelagic ecosystems

Two main sources of nitrogen supply water column primary producers: "regenerated" nitrogen, originating from the mineralization of organic matter in the euphotic zone of the water column, and "new" nitrogen transported to the euphotic zone from elsewhere or fixed there by prokaryotes (Dugdale and Goering, 1967). New nitrogen may come from the atmosphere as wet and dry deposition, from continental sources as river runoff, from N fixation or from sediments and the water column below the euphotic zone (Figure 7.14). Only "new" nitrogen, mainly in the form of NO_3^-, represents net organic matter production in the system.

The nitrogen cycle of the Kattegat and the Danish Belt Sea has been studied intensively and is used here as an example of coastal marine nitrogen cycling (Table 7.5). The annual nitrogen demand by phytoplankton in the entire area (31,000 km^2) is estimated at 1580 $\times$ 10^3 t (Richardson, 1996). Known sources of "new" nitrogen, however, account for only about 20% of this demand; the remainder is made up from nitrogen regeneration in the water column. The nitrogen budget is constructed from both nitrogen sources and sinks. Of the "new" nitrogen sources, land runoff and Kattegat bottom water are the most important, although there is a significant source from atmospheric deposition as well (Table 7.5). Sediment

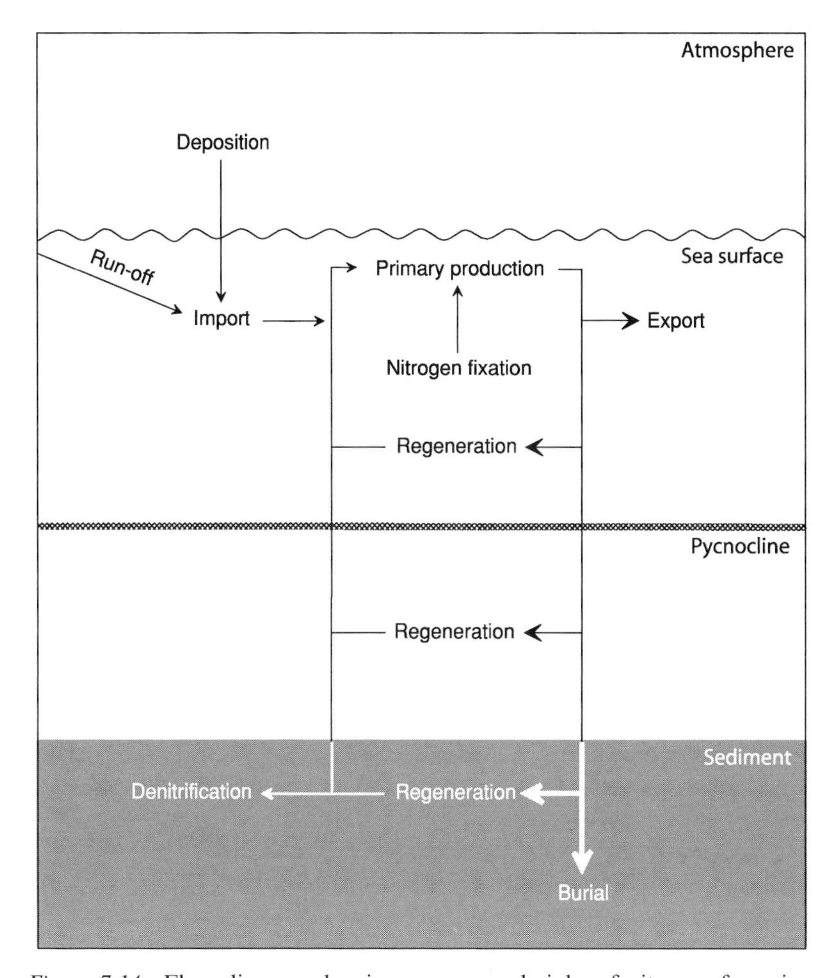

Figure 7.14 Flow diagram showing sources and sinks of nitrogen for primary producers in the photic zone of oceans. Heavy arrows represent organic nitrogen, and light arrows represent inorganic nitrogen. Processes are separated into the upper layer (including the photic zone), water masses below the pycnocline, and underlying sediments. "Deposition" represents dry and wet atmospheric inputs; "run-off" represents river, diffuse and sewage discharges from land (import = run-off + deposition); "export" represents the organic nitrogen, which is removed from the system via ocean currents; "regeneration" represents microbial conversion of organic nitrogen to inorganic nitrogen; "denitrification" represents the loss of combined nitrogen in the form of atmospheric nitrogen; and "burial" represents nitrogen, which is permanently buried in sediments. Note that nitrogen fixation is not shown.

Table 7.5 Budget for total nitrogen in the Kattegat and Danish Belt Sea

N sources and sinks	10^3 t N y^{-1}
Phytoplankton demand	1580
Regenerated in the photic zone	1270
Total "new" nitrogen	310
Atmospheric deposition	50
Run-off from land	140
From below photic zone	120
Regeneration below photic zone	144
Denitrification in sediments	196
Burial in sediments	3.4
Exported in bottom currents from system	23

Data compiled from Blackburn and Henriksen (1983), Hansen *et al.* (1994), Richardon (1996).

dentrification is by far the most important sink, nearly balancing inputs from land runoff and atmospheric deposition. The nitrogen remineralized below the photic zone, including sediments, supplies the "new" nitrogen upwelled into the photic zone. Some of this remineralized nitrogen is also exported in bottom currents out of the system.

5. NITRIFICATION

Nitrification describes the oxidation of NH_4^+ to NO_2^- and ultimately to NO_3^-. Nitrification links the most reduced and the most oxidized forms of the nitrogen cycle, and it exerts considerable influence on nitrogen dynamics in aquatic environments (Herbert, 1999). Indeed, as a result of nitrification (or uptake by other organisms), the primary product of nitrogen mineralization, NH_4^+, rarely occurs at significant concentrations in oxic environments. The NO_3^- produced by nitrification below the photic zone of lake and marine water bodies accumulates. Some of this NO_3^- may be transported upward to the photic zone as a source of "new" nitrogen for primary producers. Some may also be transported downward to the anoxic portion of the water body, if one exists, or to the sediments, where the NO_3^- can to be reduced to N_2 (denitrification) or NH_4^+ (NO_3^- ammonification). Because nitrification oxidizes NH_4^+ to NO_3^-, with a potential sink in denitrification, it is indirectly responsible for the loss of nitrogen from the system. Below, we explore the mechanisms and kinetics of nitrification and the factors regulating the rates.

5.1. Biochemistry and thermodynamics of nitrification

The complete oxidation of NH_4^+ to NO_3^- requires the transfer of eight electrons. The first step is the six-electron oxidation of NH_4^+ to NO_2^- (Equation 7.11) accomplished primarily by bacteria of the genera *Nitrosomonas* and *Nitrosospira*. These organisms are aerobic chemolithoautotrophs and function best at near neutral pH (~7 to 8).

$$NH_4^+ + 1.5O_2 \rightarrow NO_2^- + H_2O + 2H^+, \ \Delta G^0 = -272 \text{ kJ mol}^{-1} \qquad (7.11)$$

This overall oxidation reaction proceeds in at least two steps, with hydroxylamine (NH_2OH) as an intermediate. The first step, involving two electrons, is the oxidation of NH_4^+ to NH_2OH:

$$NH_4^+ + 0.5O_2 \rightarrow NH_2OH + H^+, \ \Delta G^0 = +17 \text{ kJ mol}^{-1} \qquad (7.12)$$

This step is catalyzed by the membrane-associated enzyme ammonium monooxygenase, and it does not yield biochemically useful energy. Since NH_2OH is unstable in aqueous solution and rapidly degrades to N_2, N_2O or NH_4^+, the energy-gaining second step (Equation 7.13) must be coupled closely to the first step.

$$NH_2OH + O_2 \rightarrow NO_2^- + H_2O + H^+, \ \Delta G^0 = -289 \text{ kJ mol}^{-1} \qquad (7.13)$$

This reaction involves a four-electron transfer and is catalyzed by the periplasmic enzyme hydroxylamine oxidoreductase.

The next reaction, oxidation of NO_2^- to NO_3^-, is primarily accomplished by bacteria of the genera *Nitrobacter*, *Nitrococcus Nitrospira*, and *Nitrospina*. While capable of chemolithoautotrophic metabolism, many of these organisms (e.g., *Nitrobacter*) also augment their autotrophic lifestyle with heterotrophic metabolism. Oxidation of NO_2^- to NO_3^- is a simple two-electron transfer with molecular oxygen as the terminal electron acceptor:

$$NO_2^- + 0.5O_2 \rightarrow NO_3^-, \ \Delta G^0 = -76 \text{ kJ mol}^{-1} \qquad (7.14)$$

The membrane-bound enzyme nitrite oxidase catalyzes this process. Since intermediate compounds and byproducts other than NH_2OH (e.g., N_2O) have been identified (Kaplan, 1983), it appears that the nitrification processes may involve more intermediate steps than depicted above.

The electron donors for nitrification have high E'_0 values for both the NH_2OH/NH_4^+ couple, which is close to 0 V, and the NO_3^-/NO_2^- couple, which is 0.43 V. In either case, these redox couples are too oxidized to reduce NAD^+ to NADH ($E'_0 = -0.32$ V), and so a reverse electron transport system is required to produce NADH (see Chapter 3). This energy-requiring process limits the amount of ATP produced per NH_4^+ oxidized, reducing the growth yield (Kaplan, 1983). Growth yields of nitrifying bacteria are further affected by the energy demands of CO_2 fixation. Like many other aerobic chemolithoautotrophs, nitrifiers employ the Calvin cycle (reductive pentose phosphate cycle) for CO_2 fixation, which requires 3 moles of ATP for every CO_2 fixed (Chapter 4). Nitrifiers are, therefore, extremely slow growing, and a small microbial biomass must oxidize large amounts of reduced nitrogen to maintain growth (Figure 7.15). Thus, the thermodynamic efficiency (energy conserved for growth: metabolically available energy) is only 1–4% for *Nitrosomonas* and 3–10% for *Nitrobacter*, much less than aerobic heterotrophs (Fenchel and Blackburn, 1979).

The chemolithoautotrophic oxidation of NH_4^+ might also occur under anoxic conditions. Luther *et al.* (1997) suggested the possibility of NH_4^+ oxidation with MnO_2 as an electron acceptor. Two pathways were considered, one forming N_2 (Equation 7.15), and the other forming NO_3^- (Equation 7.16).

$$2NH_4^+ + 3MnO_2 + 4H^+ \rightarrow 3Mn^{2+} + N_2 + 6H_2O, \ \Delta G^0 = -659 \text{ kJ mol}^{-1} \tag{7.15}$$

$$NH_4^+ + 4MnO_2 + 6H^+ \rightarrow 4Mn^{2+} + NO_3^- + 5H_2O, \ \Delta G^0 = -322 \text{ kJ mol}^{-1} \tag{7.16}$$

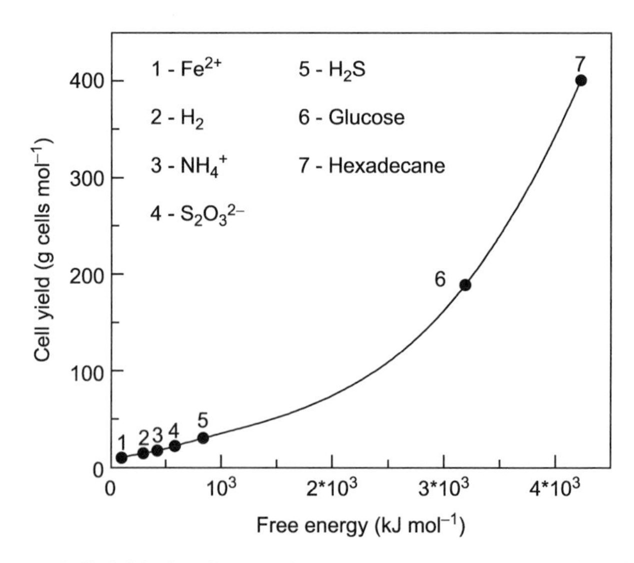

Figure 7.15 Cell yield of various prokaryotes on energy sources with different free energies of reaction. O_2 is the electron acceptor in all cases. Modified from Madigan *et al.* (2003).

Luther *et al.* (1997) further suggested that analogous cycles involving iron might be established in ecosystems with low pH, such as many lakes. However, Thamdrup and Dalsgaard (2000) could find no evidence for anaerobic NH_4^+ oxidation with Mn oxides in Mn oxide-rich marine sediments, where it might be expected, and no direct evidence exists to support an analogous cycle with Fe.

It has long been recognized that many heterotrophic bacteria and fungi can oxidize organic nitrogen by nitrification. The organisms responsible for this process include *Bacteria* of the genera *Alcaligenes* (Papen *et al.*, 1989) and *Arthrobacter* (Brierley and Wood, 2001), as well as certain fungi (Stroo *et al.*, 1986). The nitrification rate of heterotrophs is generally 10^3 to 10^4 times slower than for chemolithoautotrophs. The biochemistry of heterotrophic nitrification is not fully understood, and it appears that different organisms may employ unique enzyme pathways, which are also different from those used by chemolithoautotrophs (Jetten *et al.*, 1997). During heterotrophic nitrification, nitrogen oxidation can proceed by an inorganic pathway, an organic pathway or a combination of both (Figure 7.16). The pathway from reduced nitrogen to NO_3^- may involve intermediates such as NH_2OH, NO_2^- and various organic compounds (e.g., amides, aminopropionic acids, nitropropionic acids, nitrosoethanol). Heterotrophic nitrifiers gain no energy during nitrification, and it has been suggested that intermediates formed have specific functions as microbial growth factors or as biocidal agents. Heterotrophic nitrification has been studied mostly in terrestrial ecosystems, but it is also present in aquatic environments (Mevel and Prieur, 1998). However, our knowledge on the role of heterotrophic nitrification in aquatic ecosystems is still limited.

5.2. Phylogeny of chemolithoautotrophic nitrifiers

The numbers and diversity of organisms identified as chemolithoautotrophic nitrifiers are low compared with denitrifiers. The two broad classes of organisms, NH_4^+ oxidizers and NO_2^- oxidizers, are physiologically unrelated, and they also employ two very different enzyme systems for the energy-gaining oxidation processes. The phylogenetic relationships of several species of NH_4^+ oxidizers, based on 16S rRNA gene sequences, show two distinct groups within the proteobacteria. One group, based on the sequence of a single NH_4^+ oxidizer, *Nitrosococcus oceanus*, is deeply branching within the γ-proteobacteria (Woese *et al.*, 1985). The second group contains the majority of cultured strains and forms a tight cluster within the β-proteobacteria. This group can be subdivided into two clades, corresponding to *Nitrosomonas* spp. and *Nitrosospira* spp. (Figure 7.17) (Head *et al.*, 1993). Within each clade there are a number of closely related strains, some of which group into further clusters. Sequences from 16S rDNA amplified

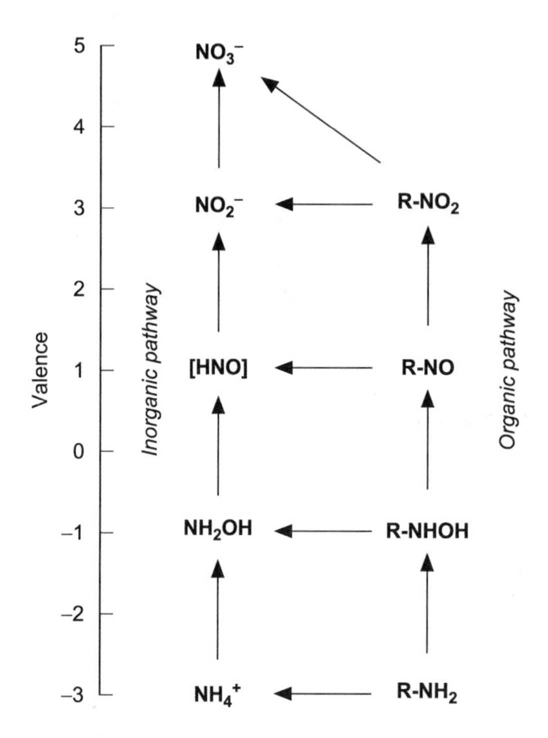

Figure 7.16 Inorganic and organic pathways of nitrification. Modified from Focht and Verstraete (1977).

directly from marine sediments and soil reveal at least seven clusters of NH_4^+ oxidizers within the β-proteobacteria, three within the *Nitrosomonas* clade and four within the *Nitrosospira* clade (Stephen *et al.*, 1998).

On the basis of ultrastructural properties, four genera of chemolithoautotrophic NO_2^- oxidizers have been described. Species of the genus *Nitrobacter* are facultative chemolithoautotrophs and are found in both terrestrial and aquatic environments. They can grow aerobically with NO_2^- as the electron donor and by the oxidation of simple organic compounds (Kaplan, 1983). Members of the genera *Nitrococcus, Nitrospina*, and *Nitrospira* are obligate chemolithoautotrophs isolated from marine habitats (Watson *et al.*, 1986). Based on 16S rRNA sequence data, the *Nitrobacter* strains constitute a closely related subcluster within the α-proteobacteria. Other NO_2^- oxidizers such as *Nitrococcus* belong to the γ-proteobacteria, whereas *Nitrospina* is a member of the δ-proteobacterial subdivision. *Nitrospira* represents its own lineage among the *Bacteria*.

The physiology and phylogenetic distribution of NO_2^--oxidizing bacteria in the α and γ subgroups of the proteobacteria suggest that they have descended from photosynthetic purple bacteria. These nitrifiers have

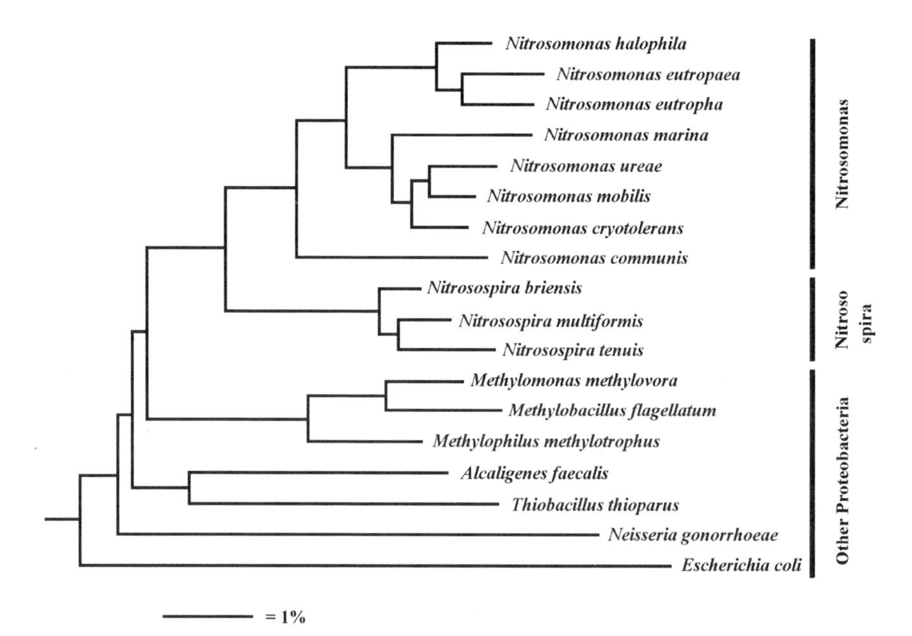

Figure 7.17 Phylogenetic tree based upon 16S rDNA sequences from NH_4^+ oxidizers, *Nitrosomonas* spp. and *Nitrosospira* spp., within the β-subdivision of proteobacteria. Scale bar indicates 1% sequence divergence. Adapted from Kowalchuk *et al.* (1997).

retained the general structural features of the ancestral photosynthetic membrane complex. NO_2^- oxidizers as a group are apparently not derived from one ancestral nitrifying phenotype. Two species within the α-proteobacteria, *Bradyrhizobium japonicum* and *Rhodopseudomonas palustris*, are closely related to *Nitrobacter* species (Figure 7.18). The main difference is that the former two fix nitrogen and the latter oxidizes NO_2^-. *B. japonicum* forms nitrogen-fixing nodules on leguminous plants, while *R. palustris* is a free-living nitrogen-fixing organism. These differences led Orso *et al.* (1994) to speculate that these three species evolved from a photosynthetic, nitrogen-fixing ancestor at about the time of differentiation of land plant families. One evolutionary line retained nitrogen fixation and lost the ability to photosynthesize, while the other evolutionary line acquired the ability to nitrify and seemingly lost the ability to fix nitrogen and to photosynthesize.

5.3. Environmental factors affecting nitrification rates

Environmental conditions control the location and magnitude of nitrification. Nitrifiers depend on the availability of O_2 and NH_4^+, and they are inhibited by extremes in pH, sulfide concentration, temperature, salinity, and light.

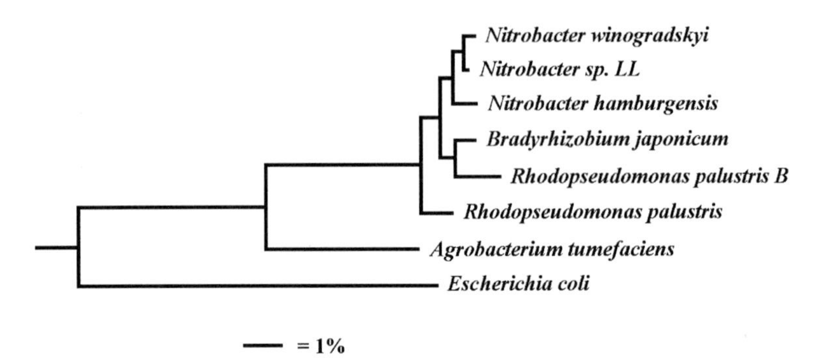

Figure 7.18 Phylogenetic tree based upon 16S rRNA sequences from NO_2^- oxidizers, *Nitrobacter* spp., within the α-subdivision of proteobacteria. *E. coli* is shown as the outgroup for comparison. Scale bar indicates 1% sequence divergence. Adapted from Teske *et al.* (1994).

Environmental effects on nitrification have been studied intensively in laboratory cultures and by *in situ* and laboratory measurements on natural materials.

5.3.1. Substrates

Because nitrifiers are aerobes, their activity is ultimately constrained by the availability of O_2, and required minimum concentration of O_2 ranges from 1 to 6 μM (Henriksen and Kemp, 1988). Surprisingly, many nitrifiers prefer relatively low O_2 conditions. Thus, the nitrifier *Nitrosomonas europaea* accumulated at low O_2 concentrations near the oxic–anoxic interface in a stratified temperate lake (Voytek and Ward, 1995), and cultures of *Nitrosomonas marina* grow best at O_2 concentrations around 5% of air saturation (Goreau *et al.*, 1980). This is not always the case, however, as Rysgaard *et al.* (1994) observed maximum potential nitrification rates of sediment slurries from a freshwater lake at O_2 concentrations between 150 and 400 μM (Figure 7.19). At higher O_2 concentrations, the nitrification rate was somewhat inhibited. Overall, the composition of the nitrifying community, and its O_2 optimum, appears to be location specific and depends on adaptations to the prevailing environmental conditions.

Indeed, different environmental adaptations can be found among closely related nitrifying species. Thus, the two *Nitrosomonas* species *N. eutropha* and *N. europaea* are closely related by 16S rDNA sequence analysis, yet *N. eutropha* dominates in oxygen-poor profundal sediments of an oligotrophic lake, while *N. europaea* primarily inhabits sediments from the oxygen-rich littoral zone (Whitby *et al.*, 1999).

Nitrifiers in sediments may survive periods of inactivity when they are periodically or persistently exposed to anoxia, and they recover their activity

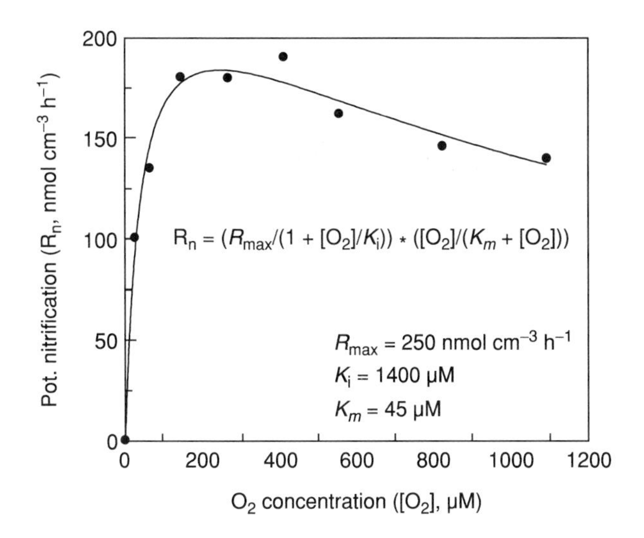

Figure 7.19 Potential nitrification in sediment from a freshwater lake as a function of O_2 concentrations in the overlying water. The equation describing O_2 as both a substrate (Michaelis-Menten kinetics) and a non-competitive inhibitor is shown. Values for the constants, R_{max}, K_i and K_S, which provide the best fit to the data points, are also given. Modified from Rysgaard *et al.* (1994).

instantly following O_2 exposure (Henriksen *et al.*, 1981). The physiological basis of this tolerance to anoxia is not well understood. However, a number of both NH_4^+ and NO_2^- oxidizers are apparently capable of heterotrophic growth using low-molecular organic substrates as electron donors while performing partial or complete dissimilatory nitrate reduction. Nitrifiers in marine sediments, by contrast, are almost completely inactivated by exposure to free sulfide (Kaplan, 1983), and they may require days to weeks to recover completely (Joye and Hollibaugh, 1995). Thus, the ability of nitrifiers to recover from environmental O_2 fluctuations may be impeded if sulfide exposure also occurs.

Ammonium oxidation is generally the rate-limiting step of nitrification when there are ample supplies of O_2 and when NO_2^- accumulation is unlikely (Kaplan, 1983). However, since NH_4^+ oxidizers generally have higher affinity for O_2 than NO_2^- oxidizers, accumulation of NO_2^- and N_2O may occur at very low O_2 concentrations (Henriksen and Kemp, 1988).

Generally, the oxidation of NH_4^+ and NO_2^- by nitrifiers follows Michaelis-Menten kinetics, and a range of K_m values have been observed. The nitrification of NH_4^+ to NO_2^- in eutrophic environments has K_m values in the range of 50–700 μM NH_4^+, and K_m values of 350–600 μM NO_2^- are observed for NO_2^- oxidation to NO_3^- (Focht and Verstraete, 1977; Koops and Pommerening-Roser, 2001). However, nitrifiers from oligotrophic open ocean and lake environments are adapted to much lower substrate concentrations, and K_m values ranging from 0.1 to 5 μM have been observed for both NH_4^+ and NO_2^- oxidation (Olson, 1981a; Koops and Pommerening-Roser, 2001). Thus, the substrate affinity of nitrifying populations adapts to the substrate availability in the environment. Marine populations of nitrifiers are rather insensitive to perturbations in NH_4^+ and NO_2^- concentrations, as their activity is usually saturated at normal environmental levels of NH_4^+ and NO_2^- (Ward, 2000). However, intense competition with phytoplankton or benthic microalgae for NH_4^+ and NO_2^- can limit the growth of nitrifiers in the upper photic zone of oligotrophic waters and in surface sediments (Jensen *et al.*, 1994). Nitrification rates in eutrophic lakes, rivers, and coastal marine areas are generally higher than in coastal oligotrophic ocean environments (Table 7.6). Not surprising, eutrophic environments receiving elevated supplies of NH_4^+ from organic compound mineralization and anthropogenic input support denser and more active nitrifier populations. The activity of nitrifiers may be lower in rivers and other environments receiving reactive organic matter poor in nitrogen. In this case, the nitrifiers must compete with heterotrophic bacteria for the available NH_4^+ (Strauss *et al.*, 2002).

Table 7.6 Range of depth-integrated nitrification and denitrification rates in sediment (mmol N m^{-2} d^{-1}), and volume-specific rates in the water column (μmol Nl^{-1} d^{-1}), from various aquatic ecosystems

Ecosystem	Nitrification	Denitrification
Lake		
Sediment	2–20	0.2–6
Water	0–1	0.2–2[a]
River		
Sediment	3–23	0.9–20
Water	0–4	0–2[a]
Coastal marine (<100 m)		
Sediment	0.2–7	0.1–12
Water	0.002–0.2	0.7–4[a]
Deep ocean (>100 m)		
Sediment	0.003–0.1	0.003–0.6
Water	0.001–0.01	0.001–0.01[a]

[a]Only valid in the rare cases when the water column or part of it is devoid of oxygen.

Data compiled from Kaplan (1983), Christensen and Rowe (1984), Koike and Sørensen (1988), Seitzinger (1988), Jensen *et al.* (1993), Omnes *et al.* (1996), Ward (1996), Sjodin *et al.* (1997), Iriarte *et al.* (1998), Lorenzen *et al.* (1998), Herbert (1999), and Pauer and Auer (2000).

5.3.2. pH and surface growth

The growth and activity of most chemolithoautotrophic nitrifiers are optimal in the neutral to slightly alkaline pH range (pH 7 to 8.5) (Focht and Verstraete, 1977). The restricted optimal pH range for nitrification is mainly due to the toxicity of free NH_3 at high pH and of nitrous acid (HNO_2) at low pH. *Nitrobacter* spp. are sensitive to both high and low pH levels, whereas *Nitrosomonas* spp. are more affected by alkaline conditions when high NH_3 concentrations are present (Figure 7.20).

Apart from acid lakes, most natural aqueous environments have pH levels within the optimal range for nitrifiers. Sedimentary environments, on the other hand, can experience pH extremes well outside of the optimal range. For example, shallow water sediments with benthic microalgae may reach pH 10 in the photic zone during illumination (Revsbech *et al.*, 1983). Also, sulfide-rich salt marsh and mangrove sediments, with intensive sulfide oxidation, may have pH <6 in the upper sediment zone where sulfide oxidation is most intensive (Kristensen *et al.*, 1991). Nevertheless, nitrification is rarely

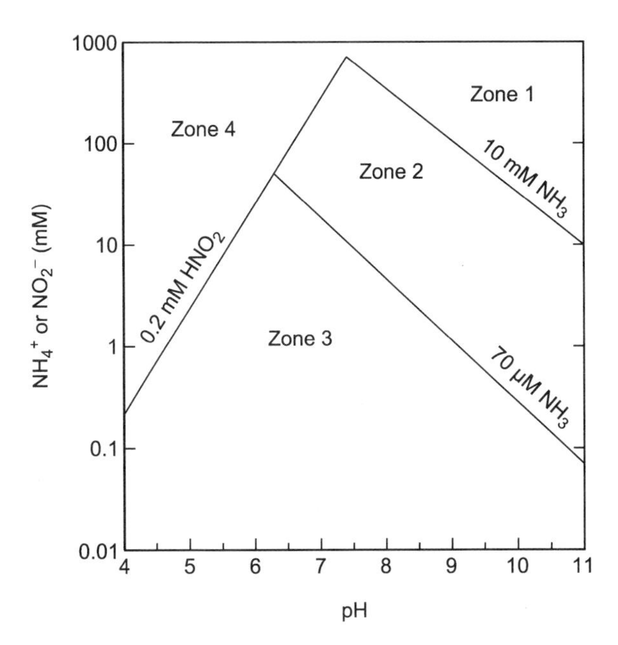

Figure 7.20 The pH tolerance of nitrification. Zone 1: Inhibition of *Nitrosomonas* and *Nitrobacter* by NH_3; zone 2: inhibition of *Nitrobacter* by NH_3; zone 3: complete nitrification; zone 4: inhibition of *Nitrobacter* by HNO_2. Lines represent the approximate tolerance levels for NH_3 and HNO_2. The *y*-axis is given as total of dissociated plus undissociated compounds. Modified from Anthonisen *et al.* (1976).

inhibited completely under these extremes of pH sometimes found in sediments. Thus, species of *Nitrosospira* and *Nitrosovibrio* can be active in acid soils and sediments at pH values around 4 (de Boer and Laanbroek, 1989), while *Nitrobacter alkalicus* has been isolated from sediments of soda lakes with pH values up to around 10 (Sorokin *et al.*, 1998). Heterotrophic nitrifiers may also contribute to NO_3^- formation under acidic conditions (Brierley and Wood, 2001).

Sediments generally host, on a per volume basis, nitrifier populations orders of magnitude greater than in the water column Thus, nitrifiers are typically present in the range of 10^5 to 10^7 cells cm^{-3} in sediments, while in the water column they range from 10^1 to 10^4 cells cm^{-3} (Focht and Verstraete, 1977; Pauer and Auer, 2000). The strong association of nitrifiers with the particulate- and organic-rich fractions of sediments is linked to the high availability of NH_4^+ derived from microbial nitrogen mineralization of the organic particles. Furthermore, the pH-buffering of clay and silt particles is likely to help retain an optimal pH for the nitrifiers.

5.3.3. Temperature

The temperature optimum for typical mesophilic nitrifiers in pure culture ranges from 25 to 35 °C, and growth usually occurs from 3 to 45 °C (Herbert, 1999). Thermophilic heterotrophic nitrifiers with temperature optima of around 65 °C have also been isolated from deep-sea hydrothermal vents (Mevel and Prieur, 1998). Heterotrophic thermophilic nitrifiers have been identified in active compost at 53 °C (Pel *et al.*, 1997). Autotrophic nitrification was not found in this mixed compost population suggesting that thermophilic chemolithoautotroph nitrification apparently is rare in nature.

When metabolizing within their tolerance range, pure cultures and natural populations of nitrifiers respond exponentially to temperature, with Q_{10} values ranging from 2 to 3 (Pomeroy and Wiebe, 2001). In nature, nitrifying communities tend to adapt to the temperature of the environment. Thus, nitrifying populations in arctic sediment from Svalbard, Norway have a temperature optimum of only 14 °C, well within the psychrophilic range. By contrast, the temperature optimum was near 40 °C for nitrifiers from warmer temperate sediments off Germany (Thamdrup and Fleischer 1998) (see also Chapter 2 for more discussion on temperature adaptations). In some instances, individual populations can also adapt their metabolic range to the temperature of the environment in which they live. For example, an NH_4^+-oxidizing strain was isolated from arctic waters with growth minimum of −5 °C, an optimum of 22 °C, and a maximum growth temperature of about 29 °C when adapted to 5 °C. Cells grown at 25 °C, on the other hand, showed an optimum temperature of 30 °C and a maximum of about 38 °C.

Nitrification rates in these warm-adapted populations exceeded the cold-adapted cells only above 25 °C (Jones and Morita, 1985).

Nitrifying bacteria in shallow-water sediments from temperate regions are subject to large seasonal changes in temperature, and they exhibit the highest cell-specific activity during the warm summer months. However, when *in situ* measurements of nitrification in sediments are compared on a seasonal basis, different patterns emerge. Distinct summer maxima for nitrification have been observed, for example, in Aarhus Bay, Denmark (Figure 7.21) (Hansen *et al.*, 1981), the Tay Estuary, Scotland (Macfarlane and Herbert, 1984), and the middle reaches of Narragansett Bay, USA (Seitzinger *et al.*, 1984). Nitrification rates in some shallow Danish fjords (Figure 7.21) (Hansen *et al.*, 1981) and the Providence River station in Narragansett Bay (Jenkins and Kemp, 1984), on the other hand, are lowest during the summer and highest during the winter. The reduced nitrification activity in the summer has been attributed primarily to reduced availability of O_2 to the nitrifying population. The warm conditions during summer reduce O_2 solubility, and high rates of benthic respiration increase O_2 demand. In concert, both of these factors reduce the depth of O_2 penetration into the sediment. Other factors contributing to reduced rates of summer nitrification include greater competition for NH_4^+ by heterotrophic bacteria and sulfide toxicity.

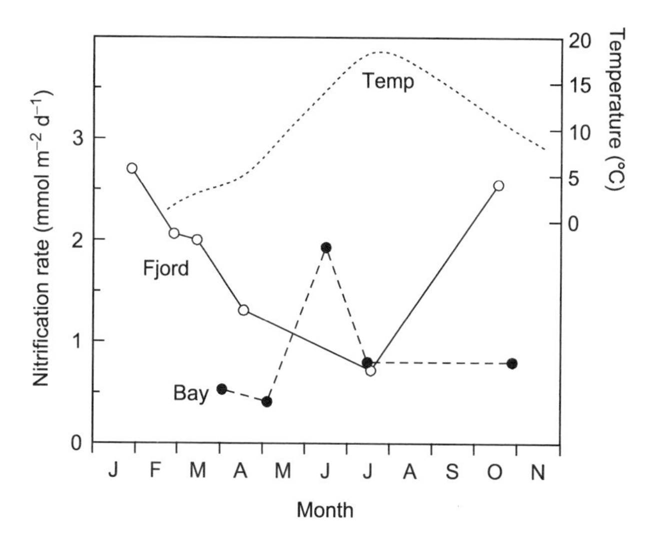

Figure 7.21 Seasonal variations in nitrification for a sediment from the estuary, Norsminde Fjord, Denmark, and at 17 m depth from Århus Bay, Denmark. Water temperature is shown as the dotted line. Modified from Hansen *et al.* (1981).

5.3.4. Salinity

Nitrifiers are adapted to the salinity prevailing in their environment, although individual species are generally able to acclimate to a broad salinity range. The tolerance to varying salinities, however, differs from species to species. Most freshwater species rapidly reduce their activity when salinity increases, and some do not grow at all in sea water, whereas the opposite pattern is observed for many marine species. NH_4^+ oxidizers from estuarine environments have optimum activity at intermediate salinities (5–10), and many of them can grow in the entire salinity range of 0–35. For example, *Nitrosomonas eutropha*, isolated from the Elbe River, is an euryhaline species exhibiting at least 75% of the maximum activity in the whole of the salinity from 0 to 30 (Stehr *et al.*, 1995) (Figure 7.22). Other nitrifiers isolated from the same river exhibit different salinity responses. Thus, *N. oligotropha* appears to be a freshwater species, which is completely inhibited at salinities above 10, and *N. europaea* has a higher salinity tolerance but cannot function at full seawater salinity. In the Schelde estuary, changes in salinity along the estuary cause community shifts of the NH_4^+ oxidizing population (de Bie *et al.*, 2001). The most dramatic change occurs in the estuarine region with the sharpest salinity gradient.

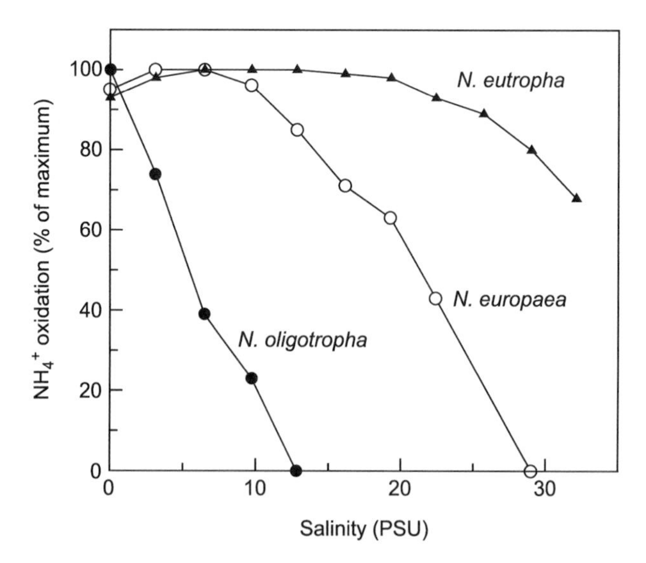

Figure 7.22 Sensitivity to increasing salinity of three species of NH_4^+-oxidizing *Nitrosomonas*, isolated from the lower River Elbe, Germany. Modified from Stehr *et al.* (1995).

5.3.5. Light

Light often inhibits nitrification in the open ocean and in lake environments (Vanzella *et al.*, 1989; Ward, 2000). Although NO_2^- oxidizers are usually considered more sensitive to sunlight than NH_4^+ oxidizers, this generalization may be obscured by species-specific as well as dose- and wavelength-dependent responses (Guerrero and Jones, 1996). However, the higher sensitivity of NO_2^- oxidizers to light supports the hypothesis of Olson (1981b) that light inhibition is responsible for the commonly observed near-surface NO_2^- maximum in the water column of many oceanic environments.

Measuring nitrification and denitrification

Specific inhibitors are widely used to determine nitrification and denitrification rates. In using these, the process of interest is blocked and the accumulation of unused substrates is determined (e.g., NH_4^+ or NO_2^- for nitrification and N_2O for denitrification). The most commonly used inhibitors of nitrification are nitrapyrin, allylthiourea, and chlorate (Henriksen and Kemp, 1988). The former two inhibit NH_4^+ oxidation and the latter inhibits NO_2^- oxidation. Inhibitor-based quantification of denitrification is usually done by acetylene inhibition (Sørensen, 1978). Acetylene inhibits N_2O reductase, with a resulting accumulation of N_2O. Problems sometimes exist with these inhibitor-based methods, however, limiting the usefulness of the results. For example, sediment denitrification rates obtained with the acetylene-block technique may be too low because coupled nitrification-denitrification is excluded due to the simultaneous inhibition of nitrification by acetylene.

Alternatively, nitrification can be estimated from inorganic nitrogen fluxes (Christensen and Rowe, 1984), whereas denitrification can be measured directly as the production of N_2 (Seitzinger, 1993). These methods, however, lack accuracy, and the experimental design can be elaborate. More recently, a variety of ^{15}N tracer methods has been proposed. The most promising of these is the relatively simple, accurate, and versatile nitrogen isotope pairing technique of Nielsen (1992), which provides simultaneous measurements of both nitrification and denitrification (see text box below).

The use of newly developed microelectrodes can overcome biases resulting from uneven dispersal of inhibitors or tracers, or unwanted side effects from inhibitors. Microelectrodes with very high spatial resolution can measure the fine-scale distributions of O_2 and NO_3^- in sediments (Figure 7.23). At steady state, reaction diffusion models can be used to estimate rates of nitrification and denitrification and to determine the location of the processes in relation to the chemical profiles (Revsbech and Jørgensen, 1986).

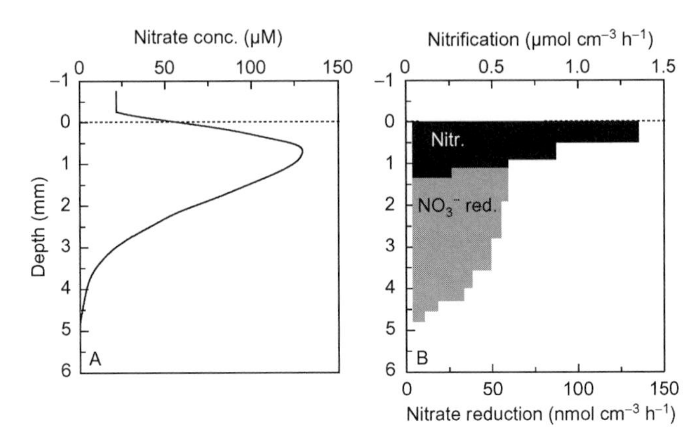

Figure 7.23 (A) Steady-state concentration profile of NO_3^- in a freshwater lake sediment measured with a NO_3^- microsensor. A concentration of 300 μM of NH_4^+ was added to the overlying water. (B) The corresponding distribution and magnitude of the volume-specific nitrification (Nitr.) and NO_3^- reduction rates (NO_3^- red.) are indicated by the black and grey areas, respectively. Modified from Jensen *et al.* (1993).

6. DISSIMILATORY NITRATE REDUCTION

Dissimilatory NO_3^- reduction is a microbial process in which, NO_3^- is reduced with various electron donors by an energy-gaining metabolism in the absence or near absence of O_2. NO_2^- is the first intermediate in this reduction, and based on the fate of NO_2^-, three different pathways can be distinguished (Figure 7.3) (Bonin *et al.*, 1998; Thamdrup and Dalsgaard, 2002): (1) reduction to gaseous products (N_2O or N_2) by denitrification, (2) reduction to NH_4^+ in a process termed NO_3^- ammonification or dissimilatory NO_3^- reduction to ammonium, and (3) reduction to N_2 coupled to the oxidation of NH_4^+ by the newly discovered anammox process. Most dissimilatory NO_3-reducing prokaryotes are facultative anaerobic heterotrophs utilizing either dissolved low-molecular-weight carbon sources (e.g., *Pseudomonas* spp.) or one-carbon compounds (e.g., *Alcaligenes* spp.) (Table 7.7). Other dissimilatory NO_3^- reducers grow as chemolithoautotrophs by oxidizing reduced inorganic compounds, such as H_2 (*Paracoccus* spp.), H_2S (*Thiobacillus* spp.). Also, the anammox bacteria of the *Planctomycetales* are chemolithoautotrophs.

NO_3^- reduction in sediments occurs below the oxic surface zone. In open waters, NO_3^- reduction may occur in O_2-depleted zones of the oceans (oxygen minimum zones) and in the anoxic hypolimnion of lakes when NO_3^- is present. River plumes rich in suspended particles support NO_3^- reduction within aggregates, where anoxic microenvironments occur in an otherwise oxygenated water body (Omnes *et al.*, 1996).

The isotope pairing technique

The nitrogen isotope pairing technique (Nielsen, 1992) was developed to measure nitrification and denitrification rates in sediments but can be applied to other environments as well. The water in the enclosed experimental system is enriched with $^{15}NO_3^-$, which mixes with the $^{14}NO_3^-$ of the naturally occurring NO_3^-. The formation of single-labeled ($^{14}N^{15}N$) and double-labeled ($^{15}N^{15}N$) dinitrogen pairs by denitrification is measured by mass spectrometry. Denitrification, including the formation of unlabeled ($^{14}N^{14}N$) dinitrogen, can be determined assuming random isotope pairing by denitrification of the uniformly mixed NO_3^- species. Rates of denitrification based on $^{15}NO_3^-$ (D_{15}) are calculated as follows:

$$D_{15} = (^{14}N^{15}N) + 2(^{15}N^{15}N)$$

Because the production of $^{14}N^{14}N$ cannot be precisely measured due to the large natural background, denitrification based on $^{14}NO_3^-$ (D_{14}) is calculated according to the following:

$$D_{14} = [(^{14}N^{15}N)/2(^{15}N^{15}N)]D_{15}$$

D_{14} thus represents the indigenous denitrification rate of the system.

To gain further insight into the sources of NO_3^-, denitrification may be divided between the activity based on NO_3^- from the overlying water (D_{14}^{w}) and that based on NO_3^- from nitrification (D_{14}^{n}):

$$D_{14}^{w} = D_{15}/e$$

$$D_{14}^{n} = D_{14} - D_{14}^{w}$$

where e is the ^{15}N-labeled fraction of the NO_3^- in the surrounding water (Figure 7.24).

NO_3^- ammonification (DNRA) can be estimated as

$$\text{DNRA} = (F_{\text{NH4}}\ x\ \Delta y)/n$$

where F_{NH4} is the formation of NH_4^+, Δy is the change in ^{15}N labeled fraction of NH_4^+, and n is the ^{15}N-labeled fraction of NO_3^- that is reduced to NH_4^+ (modified from Rysgaard *et al.*, 1993).

Total nitrification (N_t) can be determined as

$$N_t = D_{14}^{n} + N_f$$

where N_f is the measured net formation of unlabeled NO_3^-. The fundamental limitation of the isotope pairing technique is the demand for a uniform mixing of the added $^{15}NO_3^-$ with the endogenous sources of $^{14}NO_3^-$. The assumption of random isotope pairing is not valid when anammox contributes significantly to N_2 production, and further measurements are needed to determine denitrification rates.

Table 7.7 Examples of archaeal and bacterial genera harboring denitrifying species with the source of energy, electrons, and carbon indicated

	Genus	Energy/electron source	Carbon source
Archaea			
	Halobacterium	Chemoorganotrophic	Heterotrophic
	Pyrobaculum	Chemoorganotrophic	Heterotrophic
Bacteria			
	Pseudomonas	Chemoorganotrophic	Heterotrophic
	Aquaspirillum	Chemoorganotrophic	Heterotrophic
	Azospirillum	Chemoorganotrophic	Heterotrophic
	Pseudomonas	Chemoorganotrophic	Heterotrophic
	Alcaligenes	Chemoorganotrophic	Heterotrophic
	Rhodobacter	Photolithotrophic	Autotrophic
	Beggiatoa	Chemolithotrophic	Autotrophic
	Thiobacillus	Chemolithotrophic	Autotrophic
	Thioploca	Chemolithotrophic	Autotrophic
	Paracoccus	Chemolithotrophic	Autotrophic

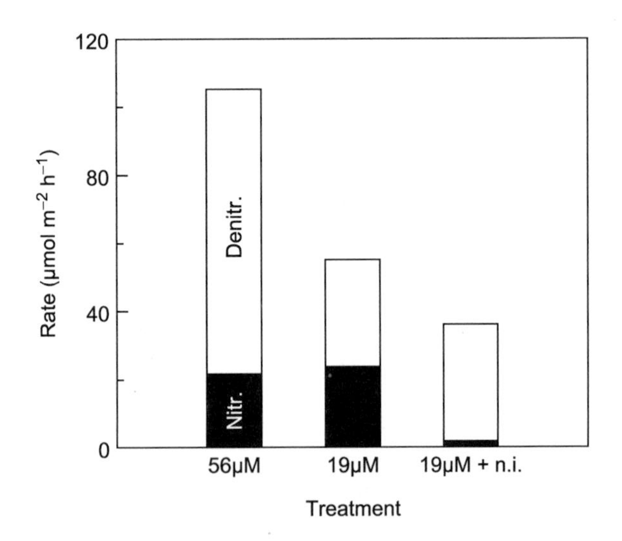

Figure 7.24 Rates of nitrification and denitrification in stream sediment as determined by the isotope pairing technique. Black bars (Nitr.) indicate coupled nitrification-denitrification (equivalent to nitrification when no NO_3^- efflux occurs across the sediment–water interface) and white bars indicate denitrification of NO_3^- diffusing from the overlying water (Denitr.). Scenarios are shown with two different NO_3^- concentrations and with the addition of the nitrification inhibitor (n.i.) thiourea. Modified from Nielsen (1992).

6.1. Biochemistry of denitrification

Denitrification is a major sink in the nitrogen cycle, converting NO_3^- to N_2 and removing fixed nitrogen from the environment. It consists of a number of respiratory reduction steps (Equation 7.17).

$$NO_3^- \rightarrow NO_2^- \rightarrow NO \rightarrow N_2O \rightarrow N_2 \tag{7.17}$$

The characteristics and function of the enzymes (reductases) catalyzing each of the reduction steps are briefly discussed in the following sections.

6.1.1. NO_3^- reductase

Respiratory NO_3^- reductases are membrane-bound complexes consisting of two to three subunits (Zumft, 1997), and they are remarkably similar among denitrifiers in subunit structure and molecular weight. The enzyme contains eight to twelve Fe-S groups and one atom of molybdenum in the catalytic subunit. The concentration of NO_3^- reductase in the cytoplasmic membrane of denitrifiers may be as high as 25% of the total membrane proteins (Stouthamer, 1988). The mechanism of coupled proton-electron transfer during NO_3^- reduction involves the interaction of Mo(IV) with NO_3^-. A fully protonated ligand (X) is attached to the Mo atom. As the Mo(IV) reduces NO_3^- by two electrons, it is oxidized to Mo(VI) and the donor ligand X transfers a proton to NO_3^-, which then splits into NO_2^- and hydroxide:

$$\begin{aligned} HX - Mo(IV)^+ + NO_3^- \rightarrow HX - Mo(IV) - NO_3 \rightarrow \\ X - Mo(VI)^{2+} + NO_2^- + OH^- \end{aligned} \tag{7.18}$$

The Mo(VI) is reduced again by the Fe-S clusters in the enzyme (Stouthamer, 1988).

6.1.2. NO_2^- reductase

Two main types of respiratory NO_2^- reductases have been isolated from denitrifiers: a tetraheme (cytochrome cd_1-Nir) enzyme and a copper-containing (CuNir) enzyme (Zumft, 1997). About 75% of the known denitrifying strains have the cd_1-Nir enzyme, and the two enzymes are not found together in the same strain (Gamble *et al.*, 1977). The cd_1-Nir enzyme consists of two identical subunits, each containing two types of prosthetic groups: heme *c*, which is covalently linked to the protein, and a non-covalently bound chlorine type, heme d_1. Electrons enter cd_1-Nir at heme *c* (Fe^{2+}) and are transferred to NO_2^- via heme d_1. Transfer of one electron is associated with protonation and removal of water and yields one molecule of NO:

$$d_1\text{–}Fe^{2+} + NO_2^- + 2H^+ \rightarrow d_1\text{–}Fe^{2+} - NO_2^- + 2H^+ \rightarrow d_1\text{–}Fe^{3+} + NO + H_2O \qquad (7.19)$$

The product of the cd_1-Nir pathway is always NO.

The Cu-Nir enzyme contains two identical subunits with one copper atom each. The transfer of electrons from Cu to NO_2^- is associated with protonation and removal of water according to the following:

$$Cu^+ + NO_2^- + 2H^+ \rightarrow Cu^+ - NO_2^- + 2H^+ \rightarrow Cu^{2+} + NO + H_2O \qquad (7.20)$$

The product of the Cu-Nir pathway is mostly NO, but N_2O may be formed under strongly reducing conditions and at high pH.

As NO is highly reactive with O_2 and transition metals. It is toxic to living organisms and rarely accumulates in cells, but is either excreted to the surroundings or rapidly reduced intracellularly.

6.1.3. NO reductase

NO reductase, which reduces NO to N_2O, was the last identified enzyme involved in denitrification. The enzyme is composed of two subunits, NorB (highly hydrophobic heme *b*-type cytochrome) and NorC (membrane-bound monoheme *c*-type cytochrome) (Zumft, 1997). Electrons are believed to enter the iron-containing reaction center in the NorB subunit via the NorC subunit and the process proceeds according to the following:

$$2NorB\text{-}Fe^{2+} + 2NO + 2H^+ \rightarrow 2NorB\text{-}Fe^{3+} + N_2O + H_2O \qquad (7.21)$$

However, the exact mechanisms of NO reduction are still not fully understood.

6.1.4. N_2O reductase

The conversion of N_2O to N_2 is the last step of the denitrification pathway. The role of N_2O as an obligatory intermediate is confirmed by the fact that many denitrifying bacteria are able to grow at the expense of N_2O as the sole electron acceptor. N_2O reductase (NosZ proteins) consists of two subunits with about four Cu atoms per subunit. Parallel pathways of electron transfer, and sometimes alternative electron donors for the enzyme, exist in different denitrifiers, but there are strong indications for the involvement of *c*- and *b*-type cytochromes:

$$2NosZ\text{-}Cu^+ + N_2O + 2H^+ \rightarrow 2NosZ\text{-}Cu^{2+} + N_2 + H_2O \qquad (7.22)$$

6.1.5. The entire denitrification process

NO_3^- reductase and NO reductase are membrane-bound enzymes in most denitrifiers, whereas NO_2^- reductase and N_2O reductase are located in the cell periplasm. Although complete denitrification of NO_3^- to N_2 is most common, a number of denitrifiers lack one or more of the reductases. For example, N_2O reductase is absent in some *Rhizobium* and *Pseudomonas* strains. In some cases, organisms that normally contain the full complement of reductase enzymes can miss one of them in defective mutants (Stouthamer, 1988). Alternatively, the environment may be deficient in one or more trace element, such as the metals Fe, Cu, and Mo, which are essential for the biosynthesis of denitrification enzymes. Since N_2O reductase is only fully operative around pH 7 and is inhibited completely at pH 5 due to conformational changes of the protein (Zumft, 1997), the end product of denitrification will gradually shift from N_2 to N_2O when pH decreases in the environment.

Complete denitrification can be viewed as the modular assemblage of four partly independent respiratory processes (Zumft, 1997). Complete denitrification is achieved only when all four modules are activated simultaneously (Figure 7.25). When there is no overlap between the modules, or only pairwise overlap, accumulation of intermediates such as NO_2^-, NO, and N_2O may occur.

6.2. Biochemistry of NO_3^- ammonification

NO_3^- ammonification, or dissimilatory nitrate reduction to ammonium, is used by organisms to detoxify NO_2^-, and in some cases as an electron sink during fermentation (see also Section 4 in Chapter 9). Some NO_3^- ammonifiers are also true respirers (Welsh *et al.*, 2001). NO_3^- ammonification is therefore defined as the dissimilatory transformation of nitrogen oxides to NH_4^+. In some cases, but not all, this process is coupled to energy conservation. The complete reduction of NO_3^- to NH_4^+ proceeds in two steps:

$$NO_3^- \rightarrow NO_2^- \rightarrow NH_4^+ \tag{7.23}$$

NO_3^- ammonification can occur under the same environmental conditions as denitrification, and because N_2 gas is not produced, this process keeps nitrogen in the ecosystem. NO_3^- ammonification gains the most prominence in highly reducing environments, particularly in the presence of free sulfide (Brunet and Garcia-Gil, 1996).

Both facultative and obligate anaerobes mediate NO_3^- ammonification, but the pathways are often different and complex (Bonin, 1996). Since the first step in both denitrification and NO_3^- ammonification is the respiratory transformation of NO_3^- to NO_2^-, catalyzed by NO_3^- reductase, these two processes separate biochemically only after NO_2^- is formed.

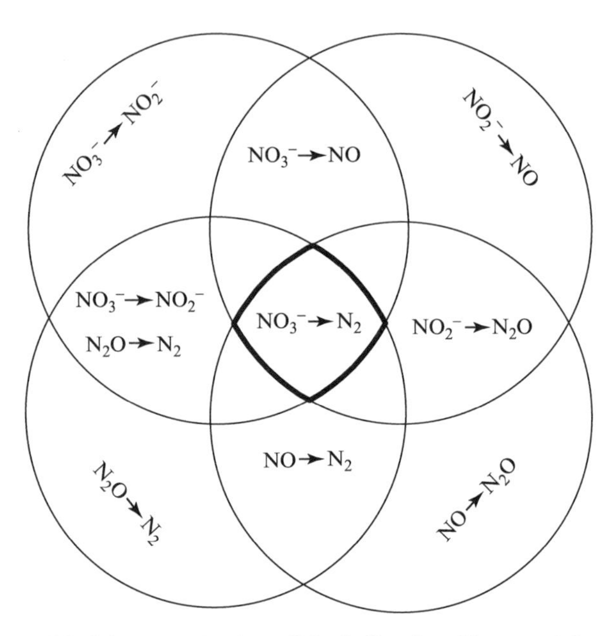

Figure 7.25 Modular organization of denitrification. Four modules representing the respiratory systems utilizing NO_3^- (upper left), NO_2^- (upper right), NO (lower right), and N_2O (lower left) constitute the overall process. Complete denitrification is achieved only when all four modules are active (zone bordered by heavy lines). Partial overlap of two or three modules occurs naturally in individual denitrifiers and in microbial communities. Modified after Zumft (1997).

NO_3^- ammonification is widespread among heterotrophic and chemoautotrophic *Bacteria*, and the reaction can be considered a short circuit in the nitrogen cycle (Figure 7.3). A number of different reductases are known to catalyze the terminal six-electron reduction of NO_2^- to NH_4^+. Common to them all is the presence of heme, which has iron and sulfur clusters in the oxidation-reduction centers. A NO_2^- reductase containing six *c*-type heme groups per molecule has been purified from *Desulfovibrio desulfuricans*, and it reduces NO_2^- by the following reaction (Liu and Peck, 1981):

$$6c\text{-}Fe^{2+} + NO_2^- + 8H^+ \rightarrow 6c\text{-}Fe^{3+} + NH_4^+ + 2H_2O \qquad (7.24)$$

Some of the most common NO_3^- ammonifiers in coastal marine sediments are fermenting bacteria within the genera *Aeromonas* and *Vibrio*. These organisms couple the reduction of NO_2^- to the oxidation of NADH produced during glycolysis (Bonin, 1996). We can follow this path as follows: pyruvate is generated together with NADH in the first step of glucose fermentation:

$$\text{glucose} + 2\ \text{NAD}^+ \rightarrow 2\ \text{pyruvate} + 2\ \text{NADH} \tag{7.25}$$

Among other products, a fraction x of the pyruvate will be fermented to ethanol with the consumption of NADH, while a fraction y will be fermented to acetate (White, 1995):

$$\text{pyruvate} + \text{NADH} \begin{cases} \nearrow x\ \text{ethanol} + x\ \text{NAD}^+ \\ \searrow y\ \text{acetate} \end{cases} \tag{7.26}$$

The acetate pathway can only proceed when reactions other than the ethanol pathway are available to oxidize NADH and replenish NAD^+ in the cells. This is where NO_2^- comes in, as it can act as an alternative electron sink for NADH:

$$NO_2^- + NADH \rightarrow NH_4^+ + NAD^+ \tag{7.27}$$

The overall outcome of glucose fermentation by NO_3^- ammonifiers depends, therefore, on the availability of NO_2^-; acetate formation occurs and $y/x > 0$ when NO_2^- is present, whereas this pathway is inhibited and y/x approaches 0 when NO_2^- is deficient.

Chemolithoautotrophic NO_3^- ammonification coupled to sulfide oxidation has been discovered in giant marine bacteria belonging to the genus *Thioploca* (Schulz and Jørgensen, 2001). *Thioploca* (see Chapter 9 for further discussion) is capable of accumulating up to 500 mM NO_3^- in its large central vacuole from sea water containing ~25 μM NO_3^- (Fossing *et al.*, 1995). The nitrate is used as an electron acceptor in oxidizing sulfide to sulfate, with the production of NH_4^+ according to Equation 7.28 (Otte *et al.*, 1999). Other phylogenetically related NO_3^- accumulating sulfide oxidizers of the genera *Beggiatoa* and *Thiomargarita* may also reduce NO_3^- to NH_4^+ during sulfide oxidation, but this still needs to be established.

$$HS^- + NO_3^- + H_2O + H^+ \rightarrow SO_4^{2-} + NH_4^+ \tag{7.28}$$

It appears that both bicarbonate and acetate may act as carbon sources, indicating that *Thioploca* is a facultative NO_3^- ammonifying chemolithoautotroph capable of mixotrophic growth.

6.3. Phylogeny and detection of denitrifiers

Nearly 130 denitrifying prokaroyte species have been isolated within more than 50 genera, and most of these are found within the α, β, and γ subdivisions of the proteobacteria (Zumft, 1997). Denitrification is, however, widespread among the *Bacteria*, including gram-positive *Bacillus* spp. and the deep-branching *Aquifex pyrophilus*. Among the *Archaea*, denitrification is described from the halophilic *Haloarcula denitrificans*, while other halophiles and hyperthermophiles such as *Pyrobaculum aerophilum* denitrify NO_2^- to N_2O.

Key enzymes in the denitrification process, NO_2^- reductase and N_2O reductase, have been the focus of phylogenetic studies (Hallin and Lindgren, 1999; Scala and Kerkhof, 1999). The two structurally different but functionally equivalent enzymes catalyzing NO_2^- reduction, the tetraheme (cd_1-Nir) enzyme and the copper-containing (Cu-Nir) enzyme, are encoded by the genes *nir*S and *nir*K. The two types of *nir* genes are mutually exclusive in a given strain, although the *nir* type may differ within the same genera and even within the same species (Coyne *et al.*, 1989). There is high diversity and near-random distribution of the two *nir* genes among denitrifying genera (Figure 7.26). For example, the genera *Alcaligenes* and *Pseudomonas* contain both *nir*S and *nir*K genes. The same *nir*-type gene among otherwise phylogenetically different groups, and the occasional occurrence of different *nir* types within the same species, could be an indication of horizontal gene transfer (Braker *et al.*, 1998).

The gene encoding N_2O reductase (*nos*Z) is largely unique to denitrifying bacteria and has been used to indicate the presence of denitrifiers in the environment (Scala and Kerkhof, 1999). Phylogenies constructed from *nos*Z gene sequences can also be compared with phylogenies from 16S rRNA to check for congruent evolution between the two genes. Indeed, comparisons show some inconsistencies. For example, the 16S rRNA unrelated *Pseudomonas denitrificans* and *Paracoccus pantotrophus* (formerly *Thiosphaera pantotropha*) have almost identical *nos*Z genes (Figure 7.27). By contrast, the 16S rRNA

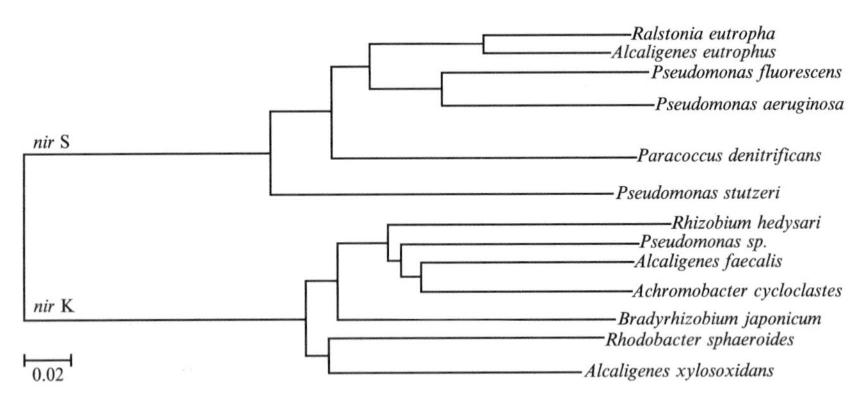

Figure 7.26 Neighbor-joining phylogenetic tree based upon *nir* gene sequences from denitrifying prokaryotes. Scale bar indicates 2% sequence divergence. Adapted from Braker *et al.* (1998, 2000).

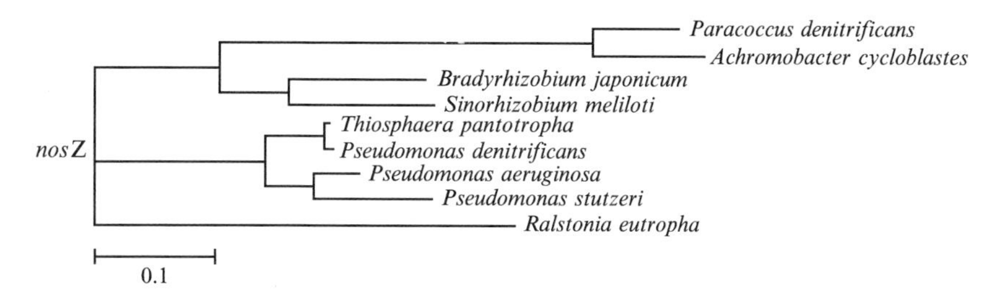

Figure 7.27 Phylogenetic tree based on *nosZ* gene sequences from different species of denitrifiers. Scale bar indicates 10% sequence divergence. Adapted from Scala and Kerkhof (1999).

closely related *Paracoccus pantotrophus* and *Paracoccus denitrificans* have only distantly related *nos*Z genes. Taken together, these results imply that *P. pantotrophus* has either acquired a 16S rRNA gene from *Paracoccus denitrificans* (or a related organism) or eliminated a functional *nos*Z gene in favor of a *nos*Z gene from *Pseudomonas denitrificans*.

There is a substantial phylogenetic diversity among denitrifiers in both freshwater and marine environments. Many of the variety of clones and strains obtained by sequence analysis of *nir* and *nos* genes from environmental samples are not represented in culture collections, suggesting that denitrification is more widespread among prokaryotes than previously anticipated (Scala and Kerkhof, 1999; Braker *et al.*, 2000). Denitrifying clones often group according to the environment from which they are obtained. For example, there are strong geographic differences among denitrifier populations in marine sediments (Scala and Kerkhof, 1999), and these are greater than the difference in population structure one might observe with depth in a sediment at a given location. Thus, mixing by burrowing animals and other advective mechanisms may limit the gradients in sedimentary population composition (Braker *et al.*, 2001).

On the other hand, changes in key environmental parameters within relatively short geographical distances exert site-specific selection pressure and could cause diversification among denitrifiers. Hence, denitrifiers are apparently well adapted to the environment where they are found. As an example, the relatively high input of refractory organic matter into near-coastal sediments off the Washington coast seems to support distinct denitrifier populations compared to continental shelf sediments receiving a greater input of labile carbon (Braker *et al.*, 2000). The long-term geographical separation of similar environments may also result in species diversification. Thus, enclosed basins such as the Baltic Sea and the Black Sea, with anoxic deep water, create an ideal environment for denitrifiers at the oxic–anoxic interface. However, many of the denitrifier species in these two similar environments are phylogenically distinct but in some cases closely related, such as *Shewanella baltica* in the Baltic and *S. oneidensis* in the Black Sea (Brettar *et al.*, 2001).

6.4. Environmental factors affecting denitrification

Denitrification is an inducible process, occurring only when O_2 is absent or nearly absent and when NO_3^- is present. Factors other than absence of O_2, however, may influence denitrification rates in aquatic environments; these include temperature and substrate availability (NO_3^- and organic matter). The following discussion concerns the principal factors controlling denitrification rates in the environment and how they are interrelated. We primarily focus on sedimentary environments, as denitrification is a ubiquitous and important process in most aquatic sediments (Table 7.6).

6.4.1. Dependence of NO_3^- *and* O_2

Rates of denitrification in cultures and in the environment follow Michaelis-Menten kinetics with respect to NO_3^- (Seitzinger, 1990). In sediments, half-saturation constants (K_m) are generally between 2 and 170 μM NO_3^- with an average of about 50 μM NO_3^-. However, values higher than 500 μM NO_3^- also have been reported (Joye *et al.*, 1996; Garcia-Ruiz *et al.*, 1998). Accordingly, denitrification in most natural and unpolluted aquatic environments responds to NO_3^- concentration because concentrations are usually lower than 20 μM in these environments. Adequate supplies of NO_3^- are therefore essential to maintain the denitrification process at high rates.

There are two major sources of NO_3^- for denitrifying communities in aquatic sediments: (1) NO_3^- diffusing into the sediment from the overlying water and (2) NO_3^- produced within the sediment by nitrification. Rates of denitrification, and the partitioning between the two sources of NO_3^-, are strongly dependent on the NO_3^- concentration in the overlying water as well as the penetration depth of O_2 into the sediment (Figure 7.28). Since denitrification is restricted to a thin anoxic layer immediately under the oxic zone, it is primarily the thickness of the oxic zone that controls the diffusional supply of NO_3^- from the overlying water to the denitrifiers below (Nielsen *et al.*, 1990). The longer the diffusion path for NO_3^- through the oxic zone, the less steep the NO_3^- gradient and the lower the diffusional supply. Often, however, NO_3^- generated within the sediment is the main source for denitrification due to a close coupling between nitrification in the oxic zone and denitrification in the anoxic zone (Jenkins and Kemp, 1984). The NO_3^- produced in the oxic zone will diffuse both into the overlying water and into the anoxic sediment to be denitrified. Even with an oxygen penetration of just 1 mm, some 30% of the NO_3^- produced by nitrification may be denitrified (Rysgaard *et al.*, 1994), and higher percentages may be reached in less active sediments with a thicker oxic zone (Seitzinger, 1988). At any given location, this percentage depends on the O_2 concentration in the overlying water. Increasing O_2 levels expand the O_2 penetration depth, which decreases the NO_3^- diffusional loss of from the overlying water and favors the diffusion of NO_3^- from nitrification into the denitrification zone (Figure 7.28).

With high NO_3^- concentrations in the overlying water (eutrophic estuaries, lakes, and streams), total sediment denitrification rates are inversely proportional to O_2 penetration depth. This is intuitive, as deep O_2 penetration depths will also decrease the NO_3^- gradient into the sediment decreasing the NO_3^- flux. This factor overrides the more modest increase in coupled nitrification-denitrification rates as O_2 penetration depth increases (Jensen *et al.*, 1994). Oxygen regulates the genes that encode the enzymes required for denitrification, and gene expression is inhibited by oxygen concentrations of $>1\,\mu$M (Zumft, 1997). Similarly, denitrification in the oceanic oxygen minimum zones requires O_2 concentrations $\leq 3\,\mu$M (Codispoti *et al.*, 2001), while somewhat higher limiting concentrations have been inferred for other waters and sediments (Seitzinger, 1988). In a few unusual cases, as with the denitrifying mixotroph, *Paracoccus (Thiosphaera) pantotrophus*, NO_3^- and O_2 can be used simultaneously as electron acceptors at O_2 concentrations up to 90% of air saturation (Robertson and Kuenen, 1984). Thus, in some instances, active denitrification occurs under well-oxygenated conditions (Robertson *et al.*, 1995). The significance of this aerobic denitrification in nature is presently unknown.

6.4.2. Influence of animals and plants

Burrow-dwelling animals and rooted plants create a three-dimensional mosaic of physico-chemical and biological microenvironments reaching deep into sediments. The surface area available for diffusive exchange and the area of the oxic–anoxic boundaries is considerably increased by the presence of these organisms (Aller, 1982) (see Chapter 4). Furthermore, the distribution of reaction rates and solutes within the sediment varies in both time and space according to the patterns of activity of the plants and animals involved (Aller, 1994a; Christensen *et al.*, 1994).

Burrowing invertebrates such as polychaetes, crustaceans, and insect larvae are known to stimulate denitrification rates, measured per area of sediment surface, by a factor of 2–6 (Pelegri *et al.*, 1994; Svensson, 1997). Irrigation of animal burrows results in advective transport of O_2 and NO_3^- from the overlying water to deeper anoxic sediment layers, where both nitrification and denitrification are stimulated (Kristensen *et al.*, 1991b).

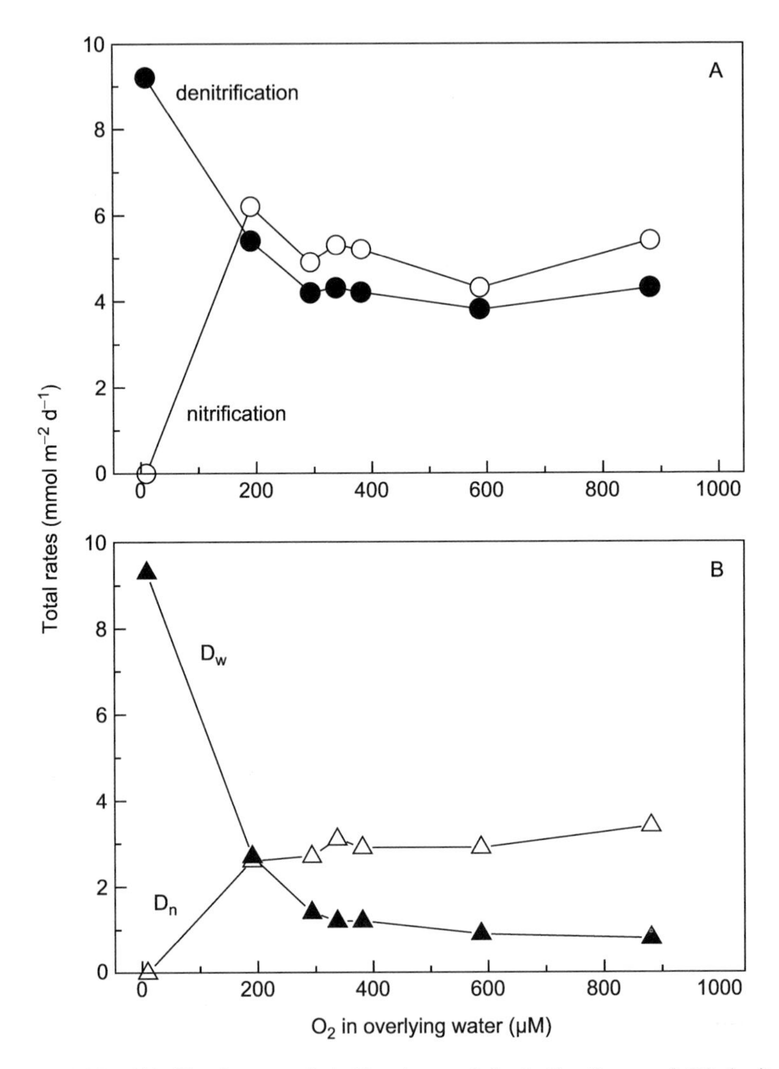

Figure 7.28 (A), Total rates of nitrification and denitrification, and (B) denitrification as determined from ^{15}N labeling in microcosm experiments on sediment collected from Lake Vilhelmsborg, Denmark. Indicated in (B) is whether denitrification is based on NO_3^- from the overlying water (D_w) or is coupled to nitrification-denitrification (D_n). Experiments were performed with different O_2 concentration in the overlying water. The NO_3^- concentration in the overlying water was 30 μM. Modified from Jensen *et al.* (1994).

However, they may not be stimulated equally, as Bartoli *et al.* (2000) reported that denitrification of water column NO_3^- (30 μM) was enhanced 3–10 times more than coupled nitrification-denitrification in sediments bioturbated by the polychaete *Nereis succinea*. Nitrogen-transforming processes are particularly enhanced when secreted mucus linings and associated organic particles along burrow walls provide labile organic carbon and nitrogen sources as well as reactive surfaces for microbial growth.

Submerged and rooted macrophyte communities also stimulate area-specific denitrification rates in sediments. The degree of stimulation varies from near zero for the marine eelgrass (*Zostera* sp.) to more than sixfold in freshwater lobelia (*Lobelia* sp.) (Risgaard-Petersen and Jensen, 1997; Ottosen *et al.*, 1999). Plants growing with roots in anoxic sediments have a well-developed lacunar system by which they establish a gas space continuum between leaves and the root tissue. Oxygen rapidly diffuses to the roots via the lacunar conduits, where it supports the aerobic metabolism of the root cells. However, O_2 may also diffuse across the epidermis of the roots and leak into the surrounding rhizosphere. Consequently, a niche is formed for microbial communities performing coupled nitrification-denitrification. Species of macrophytes differ in the amount of O_2 released from their roots, ranging from almost nothing in *Zostera* sp. to large quantities in *Lobelia* sp. (Sand-Jensen *et al.*, 1982). The denitrifier community in rhizospheres may be further supported by the concurrent release of easily degradable dissolved organic carbon from the roots.

6.4.3. Temporal variations

Denitrification rates in aquatic sediments may vary on both seasonal and diel time scales, and the rates rarely correlate with temperature, which contrasts with other sediment respiratory processes such as sulfate reduction (see Chapter 9). The metabolic activity of denitrifiers in nature does correlate with temperature, and a Q_{10} response of around 2 is normal (from 0 to 35 °C; Focht and Verstraete, 1977). However, seasonal variations in denitrification rates are primarily regulated by the availability of organic carbon and NO_3^- supplied by diffusion from the overlying water and by the intensity of nitrification in the sediment. These parameters do not all vary in concert with temperature.

Denitrification maxima are usually observed during the spring and occasionally, but to a lesser extent, during the fall in freshwater and marine sediments (Figure 7.29) (Rysgaard *et al.*, 1995; Pattison *et al.*, 1998). The distinct spring maximum is caused by a combination of increasing temperatures, ample supplies of NO_3^- from the overlying water, high rates of nitrification in the oxic sediment, and an increased supply of labile organic carbon from the microalgal spring bloom (benthic as well as pelagic). Low denitrification rates during summer are primarily related to lower NO_3^- and O_2 availability. Denitrification in sediments is frequently hampered as water

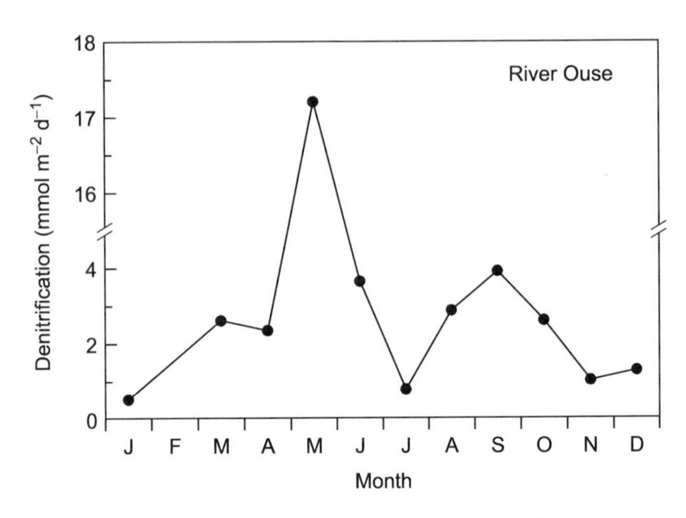

Figure 7.29 Denitrification in the sediment of the River Ouse over one seasonal cycle. Modified from Pattison *et al.* (1998).

column NO_3^- is scavenged by primary producers. Coupled nitrification-denitrification, on the other hand, may be O_2 limited during warm summer months when high rates of microbial respiration create anoxic conditions near the sediment surface. Oxygen saturation is also reduced in warm water. Nitrification may be further suppressed due to competition for NH_4^+ with a rapidly growing heterotrophic population. The reducing conditions prevailing in most organic-rich sediments during summer may also cause a partial shift from denitrification to NO_3^- ammonification. The latter process thrives best under strongly reducing conditions, when up to 80% of the reduced NO_3^- may pass via this pathway (Christensen *et al.*, 2000).

A small denitrification maximum in fall develops as a response to low temperature and reduced light conditions. Coupled nitrification-denitrification increases following the expansion of the oxic surface sediment, and the competition for NH_4^+ by heterotrophs is reduced as overall microbial respiration rates diminish. The NO_3^- supply from the overlying water increases as primary production drops and primary producers are unable to consume all the NO_3^- supplied. The low rates of denitrification prevailing during winter months are probably a direct temperature effect because NO_3^- and O_2 usually are found in sufficient amounts.

Diel variations in denitrification are observed only in shallow areas with dense populations of benthic microalgae. The high daytime primary production rates by microalgae increase O_2 penetration into the sediment (Chapter 13). The primary producers also actively consume available NO_3^- to support biomass production. Consequently, the supply of NO_3^- from the overlying water is reduced, and NO_3^- originating from nitrification within the sediment is increased (Rysgaard *et al.*, 1994). Thus, during winter and spring, when most of the NO_3^- for denitrifiers is supplied from the overlying water, the denitrification rate is lower during the day than at night. During summer, on the other hand, the rates are highest during the day when denitrification is primarily coupled to nitrification. However, competition for NH_4^+ between benthic microalgae and nitrifiers may hamper the latter process during the day when limited NH_4^+ is available.

7. ANAMMOX

Nearly four decades ago Richards (1965) noted that NH_4^+ did not accumulate in O_2 minimum zones of the ocean supporting denitrification, and he speculated that NH_4^+ might be removed biologically by reaction with NO_3^-, forming N_2 gas, although no specific proof for this process was available. Somewhat later, thermodynamic considerations led Broda (1977) to the conclusion that the anaerobic oxidation of NH_4^+ with NO_2^- (anammox) is as energetically favorable as oxic nitrification (Table 7.8), and he predicted the existence of chemolithoautotrophic organisms driving the following reaction:

$$NH_4^+ + NO_2^- \rightarrow N_2 + 2H_2O \quad (7.29)$$

Studies on the biological nature of the anammox process in wastewater bioreactors (van de Graaf *et al.*, 1995; Jetten *et al.*, 1999) have verified that N_2 is indeed formed when one nitrogen atom from NH_4^+ is paired with one from NO_2^-, as predicted by Broda (1977). Generation of N_2 from NH_4^+ and NO_3^-, on the other hand, requires more reducing power than available from NH_4^+ and is not in agreement with the stoichiometry of the anammox

Table 7.8 Physiological and kinetic parameters of bacterial cultures during aerobic ($NH_4^+ \rightarrow NO_2^-$) and anaerobic (anammox) ($NH_4^+ \rightarrow N_2$) NH_4^+ oxidation in a sequencing batch reactor

Parameter	Aerobic	Anaerobic
Free energy (kJ mol^{-1})	−272	−357
Biomass yield (mol mol^{-1} C)	0.08	0.07
Aerobic rate (nmol mg^{-1} min^{-1})	400	0
Anaerobic rate (nmol mg^{-1} min^{-1})	2	60
Growth rate (h^{-1})	0.04	0.003

Modified from Jetten *et al.* (2001).

process (Jetten *et al.*, 2001). The source of NO_2^- for anammox in NO_2^--poor environments is likely from NO_3^- reduction catalyzed by NO_3^- reductases and coupled to the oxidation of organic carbon.

The only known chemolithoautotrophs involved in the anammox process belong to the order *Planctomycetales* (Strous *et al.*, 1999a). Only a few species have yet been identified by enrichment and 16S rDNA sequencing; the first two are *Brocadia anammoxidans*, from a denitrifying pilot plant at Gist-Brocades, Delft, Holland, and *Kuenenia stuttgartiensis*, from a wastewater treatment plant in Stuttgart, Germany (Jetten, 2001). Organisms related to the wastewater species have been found in the water column of the Black Sea, where their distribution coincides with the zone of anaerobic ammonium oxidation (Kuypers *et al.*, 2003).

The metabolic pathway for anammox is not fully understood. Investigations to date suggest that the electron acceptor NO_2^- is reduced to hydroxylamine (NH_2OH), which reacts with the electron donor NH_4^+, leading to the production of N_2 via the intermediate hydrazine (N_2H_4) (Jetten *et al.*, 2001). At least three enzymes are believed to be involved in the process. The first enzyme reduces NO_2^- to NH_2OH, the second catalyzes the condensation of NH_4^+ and NH_2OH into N_2H_4, and the third enzyme catalyzes the oxidation of N_2H_4 to N_2, thereby releasing the electrons required for NO_2^- reduction. Although the main product of the reaction is N_2, about 20% of the consumed NO_2^- is recovered as NO_3^-. NO_3^- is probably formed as a byproduct when reducing equivalents are needed for CO_2 reduction (van de Graaf *et al.*, 1997). The anammox process is reversibly inhibited by O_2 (Jetten *et al.*, 2001), and no NH_4^+ is oxidized at O_2 concentrations >0.5% of air saturation (Figure 7.30) (Strous *et al.*, 1997).

The generation of N_2 by anammox may be important in many aquatic sediments. Thamdrup and Dalsgaard (2002) demonstrated the anaerobic oxidation of NH_4^+ with NO_2^- in marine sediments, presumably by the anammox reaction. They found that the relative importance of anaerobic NH_4^+ oxidation for total N_2 production in continental shelf sediments increases significantly with water depth, from 2% in a eutrophic coastal bay (16 m) to 24% at 380 m and 67% at 695 m (Figure 7.31). The low relative significance of anammox in shallow sediments conceals that the absolute rates are two to three times higher than in sediments underlying deeper waters because the depth-dependent difference in denitrification is orders of magnitude higher. The strong variation in denitrification is probably related to differences in the availability of electron donors, reflecting the decreasing input and decreasing lability of sedimentary organic matter with increasing water depth (Liu and Kaplan, 1984). Thus, anammox has the potential to consume a significant fraction of NH_4^+ produced in sediments when only NO_3^- and NO_2^- are available, whereas denitrification is strongly

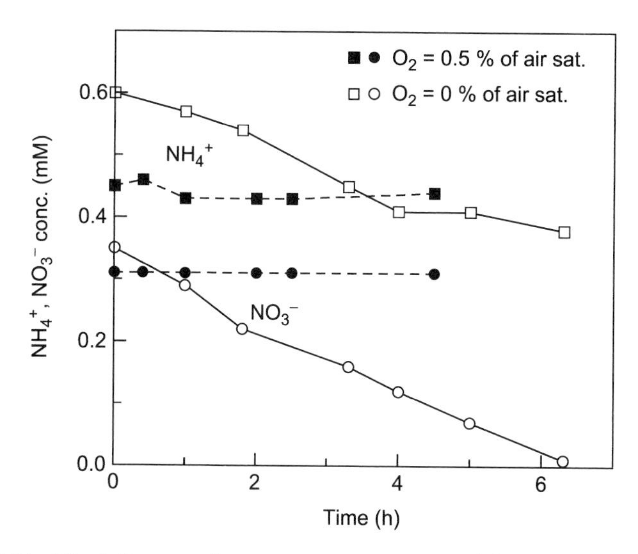

Figure 7.30 The influence of oxygen on anammox activity. Cultures were incubated at the defined oxygen concentrations in a batch reactor. Activity could only be detected when all oxygen was removed. Modified from Strous *et al.* (1997).

dependent on the availability of reactive organic matter (Thamdrup and Dalsgaard, 2002). The process may also contribute substantially to N_2 production in NO_3^--rich anoxic water columns (Dalsgaard *et al.*, 2003). Here anammox is dependent on heterotrophic NO_3^- reduction as a source of NH_4^+ through ammonification.

Anammox produces twice as much N_2 per NO_3^- and NO_2^- molecule as denitrification. The anammox process can therefore explain nitrogen deficiencies in many anoxic waters and sediments, as well as the very efficient conversion of NH_4^+ to N_2 in many shelf sediments, which has previously been attributed solely to a tight coupling of nitrification and denitrification. The process must be included in future revisions of global nitrogen budgets, as it may fill significant gaps in the aquatic N_2 production estimates.

8. ISOTOPE FRACTIONATION

There are two stable isotopes of nitrogen, the major isotope, ^{14}N (99.6337%), and the minor isotope, ^{15}N (0.3663%). As usual, the isotopic composition of a sample is expressed relative to the minor isotope (Equation 7.30), and relative to a standard, taken as the ($^{15}N/^{14}N$) ratio of the atmosphere.

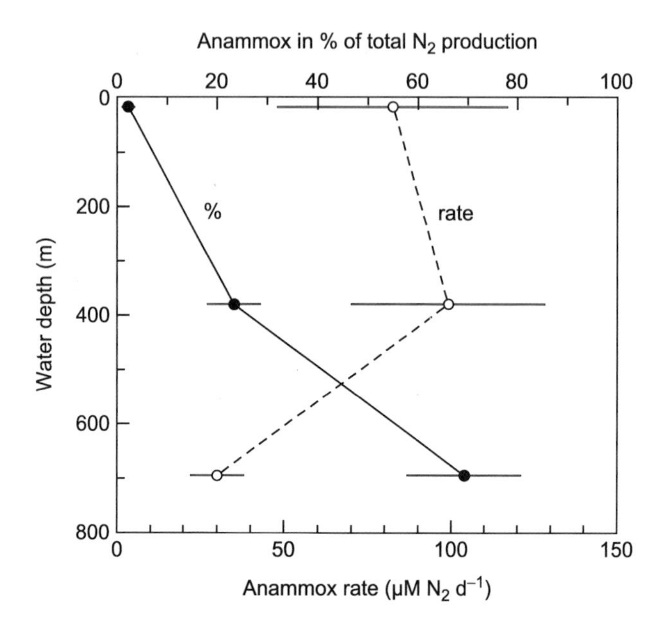

Figure 7.31 Relative contribution of anammox to the total sediment N_2 production (filled symbols), and the rate of sediment N_2 production by anammox (open symbols) as a function of water depth in Danish waters. The shallow station (16 m) is from Aarhus Bay, the intermediate station (380 m) is from the southern Skagerrak, and the deep station (695 m) is from the central Skagerrak. Modified from Thamdrup and Dalsgaard (2002).

$$\delta^{15}N_{sam} = 1000\left[\left(\frac{^{15}N/^{14}N_{sam}}{^{15}N/^{14}N_{std}}\right) - 1\right] \tag{7.30}$$

As with many of the other biologically active elements undergoing redox transformations, isotope fractionation also accompanies the microbially mediated transformations of nitrogen compounds (Owens, 1987). Thus, assimilatory nitrate reduction, N fixation, denitrification, and nitrification impart distinct fractionations for potential use in reconstructing the relative intensity of these processes regionally and globally within aquatic ecosystems (e.g., Cline and Kaplan, 1975; Altabet and Francois, 1994; Sigman *et al.*, 2000; Brandes and Devol, 2002), through the Holocene and Pleistocene (e.g., Francois *et al.*, 1992; Farrell *et al.*, 1995) and over geologic time (Beaumont and Robert, 1999). In the following sections we briefly overview the magnitudes of fractionation associated with the biological processing of nitrogen.

8.1. Denitrification

The isotope fractionation associated with denitrification has been widely studied in pure cultures and natural populations of denitrifiers, and it has been inferred from the distribution of nitrate isotopic compositions in nature. From pure culture and natural population studies, relatively large fractionations ($\varepsilon_{NO_3\text{-}N_2}$) of between 13‰ and 29‰ have been determined (Figure 7.32) (Delwiche and Steyn, 1970; Mariotti *et al.*, 1981; Barford *et al.*, 1999). Generally, these studies have been conducted under optimal growth conditions.

Bryan *et al.* (1983) observed a negative relationship between the extent of fractionation and rates of denitrification (NO_2^- reduction to N_2) for cells of *Pseudomonas stutzeri* under conditions in which NO_2^- was non-limiting and electron donor availability was controlling denitrification. A similar trend was also observed for denitrification by cell-free extracts of *P. stutzeri* with abundant NO_2^- and limiting electron donor. When electron donors were abundant and NO_2^- was limiting, both pure cultures and cell-free extracts showed a positive correlation between rate and fractionation. In this case, NO_2^- availability controlled the rate. Bryan *et al.* (1983) concluded that fractionation is not controlled by NO_2^- exchange across the cell membrane, as is partly the case for sulfate reduction (see Chapter 9). Rather, fractionation is apparently controlled by the availability of NO_2^-, as it controls the build-up of intermediates in the nitrite reduction process and the extent of back reactions. The fractionations observed during the denitrification of NO_2^- for whole cells ranged from about 5% to 25%. These fractionations are similar in range to those observed during denitrification from NO_3^- (see Figure 7.32).

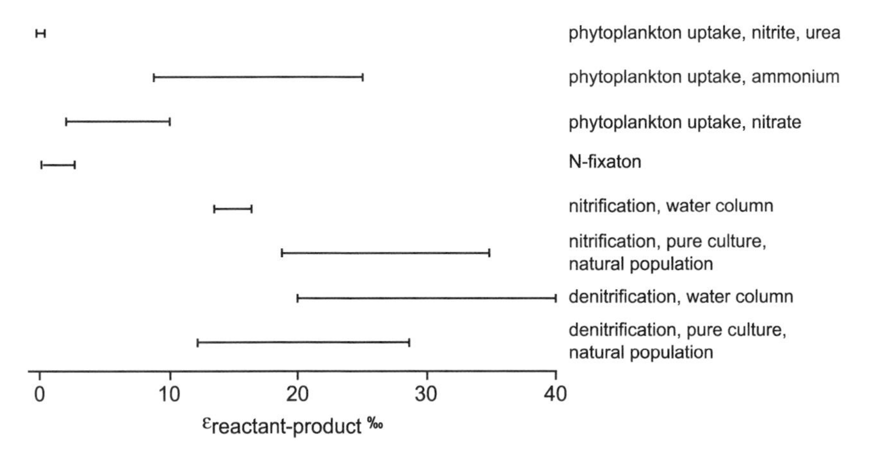

Figure 7.32 Isotope fractionations associated with N transformations by organisms. See text for details and references.

The isotopic composition of NO_3^- increases in oxygen minimum zones of the oceans (e.g., Cline and Kaplan, 1975; Brandes *et al.*, 1998; Voss *et al.*, 2001; Brandes and Devol, 2002). This isotope shift is indicative of denitrification, and modeling of the NO_3^- concentration and isotope profiles, typically with a Rayleigh distillation model, produces fractionations in the range of 20‰ to 40‰. These fractionations are similar to those found in pure culture studies (Figure 7.32). Denitrification in sediments, on the other hand, apparently produces only small net fractionations of around 3‰ (Brandes and Devol, 1997). This low fractionation is probably not due to low fractionations during the denitrification process, but rather to limiting NO_3^- availability in rapidly metabolizing sediment systems.

8.2. Nitrification

Large fractionations also accompany the oxidation of NH_4^+ with oxygen. In natural populations of soil nitrifiers, Delwiche and Steyn (1970) observed fractionations of between 18% and 36% ($\varepsilon_{NH_4\text{-}NO_3}$), while Mariotti *et al.* (1981) observed a large fractionation of 35% during nitrification by *Nitrosomonas europaea*. In a field study, the isotope and concentration profiles of NH_4^+ in the water column of the Chesapeake Bay yielded model-derived fractionations during nitrification in the range of 13‰ to 17‰ (Horrigan *et al.*, 1990). Taken together, nitrification produces substantially ^{15}N-depleted NO_3^-.

8.3. Nitrogen fixation and assimilation

Small but significant fractionations accompany the fixation of atmospheric N_2 by microorganisms. Cyanobacteria have been the best studied, and in culture experiments with *Trichodesmium* sp. (strain IMS101), fractionations ($\varepsilon_{N_2\text{-biomass}}$) of 1.3‰ to 3.6‰ were observed (Carpenter *et al.*, 1997). These fractionations are slightly higher than the fractionations of 0.5‰ to 2‰ typically seen for *Trichodesmium* spp. in nature (e.g., Wada and Hattori, 1976; Carpenter *et al.*, 1997; Montoya *et al.*, 2002). Although the fractionations during N fixation are small, the isotopic consequences can be significant. The nitrogen source for N fixation is atmospheric N_2, which, with a $\delta^{15}N$ of 0‰, can have a significantly different isotopic composition than the fixed nitrogen in aquatic systems. Nitrogen fixation, therefore, can produce isotopically distinct organic biomass. For example, the $\delta^{15}N$ of *Trichodesmium* biomass in the surface oceans is typically −1‰ to −2‰, with associated total particulate organic nitrogen (PON) in the range of −3‰ to 2‰ (Carpenter *et al.*, 1997). By contrast, surface water PON in areas of the ocean devoid of N fixers is much more ^{15}N enriched, with typical $\delta^{15}N$ values of 5‰ to 10‰ (Carpenter *et al.*, 1997).

Fractionations also accompany the uptake of nutrients by plankton and microorganisms and the subsequent processing of PON down the food chain. Phytoplankton fractionate ($\varepsilon_{NO_3\text{-biomass}}$) from 3‰ to 10‰ as they take up NO_3^- during growth (Montoya and McCarthy, 1995; Waser *et al.*, 1998), with flagellates producing lower fractionations than diatoms (Montoya and McCarthy, 1995). When growing with NH_4^+, fractionation seems to depend on the NH_4^+ concentration, and in experiments with diatoms, fractionations ranged from 8‰ at low NH_4^+ concentrations of 5 to 20 μM to 25‰ at concentrations between 50 and 100 μM (Figure 7.32) (Pennock *et al.*, 1996). High fractionations of up to 40% may also accompany microbial growth on NH_4^+, but these results arise from modeling studies in euxinic basins and are not confirmed by direct determination (Velinsky *et al.*, 1991). When grown on NO_2^- and urea, phytoplankton express only small fractionations of around 0.5‰ to 1‰ (Figure 7.32).

The extent to which high fractionations are expressed during nitrogen assimilation in nature depends on whether the nitrogen source is used to exhaustion during primary production. When the substrate is completely consumed, even large fractionations will not be expressed. The processing of PON through the food chain also influences the isotopic composition of the PON. Thus, the PON in animals is about 3.5% enriched in ^{15}N, on average, compared to their food sources (e.g., DeNiro and Epstein, 1981; Montoya *et al.*, 2002). The further up the food chain, the more ^{15}N enriched the PON becomes. Isotope balance occurs as ^{15}N-depleted NH_4^+ is excreted by the organism (Montoya *et al.*, 2002). It is believed that the ^{15}N enrichment of PON through the food chain is responsible for delivering ^{15}N-enriched PON to the deep ocean, which, through subsequent oxidation, enriches deep-water NO_3^- in ^{15}N ($\delta^{15}N$ around 4.5% on average) relative to the primary-produced PON in surface waters (Montoya *et al.*, 2002).

Chapter 8

The Iron and Manganese Cycles

1. INTRODUCTION

Iron has a high abundance of 4.3% by mass in the continental crust (Wedepohl, 1995). This, coupled with a variety of unique geochemical characteristics, allows Fe to interact significantly with the cycles of a number of other biologically important elements such as carbon, sulfur, and phosphorus. Iron redox cycling has been active and globally important throughout

0-12-026147-2
© 2005 Elsevier Inc.
All rights reserved

Earth's history as indicated by the massive deposition of iron-rich sediments, banded iron formations, during the Archean Eon (>2.5 billion years ago) and parts of the Proterozoic Eon (2.5 to 0.54 billion years ago). At a 50-fold lower crustal abundance than iron, manganese is the second most abundant redox-active metal (Wedepohl, 1995). However, manganese concentrates in certain environments, thereby enhancing its local biogeochemical significance. As there are many similarities between iron and manganese in terms of both geochemistry and microbiology, the two elements are discussed together in this Chapter, while important differences are emphasized when appropriate. Many of the microbes transforming iron and manganese also catalyze the redox transformations of other rarer transition elements (Lovley, 2000c; Lloyd, 2003), but this is not discussed in this Chapter because those elements are not known to influence major biogeochemical cycles.

In addition to redox reactions with many other major elements, oxidized and reduced manganese and iron species adsorb and coprecipitate transition metals and other compounds of environmental significance. One such interaction is the binding of phosphate to iron oxides (Chapter 11), which regulates the release of this nutrient from sediments, and which may have significantly limited primary production in the iron-rich Archean ocean (Bjerrum and Canfield, 2002).

Both the oxidation and the reduction of iron and manganese in natural environments is, to a large extent, promoted by microbial catalysis, but abiotic transformations are often important too and may compete with the biological processes. In what follows we look at the processes regulating the transformations of iron and manganese in nature and the relationship between the cycling of these and other biologically active elements. The story is part biological and part chemical, and the relative importance of, and interactions between, abiotic and microbial iron and manganese transformations are a central issue in this Chapter.

2. GLOBAL MANGANESE AND IRON CYCLES

Due to the very low solubility of oxidized manganese and iron, their global cycles are heavily influenced by physical processes. In the Earth's crust the two elements are mainly found as minor components of rock-forming silicate minerals such as olivine, pyroxenes, and amphiboles (Table 8.1). Under oxic conditions and at near neutral pH, iron and manganese released during weathering are rapidly reprecipitated as oxides and hydroxides, as expressed here by the oxidative weathering of fayalite (Fe_2SiO_4) to goethite (FeOOH):

$$Fe_2SiO_{4(s)} + 0.5O_2 + 3H_2O \rightarrow 2FeOOH_{(s)} + H_4SiO_{4(aq)} \qquad (8.1)$$

Table 8.1 Representative iron- and manganese-containing minerals[a]

Silicates	Olivines: fayalite, $Fe(II)_2SiO_4$
	Pyroxenes: augite, $Ca(Mg,Fe(II))Si_2O_6$
	Amphiboles: hornblende, $(Ca,Na)_{2-3}(Mg,Fe(II),Al)_5(Al,Si)_8O_{22}(OH)_2$
	Garnets: almandine, $Fe(II)_3Al_2(SiO_4)_3$
Sheet silicates	Micas: biotite, $K_2(Mg,Fe(II))_{6-4}(Fe(III), Al,Ti)_{0-2}(Si_{6-5}Al_{2-3}O_{20})(OH, F)_4$
	Smectites: nontronite, $Na_{0.3}Fe(III)_2(Si,Al)_4O_{10}(OH)_2 \cdot nH_2O$
Oxides	Ferrihydrite, $Fe(III)_5HO_8 \cdot 4H_2O$; lepidocrocite, γ-Fe(III)OOH; goethite, α-Fe(III)OOH; hematite, α-$Fe(III)_2O_3$; magnetite, $Fe(II)Fe(III)_2O_4$; ilmenite, $Fe(III)_2TiO_4$
Green rusts	E.g., $[Fe(II)_4Fe(III)_2(OH)]_{12}[(CO_3,SO_4) \cdot 3H_2O]$
Sulfides	Mackinawite, Fe(II)S; greigite, $Fe(II)_3S_4$; pyrite, $Fe(II)S_2$
Carbonates	Siderite, $Fe(II)CO_3$
Manganese silicates[b]	Rhodonite, $(Mn(II),Fe(II),Ca,Mg)SiO_3$
Oxides	Birnessite, $Mn(III)Mn(IV)_6O_{13} \cdot 5H_2O$; pyrolusite, β-$Mn(IV)O_2$; manganite, γ-Mn(III)OOH; hausmannite, $Mn(II)Mn(III)_2O_4$
Carbonates	Rhodocrosite, $Mn(II)CO_3$

[a]After Deer *et al.* (1966), Hurlbut and Klein (1977), Burns and Burns (1979), Cornell and Schwertmann (1996).
[b]Manganese-rich silicates are rare. Manganese occurs as a minor component in the same coordination sites as Fe(II).

Most of the manganese and iron transported to the oceans is associated with river-borne particulates, while glaciers and wind-borne dust represent minor fluxes (Figure 8.1) (Poulton and Raiswell, 2002). Some 90% of the riverine inputs deposit on the continental shelves, and the atmosphere therefore is an important source of the metals to surface waters far off shore (Duce and Tindale, 1991), providing a potentially important source of iron for primary production (Section 4). In addition to the particulate fluxes, relatively small amounts of soluble iron and manganese are delivered to the ocean basins with hydrothermal fluids. Metals from this source are oxidized microbially or abiotically and may form copious precipitates around the vents.

Once buried into anoxic sediments, manganese and iron may be mobilized through microbial respiration and abiotic reduction. However, soluble manganese and iron are efficiently trapped through oxidation in oxic surface sediment layers only a few millimeters thick, and typically only a small fraction escapes to the bottom waters (e.g., Thamdrup *et al.*, 1994b). Thus, there is an intensive redox cycling of manganese and iron around the oxic–anoxic interface in many sediments (Canfield *et al.*, 1993b). Still, some manganese can be lost from sediments with shallow oxygen penetration, leading to a redistribution toward locations with deeper oxic zones where manganese is retained (Sundby and Silverberg, 1981). A supply of soluble manganese from the water column is found in the deep sea where centimeter- to decimeter-size

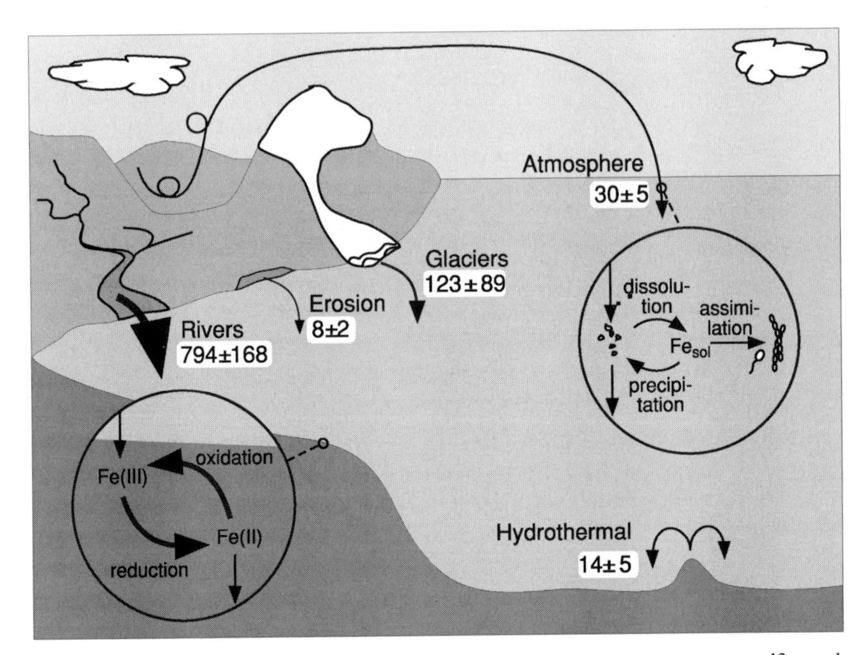

Figure 8.1 The global iron cycle with fluxes in terragrams Fe per year (10^{12} g y^{-1}). Magnified circles illustrate biogeochemical cycling in pelagic waters and sediments. These are the main topic of this Chapter. After Poulton and Raiswell (2002).

nodules of manganese and iron oxides, "manganese nodules," grow at the sediment surface by slow precipitation of metals from the bottom water (Glasby, 2000).

Benthic iron fluxes, although very small relative to the rates of iron cycling within the sediment, may cover the iron requirements of primary production in shelf waters (Berelson *et al.*, 2003). Far off shore, however, soluble concentrations of iron and manganese are below 1 nM (Landing and Bruland, 1987; Johnson *et al.*, 1997), and organisms in the surface waters may rely on scavenging of these trace metals from the particles that settle from the atmosphere. Under these conditions, solubilization is aided by organic chelators produced by the plankton and may involve photochemical reduction (see below).

3. ENVIRONMENTAL GEOCHEMISTRY

In contrast to other elements discussed in this volume, which are mostly covalently bonded in well-defined molecules, iron and manganese in the environment are ions that interact with a wide variety of other species and are mainly found in solid phases (Table 8.1). The oxidized forms, Fe^{3+}, Mn^{3+}, and Mn^{4+}, have very low solubility at neutral pH and are precipitated as hydroxides, oxyhydroxides or oxides, here collectively termed oxides. Chelation with organic compounds may keep Fe^{3+} and Mn^{3+} in solution, but their concentrations typically remain well below 1 μM in natural waters of near neutral pH. The reduced forms, Fe^{2+} and Mn^{2+}, are more soluble, reaching concentrations in the millimolar range, but they also readily precipitate, for example, with sulfides and carbonates (Table 8.1). In sediments, such precipitates are the most important sinks for these reduced ions.

Thermodynamic calculations predict the existence of single, well-defined, mineral forms depending on the environmental conditions. For iron, examples include goethite in the presence of oxygen, magnetite at intermediate redox potentials, and pyrite under sulfidic conditions (Stumm and Morgan, 1981). Kinetic inhibition, however, allows the co-occurrence and frequent dominance of metastable phases that may be poorly crystalline and hard to define analytically. Additionally, a large fraction remains bound as a minor component in silicate minerals (Figure 8.2). While most of this is only reactive on geological timescales, iron in some sheet silicates is microbially reducible (Kostka *et al.*, 1996). The kinetics of oxidation or reduction of iron and manganese, by both biological and abiotic mechanisms, depend on the mineral form, crystallinity, and grain size. In the reaction with sulfide, for example, the complex assembly of oxidized iron and manganese species found in sediments makes up a continuum of reactivities spanning several orders of magnitude from highly reactive, poorly crystalline,

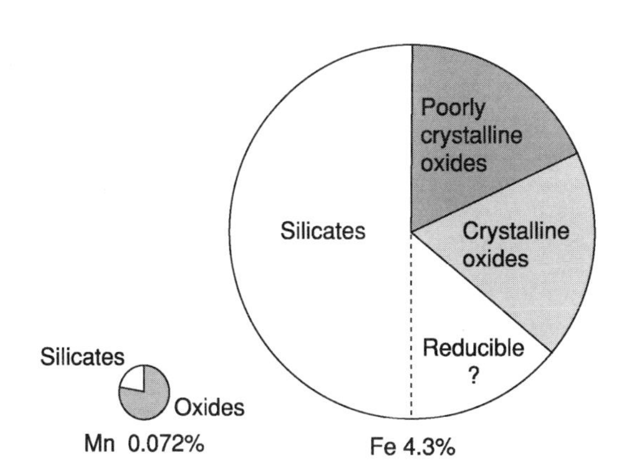

Figure 8.2 Speciation of manganese and iron in river particulates. The poorly crystalline oxide fraction corresponds roughly to the pool available to microbial reduction. After Canfield (1997); redrawn from Thamdrup (2000).

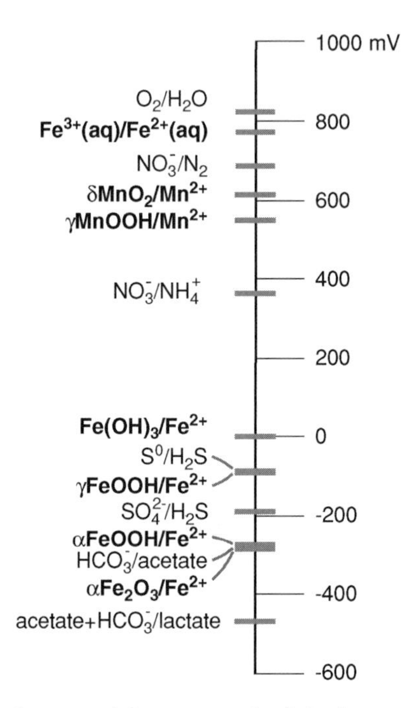

Figure 8.3 Electrode potentials vs. standard hydrogen electrode of selected iron and manganese species, compared to other important redox couples. Potentials were calculated for pH 7 and environmentally relevant concentrations of solutes (Thamdrup, 2000).

nanometer-sized oxides to silicate phases that are transformed only on million year timescales (Canfield *et al.*, 1992).

The reduction potentials of the Fe^{2+}/Fe^{3+} and $Mn^{2+}/Mn^{3/4+}$ couples depend strongly on the speciation of both the reduced and oxidized metal species (Figure 8.3). Thus, the standard potential for the Fe^{3+}/Fe^{2+} couple is 0.77 V, but this potential is only relevant at low pH, where Fe^{3+} is soluble. With poorly crystalline Fe oxide as the oxidant, the potential is around 0.0 V, and with more crystalline oxides it may even become more negative than the H_2S/SO_4^{2-} couple, rendering sulfate the thermodynamically more favorable oxidant.

Both oxidized and reduced forms of iron and manganese react spontaneously with a range of compounds, and microbes that utilize manganese or iron metabolically will often have to compete for the metals with abiotic reactions (see also Chapter 9). Figure 8.4 gives an overview of the environmentally significant abiotic redox reactions involving iron and manganese, together with the microbially catalyzed pathways discussed in detail below. While Fe^{2+} is rapidly oxidized by oxygen with a half-life on the order of

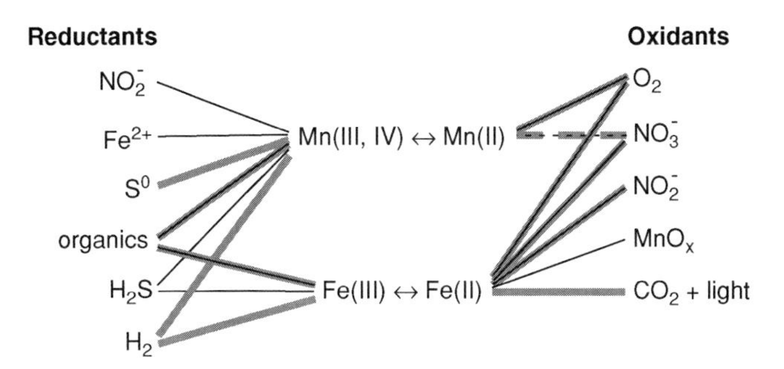

Figure 8.4 Oxidants and reductants for manganese and iron in aquatic environments. Thin lines indicate abiotic reactions; thick grey lines indicate microbially catalyzed reactions; stippled lines indicate the reaction was not experimentally documented. The oxidation of nitrite by manganese oxide requires acidic pH. See text for details.

minutes, Mn^{2+} in solution oxidizes much more slowly, with a half-life of months or more (Stumm and Morgan, 1981). Abiotic manganese oxidation is accelerated by catalytic surfaces such as manganese and iron oxides (Davies and Morgan, 1989), but the overall slower oxidations kinetics result in a much larger mobility for manganese than for iron in aquatic environments. Rapid abiotic oxidation of ferrous iron further occurs by reaction with manganese oxides (Postma, 1985), while iron oxidation with nitrate or nitrite, known as chemo-denitrification, is a relatively slow process, which is catalyzed by the surfaces of iron oxides and other minerals (Ottley *et al.*, 1997).

The environmentally most significant reductant for iron oxide in abiotic reactions is hydrogen sulfide, which also reduces manganese oxide (Yao and Millero, 1993, 1996). Certain organics may also reduce the metal oxides, but for most environmentally relevant compounds the kinetics are slow at neutral pH, and the reaction leads to only a partial oxidation of the organic molecule (Stone *et al.*, 1994). One interesting group of organics is quinones. In the reduced (quinol) form they reduce iron oxides quite rapidly. The resulting oxidized quinones can be reduced microbially, and humic substances containing quinone moieties may serve as an electron shuttle between organisms and iron oxides (Lovley *et al.*, 1996) (see also Section 6.3).

Manganese oxide reduction by ammonium has been indicated in some laboratory experiments but has not been detected by nitrogen-isotope tracer techniques in marine sediment (Thamdrup and Dalsgaard, 2002). A link to nitrogen exists through the oxidation of nitrite to nitrate with manganese oxide at acidic pH (Luther and Popp, 2002).

4. ASSIMILATION

Both manganese and iron are essential as trace elements for most living organisms (Wooldridge and Williams, 1993; Kehres and Maguire, 2003). The requirement for iron is of particular biogeochemical interest, because the availability of this element limits the growth of microorganisms in large parts of the surface ocean, thereby influencing the global carbon cycle (Price and Morel, 1998; Tortell *et al.*, 1999). Manganese, like other trace elements, is required in lower amounts than iron (one important biochemical function of manganese is in the water-splitting complex of photosystem II), and its slower oxidation kinetics make it more accessible in oxic environments (Martin and Knauer, 1973; Bruland *et al.*, 1991). Manganese availability may confer a selective pressure on phytoplankton community composition in some marine environments (Brand *et al.*, 1983), but effects on community metabolism have not been reported.

Almost all organisms take extensive advantage of the redox properties of iron by incorporating this element as an electron carrier in many parts of their biochemical machinery. The redox potential of the Fe^{3+}/Fe^{2+} couple in different enzymes spans 0.8 V as a function of its coordination environment (Braun and Killmann, 1999). Important structural classes of iron-containing proteins are cytochromes, with iron chelated by the heme prosthetic group, and Fe-S proteins such as ferredoxin, with clusters of several mutually coordinated iron and sulfur atoms. Other functionally important iron-dependent enzymes include ribonucleotide reductase, RNA and DNA polymerases, which are enzymes of the tricarboxylic acid (TCA) cycle, oxidative defense systems, and nitrogenase. Accordingly, iron is the quantitatively most important trace element in almost all organisms. Notable exceptions are the lactic acid bacterium *Lactobacillus plantarum* and the spirochaete *Borrelia burgdorferi* (Archibald, 1983; Posey and Gherardini, 2000). The former is an obligate fermenter that avoids the iron requirements of respiration and substitutes manganese for iron in other cell functions, while *Borrelia*, causing Lyme's disease, relies on obligate parasitism.

The requirement for iron relative to other cell constituents varies from species to species and as a function of iron availability, with typical values of 3 to 60 μmol Fe mol^{-1} organic C. The high values are from *E. coli* grown under aerobic, iron-replete conditions, in which almost 90% of the cellular iron is associated with NADH dehydrogenase and succinate dehydrogenase of the respiratory chain. The Fe:C ratio of heterotrophic planktonic prokaryotes from iron-poor environments is an order of magnitude lower, while eukaryotic phytoplankton appear to have somewhat lower requirements than the prokaryotes (Tortell *et al.*, 1999). As a first approximation, the elemental stoichiometry of plankton, including Fe and other trace metals, is given as the following (Martin and Knauer, 1973; Bruland *et al.*, 1991; Tortell *et al.*, 1999):

$$C:N:P:S:\mathbf{Fe}:Zn:\{Cu, Mn, Ni, Cd\} = 106:16:1:1:\mathbf{0.005}:0.002:0.0004$$

See also Whitfield (2001) for further discussion on the interactions between plankton and iron and other trace metals in the oceans.

4.1. Uptake

Microorganisms employ several different strategies in the uptake of iron from the environment. Soluble Fe^{2+} is taken up through similar, or even the same, ATP-consuming active transport systems as other divalent cations (Braun *et al.*, 1998; Braun and Killmann, 1999). The uptake of $Fe^{3+}_{(aq)}$ is hindered by its extremely low solubility at neutral pH. To overcome this problem, aerobic microorganisms synthesize a wide variety of organic chelators called siderophores. These compounds increase the solubility of Fe^{3+} by forming very strong complexes. Siderophores are typically hexadentate ligands that fill out the coordination sphere of the ferric ion. The three major classes of compounds coordinating to Fe^{3+} are hydroxamates, cathecolates, and α-hydroxycarboxylates (Albrecht-Gary and Crumbliss, 1998).

Much of the knowledge about microbial iron acquisition comes from studies of pathogenic bacteria. Due to the tight binding of iron by hemoglobin and other iron-binding proteins, bacteria in the human body are met with difficulties in obtaining iron similar to aerobic bacteria in the oceans or other oxic aquatic environments. It also appears that the strategies for obtaining Fe^{3+} are similar. The siderophore–Fe^{3+} complexes are typically taken up as a whole by the cell through specialized uptake systems (Braun *et al.*, 1998). After dissociation in the cytoplasm, the cell recycles the siderophore to the environment (Figure 8.5). In addition to the energy requirements of the uptake itself, both siderophore synthesis and the inevitable partial loss associated with excretion to the environment impose a cost to the organism. Accordingly, the expression of iron uptake systems is subject to stringent genetic regulation. In addition, siderophore design, as well as the designs of associated transporters, are species specific. This minimizes the risk of piracy, in which one species utilizes iron from siderophores excreted by another species. Still, many prokaryotes are able to utilize foreign siderophores. Furthermore, eukaryotic algae can acquire iron from some bacterial siderophores, while apparently not producing soluble siderophores themselves (Hutchins *et al.*, 1999). Since the algae are the source of DOC for the bacteria, such interactions are potentially of mutual benefit.

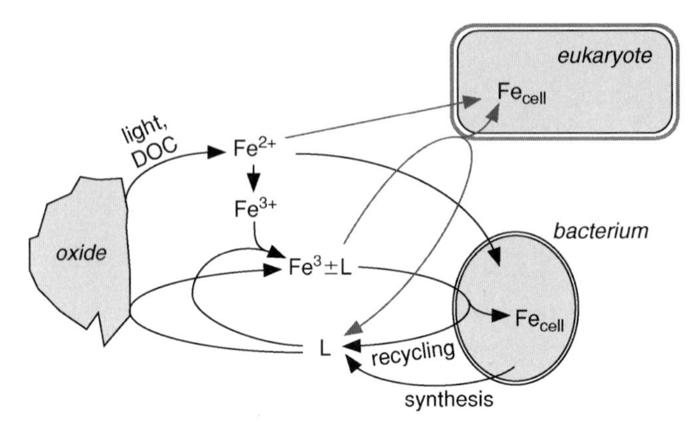

Figure 8.5 Iron cycling in iron-poor ocean waters. Particulate iron is solubilized through photochemical reduction, potentially involving organic compounds, and by strong ligands (L). Soluble Fe^{2+} can be assimilated directly, but it also oxidizes rapidly. Free $Fe^{3+}_{(aq)}$ is virtually absent due to high concentrations of strong ligands. *Bacteria* assimilate the $Fe^3 \pm L$ complex and recycle the ligand and may also synthesize new ligand molecules (siderophores). Eukaryotic algae do not synthesize siderophores but may utilize Fe complexed by bacterial siderophores.

4.2. Iron in the oceans

Dissolved iron concentrations in ocean waters average only 0.07 nM (Figure 8.6) (Johnson *et al.*, 1997), which may limit both primary production and heterotrophic activity. This is the case in parts of the oceans—often referred to as HNLC regions—characterized by *h*igh concentrations of *n*utrients (N, P, Si) and *l*ow levels of *c*hlorophyll in the surface water (see also Chapter 12 for a discussion from the Si perspective). HNLC regions include the Southern Ocean and the Equatorial Pacific. Iron limitation has been demonstrated both in batch and in open-ocean experiments in which patches of up to 225 km^2 were fertilized with iron (Martin *et al.*, 1994; Coale *et al.*, 1996, 2004; Buesseler *et al.*, 2004). The fertilization resulted in increased primary production, mainly through diatoms, and a drawdown of the CO_2 partial pressure in the water, leading to a shift in the air/sea CO_2 exchange toward the water and an increased flux of particulate organic matter toward the sea floor.

While coastal surface waters receive iron through riverine input, resuspension and possibly also emission of soluble Fe^{2+} from sediments, wind-borne dust can be a major source of iron for the surface waters of the open ocean (Duce and Tindale, 1991). Dust fluxes to the oceans are increased during glacial periods due to dryer and windier climates, possibly stimulating primary production. This would lead to increased export of carbon to the

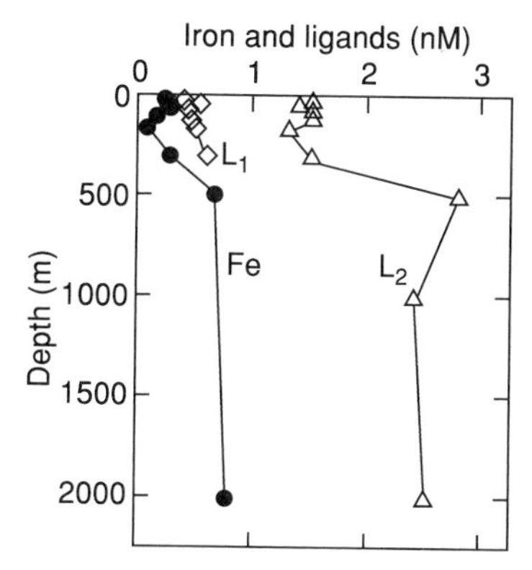

Figure 8.6 Distributions of total soluble iron and unidentified iron-binding ligands in the central North Pacific. L_1 and L_2 represent two ligand classes with conditional stability constants, K', of $10^{-13}\ M^{-1}$ and $3 \times 10^{-11}\ M^{-1}$, respectively. $K' = [FeL]/([Fe'][L'])$, where primes indicate uncomplexed species. The stability constants are similar to those of bacterial siderophores. Redrawn after Rue and Bruland (1995).

deep sea (the "biological pump") and could have contributed to decreased atmospheric CO_2 concentrations during glacial periods (Martin *et al.*, 1990).

Some 99% of the dissolved iron in ocean water is bound in strong organic complexes that have similar stability to siderophore–Fe^{3+} complexes (Figure 8.6). It appears likely that the abundant iron chelators are indeed microbial siderophores or breakdown products of these. This would add a new dimension to the significance of the microbial loop (Chapter 5) for the regulation of primary productivity in the oceans.

5. MICROBIAL OXIDATION

Microbes play an important role in the oxidation of reduced iron and manganese. A fair amount of free energy is available from the oxidation of ferrous iron with oxygen or nitrate at near neutral pH (Figure 8.3), and a number of organisms are known to grow by these processes (Section 5.1). At low pH the energy yield is much smaller, yet specialized organisms grow by aerobic iron oxidation even at pH values near zero (Section 5.1.2). Furthermore, ferrous iron functions as the electron donor for the reduction of CO_2 by anoxygenic phototrophs in what could be an evolutionarily early form of phototrophy (Section 5.2). The oxidation of Mn^{2+} yields relatively less energy than Fe^{2+} oxidation, and so far organisms that couple growth to manganese oxidation have not been found, although manganese oxidation in many aquatic environments is microbially catalyzed (Section 5.3).

A common problem associated with the oxidation of Fe^{2+} and Mn^{2+} is the rapid precipitation of oxides, which may eventually encrust the organism carrying out the process, to their own disadvantage. As we discuss below, iron oxidizers have developed different strategies to cope with this problem, and in at least one case the precipitation of manganese oxide is favorable to the organism.

5.1. Chemotrophic iron oxidation

Iron-oxidizing bacteria were already described by the 19th century, and their involvement in iron oxidation was noted (Ehrenberg, 1836; Winogradsky, 1888). Conspicuous precipitations of iron oxides, or ochre, are well known in environments such as groundwater seeps, and the presence of the classic "iron bacteria" of the genera *Gallionella* and *Leptothrix* is easily recognized through the characteristic iron oxide-rich stalks and sheaths that these organisms produce (Figure 8.7). Despite their long microbiological history, the role of iron oxidation in the energetics of these organisms still awaits definitive description (Emerson, 2000).

Today, several different groups of prokaryotes are known to catalyze the oxidation of ferrous iron with O_2 (Table 8.2):

$$4Fe^{2+} + O_2 + 4H^+ \rightarrow 4Fe^{3+} + 2H_2O \quad (8.2)$$

The ferric iron rapidly precipitates, mainly as ferrihydrite:

$$Fe^{3+} + 3H_2O \rightarrow Fe(OH)_3 + 3H^+ \quad (8.3)$$

The oxidation also proceeds spontaneously, however, and in fully aerated fresh water at near-neutral pH the half-life of Fe^{2+} is only a few minutes. Many organisms compete with the abiotic reaction by growing either at low pH or in microoxic conditions in which the spontaneous reaction is much slower. Iron oxidizers growing at neutral pH include *Gallionella ferruginea* and recently isolated strains within the α-, β-, and γ-proteobacteria (Table 8.2). These organisms grow chemolithotrophically, and at least in some cases autotrophically, through iron oxidation (Hallbeck and Pedersen, 1991; Emerson and Moyer, 1997; Edwards *et al.*, 2003). They can be cultured in gel-stabilized systems with opposing gradients of Fe^{2+} and O_2 where they position themselves at low oxygen concentrations (Figure 8.8). The organisms also exploit mineral sources of ferrous iron, including $FeCO_3$, FeS, Fe(II)-rich basaltic glass, and pyrite.

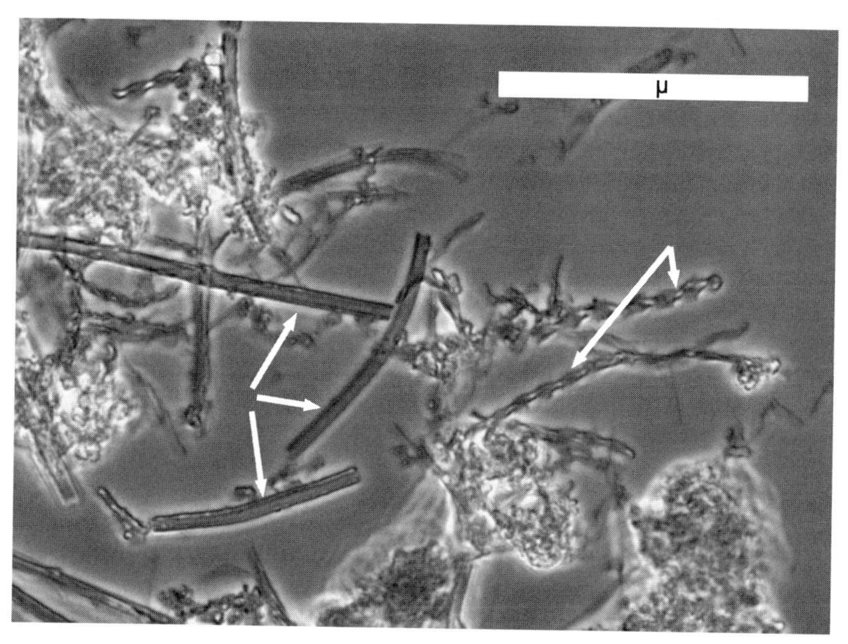

Figure 8.7 Photomicrograph of iron oxide-rich particulates from a Danish iron seep with characteristic structures produced by iron-oxidizing bacteria. *G*, stalks of *Gallionella*; *L*, sheaths of *Leptothrix*; a, amorphous iron oxide masses. Photograph by B.T.

Gallionella ferruginea is recognized by its excreted helical stalk (Figure 8.7). The stalk can be several tens of micrometers long, and it consists of an organic matrix covered and partly embedded with iron oxide (Hallberg and Ferris, 2004). A novel isolate from a submarine hydrothermal setting deposits ferrihydrite in irregularly twisted filaments (Emerson and Moyer 2002), while ferrihydrite deposited by other novel isolates forms amorphous masses.

Other microbes have been classified as "iron bacteria" based on the precipitation of iron oxide in their sheaths or capsules (Table 8.2), but it is not clear whether iron oxidation plays a role in their energy metabolism or whether it constitutes any other benefit for the organisms. Most notably, the sheath-forming morphotype *Leptothrix ochracea* is very common in environments where *Gallionella* and other iron oxidizers also are found (Figure 8.7), but it has not yet been isolated. *Leptothrix* sheaths are found at high oxygen concentrations, indicating that it may not be a microaerophile as are *Gallionella* and the γ-proteobacterial strains (Emerson and Revsbech, 1994a).

Table 8.2 Representative chemotrophic iron-oxidizing *Bacteria* and *Archaea*

Species or strain	Affiliation	Special characteristics
Neutralophilic aerobic lithotrophs		
Gallionella ferruginea	β proteobacteria	Autotroph, stalked microaerobe
Strain TW2[a]	β proteobacteria	
Strains FO1-3[b]	α proteobacteria	Obligate autotrophs, microaerobes, fac. anaerobes
Strains FO4-6,-8,-9,-15[b]	γ proteobacteria	
ES-1, ES-2, RL-1, JV-1[c]	γ proteobacteria	Autotrophs?, microaerobes
Acidophilic aerobic lithotrophs		
Acidithiobacillus ferrooxidans	γ proteobacteria	Autotroph, mesophile
Leptospirillum spp.	*Nitrospira* group	Autotrophs, mesophiles
Sulfobacillus spp.	High G+C Gram pos.	Heterotrophs, thermophiles
Acidimicrobium ferrooxidans	High G+C Gram pos.	Heterotroph, thermophile
Sulfolobus spp.	*Crenarchaeota*	Autotrophs, hyperthermophiles
Acidianus brierleyi	*Crenarchaeota*	Autotroph, hyperthermophile
Aerobic organotrophs (?)		
Leptothrix spp.	β proteobacteria	Filamentous, sheathed
Sphaerotilus natans	β proteobacteria	Filamentous, sheathed
Siderocapsaceae	?	Mucoid capsules
Pedomicrobium spp.	α proteobacteria	Hyphal, budding
Anaerobic oxidizers		
BrG2, HidR2[d]	β proteobacteria	Denitrifiers, mixotrophs
Dechlorosoma suillus	β proteobacteria	
Strains FO1-3[b]	α proteobacteria	Obligate autotrophs, facultative anaerobes
Strains FO4-6,-8,-9,-15[b]	γ proteobacteria	
BrG3[d]	γ proteobacteria	Denitrifier
Geobacter metallireducens	δ proteobacteria	NO_3^- reduction to NH_4^+
Ferroglobus placidus	*Archaeoglobales*	Hyperthermophile

[a]Sobolev and Roden (2001).
[b]Edwards *et al.* (2003).
[c]Emerson (2000); Emerson and Moyer (1997, 2002).
[d]Straub *et al.* (1996); Buchholz-Cleven *et al.* (1997).

It is not known how iron oxidizers control the precipitation of characteristic ferrihydrite structures, but such directed precipitation is advantageous because the organisms avoid self-encrustation. Likewise, the iron oxidizers precipitating amorphous oxide particles appear to avoid encrustation by delaying the precipitation of Fe^{3+}, possibly through the excretion of chelators. Thus, in gradient cultures, Fe^{3+} may precipitate several millimeters from the zone of iron oxidation (Sobolev and Roden, 2001). Increasing the

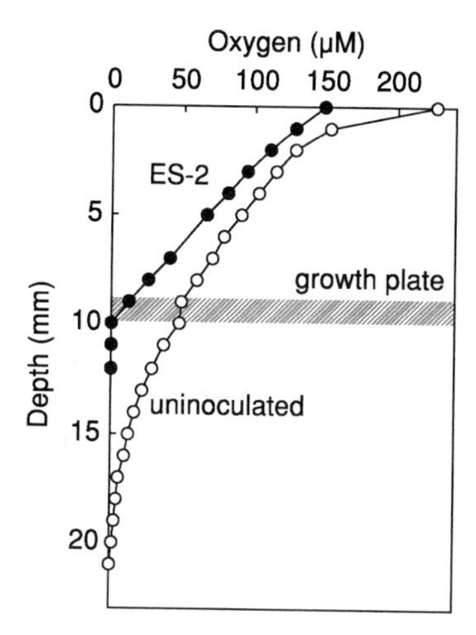

Figure 8.8 Vertical oxygen distributions in a gradient culture of the iron-oxidizing strain ES-2 compared to an uninoculated control. The culture grows in an agarose gel with supplies of oxygen from above and soluble Fe^{2+} from below. Cells are concentrated in a thin plate near the depth of oxygen depletion, and essentially all oxygen consumption (and iron oxidation) takes place at this depth at a high volume-specific rate, whereas the continuous curvature of the oxygen profile in the uninoculated control shows that abiotic iron oxidation is diffuse and specific rates are lower. Redrawn after Emerson and Moyer (1997).

mobility of ferric iron may convey an additional advantage to the iron oxidizers because it can stimulate iron reduction and thereby the production of ferrous iron in anoxic niches nearby (Sobolev and Roden, 2001).

Aerobic, neutrophilic iron oxidizers are found in both freshwater and marine environments, and they are known from mass occurrences associated with groundwater and hydrothermal sources of Fe^{2+}. They are also known from iron crusts forming on the roots of aquatic plants. In all of these environments iron oxidation is mainly biological. In some cases, *Leptothrix* sheaths and *Gallionella* stalks are the dominating type of iron oxide, but more often amorphous particulates dominate and unicellular unappendaged organisms are the main contributors to iron oxidation (Emerson and Revsbech, 1994a,b; Emerson and Moyer, 1997).

5.1.1. Iron oxidation with nitrate

There is good geochemical evidence, based on the distribution of chemical gradients, that nitrate serves as an oxidant for ferrous iron in anoxic environments such as marine sediments (e.g., Froelich *et al.*, 1979). Indeed, nitrate-reducing, iron-oxidizing bacteria were recently described, and this group now includes a diverse assembly of organisms (Table 8.2). The first organisms identified were denitrifying bacteria conserving energy from the following reaction (Straub *et al.*, 1996; Buchholz-Cleven *et al.*, 1997; Benz *et al.*, 1998):

$$10Fe^{2+} + 2NO_3^- + 6H_2O \rightarrow 10Fe^{3+} + N_2 + 12OH^- \qquad (8.4)$$

Synthetic $FeCO_3$ can be used as a source of Fe^{2+}, and the ferric iron produced precipitates as ferrihydrite.

Nitrate-reducing iron oxidizers also catalyze the oxidation of ferrous iron in various mineral phases (Chaudhuri *et al.*, 2001; Weber *et al.*, 2001). Iron-based denitrification has been observed in new microbial isolates enriched in ferrous iron/nitrate medium and in previously described denitrifiers, including *Thiobacillus denitrificans* and *Pseudomonas stutzeri*. In all of these strains iron oxidation is most rapid during mixotrophic growth with an organic co-substrate such as acetate, but one new strain, BrG2, and *T. denitrificans* and *P. stutzeri* appear to grow chemolithoautotrophically by iron oxidation in the absence of organics. These iron oxidizers also grow organotrophically on nitrate or oxygen and are affiliated with the β- and γ-proteobacteria. Other isolates including α- and γ-proteobacteria are obligate chemolithoautotrophs using only Fe^{2+} as electron donor and either oxygen or nitrate as electron acceptors (Edwards *et al.*, 2003). The product of nitrate reduction for these strains was not reported. The δ-proteobacterium *Geobacter metallireducens*, known primarily as an iron reducer, also oxidizes iron while reducing nitrate to ammonium (Finneran *et al.*, 2002b). It is not known whether energy for growth is conserved through this process.

Nitrate-dependent iron oxidation is furthermore the basis for growth in the archaeal hyperthermophile *Ferroglobus placidus*. This organism was isolated from a submarine hydrothermal vent, is obligate anaerobic and lithotrophic, and reduces nitrate to nitrite and nitrous gases (Hafenbradl *et al.*, 1996).

Iron-oxidizing denitrifiers are widespread in freshwater and marine sediments, but they typically make up only a small fraction ($\leq$1%) of the total denitrifying community (Straub and Buchholz-Cleven, 1998; Hauck *et al.*, 2001; Straub *et al.*, 2001). Nitrate-dependent iron oxidation has also been demonstrated in activated sludge (Nielsen and Nielsen, 1998). As noted previously, distributions of ferrous iron and nitrate in deep-sea sediments and subsurface environments indicate that in these environments nitrate is an important oxidant for iron (Froelich *et al.*, 1979; Ernstsen *et al.*, 1998), and microbial catalysis is a likely mechanism for this reaction. It is not quite clear, however, to which extent abiotic reactions also contribute. Although the oxidation of Fe^{2+} by nitrate is extremely slow at low temperature, it can be catalyzed by various particulates (Ottley *et al.*, 1997).

Nitrate also spontaneously oxidizes ferrous iron in some particulate phases, with green rusts being particularly reactive (Postma, 1990; Hansen *et al.*, 1996). Nitrite may substitute for nitrate in this type of reaction, and the kinetics of nitrite-coupled iron oxidation are generally faster than with nitrate (van Cleemput and Baert, 1983; Sørensen and Thorling, 1991; Weber *et al.*, 2001). Nitrite is a free intermediate in both denitrification and nitrification (Chapter 7), but low concentrations in the environment could mitigate against the rapid reaction kinetics.

5.1.2. Acidophilic iron oxidizers

While the microorganisms mentioned above generally require pH values between 6 and 8, a broad variety of acidophilic iron-oxidizing prokaryotes are also known (Table 8.2), including hyperacidophiles that grow at pH values below 1. At pH ≤ 2, the organisms face no competition from the abiotic oxidation of Fe^{2+}. However, due to the high solubility of ferric iron in acidic waters, the energy yield from iron oxidation can be much smaller than at neutral pH.

Acidophilic iron oxidizers are typically associated with the oxidation of metal sulfide ores and contribute to the formation of acidic runoff known as acid mine drainage (see also Chapter 9). Sulfides such as pyrite react much more rapidly with ferric iron than with O_2:

$$FeS_2 + 14Fe^{3+} + 8H_2O \rightarrow 15Fe^{2+} + 2SO_4^{2-} + 16H^+ \quad (8.5)$$

and the acidophilic iron oxidizers stimulate this reaction by reoxidizing the ferrous iron.

The acidophilic iron oxidizers are phylogenetically diverse and include members of both *Bacteria* and *Archaea*. The archaeon *Ferroplasma acidarmanus* dominates biofilms growing at extremely low pH values of 0 to 1 and at high metal concentrations (Edwards *et al.*, 2000), while the bacteria *Leptospirillum ferrooxidans* and *Acidithiobacillus ferrooxidans* are generally most abundant in more dilute solutions (Rawlings *et al.*, 1999). Acidophilic iron oxidizers are also found in hydrothermal environments, which house moderately thermophilic gram-positive bacteria and archaeal thermophiles.

Electron transport from ferrous iron has mainly been studied in *Acidithiobacillus ferrooxidans*. In this organism, Fe^{2+} is oxidized either at the outer surface of the outer membrane or in the periplasm (Ingledew, 1982), and electrons are transported via a periplasmic rusticyanin–cytochrome *c* complex to an oxygen-reducing cytochrome *c* oxidase in the inner membrane (Giudici-Orticoni *et al.*, 2000). The intracellular reduction of O_2 consumes protons and thus maintains a proton gradient from the acidic environment to the cytoplasm, where the pH is around 6. This proton gradient drives ATP formation. Like most other acidophilic iron oxidizers, *A. ferrooxidans* is an autotroph. CO_2 fixation is through the reductive pentose phosphate cycle (utilizing Rubisco; see Chapter 4), and it requires reverse electron transport (see also Chapter 4) because the mid-point potentials of the redox couples involved in electron transport from Fe^{2+} are too high for the reduction of NAD(P).

Many of the acidophilic iron oxidizers also grow by the oxidation of reduced sulfur species such as elemental sulfur (the common acidophile *Leptospirillum ferrooxidans* is a notable exception), though they do not directly oxidize sulfide minerals (see also Chapter 9). In addition, both *A. ferrooxidans* and *Sulfolobus acidocaldarius* grow by dissimilatory ferric iron reduction coupled to the oxidation of elemental sulfur (see also below).

5.2. Anoxygenic iron-oxidizing phototrophs and a note on evolution

Iron-oxidizing phototrophs grow photolithoautotrophically by the following reaction:

$$4Fe^{2+} + CO_2 + 4H^+ + \text{light} \rightarrow (CH_2O) + 4Fe^{3+} + H_2O \qquad (8.6)$$

Ferrous iron can be supplied in solid form as $FeCO_3$ or FeS, and the ferric iron rapidly precipitates, mainly as poorly crystalline oxides (Widdel *et al.*, 1993; Kappler and Newman, 2004). The discovery of iron-oxidizing phototrophs is quite recent (Widdel *et al.*, 1993) and so far only about six strains have been described (Table 8.3). All strains are closely related to other phototrophic purple and green bacteria (see Chapter 9), and the biochemistry is most likely quite similar to phototrophic sulfide oxidation (see Chapters 4 and 9). With a redox potential of around 0.0 V at pH 7 (Figure 8.3), the $Fe(OH)_3/Fe^{2+}$ couple is well suited as an electron donor for the photosystems of both purple and green bacteria, which have midpoint potentials of 0.45 and 0.3 V, respectively (Widdel *et al.*, 1993). Conversely, the MnO_2/Mn^{2+} couple, with redox potential 0.6 V, is an unlikely donor for the known anoxygenic photosystems (Straub *et al.*, 1999).

It is believed that Fe^{2+} delivers electrons either directly to the photosystem located in the cytoplasmic membrane or to another carrier that transports the electrons across the outer membrane and periplasm (Ehrenreich and Widdel, 1994). For most strains of iron oxidizers the resulting oxide is separated from the cells, while *Rhodomicrobium vanielii* becomes encrusted in the oxide, which impedes further metabolism (Straub *et al.*, 2001). Iron chelators were not detected in cultures of phototrophic iron oxidizers, which

Table 8.3 Anoxygenic iron-oxidizing phototrophic bacteria[a] and their general characteristics

Name	Affiliation	H_2S oxid.	Aerobe	Source
Rhodovulum iodosum, R. robiginosum[b]	*α Proteob.*	+	+	Marine
Strain SW2,[c] *Rhodomicrobium vanielii*[d]	*α Proteob.*	–	?	Fresh
Strain L7[c]	*γ Proteob.*	–	?	Fresh
Chlorobium ferrooxidans[e]	Green non-S	–	–	Fresh

[a]Iron oxidation has also been observed in *Rhodobacter capsulatus, Rhodopseudomonas palustris,* and *Rhodospirillum rubrum.*[d]
[b]Straub *et al.* (1999).
[c]Ehrenreich and Widdel (1994).
[d]Heising and Schink (1998).
[e]Heising *et al.* (1999).

suggests that oxide precipitation is directed away from the cells (Kappler and Newman, 2004).

Like other anoxygenic phototrophs, the iron oxidizers also grow photoautotrophically with H_2, and photoheterotrophically with organic compounds. The most versatile marine *Rhodovulum* strains also utilize reduced sulfur compounds for photoautotrophic growth and grow chemoorganoheterotrophically with oxygen. The organisms have been isolated from shallow freshwater and marine surface sediments, but their general distribution, abundance, and biogeochemical impact in nature remain to be explored. Photic, anoxic, ferrous iron-rich environments are not extensive on the modern Earth, but the requirements for the process may be met in the lower hypolimnion of some anoxic lakes and iron-rich microbial mats (Emerson and Revsbech, 1994a; Pierson *et al.*, 1999; Rodrigo *et al.*, 2001).

Extensive iron-rich sedimentary deposits indicate that the ocean was rich in dissolved ferrous iron during the Archean Eon from 3.8 to 2.5 billion years ago, and the oceans could have been an important habitat for iron-dependent photosynthesis. Evolutionarily, anoxygenic photosynthesis is expected to predate the oxygenic type (see Chapters 4 and 9), and since cyanobacteria appear to have existed by 2.7 billion years ago (Brocks *et al.*, 1999; Chapter 6), it is plausible that the iron oxidizers had also evolved by then. To evaluate their possible role in primary production and the early carbon cycle, we need detailed information about environmental requirements and competition for nutrients with cyanobacteria. As we have seen in Chapter 4, the photosynthetic potential of oxygenic photosynthesis at the community level is far greater than that of anoxygenic phototrophs, because the latter will be limited by the availability of the electron donor (here Fe^{2+}) while water is always available for oxygenic photosynthesis.

5.3. Microbial manganese oxidation

At pH values below 8, Mn^{2+} is kinetically inert toward oxidation, and sluggish spontaneous oxidation is achieved only in the presence of catalytic surfaces such as manganese oxides. The microbial catalysis of Mn^{2+} oxidation, with the precipitation of manganese oxide, has been observed in a wide variety of cultures (Ghiorse, 1984; Tebo *et al.*, 1997). Manganese oxidizers have been found in the α-, β-, and γ-proteobacteria and among the high-GC gram-positive bacteria (Table 8.4). The oxidation product can be insoluble Mn(IV), which is thermodynamically stable under oxic, neutral conditions:

$$2Mn^{2+} + O_2 + 2H_2O \rightarrow 2MnO_2 + 4H^+ \tag{8.7}$$

or Mn(III), which, if not oxidized further, may either precipitate (Equation 8.9) or disproportionate (Equation 8.10):

$$4Mn^{2+} + O_2 + 4H^+ \rightarrow 4Mn^{3+} + 2H_2O \tag{8.8}$$

$$Mn^{3+} + 2H_2O \rightarrow MnOOH + 3H^+ \tag{8.9}$$

$$2Mn^{3+} + 2H_2O \rightarrow Mn^{2+} + MnO_2 + 2H^+ \tag{8.10}$$

Microbially precipitated manganese oxides include a range of mineralogies, with average manganese oxidation states ranging from 2.67 to 4 (Mandernack *et al.*, 1995). However, in oxic environments Mn(IV) manganates prevail (Kalhorn and Emerson, 1984; Murray *et al.*, 1984; Friedl *et al.*, 1997). In partial analogy to iron oxidizers, manganese oxides may be precipitated as characteristic structures including sheaths and appendages, or they may simply precipitate outside the cell wall.

The manganese oxidizers are all organoheterotrophs, and there is no compelling evidence that any of them gain energy from manganese oxidation, nor is it clear that the process confers any particular metabolic advantage to the cells. Other possible advantages include protection against oxygen radical species, UV radiation as well as predation. In addition, Mn

Table 8.4 Model organisms for microbial manganese oxidation

Species	Affiliation	Site of Mn oxide deposition
Bacillus sp. SG-1	High G + C gram pos.	Spore coat
Leptothrix discophora	β proteobacteria	Sheath
Pedomicrobium manganoxidans	α proteobacteria	Appendages
Pseudomonas putida	γ proteobacteria	Cell surface
Strain SD-21[a]	α proteobacteria	Cell surface

[a]Francis *et al.* (2001).

oxides strongly adsorb trace metals required for growth as well as toxic trace metals (Tebo *et al.*, 1997; Brouwers *et al.*, 2000). A rather curious case of manganese oxidation is represented by *Bacillus* strain SG-1. This organism oxidizes and deposits manganese on the surface of the resting spore. Because the spore is metabolically inactive it is clear that manganese oxidation is not coupled to energy conservation or to the germination processes (Rosson and Nealson, 1982). *Bacillus* strain SG-1 is also capable of reducing Mn oxides during vegetative growth, possibly in a respiratory process (de Vrind *et al.*, 1986). Therefore, in this case, manganese oxides could fuel manganese reduction for cells newly germinated from spores in anoxic sediments.

Recent evidence suggests that a particular class of enzymes, the multicopper oxidases, is involved in microbial manganese oxidation in many of the known manganese oxidizers (Brouwers *et al.*, 2000; Francis *et al.*, 2001; Francis and Tebo, 2001, 2002). Multicopper oxidase is the only type of enzyme, other than the terminal respiratory heme-copper and *bd*-type (copper-free) oxidases, known to catalyze the four-electron reduction of O_2 to water (Chapter 6). Manganese oxidation by a multicopper oxidase, laccase, is known from fungi, which use the enzyme extracellularly to generate strongly oxidizing Mn^{3+} complexes, which in turn attack and oxidize lignin compounds (Archibald and Roy, 1992; Thurston, 1994; Schlosser and Höfer, 2002). Although Mn is here reduced again, it is conceivable that fungi also contribute to (net) microbial manganese oxidation in some environments.

Despite the lack of an obvious metabolic advantage, most manganese oxidation in aquatic environments is apparently accomplished by microbes (Emerson *et al.*, 1982; Thamdrup *et al.*, 1994b). Thus, the reaction kinetics of Mn oxidation are too fast to be explained by abiotic mechanisms. For example, Thamdrup *et al.* (1994b) noted that if manganese oxidation were chemical, most of the Mn^{2+} formed in the anoxic zone of sediments would escape unoxidized to the water column. In another sign of microbial involvement, manganese oxidation is sensitive to biological inhibitors such as azide (Figure 8.9). It is not known, however, which organisms are involved in the oxidation of manganese in any particular environment.

The high rates of Mn oxidation influence overall sediment biogeochemistry. In marine sediments, for example, microbially catalyzed manganese oxidation results in the retention of Mn within the sediment, where the manganese can play an important role as a redox intermediate (see below). Manganese oxides may also play a catalytic role in the abiotic cracking of refractory organic matter in marine sediments (Sunda and Kieber, 1994).

We normally think of oxygen as the oxidant for Mn^{2+} in aquatic environments. However, thermodynamically, nitrate could also serve as an oxidant for Mn^{2+} under the conditions found in many anoxic marine settings (Murray *et al.*, 1995; Aller *et al.*, 1998). This reaction, and a possible microbial involvement, awaits experimental demonstration.

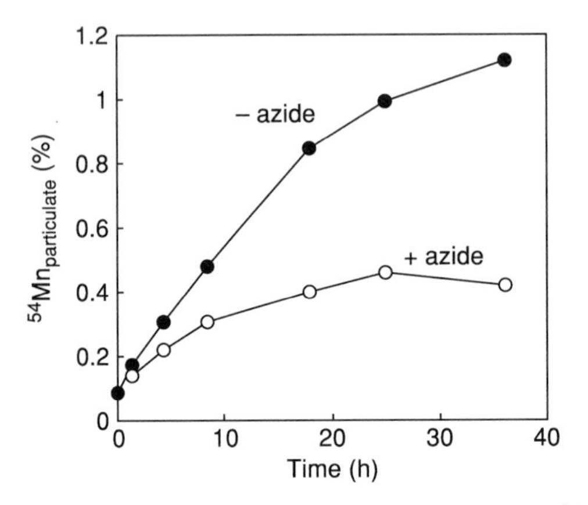

Figure 8.9 Microbial scavenging and oxidation of soluble Mn^{2+} in low-temperature waters below a manganese-rich hydrothermal Galapagos vent plume. The contribution of microbes is illustrated by the inhibitory effect of sodium azide on the production of particulate radiolabeled ^{54}Mn. Approximately half of the scavenging was attributed to oxidation, with the rest due to adsorption of Mn^{2+}. Redrawn after Mandernack and Tebo (1993).

6. MICROBIAL REDUCTION

Dissimilatory iron- and manganese-reducing microorganisms catalyze the reduction of Fe(III) to Fe(II), and of Mn(III) or Mn(IV) to Mn(II). Examples with acetate as the electron donor are given below:

$$CH_3COO^- + 8FeOOH + 3H_2O \rightarrow 2HCO_3^- + 8Fe^{2+} + 15OH^- \qquad (8.11)$$

$$CH_3COO^- + 4MnO_2 + 3H_2O \rightarrow 2HCO_3^- + 4Mn^{2+} + 7OH^- \qquad (8.12)$$

The oxidized metals are here formulated as solid oxyhydroxide and oxide, but microbes reduce iron and manganese in a variety of forms. These include soluble $Fe^{3+}_{(aq)}$ at low pH, soluble Fe^{3+} and Mn^{3+} complexes, Fe(III) and Mn(III)/Mn(IV) in oxides and hydroxides with several different crystal structures and even Fe(III) in the lattices of some clays (Lovley, 1991; Kostka and Nealson, 1995; Kostka *et al.*, 1995, 1996; Roden and Zachara, 1996). The ability to reduce soluble iron species is more widespread among microbes than the ability to reduce solid phases. Although the organisms are all classified as dissimilatory iron reducers, there are fundamental differences between the use of solids and the use of soluble Fe^{3+} or Mn^{3+} as oxidants. Thus, the standard redox potential of $Fe^{3+}_{(aq)}$ reduction at a pH of zero is 0.77 V, approaching that of oxygen reduction, while the reduction of hydrous ferric oxide, the most favorable form of particulate Fe(III) at neutral pH, has a redox potential close to 0.0 V (Figure 8.3; see also Chapter 4 on the energetics of Fe reduction with different forms of Fe). The reduction of solid phases requires electron transport to the cell surface, which may constrain the organism to the vicinity of oxide particles, whereas soluble electron acceptors are transported through molecular diffusion and may be reduced inside the cell. There are several different biochemical strategies for the microbial reduction of solid phases, as we shall see in the following Sections.

6.1. Diversity and phylogeny

About 100 dissimilatory iron-reducing prokaryotes are presently in culture (Lovley, 2000a). When tested, most of these organisms can reduce manganese as well. Similarly, organisms that were originally isolated as manganese reducers generally also reduce iron, but relatively few organisms have been obtained in this way. Thus, the existence of specialized manganese reducers cannot be excluded. As the same organisms generally reduce both iron and manganese, and because the same special considerations concerning the use of solid electron acceptors apply to both processes, we here treat these two types of metabolism together.

About half of the known iron reducers do not couple growth to iron reduction, but grow only through other types of metabolism. This group includes *Bacteria* such as *Bacillus* spp., *Clostridium* spp. and *Escherichia coli*, which during fermentation transfer a minor part, typically less than 5%, of the electrons from, for example, glucose to Fe(III) (Lovley, 2000b). Some sulfate reducers and methanogenic *Archaea* reduce iron with the electron donors they use in their normal metabolism, but the presence of ferric iron can inhibit their normal metabolism. These organisms do not apparently conserve energy for growth during iron reduction (Jones *et al.*, 1984; Lovley *et al.*, 1993; Bond and Lovley, 2002). The small electron flow to iron from the fermentative iron reducers probably limits their importance in iron reduction in natural environments. This is because most of the electron equivalents contained in the fermented substrates will be retained in the fermentation products and therefore will be available for respiratory iron reducers. By contrast, the iron-reducing sulfate reducers and methanogens could play a role in environments in which fluctuating conditions alternatingly favor their normal metabolism and iron reduction.

The first detailed descriptions of microorganisms coupling growth to dissimilatory iron and manganese reduction were reported in 1988 (Lovley and Phillips, 1988; Myers and Nealson, 1988), and since then the known diversity of such organisms in culture has grown rapidly. A large number of the known iron and manganese reducers group into two phylogenetically distinct clusters,

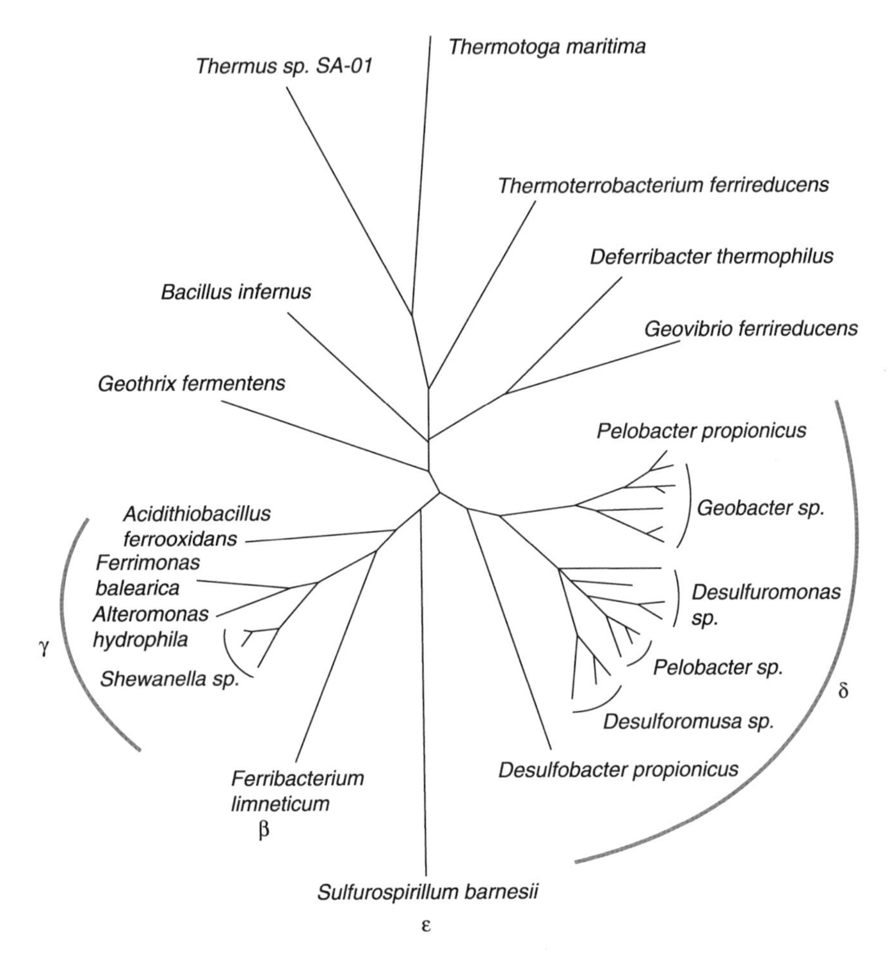

Figure 8.10 Phylogeny of dissimilatory iron-reducing bacteria based on 16S rRNA sequences. Proteobacterial subdivisions are located in the lower half and deep-branching lineages in the upper half. Redrawn after Lovley (2002).

the *Geobacteraceae* family, and the *Shewanella-Ferrimonas-Aeromonas* group (Figure 8.10; Table 8.5). Members of the *Geobacteraceae* are obligate anaerobes or microaerophiles, and in most cases, they couple iron and manganese reduction to the complete oxidation organic substrates such as acetate to CO_2. Most of the iron and manganese reducers in the *Shewanella-Ferromonas-Aeromonas* group are facultative anaerobes coupling metal oxide reduction to hydrogen oxidation or to the incomplete oxidation of lactate to acetate.

Table 8.5 Comparison of the main characteristics of the dissimilatory manganese and iron-reducing genera *Geobacter* and *Shewanella*

	Geobacter	*Shewanella*
Affiliation	δ-proteobacteria	γ-proteobacteria
Number of species	~8	~5
Morphology	Rod	Rod
Motility	Flagella and pili when metal oxides are present	Flagella
e^- donors with Fe(III)[a]	Acetate, volatile fatty acids, H_2, lactate, pyruvate, ethanol, toluene, phenol, p-cresol, benzaldehyde, benzoate	H_2, lactate, formate, glucose, casamino acids, yeast extract, pyruvate
Carbon oxidation with Fe(III)	Complete	Incomplete
e^- acceptors for growth[b]	$O_2 < 10\%$ air sat., nitrate, ferrihydrite, complexed Fe^{3+}, δ MnO_2, S^0, U(VI), Co(III), humics, AQDS, fumarate, malate, graphite anode	O_2, nitrate, ferrihydrite, magnetite, goethite, complexed Fe^{3+}, δ MnO_2, complexed Mn^{3+}, nitrate, S^0, thiosulfate, sulfite, humics, AQDS, TMAO, fumarate
Utilization of solid e^- acceptors	Contact	Contact, melanin shuttle
Cytochromes linked to oxide reduction	*c* in outer membrane	*c* in outer membrane

[a]Acetate is the only electron donor used by all species of *Geobacter*. Hydrogen and lactate are the only donors used by all species of *Shewanella*.
[b]Not all acceptors used by all species.

Phylogenetically, the ability to grow through dissimilatory iron or manganese reduction is spread over both prokaryote domains. Most known mesophilic strains belong to the proteobacteria, with representatives in the α, β, γ, δ, and ϵ subdivisions (Figure 8.10) (Küsel *et al.*, 1999; Lovley, 2002). Mesophiles in separate lineages include *Geothrix fermentens* and *Geovibrio ferrireducens*, as well as two gram-positive bacteria that couple growth to the reduction of soluble iron (Blum *et al.*, 1998; Finneran *et al.*, 2002a). Thermophilic metal reducers are found in several prokaryotic lineages. Such metal reducers include *Bacillus infernus*, a gram-positive bacterium, *Thermotoga maritima* and *Thermus* sp., both in deeply branching bacterial lineages, as well as members of the archaeal genera *Pyrobaculum* of the *Thermoproteales* and *Geoglobus* and *Ferroglobus* of the *Archaeoglobales*.

All hyperthermophiles tested so far have the ability to reduce Fe(III) with hydrogen, although only those mentioned above conserve energy for growth from the reaction (Lovley, 2002). The great phylogenetic diversity of dissimilatory iron and manganese reduction, and particularly its occurrence in all deeply branching hyperthermophilic clades of the Tree of Life, supports the idea that the last common ancestor of all living organisms may have had the ability for dissimilatory iron reduction. Therefore, this metabolism could have been one of the first to support life on Earth (Lovley, 2000b). It remains to be shown, however, that the biochemical pathway of iron reduction is monophyletic in its origin and that it has not evolved repeatedly in separate lineages.

6.2. Metabolic diversity

6.2.1. Electron donors

Iron- and manganese-reducing microbes oxidize hydrogen as well as a wide variety of organic substrates, including short- and long-chain fatty acids, alcohols, aromatic compounds, and sugars (Table 8.5) (Lovley, 2000a). As we have seen previously in this Chapter, some organisms, such as *Shewanella* spp., perform only an incomplete oxidation such as the oxidation of lactate to acetate, while others, such as *Geobacter* spp., completely oxidize their substrate to CO_2. The metal reducers are not known to oxidize complex organic matter on their own, but together with fermentative organisms, complete oxidation can be accomplished (see Chapter 5). Thus, in terms of carbon oxidation, the iron and manganese reducers occupy an ecological niche similar to that of the sulfate reducers (Chapter 9). Unlike the sulfate reducers, however, there are no indications yet of iron or manganese reduction coupled to the anaerobic oxidation of methane (AOM) (see Chapter 10).

Hydrogen is the only inorganic substrate known to support growth in iron and manganese reducers at normal environmental pH. Both *Desulfuromonas acetoxidans* and *Geobacter metallireducens*, as well as some sulfate reducers, can couple manganese reduction to the oxidation of elemental sulfur to sulfate, but they do not conserve energy for growth from this process (Lovley and Phillips, 1994). No Fe reducers are known to couple iron reduction to elemental sulfur oxidation, and attempts to enrich for such organisms resulted in sulfur disproportionators (Chapter 9) (Thamdrup *et al.*, 1993)—a negative result, but nonetheless important. Numerous attempts to isolate iron reducers utilizing either FeS or pyrite as electron donor also have been unsuccessful (Schippers and Jørgensen, 2002). Ammonium has been suggested as an electron donor for manganese reduction in sea water and sediments with a possible microbial catalysis (Luther *et al.*, 1997; Hulth *et al.*, 1999). Despite the use of sensitive assays, however, this process has not been detected at several locations, including some very Mn oxide-rich sites, where it might best be expected (Thamdrup and Dalsgaard,

2000, 2002; Dalsgaard *et al.*, 2003) (see Chapter 7). The acidophiles *Acidithiobacillus ferrooxidans* and *Sulfolobus acidocaldarius* couple the oxidation of elemental sulfur with the reduction of soluble Fe^{3+}, and the former grows through this process (Das *et al.*, 1992; Pronk *et al.*, 1992). In addition to the obligately autotrophic *A. ferrooxidans*, autotrophy coupled to iron or manganese reduction is only known from the hyperthermophile *Geoglobus ahangari* growing on hydrogen (Kashefi *et al.*, 2002).

6.2.2. Electron acceptors

The ability to reduce soluble iron complexed, for example, by NTA (nitrilotriacetic acid) appears to be common to all dissimilatory iron reducers (Lovley, 2000a). Organisms growing with solid oxides are typically cultivated on ferrihydrite (occasionally referred to as amorphous or hydrous ferric oxide), which is thermodynamically metastable and has a very high specific surface area. Species of *Shewanella* also grow with the crystalline oxides magnetite (at pH $\leq$ 6) and goethite, as well as with ferric iron in smectite clays; such growth has not been found with *Geobacter* spp. (Table 8.5) (Kostka and Nealson, 1995; Kostka *et al.*, 1996; Roden and Zachara, 1996). Since solid oxide phases dominate in natural environments at near neutral pH, microbes reducing such solids are of particular interest from a biogeochemical perspective. We further discuss how oxide forms influence Fe availability to microbial reduction in Section 7.2.3.

Characteristically, iron- and manganese-reducing organisms are also able to utilize a wide range of other electron acceptors (Table 8.5). Thus, most species grow by the reduction of elemental sulfur, nitrate, humic substances, and the humic analog anthraquinone disulfate (AQDS). In addition, some, including species of *Shewanella*, are facultative anaerobes. These abilities expand the niche of iron and manganese reducers in many natural environments, permitting growth also outside the often narrow zone in which iron and manganese reduction are possible (see below).

In addition to the environmentally abundant compounds, several rarer metals and metalloids are reduced directly by Fe and Mn reducers, including uranium(VI), chromium(VI), cobalt(III), gold(III), selenate, and arsenate (Lovley, 2000a; Lloyd *et al.*, 2003). Microbial reduction influences the geochemical behavior of such elements, and microbes may be employed, for example, in bioremediation of uranium-contaminated ground water through the precipitation of soluble U(VI) as particulate U(IV).

The versatility of iron and manganese reducers is further demonstrated by their unique ability to transfer electrons to a solid graphite electrode in an electrical circuit (Bond *et al.*, 2002). *Desulfuromonas acetoxidans* grows by the oxidation of acetate in a medium containing no electron acceptors, when

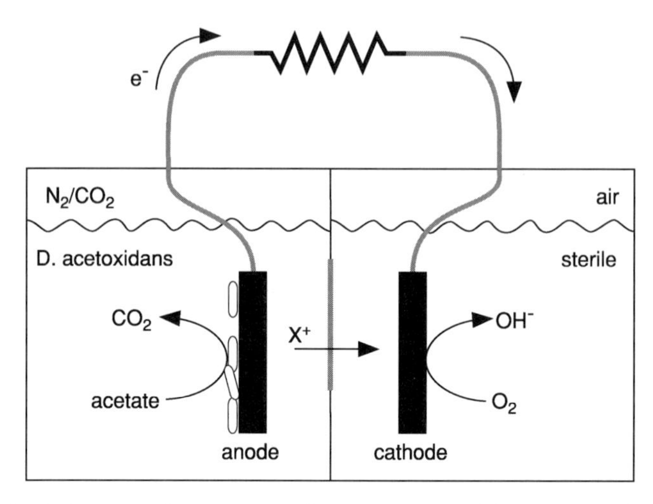

Figure 8.11 Bacterial battery. *Desulfuromonas acetoxidans* in anaerobic medium oxidizes acetate and delivers the electrons to a graphite anode, thereby generating a current through a resistor to a cathode in sterile, oxic solution, where oxygen is reduced abiotically. Charge balance is maintained through a cation-selective membrane between the anoxic and oxic compartments.

it is in contact with a graphite electrode (anode) connected electrically to another electrode (cathode) submerged in oxic medium (Figure 8.11). By transferring electron equivalents from acetate to the anode through their respiration, the bacteria generate an electrical current that can be harvested; the system constitutes a "bacterial battery." Electricity can similarly be harvested from natural anaerobic microbial communities by burying an electrode in anoxic sediment and connecting it to another electrode in the overlying oxic water (Reimers *et al.*, 2001). Iron and manganese reducers also appear to generate a current in such environmental bacterial batteries, since the anodes of such systems after prolonged deployment are predominantly colonized by microbes of the *Geobacteraceae* (Bond *et al.*, 2002).

6.3. Mechanisms of iron and manganese reduction

Iron- and manganese-reducing microbes in most aquatic environments are faced with the special task of exploiting a solid electron acceptor. In principle, an iron reducer has two possibilities: (1) to reduce Fe^{3+} after its dissolution from the solid phase or (2) to deliver the electrons to Fe(III) at the oxide surface. Compared to the use of iron oxides, iron reducers generally grow and metabolize much faster when supplied with soluble ferric iron, such as in the form of ferric citrate. However, both theoretical considerations and experimental evidence indicate that oxide reduction dominates in natural environments. Thus, the non-reductive dissolution of iron oxides into soluble Fe^{3+}, the first step in soluble Fe^{3+} reduction, is an inherently slow process. Dissolution is accelerated in the presence of Fe^{3+} chelators, but even chelator-aided dissolution is much slower than reduction of the solid oxide by simple reductants (Sulzberger *et al.*, 1989; Childers *et al.*, 2002). Also, the location of iron reductase activity in the outer membrane of species of both *Shewanella* and *Geobacter* (see below) suggests an ability to reduce particle-bound iron. Furthermore, electron transfer to the solid phase appears to be the only way to reduce manganese(IV) oxide, which is practically insoluble.

So far, two different general strategies have been identified for the transfer of electrons to the metal oxides (Figure 8.12): (1) reduction through direct cell–oxide contact and (2) reduction through soluble electron-shuttling compounds. *Geobacter metallireducens*, and probably also other *Geobacter* species, requires physical attachment to the manganese or iron oxide surface (Childers *et al.*, 2002). When grown in the presence of metal oxides, *G. metallireducens* is motile and actively seeks the oxide by chemotaxis toward Mn^{2+} and Fe^{2+} liberated from the oxide after reduction. The cells develop pili, flagellum-like appendages that are used for the attachment to and movement on solid surfaces (Skerker and Berg, 2001).

Species of *Shewanella* as well as *Geothrix fermentans* do not require direct contact with the oxide because they excrete electron-shuttling compounds. These are organic molecules reduced by the cell and spontaneously reoxidized at the oxide surface. After reoxidation, the shuttle can be used repeatedly as an electron acceptor by the organism. Through the use of electron shuttles, the organisms can oxidize their electron donor while spatially separated from the metal oxides. The cost of this freedom includes the extra energy used for shuttle synthesis. Also, in most natural environments, there is a continuous loss of the shuttle through diffusion, advection, microbial degradation, and piracy. One such compound produced by *Shewanella algae* has been identified as melanin (Turick *et al.*, 2002). Melanin is a quinone-containing polymer with similarities to humic substances, and the type produced by *S. algae* has a molecular weight of 12,000–14,000. However, it is not completely clear that melanin serves as a soluble shuttle, because it may also bind to the cell surface and facilitate electron transfer over a short distance through semi-conductor properties (Turick *et al.*, 2003).

A number of naturally occurring compounds may also serve as electron shuttles between cells and iron and manganese oxides. Humic substances appear to be particularly important in this respect. Humics contain quinone functional groups that carry electrons in a similar fashion to the melanin produced by *Shewanella*. Many iron and manganese reducers are also able to use humics as electron acceptors without the reoxidation of the humics by iron oxide (Coates *et al.*, 1998). Some soluble iron complexes also function as electron shuttles. For example, ferrous citrate and EDTA complexes can

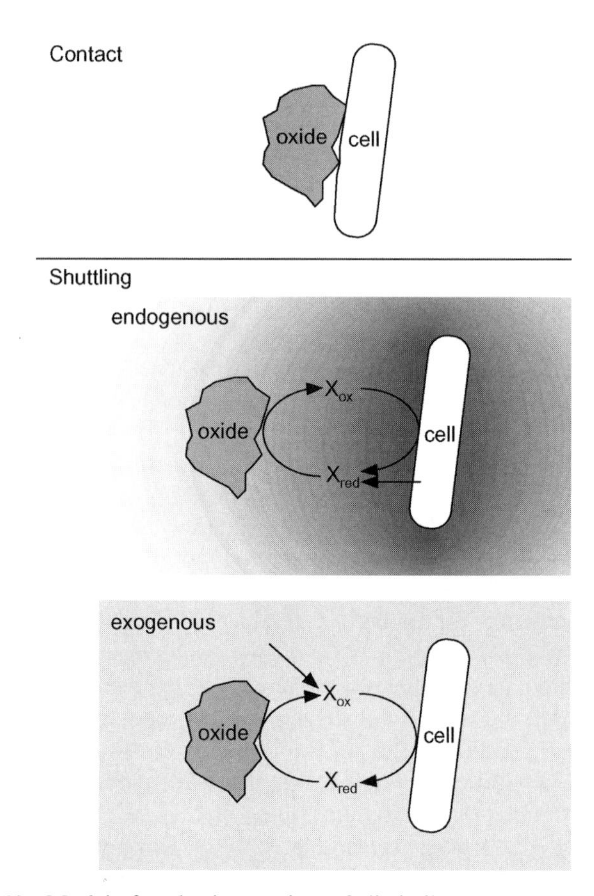

Figure 8.12 Models for the interaction of dissimilatory manganese- and iron-reducing bacteria with solid electron acceptors. (Top) Contact between the cell surface and the solid. (Middle and bottom) Solids are reduced through electron-shuttling compounds that are either produced by the organism (middle) or naturally present in the environment (bottom).

reduce iron oxides with the oxidation of the complexed iron to Fe^{3+} (Segal and Sellers, 1984; Suter *et al.*, 1988), and the ferric complexes are good electron acceptors for iron and manganese reducers.

6.4. Intracellular electron transport

The cellular biochemistry involved in iron and manganese reduction is not well understood. Several enzymes involved in dissimilatory iron reduction have been found to display iron reductase activity *in vitro*. However, it appears that iron reduction is not the main function of many of these enzymes, as they do not contact the Fe(III) *in vivo*. Thus, the biochemical specialization required for respiration with metal oxides probably is related to the spatial configuration of the electron transport system, which has to deliver electrons to the outside of the cell wall rather than to the inner or outer surface of the cytoplasmic membrane as in other types of respiration (e.g., Chapter 6).

Iron and manganese oxide reduction in both *Shewanella* spp. and *Geobacter* spp. probably occurs through *c*-type cytochromes located in the outer membrane (Myers and Myers, 1992, 1997; Lovley, 2000b; Myers and Myers, 2001; DiChristina *et al.*, 2002). Current models suggest that the electrons used in iron and manganese reduction are delivered to the cell membrane by NADH or NADPH and are transported across the cell membrane, periplasm/cell wall, and outer membrane through a chain of cytochrome *c*-type proteins (Lovley, 2000b) (Figure 8.13). It is not clear how the membranes are energized for ATP synthesis with this unique configuration of the electron transport chain. The key role for *c*-type cytochromes cannot be common to all iron reducers, since both species of *Pelobacter* from the *Geobacteraceae* and the archaeon *Pyrobaculum islandicum* are capable of dissimilatory iron reduction yet do not contain cytochrome *c* (Schink, 1992; Childers and Lovley, 2001).

7. MICROBIAL MANGANESE AND IRON REDUCTION IN AQUATIC ENVIRONMENTS

7.1. Environmentally important manganese and iron reducers

Culture-independent techniques are, as pointed out many places in this book, gaining increasing popularity in determining microbial population structure. However, the phylogenetic diversity of known manganese and

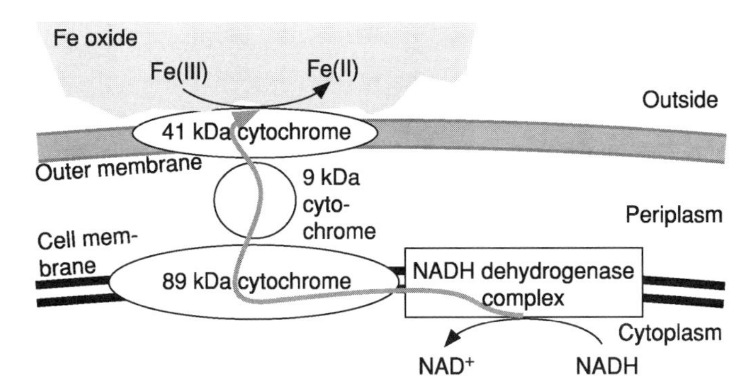

Figure 8.13 Model for electron transport to extracellular electron acceptors (here iron oxide) in *Geobacter sulfurreducens*. The fat grey curve symbolizes electron flow. Redrawn after Lovley (2000b).

iron reducers is large and growing, and therefore, for a faithful survey of natural abundance, one must search for several highly divergent genotypes. In addition to the inherent difficulties in conducting such a broad-based search, such a screening may still miss important organisms for which the capability of oxide respiration has not yet been recognized. Nonetheless, it appears that the composition of dominant metal-reducing populations in nature may not be as phylogenetically complex as the cultured diversity might suggest. Thus, members of the iron-reducing genera *Geobacter* and *Desulfuromonas* repeatedly show up as important components of microbial communities in aquifers in which iron reduction is the dominating electron-accepting process (Rooney-Varga *et al.*, 1999; Snoeyenbos-West *et al.*, 2000; Röling *et al.*, 2001; Holmes *et al.*, 2002). *Shewanella*-related genotypes have been found in aquatic sediments (DiChristina and DeLong, 1993; Li *et al.*, 1999), but there is no evidence yet that this group is quantitatively important for iron reduction in sediments or aquifers. Organisms related to *Geothrix fermentans* have been identified in the early stages of iron reduction in an aquifer, but these organisms were 1000-fold less abundant than the *Geobacteraceae* (Rooney-Varga *et al.*, 1999).

So far, studies of iron-reducing communities have focused on contaminated aquifers, and it remains to be shown whether *Geobacter*/*Desulfuromonas* affiliates also dominate iron reduction in pristine environments. In one Arctic marine sediment, 13% of clones in a general bacterial rDNA clone library were most closely related to *Desulfuromonas* species (Ravenschlag *et al.*, 1999), which might indicate that such organisms are also important for carbon oxidation there. Still, carbon oxidation could be supported by nitrate and elemental sulfur, which are both utilized by members of this genus. Thus far there has been little attention paid to the structure of microbial communities in which manganese reduction dominates carbon oxidation; this represents a field ripe for exploration.

7.2. Controls of microbial iron and manganese reduction

Three basic conditions must be fulfilled for microbial iron or manganese reduction to thrive in normal aquatic environments of near neutral pH. Oxygen must be absent or nearly so, suitable electron donors must be present, and oxidized manganese or iron must be available in an appropriate form. Before examining in greater detail the rates and kinetics of Fe and Mn reduction in natural environments, we briefly explore each of these necessary conditions.

7.2.1. Oxygen

Oxygen inhibits metabolism in strictly anaerobic iron and manganese reducers, whereas in facultative anaerobes, such as *Shewanella*, oxygen is used for respiration when it is present (Myers and Nealson, 1988; Arnold *et al.*, 1990; Lovley, 1991). By contrast, acidophilic iron reducers can reduce soluble Fe^{3+} in the presence of oxygen (Küsel *et al.*, 1999).

7.2.2. Electron donors

Iron and manganese reducers can use a range of simple organic compounds and hydrogen as electron donors (Table 8.5). Since the organisms do not degrade organic polymers nor, generally, oxidize carbohydrates, cooperation with fermenters is required for an efficient oxidation of the complex organic matter typically found in natural environments. By analogy with sulfate reduction (Chapter 9), we may assume that the fermentation product acetate is the most important carbon source for microbial iron and manganese reduction in aquatic sediments, though this remains to be tested. Iron and manganese reducers will, however, have to compete with other microbes for their substrates. As discussed in Chapter 3, this competition is normally decisive for the relative importance of the various pathways of carbon mineralization, with pathways using more energetically favorable electron acceptors prevailing over the less favorable ones. Thus, with nitrate or manganese oxide as the terminal electron acceptor, hydrogen concentrations can be kept so low that it cannot be used for microbial iron reduction, while iron reduction, in turn, competitively excludes sulfate reduction and methanogenesis when sufficient iron oxide is available (Lovley and Phillips, 1987a; Lovley and Goodwin, 1988; Thamdrup, 2000). As a further illustration of this, Roden (2003) found that the K_m for acetate mineralization in freshwater wetland sediment was $0.83\,\mu M$ when iron reduction was the terminal electron-accepting process, while it increased to $11.5\,\mu M$ when mineralization was coupled to methanogenesis. The competitive relationship between nitrate and manganese reduction has not been explored in detail, but in manganese oxide-rich sediment the two processes occur simultaneously (Thamdrup and Dalsgaard, 2000; Dalsgaard and Thamdrup, 2002).

7.2.3. Oxide speciation

The availability of manganese and iron oxides is a critical factor regulating the activity of manganese and iron reducers. This regulation is multifaceted, involving mineral crystallinity, mineral surface area, the presence of adsorbants and precipitates on the mineral surfaces, and the spatial distribution of the oxides. These effects have mainly been explored for iron reduction. Rates of microbial manganese reduction have also been found to depend on the oxide form (Burdige *et al.*, 1992), but it is not clear to which extent manganese mineralogy affects the process in nature.

Species of *Shewanella* using energy-rich substrates such as lactate or H_2 can grow with crystalline iron oxide phases such as goethite and magnetite. By contrast, iron-reducing communities in natural sediments, subject to electron donor limitation, have only been found to compete successfully with sulfate reducers or methanogens in the presence of ferrihydrite or similar poorly crystalline iron oxide forms (Lovley and Phillips, 1987b; Thamdrup, 2000; Roden and Wetzel, 2002). Recently, however, Roden (2003) showed that in freshwater sediment iron reducers could also outcompete methanogens through the reduction of crystalline goethite or hematite, when these oxides were added in large amounts. The same suppression of methanogenesis was obtained with either goethite or ferrihydrite when the dosage of oxide was the same in terms of oxide surface area. This required the addition of ~10 times more goethite than ferrihydrite, reflecting a 1:10 ratio of their specific surface area. Based on these observations, Roden suggested that it is the available surface area of iron oxides that regulates the competition of iron reducers and methanogens in natural environments. In aquatic sediments, concentrations of crystalline oxides and ferrihydrite/poorly crystalline oxides are often of similar magnitude (Figure 8.2). Crystalline oxides, therefore, probably contribute little to the bulk surface area, and, according to the interpretation above, to the competitive success of iron reducers.

The different thermodynamic stabilities of the iron oxides affect the energy that may be gained from iron reduction. This, in turn, should influence metabolic rates and the competition between iron reducers and sulfate reducers or methanogens, at least at low substrate concentrations when the energy yield approaches the minimum required for energy conservation (Chapter 3). Thus, at typical environmental concentrations of reactants and products, the redox potential for the goethite/Fe^{2+} couple can be lower than for sulfate/hydrogen sulfide, and the energetic advantage of sulfate reduction would be expected to result in metabolic dominance of sulfate reducers (Figure 8.3) (see also Postma and Jakobsen, 1996). We still, however, lack a complete picture of how surface area and energetics interact to control microbial iron reduction in natural systems.

In a further complication, the sorption of Fe^{2+} to iron oxide surfaces affects both the rate and extent of microbial iron reduction (Roden and Zachara, 1996; Roden and Urrutia, 2002). Thus, in batch systems with *Shewanella alga*, crystalline oxides with low specific surface areas are only reduced to a minor extent before reduction ceases, despite the presence of ample electron donor and iron-reducing organisms. Iron reduction resumes if Fe^{2+} is removed from the oxides by washing, and almost complete reduction is possible if incubations are made in flow-through systems in which Fe^{2+} is continuously removed (Roden *et al.*, 2000). Sorption of Fe^{2+} to oxide surfaces may involve both reversible surface complexation and the precipitation of ferrous or mixed ferrous–ferric phases (Roden and Urrutia, 2002). Ferric iron bound in some clay minerals may also support iron reduction in nature (Kostka *et al.*, 1996, 2002a), but its significance has not been well studied.

7.3. Quantification of microbial manganese and iron reduction

Microbially reducible ferric iron in nature is present in a range of forms, from freshly precipitated hydrous ferric oxides to crystalline oxides, as well as in clay minerals. These all have different specific surface areas, and they may incorporate foreign species such as Al, Si, and phosphate. In addition, they may be associated with other mineral phases and organic matter, and may have solutes adsorbed to their surfaces. A similar diversity is likely for oxidized manganese. These iron phases have reactivities toward abiotic reduction spanning several orders of magnitude (Canfield *et al.*, 1992; Postma, 1993). One would expect a similar variation in the susceptibility of the iron phases to microbial reduction (Roden and Wetzel, 2002). It is generally not possible to capture the details of this variability, and studies of microbial iron and manganese reduction in natural environments typically rely on a simple classification of the oxide phases into one or two pools, such as "reactive–non-reactive" or "poorly crystalline–crystalline", based on the selectivity of chemical extractions (e.g., Canfield, 1989a; Aller, 1994b).

Two experimental approaches have been used to determine rates of microbial Fe reduction in aquatic sediments. The first is a direct quantification of changes in Fe(III) or Fe(II) pools during sediment incubations (Roden and Wetzel, 1996). This method has been validated by monitoring the iron pools in sulfate-depleted freshwater sediments in which abiotic iron reduction was insignificant. In marine sediments, however, hydrogen sulfide produced from sulfate reduction will reduce iron abiotically, and it is therefore not certain which process reduces the Fe(III). Analogous considerations apply to the reduction of manganese oxide. Thus, in a second commonly used approach, rates of dissimilatory iron and manganese reduction are determined by comparing the depth distribution of total carbon oxidation, based on production of dissolved inorganic carbon, to measured rates of sulfate reduction. The carbon oxidation not accounted for by sulfate reduction is attributed to the other processes of carbon mineralization, including oxic respiration, denitrification, manganese reduction, and iron reduction.

The contribution of each of these is then inferred from the depth distributions of oxygen, nitrate, manganese, and iron species. For further discussion of strengths and weaknesses of these and other techniques, see Thamdrup (2000) and Thamdrup and Canfield (2000).

Some examples of microbial iron and manganese reduction rates in sediments are shown in Figure 8.14. Rates of iron and manganese reduction are assumed to be zero in the oxygen- and nitrate-containing surface layer (see above). In the freshwater and coastal marine sediment shown in Figure 8.14, as in these types of sediments in general, manganese concentrations are very low, and reactive manganese oxides are not detected below the oxygen/nitrate zone. In contrast, poorly crystalline ferric iron reaches several centimeters into the sediment, and microbial iron reduction is the dominant pathway of carbon oxidation in this zone, inhibiting sulfate reduction and methanogenesis. A rather unusual marine site holds very high concentrations of manganese oxide, and accordingly microbial manganese reduction accounts for nearly all of the anaerobic carbon oxidation (Figure 8.14). As manganese reduction is active in carbon mineralization, ferric iron concentrations do not change with depth. In all cases, the disappearance of ferric iron or manganese oxide with depth is mirrored in an accumulation of particulate ferrous and manganous phases (not shown).

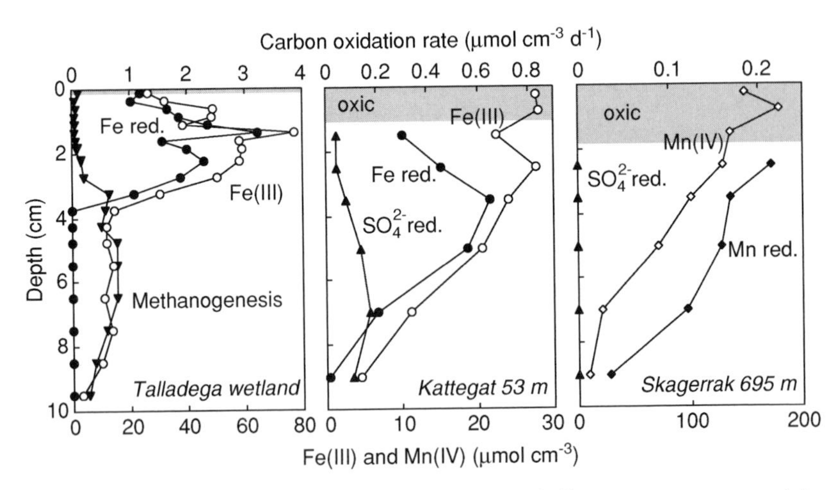

Figure 8.14 Examples of the distribution of dissimilatory manganese and iron reduction and competing methanogenesis (freshwater) or sulfate reduction (marine) relative to manganese oxide and poorly crystalline iron oxide concentrations in sediments. (Left) Talladega freshwater wetland, Alabama. (Center and right) Marine sites of the Baltic–North Sea transition. Manganese reduction was only significant in the Skagerrak deepsite, where iron reduction was not detected. Grey shading indicates the oxic, nitrate-containing surface layers. Redrawn after Canfield *et al.* (1993b), Roden and Wetzel (1996), and Jensen *et al.* (2003).

7.4. Kinetics

Compilations of data from marine sediments such as those in Figure 8.14, as well as detailed investigation in a single freshwater sediment, have shown that rates of dissimilatory iron reduction approximate Michaelis-Menten-type kinetics (Chapter 2) with respect to ferric iron (Figure 8.15). For a wide range of continental shelf and slope sediments, the half-saturation concentration was $\sim$10 μmol cm^{-3} of poorly crystalline iron oxide (Figure 8.15), and a value of 25–50 μmol cm^{-3} was indicated in sediment from a freshwater wetland (Roden and Wetzel, 2002). These concentrations are high compared to the concentrations of poorly crystalline Fe(III) in many sediments (Figure 8.14), and Fe(III) availability therefore exerts a strong control on dissimilatory iron reduction rates. From a much more limited data base, a similar regulation has been indicated for dissimilatory manganese reduction with a slightly lower half saturation concentration (Thamdrup *et al.*, 2000).

Since iron and manganese oxides are solids, it is their surface area rather than their concentration that should be considered in kinetic analysis. Surface areas of poorly crystalline oxides in marine sediments are not well known but probably scale, at least approximately, with concentration. If the microbes require contact with the oxides, the observed dependence of rates on Fe(III) concentration/surface area could correspond to complete occupation of oxide surfaces by organisms. However, estimates suggest that only a small fraction of the oxide surface is likely occupied by iron reducers (Thamdrup, 2000). Other factors such as steric hindrance and the partial occlusion of surfaces could also play a role.

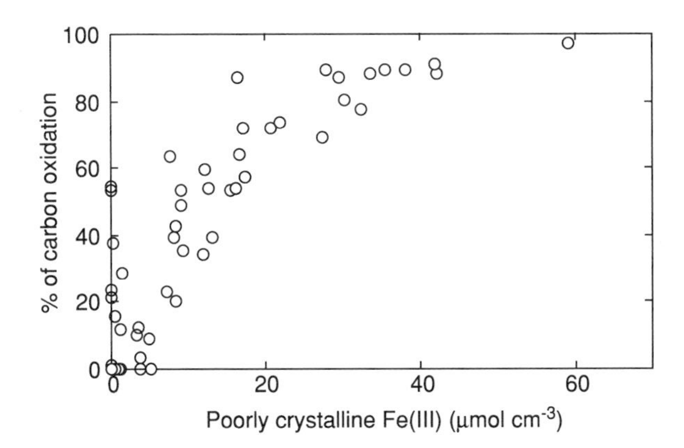

Figure 8.15 Relative contribution of dissimilatory iron reduction to anaerobic carbon mineralization (iron reduction plus sulfate reduction) in marine sediments as a function of the concentration of poorly crystalline iron oxide. Data from eight fine-grained shelf sediments from the Atlantic Arctic, Baltic–North Sea transition, and off central Chile. Redrawn after Jensen *et al.* (2003).

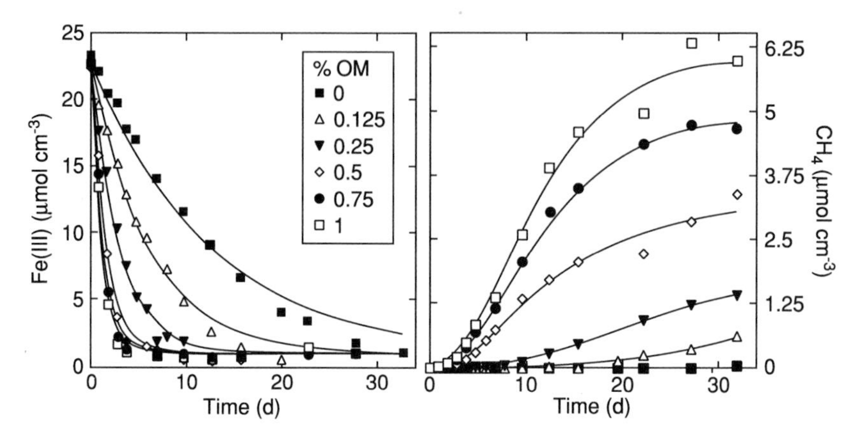

Figure 8.16 Dissimilatory iron reduction (left) and methanogenesis (right) in a freshwater sediment amended with different amounts of organic matter. Redrawn after Roden and Wetzel (2002).

As microbial iron reduction rates decrease with decreasing ferric iron availability, an increasing fraction of the carbon mineralization is channeled through sulfate reduction or, in sulfate-depleted environments, methanogenesis (Figures 8.14 and 8.16). When iron oxide concentrations are varied independently of carbon availability, changes in rates of dissimilatory iron reduction are accompanied by equivalent opposite changes in the rates of the alternative processes, with the total rate of carbon mineralization remaining the same (Figure 8.16) (Lovley and Phillips, 1987a). This indicates a similar efficiency of carbon oxidation through the different pathways of mineralization. This also indicates, as generally assumed, that rates of anaerobic carbon mineralization are governed by the concentrations and composition of the organic matter, because these factors control the initial rates of anaerobic carbon mineralization through hydrolysis and fermentation (Chapter 5). Overall, when ferric iron is limiting, its concentration controls the fraction of carbon oxidation coupled to iron reduction, while the concentration and reactivity of the organic matter control the rate. The rate of dissimilatory iron reduction, *FeRR*, may thus be approximated by "linearized Michaelis-Menten kinetics" (Roden and Wetzel, 2002):

$$
\begin{aligned}
&[\mathrm{Fe(III)}] < [\mathrm{Fe(III)}]_{\mathrm{lim}}^{-1} : FeRR = \alpha\, CMR[\mathrm{Fe(III)}][\mathrm{Fe(III)}]_{\mathrm{lim}}^{-1} \\
&[\mathrm{Fe(III)}] \geq [\mathrm{Fe(III)}]_{\mathrm{lim}}^{-1} : FeRR = \alpha\, CMR \qquad (8.13)
\end{aligned}
$$

where α represents the stoichiometric ratio of iron reduced to carbon oxidized during iron reduction (expectedly $\alpha = 4$; Equation 8.11), *CMR* is the total rate of carbon mineralization (production of CO_2 plus CH_4) and $[Fe(III)]_{lim}$ is the concentration below which inhibition of iron reduction commences. Here, at constant *CMR*, the rate of iron reduction increases linearly with [Fe(III)] until the limiting concentration is reached. At higher [Fe(III)], the rate is set exclusively by the total rate of carbon mineralization. These expressions apply to systems in which microbial population size has adjusted to the substrate supply. Deviations from these equations have been observed, for example, in freshly wetted rice paddy soil in which the sudden release of fermentation products relieved the competitive inhibition of sulfate reduction and methanogenesis. In this case, the population size of iron reducers was not large enough to consume all the electron donors (Achtnich *et al.*, 1995).

The coexistence of iron reduction with sulfate reduction or methanogenesis over a broad range of ferric iron concentrations (Figures 8.14 and 8.16) occurs under to competition for electron donors (Roden and Wetzel, 2003). The high concentrations of ferric iron needed for iron reducers to outcompete the other pathways are probably related to the necessary association of iron reducers with the surfaces of the iron oxides. Although iron reducers may locally deplete electron donors to concentrations too low for use by the competitors (see above and Chapter 3), concentration gradients between the iron reducer and the substrate source (particulate organic matter subject to hydrolysis and fermentation) may allow active sulfate reduction and methanogens closer to the source (Hoehler *et al.*, 1998). As iron oxide particles are consumed, the average distance between the remaining particles will increase, as will the volume of sediment where substrate concentrations are sufficient for the competing processes.

7.5. Abiotic reduction

In addition to the direct competition for electron donors, the interactions between the iron- and sulfate-reducing microbial communities include the reaction of hydrogen sulfide with iron oxides (Yao and Millero, 1996):

$$H_2S + 2FeOOH + 4H^+ \rightarrow S^0 + 2Fe^{2+} + 4H_2O \tag{8.14}$$

This abiotic reaction is very fast with poorly crystalline iron oxides, and because of this reaction, in conjunction with the precipitation of ferrous sulfide,

$$Fe^{2+} + H_2S \rightarrow FeS + 2H^+ \tag{8.15}$$

submicromolar hydrogen sulfide concentrations are often maintained in marine porewaters, despite active production of hydrogen sulfide through sulfate reduction (Canfield, 1989a; Thamdrup *et al.*, 1994a). By adding Equations 8.14 and 8.15 we find that three moles of H_2S molecules, equivalent to the oxidation of six moles of organic carbon, are consumed by two moles of FeOOH. In contrast, with dissimilatory iron reduction, six moles of organic carbon reduce 24 moles of FeOOH. Thus, the sulfide-based demand for ferric iron relative to the amount of carbon oxidized is 12 times lower than the demand through direct carbon oxidation with iron. So, hydrogen sulfide constitutes a minor iron sink in sediment horizons in which microbial iron and sulfate reduction are of similar importance.

By analogy with abiotic iron reduction, manganese oxides react rapidly with both hydrogen sulfide and ferrous iron (Postma, 1985; Yao and Millero, 1993):

$$H_2S + MnO_2 \rightarrow S^0 + Mn^{2+} + 2OH^- \ (\mathrm{high}[H_2S]) \tag{8.16}$$

$$H_2S + 4MnO_2 + 2H_2O \rightarrow SO_4^{2-} + 4Mn^{2+} + 6OH^- \ (\mathrm{low}[H_2S]) \tag{8.17}$$

$$2Fe^{2+} + MnO_2 + 2OH^- \rightarrow 2FeOOH + Mn^{2+} \tag{8.18}$$

Microbes may also contribute to the oxidation of elemental sulfur and iron monosulfides through the reduction of manganese oxides (Aller and Rude, 1988; Lovley and Phillips, 1994). It is difficult to unravel the relative significance of manganese-reducing pathways in manganese-poor environments due to the thinness of the zone in which the oxides are consumed. However, in some cases, the reoxidation of reduced iron or sulfur compounds appears to dominate the consumption of manganese oxide, leaving little oxide for dissimilatory manganese reduction (Canfield *et al.*, 1993b; Aller, 1994b). This conclusion was reached through comparison of the gradients of solid phase species, which also indicated that most of the iron and sulfur reoxidation could be coupled to manganese reduction.

7.6. Recycling

The allochthonous supply of iron and manganese oxides to aquatic sediments is generally small relative to the input of organic matter. Therefore, quantitatively significant contributions of manganese and iron reduction to benthic biogeochemistry depend on efficient recycling of the reduced metals. An iron or manganese atom may be reduced and reoxidized 100 times before it is permanently buried (Canfield *et al.*, 1993b). Due to the strong association of reduced iron and manganese with the solid phase, such recycling requires that these phases are transported to the oxidizing environment through physical mixing of sediment particles (Aller, 1990; Canfield *et al.*, 1993b). The most important agents for mixing are bioturbation by the benthic fauna and waves and currents causing sediment resuspension. The effect of bioturbation is exemplified to the extreme in salt marsh sediments, where dissimilatory iron reduction accounted for practically all carbon oxidation at sites inhabited by actively burrowing fiddler crabs, while the process was insignificant at neighboring, uninhabited sites (Kostka *et al.*, 2002b).

Sediments are often heavily reworked by animals, and each organism has its own characteristic way of moving sediment particles around depending on its size, lifestyle, and foraging habit. However, sediment mixing by infaunal assemblages is generally treated as a diffusional process, in which fluxes of a particulate species are proportional to its concentration gradient, with the proportionality constant referred to as the biodiffusion coefficient, D_b (Boudreau, 1997):

$$\text{Flux} = D_b \, dC/dx \tag{8.19}$$

Thus, if the biodiffusion coefficient is known, rates of manganese and iron reduction (but not of individual manganese- and iron-reducing processes) in a sediment can be estimated from the vertical gradients of manganese or iron oxide concentration into the zones of reduction.

The dependence of manganese and iron cycling on bioturbation complicates our understanding of the controls on these processes. Bioturbation typically affects marine sediments to depths on the order of 10 cm, with no obvious trends from shallow to deep waters (Boudreau, 1994). Bioturbation coefficients, however, vary by several orders of magnitude. D_b correlates positively with sediment accumulation rates, but a variation of two orders of magnitude is observed between sediments with the same accumulation rate (Boudreau, 1994). Thus, although it is expected that the availability of iron and manganese oxides will vary in relationship to the style and intensity of benthic faunal activity, there is no easy way to quantitatively predict this variability.

Sediment resuspension is particularly important in stimulating iron and manganese cycling in mobile mud belts. Mobile muds are found on continental shelves off major rivers carrying a high particle load, such as the Amazon River. Through cycles of resuspension and sedimentation, these muds are reworked to a depth of up to 1 m or more. Iron oxides are the major terminal electron acceptor for carbon oxidation in these muds, but it is not clear whether iron reduction occurs through direct organotrophic microbial reduction or through reoxidation of reduced sulfur species, with sulfate reduction as the dominating respiratory process (Aller *et al.*, 1996; Madrid *et al.*, 2001). Mobile muds are of great biogeochemical interest because they receive a large fraction of the particle flux to the oceans.

In freshwater sediments, mixing linked to the ebullition of methane may also stimulate iron reduction (Roden and Wetzel, 1996). Furthermore, in both freshwater and shallow marine environments, plant roots may stimulate iron and manganese cycling by leaking oxygen to the sediment and thereby increasing the area of the oxic–anoxic interface (Kristensen *et al.*, 2000; Ratering and Schnell, 2000).

7.7. Role in benthic carbon mineralization

Microbial manganese and iron reduction depend on oxide concentrations, which are influenced by competing abiotic reduction processes and by sediment reworking. Variability in the intensity and styles of reworking and in the importance of abiotic reduction ensures a large variability in the oxide reduction rates and their significance for carbon oxidation in aquatic sediments (Figure 8.17; see also Chapter 5). Generally, high rates and large relative contributions from iron reduction are found in fine-grained marine sediments on the open shelf. These environments have a high content of diagenetically active iron and well-oxygenated bottom water, and a moderately high input of reactive organic matter support a large infauna. Active infauna maintain high concentrations of iron oxide through continuous sediment reworking. At such sites, iron reduction may account for up to 50% of carbon oxidation

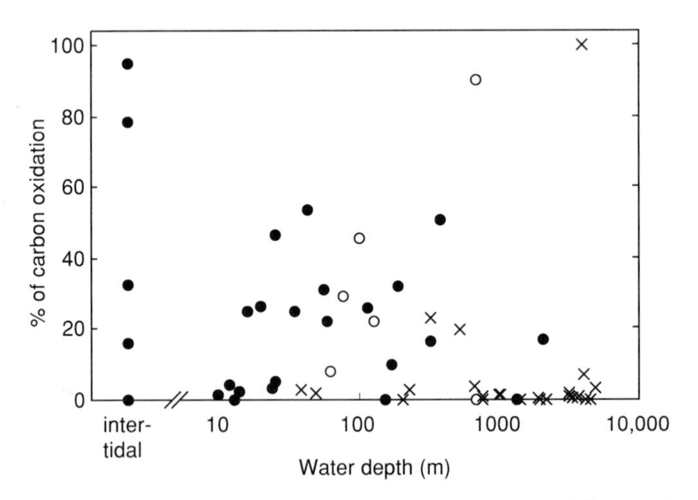

Figure 8.17 Contributions of dissimilatory manganese and iron reduction to carbon oxidation in marine sediments as a function of water depth. Circles, results from sediment incubations; filled circles, iron reduction (manganese reduction insignificant); open circles, manganese reduction (iron reduction insignificant); crosses, results from diagenetic modeling presented as the sum of manganese and iron reduction. Sources: Thamdrup (2000) and references therein, Bender and Heggie (1984), Slomp *et al.* (1997), Boudreau *et al.* (1998), Lohse *et al.* (1998), Kristensen *et al.* (2000), Thamdrup *et al.* (2000), Kostka *et al.* (2002b), and Jensen *et al.* (2003).

and may be the dominating anaerobic mineralization pathway, though the contribution is not always so high. High contributions are also found in vegetated intertidal sediments, where crab bioturbation and possibly oxygen transport through plant roots stimulate iron recycling. Conversely, iron reduction contributes little in sediments underlying permanently or seasonally anoxic waters. In these environments, infaunal activity is reduced. Iron reduction is also not important in sandy sediments, where low concentrations of reactive iron may inhibit the process. In sands and in sediments underlying anoxic water columns, sulfate is the dominating anaerobic electron acceptor. At the sites where rates of dissimilatory iron reduction have been quantified experimentally (Figure 8.17), its average contribution was 17% of the total carbon mineralization and 26% of the anaerobic carbon oxidation.

Microbial manganese reduction generally contributes little to carbon oxidation in continental shelf sediments due to their low manganese content. In these environments, manganese oxides are depleted just below the oxic–anoxic interface (Figure 8.17). Reduced Mn^{2+}, however, is mobile, frequently escaping the sediment (see above). In some instances, this manganese is concentrated in well-ventilated sedimentary basins such as the Norwegian Trough (Figure 8.14) and in oxic sediments bordering anoxic basins. In such environments, manganese concentrations can reach several percent by weight, and microbial manganese reduction may be the dominating pathway of carbon oxidation (Aller, 1990; Canfield *et al.*, 1993a).

The significance of dissimilatory iron and manganese reduction on the continental slopes and in the deep sea has not been extensively explored; most estimates are from diagenetic modeling (Figure 8.17). Some models include only dissolved Mn^{2+} and Fe^{2+} in the porewater, neglecting the recycling of particulate Fe(II), in which case they likely underestimate the gross rates of iron cycling (Thamdrup, 2000). Nevertheless, the general increase in oxygen penetration depths (Chapter 6) and decrease in bioturbation intensities with increasing water depth should favor aerobic over anaerobic mineralization.

In freshwater environments, high rates of microbial iron reduction have been found in river and wetland sediments, with the process accounting for up to 50% of carbon oxidation in the latter (Roden and Wetzel, 1996). Likewise, iron is the dominating electron acceptor for carbon oxidation in parts of many aquifers (Lovley, 2000b).

7.8. Manganese and iron reduction in water columns

The chemoclines of stratified water bodies often exhibit qualitatively similar redox stratification to sediments, and they therefore provide potential niches for manganese- and iron-reducing microbes. The concentrations of particulate manganese and iron phases in anoxic water columns are, however, several orders of magnitude lower than found in sediments. Microbial metal oxide reduction is potentially significant in non-sulfidic anoxic hypolimnia of lakes, where soluble Mn^{2+} and Fe^{2+} accumulate. Conditions favoring these processes, such as anoxia in the absence of nitrate and hydrogen sulfide, are not found extensively in marine water columns. However, organotrophic manganese reduction, as well as manganese reduction coupled to ammonium oxidation, has been suggested to play a role in the some anoxic waters, including the non-sulfidic intermediate layer in the Black Sea and in the chemocline above the hypersaline Orca Basin (Nealson *et al.*, 1991; Murray *et al.*, 1995; Van Cappellen *et al.*, 1998).

8. STABLE IRON ISOTOPES AND IRON CYCLING

Whereas manganese has only one stable isotope, iron has four stable forms, ^{54}Fe, ^{56}Fe, ^{57}Fe, and ^{58}Fe, accounting for 5.8, 91.7, 2.2, and 0.3%, respectively, of the total mass. Isotope ratios may, therefore, potentially serve as tracers of biogeochemical iron transformations. Isotopic shifts up to, typically, a few per mil. are imparted by both abiotic and biological pathways during both the oxidation and the reduction of iron (Beard *et al.*, 1999; Anbar *et al.*, 2000; Bullen *et al.*, 2001; Beard *et al.*, 2003; Croal *et al.*, 2004). Results from this emerging field of research suggest that variations in iron isotope abundances are to a large extent related to equilibrium fractionation effects between different iron species, although kinetic effects may also play a role (Roe *et al.*, 2003; Croal *et al.*, 2004). Fractionation arises between iron species with different bonding energies, with the heavier isotopes enriched in the more strongly coordinated species. For example, at equilibrium, octahedrally coordinated iron will be isotopically heavier than coexisting tetrahedrally coordinated iron. Phototrophic iron oxidation, abiotic iron oxidation, and bacterial iron reduction all show initial isotopic fractionations of $\sim$1.5‰ in $^{56}Fe.^{54}Fe$ ratios ($\delta^{56}Fe_{oxide}$-$\delta^{56}Fe_{Fe^{2+}}$) between oxide phases and soluble Fe^{2+} (Croal *et al.*, 2004). The similarity in signatures imparted by these different processes may complicate the use of Fe isotope systematics to unravel the details of biogeochemical cycling. This field is, however, still in its infancy.

Chapter 9

The Sulfur Cycle

ADVANCES IN MARINE BIOLOGY VOL 48
0-12-026147-2

© 2005 Elsevier Inc.
All rights reserved

1. INTRODUCTION

One's concept of the sulfur cycle depends greatly upon discipline. For example, to an atmospheric chemist, the sulfur cycle involves gas-phase reactions of sulfur compounds and the formation and subsequent involvement of sulfate aerosols as cloud-forming nuclei. To a geochemist, the sulfur cycle involves all processes influencing the transfer of sulfur compounds to and from the various Earth surface reservoirs (Figure 9.1). Thus, the sulfur cycle involves the weathering of sulfide and sulfate minerals from rocks on land, the transfer of sulfate to the oceans, the removal of sulfate as both

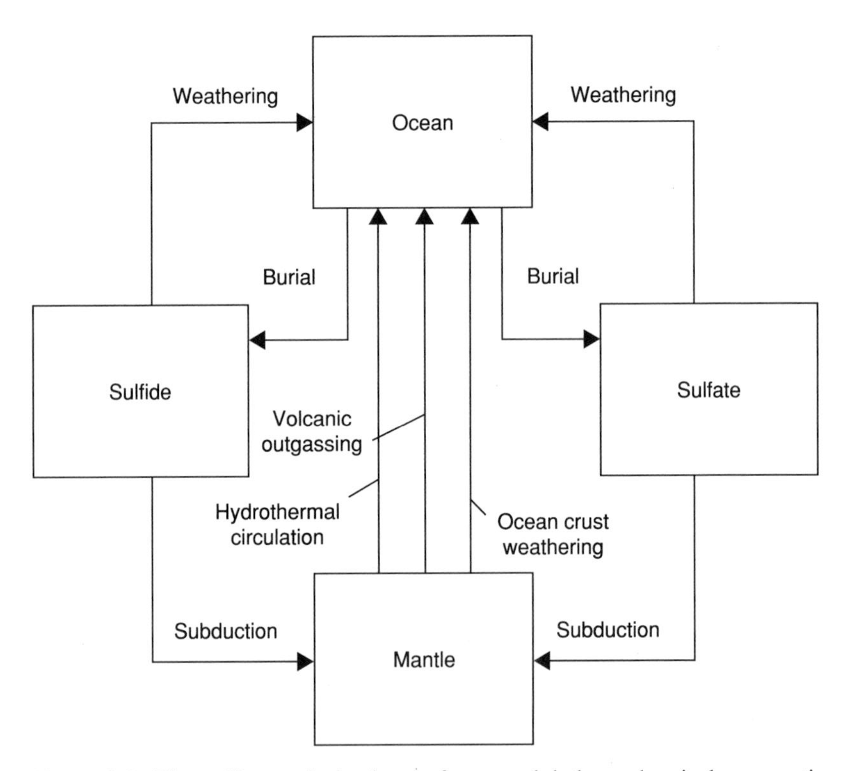

Figure 9.1 The sulfur cycle is shown from a global geochemical perspective. Sulfide and sulfate minerals are weathered on land, delivering sulfate to the oceans. Marine sulfate may be formed into sulfide during sulfate reduction, returning sulfur back to the sulfide pool, or sulfate may be precipitated in evaporitic basins. The mantle also represents a source of sulfur for the Earth's surface. Mantle sources include volcanic outgasing of, mainly, SO_2 gas, the delivery of sulfide to the oceans through hydrothermal circulation at mid ocean ridges and the weathering of sulfide in ocean crust. The subduction of sediments containing sulfide and sulfate represents return fluxes of sulfur back into the mantle. Modified from Canfield (2004).

pyrite in marine sediments and evaporitic gypsum deposits, and subsequent delivery of these marine deposits back on to land. Also important could be the exchange of sulfur compounds between the mantle and the Earth surface reservoirs. Our perspective here encompasses some of the geochemist's view of the sulfur cycle but is decidedly more organismal in its emphasis. Thus, we wish to explore the relationship between the different organisms metabolizing sulfur compounds and the environment. We therefore pay attention to the organisms themselves, their diversity, ecology, and phylogeny, as well as the biogeochemical processes influencing their activity.

From our perspective then, the sulfur cycle is better represented in Figure 9.2. Listed are the most important compounds involved in the biological transformation of sulfur, ranging in oxidation level from +6 (SO_4^{2-}) to −2 (H_2S), as well as an indication of the processes of transformation. In a brief overview, there are two ways by which sulfate can be reduced to sulfide: assimilatory sulfate reduction, in which sulfate is reduced to form

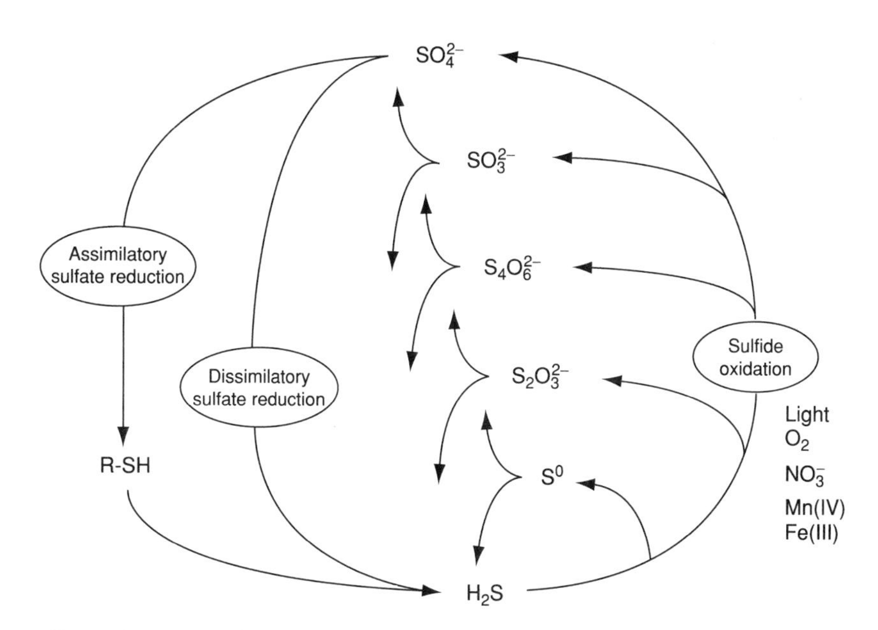

Figure 9.2 The sulfur cycle from a microbial ecological perspective. Ultimately, sulfide is formed from sulfate either through assimilatory sulfate reduction or through dissimilatory sulfate reduction (the major sulfide formation pathway). Most of the sulfide formed in nature is reoxidized again through light-mediated photosynthetic pathways, biologically mediated non-photosynthetic pathways or through inorganic reactions. Numerous sulfur intermediate compounds may be formed, and these may be oxidized, reduced or in many cases disproportionated to sulfide and sulfate.

bioessential compounds at an energy cost to the organism, and dissimilatory sulfate reduction, in which the organism gains energy for growth and maintenance from the reduction of sulfate. Once sulfide is produced, a portion can form with reactive Fe(III) and Fe(II) compounds to form various iron sulfide phases and eventually pyrite. We do not yet know if organisms are directly involved in this transformation. However, most sulfide is reoxidized to a variety of sulfur compounds of intermediate oxidation state (Jørgensen, 1982b) and eventually to sulfate. Reoxidation may occur by strictly inorganic reactions, as in, for example, the reaction between reactive Fe oxides and dissolved H_2S (Pyzik and Sommer, 1981; Canfield, 1989a) (see Chapter 8), or by biologically mediated processes utilizing light energy, as in the case of anoxygenic phototrophs, or for nonphototrophs, with oxygen and nitrate as electron acceptors.

Once intermediate compounds are formed, a variety of fates is possible, including biological oxidation, reduction, and disproportionation (Bak and Pfennig, 1987; Thamdrup *et al.*, 1993). The last is analogous to the fermentation of organic compounds; the sulfur compound forms into both sulfate and sulfide. We still have a poor understanding of the quantitative significance of the different biological processes involved in the turnover of sulfur of intermediate oxidation levels. However, a great number of organisms have been isolated from nature that can accomplish these turnover reactions, and overall, organisms of the sulfur cycle have perhaps the greatest diversity of any group of microorganisms metabolizing a single element. Table 9.1 summarizes the metabolically active compounds of the S cycle and their possible involvement in different biological metabolisms and inorganic transformations.

We consider below each of the biologically mediated sulfur metabolisms. We first consider those metabolisms involved in the reduction of sulfur compounds, then the oxidative processes, and last the disproportionation of sulfur compounds. We consider the phylogenies and ecology of sulfur-metabolizing organisms, and we take a close look at the global significance of sulfate reduction in the mineralization organic matter.

2. ASSIMILATORY SULFUR METABOLISMS

Sulfur is an essential nutrient element and constitutes around 0.5 to 1.0% of the dry weight of prokaryotic organisms (Zehnder and Zinder, 1986). Sulfur is most abundant in the amino acids cysteine and methionine (Table 9.2), key building blocks of proteins. No less important is the fact that sulfur is ubiquitous among organisms in numerous co-enzymes, vitamins, and electron carriers (cofactors) crucial to cellular metabolism (Table 9.1). In

Table 9.1 Metabolically active compounds of the S cycle

Formula	Name	Oxidation level	Dissimilatory reduction	Assimilatory reduction	Chemotrophic oxidation	Phototrophic oxidation	Dispropor-tionation	Chemical oxidation[a]	Chemical reduction[a]
SO_4^{2-}	Sulfate	+6	+	+	–	–	–	–	–
SO_3^{2-}	Sulfite	+4	+	+	+	+	+	+	+
$S_3O_6^{2-}$	Trithionate	+3.33	+	?	+	+	?	?	+
$S_4O_6^{2-}$	Tetrathionate	+2.5	+	?	+	?	+	?	+
$S_2O_3^{2-}$	Thiosulfate	–1, +5	+	+	+	+	+	+	–
S^0	Elemental Sulfur	0	+	+	+	+	+	+	+
H_2S	Hydrogen Sulfide	–2	–	–	+	+	–	+	–
FeS_2	Pyrite Marcasite	–1	–	–	+	–	–	+	(+)

+, yes; – no; ?, uncertain.

[a]Conditions with normal environmental oxidants and reductants such as O_2, NO_3^-, metal oxides, Fe^{2+}, Mn^{2+}, and H_2S, and at near neutral pH.

Table 9.2 Some sulfur-containing compounds of importance to microorganisms

Name and occurrence
1. Common amino acids
Cysteine in all organisms
Methionine in all organisms
2. Cofactors (coenzymes, vitamins, electron couriers)
Thiamine (vitamin B_1) in all organisms
Biotin (vitamin H) in all organisms
Coenzyme A (coenzyme) in all organisms
Lipoic acid (vitamin) in all organisms
Ferredoxin (electron center) in all organisms
Coenzyme M (coenzyme) methane-forming bacteria
3. Sulfate esters
Choline sulfate in fungi and bacteria
4. Sulfonates
Sulfolipid in plants and photosynthetic bacteria

Adapted from Zahnder and Zinder (1980).

addition, sulfate esters and sulfonates, representing oxidized sulfur forms, have an important structural role in cell walls, photosynthetic membranes, and the connective tissues of plants and animals (not shown in Table 9.2).

Most prokaryotic organisms obtain a large part of their organic sulfur by internal cellular metabolisms (Schiff and Fankhauser, 1981; Widdel, 1988). As many of the biologically essential organic sulfur compounds contain sulfur in a reduced oxidation level, organisms have developed a biochemical pathway of sulfide formation based on the reduction of oxidized sulfur species. The most widely used assimilatory reduction process involves the reduction of sulfate, although many prokaryotes may also reduce thiosulfate, polythionates, and elemental sulfur for their assimilatory sulfur needs by lesser-known pathways (Le Faou *et al.*, 1990; Thauer and Kunow, 1995). Here we focus on the process of assimilatory sulfate reduction. This process, by contrast to dissimilatory sulfate reduction, is an energy-requiring process.

The first step in assimilatory sulfate reduction is the active cellular uptake of sulfate by specific sulfate-binding proteins. This process requires energy and is unidirectional; no sulfate is lost from the cell after uptake (Cypionka, 1995). Following this, there are two main paths for reduced sulfur formation (Figure 9.3). The first pathway (the APS pathway) is utilized by most oxygenic phototrophs, including many cyanobacteria, all eukaryotic algae, and higher plants (Schiff and Fankhauser, 1981). In this pathway sulfate is combined with ATP, mediated by the enzyme ATP sulfurylase, to form APS (adenosine-5′-phosphosulfate). From here, two branch points are possible. In one, leading to the formation of sulfate esters, APS is phosphatized with ATP and the enzyme APS kinase to form PAPS (adenosine 3′-phosphate-5′-phosphosulfate) (Figure 9.3). In the other branch, APS is converted to an

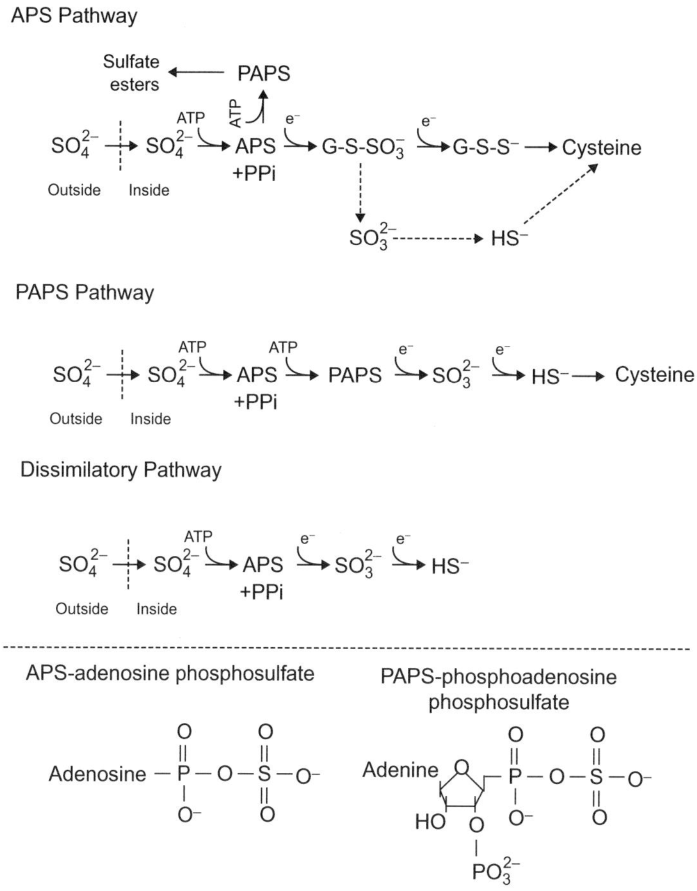

Figure 9.3 The major biochemical pathways by which sulfate is reduced. During assimilatory sulfate reduction, either the APS pathway or the PAPS pathway is used. The APS pathway is used by oxygen-evolving eukaryotes and some cyanobacteria, while the PAPS pathway is used by some cyanobacteria, but mostly by microorganisms whose metabolism does not produce oxygen. Also shown are the principal steps during dissimilatory sulfate reduction.

organic thiosulfate with the enzyme APS sulfotransferase. A further reduction step, promoted by the enzyme organic thiosulfate reductase, produces reduced organic sulfur bonds, ultimately leading to cysteine. In a minor pathway, the organic thiosulfate may decompose to sulfite, which can be converted to hydrogen sulfide and eventually to cysteine through the enzyme sulfite reductase (Schiff and Fankhauser, 1981).

In the second pathway (the PAPS pathway) of assimilatory sulfate reduction, utilized by some cyanobacteria but mostly by microorganisms not producing oxygen, PAPS is the main intermediate compound from which the initial reduction of sulfate occurs. After PAPS formation, the reduction to sulfite occurs through the enzyme PAPS sulfotransferase. Sulfite is reduced to hydrogen sulfide through an assimilatory sulfite reductase enzyme, and the sulfide produced is utilized in organic sulfur formation.

In both pathways of assimilatory sulfate reduction, no proton motive force is generated, no oxidative phosphorylation occurs and, therefore, no energy is conserved. During assimilatory sulfate reduction, energy in the form of ATP, as well as reducing power, must be supplied from other respiratory or fermentative processes. The assimilatory reduction of sulfate is, therefore, energy expensive.

3. DISSIMILATORY SULFATE REDUCTION

In the following Sections we review the ecology and phylogeny of sulfate reducers in nature, and we look at the principal factors regulating the process. Many sulfate reducers can also reduce elemental sulfur, and elemental sulfur reduction is considered in a separate Section. Sulfate reducers also commonly reduce thiosulfate and sulfite, and although present in nature, obligate thiosulfate and sulfite-reducing organisms seem to be rare. We say little about thiosulfate- and sulfite reduction in the following pages. We acknowledge that thiosulfate and sulfite reduction occur in nature, but we view these processes, rather, as a subset of sulfate reduction.

3.1. Environmental significance and distribution

Dissimilatory sulfate reduction is a widespread anaerobic mineralization process. To assess its significance in the marine realm, we have compiled sulfate reduction rates and oxygen uptake rates from the whole range of marine sedimentary environments where data are available. In making our compilation, depositional environments have been subdivided based on water depth. Considered also is whether the sediment underlies an upwelling zone, and for the shelf environment, whether sediment is depositing or not. Several different coastal depositional environments are recognized, including special high deposition rate shallow water sites, salt marshes, mangroves, intertidal zones, seagrass beds, and microbial mats. Individual results for total carbon mineralization and sulfate reduction rates are plotted as a function of sediment deposition rate in Figure 9.4, with the compiled rate averages shown in Table 9.3.

Measuring rates of sulfate reduction

Various methods have been used to quantify rates of sulfate reduction in the environment; these include closed jar incubations in which rates of sulfate depletion are monitored, diagenetic modeling of sulfate profiles in sediments, and mass transport modeling of sulfate or sulfide gradients in water columns. However, the most widely used, and probably most accurate, method involves the injection of radiolabeled $^{35}SO_4^{2-}$ into sediment or water column samples and determining the distribution of radioisotope in the sulfate and sulfide pools after a known incubation period. The following equation is used to quantify rates of sulfate reduction (Jørgensen, 1978a):

$$\mathrm{Rate} = \alpha * A_{\mathrm{sulfide}}[\mathrm{SO_4^{2-}}]/(A_{\mathrm{sulfate}} * t)$$

where $A_{sulfide}$ is the total activity of radioisotope in the sulfide pool, $A_{sulfate}$ is the total activity of radioisotope in the sulfate pool, $[SO_4^{2-}]$ is the concentration of sulfate, t is time, and the factor α expresses the fractionation between $^{32}SO_4^{2-}$ (the most abundant form of sulfate in the environment; see Section 10) and the $^{35}H_2S$ produced. This factor is normally taken as 1.06, but should be adjusted accordingly in situations, such as in low sulfate lake environments, in which reduced fractionations might be expected.

In calculating sulfate reduction rates with this method, various assumptions apply. The radiolabeled sulfate is assumed to be present in tracer amounts, and during the course of the incubation, only a small fraction of the added tracer is assumed to be reduced to sulfide. For the most accurate determination of sulfate reduction rate in sediments, radiotracer should be injected into intact cores (Jørgensen, 1978a). However, because most sulfide produced in sediments is ultimately reoxidized to sulfate, incubations should be kept as short as possible to minimize reoxidation. In active systems such as microbial mats, this could mean incubations as short as 10 minutes (Canfield and Des Marais, 1993), while for typical coastal marine sediments incubations should probably be 1 h or less (although incubation times have normally been longer). A special problem in sediments is the rapid incorporation of radiolabeled sulfide into pyrite (Howarth, 1979), which is inaccessible using standard HCl acid digestions. Therefore, to accurately quantify the size of the radiolabeled sulfide pool, sulfide is distilled from sediments with hot acidic Cr^{2+}, which effectively liberates all the inorganic sulfide from the sediment (Canfield *et al.*, 1986; Fossing and Jørgensen, 1989). Compared to sediments, there have been relatively few determinations of sulfate reduction rates in anoxic water columns. Rates are frequently low compared to sediment rates, requiring incubation with high radioactivity for long periods of time. This is necessary to generate activity in the sulfide pool significantly larger than normal procedure blanks.

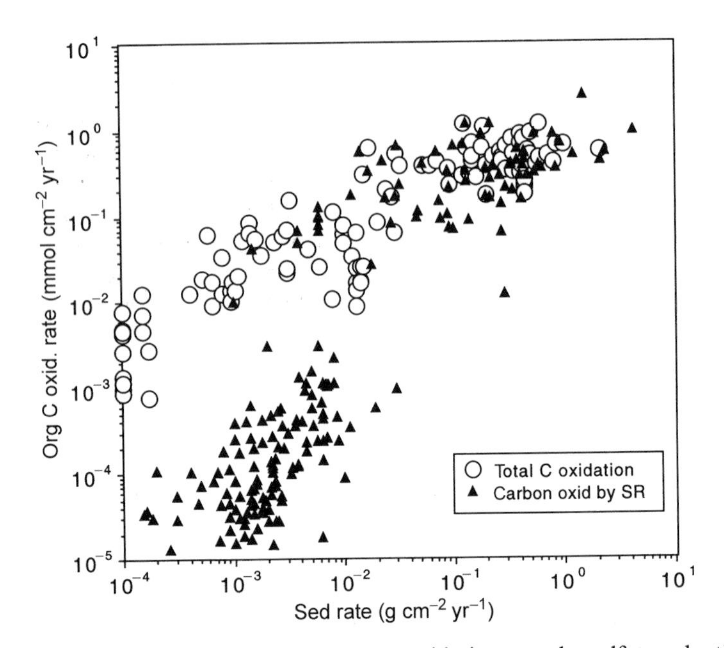

Figure 9.4 A global compilation of carbon oxidation rates by sulfate reduction, as well as total carbon mineralization rates, as obtained from rates of sediment oxygen uptake. In environments with relatively high sedimentation rates, sulfate reduction may account for half or more of the total carbon mineralization. However, as sedimentation rate drops, sulfate reduction accounts for a progressively smaller fraction of total carbon oxidation. Updated from Canfield (1993).

From Figure 9.4, and consistent with previous discussions (Jørgensen, 1982b; Canfield 1993), it can be seen that sulfate reduction plays a prominent or even dominant role in carbon mineralization in high deposition rate settings characterizing the continental margin. The significance of sulfate reduction drops in moving to slower deposition rate settings (Canfield, 1993; Ferdelman *et al.*, 1999). Nonetheless, sulfate reducers seem to be active, albeit at very much reduced rates, under virtually the whole deep ocean floor (Canfield, 1991; D'Hondt *et al.*, 2002).

Taken individually, most shallow water environments demonstrate a dominant role for sulfate reduction in carbon mineralization. This means that aerobic respiration is suppressed, and, on average, other carbon mineralization pathways such as metal oxide reduction are sub-prominent. Of course, some variability exists, and high rates of metal oxide reduction have been demonstrated in certain marine settings (Canfield *et al.*, 1993; Rysgaard *et al.*, 2001; see Chapter 8). Also, sulfate reduction seems to be

Table 9.3 Summary of average sulfate reduction rates and oxygen uptake rates in different marine depositional settings

Region	Area (km^2)	Sulfate reduction rate (mmol cm^{-2} y^{-1})	O_2 uptake (mmol cm^{-2} y^{-1})
Shallow, high deposition	1×10^4	1.3	2.6
Salt marsh	4×10^5	2.8	8.5
Mangroves	1×10^5	0.72	6.1
Intertidal	5×10^4	0.42	1.31
Seagrass beds	1.1×10^6	0.74	1.48
Estuaries and embayments	2×10^6	0.26	0.94
Shelf-depositional	1.1×10^7	0.17	0.43
Shelf-non depositional	1.4×10^7	0.04	0.11
Upwelling	1.3×10^5	0.27	0.7
Upper slope (200–1000 m)	1.6×10^7	7.4×10^{-2}	0.13
Lower slope (1000–2000 m)	1.5×10^7	1.5×10^{-2}	5.3×10^{-2}
Rise (2000–3000 m)	2.2×10^7	2×10^{-3}	6.3×10^{-2}
Abyss (3000–4000 m)	7.1×10^7	1×10^{-3}	4×10^{-2}
Abyss (4000–5000 m)	1.2×10^8	1.5×10^{-4}	1.6×10^{-2}
Abyss (>5000 m)	8.8×10^7	1.2×10^{-4}	3.6×10^{-3}

Data are compiled from over 100 literature sources.

suppressed in mangrove environments, where, as shown in Chapter 13, active burrowing by animals, as well as sediment oxidation by mangrove plant roots and rhizomes, provides the opportunity for alternative pathways of carbon mineralization, including Fe reduction. It would appear that sulfate reduction is an important mineralization process in sediments as deep as 2000 m (Table 9.3). However, the results from slope environments should be met with some suspicion because the database for sulfate reduction rates is small, and generally, oxygen uptake rates (from which total carbon mineralization is determined) and sulfate reduction rates have been measured from different locations. Sulfate reduction is clearly of minor significance in deep-sea sediments, as is also evident from Figure 9.4. In total, sulfate reduction is globally responsible for the mineralization of about 1.3×10^{14} mol C yr^{-1}, which accounts for about 55% of the total mineralization of carbon in marine sediments. These numbers should be adjusted as new results appear on sulfate reduction rates in marine sediments, particularly in slope environments (and deeper) where data is lacking.

Sulfate reducers can also be active in lake sediments (Capone and Kiene, 1988; Bak and Pfennig, 1991), as well as in the water columns of anoxic lakes and marine basins (Albert *et al.*, 1995). Sulfate reduction rates in lake sediments are sometimes as high as in coastal marine sediments (Bak and

Pfennig, 1991), and sulfate reduction can account for a significant fraction (10 to 80%) of the anaerobic carbon mineralization in the sediment (Holmer and Storkholm, 2001). Generally, though, sulfate reduction is thought to account for only a small amount of the total carbon mineralization in lake sediments (Capone and Kiene, 1988; Holmer and Storkholm, 2001). In most instances, however, total sediment metabolism is not measured in lake sediments, so the significance of sulfate reduction in total carbon mineralization is difficult to quantify. In one recent study of oligotrophic Lake Michigan sediments, Thomsen *et al.* (2004) found that sulfate reduction accounted for a modest 19% of the total organic matter mineralization.

There are very few measurements of sulfate reduction rate in anoxic water columns. Part of the problem is that the rates are often extremely low and difficult to measure with standard radiotracer techniques (see text box). Nonetheless, Albert *et al.* (1995) have measured sulfate reduction rates in the water column of the Black Sea that are apparently comparable, on an aerial basis, to sediment rates (Table 9.4). The database, however, is still very limited, and hard conclusions about the relative significance of water column sulfate reduction in the Black Sea are probably premature. In another study, Indrebø *et al.* (1979) measured relatively high water column rates of sulfate reduction in Saelenvann, a Norwegian fjord, exceeding sediment rates (Table 9.4). By contrast, in our studies of Mariager Fjord, an anoxic Danish Fjord on the East coast of Jutland, water column sulfate reduction is <10% of the rates of sulfate reduction measured in the sediment (Sørensen and Canfield, 2004) (Table 9.4).

Sulfate reducers will also be active in environments other than lakes and marine sediments in which a suitable electron donor, such as organic matter or hydrogen, is available along with sulfate. These environments include

Table 9.4 Sulfate reduction rates in anoxic water bodies

Site	Max. rate (nmol cm^{-3} d^{-1})	Integrated rate (mmol m^{-2} d^{-1})	% of total sulfate reduction in basin
Black Sea	3.5×10^{-3}	0.22 to 1.2	~50
Mariager Fjord, Denmark	0.14	<1.5	<25
Saelenvann, Norway	1.0	2.1 to 3.8	68 to 86
Mahoney Lake	0.3	1 to 13	–
Big Soda Lake	1 to 3	24 to 80	>50
Solar Lake	6	2.4 to 4.6	3 to 6
Lake Mogil'noe	13.6	10.9	–

References: Indrebø *et al.* (1979); Jørgensen *et al.* (1979); Smith and Oremland (1987); Albert *et al.* (1995); Overmann *et al.* (1996b); Ivanov *et al.* (2001); Sørensen and Canfield (2004).

hydrothermal springs such as those found at mid-ocean spreading centers in the deep sea or in terrestrial hydrothermal environments such as Yellowstone Park (Jørgensen *et al.*, 1992; Weber and Jørgensen, 2002). Sulfate-reducing populations are also well developed in cyanobacterial mats, where the highest recorded rates have been found (Jørgensen and Cohen, 1977; Canfield and Des Marais, 1993), and in stromatolites (Visscher *et al.*, 2000), and they may form an important component of a subsurface biosphere believed to be driven by the autotrophic oxidation of hydrogen and possibly methane (Stevens, 1997) (see Chapter 4).

One reason for the wide environmental distribution of sulfate reducers is their wide environmental tolerance. For example, they have been reported in environments with temperatures ranging from less than 0 °C to just over 100 °C (Jørgensen *et al.*, 1992; Sagemann *et al.*, 1998; Kostka *et al.*, 1999), and at salt concentrations up to 28% NaCl, or about eight times the salinity of sea water (Brandt *et al.*, 2001; Sørensen *et al.*, 2004). Finally, although all known pure cultures of sulfate reducers must have strictly anaerobic conditions to reduce sulfate by dissimilatory metabolism, sulfate reduction has been detected in the oxic zones of cyanobacterial mats and marine sediments (Canfield and Des Marais, 1991; Jørgensen and Bak, 1991). The significance of "aerobic" sulfate reduction in nature is currently unknown, and a more complete discussion of this process is provided in Section 3.6.1.

3.2. Phylogeny

Organisms from both prokaryote domains are engaged in dissimilatory sulfate reduction (Widdel, 1988; Stackebrandt *et al.*, 1995; Castro *et al.*, 2000) (Figure 9.5). Deep branching within the *Bacteria* are sulfate reducers of the genera *Thermodesulfobacterium* and the newly described *Thermodesulfatator* (not shown; Moussard *et al.*, 2003). Both genera display maximum growth temperatures of 80 to 85 °C, and this range is among the highest of any member of the bacterial domain. Another thermophilic genus, *Thermodesulfovibrio* (growth maximum of 70 °C; Henry *et al.*, 1994; Sonne-Hansen and Ahring, 1999), is located within the *Nitrospira* group (known mainly for housing nitrifying bacteria). Most known sulfate reducers fall within the δ-subdivision of the proteobacteria (Figure 9.5). A relatively large group of spore-forming sulfate reducers is also found among the gram-positive bacteria. Most gram-positive sulfate reducers are housed within the genus *Desulfotomaculum*, while a separate, phylogenetically distinct genus, *Desulfosporosinus* (not shown in Figure 9.5), has also been proposed (Castro *et al.*, 2000). Many gram-positive sulfate reducers are thermophilic.

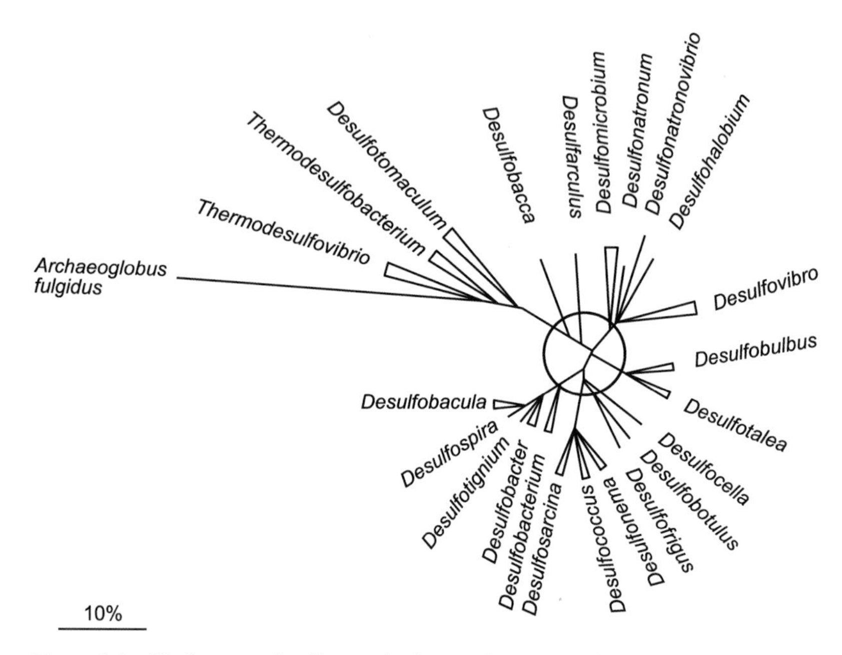

Figure 9.5 Phylogeny of sulfate-reducing prokaryotes with the archaeon *Archaeoglobus fulgidus* as the outgroup. The phylogeny is based on sequences from the 16S rRNA gene. Simplified from Detmers *et al.* (2001). The circle outlines branching within the δ proteobacteria.

Located within the *Archaea* is a single genus of sulfate reducers, *Archaeoglobus* (Stetter, 1992; Stackebrandt *et al.*, 1995), which is distinguished by high maximum growth temperatures of up to 92 °C. These are the highest known growth temperatures for any pure culture of sulfate reducer. These high temperatures, however, do not quite match the highest temperature (about 105 °C; Jørgensen *et al.*, 1992) for which sulfate reduction has been detected in nature, suggesting that perhaps other hyperthermophilic species of sulfate reducer are present in the environment but have yet to be cultured.

Archaeoglobus is located among a group of acetotrophic (split acetate into CH_4 and CO_2) methanogenic bacteria (*Methanomicrobiales*), and it shares certain biochemical similarities with the methanogens (Thauer and Kunow, 1995). Many speculate that *Archaeoglobus* was originally a methanogenic species, having received the ability to reduce sulfate by lateral gene transfer (Stackebrandt *et al.*, 1995; Stetter, 1996). Such a conclusion is at least loosely supported by comparisons between phylogenies constructed from 16S rRNA and the dissimilatory sulfite reductase (DSR) gene, for which sulfite reductase is a key enzyme in the sulfate reduction path (Wagner *et al.*, 1998; Klein *et al.*, 2001) (see Section 3.3). Importantly, the branching order for the deeply rooted genera *Thermodesulfovibrio* and *Archaeoglobus* differs between the 16S rRNA and DSR-based trees, indicating the possibility for gene transfer (Figure 9.6). Furthermore, when the two separate subunits of the DSR gene (DsrA and DsrB) are aligned and compared, *Thermodesulfovibrio* forms the root of the tree, allowing the possibility that *Archaeoglobus* obtained its DSR by later gene transfer from the bacterial domain. A root between *Archaeoglobus* and *Thermodesulfovibrio* would have been more consistent with an early evolved DSR gene having transcended both prokaryote domains.

Comparisons between the 16S rRNA and DSR trees reveal several other instances of lateral gene transfer of the sulfite reductase gene (or complex of genes controlling sulfate reduction). Thus, multiple episodes of DSR gene transfer place numerous species of gram-positive sulfate reducers of the genus *Desulfotomaculum* clearly within the δ-proteobacterial subdivision. Only a handful of *Desulfotomaculum* DSR sequences branch outside of the proteobacteria, and this placement would be expected for gram-positive organisms in the absence of gene transfer. Surprisingly, the DSR genes of the deep-branching genus *Thermodesulfobacterium* fall well within the δ-subdivision of the proteobacteria. This phylogenetic association indicates that *Thermodesulfobacterium* obtained its DSR gene from lateral gene transfer from the proteobacteria. Thus, unless there is another unidentified DSR gene within its genome with a different phylogenetic association, *Thermodesulfobacterium* apparently obtained its DSR gene, and ability to reduce sulfate, long after the initial emergence of the lineage.

3.3. Principal steps in dissimilatory sulfate reduction

Dissimilatory sulfate reduction is an energy-gaining respiratory process in which the reduction of sulfate to sulfide is linked to the oxidation of an electron donor (either organic compounds or hydrogen). If the organic substrate is complex, intracellular fermentations break the substrate into simple compounds such as acetate and H_2. Some energy is gained during these fermentation steps by substrate-level phosphorylation, but most energy is gained as acetate (in complete oxidizers) or H_2 (in most incomplete oxidizers and some complete oxidizers; see Section 3.4) oxidation is coupled to electron-transport-linked phosphorylation (Figure 9.7) (see Chapter 3).

The main steps in the reduction of sulfate are reminiscent of those in assimilatory sulfate reduction but with some major differences. One difference is that the transport of sulfate into the cell is coupled to the transport of either protons or sodium ions and is highly reversible (Cypionka, 1995). The driving force behind sulfate transport is concentration gradients of protons,

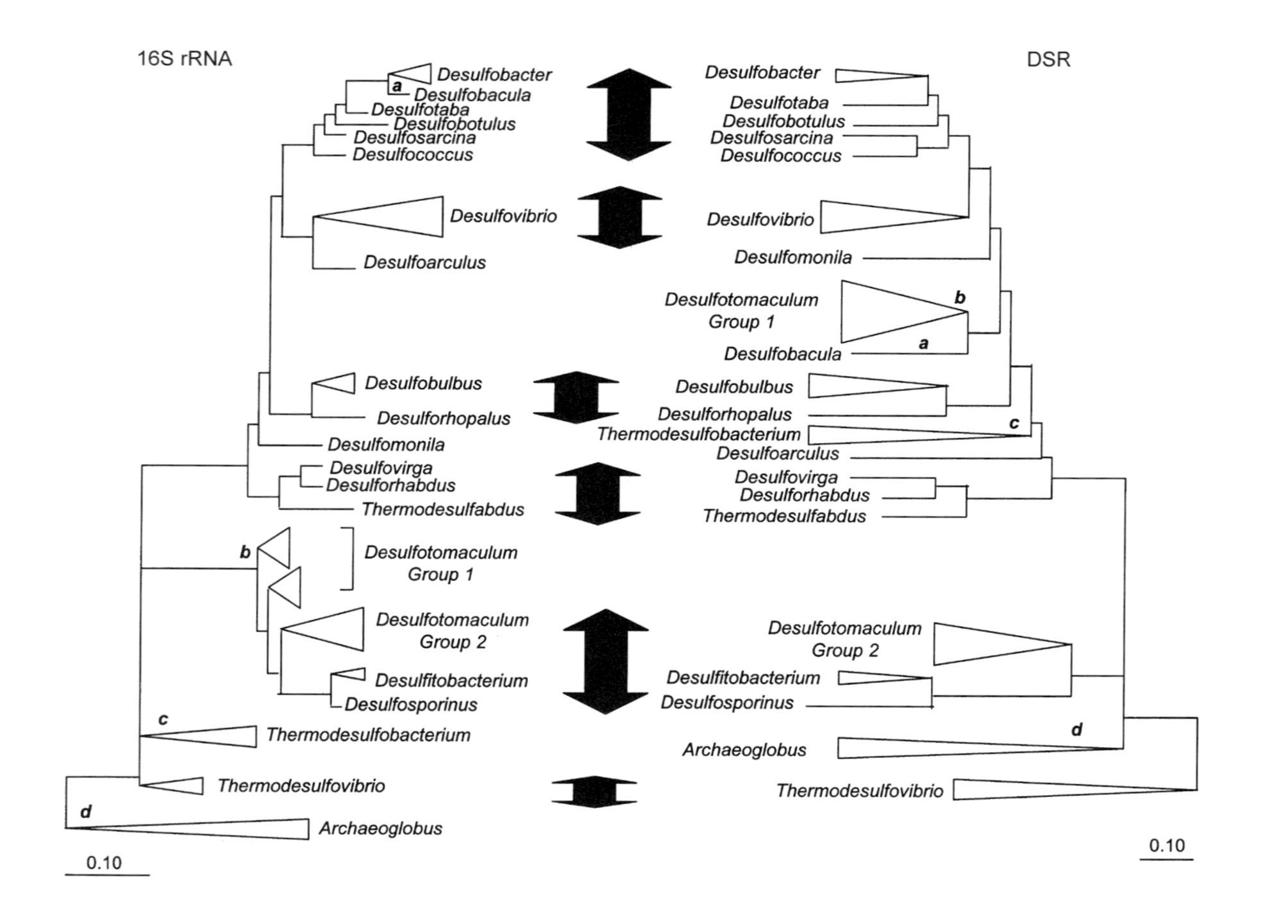
16S rRNA
DSR
Desulfobacter
a
Desulfobacula
Desulfotaba
Desulfobotulus
Desulfosarcina
Desulfococcus
Desulfovibrio
Desulfoarculus
Desulfobulbus
Desulforhopalus
Desulfomonila
Desulfovirga
Desulforhabdus
Thermodesulfabdus
b
Desulfotomaculum Group 1
Desulfotomaculum Group 2
Desulfitobacterium
Desulfosporinus
c
Thermodesulfobacterium
Thermodesulfovibrio
d
Archaeoglobus
0.10
Desulfobacter
Desulfotaba
Desulfobotulus
Desulfosarcina
Desulfococcus
Desulfovibrio
Desulfomonila
Desulfotomaculum Group 1
b
a
Desulfobacula
Desulfobulbus
Desulforhopalus
Thermodesulfobacterium
c
Desulfoarculus
Desulfovirga
Desulforhabdus
Thermodesulfabdus
Desulfotomaculum Group 2
Desulfitobacterium
Desulfosporinus
d
Archaeoglobus
Thermodesulfovibrio
0.10

or sodium ions, and electrical potential across the cell membrane. In freshwater sulfate reducers, proton gradients across the membrane can result in a 1000-fold accumulation of sulfate into the cell (Cypionka, 1995). Sulfate accumulation is likely much less extreme for marine sulfate reducers, but has not been well explored.

Once into the cell, ATP sulfurylase activates the sulfate to APS (Figure 9.3). Following this, APS is reduced to sulfite with the enzyme APS reductase. This step is not shared with assimilatory sulfate reduction. Once formed, sulfite is reduced to sulfide with the enzyme dissimilatory sulfite reductase (DSR). Sulfide is the only significant product of dissimilatory sulfate reduction, and once formed, it is quickly lost from the cell. Active debate surrounds the possibility of a more complex biochemistry for sulfate reduction, including involvement of the intermediates thiosulfate and trithionate (Chambers and Trudinger, 1975; Widdel and Hansen, 1992; Akagi, 1995). The details and significance of such intermediates in the dissimilatory sulfate reduction process are, however, currently unclear.

3.4. Substrate utilization

Complex macro biomolecules are inaccessible to sulfate reducers for direct mineralization (Widdel, 1988). In large part this is because prokaryotes can only ingest macromolecules with an upper molecular weight limit of about 600 Daltons (one Dalton is 1/16 of the mass of an oxygen atom) (Weiss *et al.*, 1991). Furthermore, anaerobes do not generate active oxygen species such as superoxide anion ($^{\bullet}O_2^-$), hydroxyl radicals ($^{\bullet}OH$), and hydrogen peroxide (H_2O_2), used by aerobes to decompose complex biomolecules into simpler moieties (see Chapters 5 and 6). Thus, in anaerobic decomposition smaller compounds must first be made available by the hydrolysis and fermentation of complex organic material (see Chapters 3 and 5). Nevertheless, the spectrum of substrates potentially utilized by sulfate reducers is impressive (Table 9.5), including inorganic electron donors, mono- and dicarboxylic acids, alcohols, amino acids, sugars, a wide variety of aromatic compounds

---

Figure 9.6 Comparison of phylogenies for sulfate reducers based on sequences derived from the 16S rRNA gene and the dissimilatory sulfite reductase (DSR) gene. This comparison documents similarities as well as large differences between the two phylogenies. Regions of similarity are shown by the double-sided arrows. Differences likely represent episodes of lateral gene transfer of the sulfite reductase gene between sulfate-reducing lineages. Likely gene transfer events are designated by boldfaced letters, and particularly among the gram-positive sulfate reducers (*Desulfotomaculum*), more instances than indicated of gene transfer are probable. Lines at the bottom of the figure indicate 10% differences in nucleotide (16S rRNA) or amino acid sequence (DSR gene). Redrawn and simplified from Klein *et al.* (2001).

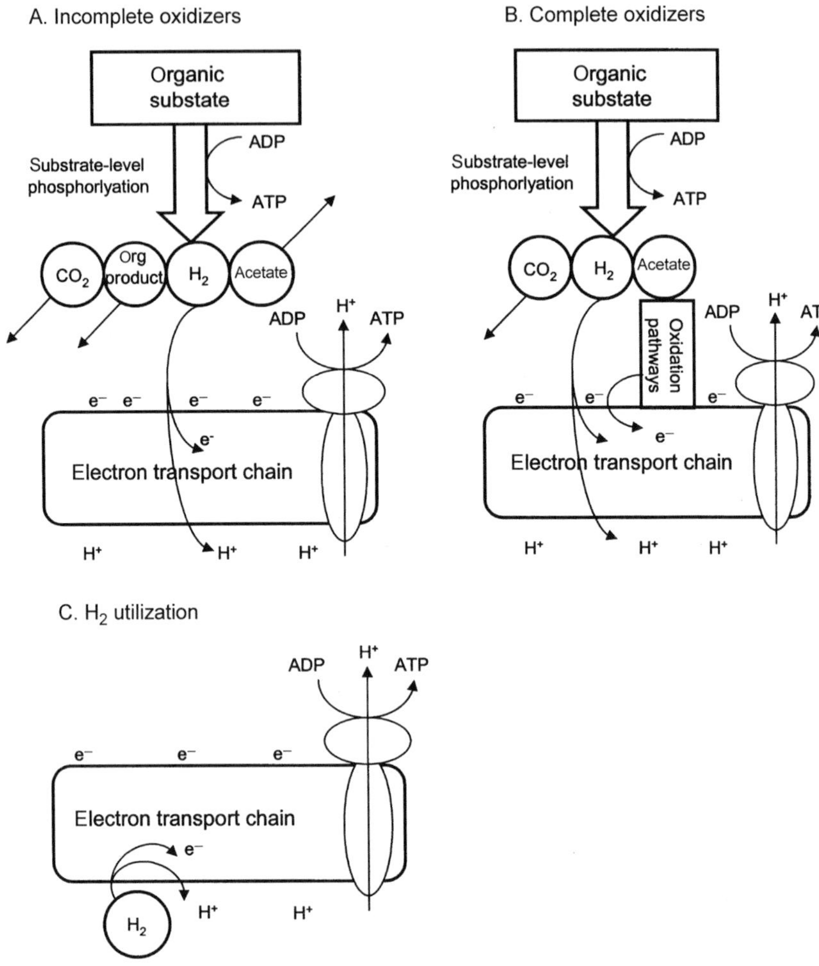

Figure 9.7 Illustration of how different sulfate-reducing organisms metabolize organic substrate, or hydrogen, leading ultimately to the generation of proton motive force across the cell membrane and ATP generation. For incomplete oxidizers (A), organic substrate is broken down by substrate-level phosphorylation to organic products, including acetate, CO_2, and H_2. The H_2 is used in generating a proton motive force. Thus, ATP is formed both through substrate-level phosphorylation and through oxidative phosphorylation. For complete oxidizers (B), substrate-level phosphorylation produces organics that are oxidized completely to CO_2. When utilizing H_2 (C), the hydrogenase enzyme strips electrons from the H_2, setting them into an electron transport chain, ultimately generating a proton motive force leading to ATP formation from oxidative phosphorylation.

Table 9.5 Compounds that can be used as substrates for dissimilatory sulfate reduction

Class of compound	Specific compounds utilized
Inorganic	Hydrogen, carbon monoxide
Hydrocarbons	Straight alkanes (C_{12} to C_{20})
Monocarboxylic acids (aliphatic)	Formate, acetate, propionate, butyrate, higher fatty acids up to C_{20}, isobutyrate, 2-methylbutyrate, 3-methylbutyrate, 3-methylvalerate, pyruvate, lactate
Dicarboxylic acids	Succinate, fumarate, malate, oxalate, maleinate, glutarate pimelate
Alcohols	Methanol, ethanol, 1-propanol, 1-butanol, 1-pentanol, isobutanol, 2-propanol, 2-butanol, ethylene glycol (mono-, di-, tri- and tetra-), 1-,2-propanediol, 1,3-propanediol, glycerol
Amino acids	Glycine, serine, alanine, cysteine, cystine, threonine, valine, leucine, isoleucine, aspartate, glutamate, phenylalanine
Sugars	Fructose, glucose, mannose, xylose, rhamnose
Aromatic compounds	Benzoate, 2-,3-, and 4-hydroxybenzoate, phenol, p-cresol, catechol, resorcinol, hydroquinone, phloroglucinal, pyrogallol, 2-,4-dihydroxybenzoate, 3-,4-dihydroxybenzoate (protocatechuate), 2-aminobenzoate (anthranilate), 4-aminobenzoate, aminobenzene (aniline), hippurate, nicotinic acid, indole, quinoline, phenylacetate, 4-hydroxyphenylacetate, indolylacetate, 3-phenylpropionate, trimethoxybenzoate, *m*-anisate, *p*-anisate, vanillate, 3-methoxysalicylate, vanillin, syringaldehyde, *p*-anisaldehyde, 2- and 3-hydroxybenzaldehyde, toluene, benzene
Miscellaneous	Choline, betaine, oxamate, acetone, dihydroxyacetone, furfural, cyclohexanecarboxylate

Modified from Hansen (1993).

including, apparently, benzene (Lovley, 2000a), straight chain alkanes, (Aeckersberg *et al.*, 1991) and various other compounds, including some man-made xenobiotics (Widdel, 1988; Hansen, 1993). In a novel metabolism, the sulfate reducer *Desulfotignum phosphitoxidans* can, in addition to completely oxidizing a variety of organic substrates (see below) to CO_2, also oxidize phosphite (PO_3^{3-}) to phosphate (PO_4^{3-}) while reducing sulfate (Schink *et al.*, 2002) (see also Chapter 11).

Historically, sulfate reducers were thought capable of utilizing only a limited range of organic substrates and of being able to oxidize these only incompletely, with acetate as the most important product. The best-studied sulfate reducer, *Desulfovibrio desulfuricans*, falls into this class of incomplete oxidizers, as do numerous others (Postgate, 1984; Widdel, 1988) (Table 9.6).

Table 9.6 Basic physiological properties of sulfate-reducing prokaryotes[a]

Genus	Temp optima (°C)	Oxidation of organic electron donors	H_2	Acetate	Lactate	Propionate	Higher fatty acids	Ethanol	Reference
Bacteria									
Desulfovibrio	30–38	i	+	–	+	–	–	+	1
Desulfomicrobium	28–37	i[b]	+	–	+	–	–	±	1
Desulfobulbus	28–39	i	+	–	+	+	–	+	1
Desulfobotulus	34	i	–	–	+	–	+	–	1
Desulfohalobium		i	+	–	+			+	2
Thermodesulfovibrio	65	i	+	–	+	–	–	–	3
Thermodesulfobacterium	65–70	i	+	–	+	–	–	–	1
Desulfotomaculum	30–65	i, c[b]	±	±	±	±	±	±	1
Desulfobacter	28–32	c[b]	±	+	–	–	–	+	1
Desulfobacterium	20–35	c[b]	±	(+)[c]	±	(±)	±	±	1
Desulfococcus	28–35	c	–	(+)	+	+	+	+	1
Desulfosarcina	33	c[b]	+	(+)	+	+	+	+	1
Desulfomonile	37	c	+	–[d]	–			–	1
Desulfonema	30–32	c[b]	±	(+)	±	+	+	–	1
Desulfobacula	28	c	–	–[d]	–	–	+	+	4
Desulfacinum	60	c[b]	+	+	+	+	+	+	5
Desulfofrigus	10–18	c	+	+	+	+	±	+	6
Desulfofaba	7	c	+	+	+	+	–	+	6
Desulfotalea	10–18	c	+	+	+	+	–	+	6
Desulfoarculus	35–39	c	–	(+)	–	(+)	+	–	1
Archaea									
Archaeoglobus	75–83	c	+	–	+				1

1, Widdel and Hansen (1992); 2, Ollivier *et al.* (1991); 3, Henry *et al.* (1994); 4, Rabus *et al.* (1993); 5, Rees *et al.* (1995); 6, Knoblauch *et al.* (1999).

[a]While this list is comprehensive, it is likely not complete with new genera being regularly described. Only a subset of possible electron donors is indicated.
[b]Autotrophic growth in some members.
[c](+) poorly utilized.
[d]While acetate is not oxidized, other organic electron donors are completely oxidized.
i, incomplete oxidizer; c, complete oxidizer.

Table 9.7 Some oxidation reactions during sulfate reduction

Incomplete oxidation of lactate:
$2CH_3CHOHCOO^- + SO_4^{2-} \rightarrow 2CH_3COO^- + 2HCO_3^- + H_2S$
Incomplete oxidation of valerate:
$2CH_3(CH_2)_3CO_2^- + SO_4^{2-} \rightarrow 2CH_3COO^- + 2CH_3CH_2COO^- + H_2S$
Complete oxidation of lactate:
$2CH_3CHOHCOO^- + 3SO_4^{2-} \rightarrow 6HCO_3^- + 3HS^- + H^+$
Oxidation of acetate:
$CH_3COO^- + SO_4^{2-} \rightarrow 2HCO_3^- + HS^-$
Oxidation of H_2[a]:
$4H_2 + SO_4^{2-} + H^+ \rightarrow 4H_2O + HS^-$

[a]Some grow autotrophically.

Some examples of incomplete (and complete) oxidation reactions are presented in Table 9.7. Presumably, incomplete oxidizers lack the necessary biochemical machinery (such as a complete citric acid cycle or the acetyl CoA pathway) to oxidize acetate (Figure 9.7). Incomplete substrate oxidation, however, is not necessarily a disadvantage, as these organisms tend to grow very rapidly, with doubling times under optimal conditions of less than 4 h (Widdel, 1988). The rapid growth of incomplete oxidizers explains why they are often so easily enriched. Rapid growth could give them a competitive advantage in nature where organic matter input to surface sediments can be quite episodic, and coupled, for example, to phytoplankton blooms.

Despite the easy enrichment of incomplete oxidizing sulfate reducers from nature, their main product, acetate, does not accumulate to high concentrations in anoxic environments in which sulfate reduction is the dominant mineralization process. This apparent paradox was solved when Widdel and Pfennig (1977) reported the first acetate-oxidizing sulfate reducer capable of complete substrate oxidation (Tables 9.6 and 9.7). Numerous other reports have followed, including the report that sulfate reducers completely oxidize organic substrates other than acetate (Widdel, 1988). Complete oxidizers, therefore, represent a second major metabolic type of sulfate reducer. These organisms tend to grow slowly and are easily outcompeted by incomplete-oxidizing sulfate reducers in enrichments with substrates such as lactate.

A third major metabolism for sulfate-reducing bacteria is hydrogen utilization. Most incomplete oxidizers, and a large number of complete oxidizers, are able to use H_2 as an electron donor during sulfate reduction (Figure 9.7; Table 9.6). Despite the prospects for chemolithoautotrophic growth on H_2, for most species, acetate must be supplied as a carbon source. Therefore, metabolism with H_2 is normally chemolithoheterotrophic. True autotrophic growth with H_2 is, however, known among some sulfate reducers (Table 9.6), and many of these also completely oxidize organic substrate. This association may not be chance. Complete oxidizers have associated with them carbon oxidation pathways suited for the oxidation of acetate, including

Table 9.8 Substrates used by natural populations of sulfate reducers

Substrate	% of total substrate used			
	Loch Etire	Loch Eil	Tay Estuary	Coastal Denmark
Acetate	98	63.8	35.5	40 to 50
Lactate	—	—	43.4	ND
Propinate	—	12.5	6.0	10 to 20
H_2	1.9	0.1	—	5 to 10
Arginine	—	2.9	—	ND
Serine	—	2.3	0.8	ND
Glutamate	—	1.7	0.5	ND
Alanine	—	0.8	0.4	ND
Aspartate	—	0.6	0.3	ND
Leucine	—	0.3	0.2	ND
Lysine	—	0.3	0.2	ND
Isoleucine	—	—	0.1	ND
Phenylalanine	—	0.3	—	ND
Butyrate	ND	ND	ND	10 to 20

–, not detected; ND, not determined.
References: Sørensen *et al.* (1981); Parkes *et al.* (1989).

modifications of the citric acid cycle and the acetyl-CoA pathway (Widdel and Hansen, 1992). These pathways, operating in reverse, are the basis for two of the four known autotrophic carbon fixation pathways (see Chapter 4).

While sulfate reducers can utilize a wide range of organic substrates, what are the substrates most available to them in nature? Information is not extensive, but in experiments with marine sediments substrate availability has been assessed by blocking sulfate reduction with the addition of molybdate and monitoring the accumulation of organic compounds and H_2 (Sørensen *et al.*, 1981; Parkes *et al.*, 1989). The idea is that the substrates accumulating after molybdate addition are the same as those utilized during active sulfate reduction before inhibition. This may not be strictly true, as other mineralization pathways such as Fe reduction or methanogenesis could utilize accumulating substrates. Nevertheless, the wide variety of organic compounds and H_2 accumulating in these experiments are potential substrates for sulfate reducers in nature (Table 9.8). The most important substrates seem to be acetate, lactate, H_2, butyrate, and propionate.

3.5. Natural diversity of sulfate-reducing organisms

More than 100 species of sulfate reducers have been described, which must represent only a small fraction of their real diversity in nature. Even if we consider individual environments, such as a lake or marine sediment or an anoxic water column, the diversity of sulfate reducers may be large. For example, in the sediments of Lake Stechlin, Germany, a total of 34 different pure cultures of sulfate reducers were isolated from various sediment locations within the lake, and up to 20 from a single sediment location (Sass *et al.*, 1997). The real diversity of sulfate reducers in these sediments was probably much higher, as culturing techniques can only enrich for organisms that happen to grow under the specified growth conditions of the culture media. Ultimately, extraction and amplification of genetic material may be required to appreciate the diversity of sulfate reducers in natural habitats (Devereux and Mundfrom, 1994). Even with molecular techniques, however, it is not certain that species present in low abundance, dormant or unreactive to the amplification primers used will be easily identified.

The diversity of sulfate-reducing populations would seem to be overly large, particularly as these organisms all have a similar overall metabolism. The large natural diversity must relate to subtle differences in physiology, including differences in growth rate, substrate preference, substrate threshold requirements, optimal temperature range, and possibly oxygen tolerance. These subtle (and not so subtle) differences probably allow certain species to capitalize on dynamic changes in the environment, of which temperature, substrate availability, and periodic oxygen exposure may be among the most important. Recent work indicates that sulfate-reducing populations stratify with depth in sedimentary environments and that this stratification may reflect depth-dependent differences in the nature of the substrates available (Risatti *et al.*, 1994; Sass *et al.*, 1997). Thus, in Lake Stechlin, acetate-utilizing sulfate-reducing bacteria occupy the upper sediment layers, while the maximum population size of lactate-utilizing sulfate reducers is located deeper (Sass *et al.*, 1997) (Figure 9.8). Based on most probable number (MPN) counting (see Chapter 2), lactate-utilizing sulfate reducers appear, overall, to dominate the sulfate-reducing population in the sediments of this lake. From a microbial mat in Baja California, Mexico, complete-oxidizing, acetate-utilizing genera of sulfate-reducing bacteria were concentrated in both the upper and lower regions of active sulfate reduction, whereas incomplete oxidizers of the genus *Desulfovibrio* were dominant in between (Risatti *et al.*, 1994).

At the genus level, RNA oligonucleotide probes have revealed profound seasonal variations in the relative abundance of sulfate-reducing bacteria in a lake sediment from Japan. In this case, the population size of *Desulfobulbus* sp. followed seasonal patterns in rates of sulfate reduction, whereas *Desulfobacterium* sp. and *Desulfovibrio* sp. showed little seasonal variation in population size (Li *et al.*, 1999a). Species-level dynamics could not be addressed with the specific probing technique utilized. Although progress is being made in understanding the structure of sulfate-reducing populations, our understanding is still very much at a rudimentary level.

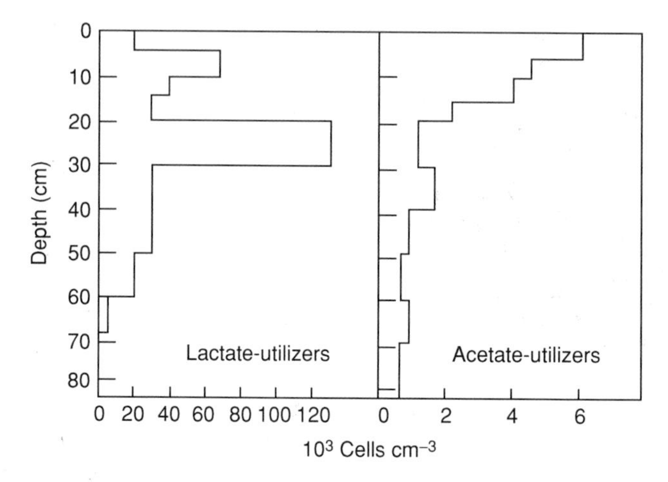

Figure 9.8 The depth distribution of lactate-utilizing and acetate-utilizing sulfate reducers as determined from MPN counting of lake sediment from Lake Stechlin, Germany. Redrawn from Sass *et al.* (1997).

3.6. Factors controlling rates of sulfate reduction

The most important factors controlling rates of sulfate reduction in natural environments include the presence or absence of oxygen, the availability of sulfate, the availability of electron donor (hydrogen or organic matter), and temperature. We take a brief look at each of these in turn.

3.6.1. Oxygen

The absence of oxygen is often considered a primary requisite for the establishment of sulfate-reducing populations, and no pure culture of sulfate reducers is known to reduce sulfate in the presence of oxygen. However, as discussed previously, active sulfate reduction has been measured in the oxic zone of microbial mats and oxic marine sediments (Canfield and Des Marais, 1991; Jørgensen and Bak, 1991). Sigalevich *et al.* (2000) offer a possible explanation for this paradox by observing that the sulfate reducer *Desulfovibrio oxyclinae* can apparently reduce sulfate in the presence of at least 87 μM oxygen when grown in co-culture with a facultative aerobic heterotroph tentatively identified as a *Marinobacter*. This consortium was isolated from the same surface portions of a microbial mat from Solar Lake, Sinai, where oscillating sulfidic and oxic conditions are found, and where "aerobic" sulfate reduction has been postulated (Fründ and Cohen, 1992). Oddly, *Desulfovibrio oxyclinae* cannot alone reduce sulfate in the presence of oxygen. It is obvious that *Marinobacter* sp., as an aerobic heterotroph, consumes oxygen and thereby reduces the oxygen level. However, the two organisms were not observed to form aggregates together, and therefore *D. oxyclinae* was apparently exposed to the same oxygen level as *Marinobacter* sp. It is therefore unclear how *Marinobacter* sp. encourages aerobic sulfate reduction by *Desulfovibrio oxyclinae*.

Many sulfate-reducing strains employ oxidative metabolisms in the presence of oxygen to actively oxidize hydrogen, sulfide, sulfite, and thiosulfate (Krekeler and Cypionka, 1995; Fuseler *et al.*, 1996). Numerous sulfate-reducing strains can also oxidize a variety of organic compounds with oxygen. Sulfate reducers best use oxygen under microoxic conditions, with metabolic rates increasing in response to reduced oxygen concentrations (Cypionka, 1994). Sulfate reducers, however, grow only poorly, if at all, while metabolizing oxygen and probably use oxidative metabolisms as a mode of survival under fluctuating redox conditions as often occurs in the environment. Thus, although, sulfate reducers demonstrate certain adaptations to oxic conditions in nature, their activity is restricted, for the most part, to anoxic environments.

3.6.2. Sulfate

The presence of sulfate is a mandatory prerequisite for sulfate reduction. However, not much sulfate is needed, particularly among freshwater strains of sulfate reducers, in which half-saturation constants (K_m) as low as 4.8 μM have been measured (Ingvorsen and Jørgensen, 1984). Active metabolism of organic substrate may occur even at sulfate concentrations of less than 1 μM. As freshwater environments typically contain sulfate at low concentrations of 10 to 500 μM, the ability to utilize low sulfate concentrations by freshwater sulfate reducers is a decided advantage. However, low water column sulfate concentrations nearly guarantee that sulfate becomes rapidly depleted in anoxic freshwater sediments. Therefore, even though rapid rates of sulfate reduction may be encountered in anoxic sediment layers where sulfate is available, depth-integrated rates of sulfate reduction in freshwater sediments are typically low (see Section 3.1).

The average concentration of sulfate in the ocean is 28 mM. Perhaps owing to such high sulfate concentrations, marine strains of sulfate reducers have higher half-saturation constants (K_m) of around 70 to 200 μM sulfate (Ingvorsen and Jørgensen, 1984; Ingvorsen *et al.*, 1984). However, these half-saturation constants are still considerably lower than the concentration of sulfate in sea water, ensuring that sulfate does not limit sulfate reduction until deeper in marine sediments where sulfate becomes nearly exhausted. The high concentration of seawater sulfate, coupled with the mixing and pumping activities of benthic animals, ensures that sulfate is abundant to sometimes considerable depths (typically tens of centimeters to well over 1 m) in coastal sediments. The rapid exhaustion of oxygen in coastal sediments (see Chapter 6), coupled with sulfate availability to depth, explains why sulfate reduction is often so dominant a carbon mineralization pathway, as discussed in Section 3.1.

3.6.3. Carbon availability

There is a clear relationship between carbon availability and rates of sulfate reduction, particularly in marine environments where sulfate is abundant (Canfield, 1989b). Not only the concentration of organic matter, but also its degradability or "quality," is important. As outlined in Chapter 5, freshly produced organic biomass can be viewed as an assemblage of organic compounds with different degradabilities. During decomposition, the most biologically "labile" components of cellular biomass will be preferentially decomposed, leaving more refractory organic components behind. These more refractory organics may also be degradable, but on a slower timescale. In a series of laboratory and field experiments Westrich (1983) and Westrich and Berner (1984) elegantly demonstrated that under sulfate-reducing conditions, rates of sulfate reduction will depend on both the quantity and quality of available organic matter. This was quantified with the following kinetic relationship (Equation 9.1; see also Chapter 5) (the "multi-G" model), in which the overall rate of organic carbon oxidation by sulfate reduction is the sum of the rates of decomposition (k_iG_i, where k_i is a first order rate constant and G_i is the concentration of organic carbon of a particular degradability).

$$\mathrm{d}G/\mathrm{d}t = -(k_1G_1 + k_2G_2 + k_3G_3 + \ldots) \tag{9.1}$$

Organic matter may well degrade with continuously decreasing reactivity, rather than the step functions in reactivity implied by the multi-G model (Boudreau and Ruddick, 1991). Nevertheless, the multi-G is a useful, and mathematically expedient, approximation for organic matter decomposition under sulfate-reducing conditions (see also discussion in Chapter 5). Westrich (1983) has compiled likely k_i values for organic matter reactivity in the environment, and these are summarized in Table 9.9.

3.6.4. Temperature

When rates of sulfate reduction are determined over the course of a season, and are compiled on an aerial basis (rates underlying a given area of anoxic water column or sediment), they generally correlate with seasonal changes in

Table 9.9 First order rate constants (k_i) for organic matter reactivity under sulfate-reducing conditions

Organic matter type	k (yr^{-1})	$t_{1/2}$ (yr)
G_1	3.3 ± 0.8	0.21 ± 0.05
G_2	$4.0 \pm 1.3 \times 10^{-1}$	1.7 ± 0.4
G_3	$1.5 \pm 0.9 \times 10^{-2}$	$4.6 \pm 2.7 \times 10^{1}$
G_4	$1.6 \pm 0.8 \times 10^{-3}$	$4.3 \pm 2.2 \times 10^{2}$
G_5	$7.0 \pm 3.0 \times 10^{-5}$	$9.9 \pm 4.2 \times 10^{3}$
G_6	$5.0 \pm 2.0 \times 10^{-6}$	$1.4 \pm 0.6 \times 10^{5}$
G_7	$3.0 \pm 1.5 \times 10^{-7}$	$2.3 \pm 1.2 \times 10^{6}$
G_8	0	1

Data from Westrich (1983).

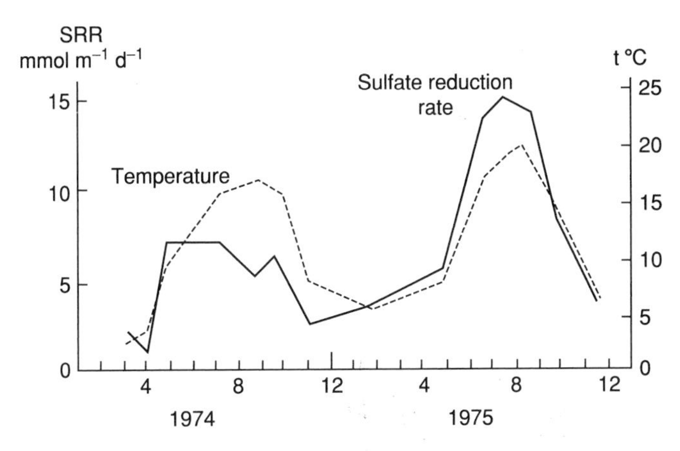

Figure 9.9 Seasonal trends in temperature and sulfate reduction rates from a shallow-water site in the Limfjorden, Denmark. Redrawn from Jørgensen (1997).

temperature. An example of such a correlation for sediments sampled from Limfjorden in Denmark is shown in Figure 9.9 (Jørgensen, 1977c).

Correlations such as these are thought to represent, for the most part, the response of the metabolic activity of the sulfate-reducing community to temperature. It is well known that within their temperature range of active growth, the metabolic rate of sulfate-reducing organisms, and indeed all prokaryotes, responds to temperature. Typically, pure cultures of microorganisms increase their metabolic rate by a factor of two to four for a 10 °C increase in temperature (also known as the Q_{10} response; see Chapter 2). Similar Q_{10} responses are also found for populations of sulfate reducers in nature. Therefore, correlations as observed in Figure 9.9 would seem to be well justified.

However, as described above (see also Table 9.6), sulfate reducers are present in nature with a wide range of metabolic adaptations to temperature. Thus, while an individual species of sulfate reducer might metabolize between 10 and 36 °C, with a Q_{10} response of 3, a different species can metabolize between −1 and 15 °C with a similar Q_{10} response. Sulfate reducers are well adapted to the temperature range in which they grow, with no significant differences between the specific rates of sulfate reduction (rate cell^{-1}) over the broad temperature range in which sulfate reducers are active (Figure 9.10). Therefore, to state the matter simply, why do sulfate reduction rates fall so decisively in Limfjorden sediments (and indeed most marine sediments) as temperature drops, when an efficiently metabolizing, low-temperature-adapted population can in principle prosper under these conditions? We have no solid answer for this apparent paradox, but apparently, low-temperature-adapted sulfate reducers grow too slowly to take advantage of winter-long periods of cold sediment temperatures. Because of this, the higher temperature-adapted organisms seem to control the temperature response of sulfate reduction.

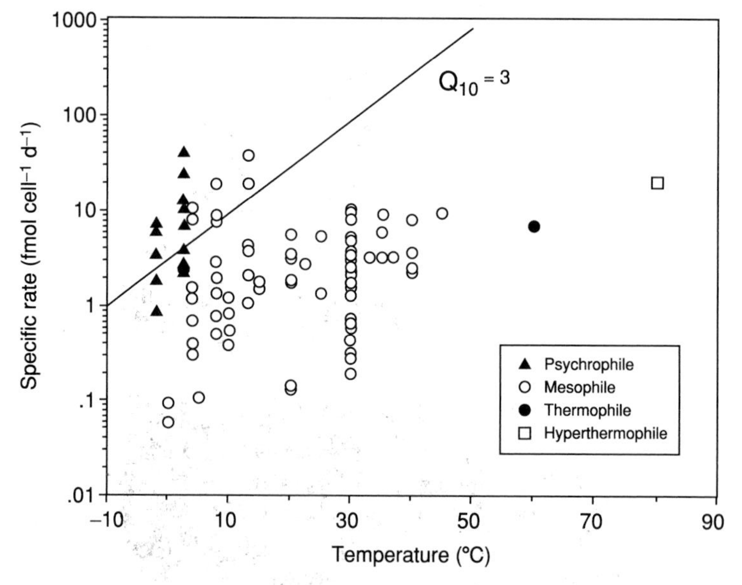

Figure 9.10 Specific rates of sulfate reduction are shown as a function of temperature for a variety of sulfate reducers that are categorized, based on their metabolic temperature range, as psychrophiles, mesophiles, thermophiles or hyperthermophiles. There is not a strong tendency for specific rates of sulfate reduction to increase with increasing temperature. Shown for comparison is the trend expected if specific rates of sulfate reduction increased with a Q_{10} of 3. Adapted from Canfield *et al.* (2001).

4. ELEMENTAL SULFUR REDUCTION

In the following Sections we consider aspects of the ecology, phylogeny, and environmental distribution of organisms reducing elemental sulfur in nature.

4.1. Elemental sulfur in nature

Elemental sulfur is a common and sometimes abundant intermediate of sulfide oxidation. In marine sediments, elemental sulfur often accumulates to concentrations of 1 to 20 μmol cm^{-3} (Table 9.10) near interfaces between sulfide and oxidants such as oxygen, nitrate, and Fe and Mn oxides. As we discuss below, sulfur sometimes forms as a result of biologically mediated oxidation processes, and it may accumulate within the cells of sulfide-oxidizing organisms. In other cases, elemental sulfur forms from strictly inorganic reactions, such as those between sulfide and Fe and Mn oxides (Pyzik and Sommer, 1981; Burdige and Nealson, 1986) (see also Chapter 8).

Table 9.10 Concentrations of elemental sulfur in various anoxic environments

Sediment	Max concentrations (μmol cm^{-3})	Reference
Chilean Coast		
C6	8	1
C7	8	1
C18	8	1
C26	25	1
C40	2	1
C41	1	1
Danish Coast		
Kysing Fjord	3	2
Aarhus Bay	17	2
Microbial Mat		
Solar Lake	4	5
Water column	(nM)	
Black Sea		
Sta 1	60	3
Sta 2	60	3
Sta 6	22	
Mariager Fjord	10,000	4

1, Thamdrup and Canfield (1996); 2, Troelsen and Jørgensen (1982); 3, Luther *et al.* (1991); 4, Zopfi *et al.* (2001); 5, Habicht and Canfield (1997).

$$2FeOOH + 3H_2S \rightarrow S^0 + 2FeS + 4H_2O \quad (9.2)$$

$$2H^+ + MnO_2 + H_2S \rightarrow Mn^{2+} + S^0 + 2H_2O \quad (9.3)$$

Elemental sulfur can also be found in anoxic water columns (Table 9.10), and it may be particularly abundant in hot springs with substantial hydrothermal or volcanic sources of sulfide.

4.2. Organisms reducing elemental sulfur, substrates, and phylogeny

The ability to reduce elemental sulfur among prokaryotes is widespread (Figure 9.11) (Widdel and Hansen, 1992). Within the bacterial domain, elemental sulfur reduction is conducted as an alternative metabolism by numerous members of the proteobacteria, including the sulfate reducers *Desulfovibrio gigas, Desulfomicrobium* spp., and the newly discovered *Desulfospira joergensenii* (Finster *et al.*, 1997a). Also, some facultative anaerobic organisms with wide metabolic capabilities (although they are not sulfate reducers) such as *Wolinella succinogenes, Pseudomonas mendocina*, and *Shewanella putrefaciens* can also reduce elemental sulfur. Elemental sulfur

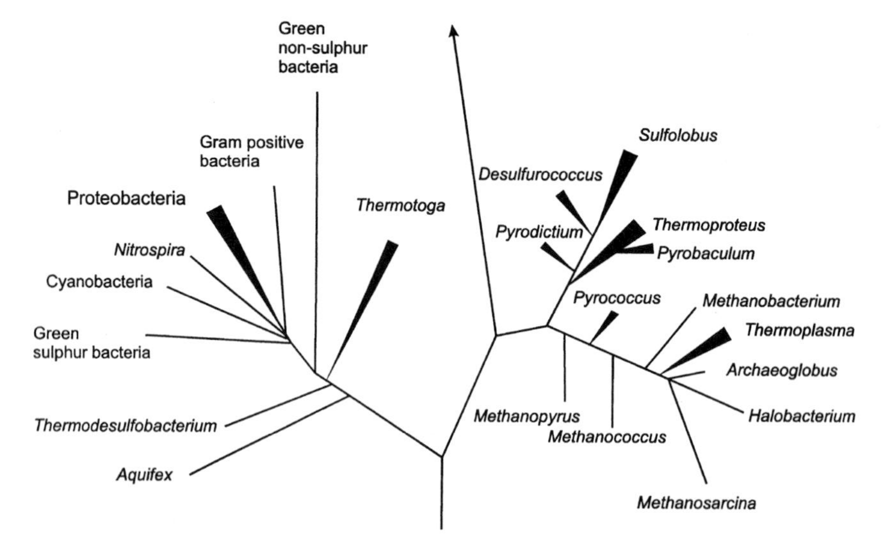

Figure 9.11 The lineages within the two prokaryote domains housing elemental sulfur-reducing organisms. Redrawn from Canfield and Raiswell (1999).

reduction, however, is probably not an important metabolism for any of these organisms.

By contrast, members of the genus *Desulfuromonas*, the thermophilic *Desulfurella acetivorans* (Widdel and Pfennig, 1982), and the newly discovered *Sulfurospirillum arcachonense* (Finster *et al.*, 1997b), all from the δ-subdivision of the proteobacteria, conduct elemental sulfur reduction as an important, or even principal, metabolism. In some cases, alternative inorganic electron acceptors such as nitrate, Fe(III) (see Chapter 8) or oxygen may be used, but in general, the variety of possible electron acceptors is limited, and some members of *Desulfuromonas* are obligate S^0 reducers.

Both incomplete and complete oxidation of organic substrates is demonstrated by sulfur reducers. Members of the genus *Desulfuromonas,* as well as *Desulfospira joergensenii,* are complete oxidizers, metabolizing in some cases only acetate. In other cases, a broader but still limited range of organic substrates is used, especially compared to many complete-oxidizing sulfate reducers (Widdel and Hansen, 1992; Finster *et al.*, 1994, 1997a). Sulfate reducers able to reduce S^0 appear to be incomplete oxidizers of organic compunds, as is *Shewanella putrefaciens*, while *Wolinella succinogenes* and *Pseudomonas* spp. use only H_2, and in some cases formate, as electron donors.

Several members of the deep-branching bacterial hyperthermophilic genus *Thermotoga* can also reduce elemental sulfur (Huber *et al.*, 1986; Schönheit and Schäfer, 1995), although this is not thought to be an energy-gaining metabolism. Rather, for *Thermotoga*, as well as other hyperthermophilic organisms capable of reducing elemental sulfur (see below), elemental sulfur probably acts as an H_2 sink during fermentative metabolism. In this way, no ATP is generated during S^0 reduction, but the organism benefits by maintaining low partial pressures of H_2, which favors fermentation (Huber *et al.*, 1986) (see Chapter 3).

The reduction of S^0 is widespread among members of the *Archaea*, including deep-branching hyperthermophilic members. There are, however, considerable differences among organisms as to how elemental sulfur is used, and at least four different types of metabolism with sulfur can be identified:

1. The most widespread is the chemolithoautotrophic reduction of sulfur with H_2, accomplished by many hyperthermophiles from the *Crenarchaeota* (Widdel and Hansen, 1992) (see Figure 9.11). A representative, though not exhaustive, list of organisms with this physiology includes members of the genera *Pyrodictium, Pyrobaculum, Sulfolobus, Acidianus*, and *Thermoproteus.*
2. Some members of the order *Thermoproteales*, and *Pyrobaculum islandicum*, can respire heterotrophically with elemental sulfur; this is apparently an energy-gaining metabolism (Schönheit and Schäfer, 1995).

3. For some members of the *Archaea*, including representatives from the genera *Pyrococcus, Thermococcus, Desulfurococcus, Pyrodictium*, and *Thermoproteus*, respiration with elemental sulfur is only apparent and, similar to *Thermotoga* (see above), is coupled to fermentation with sulfur reduction acting as an H_2 sink.
4. Finally, numerous H_2-oxidizing methanogenic *Archaea* can also reduce elemental sulfur with H_2 or methanol (Stetter and Gaag, 1983). Sulfide may be produced in large amounts, and in some instances, sulfide accumulates simultaneously with methane. Methanogens do not seem to grow while reducing sulfur, and in some cases can survive for only a matter of days with this metabolism. Elemental sulfur reduction by methanogens, therefore, seems to be a secondary metabolism and may benefit the organism by quickly establishing reducing conditions in otherwise oxic, elemental sulfur-containing environments (Stetter and Gaag, 1983).

4.3. Environmental distribution and a note on evolution

Compared with sulfate reducers, the environmental distribution of elemental sulfur-reducing organisms, and their quantitative role in carbon cycling, is poorly understood. The distribution of sulfur reducers will obviously be limited by the availability of elemental sulfur and electron donors, as discussed above. Not surprisingly, prokaryotes capable of reducing elemental sulfur are found in environments ranging from marine and freshwater sediments to submarine and terrestrial hydrothermal settings (Widdel and Hansen, 1992). Collectively, sulfur reducers can tolerate temperatures probably down to near 0 °C (although no psychrophilic S^0 reducers have yet been reported), and they can metabolize at temperatures up to 110 °C (Stetter *et al.*, 1986), near the upper demonstrated maximum for life.

Many sulfur-reducing archaeans are deep-branching, hyperthermophilic, and are of probable importance in the early history of life on Earth (Stetter, 1996). One difficulty, however, is to understand the source of elemental sulfur in an early Earth devoid of oxygen. One possible source may have been the rapid cooling of magma-derived volcanic gases containing SO_2 and H_2S, in which, during cooling, equilibrium shifts in favor of elemental sulfur formation (Grinenko and Thode, 1970):

$$SO_2 + 2H_2S \leftrightarrow 3S^0 + 2H_2O \qquad (9.4)$$

The involvement of volcanic emissions in S^0 formation might also explain the hyperthermophilic nature of the deepest branching elemental sulfur-reducing organisms. Another possible source for S^0 would have been the

UV photolysis of volcanically derived SO_2 gas in the atmosphere (Farquhar *et al.*, 2002). This would have been an ancient source of elemental sulfur, also consistent with the apparent early emergence of microbial elemental sulfur reduction.

4.4. Principal steps in the reduction of sulfur

Compared to sulfate reducers, little is known about the biochemistry of sulfur reduction. The best-studied pathway is the reduction of sulfur with formate by *Wolinella succinogenes* (Widdel and Hansen, 1992). Key aspects of the reduction process include the solubilization of S^0 with sulfide to form polysulfides, which are subsequently brought into the cell. The polysulfides are reduced to sulfide with a membrane-bound sulfur reductase (also known as sulfur oxidoreductase). Electrons driving the reduction process are supplied by H_2 through the enzyme hydrogenase or by formate (in this particular example) through the enzyme formate dehydrogenase. Both hydrogenase and formate dehydrogenase promote an electron motive force to drive ATP formation with ATPase. It is not known if these steps are common to sulfur reducers in general.

5. INORGANIC SULFIDE OXIDATION

Sulfide is abundant in the environment. It is delivered in copious quantities in terrestrial and marine hydrothermal areas and is produced in large amounts in marine sediments of the continental margin. When physical processes promote stagnation, sulfide can also accumulate into the water columns of lakes and marine basins. The Black Sea, the Cariaco Basin, and Fayetteville Green County Lake in New York are just a few examples of basins containing sulfidic water columns (also called euxinic basins, after *Euxinos pontos*, the Greek name for the Black Sea).

Much of this sulfide is available for oxidation (more on this below), and inorganic reactions are possibly significant oxidation pathways in nature. In fact, the reactions between dissolved sulfide and poorly crystalline Fe and Mn oxides occur with no known biological mediation and can be fast. For example, the half-life for sulfide reacting with Fe oxyhydroxide phases can be as short as 25 min, and sulfide has a half-life of 50 s when in contact with colloidal Mn oxides (Millero, 1991). In sediments, reactive Fe oxyhydroxides can buffer sulfide concentrations to less than 1 μM, even with active sulfide production by sulfate reduction (Canfield, 1989a). It is possible that organisms cannot compete with the rapid kinetics of the inorganic reactions and the low sulfide concentrations that result. It is also possible that organisms conducting these reactions exist, but have not yet been isolated.

The reaction products of these inorganic reactions vary. For example, the reaction between poorly crystalline Fe oxides (ferrihydrite) and sulfide produces mostly elemental sulfur at near-neutral pH, as does the reaction between Mn oxides and sulfide (Burdige and Nealson, 1986). By contrast, the reaction between sulfide and oxygen at near-neutral pH produces, initially, a mixture of reaction products, with sulfate the most important (Zhang and Millero, 1993). However, 40 to 50% of the initial reaction products also consist of sulfite and thiosulfate. Over time, the sulfite is further oxidized to sulfate (Zhang and Millero, 1993), although thiosulfate seems to be stable in the absence of prokaryotes.

In nature, the reaction rate between oxygen and sulfide varies by a millionfold (Jørgensen, 1982a). In some instances, such as in the water column of the Black Sea, sulfide persists in the presence of oxygen for up to 5 days (Jørgensen, 1982a). In this case, the reaction between oxygen and sulfide may be partly inorganic (a well-defined O_2–H_2S interface is not a persistent feature of the Black Sea water column, and recent studies demonstrate that O_2 and H_2S can be well separated, with NO_3^- and H_2S overlapping instead [Murray *et al.*, 1989] [see Chapter 13]).

The inorganic oxidation rate between O_2 and H_2S is given by the following (Millero *et al.*, 1987):

$$d[H_2S]_T/dt = -k[O_2][H_2S]_T \tag{9.5}$$

where $[H_2S]_T$ is the total concentration of dissolved sulfide (M), $[O_2]$ is the concentration of dissolved oxygen (M), and the rate constant k ($M^{-1}\ h^{-1}$), at a pH of 8, is given by

$$\log(k) = 11.78 - 3.0 \times 10^3/T + 0.44I^{1/2} \tag{9.6}$$

At a temperature (T) of 288.15 K (15 °C), and at the ionic strength (I) of sea water (0.7), k has a value of 54 M h^{-1}. Trace metals catalyze the oxidation rate, and Fe(II) is the strongest catalyst, whose influence on k is expressed as

$$\log(k_{Fe}/k) = a + b\log[\mathrm{Fe(II)}] \tag{9.7}$$

where a and b are constants and k_{Fe} is the Fe(II)-enhanced k value. A value for k_{Fe} of 351 M h^{-1} is obtained for $a = 6.55$, $b = 0.820$ and a representative Fe(II) concentration of 1×10^{-7} M.

We can use the above equations and our calculated value for k_{Fe} to assess the inorganic reaction rate between H_2S and O_2 in the Black Sea. We assume O_2 and H_2S overlap with concentrations of 3 μM each, as has been observed in the past (Jørgensen, 1982a), and calculate an inorganic H_2S oxidation rate of 3.5×10^{-8} M h^{-1}. This reaction rate yields a half-life for H_2S of 440 h, or a little over 18 days. Such a time frame is considerably longer than the observed oxidation rates as reported by Jørgensen (1982a; see above) and suggests that although the H_2S oxidation rate in the Black Sea is slow, there may still be a significant biological component.

6. INORGANIC VS. BIOLOGICAL CONTROL OF SULFIDE OXIDATION

In microbial mats, the residence time for sulfide in the overlapping reaction zone containing oxygen is only 0.6 s (Jørgensen, 1982a), and the reaction is surely mediated by microorganisms. The cell density of sulfide oxidizers is a principal factor contributing to the rapid rates of sulfide oxidation in microbial mats, and generally, cell density seems to control rates of biologically mediated sulfide oxidation in nature (Jannasch *et al.*, 1991). Thus, in microbial mats, sulfide-oxidizing organisms are frequently found in high density in well-developed layers (Jørgensen, 1982a) (see Section 7). By contrast, in euxinic water columns such as the Black Sea, cell densities are much lower, being limited, probably, by ciliate grazing, turbulent dispersion, and lower inherent available energy due to shallower gradients in O_2 and sulfide.

We can explore quantitatively how population size acts to regulate biological sulfide oxidation rates. We assume that rates of biological oxidation are approximated by Michaelis-Menten kinetics (see Chapter 2) and that rates respond to both H_2S and O_2 concentrations. We express rates of biologically mediated sulfide oxidation as the following:

$$\frac{d[H_2S]_T}{dt} = \frac{-V'_{max}[H_2S][O_2]}{(K_{m-H_2S}[H_2S] + K_{m-O_2}[O_2] + [H_2S][O_2])} \qquad (9.8)$$

where, in addition to the terms already defined above, V'_{max} is the maximum sulfide oxidation rate in the environment (mol l^{-1} h^{-1}), and $K_{m\text{-}H_2S}$ and $K_{m\text{-}O_2}$ are the respective half saturation constants (M) for H_2S and O_2 in sulfide oxidation. The term V'_{max} is derived from V_{max}, which is the actual measured protein-specific maximum oxidation rate, with the following expression:

$$V'_{max} = V_{max} \times (\text{protein content}) \times (\text{cell density}) \qquad (9.9)$$

Therefore, to calculate V'_{max} we need to know the protein content of the sulfide-oxidizing cells (mg protein $cell^{-1}$) and the cell density (cells l^{-1}).

Values of V_{max}, in particular, vary widely both within and among different genera of sulfide oxidizers (Table 9.11). The values for half saturation constants appear less variable, although there is little information for $K_{m\text{-}O_2}$ (Table 9.11). We use a value for protein content of 25 fg protein cell^{-1}, appropriate for a *Thiobacillus*-sized organism (van den Ende, 1997), and calculate V'_{max}, and subsequently sulfide oxidation rates, with $K_{m\text{-}H_2S}$ of 5×10^{-6} M and $K_{m\text{-}O_2}$ of 1.5×10^{-6} M (Table 9.11). We also use H_2S (total) and O_2 concentrations of 3×10^{-6} M. Comparing these biological rates (Table 9.12) to the inorganic rate of sulfide oxidation in the Black Sea, as calculated above, biological oxidation becomes faster than inorganic sulfide oxidation for cell densities between 10^4 and 10^5 cells cm^{-3}. These calculations confirm the experimental observations of Jannasch *et al.* (1991), who found that biological oxidation could only compete with inorganic sulfide oxidation rates for sulfide oxidizers at cell densities greater than 10^4 cells cm^{-3}. The sulfide oxidizers from the Jannasch *et al.* (1991) study were enriched from the Black Sea water column.

Table 9.11 Kinetic parameters used to calculate biologically mediated rates of sulfide oxidation[a]

Genus	V_{max}[b] (mol $mg_{protein^{-1}}$ h^{-1})	$K_{m\text{-}H_2S}$ (M)	$K_{m\text{-}O_2}$ (M)
Thiothrix	1.2 to 18 $\times 10^{-5}$	2.4 to 12.4 $\times 10^{-6}$	1.5×10^{-6}
Beggratoa	0.6 to 6 $\times 10^{-5}$	–	–
Thiobacillus	60×10^{-5}	2.3×10^{-6}	–

[a]Data from van den Ende (1997).
[b]These V_{max} values for sulfide oxidation are estimated from V_{max} oxygen values assuming V_{max} sulfide is 2.5 greater than V_{max} oxygen, as is true for *Thiothrix* species (van den Ende, 1997).

Table 9.12 Calculated rates of biological sulfide oxidation compared to the abiological rate[a]

Cell density (cells cm^{-3})	V'_{max} (M h^{-1})	d[H_2S]/dt (M h^{-1})
10	1.5×10^{-11}	9×10^{-12}
10^2	1.5×10^{-10}	9×10^{-11}
10^3	1.5×10^{-9}	9×10^{-10}
10^4	1.5×10^{-8}	9×10^{-9}
10^5	1.5×10^{-7}	9×10^{-8}
10^6	1.5×10^{-6}	9×10^{-7}
10^7	1.5×10^{-5}	9×10^{-6}
Abiological rate		3.2×10^{-9}

[a]$[H_2S] = [O_2] = 3 \times 10^{-6}$ M.

7. NON-PHOTOTROPHIC BIOLOGICALLY MEDIATED SULFIDE OXIDATION

7.1. Environments

In the absence of light, and when cell numbers are sufficiently high (see above), significant biological oxidation of sulfide can occur, provided an appropriate oxidant is available. Conspicuous examples of non-phototrophic biological sulfide oxidation are found in anoxic, sulfidic water columns, marine and lake sediments, microbial mats, and even the carcasses of decomposing animals (Bennett *et al.*, 1994). The most extensive populations of sulfide oxidizers yet described are the centimeter-thick layers of *Thioploca* spp. developed over thousands of square kilometers in the oxygen minimum zone off the Chilean and Peruvian coasts (Fossing *et al.*, 1995). Less dramatic, but nonetheless impressive, millimeter-thick populations of *Beggiatoa* spp. can be found as distinctive white or yellow layers at the surface of sulfidic sediments.

Other environments with active non-phototrophic biologically mediated sulfide oxidation include hydrothermal areas where sulfide oxidizers may be free living, bound to sediments or living in symbiotic association with a variety of animals. In the latter case (see Chapter 4), the sulfide oxidizers obtain a refuge against predation and provide the animal with nutrition for growth. In sulfidic sediments well removed from hydrothermal areas, similar symbiotic associations are found between sulfide oxidizers and a variety of animals, including nematodes, annelid worms, and a host of mollusks (Canfield and Teske, 1996; Hentschel *et al.*, 1999; Dubilier *et al.*, 2001). Solid phase sulfides are also, in principle, subject to biological oxidation in environments ranging from sulfide-containing ore deposits and mine tailings to anoxic pyrite-containing sediments.

7.2. Organisms oxidizing sulfide

A wide variety of non-phototrophic organisms (also known as colorless sulfur bacteria) oxidize sulfide in nature (Robertson and Kuenen, 1992). Most of these are chemolithoautotrophs coupling sulfide oxidation with O_2 or NO_3^- reduction (Table 9.13). In some cases, such as for a number of freshwater strains of *Beggiatoa*, for example, chemolithoheterotrophic metabolism also occurs (Nelson *et al.*, 1986) (Table 9.13). Sulfide oxidizers also commonly oxidize sulfur compounds of intermediate oxidation states such as elemental sulfur, thiosulfate, and tetrathionate, and furthermore, they can reduce elemental sulfur with various organic compounds or H_2 as electron

Table 9.13 Basic properties of selected sulfide-oxidizing organisms

Genus	C source	e acceptor (O_2 or NO_3^-)	Phylogenic position	Comments
Thiobacillus	A/H	O_2, NO_3^-	α, β, γ	Straight rods. Environmentally diverse and widely dispersed. Varied metabolism. As a genus, wide pH tolerance.
Thiomicrospira	A	O_2, NO_3^-	γ, ε	Common in anoxic water columns, marine sediments, and hydrothermal vents. Vibrio or bent rod morphology.
Paracoccus (Thiosphaera)	A/H	O_2, NO_3^-	α	Isolated from fluidized bed reactors, coccoidal. Known also as aerobic denitrifier. Genus recently renamed.
Sulfolobus/ Acidianus	A	O_2	*Archaea*	Hyperthermophilic, acidophile.
Thiovulum	A	O_2	ε	Highly motile cells swarming in conspicuous viels within O_2/H_2S interface above a solid surface.
Beggiatoa	A/H	O_2, NO_3^-	γ	Long slender motile filaments of greatly variable size. Typical in marine sediments and microbial mats. Some species concentrate NO_3^- in a central vacuole.
Thioploca	ND	NO_3^-	γ	Forms trichomes in sheathed filaments. Stores NO_3^- in a central vacuole. Exhibits phobic response to O_2. Closely related to *Beggiatoa*.
Thiomargarita	ND	NO_3^-, $O_2^?$	γ	Giant spheres up to 750 μm in diameter. Abundant in oxygen-minimum sediments of Namibia. Concentrates NO_3^- into a central vacuole.
Thiothrix	A/H	O_2, NO_3^-	γ	Filamentous bacteria. Can occupy H_2S–O_2 interface. Common as sulfide oxidizer in waste water treatment systems.
Sulfurospirillum	H	O_2, NO_3^-	ε	Reduces nitrate to ammonia, oxidizing sulfide to elemental S.
Desulfovibrio	–	O_2, NO_3^-	γ	Can oxidize sulfide with O_2 and nitrate but does not grow by this metabolism.

A, autotroph; H, heterotroph; ND, not determined; ?, not sure if oxygen is used.

donors. Some examples of the sulfide-oxidizing reactions conducted by sulfide oxidizers are presented in Equations 9.10 to 9.13.

$$H_2S + 2O_2 \rightarrow H_2SO_4 \quad (9.10)$$

$$2H_2S + O_2 \rightarrow 2S^0 + 2H_2O \quad (9.11)$$

$$5H_2S + 8NO_3^- \rightarrow 4N_2 + 5SO_4^{2-} + 2H^+ + 4H_2O \quad (9.12)$$

$$H_2S + NO_3^- + H_2O \rightarrow SO_4^{2-} + NH_4^+ \quad (9.13)$$

Most sulfide oxidizers are either obligate or facultative aerobes, although in some cases, as for members of the genera *Beggiatoa* and *Thiovulum*, for example, microoxic conditions are preferred with phobic responses to oxygen levels above 5 to 10% air saturation (see Section 7.4) (Jørgensen and Revsbech, 1983; Nelson *et al.*, 1986; Fenchel, 1994). *Thioploca* spp. demonstrates a strong phobic response to oxygen and a strong positive response to NO_3^- (Huettel *et al.*, 1996; Jørgensen and Gallardo, 1999). Together with some strains of *Beggiatoa* and the newly discovered giant sulfide oxidizer *Thiomargarita namibiensis* (found in sediments of the Benguela upwelling zone; Schulz *et al.*, 1999), *Thioploca* spp. can concentrate up to 500 mM nitrate in a central vacuole. A nitrate store enhances the capability of these organisms to use nitrate in sulfide oxidation (Schultz and Jørgensen, 2001) (see Chapter 2).

For *Thioploca* spp. and *Sulfurospirillum deleyianum* (Eisenmann *et al.*, 1995), nitrate is reduced to ammonia during sulfide oxidation, while for *Thiobacillus denitrificans, Thermothrix thiopara*, some species of the genus *Paracoccus* and *Thiomicrospira denitrificans,* nitrate is denitrified to N_2 (Robertsen and Kuenen, 1992). The products of nitrate reduction by nitrate-utilizing *Beggiatoa* and *Thiomargarita* are not yet known. Some sulfate reducers of the genus *Desulfovibrio* can also reduce nitrate to ammonia during sulfide oxidation, but they do not apparently gain energy for growth from this metabolism (Cypionka, 1994).

The oxidation of sulfide may proceed all the way to sulfate, or part way to elemental sulfur. Indeed, sulfide-oxidizing strains commonly precipitate elemental sulfur within the cell (Figure 9.12). This elemental sulfur is an available energy store organisms can either oxidize, as in the case of *Thiothrix nivea* for example, or reduce, as in the case of *Beggiatoa alba*, to gain energy when a ready interface between oxidant and sulfide cannot be found (Schmidt *et al.*, 1987).

Most of the described sulfide-oxidizing organisms are active within the mesophilic temperature range; however, thermophilic and hyperthermophilic representatives are also known. For example, members of the archaeal family *Sulfolobales* (including the genera *Sulfolobus* and *Acidianus*) have temperature maxima between 80 and 95 °C (Segerer and Stetter, 1992), while the bacterial *Thermothrix thiopara* has a temperature maximum of 72 °C and some strains

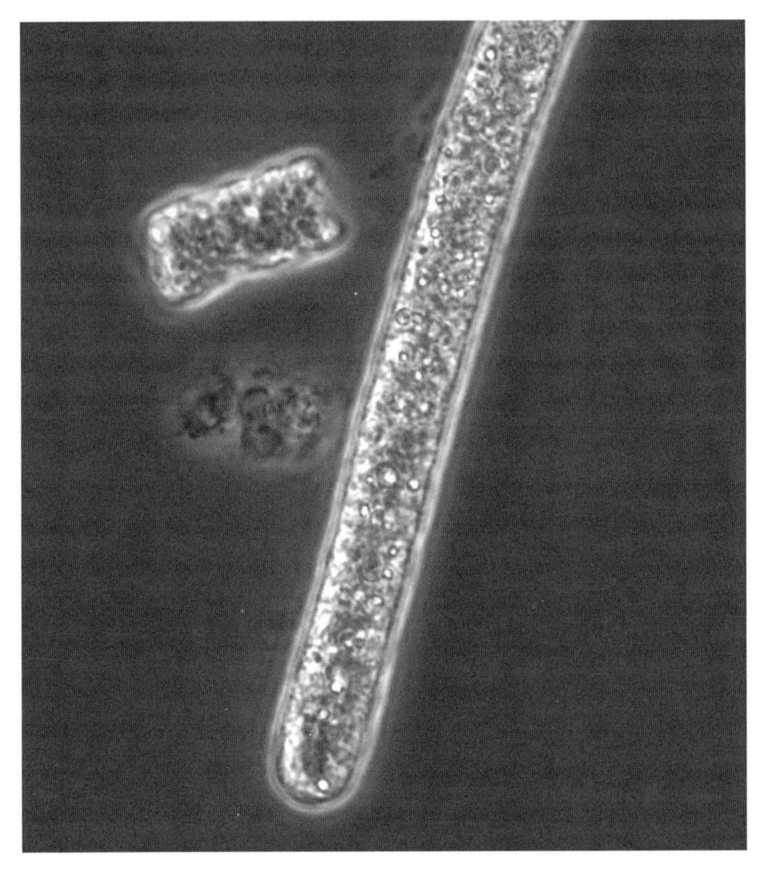

Figure 9.12 Photomicrograph of a sulfide-oxidizing *Beggiatoa* sp. in which the included sulfur grains are easily seen. Photo by Jacob Zopfi.

of *Thiobacillus*, also *Bacteria*, can metabolize up to 55 °C (Robertson and Kuenen, 1992). Psychrophilic sulfide oxidizers are not well known, but some strains of the bacterial *Thiobacillus* can apparently grow down to 2 °C (Nordstrom and Southam, 1997). Given the vast regions of cold sedimentary environments on Earth, psychrophilic sulfide oxidizers are probably well represented in nature, but to this point, are understudied.

7.3. Phylogeny

Sulfide oxidizers are found among the *Archaea*, within the family *Sulfolobales*, and are widely spread among the α-, β-, γ-, and ε-subdivisions of the proteobacteria (Robertson and Kuenen, 1992) (Table 9.13). The phylogeny of sulfide oxidizers from the γ- and β-subdivisions of the proteobacteria is shown in Figure 9.13. As mentioned above, some members of the sulfate-reducing genus *Desulfovibrio* (δ-subdivision, also shown in Figure 9.13) can oxidize sulfide but do not grow by this process. Members of the gram-positive *Hydrogenobacter* can oxidize reduced sulfur compounds (elemental sulfur and thiosulfate) but apparently not sulfide. Deep-branching sulfide oxidizers have not yet been identified.

Genus-level classification of sulfide oxidizers has traditionally been based on the organism's morphology and physiology. In many instances this has proven satisfactory, as, for example, different strains of the filamentous genus *Beggiatoa* are identifiable by these criteria. By contrast, species of the genus *Thiobacillus* are spread among the α-, β-, and γ-subdivisions of the proteobacteria (see Figure 9.13). They are all gram-negative rods able to

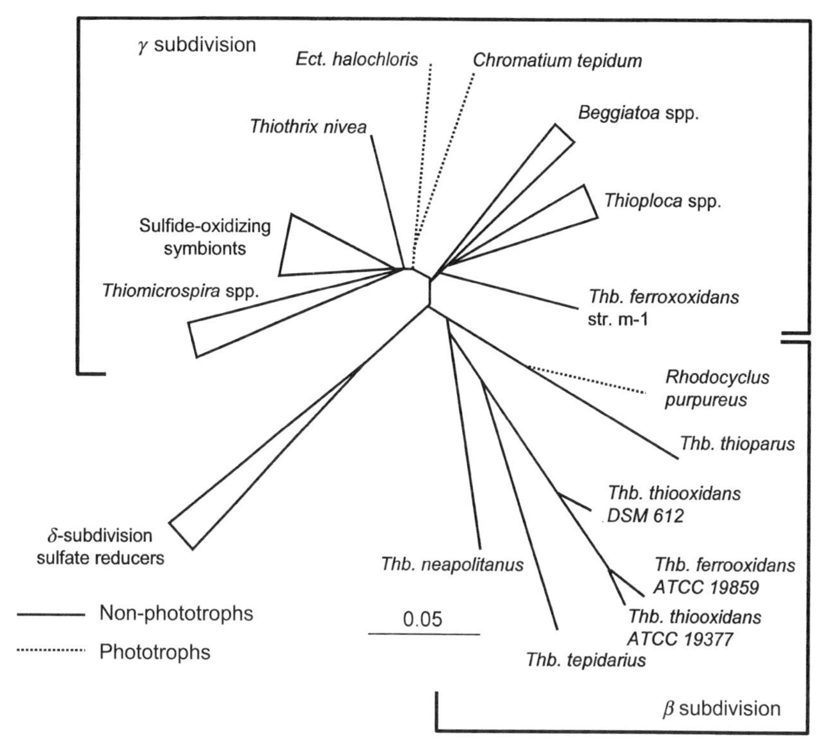

Figure 9.13 The phylogeny of selected sulfide-oxidizing bacteria, including some phototrophs, from the γ- and β-subdivisions of the proteobacteria. Sulfate reducers from the δ-subdivision are shown as an outgroup. Redrawn and simplified from Teske *et al.* (1996a).

oxidize reduced sulfur compounds but have widely different ancestral relationships. For this reason, the genus is currently under reclassification, with *T. thioparus*, from the β-subdivision of the proteobacteria, taken as the "type" species (McDonald *et al.*, 1996). Those species with close association (based on 16S rRNA) with *T. thioparus* remain as members of *Thiobacillus*, whereas distantly related species are being actively reassigned to other genera, including *Acidithiobacillus, Halothiobacillus*, and *Thermithiobacillus* of the γ-subdivision of the proteobacteria and *Starkeya* from the α-subdivision (Kelly and Wood, 2000; Kelly *et al.*, 2000).

7.4. Notes on behavior

We highlight the behavior of three different genera of sulfide-oxidizing organisms, each with unique strategies for maximizing their potential for growth, starting with *Beggiatoa*. Members of this genus may be either chemolithoautotrophic (using the reductive pentose phosphate cycle for CO_2 fixation; see Chapter 4) or chemolithoheterotrophic and typically, although not exclusively, occupy the interface between steep gradients of oxygen and sulfide. They are sediment bound, continuously gliding filaments, with strong phobic responses to oxygen and light and a weak phobic response to sulfide (Schultz and Jørgensen, 2001). The phobic response to oxygen occurs at oxygen levels greater than 5% air saturation. In microbial mats, oxygen and sulfide frequently overlap in a zone around 50 μm thick (Figure 9.14), and within this zone is often found a layer of *Beggiatoa* filaments 0.5 mm thick or greater (Jørgensen and Revsbech, 1983). Thus, the lower part of the *Beggiatoa* layer is under constant anoxic conditions, but due to constant gliding, the individual cells of a filament will occasionally migrate into the oxygen–sulfide interface. Presumably the reduction of accumulated S^0 under anoxic conditions can also help to maintain the cells.

Phobic responses to oxygen (mostly) and sulfide, and the change in gliding direction that results, assures that the filaments position themselves in their optimal microenvironment. In a dynamic photosynthetic system such as a microbial mat, this could mean that the *Beggiatoa* population migrates up and down several millimeters within the mat during the course of a day along with the oxygen–sulfide interface (see Chapter 13). In a microbial mat, the phobic response to light is advantageous, as in the early morning hours the oxygen produced by cyanobacteria may quickly accumulate below a *Beggiatoa* layer, which may be positioned at or near the mat surface over the night. Thus, the phobic response to light drives the *Beggiatoa* through the accumulating oxygen layer into its preferred interface between oxygen

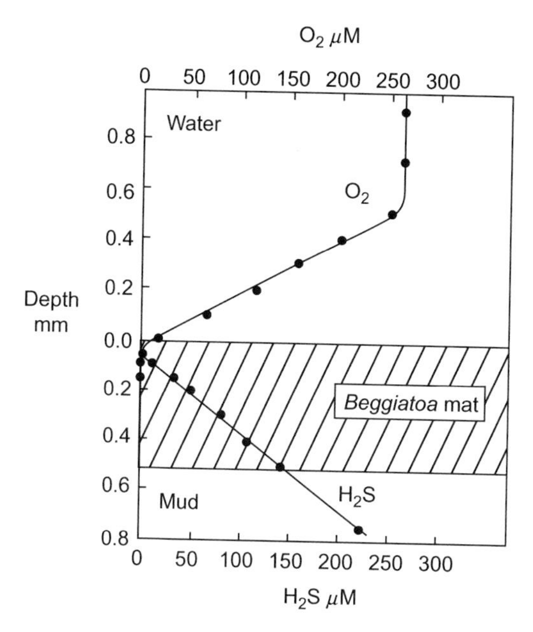

Figure 9.14 The distribution of oxygen and sulfide in sediment containing a mat of *Beggiatoa* at the surface. Note that oxygen and sulfide only overlap by about 50 μm. Redrawn from Jørgensen and Revsbech (1983).

and sulfide (Schultz and Jørgensen, 2001). Some species of *Beggiatoa* use nitrate rather than oxygen as an electron acceptor (McHatton *et al.*, 1996; Schultz and Jørgensen, 2001). When using nitrate, *Beggiatoa* can be more broadly distributed within the sediment and not obviously occupying a strict interface between electron acceptor and sulfide. The ecology and behavior of these organisms are not yet well studied.

Like many species of *Beggiatoa, Thiovulum* spp. are chemolithoautotrophic microaerophiles occupying steep gradients between oxygen and sulfide. However, unlike *Beggiatoa, Thiovulum* are fast swimmers (for a bacterium), are water column bound and have their own special strategy for maintaining an optimal environment. Specifically, *Thiovulum* have a strong phobic response to oxygen and adjust their swimming direction to concentrate themselves within the oxygen–sulfide interface (Fenchel, 1994). Once there, individual organisms secrete slime threads, which stick together, forming a two-dimensional veil. The veil effectively separates turbulent oxygen-containing water above from stagnant sulfide-containing water below. Thus, the veil influences the hydrodynamics of the chemocline, and the organisms can even move the veil up or down to optimize the concentration gradients of oxygen and sulfide at the chemocline (Schultz and Jørgensen, 2001).

Thiovulum may also attach to solid surfaces with their slime threads. By rapidly spinning around this tether, they can actively pump oxygenated water to the oxygen–sulfide interface, thereby dramatically increasing oxygen and sulfide transport rates and hence their metabolic activity (Fenchel and Glud, 1998). The resulting veil architecture is a delicate patchwork of inflow regions, where the veil is concentrated, and outflow regions existing as holes within the veil (Figure 9.15).

Thioploca spp. are most conspicuous as extensive mats at the sediment surface of oxygen-minimum regions off the coast of Chile and Peru, but are also widely distributed in other marine and freshwater environments. Individual *Thioploca* filaments, consisting of a string of up to 1000 connected cells, may reach several centimeters in length (Jørgensen and Gallardo, 1999). Filaments are typically arranged as bundles sharing a common sheath. The sheaths act as mini-highways, facilitating migration of the filaments up and down in the sediment. Due to their positive nitrate response (see above), *Thioploca* filaments extend well out of their sheath when exposed to nitrate, accumulating it in their central vacuole. They then

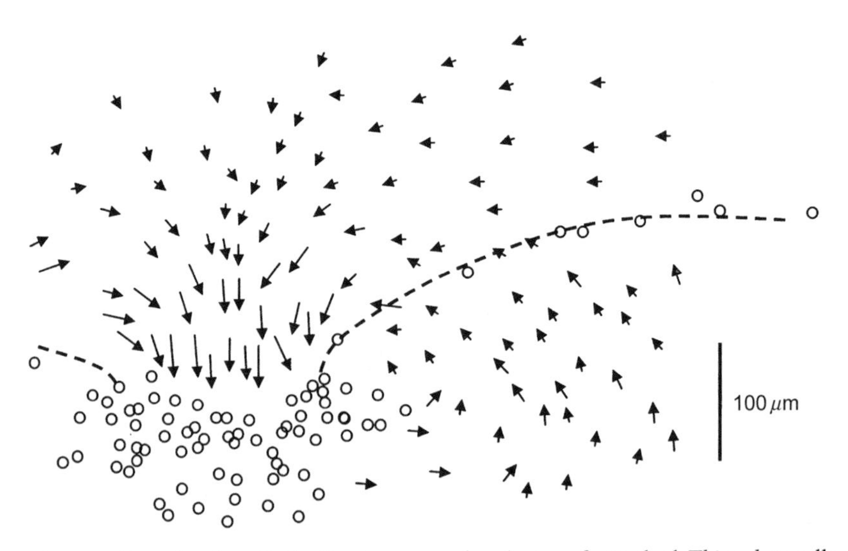

Figure 9.15 The flow field of water around a cluster of attached *Thiovulum* cells. Due to rapid rotation around their attached threads, oxygenated water is brought into the cluster with deoxygenated water displaced outward and upward outside of the cell cluster. The dashed line traces water with an oxygen concentration of about 4% atmospheric saturation, and some free-swimming *Thiovulum* cells may be found at this oxygen tension. Redrawn from Fenchel and Glud (1998).

migrate down into the sediment using the nitrate to oxidize sulfide at depth or the supply of elemental sulfur concentrated within the cells. Another potential source of sulfide is from sulfate reducers of the genus *Desulfonema* found concentrated within the *Thioploca* sheaths (Fossing *et al.*, 1995). *Thioploca* apparently utilizes only NO_3^- as an electron acceptor in sulfide oxidation, with no indications yet of oxygen utilization.

7.5. Solid phase sulfide oxidation

In low pH environments, such as those found in abandoned sulfide mines and sulfide mine tailings, the oxidation of solid phase sulfides is promoted by microorganisms. However, the organisms apparently do not oxidize sulfide directly and are therefore not true sulfide oxidizers. Rather, with *Thiobacillus ferrooxidans* as an example, Fe^{2+} in the environment is microbially oxidized to Fe^{3+} (see also Chapter 8), which at low pH is a potent chemical oxidant for sulfide minerals. During the oxidation of pyrite (FeS_2), there are the following stages in sulfide oxidation:

$$\text{biological}: 4Fe^{2+} + O_2 + 4H^+ \rightarrow 4Fe^{3+} + 2H_2O \quad (9.14)$$

$$\text{abiological}: FeS_2 + 14Fe^{3+} + 8H_2O \rightarrow 15Fe^{2+} + 2SO_4^{2-} + 16H^+ \quad (9.15)$$

Pyrite is also oxidized in marine and freshwater sediments of near neutral pH. When oxygen is present, pyrite oxidation also apparently proceeds through Fe^{3+} (Moses and Herman, 1991), and the production of Fe^{3+} is apparently biological as shown in Equation 9.14. In the absence of oxygen, pyrite and FeS can be oxidized chemically with MnO_2, but not with NO_3^- or Fe oxides (Schippers and Jørgensen, 2002). Although the data are rather limited, there is no apparent microbial involvement in anaerobic pyrite oxidation in sediments (Schippers and Jørgensen, 2002). By contrast, FeS oxidation with NO_3^- can be mediated by microbes.

8. PHOTOTROPHIC SULFIDE OXIDATION

8.1. General considerations

The phototrophic oxidation of sulfide is a type of anoxygenic photosynthesis that occurs when sulfide is found in the presence of light. Different reduced sulfur compounds, including sulfide, elemental sulfur, and thiosulfate, may be used as electron donors in the process, and the oxidation of reduced sulfur is generally coupled to the reduction of CO_2 to cell biomass (Equations 9.16–9.20). Elemental sulfur is a common oxidation intermediate (Equations 9.16 and 9.19), accumulating inside the cell in some genera of sulfide oxidizers and outside the cell in others. In most cases, but not all, elemental sulfur is further oxidized to sulfate (Equation 9.17). In some cases, reduced sulfur is oxidized completely to sulfate in a single step (Equations 9.18 and 9.20).

$$CO_2 + 2H_2S \rightarrow CH_2O + H_2O + 2S^0 \quad (9.16)$$

$$3CO_2 + 2S^0 + 5H_2O \rightarrow 3CH_2O + 2SO_4^{2-} + 4H^+ \quad (9.17)$$

$$2CO_2 + H_2S + 2H_2O \rightarrow 2CH_2O + SO_4^{2-} + 2H^+ \quad (9.18)$$

$$CO_2 + 2S_2O_3^{2-} + H_2O \rightarrow CH_2O + 2S^0 + 2SO_4^{2-} \quad (9.19)$$

$$2CO_2 + S_2O_3^{2-} + 3H_2O \rightarrow 2CH_2O + 2SO_4^{2-} + 2H^+ \quad (9.20)$$

Many sulfide oxidizers act as photoorganoheterotrophs assimilating or oxidizing organic compounds during their photosynthesis. Hydrogen gas (H_2) is also a common electron donor for sulfide-oxidizing phototrophs.

Environments housing phototrophic sulfide oxidizers are rather common. One might notice purple layers atop black sulfidic sediments in a stagnant pool or a salt marsh, or one might also find them as distinctive purple or green layers just below the surface of waterlogged sands. Indeed, phototrophic sulfide oxidizers are common in cyanobacterial mats, many types of waterlogged soils, the water columns of anoxic lakes and marine basins, and even in wastewater treatment facilities. Dense populations may also develop in effluents of sulfidic hydrothermal springs. In a typical cyanobacterial mat, sulfide-oxidizing phototrophs may be found immediately below an active population of cyanobacteria and immediately above anoxic sulfide-containing porewater. Similarly, in sulfidic lakes, dense bacterial "plates" of sulfide-oxidizing phototrophs may be found populating the oxygen–sulfide interface. More details on the ecology of sulfide-oxidizing anoxygenic phototrophs will follow.

8.2. Major groups and a historical note on taxonomy

Five major groups of anoxygenic phototrophs have been identified with the ability, of varying degrees, to oxidize sulfide during photosynthesis. These groups are the purple sulfur bacteria, the purple nonsulfur bacteria, the green sulfur bacteria, the green nonsulfur bacteria, and the heliobacteria (Table 9.14). Historically, these groups have been distinguished based on combined considerations of physiology, morphology, pigment composition,

Table 9.14 Some ecophysiological characteristics of sulfide-oxidizing anoxygenic phototrophs

Photosynthetic group	Phylogenetic position	Reduced substrates	Pigments	Motility (G, F)[a]	Autotrophy (L, P)[b]	Aerobic metabolism	Chlorosomes	Gas vescicles	S storage in cell	C fixation pathway[c]	RC[d]
Purple S	α proteobacteria	H_2S, S^o, $S_2O_3^{2-}$, H_2	Bchl *a,b*	F	L, P	yes	no	yes	yes	RPP	Ph-Q
Purple non-S	α, β proteobacteria	H_2S, S^o, $S_2O_3^{2-}$, H_2	Bchl *a,b*	F	L, P	yes	no	no	no	RPP	Ph-Q
Green S	own lineage	H_2S, S^o, $S_2O_3^{2-}$, H_2?	Bchl *a, c,d,e*	G	P	no	yes	yes	no	RCA	Fe-S
Green non-S	own lineage	H_2S, H_2	Bchl *a, c,d*	G	P	yes	yes	no	no	3HP, RPP	Ph-Q
Heliobacteria	gram positive	H_2S	Bchl *g*	G, F	–	no	no	no	no	–	Fe-S

[a] G, gliding; F, flagellar.
[b] L, lithoautotrophy; P, photoautotrophy.
[c] RPP, reductive pentose phosphate cycle; RCA, reductive citric acid cycle; 3HP, 3-hydroxypropionate cycle.
[d] RC, reaction center; Ph-Q, pheophytin-quinone, Fe-S, iron sulfide.

and membrane structure. The nomenclature has developed such that groups of organisms are broadly identified as "sulfur" or "nonsulfur" bacteria. As originally conceived, sulfur bacteria oxidized reduced sulfur compounds during photosynthesis, depositing elemental sulfur within or on the cell (Pfennig, 1977), with the elemental sulfur subsequently oxidized to sulfate. The nonsulfur bacteria were believed not to use sulfide during photosynthesis, instead oxidizing or assimilating organic compounds as photoorganoheterotrophs. Subsequently, representatives within the nonsulfur clades have been found to oxidize sulfide to elemental sulfur (though generally outside of the cell), with further oxidation of the sulfur to sulfate in many cases (Pfennig, 1975, 1977; Imhoff and Trüper, 1992). Thus, although the original intentions of the sulfur and nonsulfur designations have been somewhat blurred, the groupings nevertheless have phylogenetic significance and have been retained. In the following Sections we consider each of these groups in more detail.

8.2.1. Purple sulfur bacteria

The purple sulfur bacteria are all found within the γ-subdivision of the proteobacteria. As a group they synthesize both BChl *a* and BChl *b*, with BChl *b* being especially noteworthy because it has absorption maxima well into the infrared range (Figure 9.16). The purple sulfur bacteria are divided into two families, the *Chromatiaceae* and the *Ectothiorhodospiraceae* (Pfennig, 1989), and the genera within these families are summarized in Table 9.15. The *Chromatiaceae*, some of which contain gas vesicles, commonly store elemental sulfur in their cells during anoxygenic photosynthesis. They are widely distributed where light is available at the interface between oxygen and sulfide in freshwater and saline environments, including anoxic lakes, anoxic marine basins, sediments, and microbial mats and sulfur springs (Pfennig and Trüper, 1992). In microbial mats, *Chromatium*-like species can be found at depths where the visible light has been fully adsorbed by the overlying oxygenic phototrophs, yet sufficient near-IR radiation penetrates to support the photosynthesis of these BChl *b*-containing anoxygenic phototrophs (Figure 9.17) (Jørgensen and Des Marais, 1986a). This is a beautiful example of how the absorption properties of specific BChl molecules are used to capitalize on a specific environmental niche.

The *Chromatiaceae* are especially noteworthy for forming dense plates in many stratified, sulfidic lakes (Van Gemerden and Mas, 1995). In some cases, carbon and sulfur cycling in the lake monimolimnion (the lower stagnant portion) is so efficient that carbon fixation by the anoxygenic

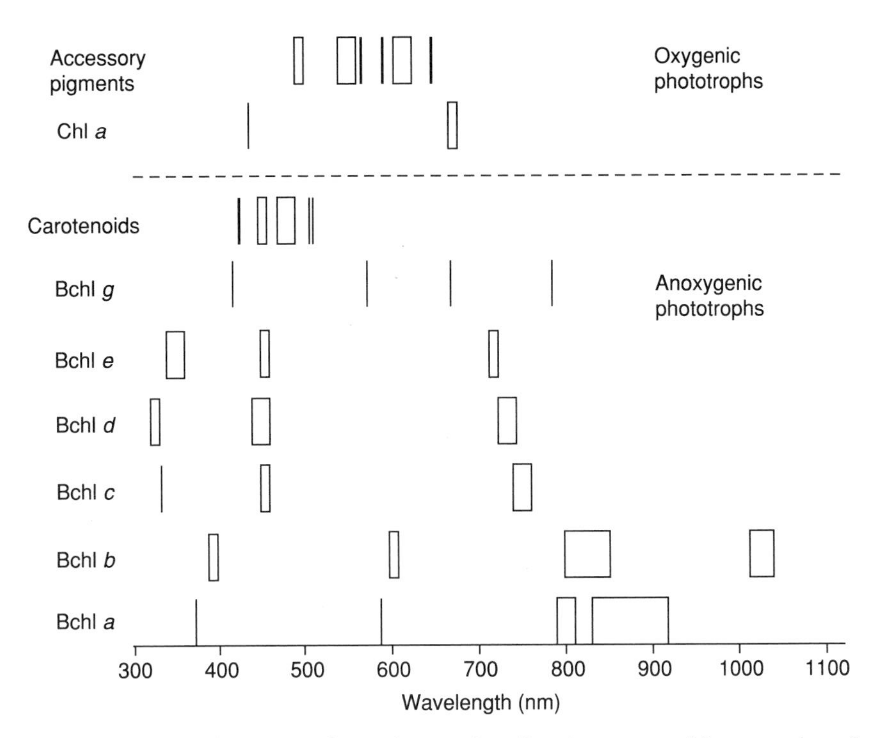

Figure 9.16 Major *in vivo* absorption maxima for pigments used in oxygenic and anoxygenic photosynthesis. Compiled from data in Castenholz *et al.* (1992).

Table 9.15 Genera of anoxygenic phototrophs

Purple sulfur bacteria	Purple nonsulfur bacteria	Green sulfur bacteria	Green nonsulfur bacteria	Heliobacteria
Chromatium	*Rhodospirillum*	*Chlorobium*	*Chloroflexus*	*Heliobacterium*
Thiospirillum	*Rhodobacter*	*Prosthecochloris*	*Chloronema*	*Heliobacillus*
Thiocapsa	*Rhodomicrobium*	*Ancalochloris*	*Oscillochloris*	
Thiocystis	*Rhodopseudomonas*	*Pelodictyon*	*Heliothrix*	
Lamprocystis	*Blastochloris*	*Chloroherpeton*		
Thiopedia	*Rhodocyclus*	*Clathrochloris*		
Thiodictyon	*Ribrivirav*			
Amaebobacter	*Acidiphilium*			
Lamprobacter				
Ectothiorhodospira				

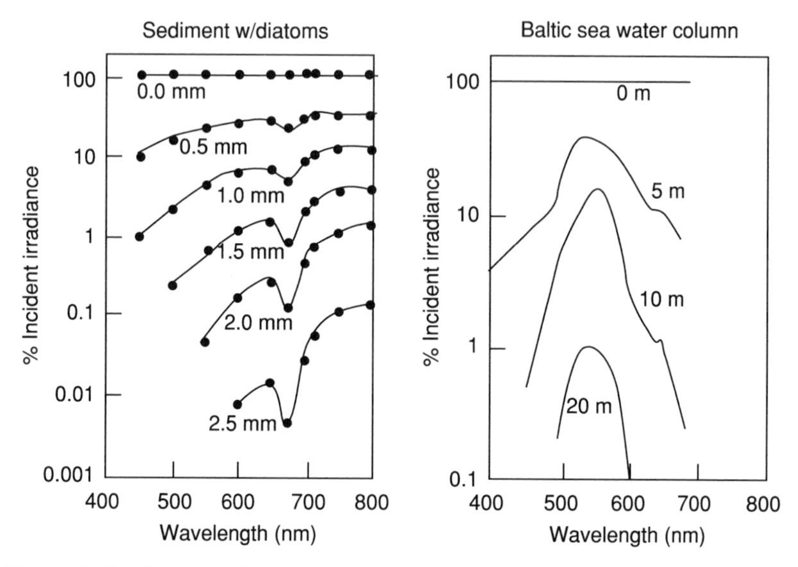

Figure 9.17 Spectra of scalar irradiance, normalized to incident downwelling irradiance, for a sandy sediment containing diatoms and for the water column of the Baltic Sea. Note that near-infrared radiation is quickly absorbed with depth in the marine water column but persists through shallow sediment layers supporting oxygenic photosynthesis. This near-infrared radiation can fuel anoxygenic phototrophs below the oxic surface layer. Redrawn from Kühl and Jørgensen (1994) and from data presented in Jerlov (1976).

phototrophs equals or even exceeds primary production by the oxygenic phototrophs (Van Gemerden and Mas, 1995) (see Chapter 13). The *Ectothiorhodospiraceae* favor saline and hypersaline alkaline environments. They are further distinguished from *Chromatiaceae* by precipitating sulfur outside of the cell during anoxygenic photosynthesis. This sulfur is further oxidized to sulfate. All *Ectothiorhodospiraceae* are motile with polar flagella.

As a group, the purple sulfur bacteria prefer anoxic habitats and most are obligate anaerobes, although some can tolerate low concentrations of oxygen. Most can grow photolithoautotrophically, utilizing the reductive pentose phosphate pathway (see Chapter 4) for carbon fixation. Some oxygen-tolerant species can grow chemolithoautotrophically in the absence of light, oxidizing sulfide with oxygen (Pfennig, 1989), while some are also chemoorganoheterotrophic. This ability to respire organic compounds, however, seems to be quite limited (Pfennig, 1989). Some purple sulfur bacteria have the ability to assimilate organic compounds during photosynthesis, but only a limited range of organic compounds can be used. In summary, the purple bacteria are mainly anaerobes living by the photolithoautotrophic oxidation of reduced sulfur compounds.

8.2.2. Purple nonsulfur bacteria

In terms of cell morphology, phylogeny, and physiology, the purple nonsulfur bacteria are the most diverse group of anoxygenic phototrophs (Imhoff, 1995). Most are located within α-subdivision of the proteobacteria, while two genera (*Rhodocyclus* and *Rubrivirax*) are found within the β-subdivision. As a group, all contain BChl *a*, while some also have BChl *b*. Purple nonsulfur bacteria are impressive physiological generalists. They are all photoorganoheterotrophs, oxidizing and assimilating a wide variety of organic compounds in the presence of light, including a number of aromatic compounds (Gibson and Harwood, 1995). Photoorganotrophy is their preferred metabolism. They can typically withstand low oxygen concentrations, and some species can even tolerate fully oxic conditions. In the absence of light, many respire organic matter with oxygen. Without both light and oxygen, many species ferment, and some, such as *Rhodobacter sphaeroides* denitrify when nitrate is available (Imhoff and Trüper, 1992). Some may also grow chemolithoautotrophically with H_2 or H_2S as electron donors (Imhoff and Trüper, 1992). Photolithoautotrophic growth is possible in many species using H_2S, thiosulfate or H_2 as electron donors. When oxidizing reduced sulfur compounds the end product is usually elemental sulfur, which is deposited outside of the cell. However, in some cases sulfide may be completely oxidized to sulfate (Pfennig, 1975; Imhoff and Trüper, 1992). During autotrophic growth, CO_2 is fixed with the reductive pentose phosphate pathway (Imhoff and Trüper, 1992).

Purple nonsulfur bacteria are widely distributed in stagnant lakes, ponds, lagoons, sediments, soils, and rice paddy fields with low oxygen concentrations and high concentrations of dissolved organic carbon (DOC). They may be found on rotting leaves and are common at all stages of sewage plant operation (Imhoff and Trüper, 1992). When anaerobic sewage sludge is incubated in the light, a red-brown color often develops, indicating the presence of purple nonsulfur bacteria. Purple nonsulfur bacteria are easily enriched under low-oxygen conditions in the light with organic substrates. Unlike purple sulfur bacteria, they rarely form blooms in the environment.

In summary, purple nonsulfur bacteria prefer photoorganoheterotrophy as a way of life. They are typically found in organic-rich areas where light is present, oxygen is low or absent, and sulfide may not be available. While purple nonsulfur bacteria may be found together with purple sulfur bacteria in light-exposed sulfidic settings, they do not apparently compete well under these circumstances.

8.2.3. Green sulfur bacteria

The green sulfur bacteria are organized into five genera (Table 9.15) under the family name *Chlorobiaceae*. They are a monophyletic, phylogenetically distinct group of anoxygenic phototrophs forming their own lineage on the 16S rRNA Tree of Life (Overmann, 2001) (see Chapter 1). The green sulfur bacteria are obligate anaerobes and obligate phototrophs, and many are mixotrophic, assimilating simple organic compounds such as acetate and pyruvate during photosynthetic sulfide oxidation. Some are also capable of strict photolithoautotrophic growth, using H_2S, H_2, and $S_2O_3^{2-}$ (for some) as electron donors. Fixation of CO_2 in the green sulfur bacteria is accomplished by the reverse citric acid cycle (see Chapter 4). Elemental sulfur is a common oxidation product during sulfide oxidation, and it is deposited outside of the cell. In some situations, particularly when H_2S is limiting, sulfide may be directly oxidized to sulfate (Pfennig, 1975). All green sulfur bacteria contain BChl *a* in their reaction centers but differ in their major antenna pigments, with the green-colored species containing BChl *c* or *d* and the brown-colored species containing BChl *e* (Trüper and Pfennig, 1992; Overmann, 2001). The major antenna pigments are contained in chlorosomes, special spherical-to-oblong bodies attached to the inside of the cytoplasmic membrane. Chlorosomes are lacking in the anoxygenic photosynthetic purple bacteria and the heliobacteria.

Members of the green sulfur bacteria have special adaptations to low light, and they grow at rates comparable to the purple sulfur bacteria at only one-fourth of the light intensity (Trüper and Pfennig, 1992). Indeed, the lowest light intensity supporting the growth of any photosynthetic organism is found for *Chlorobium phaeobacteroides*. This organism was isolated from the chemocline of the Black Sea with an ambient light level of <0.25 μmol photons m^{-2} s^{-1} (Overmann, 1992). Part of this adaptation to low light may result from the low energy requirement of the reverse citric acid cycle used in carbon fixation (Overmann, 2001) (see Chapter 4). Among the *Chlorobiaceae*, the brown-colored species, containing BChl *e*, tend to have the lowest light adaptation.

Green sulfur bacteria can be readily enriched from many shallow fresh-water and marine sedimentary environments but rarely show the same mass accumulations in these environments as is common with the purple sulfur bacteria. Conspicuous populations of green sulfur bacteria, however, have been found in sulfur springs in New Zealand and in the deeper photosynthetic layers of microbial mats in Great Sippewissett Marsh, Cape Cod, USA (Nicholson *et al.*, 1987; Trüper and Pfennig, 1992). Green sulfur bacteria form prominent, and even dominant, populations in many meromictic lakes, particularly when the anoxic zone is deep (Pfennig, 1989; Overmann, 2001).

An example is Fayetteville Green Lake, New York, where a dense population of *Chlorobium phaeobacteroides* supports high rates of carbon fixation at around 18 m water depth (Culver and Brunskill, 1969). The prominence of these, and other, deep-water examples of green sulfur bacteria, is due mostly to their adaptation to low light. However, as blue-green light with wavelengths between 500 and 600 nm penetrates deepest in the water column (Figure 9.17), deep-water green sulfur bacteria must rely on accessory carotenoid pigments for light absorption in these wavelengths (Figure 9.16).

Both green and brown members of the *Chlorobiaceae* are often found as well-organized epibionts surrounding a single nonphotosynthetic central prokaryote. These microbial consortia are given species names such as *Chlorochromatium aggregatum* and *Pelochromatium roseum*, and they can account for the major part of the photosynthetic population in some sulfidic lakes (Overmann and Schubert, 2002). Previously, the consortium partners were believed to be bound in a tight sulfur cycle where the central microbe was thought to be a sulfate reducer supplying sulfide to the sulfide-oxidizing phototrophs (e.g., Overmann *et al.*, 1998). This, however, is not likely, as the central organism is now known to be a member of the β-subdivision of the proteobacteria where sulfate reducers are not found (Fröstl and Overmann, 2000; Overmann and Schubert, 2002). Thus, the function of the consortium remains unclear. Nevertheless, interesting signaling communication between the consortium partners has been demonstrated. For example, the consortia are motile with a polar flagellum located on the central bacterium, yet the consortia shows a tactic response to wavelengths of light identical to the absorption spectra of the BChl's of the phototrophic partner (Overmann and Schubert, 2002). Positive tactic response to sulfide, which benefits the sulfide oxidizing phototroph, would also seem to require signaling between consortium partners.

8.2.4. Green nonsulfur bacteria (filamentous anoxygenic phototrophic bacteria)

The family *Chloroflexaceae* houses a variety of phototrophic organisms that have been collectively termed the green nonsulfur bacteria. The reason for this designation is that, by comparison with the purple nonsulfur bacteria, the green nonsulfur bacteria have photoorganoheterotrophy as their principal metabolism. Recently, Pierson (2001) advocated renaming the green nonsulfur bacteria the "filamentous anoxygenic phototrophic bacteria." This change in nomenclature is driven in part by the recognition that the green nonsulfur bacteria stand phylogenetically distinct from the green sulfur bacteria, occupying one of the deepest branches within the bacterial domain of the Tree of Life (see Chapter 1). As we shall see, they also display a wider range of physiological flexibility than originally conceived when the nonsulfur designation was first adopted. We retain the green nonsulfur bacterial designation here, however, to provide consistency with most of the available literature discussions on anoxygenic phototrophs, but we recognize that an important change in nomenclature may be underway.

There are currently four recognized genera of the green nonsulfur bacteria (Table 9.15), and these organisms group together with other nonphototrophic bacteria, including some filamentous representatives (Pierson, 2001). Members of the genus *Chloroflexus* were the first isolated green nonsulfur bacteria. These organisms are filamentous, possess gliding motility, and are predominantly chemoorganoheterotrophs (Pierson and Castenholz, 1995). A typical environment for *Chloroflexus* is hot springs, where some strains have an upper temperature limit of 70 °C (Pierson and Castenholz, 1992). However, mesophilic strains of *Chloroflexus* are known that can be isolated from sulfidic lake sediments and microbial mats. Some strains of *Chloroflexus* can grow chemolithoautotrophically with H_2S or H_2, and they use the 3-hydroxypropionate pathway for carbon fixation (see Chapter 4). *Chloroflexus* spp. are also aerotolerant and can respire with oxygen in the dark, although atmospheric oxygen levels completely inhibit growth (Pierson and Castenholz, 1992). Similar to the green sulfur bacteria, members of the genera *Chloroflexus* house antenna pigments in chlorosomes. *Chloroflexus* contains BChl *a* and BChl *c*.

Members of the genera *Chloronema* and *Heliothrix* have not yet been grown in pure culture, and their physiology has not, therefore, been studied in detail. However, it is doubtful whether the known species of either of these genera oxidize sulfide in nature. *Chloronema* spp. are found in the water column of meromictic lakes with high iron contents and limited sulfide (Pierson, 2001), while *Heliothrix* spp. are typically found in hot springs environments and are probably aerobic photoorganoheterotrophs (Pierson, 2001).

Organisms from the genus *Oscillochloris* are filamentous and have gliding motility, and some may have a sheath. These organisms contain mainly BChl *c*, with some BChl *a*, and the antenna pigments are contained in chlorosomes. *Oscillochloris* spp. prefer photolithoheterotrophic growth on sulfide under anaerobic conditions, but they can also grow, though poorly, as photolithoautotrophs on H_2S and H_2 (Pierson, 2001). Sulfide oxidized during phototrophic growth is deposited as elemental sulfur outside of the cell. Organisms from this genus do not apparently oxidize elemental sulfur further to sulfate (Pierson, 2001). *Oscillochloris* spp. fix carbon during autotrophic growth with the reductive pentose phosphate cycle (see Chapter 4). This is by contrast with the 3-hydroxypropionate pathway used by *Chloroflexus* spp. (see above).

Oscillochloris spp. can be found in sulfidic freshwater sediments and algal mats, as well as in sulfidic springs at temperatures up to 45 °C (Pierson, 2001). They may be found in association with other anoxygenic phototrophs and cyanobacteria.

8.2.5. Heliobacteria

The heliobacteria form the family *Heliobacteriaceae* with two recognized genera (Table 15), *Heliobacterium* and *Heliobacillus*. These organisms are found among the gram-positive group of prokaryotes (although their cells stain gram negative) and are unique among the other anoxygenic phototrophs in containing BChl *g*. This bacteriochlorophyll is rather unusual, with structural relationships to both Chl *a* and the other bacteriochlorophylls (Madigan, 1992). The heliobacteria are known mostly as nonsulfur bacteria, with photoorganoheterotrophy as their primary metabolism in the light (utilizing simple organic substrates such as acetate, pyruvate, lactate, malate, and propionate). Many can ferment in the dark (Stevenson *et al.*, 1997). They are also strict anaerobes in their metabolism, although through endospore formation some can withstand exposure to oxygen and even boiling temperatures (Madigan, 1992). The heliobacteria have been isolated from a variety of soil types, including waterlogged rice paddies (Madigan, 1992; Stevenson *et al.*, 1997). They seem to be rare in marine and lake settings, but they have been isolated from hydrothermal spring environments. Indeed, from cyanobacterial mats growing in alkaline hot springs of the Buryatia (Russia) new strains of heliobacteria that can phototrophically oxidize sulfide to elemental sulfur were isolated (Bryantseva *et al.*, 2000). However, no carbon fixation could be demonstrated during phototrophic sulfide oxidation, and even these sulfide-oxidizing heliobacteria were mainly photoorganoheterotrophs. In summary, although some heliobacteria apparently possess the ability to oxidize sulfide during their phototrophic metabolism, they are, as a group, photoorganoheterotrophs living mostly in terrestrial soil environments.

8.3. Pathways of sulfide oxidation

We briefly outline some of the main enzymatic steps involved in the phototrophic oxidation of sulfide. We concentrate here on sulfide oxidation. A more complete discussion of the enzymatics of phototrophic sulfur compound oxidation can be found in Dahl and Trüper (1994), from which the following discussion is abstracted. The main steps in sulfide oxidation are the oxidation of sulfide to elemental sulfur or sulfite, the oxidation of elemental sulfur to sulfite, and the oxidation of sulfite to sulfate (Figure 9.18).

To follow these steps more closely, in the initial oxidation step, sulfide is oxidized to polysulfides and elemental sulfur with *c*-type cytochromes, in which the sulfur and polysulfides are in thermodynamic equilibrium. The formation of sulfite from polysulfides/elemental sulfur is accomplished with a "reverse" siroheme-containing sulfite reductase, which is also responsible for the direct oxidation of sulfide to sulfite. In a step essentially reverse to what happens during sulfate reduction (see Figure 9.3), sulfite reacts with AMP to form APS, catalyzed by the enzyme APS reductase. The sulfate in APS is liberated in two different ways: either by replacing sulfate with a phosphate group, forming ADP, with mediation by the enzyme APAT (adenylylsulfate:phosphate adeylyltransferase), or by replacing the phosphate with pyrophosphate, liberating ATP and sulfate, with mediation by the enzyme ATP sulfurylase. Again, the step mediated by ATP sulfurylase has analogies with sulfate reduction, only operating in reverse. Sulfide may also be oxidized directly to sulfate with the enzyme sulfite:acceptor oxidoreductase. There are no analogies with this step in sulfate reduction.

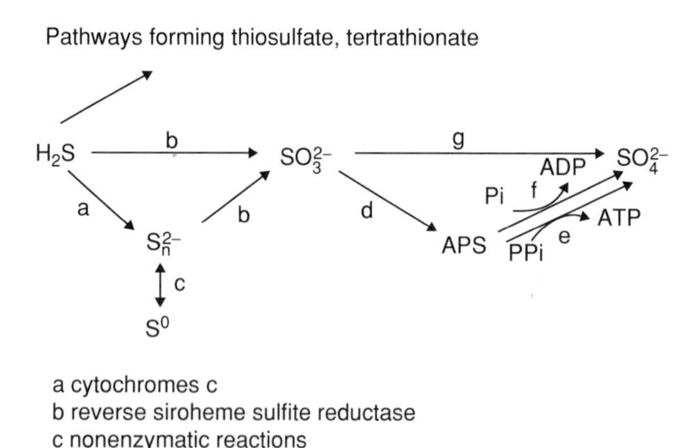

Figure 9.18 Pathways of sulfur metabolism in sulfide-oxidizing phototrophs. Additional pathways also lead to the formation of thiosulfate and tetrathionate, which are not shown here. Abstracted from Dahl and Trüper (1994).

8.4. A note on evolution

The SSU rRNA-based Tree of Life (Chapter 1), if read literally, suggests the early emergence of anoxygenic photosynthesis within the green nonsulfur bacteria. This was then followed by the much later development of oxygenic photosynthesis, and the near contemporaneous proliferation of other anoxygenic photosynthetic lineages in a pattern not easy to distinguish based on the vagaries in tree topology (Knoll and Bauld, 1989; Canfield and Raiswell, 1999). An early development of anoxygenic photosynthesis is also reasonable based on the much more complex coupled photosystems used in oxygenic photosynthesis, which was apparently derived from preexisting anoxygenic photosynthetic reaction centers (Blankenship, 1992) (see Chapter 4).

Still, some argue otherwise. The Granick hypothesis (Granick, 1965) quite reasonably states that the pathway of pigment biosynthesis recapitulates the evolutionary appearance of the pigment. Therefore, since bacteriochlorophyll is synthesized through a chlorophyll-like intermediate step (Beale, 1999), the Granick hypothesis predicts that the evolution of chlorophyll predated the evolution of bacteriochlorophyll. This would imply that the capacity for oxygen-evolving photosynthesis predated anoxygenic photosynthesis.

As the Granick hypothesis deals specifically with the evolution of photosynthetic pigments, it would seem most pertinent to explore the phylogeny of genes involved in photosynthetic pigment synthesis in order to resolve this issue. This was accomplished by Ziong *et al.* (2000), who compiled, aligned, and phylogenetically analyzed a number of genes involved in the synthesis of chlorophyll and bacteriochlorophyll from the major groups of anoxygenic phototrophs, cyanobacteria, and algae. Using a variety of different analytical protocols, a number of different phylogenetic trees were constructed, with the consensus conclusion that the pigment synthesis system of purple bacteria emerged first among the phototrophs. This was followed by the synthesis systems of the green sulfur bacteria and the green nonsulfur bacteria, which cluster as groups. The pigment systems of the heliobacteria were the latest emerging among the anoxygenic phototrophs, coming just before the cyanobacteria, which clearly predate photosynthetic eukaryotes.

These results reject the Granick hypothesis and demonstrate that anoxygenic photosynthesis emerged prior to oxygenic photosynthesis. However, the results do provide some surprises, as, based on SSU rRNA phylogenies, the purple bacteria emerge well after the green sulfur and the green nonsulfur bacteria (see Chapter 1). Furthermore, from SSU rRNA, the green sulfur bacteria and the green nonsulfur bacteria are well-separated groups. This divergence in tree topologies forces us, again, to conclude that metabolic genes are subject to lateral transfer and provide important, independent, though often conflicting, phylogenies compared to SSU rRNA. As a further example of this, heliobacteria emerge first among the phototrophs in phylogenies based on heat shock protein genes (Hsp 60 and Hsp70), followed by the green nonsulfur bacteria, the cyanobacteria, the green sulfur bacteria, and finally photosynthetic purple bacteria (Gupta *et al.*, 1999). It is clear that the heat shock proteins have experienced a different evolutionary path than either the photosynthetic genes or the SSU rRNA. It is equally clear that a full understanding of the divergent phylogenies obtained from different genes of photosynthetic organisms is still some time away.

9. DISPROPORTIONATION

9.1. General considerations

The disproportionation of sulfur compounds of intermediate oxidation state is now recognized as an environmentally significant process (Thamdrup *et al.*, 1993; Canfield and Thamdrup, 1994; Habicht and Canfield, 2001). The stoichiometries of the known microbially mediated disproportionation reactions are presented in Equations 9.21–9.24.

$$4H_2O + 4S^0 \rightarrow 3H_2S + SO_4^{2-} + 2H^+ \tag{9.21}$$

$$4SO_3^{2-} + 2H^+ \rightarrow H_2S + 3SO_4^{2-} \tag{9.22}$$

$$S_2O_3^{2-} + H_2O \rightarrow H_2S + SO_4^{2-} \tag{9.23}$$

$$4S^0 + 3H_2O \rightarrow 2H_2S + S_2O_3^{2-} + 2H^+ \tag{9.24}$$

Sulfur intermediate compounds are mainly formed during incomplete sulfide oxidation by either biologically mediated pathways or by inorganic reactions between sulfide and various oxidants (see Sections 5, 6, and 7). The potential for sulfide intermediate compound formation in nature is enormous.

In anoxic marine sediments supporting sulfate reduction, typically 90% or more of the sulfide formed during sulfate reduction is reoxidized again (Jørgensen, 1982b; Canfield, 1993). As discussed previously, pathways of reoxidation are numerous and quantitatively not well understood. Nevertheless, there is the capacity for significant formation of sulfur intermediate compounds, and in radiotracer experiments Jørgensen (1990a,b) found thiosulfate to make up over 50% of the immediate sulfide oxidation product in both lake and marine sediments. The sulfide oxidation products include elemental sulfur, sulfite, and thiosulfate, and these may be all found in small concentrations in anoxic and microoxic environments (see Table 9.10) (Thamdrup *et al.*, 1994a).

Most of the sulfide formed in sulfidic lakes and marine basins is also reoxidized again at the oxic–anoxic interface with the potential for intermediate sulfur compound formation. Indeed, elemental sulfur (Table 9.10), sulfite, and thiosulfate may all be found in small concentrations (micromolar to sub-micromolar range) in the water columns of anoxic marine basins (Luther *et al.*, 1991; Zopfi *et al.*, 2001). Small concentrations, however, should not be confused with a lack of significance but rather rapid biological utilization (see Jørgensen, 1990a,b).

9.2. Organisms disproportionating sulfur compounds

The recognition of sulfur compound disproportionation as an environmentally significant process is relatively new, and the numbers of organisms identified as conducting disproportionation reactions are still quite limited. Nevertheless, we recognize four different classes of organisms involved in the disproportionation of sulfur intermediate compounds (Table 9.16). The first class of organisms (referred to as type 1), represented by *Chlorobium thiosulfatophilum* and *Chlorobium limicola,* disproportionates elemental sulfur

Table 9.16 Types of organisms disproportionating sulfur compounds

Type	Examples	Comment
1	*Chlorobium limicola*	Anoxygenic phototroph, disproportionates elemental sulfur to H_2S and thiosulfate when phototrophically oxidizing sulfur in the absence of CO_2
2	*Desulfocapsa sulfoexigens*	Obligate anaerobe with disproportionation of sulfur compounds as its sole metabolism
3	*Desulfovibrio sulfodismutans* *Desulfocapsa thiozymogenes*	Obligate anaerobe growing by sulfate reduction and sulfur compound disproportionation
4	*Desulfovibrio desulfuricans* *Desulfobulbus propionicus*	Sulfate reducers, some oxygen tolerant, little if any growth during sulfur compound disproportionation. Some disproportionate elemental sulfur as the final step in sulfide oxidation with oxygen

during phototrophic sulfur oxidation in the absence of CO_2 (Paschinger *et al.*, 1974). A continuous removal of the product H_2S gas is required for the reaction to proceed. The disproportionation products are H_2S and thiosulfate (Equation 9.24). The initial biological product of the disproportionation may be sulfite, which reacts inorganically with elemental sulfur to form thiosulfate (Trüper and Fischer, 1982). This type of disproportionation has been described only for the two organisms named above, and its environmental significance is unknown.

The second class of organisms (type 2) involved in sulfur compound disproportionation is comprised of obligate anaerobes that perform disproportionation as their sole means of metabolism (Table 9.16). This class of organisms is represented by *Desulfocapsa sulfoexigens* (Finster *et al.*, 1998), which was isolated from nature as an elemental sulfur disproportionator; it grows obligately by the disproportionation of elemental sulfur, sulfite, and thiosulfate. *Desulfocapsa sulfoexigens* cannot reduce sulfate, and so far it is the only described obligate sulfur compound disproportionator. However, it is quite easy to enrich for sulfur-disproportionating organisms from nature, and a variety of different morphotypes have been found from a wide range of natural environments (Canfield *et al.*, 1998). It seems likely, therefore, that once brought into pure culture, a number of these different enrichment cultures will also yield organisms of type 2. The growth of *Desulfocapsa sulfoexigens*, as well as enrichment cultures of sulfur disproportionators isolated from nature, requires low sulfide concentrations. This is because sulfur disproportionation is thermodynamically favorable only with sulfide levels less than about 1 mM (Thamdrup *et al.*, 1993) (see Chapter 3). In nature, sulfide is typically removed from surficial sedimentary environments by reaction with iron oxides, (e.g., Canfield, 1989a), providing the chemical conditions for active sulfur disproportionation.

The third class of organisms (type 3) disproportionating sulfur compounds is comprised of strict anaerobes that grow both by sulfate reduction and by sulfur compound disproportionation (Table 9.16). This class of disproportionators is represented by the organisms *Desulfovibrio sulfodismutans* (Bak and Pfennig, 1987) and *Desulfocapsa thiozymogenes* (Janssen *et al.*, 1996). *Desulfovibrio sulfodismutans* is a typical *Desulfovibrio*, oxidizing lactate, ethanol, propanol, and butanol during sulfate reduction, but it also grows while disproportionating thiosulfate and sulfite, with the necessity of acetate as a carbon source. *Desulfocapsa thiozymogenes* grows as a sulfate reducer on short-chain alcohols, but grows best chemolithoautotrophically during the disproportionation of sulfite, thiosulfate, and elemental sulfur. Other *Desulfovibrio*-type sulfate reducers also fall into this class (Krämer and Cypionka, 1989).

The fourth class of sulfur compound disproportionating organisms (type 4) is represented by a number of sulfate reducers, some oxygen tolerant, which can disproportionate sulfur compounds but do not grow, or grow only poorly, by this process. Organisms in this class include *Desulfobulbus propionicus*, which can disproportionate elemental sulfur and thiosulfate (Lovley and Phillips, 1994; Canfield *et al.*, 1998), and a variety of *Desulfovibrio*-like sulfate reducers disproportionating sulfite and/or thiosulfate. Some of these organisms, such as *Desulfovibrio desulfuricans* strain CSN (Fuseler *et al.*, 1996), conduct a very interesting metabolism whereby they oxidize sulfide to elemental sulfur with low oxygen concentrations and subsequently disproportionate the sulfur to sulfide and sulfate.

There are obviously a wide variety of different physiological adaptations to sulfur compound disproportionation, allowing disproportionation to potentially occur in a broad range of different environments. Thus, in stable anoxic settings with ample production of sulfur-intermediate compounds, disproportionators of type 2 or type 3 could thrive, and if sulfide levels are below 1 mM, elemental sulfur disproportionators could be of particular significance. These conditions are met, as mentioned above, in surficial sedimentary environments. At the interface between oxygen and sulfide in anoxic water bodies, one could expect abundant sulfur intermediate compound production but a general lack of organic carbon. In this situation disproportionators of type 2 might flourish. By MPN counting we have found high numbers (10^4 cells ml^{-1}) of sulfur disproportionators in the anoxic water column of Mariager Fjord, Denmark (Sørensen and Canfield, 2004), although the physiology of these organisms has not yet been explored. Organisms of type 4 might use the disproportionation of sulfur intermediate compounds as a survival strategy when exposed to low oxygen concetrations near the oxygen–sulfide interface or when organic substrate is scarce but sulfur intermediate compounds can be found. Because they do not grow through sulfur compound disproportionation, sulfate reduction must be their primary metabolism. Finally, sulfur disproportionators of type 1 could be active in microbial mats during the day when CO_2 becomes limiting (Chapter 13), although there is no information yet to confirm this.

9.3. Biochemistry of sulfur compound disproportionation

The biochemistry of elemental sulfur disproportionation has been studied with *Desulfocapsa sulfoexigens*, and that of sulfite disproportionation with *Desulfocapsa sulfoexigens, Desulfovibrio sulfodismutans*, and *Desulfovibrio desulfuricans* (Krämer and Cypionka, 1989; Frederiksen and Finster, 2003). All of the organism studied thus far seem to share a common biochemical pathway during thiosulfate disproportionation. Thus, thiosulfate is split into sulfide and sulfite with the enzyme thiosulfate reductase (Figure 9.19). The sulfite is further transformed into sulfate following the same

Thiousulfate disproportionation

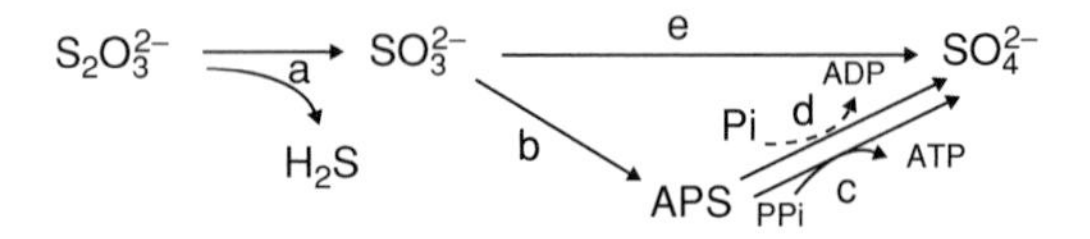

Elemental S disproportionation

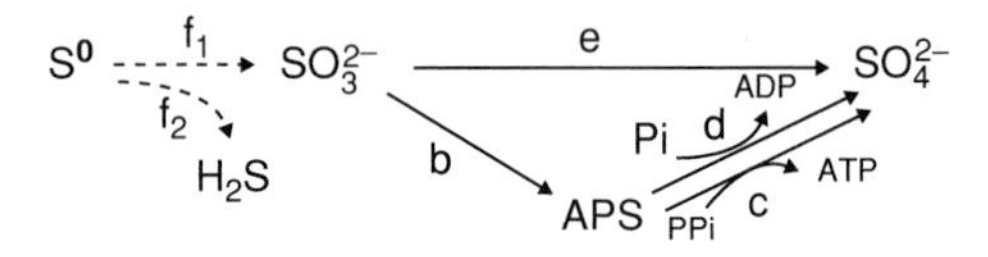

------- activity not detected in cultures
a thiosulfate reductase
b APS reductase
c ATP sulfurylase
d APAT
e sulfite oxidoreductase
f_1, f_2 unknown reactions

Figure 9.19 Pathways of sulfur metabolism during the disproportionation of thiosulfate and elemental sulfur. Note that after the formation of sulfite, both thiosulfate and elemental sulfur disproportionation follow a similar path, which is also similar to sulfide oxidation by phototrophs (see Figure 9.18). Abstracted from Frederiksen and Finster (2003).

reaction scheme proposed for anoxygenic photosynthetic sulfide oxidation (see above) (Figure 9.18). For elemental sulfur disproportionation, the sulfur is transformed into sulfite and sulfide through a yet-unknown reaction pathway (Figure 9.19). There are apparently two pathways for the further transformation of sulfite (Figure 9.19). In one, apparently minor, pathway, sulfite may be reduced to sulfide with the enzyme sulfite reductase (not shown). In the other pathway, which is presumably the most significant, sulfite follows the same path as in anoxygenic photosynthesis and sulfite disproportionation (Figures 9.18 and 9.19).

10. STABLE ISOTOPE GEOCHEMISTRY

The four stable isotopes of sulfur include the major isotopes ^{32}S (95.04%) and ^{34}S (4.20%) and the minor isotopes ^{33}S (0.749%) and ^{36}S (0.0156%). Organisms metabolizing sulfur compounds frequently fractionate these isotopes during their metabolism, and in principle, the isotopic composition of sulfur compounds in nature can provide an indication of the microbial pathways by which the compounds were formed. This is true both in modern environments and in the geologic past, where sulfur is preserved as pyrite in sedimentary rocks and as sulfate minerals in evaporite deposits (Canfield and Raiswell, 1999; Canfield, 2001).

Isotopic compositions in the sulfur system are always expressed relative to the major isotope of sulfur, ^{34}S with the following δ notation:

$$\delta^{3i}S_{sam} = 1000\left[\left(\frac{^{3i}S/^{32}S_{sam}}{^{3i}S/^{32}S_{std}}\right) - 1\right] \tag{9.25}$$

where i is either 2, 3 or 6, while "sam" refers to sample and "std" refers to standard, which is normally taken as the Cañon Diablo troilite standard. In most cases the isotope compositions of ^{34}S are reported, although there is increasing interest in the isotopic compositions of the minor isotopes, ^{33}S and ^{36}S. These minor isotopes of sulfur should follow the fractionations imparted on the more abundant ^{34}S by predictable amounts depending on the mass of the isotope. This is known as mass-dependent fractionation, and for mass-dependent processes ^{33}S fractionates about half as much (0.515 times) as ^{34}S, while ^{36}S fractionates about twice as much (1.91 times) as ^{34}S, when compared to fractionations relative to ^{32}S (Hulston and Thode, 1965). Mass-dependent fractionations follow the normal biogeochemical cycling of sulfur compounds (Farquhar *et al.*, 2000), although very interesting mass-independent fractionations should accompany the atmospheric processing of sulfur with near-UV radiation. This might have been particularly important early in Earth's history when oxygen levels were low and the UV-shielding ozone layer had not yet developed (Farquhar *et al.*, 2000).

At equilibrium, isotopes will partition themselves into different chemical bonds in variable ratios, which is the basis for isotope fractionation. However, for isotopes of sulfur, the expected thermodynamic distribution of isotopes between different compounds is rarely attained; instead, the extent of fractionation is largely controlled by kinetic factors. Fractionations generally occur as sulfur bonds are broken or formed, and these are governed largely by enzymatic processes. Therefore, the isotope differences imparted during the metabolic processing of sulfur compounds is influenced, in part, by the ability of particular enzymatic processes to fractionate. Fractionation will also depend on the extent to which reactions are reversible and the extent to which products and reactants exchange across the cell membrane.

An example from sulfate reduction illustrates these points. The important steps in the sulfate reduction process (Figure 9.3) are reproduced below in Equation 9.26:

$$SO_4^{2-}(\text{out}) \underset{}{\overset{\text{Step 1}}{\rightleftharpoons}} SO_4^{2-}(\text{in}) \overset{\text{ATP}}{\underset{}{\overset{\text{Step 2}}{\rightleftharpoons}}} \text{APS} \overset{e^-}{\underset{}{\overset{\text{Step 3}}{\rightleftharpoons}}} SO_2^{3-} \overset{e^-}{\underset{}{\overset{\text{Step 4}}{\longrightarrow}}} H_2S \tag{9.26}$$

Isotope fractionation occurs where S-O bonds are broken, during steps 3 and 4, and although poorly studied, each of the steps is believed to produce fractionations ($\varepsilon = 10^3\ (\alpha_{\text{react-prod}}-1) \approx \delta^{18}O_{\text{react}}\text{-}\delta^{18}O_{\text{prod}}$, where $\alpha_{\text{react-prod}}$ is the fractionation factor) of around 20 to 25‰. The full expression of these fractionations will only occur when steps 2 and 3 are in exchange equilibrium, and importantly, when sulfate is rapidly exchanged both in and out of the cell (step 1). If sulfate is not exchanged back out of the cell, then no fractionation will be observed despite significant fractionations by intracellular processes. Simply put, the isotopic composition of what goes into the cell as sulfate must come out again as sulfide. If steps 1 and 2 are in exchange equilibrium but step 3 is rapid with little back exchange, then the fractionation during step 2 should be expressed, but not the fractionation during step 3. Thus, all of the APS formed becomes reduced to sulfide, and despite fractionations during step 3, the isotopic composition of sulfide should be the same as the APS.

The degree of fractionation during sulfate reduction, therefore, depends on the balance between the exchange of sulfate in and out of the cell and the extent to which internal biochemical pathways are in exchange equilibrium. It is generally believed that when sulfate reducers metabolize at a high specific rate (rate cell^{-1}) fractionations become suppressed because the demand for sulfate is high within the cell, and this high demand limits the reverse reactions of steps 1–3. High specific rates of sulfate reduction might be expected with abundant electron donor (either hydrogen or organic compounds), particularly when the organism is growing near its temperature optimum. For *Desulfovibrio desulfuricans,* decreasing fractionations are clearly observed as specific rates of sulfate reduction increase (Figure 9.20), although the spread among the data at a given specific rate is generally quite large (Canfield, 2001).

High sulfate demand is also expected when sulfate concentrations become sufficiently low that the intracellular processing of sulfate is limited by sulfate transport into the cell. In this case, sulfate is not readily exchanged back out of the cell. We have recently found significant suppression of isotope fractionation when sulfate concentrations fall below around 200 μM, and this is true for both freshwater and marine natural populations of sulfate reducers (Habicht *et al.*, 2002). Fractionation apparently becomes limiting in the same range of sulfate concentrations also limiting sulfate reduction rate. This is not a surprise, as suppressed rates of sulfate reduction at low sulfate concentrations is a clear expression of sulfate limitation by the organism.

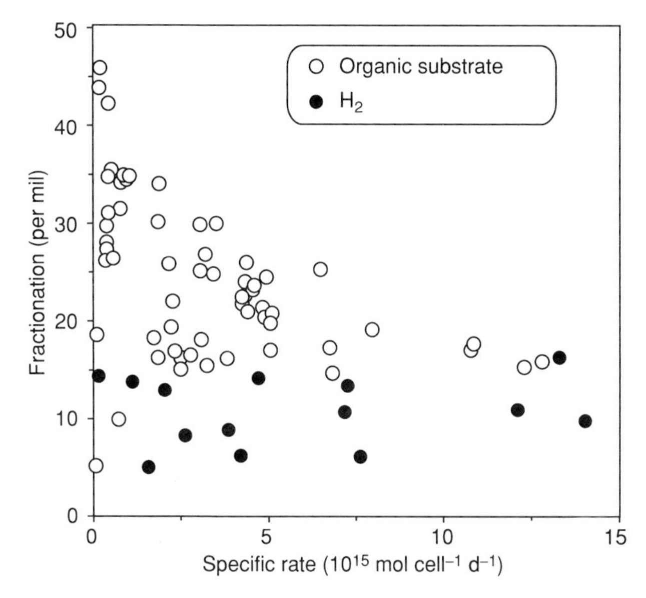

Figure 9.20 Isotope fractionation is shown as a function of specific rates of sulfate reduction for *Desulfovibrio desulfuricans* utilizing both organics substrate and H_2 as electron donors. Modified from Canfield (2001).

For pure cultures of sulfate reducers, a large range of fractionations has been observed during sulfate reduction with non-limiting sulfate concentrations (Figure 9.21), from about 3 to 46‰ (Canfield, 2001). Some of this variability expresses differences in specific rates of sulfate reduction, as discussed above, but some also probably reflects species and strain differences in the ability of organisms to fractionate. For example, Detmers *et al.* (2001) explored the ability of 32 different strains of sulfate reducers to fractionate under ideal growth conditions for each of the organisms. The organisms spanned the complete phylogenetic spectrum of known sulfate reducers. A wide range of fractionations, from 2 to 42‰, was found. Such a spread in fractionations possibly reflects differences in cell membrane properties and the ability of the membrane to promote sulfate exchange out of the cell. Metabolic differences also have some bearing on fractionation. For example, when utilizing H_2, sulfate reducers generally have lower fractionations than when utilizing organic substrate (Figure 9.20). It is possible that when utilizing hydrogen the reduction steps (steps 3 and 4) are rapid compared to the reverse reactions due to enhanced electron availability through the hydrogenase enzyme.

Fractionations by natural populations of sulfate reducers have also recently been explored. Natural populations can be viewed as intact and

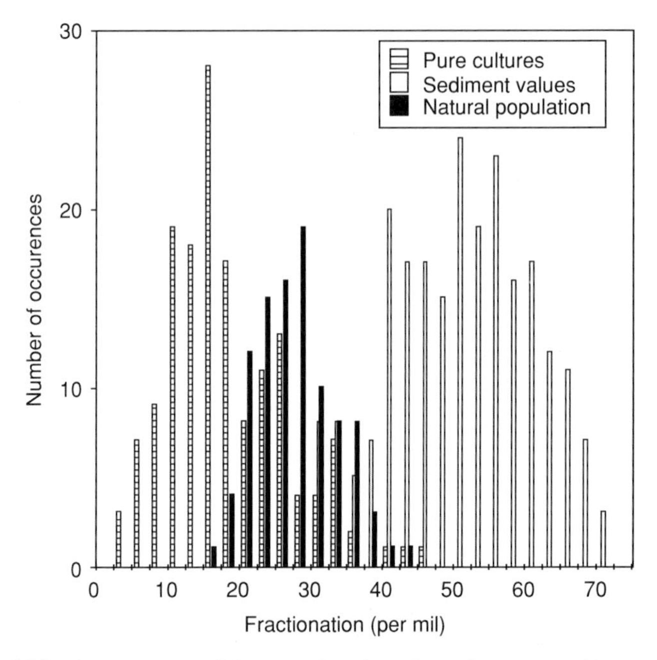

Figure 9.21 A summary of isotope fractionations for pure cultures of sulfate reducers, natural populations of sulfate reducers and the fractionations associated with pyrite formation in sediments. Note that the natural populations generally show greater fractionations than the pure cultures, with similar maximum fractionations. The isotopic composition of sedimentary pyrites shows greater fractionations than those produced by either pure cultures or natural populations of sulfate reducers.

diverse populations metabolizing indigenous organic substrate. The upper range of fractionations, around 42‰, is comparable to the pure cultures, but so far the minimum fractionations are much greater at around 16‰. The generally higher fractionations (on average) expressed by the natural populations could be due to generally lower specific rates of sulfate reduction in nature, where substrate limitation and lower specific rates of sulfate reduction are likely to be more important. It can also be that the species represented in the natural populations tend to fractionate greater than those populations studied so far in pure culture.

In principle, fractionation can also accompany oxidative sulfur metabolisms, although generally these fractionations are small (Table 9.17). Significant fractionation can sometimes be found in minor reaction products, such as when polythionates ($S_xO_6^{2-}$) are formed during sulfide oxidation, and where sulfate is the major reaction product (Table 9.17). But even in these circumstances, there is little fractionation into the major reaction product.

Table 9.17 Fractionations during the oxidation of sulfur compounds

Process	Generalized reaction	Fractionation (‰)[a]
Phototrophic		
H_2S	$H_2S \rightarrow S^0$	−2 to 0
S^0	$S^0 \rightarrow SO_4^{2-}$	0
Nonphototrophic		
H_2S	$H_2S \rightarrow S^0$	−1 to 1
	$H_2S \rightarrow S_xO_6^{2-}$[b]	0 to −19
	$H_2S \rightarrow SO_4^{2-}$[c]	10 to 18
S^0	$S^0 \rightarrow SO_4^{2-}$	0
$S_2O_3^{2-}$	$S_2O_3^{2-} \rightarrow SO_4^{2-}$	−0.4
SO_3^{2-}	$SO_3^{2-} \rightarrow SO_4^{2-}$	0
Nonbiological		
H_2S	$H_2S \rightarrow S^0, S_2O_3^2, SO_4^{2-}$	4 to 5
SO_3^{2-}	$SO_3^{2-} \rightarrow SO_4^{2-}$	0.4

Adapted from Canfield (2001).

[a]Fractionations are presented as both $\varepsilon_{A\text{-}B}$ and $\Delta_{A\text{-}B}$, where A is the reactant and B is the product.
[b]$S_xO_6^{2-}$ is a minor reaction product.
[c]SO_4^{2-} is a minor reaction product.

The reasons for small fractionations during the oxidation of reduced sulfur compounds are not clear but could be related to small kinetic fractionations during the oxidation steps, although these fractionations have not yet been studied in detail.

Significant fractionation accompanies the disproportionation of sulfur compounds. Thus, during the disproportionation of elemental sulfur, sulfide is generally depleted in ^{34}S by 5 to 7‰, while the sulfate is enriched in ^{34}S by 17 to 21‰. These extents of fractionation are remarkably consistent through a wide range of sulfur disproportionating organisms (type 2 and 3) (Table 9.18), including marine and freshwater strains isolated from diverse environments around the world (Canfield *et al.*, 1998). An exception is the sulfate reducer *Desulfobulbus propionicus* (type 4) (Table 9.18), which grows only poorly, if at all, during sulfur disproportionation. For this organism, much higher fractionations are found, indicating a possible alternative biochemical pathway during sulfur disproportionation.

The disproportionation of sulfite produces reasonably large depletions of ^{34}S into sulfide of 20 to 37‰, with corresponding enrichments of 7 to 12‰ into sulfate. The magnitude of fractionation is apparently related to specific rates of sulfite disproportionation (Habicht *et al.*, 1998). The disproportionation of thiosulfate produces ^{34}S-depleted sulfide and ^{34}S-enriched sulfate. However, the factors leading to the final isotopic compositions of sulfide and sulfate are complex. Thiosulfate ($S_2O_3^{2-}$) is composed of an inner sulfonate

Table 9.18 Isotope fractionation (‰) during elemental sulfur disproportionation by pure and enrichment cultures

Culture	Freshwater/ marine	$\Delta_{(S^0-AVS)}$ measured[a]	$\Delta_{(S^0-SO_4)}$ measured	$\Delta_{(S^0-AVS)}$ cellular[b]	$\Delta_{(S^0-SO_4)}$ cellular[b]
Desulfocapsa thiozymogenes	Fresh	5.9	−17.3	5.8	−17.4
Desulfobulbus propionicus	Fresh	15.5	−30.9	11.3	−33.9
Desulfocapsa sulfoexigens	Marine	5.8	−16	5.5	−16.4
Dangast 1	Marine	6.6	−18.2	6.2	−18.7
Dangast 2	Marine	6.5	−16.7	5.8	−17.4
Weddewarden	Marine	6.2	−17.9	6.9	−20.6
Gulfo Dulce S1	Marine	7.9	−19.7	6.9	−20.6
Gulfo Dulce S160	Marine	8.0	−17.1	6.2	−18.5
Teich 1st	Fresh	6.4	−16.4	5.7	−17.1
Teich 2nd	Fresh	6.2	−22	6.2	−18.6
Kugraben	Fresh	7.0	−19.9	6.7	−20.2

Adapted from Canfield (2001).

[a]AVS, acid volatile sulfide.

[b]A small correction is made to the observed isotope differences to account for elemental sulfur formed from reaction between sulfide and iron oxides, and subsequently disproportionated. These fractionations should, therefore, be the cellular level fractionations. See Canfield *et al.* (1998) for details.

sulfur (-SO_3^-) and an outer sulfane sulfur (-S^-). Through disproportionation, the sulfane sulfur should form into sulfide and the sulfonate sulfur should form into sulfate. Therefore, the isotopic composition of these individual sulfurs in thiosulfate should, in part, influence the final isotopic compositions of sulfide and sulfate. However, if sulfide accumulates during disproportionation, isotope exchange is apparently promoted between the sulfane sulfur and sulfide as well as between sulfonate sulfur and sulfate. This isotope exchange produces greater isotope differences between sulfide and sulfate than was apparent in the two different sulfur atoms of thiosulfate, and the extent of this isotope difference varies greatly (Habicht *et al.*, 1998) (Table 9.19). When sulfide does not accumulate, no isotope exchange occurs, and kinetic isotope fractionations seem to operate (Cypionka *et al.*, 1998).

Isotope fractionations during the disproportionation of sulfur compounds of intermediate oxidation state have a large influence on the final isotopic composition of mineral sulfides preserved in sedimentary environments. Thus, the isotopic fractionations associated with sulfate reduction alone do not produce sufficiently ^{34}S-depleted sulfide to explain the isotopic

Table 9.19 Isotope fractionation (‰) during the disproportionation of thiosulfate by pure and enrichment cultures

Culture	$\Delta S_{(sulfane\text{-}H_2S)}$ [a]	$\Delta S_{(sulfonate\text{-}SO_4)}$ [b]
Desulfovibrio sulfodismutans		
I	0.7	−1.1
II	1.6	−0.6
III	2.1	−2.0
Desulfocapsa thiozymogenes		
I	0.7	−0.9
II	3.2	−2.5
Desulfovibrio desulfuricans	20	4
Løgten Lagoon	11.4	−11.3
Solar Lake	11.8	−12.7
Weddewarden	4.7	−5.0

Adapted from Canfield (2001).

[a]Calculated as the isotope difference between the initial sulfane sulfur and the H_2S produced during disproportionation.

[b]Calculated as the isotope difference between the initial sulfonate sulfur and the SO_4^{2-} produced during disproportionation.

composition of pyrites in sediments (Canfield and Thamdrup, 1994; Habicht and Canfield, 2001). Additional fractionations imparted during the disproportionation of sulfur compounds can explain this difference. If one imagines, for example, sulfide oxidation to elemental sulfur and subsequent disproportionation of elemental sulfur to sulfate and sulfide, the sulfide produced will be more ^{34}S-depleted than the original sulfide. If the sulfide is oxidized again to elemental sulfur and further disproportionated, an even greater ^{34}S depletion can be produced. Several such cycles can produce significant ^{34}S depletions in sedimentary and water column sulfides (see Sørensen and Canfield, 2004).

Chapter 10

The Methane Cycle

1. INTRODUCTION

Methane (CH_4) is very much the biological compound of the moment. It is a greenhouse gas of possible significance in the temperature regulation of the early Earth (Pavlov *et al.*, 2000), and in the modern Earth as concentrations rise as a result of anthropogenic activity. The occasional release of methane from a vast sedimentary reservoir of methane clathrate hydrates has also been implicated in several episodes of rapid climate change through Earth's history (e.g., Dickens *et al.*, 1997; Kennett *et al.*, 2003).

Methane is produced either through the thermochemical breakdown of organic matter or through biological processes, and here we focus on the latter. Indeed, methanogenesis can contribute greatly to the microbial oxidation of organic matter in anoxic settings, particularly if sulfate concentrations are low, as the two processes of sulfate reduction and methanogenesis actively compete for electron donors in the environment (see Chapter 3). The processes of methane oxidation are also of great interest, and there has been much progress lately in our understanding of the pathways of anaerobic

ADVANCES IN MARINE BIOLOGY VOL 48
0-12-026147-2

© 2005 Elsevier Inc.
All rights reserved

methane oxidation (Hinrichs *et al.*, 1999; Boetius *et al.*, 2000; Orphan *et al.*, 2001a). This problem of the anaerobic oxidation of methane (AOM) has stood as a durable, and long-standing, microbiological puzzle. The geochemical evidence for AOM has pervaded the literature for decades and is quite compelling (e.g., Martens and Berner, 1977; Reeburgh 1980; Alperin and Reeburgh, 1984), but progress on the microbiology of the problem is only recent.

In this Chapter, we explore the biogeochemistry and microbial ecology of the methane cycle. We begin by considering some of the dynamics of methane as a greenhouse gas and the role of clathrate hydrates in climate change. We explore the processes of methane formation, paying particular attention to the various microbial pathways involved. We look at the pathways of methane oxidation and consider anaerobic methane oxidation in some detail. Finally, we look at the isotope fractionations accompanying methane formation and oxidation.

2. METHANE AND CLIMATE

Methane is a strong greenhouse gas. Indeed, per unit mass, methane is 21 times more potent a greenhouse gas than CO_2, as determined by the Intergovernmental Panel on Climate Change (IPCC). Over the last 470,000 years, the later Pleistocene, air trapped in polar ice has revealed that atmospheric methane concentrations have varied between about 350 and 700 ppbv (parts per billion by volume, equivalent to nmol mol^{-1}), in pace with changes in atmospheric CO_2 (Petit *et al.*, 1999). During the last 300 years atmospheric methane concentrations have risen sharply from preanthropogenic values of around 700 ppbv to present values of around 1760 ppbv. Over the last 5 years methane concentrations have been relatively stable (Figure 10.1) (Steele *et al.*, 2002), with an apparent balance between rates of production and consumption. The main anthropogenic sources of methane, which probably exceed natural sources by a factor of two, are digestive fermentation by livestock, waste management, rice paddies, and methane release during fossil fuel mining and use (Prather *et al.*, 1995) (Table 10.1). The most important natural sources of methane are release from wetland soils and lake sediments and digestive fermentation by termites. The residence time of methane in the atmosphere is about 9 years, where the main sink is the tropospheric reaction with OH radicals (Table 10.1) producing H_2O and CO_2.

Even with increasing atmospheric methane concentrations, increasing CO_2 has been the most important driver of contemporary climate change (Schlesinger, 1997). There is mounting evidence, however, that during some periods in Earth's history, methane, and not CO_2, has driven climate change.

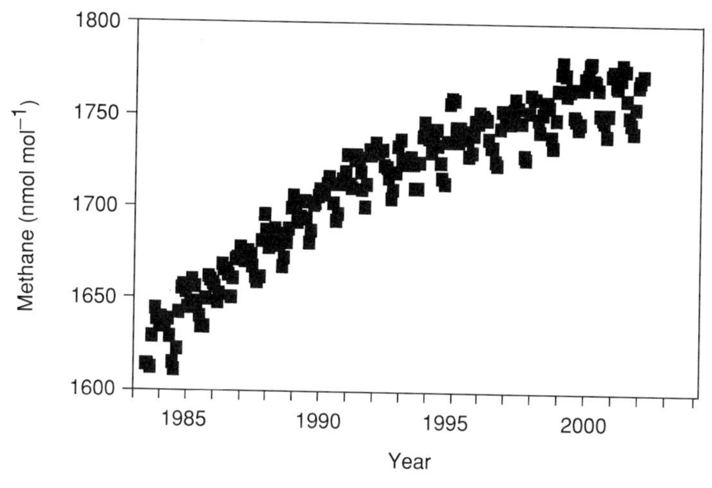

Figure 10.1 Monthly average atmospheric methane concentrations taken from flask samples at the Mona Loa, Hawaii, observatory. Data from Steele *et al.* (2002).

As mentioned previously, methane might have kept the early Earth from freezing despite reduced solar luminosity at the time (Pavlov *et al.*, 2000). Likewise, methane may have been an important greenhouse gas just prior to global glaciation around 700 million years ago, the so-called "snowball Earth" (Schrag *et al.*, 2002). Of more interest here is the occasional rapid release of methane during the dissolution of massive reservoirs of sediment-hosted clathrate hydrates and its influence on climate.

Under appropriate conditions of temperature and pressure, methane can form into an ice-like solid known as clathrate hydrate. In clathrate hydrates, methane (and other gases such as ethane and propane) is found in the hollow cavities of well-ordered lattices of water molecules (Buffet, 2000). The methane not only is trapped within the cavities, but also acts to stabilize the structure. The phase diagram for a methane–water mixture is shown in Figure 10.2 as a function of temperature and pressure (converted to water depth assuming a hydrostatic gradient of 10^4 Pa m^{-1}). The upper boundary for the stability field for methane clathrate hydrate is labeled $T_3(P)$. With increasing temperature, methane clathrate hydrates are stable with progressively increasing pressure. It is interesting that for water depths deeper than 200 m, clathrate hydrates are stable above the melting temperature of water ice.

It is obvious from Figure 10.2 that methane clathrate hydrates are stable at relatively shallow depths at low temperatures, which explains their occurrence in organic-rich permafrost areas such as northern Siberia and the North Slope of Alaska (Kvenvolden, 1988a, 1993). They also occur in relatively fast-depositing continental margin sediments when there is the

Table 10.1 Estimated atmospheric sources and sinks of methane (10^{12} gCH_4 y^{-1})

Sources	Range	Likely
Natural		
Wetlands		
Tropics	30 to 80	65
Northern latitudes	20 to 60	40
Other	5 to 15	10
Termites	10 to 50	20
Ocean	5 to 50	10
Freshwater	1 to 25	5
Total		160
Anthropogenic		
Fossil fuel related		
Coal mines	15 to 45	30
Natural gas	25 to 50	40
Petroleum industry	5 to 30	15
Coal combustion	5 to 30	15
Waste management		
Landfills	20 to 70	
Animal waste	20 to 30	
Sewage treatment	15 to 80	25
Enteric fermentation	65 to 100	85
Biomass burning	20 to 80	40
Rice paddies	20 to 100	60
Total		375
Total Sources		535
Sinks		
Reaction with OH	330 to 560	445
Stratosphere removal	24 to 55	40
Soil removal	15 to 45	30
Total sinks		515

Data from Prather *et al.* (1995).

appropriate combination of low enough temperature and active methane formation by methanogenesis. [Indeed, the source of methane for clathrate hydrates seems to be mostly biogenic (Buffett, 2000).] This confluence of conditions is apparently rather common, as methane clathrate hydrates are widely distributed in continental margin sediments (Kvenvolden, 1993). Temperature increases with depth in marine sediments and permafrost areas, defining a base of methane clathrate hydrate stability below which hydrate cannot form (Figure 10.3). Overall, methane clathrate hydrates represent an enormous

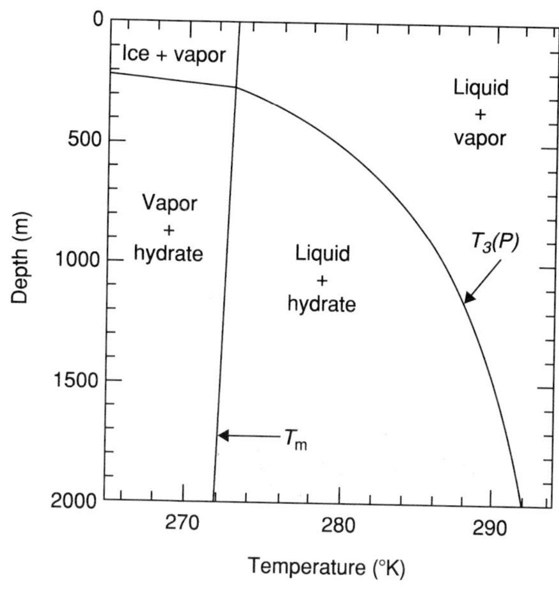

Figure 10.2 Phase diagram for water–methane mixtures as a function of temperature and pressure. T_m is the melting temperature of water ice and $T_3(P)$ defines the upper temperature boundary where methane clathrate hydrates are stable. The diagram is drawn assuming a hydrostatic gradient of 10^4 Pa m^{-1}. Redrawn after Buffet (2000).

sink for methane, trapping an estimated 2×10^{16} kg of carbon (Kvenvolden, 1988b). To put this number in perspective, there are currently 3.8×10^{12} kg of methane carbon in the atmosphere (Mackenzie, 1995), so only small amounts of methane clathrate hydrate dissolution can greatly influence the radiative balance and thus the temperature of the Earth surface.

Methane clathrate hydrates can dissolve when they no longer reside within their stability field. This can happen in response to sediment warming or to a drop in pressure, as what might occur with falling sea level. Kennett *et al.* (2003) have argued that over the last 800 kyr (thousands of years) temperature, and not pressure, changes have resulted in substantial methane clathrate hydrate dissolution from continental margin sediments during glacial terminations. The temperature of waters impinging on upper continental slope sediments varies as changes occur in the source and magnitude of intermediate water formation at the end of glacial periods. Indeed, the release of methane, which tracks CO_2 accumulation (e.g., Petit *et al.*, 1999), could amplify climate change, accelerating the warming following glacial terminations. This is known as the "methane gun hypothesis"; the "methane gun" is loaded by clathrate hydrate formation during cold periods, to be subsequently "fired" as intermediate waters warm (Kennett *et al.*, 2003). Massive methane clathrate hydrate dissolution has also been

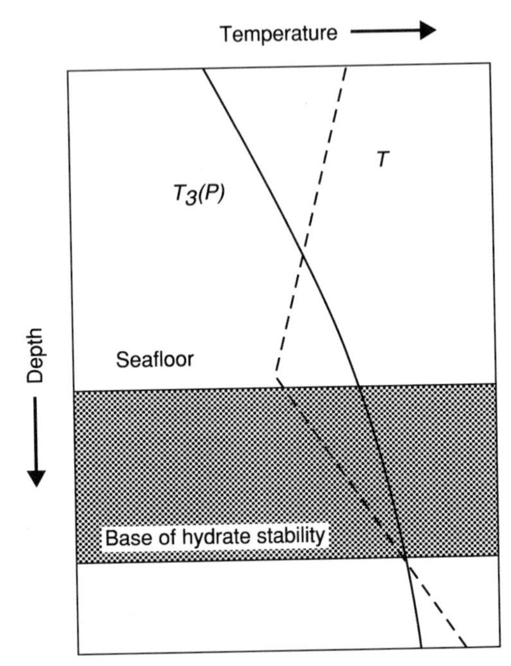

Figure 10.3 Illustration of the zone of clathrate hydrate stability in hypothetical marine sediments. In this example, methane clathrate hydrates are within their stability field (defined by $T_3(P)$; see Figure 10.2) at the sea floor (assuming an ample supply of methane), but due to increasing temperature with sediment depth (the geothermal gradient) the upper temperature (T) boundary of clathrate hydrate stability is exceeded. Redrawn after Buffet (2000).

implicated in other episodes of rapid climate change, including the Paleocene thermal maximum (Dickens *et al.*, 1997) and the end Permian, in which a link between global climate changes and mass extinction has been made (e.g., Krull and Retallack, 2000). Furthermore, Halverson *et al.* (2002) have suggested that massive methane hydrate release in the late Neoproterozoic created an unstable atmospheric greenhouse that finally collapsed, resulting in the so-called Neoproterozoic "snowball Earth" (Hoffman *et al.*, 1998).

3. METHANOGENESIS

3.1. Substrates

Some specialized groups of prokaryotes, all members of the *Archaea* and all strict anaerobes, grow through the production of methane as a major part of their metabolism. These groups are known as the methanogens. While

Table 10.2 Substrates for methanogenesis

C_1 compounds reduced with H_2 or alcohols	Disproportionated substrates
CO_2	$HCOO^-$ (formate)
CH_3OH (methanol)	CO
	CH_3OH (methanol)
	$CH_3NH_3^+$ (methylamine)
	$(CH_3)_2NH_2^+$ (dimethylamine)
	$(CH_3)_3NH^+$ (trimethylamine)
	CH_3SH (methylmercapton)
	$(CH_3)_2S$ (dimethylsulfide)
	CH_3COO^- (acetate)

methane is a principal metabolic product, a small collection of different substrates is used in methane formation. The substrates used are logically divided into two groups (Whitman *et al.*, 1999; Madigan *et al.*, 2003). The first group of substrates consists of C_1 compounds (Table 10.2) reduced to methane with H_2, or alcohols (also pyruvate) with two carbon atoms or more. The most important of these C_1 compounds is CO_2. An example of CO_2 reduction with H_2 is given in Equation 10.1, while CO_2 reduction with ethanol is given in Equation 10.2.

$$CO_2 + 4H_2 \rightarrow CH_4 + 2H_2O \quad (10.1)$$

$$CO_2 + 2CH_3CH_2OH \rightarrow CH_4 + 2CH_3COO^-(\text{acetate}) + 2H^+ \quad (10.2)$$

The free energy yields associated with these types of metabolisms tend to be rather large, with ΔG° values ranging from -136 kJ mol^{-1} CH_4 for Equation 10.1 to -116 kJ mol^{-1} CH_4 for Equation 10.2 (Garcia *et al.*, 2000). Of course, the free energy change associated with methanogenesis in the environment will depend on the chemistry of the environment and may differ substantially from the values given under standard conditions (Chapter 3).

The second group of compounds used by methanogens is disproportionated, and these compounds include formate ($HCOO^-$; Equation 10.3), carbon monoxide (CO; Equation 10.4) and a variety of compounds in which carbon is in methyl form (Table 10.2), including methanol, a variety of methylamines, and methylated sulfides (Table 10.2). The disproportionation of dimethylamine is shown in Equation 10.5.

$$4HCOO^- + 4H^+ \rightarrow CH_4 + 3CO_2 + 2H_2O \quad (10.3)$$

$$4CO + 2H_2O \rightarrow CH_4 + 3CO_2 \quad (10.4)$$

$$2(CH_3)_2NH_2^+ + 2H_2O \rightarrow 3CH_4 + CO_2 + 2NH_4^+ \quad (10.5)$$

Methylamine compounds are considered non-competitive substrates for methanogens, as these are apparently not metabolized by other groups of anaerobes. The free energy changes, ΔG^{o}, associated with methylotrophic methanogenesis (methane formation from compounds with a methyl group) are intermediate and vary from around -73 kJ mol^{-1} CH_4 to -113 kJ mol^{-1} CH_4.

Acetate is a special, additional substrate, disproportionated by a small group of environmentally significant methanogens. In this process, acetate is cleaved into methane and CO_2:

$$CH_3COO^- + H^+ \rightarrow CH_4 + CO_2 \quad (10.6)$$

This process is alternately referred to as acetotrophic methanogenesis, acetoclastic methanogenesis, acetate disproportionation or acetate fermentation. As we see below, this is not a true fermentation but rather involves a novel and interesting pathway of energy conservation. The free energy value, ΔG^{o}, associated with acetotrophic methanogenesis is low at -31 kJ mol^{-1} CH_4 (Garcia *et al.*, 2000).

3.2. Taxonomy, phylogeny, and a note on evolution

Five different orders of methanogens have been identified; they are subdivided into 11 families and 25 genera (Boone *et al.*, 1993). An outline of order, family, and genus is provided in Table 10.3, along with a brief accounting of substrate utilization. To summarize, the order *Methanobacteriales* contains non-motile methanogens, both rods and cocci, using CO_2, formate and methanol as substrates. Both thermophilic and hyperthermophilic representatives of this order are known. Members of the order *Methanococcales* are all cocci or irregular cocci, and all use CO_2 as a substrate, with some also disproportionating formate. Some members of the genus *Methanococcus* reduce CO_2 with pyruvate. Members of the order *Methanomicrobiales* also all use CO_2, and most also disproportionate formate. In some cases, alcohols are used instead of H_2 as an electron donor. A variety of morphologies, including short rods, irregular cocci, spirilla, and plate-shaped cells, are represented. The *Methanosarcinales* are mostly irregular cocci, with some long rods and filaments, and they mostly use methyl substrates. However, some members of the genus *Methanosarcina* reduce CO_2 with H_2, and some are acetotrophic methanogens, while members of the genus *Methanosaeta* (filamentous) are all acetotrophic methanogens. Finally, the order *Methanopyrus* contains a single hyperthermophilic member, *Methanopyrus kandleri*, with a growth maximum of 110 °C. This organism reduces CO_2 with H_2.

Methanogenesis is unique to the *Archaea*, and methanogens are widely distributed among, and are restricted to, the kingdom *Euryarchaeota*.

Table 10.3 Methanogen taxonomy: Order, family, and genus, with a note on substrate utilization

Order	Family	Genus	Note on substrates
Methanobacteriales	*Methanobacteriaceae*	*Methanobacterium*	H_2/CO_2, formate
		Methanothermobacter	H_2/CO_2, formate
		Methanobrevibacter	H_2/CO_2, formate
		Methanosphaera	Methanol + H_2
	Methanothermaceae	*Mathanothermus*	H_2/CO_2, formate
Methanococcales	*Methanococcaceae*	*Methanococcus*	H_2/CO_2, formate, pyruvate + CO_2
		Methanothermococcus	H_2/CO_2, formate
	Methanocaldococcaceae	*Methanocaldococcus*	H_2/CO_2
		Methanoignis	H_2/CO_2
Methanomicrobiales	*Methanomicrobiaceae*	*Methanomicrobium*	H_2/CO_2, formate
		Methanolacinia	H_2/CO_2, formate, alcohols
		Methanogenium	H_2/CO_2, formate
	Methanomicrobiaceae	*Methonoplanus*	H_2/CO_2, formate
		Methanocullens	H_2/CO_2, formate, alcohols
		Methanofollis	H_2/CO_2, formate
	Methanocorpusculaceae	*Methanocorpusculum*	H_2/CO_2, formate, alcohols
	Methanospirillaceae	*Methanospirillium*	H_2/CO_2, formate
Methanosarcinales	*Methanosarcinaceae*	*Methanosarcina*	H_2/CO_2, acetate, methanol, methylamines
		Methanolobus	Methanol, methylamines
		Methanococcoides	Methanol, methylamines
		Methanohalophilus	Methanol, methylamines, methyl sulfides
		Methanohalobium	Methanol, methylamines
		Methanosalsus	Methanol, methyamines, dimethylsulfide
	Methanosaetaceae	*Methanosaeta*	Acetate
Methanopyrales	*Methanopyraceae*	*Methanopyrus*	H_2/CO_2, formate

Data from Boone (1993) and Madigan *et al.* (2003).

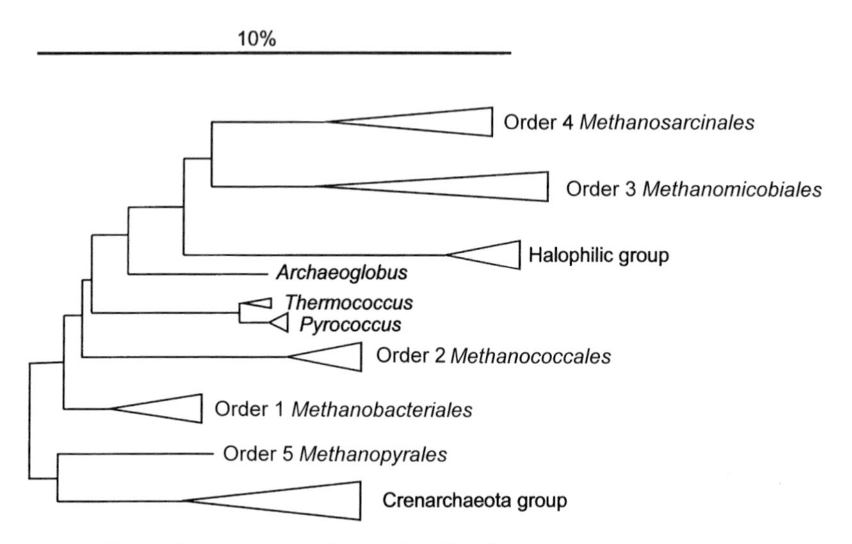

Figure 10.4 Methanogen phylogeny highlighting the placement of the five methanogen orders. See also Table 10.3. Abstracted from Garcia *et al.* (2000).

Individually, each of the five orders clusters together based on SSU rRNA comparisons (Figure 10.4). Among these orders, the hyperthermophilic *Methanopyrales* is the most deeply branching. Indeed, *Methanopyrales* branch deeply within the whole of the domain *Archaea* (see Chapter 1). This deep branching, combined with their hyperthermophilic nature and their metabolism (reduction of CO_2 with H_2), has led to speculations that methanogens may have evolved early in the history of life around hydrothermal springs (e.g., Stetter, 1996). In such an environment, both CO_2 and H_2 would have been abundant, so methanogens could have served as early primary producers on Earth. This idea has recently been challenged by House *et al.* (2003), who find little support for an early evolution of methanogens from whole genome sequence analysis. This is an interesting finding indeed. If substantiated, it demonstrates that in the history of life long time lapses may be needed for metabolic innovation to respond to metabolic potential.

3.3. Biochemistry

In this Section we outline some of the basic features of the biochemistry of methanogenesis as extracted from presentations in Whitman *et al.* (1999), Madigan *et al.* (2003), and Buckel (1999). The process of methanogenesis is complex and involves a variety of coenzymes, some of which are unique to the *Archaea*. One of the coenzymes, F_{420}, fluoresces blue-green when absorbing light at 420 nm, which is often used as a diagnostic probe for the presence of methanogens. One should be careful, however, as F_{420} has also been found among non-methanogenic *Archaea* and some groups of *Bacteria*, including the mycobacteria (Purwantini *et al.*, 1997).

We start by considering methanogenesis from CO_2 reduction with H_2. The process begins (Figure 10.5) with the activation of CO_2 with a methanofuran and the reduction of the carbon to the level of formyl. From here, the formyl carbon is transferred to methanopterin, where it is reduced first to a methylene carbon and next to methyl carbon, with an enzyme containing coenzyme F_{420}. From here, the methyl carbon is transferred to an enzyme containing coenzyme M (CoM), forming CoM-S-CH_3.

What follows is of particular significance. There are two major steps from the formation of CoM-S-CH_3 to the formation of CH_4, and to the subsequent conservation of energy as ATP (following Thauer, 1998). First is the oxidative part of the process, in which the CoM-S-CH_3 complex combines with HS-CoB. In this step the methyl group in CoM-S-CH_3 is reduced to methane (catalyzed by the enzyme methyl-coenzyme M reductase), while an oxidation accompanies the formation of a heterodisulfide bridge (this is why it is referred to as the oxidative part), bonding the two coenzymes together (Figure 10.5). No energy conservation occurs during this step. However, there is a complementary reductive part of the process in which the heterodisulfide compound CoM-S-S-CoB is reduced back to the constituent coenzymes. This step is promoted by a dehydrogenase enzyme and a heterodisulfide reductase. Energy conservation occurs here, as electrons are transferred through the two enzymes, generating a proton motive force and promoting ATP formation (see Thauer, 1998).

From methanol and other methyl-containing compounds (Figure 10.6A), the methyl group is first donated to a corrinoid protein (a protein containing a porphyrin-like ring and the parent structure for vitamin B_{12}). The methyl carbon is then transferred to Co-M, and the remaining steps in methane formation are as described above. If H_2 is not available to reduce the methyl group to methane, then some reducing power must be generated by oxidizing methanol to CO_2, which, from the methyl corrinoid complex, is the reverse of methanogenesis as shown in Figure 10.5. During methanogenesis from acetate, the acetate is first activated to acetyl CoA (Figure 10.6B) and subsequently interacts with carbon monoxide dehydrogenase (CODH), from the reductive acetyl CoA carbon fixation pathway (see Chapter 4). From here, the methyl group is transferred to the corrinoid protein, and then to coenzyme M (CoM), where methane production follows a path similar to what we have seen before. The carboxyl carbon, still attached to CO dehydrogenase, is oxidized to CO_2 (Figure 10.6B), liberating electrons used in the reduction of the methyl carbon to methane.

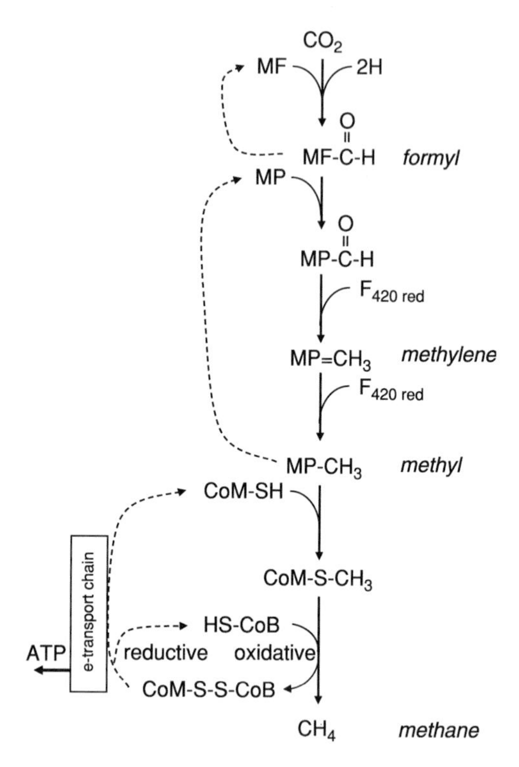

Figure 10.5 Outline of the biochemistry of methane formation from CO_2/H_2. CO_2 is activated with a methanofuran-containing enzyme (MF) with the reduction of the carbon to the level of formyl. The formyl carbon is transferred to an enzyme containing methanopterin (MP) with reduction steps to the methylene and methyl levels. The methyl carbon is transferred to a coenzyme M (CoM)-containing enzyme, forming CoM-S-CH_3. Methane is formed as this complex combines with HS-CoB, forming also the heterodisulfide CoM-S-S-CoB. ATP is conserved as this complex is reduced back to the sulfide-bonded coenzymes. Modified from Madigan *et al.* (2003), with inspiration from Thauer (1998).

When methane is formed from methyl compounds without an input of H_2, energy is required. This energy comes from a Na^+ ion pump, which works in a manner similar to a proton pump. In a Na^+ ion pump, Na^+, rather than H^+, is extruded to the outer cell membrane, and ATP is generated through a Na^+ ATPase (see Madigan *et al.*, 2003, for details). Finally, we highlight that while some of the methane formation pathways, such as acetotrophic methanogenesis, resemble fermentation reactions in their overall chemistry, ATP formation is not coupled to substrate-level phosphorylation. Rather, oxidative phosphorylation is used.

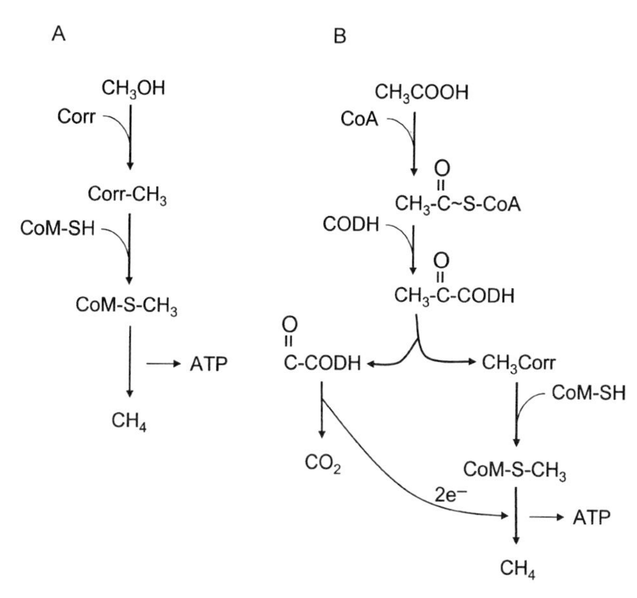

Figure 10.6 (A) Biochemistry of methane formation from methanol and other methyl-containing compounds. The methyl group of the substrate is transferred to a corrinoid protein (Corr) and then to conenzyme M (CoM), with the remaining steps in methane production as described in Figure 10.5. (B) The biochemistry of methane formation from acetate, in which the acetate is first activated with acetyl CoA (CoA). This complex then interacts with the enzyme carbon monoxide dehydrogenase (CODH), and from there the original carbons in the acetate are split. The methyl carbon is transferred to a corrinoid protein (Corr), then coenzyme M (CoM), with subsequent steps in methane formation as was shown in Figure 10.5. The carbonyl carbon is oxidized to CO_2, liberating the electrons needed to reduce the methyl carbon to methane. In both pathways of methane formation, energy conservation occurs as outlined in Figure 10.5. See text for further details. Abstracted from Madigan *et al.* (2003).

3.4. Environmental significance

As we discussed in Chapter 3, methanogens have a rather low energy yield compared to other anaerobic metabolic processes and have a difficult time competing for electron donors such as H_2 and acetate when other, more energetically favorable, anaerobic populations are active. Thus, sulfate reducers, for example, can maintain concentrations of hydrogen and acetate at energetically favorable levels for their own metabolism, but these same levels are too low to fuel methanogenesis. For this reason, methanogens dominate in anoxic environments where electron acceptors such as nitrate, metal oxides, and sulfate are exhausted, or nearly so. This is not always true, however, as in some instances active methanogenesis can be found where it might not be expected. For example, Hoehler *et al.* (2001) measured high rates of methane production in the upper anoxic portions of sulfate-rich (around 80 mM) microbial mats. Normally one thinks of high sulfate concentrations, and active sulfate reduction, as being detrimental to methanogenesis, as the sulfate reducers should actively consume the available electron donor. However, in this case, unexpectedly high H_2 concentrations were measured, particularly at night, when active H_2 production presumably occurred via nitrogen fixation (see Chapter 7).

Generally, however, methane production is limited when H_2 and acetate are used for other more energetically favorable anaerobic metabolisms, and it becomes active only when the electron acceptors driving these other metabolisms are reduced to low concentrations. In the marine realm, sulfate reduction is the main process competing with methanogenesis for electron donor. Methanogenesis, therefore, becomes active when sulfate is exhausted, or at least significantly reduced in concentration. For example, in the classic study of Martens and Berner (1974), methane did not accumulate until sulfate was depleted in homogenized marine sediments (Figure 10.7). In another example, Kuivila *et al.* (1990) measured active methane production in sediments from Saanich Inlet only below the zone of sulfate depletion. In other instances, such as at Cape Lookout Bight (Crill and Martens, 1986) and at Princess Louise Inlet in British Columbia, Canada, methanogenesis

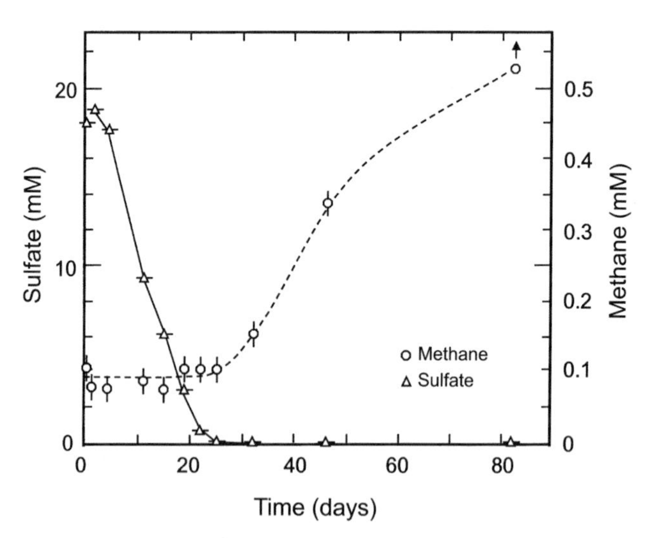

Figure 10.7 Illustration of how in marine sediment incubation experiments, methane formation does not occur until sulfate is depleted. Redrawn from Martens and Berner (1974).

occurs in the presence of 2 to 4 mM sulfate, and even where sulfate reduction is still active, but at rates significantly less than maximum rates. In these instances, methanogenesis occurs (Hoehler *et al.*, 1998) because the combination of low sulfate concentrations and relatively high electron donor concentrations allows for a thermodynamically favorable methanogenic reaction (see Chapter 3).

Methanogens and sulfate reducers co-occur in small melt-water ponds located on the McMurdo Ice Shelf, Antarctica (Mountfort *et al.*, 2003). Not surprisingly, the relative significance of methanogenesis increased as sulfate concentrations within the pond waters decreased. Generally, and regardless of the sulfate concentration, H_2 drove methanogenesis, and acetate was the most important electron donor for sulfate reducers. Indeed, increasing H_2 concentration in slurry experiments stimulated methanogenesis but had little effect on rates of sulfate reduction. In these sediments, acetate was seemingly the most important available electron donor in ponds with high sulfate concentrations, and the sulfate-reducing population was well adapted to this abundant electron donor. Surprisingly, at low sulfate concentrations H_2 was apparently the most important electron donor, driving high rates of methanogenesis but supporting only limited sulfate reduction. In this case, the sulfate-reducing population may not have been able to compete based on thermodynamic considerations because of the low sulfate concentration. Overall, the control of sulfate concentration on the relative magnitudes of sulfate reduction and methanogenesis meets with expectations, but the exact features of the competitive interactions are complex at the substrate level.

In marine environments, little reactive organic matter is usually available to fuel methanogenesis because, as discussed above, methanogenesis normally occurs at depths where other electron acceptors (in particular sulfate) are exhausted and where considerable amounts of reactive organic carbon have already been consumed. For these reasons, methanogenesis usually accounts for only about 5 to 10% as much carbon mineralization as sulfate reduction (Canfield, 1993) (Table 10.4). Because sulfate reduction accounts for, typically, half of the total carbon mineralization in coastal marine sediments, methanogenesis, therefore, is responsible for about 2 to 5% of the total carbon mineralization. In some cases in which sediment deposition rates are particularly high, a significant amount of reactive carbon can be passed through the zone of sulfate depletion, thus increasing the relative contribution of methanogenesis in carbon mineralization. This occurs at Cape Lookout Bight, where high sedimentation rates of 10 cm yr^{-1} rapidly deliver organic carbon to the zone of active methanogenesis; here methanogenesis rates are about 40% as great as rates of sulfate reduction and methanogenesis accounts for about one-third of the total carbon mineralization (Crill and Martens, 1986).

Table 10.4 Rates and pathways of methanogenesis in marine and freshwater sediments compared to rates of sulfate reduction

Site	Methane prod. rate[a]	% Methane from acetate	% Methane from H_2/CO_2	Sulf red rate[a]	Methane/SR[c]	Reference
Marine						
Skan Bay, Alaska	54	—[b]	—[b]	700	0.08	1
Santa Barbara Basin	20	2 to 11	70 to 85	—[b]		2
Cape Lookout Bight	870 to 1000	20 to 29	71 to 80	2300	0.4	3, 4, 5
Sachem Head	9 to 30	—[b]	—[b]	360	0.05	6
Saanich Inlet	53	42	58	610	0.09	7
Princess Louise Inlet	20	43 to 48	52 to 57	—[b]		8
Kattegat/Skagerrak						
Sta A	0.4	—[b]	—[b]	340	0.001	9
Sta B	30	—[b]	—[b]	320	0.09	9
Sta C	42	—[b]	—[b]	450	0.09	9
Freshwater						
White Oak estuary	2130	72	28	—[b]		10
Buck Hollow	270	51	49	—[b]		10
Orange Pond, Ant.	8.8	3	97	—[b]		11
Lake Constance	—[b]	100	0	—[b]		12
Kings Lake Bog	54 to 142	0.5	99.5	—[b]		13
Lawrence Lake	2008	14	86	803	2.5	14
Wintergreen A	17,500	80	20	2700	6.5	14
Wintergreen B	11,300	67	33	8,400	1.3	14
Lake Mendota	—[b]	76	24	—[b]		14
Rømø Site 3	68	10	90	23.4	2.9	15
Rømø Site 8	72	10	90	15	4.8	15
Rømø Site 10	11	30	70	0.12	91	15
Lake Washington	14.6	61 to 85	15 to 39	4.3	3.4	16
Rice paddy soil	—[b]	69 to 83	17 to 31	—[b]		17
Lake Michigan	0[d]	—[b]	—[b]	570	0	18

In freshwater environments, low sulfate concentrations generally restrict the activity of sulfate reducers, and rates of methanogenesis are typically two to five times greater than rates of sulfate reduction (Table 10.4). Methanogens, therefore, usually dominate anaerobic carbon mineralization. Still, sulfate reduction (or iron reduction) may be the most important carbon mineralization process in surface lake sediments, with methanogenesis most significant below (e.g., Schulz and Conrad, 1996). Furthermore, in some freshwater systems, sulfate reduction may dominate carbon mineralization altogether. Recent work by Thomsen *et al.* (2004) shows that methanogenesis could not be detected within the upper 18 cm of sediment collected from West Lake Michigan and that sulfate reduction accounted for 30% of the total anaerobic carbon mineralization (Fe reduction making up most of the rest). Unlike most of the freshwater sites studied to date, Lake Michigan is a very oligotrophic system, with a relatively slow overall sediment metabolic rate (2200 μmol C cm^{-2} yr^{-1}). This, combined with a relatively slow rate of sediment deposition (0.05 cm yr^{-1}), ensures that there is ample time for the limited sulfate in the sediment porewaters to oxidize organic matter, allowing only small amounts of refractory carbon burial into the zone of methanogenesis. Additional study of oligotrophic freshwater systems should show whether the Lake Michigan results are anomalous or characteristic.

Returning to the problem of regulation, we have discussed above, and in, Chapter 3, how energetics can control the predominance of specific microbial metabolisms in nature. We know, however, that sulfate reducers, for example, do not always completely exclude methanogens (see above), and the competition between these two populations for H_2 has been explained in kinetic terms with the following expression (Lovely *et al.*, 1982) (see also Chapter 2):

$$V_{total} = V_{max\ SRB} \frac{S}{K_{mSRB} + S} + V_{max\ MB} \frac{S}{K_{mMB} + S} \tag{10.7}$$

where SRB refers to sulfate reducers and MB refers to methanogens. The proportion of the total H_2 uptake (V_{total}) is the sum of H_2 uptake by each of

---

1, Alperin (1988); 2, Warford *et al.* (1979), 3, Crill and Martens (1983); 4, Crill and Martens (1986); 5, Chanton, (1985); 7, Devol *et al.* (1984); 8, Kuivila *et al.* (1990); 9, Iverson and Jørgensen (1985); 10, Avery *et al.* (2002); 11, Mountfont *et al.* (1999); 12, Schulz and Conrad (1996); 13, Sansdown *et al.* (1992); 14, rates summarized in Holmer and Storkholm (2001) and Lovley and Klug (1886); 15, Hansen *et al.* (2001); 16, Kuivila *et al.* (1990); 17, Rothfuss and Conrad (1993); 18, Thomsen *et al.* (2004).

[a]Rates in μmol cm^{-2} y^{-1}.

[b]Not determined.

[c]Rates of carbon mineralization by methanogenesis ratioed to rates by sulfate reduction.

[d]Down to 18 cm depth.

the two populations, and as K_m tends to be lower for natural populations of methanogens than for sulfate reducers (Lovley *et al.*, 1982) and V_{max} is comparable, sulfate reducers should out-compete methanogens at the same substrate concentration. The same analysis also holds true for acetate (Lovley and Klug, 1986).

This kinetic analysis helps explain why methanogens do poorly under sulfate-reducing conditions. However, the model would also predict sustained methanogenesis, albeit at reduced rates, when substrate levels were reduced to the point at which methanogenesis is thermodynamically unfavorable and clearly cannot occur. Thus, the Michaelis-Menten treatment cannot deal with thermodynamic thresholds (Hoh and Cord-Ruwisch, 1996; Conrad, 1999). To address this issue, the Michaelis-Menten expression has been amended:

$$V = V_{max} \frac{S(1 - K_{sp}/AP)}{K_m + S(1 + K_{sp}/AP)} \tag{10.8}$$

In this expression, AP refers to the activity product of reaction products and reactants and K_{sp} is the solubility product expressing the activity product of products and reactants at equilibrium (see Chapter 2) (Hoh and Cord-Ruwisch, 1996; Conrad, 1999). Therefore, V is zero at equilibrium, and as AP becomes large (the system becomes more supersaturated and thermodynamically favorable), the expression approaches the traditional Michaelis-Menten expression. In this model, both thermodynamic and kinetic factors control metabolic rate. A recent, more comprehensive treatment is also provided in Jin and Bethke (2003).

Acetotrophic methanogenesis and methane production through CO_2 reduction with H_2 are by far the two most significant pathways of methane production in the environment. By comparison, methane production from methylamines, methyl sulfides, methanol, CO, and formate is thought to be relatively minor (Lovley and Klug, 1986). Although the database is rather limited, CO_2/H_2 methanogenesis seems to be the most important methane production pathway in marine environments, but acetate can be nearly as important in some cases (Table 10.4). In freshwater systems, acetate is traditionally thought to be the most important methane production pathway (e.g., Conrad, 1999). However, a broad range of relative contributions of acetate and CO_2/H_2 is observed, and in many instances CO_2/H_2 dominates methane production. Conrad (1999) has overviewed the likely pathways of fermentative organic matter degradation and has concluded that, starting with glucose, a maximum of 4 mol of H_2 are formed per 2 mol acetate produced, assuming that H_2, acetate, and CO_2 are the final fermentation products. This leads to a H_2-to-acetate production ratio of 2:1. As 4 mol H_2 are required to form 1 mol methane, but only 1 mol of acetate is needed to form 1 mol of methane, a theoretical contribution from CO_2/H_2 to methanogenesis should be 1/3 (33%) or less.

The contribution of CO_2/H_2 to methane production is reduced below the theoretical 33% maximum if homoacetogenesis (Equation 10.9) is important, as seems to be the case in some environments (Conrad, 1999), or if the methanogens are out-competed by sulfate reducers for H_2 (Conrad, 1999).

$$2CO_2 + 4H_2 \rightarrow CH_3COOH + 2H_2O \tag{10.9}$$

A higher contribution of CO_2/H_2 to methane production might occur if sulfate reducers out-compete the methanogens for acetate but not H_2. This could be the case in Cape Lookout Bight (Crill and Martens, 1986), where CO_2/H_2 methanogenesis dominates methane production in the zone of sulfate reduction, and this is also apparently the case in the Antarctic melt-water ponds described above (Mountfort *et al.*, 2003). This still, however, does not explain the dominance of CO_2/H_2 methanogenesis at Cape Lookout Bight outside of the zone of sulfate reduction (Crill and Martens, 1986) and in other marine settings below the zone of sulfate reduction, where methanogenesis is most important (Table 10.4). As we saw above for Antarctic melt-water ponds, under some circumstances H_2 is apparently the most important electron donor supplied through the initial steps of hydrolysis/fermentation. These observations do not fit with the prediction of Conrad (1999) as to the maximum contribution of CO_2/H_2 to methanogenesis. Thus, production ratios of H_2 to acetate seem to differ from the predicted values in some cases, although the reasons for this are presently unknown.

The pathways to methane may, occasionally, be more complex than the discussion so far would suggest. Thus, in subtropical Lake Kinneret (Israel), radiolabel experiments show that nearly all methane is ultimately produced from acetate, yet there is also a substantial role for CO_2/H_2 in methane production (Nüsslein *et al.*, 2001). Also, through radiolabel experiments, a large portion of the methyl carbon of acetate is oxidized to CO_2, despite the absence of electron acceptors such as sulfate or iron, which would normally be coupled to acetate oxidation. These apparently contradictory observations are explained if much of the acetate is first converted to CO_2 and H_2 by acetate-oxidizing bacteria (Equation 10.10), and the H_2 is subsequently used by H_2-utilizing methanogens (Equation 10.1).

$$CH_3COOH + 2H_2O \rightarrow 4H_2 + 2CO_2 \tag{10.10}$$

In this case there is a syntrophic relationship between the acetate oxidizers and the methanogens, and thermodynamic calculations support the plausibility of this mechanism in Lake Kinneret sediments (Nüsslein *et al.*, 2001). Note that Equation 10.10 is the reverse of homoacetogenesis (Equation 10.9), and some homoacetogenic organisms known as the "reversibacteria" are capable of this metabolism (Zinder and Koch, 1984). The general significance of this syntrophic pathway to methane production in nature has not been well studied.

4. METHANOTROPHY

Methanotrophy refers to the microbial oxidation of methane. The best known methanotrophic process is conducted by aerobic organisms. This process has received broad attention in recent years due to the increases in atmospheric methane concentrations and the potential regulating role of methanotrophy on this increase. Methanotrophy also occurs under anoxic conditions, when methane oxidation is coupled to sulfate reduction. The anaerobic oxidation of methane (AOM) has also received a great deal of attention as of late. This is partly because the process has stood for so long as a microbiological enigma, with 30 years of failed attempts to cultivate the regulating organisms, and partly because the range of environments supporting this process has turned out to be fascinatingly diverse. In the following Sections we consider aspects of the microbiology, phylogeny, biogeochemistry, and environmental significance of methanotrophic processes.

4.1. Aerobic methanotrophs

4.1.1. Organisms, phylogeny, and carbon fixation pathways

Aerobic methanotrophs are prokaryotes that grow by the oxidation of methane. As such, they are included among a broader class of aerobes, called methylotrophs, which oxidize C_1 compounds such as methanol, methylamine, formate, carbon monoxide, and dimethylsulfide (Bowman, 2000). Many methanotrophs can grow by the oxidation of methanol in addition to methane, but their use of other C_1 compounds for growth is very limited (Bowman, 2000). Many methanotrophs are also capable of oxidizing a wide variety of other substrates, including some aromatic compounds (Bédard and Knowles, 1989), but they do not grow on these.

Two broad groups of methanotrophs are recognized. The first group is represented by the family *Methylococcaceae*, located within the γ-subdivision of the proteobacteria. Within this family there are two recognized subgroups. The first, referred to as "type I," comprises four phylogenetically affiliated genera (Table 10.5). Type I methanotrophs are mostly mesophiles and psychrophiles, and they branch closely to a number of methane-oxidizing endosymbionts (Bowman, 2000). The second cluster, referred to as "type X"

Table 10.5 Methanotroph types

	Type I	Type X	Type II
Family	*Methylococcaceae*		*Methylocystaceae*
Genera	*Methylosphaera* *Methylobacter* *Methylomicrobium* *Methylomonas*	*Methylococcus* *Methylocaldum*	*Methylosinus* *Methylocystis* *Methylocella* *Methylocapsa*
Pylogenetic affiliation	γ proteobacteria	γ proteobacteria	α proteobacteria
Intracytoplasmic membranes	Flat disc-shaped vesicles	Flat disc-shaped vesicles	Vesicles running parallel to cell margin
Carbon assimilation pathway	RuMP	RuMP	Serine
N fixation	No	Yes	Yes

Abstracted from Bowman (2000), with additional information from Dedysh *et al.* (2000, 2002).

methanotrophs, contains two genera and is more deeply branching within the γ-proteobacteria. Type X methanotrophs contain thermotolerant and thermophilic members (Bowman, 2000). These organisms branch near to the anoxygenic phototrophic families *Ectothiorhodospiraceae* and *Chromatiaceae*. Type I and type X methanotrophs are further distinguished by the arrangement of internal membranes, presumably used in the methane oxidation process. These membranes are arranged as a collection of flat disc-shaped vesicles within the cells.

Type II methanotrophs are all members of the family *Methylocystaceae*, and they are located within the α-subdivision of the proteobacteria, with a rather distant relationship to a variety of facultative methylotrophs and a variety of anoxygenic phototrophs. Most members contain internal membrane vesicles that run parallel to the cell margin. However, the recently isolated acidophilic methanotroph *Methylocapsa acidiphila* contains parallel vesicles on only one side of the cell (Dedysh *et al.*, 2002).

A variety of aerobic ammonia oxidizers, including members of the genera *Nitrosococcus* and *Nitrosomonas* from the β-subdivision of the proteobacteria, can also oxidize methane (see review by Bédard and Knowles, 1989). The importance of these organisms as methane oxidizers in nature, however, is currently unknown.

All aerobic methanotrophs are united in the initial step of methane oxidation involving the oxidation of methane to methanol with the enzyme methane monooxygenase (MMO) (both soluble and membrane-bound versions are known) and the conversion of methanol to formaldehyde (Figure 10.8). However, aspects of carbon metabolism are unique to each methanotroph family. Thus, all type I and type X methanotrophs use the ribulose

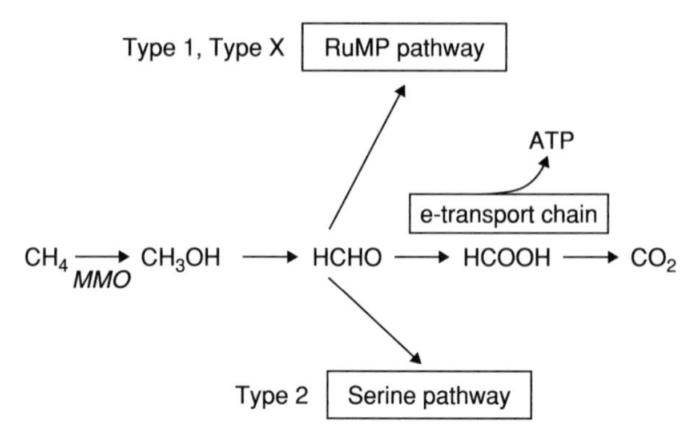

Figure 10.8 The two main pathways of carbon assimilation in methanotrophs. Formaldehyde formation through methanol is common to all methanotrophs. From here type 1 and type X methanotrophs use the ribulose monophosphate pathway (RuMP) for carbon assimilation, whereas the type 2 methanotrophs use the serine pathway. See text and Table 10.6 for more details on methanotroph types. For all methanotrophs energy conservation occurs as formaldehyde is oxidized to CO_2. Inspired by Hanson and Hanson (1996).

monophosphate pathway (RuMP) for carbon assimilation (Hanson and Hanson, 1996). In this pathway, all cell carbon is assimilated from formaldehyde, and the initial step is the incorporation of three formaldehydes into three ribulose-5-phosphate molecules, forming three molecules of hexulose-6-phosphate. Since formaldehyde is at the oxidation level of cell carbon, no reduction steps are necessary, making the pathway particularly energy efficient. One ATP molecule is used for every three molecules of formaldehyde assimilated. The three molecules of formaldehyde ultimately form one molecule of glyceraldehyde-3-phosphate, which is channeled into biosynthesis. In the type X methanogen *Methylococcus capsulatus*, some carbon may also be fixed autotrophically into biomass, as Rubisco activity has been found in this organism (Taylor *et al.*, 1981; Baxter *et al.*, 2002).

The serine pathway is used by type II methanotrophs (Bowman, 2000; Lidstrom, 2001; Madigan *et al.*, 2003). In this pathway, both formaldehyde and CO_2 (fixed with PEP carboxylase) are incorporated to form, ultimately, acetyl-CoA, some of which is further directed to biosynthesis. The initial step in the process is the incorporation of formaldehyde into glycine, forming serine. The CO_2 is incorporated somewhat later into phosphoenol pyruvate, which is converted to oxaloacetate. Overall, this pathway requires the utilization of two ATP and two NADH for each molecule of acetyl-CoA formed.

Finally, for all methanotrophs, energy conservation occurs as we have discussed earlier (Figure 10.8) (see Chapter 2). ATP is formed when methanol is stepwise oxidized to CO_2 via a membrane-bound electron transport chain, generating a proton motive force that is relaxed by the controlled transport of H^+ across the cell membrane through ATPase.

4.1.2. Environmental distribution and significance

Methanotrophs are widely distributed in nature. They are found in environments where methane and oxygen meet, including tundra, peat bogs, rice paddy soils, ponds, lakes, marine sediments, the marine water column, hydrothermal, vent areas, and even desert soils (for an excellent review, see Hanson and Hanson, 1996). The intensity of methane oxidation varies greatly among these different environments, and it is highest where associated methane production rates are high. For example, in peat-rich sediments from the Florida Everglades methane production rates are in the range of 1.4 to 2.8 mmol m^{-2} d^{-1}, with 60 to 90% reoxidized, yielding methane oxidation rates in the range of 1.1 to 2.2 mmol m^{-2} d^{-1} (King *et al.*, 1990). Similarly, in a freshwater tidal swamp from North Carolina, from 50 to 80% of the seasonally averaged methane production (rates of 1.1 to 1.4 mmol m^{-2} d^{-1}) was oxidized by methanotrophs, yielding methane oxidation rates in the range of 0.8 to 1.0 mmol m^{-2} d^{-1} (Megonigal and Schlesinger, 2002). By contrast, in soils of the Mojave Desert, the source of methane is the atmosphere, and when the soils are dry, average methane oxidation rates are much lower, in the range of 6 to 60 μmol m^{-2} d^{-1} (Striegl *et al.*, 1992). After precipitation, however, average rates increase to 0.1 mmol m^{-2} d^{-1}, with relatively high maximum rates of 0.25 mmol m^{-2} d^{-1}. Unsaturated soils including forest soils, agricultural soils, grasslands, and tundra also consume atmospheric methane, and at rates comparable to the range reported for desert soils (Whalen and Reeburgh, 1990; Hanson and Hanson, 1996).

Rice paddy soils provide a typical example of an environment supporting significant methanotrophy (Gilbert and Frenzel, 1998; Damgaard *et al.*, 1998) (Figure 10.9). Here, active methanogenesis produces methane that diffuses to the soil surface, where it is oxidized on interaction with oxygen either diffusing into (Figure 10.9B) or produced within the soil (Figure 10.9A). The overlapping zone between methane and oxygen, as revealed by the utilization of gas diffusion probes (Gilbret and Frenzel, 1998) or newly developed biosensors for methane (Damgaard *et al.*, 1998), is only a few millimeters, demonstrating rapid and efficient methane oxidation within the soil. Even at night, with no oxygen production and very shallow oxygen penetration, little, if any, methane diffuses from the soil. Such a situation might be typical, but not universal, for wetland sediments (King *et al.*, 1990). For example, in marl sediments from the Florida Everglades, methanotrophy

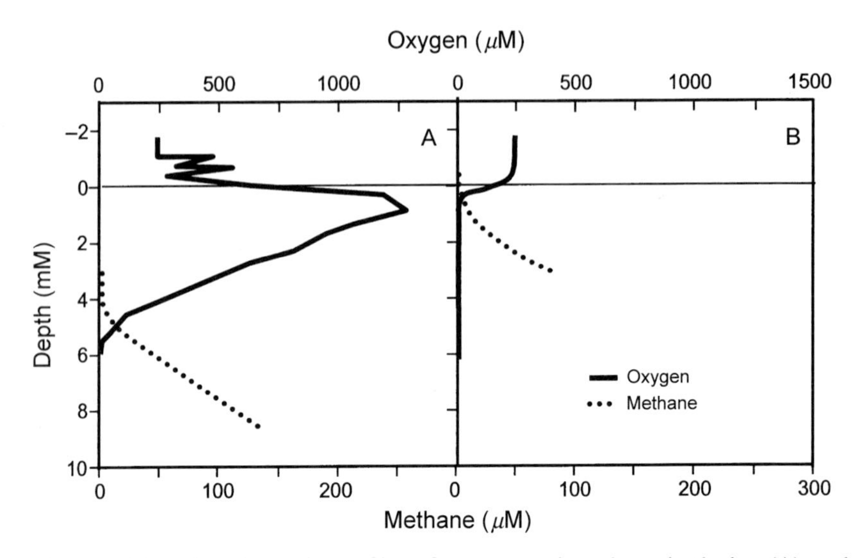

Figure 10.9 Microelectrode profiles of oxygen and methane both day (A) and night (B) from a rice paddy soil. Redrawn from Damgaard *et al.* (1998).

is inhibited and methane diffuses through the zone containing oxygen (King *et al.*, 1990). Ammonium is known to inhibit methanotrophy (Conrad and Rothfuss, 1991), but it is not known if this was the reason for the inhibition in these sediments.

Even in situations where methanotrophy in methane-rich environments is active and efficient, considerable amounts of methane can escape to the atmosphere. Thus, the amount of methane escaping wetland soils can comprise from 10 to 100% of the methane production (see summary in Megonigal and Schlesinger, 2002) with an average of about 50%. A significant mode of methane loss in rice fields is diffusion through rice plants from the soil to the atmosphere (e.g., Le Mer and Roger, 2001). It appears that rice plants absorb sediment methane into their roots, transport it by diffusion through the aerenchyma, and release it to the atmosphere through pores on the leaf sheaths (Nouchi *et al.*, 1990). Aquatic plants including water lilies (Dacey and Klug, 1979) also facilitate methane loss from sediments, as do a variety of vascular wetland plants (Sebacher *et al.*, 1985; Torn and Chapin, 1993; review in Hanson and Hanson, 1996). Methane bubble ebullition is another significant mode of methane release to the atmosphere, particularly in rapidly depositing marine sediments (e.g., Martens and Valklump, 1980) and non-vegetated, anoxic lake sediments and wetland soils.

Methanotrophs form important symbiotic associations with a variety of different invertebrate animals. For example, in the Florida Escarpment in the Gulf of Mexico high concentrations of methane, originating from deep in the Florida platform, support a symbiotic association between mytilid mussels and methanotrophs growing in the gill tissue of the mussels (Cavanaugh *et al.*, 1987; Cavanaugh, 1993). The methanotrophs gain a stable support and plentiful access to oxygen and methane, while the mussels farm the methanotrophs for nutrition. The mussels, however, are also capable of filter feeding, and at least some of their nutrition comes from detritus as well (Cavanaugh, 1993). A similar association between a different mytilid species and methanotrophs is found in a gas hydrate area off the Louisiana coast, where methane from dissolving gas hydrates supports the symbiosis (Childress *et al.*, 1986). On the Danish slope of the Skagerrak, active methane venting supports a symbiotic association between the pogonophoran tubeworm *Siboglinum poseidoni* and an active methanotroph population (Schmaljohann, 1991). As with other pogonophorans, *S. poseidoni* lacks a mouth and gut, and the endosymbionts are housed within the animal's trophosome. In this association, apparently all of the nutrition for the animal comes from the symbiotic prokaryotes. Finally, methane and sulfide fluxes from hydrothermal vent areas along the Mid-Atlantic Ridge support dual symbiotic associations between vent mussels and both methanotrophic and sulfide-oxidizing symbionts (e.g., Pond *et al.*, 1998; van Dover, 2000).

4.1.3. Kinetics

Because both oxygen and methane are required for methanotrophic metabolism, the availability of either of these gases can, in principle, limit methanotrophy in nature. Most attention has focused on methane. In a series of particularly important experiments, Bender and Conrad (1992, 1993) explored the kinetics of methane oxidation for natural populations of methanotrophs in oxic soils that would normally be viewed as atmospheric methane sinks. Noteworthy in these studies was the exploration of oxidation kinetics both with low methane concentrations, as would be experienced by the soils in nature, and after preincubation with high methane concentrations (methane mixing ratio of 20%) as might be more appropriate for methane-saturated wetland soils. The results are summarized in Table 10.6. For natural soils incubated with low methane concentrations, K_m values are low, in the range of 30 to 50 nM (Bender and Conrad, 1992; Czepiel *et al.*, 1995). These concentrations correspond to atmospheric methane levels of 20 to 34 ppmv (part per million by volume), which are 10 to 20 times higher than present atmospheric levels (see Figure 10.1). Importantly, methanotrophic activity continued down to detection-level methane concentrations of around 0.02 nM (Th_m; Table 10.6), demonstrating that these soils can

Table 10.6 Kinetic parameters for methane oxidation by aerobic methanotrophs

Sample	K_m (nM)	$V_{max}(h^{-1} g_{dw}^{-1})$ (nmol)	$^{a,c}Th_m$ (nM)	$^{a,d}Th_a$ (nM)	Reference
Temperate soil	27.7 ± 1.5^a	7.1	—[b]	—[b]	1
Cultivated cambisol	50.6 ± 13	0.7 ± 7	0.02	0.3 ± 19	2
Meadow cambisol	49.9 ± 7	0.9 ± 5	0.02	0.3 ± 1	2
Forest luvisol	29.7 ± 35	3.6 ± 8	0.02	4.0 ± 24	2
Cultivated cambisol high methane	1740 ± 34	270 ± 11	0.02	62 ± 5	2
Meadow cambisol high methane	4560 ± 33	410 ± 18	0.02	16.7 ± 27	2
Forest luvisol high methane	$27{,}900 \pm 24$	450 ± 8	0.02	18.9 ± 15	2
Landfill cover	$2{,}650 \pm 41^a$	227 ± 18	—[b]	—[b]	3
Preincubated humisol	50,000	15,320	—[b]	—[b]	4
Peat	57,000	110	—[b]	—[b]	5
Prokaryotes	580 to 54,100				6,7

1, Czepiel *et al.* (1995); 2, Bender and Conrad (1992); 3, Whalen *et al.* (1990); 4, Megraw and Knowles (1987); 5, Watson *et al.* (1997); 6, Bédard and Knowles (1989); 7, Carlsen *et al.* (1991).
[a]Original data in ppmv, converted to nM assuming 1 ppmv methane equals 1.48 nM methane.
[b]Not determined.
[c]Th_m, minimum methane concentration supporting methanotrophy.
[d]Th_a minimum methane content concentration where methane oxidation rate is first order with respect to methane concentration.

sustain appreciable rates of methane oxidation at concentrations well below atmospheric saturation. The rates of methane oxidation were first order with respect to methane down to methane concentrations of around 0.3 to 4.0 nM (Th_a; Table 10.6), with a higher order dependency below this (Bender and Conrad, 1992, 1993). For these low-methane incubated soils, V_{max} values were also low, in the range of 0.7 to 3.6 nmol h^{-1} g_{dw}^{-1} (Table 10.6).

When pre-incubated with high methane levels (methane mixing ratio of 20%) a second population emerged with much higher K_m values, in the range of 1700 to 27,000 nM, and high V_{max} values of 270 to 450 nmol h^{-1} g_{dw}^{-1} (Table 10.6). This population, with low methane affinity, displayed kinetic properties similar to populations in soils normally exposed to high methane concentrations (Table 10.6) and to pure bacterial cultures. It appears, then, that the methanotrophs brought into pure culture represent those active in high methane environments, whereas the methanogens in oxic soils are of a different sort and are not yet well characterized.

As for oxygen, natural populations of methanotrophs from a peat in Ellergower Moss, Scotland, showed an average K_m value for oxygen of 32,000 nM, with non-limiting methane and a V_{max} of 209 nmol O_2 ml_{peat}^{-1} h^{-1} (Watson *et al.*, 1997). These values are similar to those for methane

(Table 10.6). Furthermore, high rates of methanotrophy were maintained at oxygen concentrations up to air saturation, demonstrating that these methanotrophs were full aerobes and not microoxic, as is sometimes assumed. Experimental results on pure cultures of type I and type II methanotrophs also demonstrate high rates of methanotrophy at high oxygen concentrations (Ren *et al.*, 1997). In these experiments, O_2 began to limit methanotrophy only at levels below 5.7 μM.

In summary, methane limits methanotrophy in oxic soils and other methane-poor oxic environments. In methane-rich wetland soils, the kinetic properties describing O_2 and CH_4 uptake are apparently similar, as are the concentrations of O_2 and methane in the zone of active methanotrophy (King *et al.*, 1990; Damgaard *et al.*, 1998). In these cases, rates of methanotrophy are ultimately limited by the methane production rate and the extent to which other methane transport avenues, such as plant transport and bubble ebullition, are active. In situations with an active methane supply, such as in cold methane seep areas, regions of clathrate hydrate decomposition and perhaps some very active lake sediments, methane supply overwhelms the transport flux of oxygen into the sediment. Here methanotrophy, at least locally, is limited by the transport flux of oxygen along concentration gradients established through the diffusive sediment boundary layer (see Chapter 6).

4.2. Anaerobic methane oxidation

The anaerobic oxidation of methane with sulfate has been recognized by geochemists as a significant pathway of methane oxidation in marine sediments for decades:

$$CH_4 + SO_4^{2-} \rightarrow HS^- + HCO_3^- + H_2O \qquad (10.11)$$

The microbiological controls of the process, however, remained illusive, forcing many microbial ecologists to question (at least privately) whether the process occurred at all. AOM has, therefore, stood as one of the great durable problems in microbial ecology. As we discuss below, there has been substantial recent progress on the microbiology of this enigmatic process. However, before getting to this, we look at the problem historically and consider the geochemical evidence for AOM.

4.2.1. Geochemical evidence for AOM

The depth distributions of methane and sulfate in anoxic marine sediments provided the initial evidence for AOM in anoxic marine sediments (e.g., Reeburgh, 1969; Martens and Berner, 1974, 1977; Reeburgh, 1976; Alperin

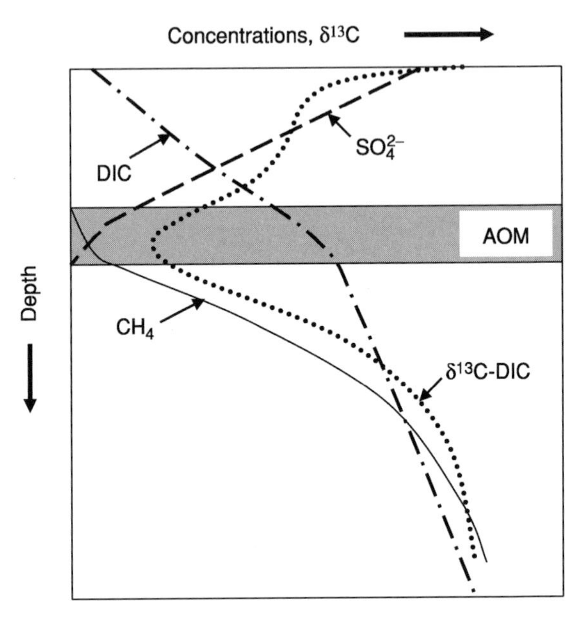

Figure 10.10 General chemical features of a marine sediment supporting active anaerobic oxidation of methane (AOM). The zone of most active AOM is found at the sulfate–methane interface. Frequently also found is a shift to ^{13}C-depleted DIC at this interface, demonstrating the oxidation of ^{13}C-depleted methane to DIC. Inspired by Reeburgh (1980).

and Reeburgh, 1984). Typically, methane does not accumulate until after sulfate is nearly exhausted at depth in sediments (e.g., Figure 10.10). Because methane is produced in the sediment but little escapes (except in some cases by bubble ebullition) it requires methane oxidation, and the depth distributions of methane and sulfate suggest a reaction front where methane is removed by sulfate reduction (Figure 10.10). The isotopic composition of DIC (total dissolved inorganic carbon) frequently shows especially ^{13}C-depleted values at the reaction front between sulfate and methane (Figure 10.10), indicating the input of isotopically depleted DIC from the oxidation of methane (see Section 5.2 on isotope systematics). From these observations, it can be concluded that the anaerobic oxidation of methane is inescapable and the coupling to sulfate reduction is highly suggestive.

Further evidence for AOM was provided by radiolabeling experiments in which peaks in methane oxidation rate (from ^{14}C-CH_4 labeling experiments) were highest at the sulfate–methane transition, consistent with the shapes and locations of the sulfate and methane profiles (e.g., Devol, 1983; Iversen and Jørgensen, 1985). Furthermore, sulfate reduction rates also often peaked where methane oxidation rates were highest (Figure 10.11), providing concrete evidence for the link between sulfate reduction and methane oxidation. From these observations, the biogeochemical community was largely convinced that anaerobic methane oxidation was coupled to sulfate reduction and that it was a widespread and important phenomenon in marine sediments. The microbiological community, however, was not yet so certain.

4.2.2. Microbiological evidence for AOM

The problem was that sulfate reducers conducting AOM could not be brought into culture. Some sulfate reducers can oxidize minor amounts of methane when they are metabolizing other electron donors, but this would not account for the AOM observed in nature, as CH_4 is the main electron donor during AOM. The dominance of AOM at the reaction front between methane and sulfate is, in many cases, substantiated by the near stoichiometric match in sulfate reduction rates and rates of methane oxidation at this interface (e.g., Iversen and Jørgensen, 1984) (Figure 10.11). Some progress was made when Zehnder and Brock (1979) reported the oxidation of

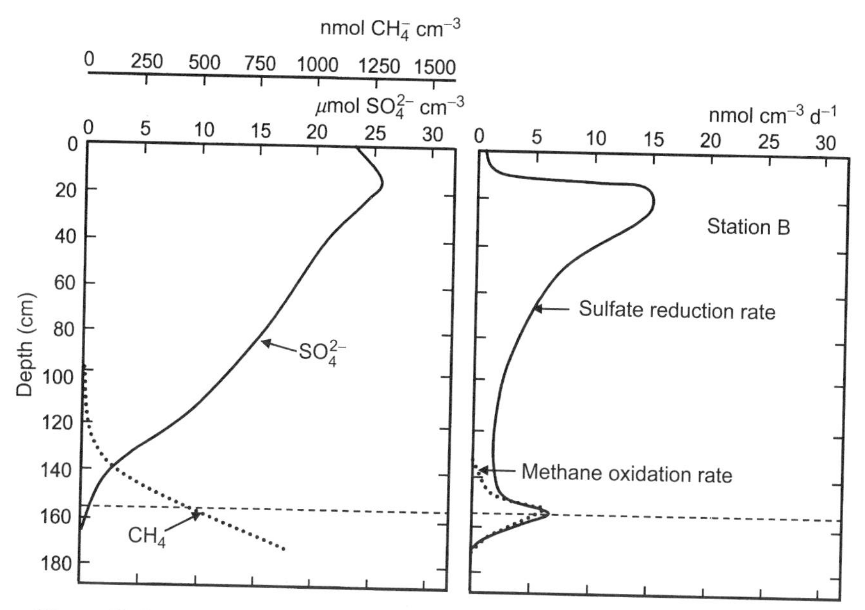

Figure 10.11 Measured rates of sulfate reduction and methane oxidation in a sediment from the Kattegat/Skagerrak transition off the coast of Denmark. Maximum rates of methane oxidation occur at the sulfate–methane transition and coincide with elevated rates of sulfate reduction, establishing the link between the two processes. Redrawn from Iversen and Jørgensen (1985).

methane by several strains of methanogens, although the amount oxidized was a small percentage of the amount produced, and no net methane oxidation was observed. Nevertheless, they proposed that AOM in nature could occur through a consortium in which a methanogen operating in reverse supplied H_2, acetate or ethanol (the reverse of methanogenesis) to a population able to use these substrates, possibly a sulfate reducer. Some support for the consortium idea came from the inhibition experiments of Alperin and Reeburgh (1985). In these experiments on sediment from the zone of AOM in Skan Bay, Alaska, the addition of molybdate inhibited sulfate reduction but not methane oxidation. This observation suggested that the two processes were decoupled, with methane oxidation and sulfate reduction likely conducted by two different sets of organisms.

Experimental work by Hoehler *et al.* (1994) demonstrated a general coupling between methanogenesis and methane oxidation below the zone of sulfate depletion in the sediments of Cape Lookout Bight, implicating methanogens in the process of methane oxidation. In this zone, however, rates of methane oxidation were only a fraction (about 10%) of the methane production rates, and net methane oxidation occurred only in the presence of sulfate. This further established the link between AOM and sulfate reduction. Based on these results, Hoehler *et al.* (1994) reinvigorated the idea of AOM fueled by a consortium in which methanogens, operating in reverse, oxidize methane, producing an electron donor, possibly H_2, used by sulfate reducers. This consortium is an example of syntrophy, where the sulfate reducer lowers the H_2 (or whatever the electron donor might be) concentration to the point where reverse methanogenesis becomes thermodynamically favorable (Hoehler *et al.*, 1994).

The first concrete evidence for a consortium conducting AOM came from the study of Hinrichs *et al.* (1999), who found archaeal lipid biomarkers extremely depleted in ^{13}C from sediments of a methane seep area in the Eel River Basin in northern California. The most plausible explanation was that methane was the carbon source for the lipids, suggesting that the *Archaea* owning these lipids derived carbon from the incorporation and oxidation of methane. Phylogenetic analysis of 16S rRNA extracted from the sediment revealed a new cluster of organisms, named ANME-1 (see Figure 10.12), located close to, but distinct from, methanogens of the orders *Methanomicrobiales* and *Methanosarcinales.* Closely following was the now-famous discovery of microbial aggregates from a methane hydrate area off the Oregon coast (Boetius *et al.* 2000). In these aggregates an inner cluster of *Archaea*, forming a separate phylogenetic cluster among the methanogens, the ANME-2 group (closely related to the *Methanosarcinales*) (Figure 10.12), was surrounded by a layer of sulfate reducers related to *Desulfosarcina/Desulfococcus*. In the study of Orphan *et al.* (2001a), similar aggregates displayed highly ^{13}C-depleted organic carbon based on direct determination

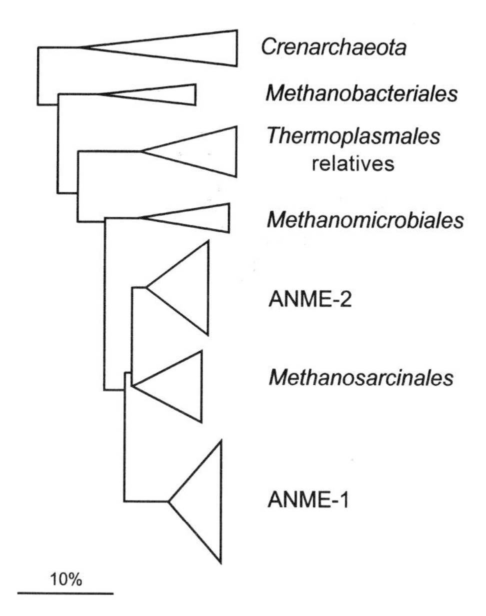

Figure 10.12 Phylogeny of methanogens and related *Archaea* highlighting the placement of the ANME-1 and ANME-2 groups. These groups are believed to contain members related to methanogens, and an active consortia with sulfate reducers, in the process of anaerobic methane oxidation (AOM). Abstracted from Orphan *et al.* (2001b).

with SIMS (secondary ion mass spectrometry), again showing evidence that the archaeal member of the aggregate is obtaining carbon from methane.

In total, these observations strongly suggest that the activities of a methanogen/sulfate reducer consortium are responsible for AOM. More evidence comes from *in vitro* experiments on Hydrate Ridge surface sediment highly enriched in methanogen-sulfate reducer aggregates (Nauhaus *et al.*, 2002). In these experiments, sulfate reduction is directly coupled to methane oxidation with the expected stoichiometry, and it does not occur in the absence of added methane. This study, however, also provided some surprises. For example, sediment free of methane and with added H_2, formate or ethanol did not display sulfate reduction rates as rapid as with added methane. If one of these electron donors was the product of methane oxidation, it would be expected to stimulate sulfate reduction; it should not matter if the source of the intermediate is from methane oxidation or experimental manipulation. Thus, the nature of the electron donor for the sulfate reducer in this AOM consortium is not clear. Also, the aggregates associated with the ANME-2 *Archaea* are unusually organized if the *Archaea* (methanogen) should be transferring electron donor to the sulfate reducer. These aggregates would require a long transfer distance for the intermediate electron donor between the inner part of the aggregate and the outer sulfate reducer population. This would increase the concentration of the electron donor in the aggregate, reducing the thermodynamic efficiency of the process. The most efficient arrangement would be the juxtaposition and direct cell-to-cell contact of sulfate reducers and methane oxidizers, minimizing the transfer distance of electron donor intermediates. It is possible, of course, that AOM might be conducted in some environments by organisms unassociated with clusters (Orphan *et al.*, 2002) and by organisms not yet identified. There are obviously critical aspects of the process that are still not understood.

5. ISOTOPE FRACTIONATION

5.1. Methanogenesis

Large isotope fractionations accompany the biological production of methane. Indeed, the isotope differences between the reactants and products of methanogenesis are large enough that fractionations are best calculated exactly, and not approximated as the isotope difference between the product and reactant. Therefore, fractionations, ε, are given as

$$\varepsilon_{A-B} = 1000(\alpha_{A-B} - 1) \tag{10.12}$$

and the fractionation factor, α, is

$$\alpha_{A-B} = \frac{(^{13}C/^{12}C)_A}{(^{13}C/^{12}C)_B} = \frac{(\delta_A + 1000)}{(\delta_B + 1000)} \tag{10.13}$$

where A is the reactant (DIC, for example), and B is the product (methane). In situations such as batch experiments, or in nature when Rayleigh distillation effects can clearly be identified, the isotopic composition of the reactants and products does not readily yield fractionation without considering the influence of the Rayleigh distillation. In this case, we can relate the initial isotope ratio of the reactant, $R_{r,0}$ to the isotope ratio after some extent of reaction $R_{r,f}$. The extent of the reaction is given by f, which is defined as the fraction of the initial substrate remaining:

$$R_{r,f} = R_{r,0} f^{(\alpha-1)} \tag{10.14}$$

or, substituting Equations 10.12 and 10.13 and solving for ε:

$$\varepsilon = \frac{1000 \ln\left(\frac{\delta_f - 1000}{\delta_0 - 1000}\right)}{\ln f} \tag{10.15}$$

If sediment porewaters are being explored, and if the system is highly evolved such that a large proportion of the original substrate (like DIC) has been converted to methane, then diagenetic modeling is best used to extract fractionation factors. In this case the fractionation is not simply the isotopic difference between DIC and methane, and the differential transport of $^{12}CH_4$ and $^{13}CH_4$ along their individual isotope gradients make a Rayleigh distillation model inappropriate (e.g., Alperin and Reeburgh, 1988). Diagenetic models might also be required to obtain fractionations associated with methane oxidation from environmental data (Alperin and Reeburgh, 1988) (see below).

A survey of the isotopic composition of methane in marine sediments ($\delta^{13}C_{CH4}$) reveals a typical range from −50 to −100‰ relative to the PDB standard (Whiticar *et al.*, 1986; Blair, 1998). Methane is similarly ^{13}C-depleted relative to the initial substrate (mostly DIC or acetate) from which the methane was formed, which typically has an isotopic composition ranging from +10 to −20‰ (Whiticar *et al.*, 1986; Blair, 1998). For lake sediments, large depletions of methane in ^{13}C are also observed, but generally less so than in marine environments, with $\delta^{13}C_{CH4}$ values mostly between −50 and −70‰ compared to a range in substrate $\delta^{13}C$ similar to the range in marine environments. The differences between the isotopic composition of methane in lake environments, compared to marine environments, have frequently been ascribed to differences in methane production pathway, with the CO_2/H_2 pathway providing the greatest fractionations (see below) and dominating in marine environments (Whiticar *et al.*, 1986). By contrast, acetoclastic methanogenesis is argued to produce lower fractionations and is viewed as the dominating methane production pathway in freshwater environments (Whiticar *et al.*, 1986).

As we have seen in Section 3.4, there is indeed a tendency for acetoclastic methanogenesis to dominate in freshwater environments and for the CO_2/H_2 pathway to dominate in the marine realm (Table 7.5). However, both pathways are operative in each environment, and in quite variable proportions. Thus, it is not proven that differences in methane production pathway account for the differences in the $\delta^{13}C_{CH4}$ between freshwater and marine environments. Other factors such as strain differences in fractionation and the intensity and pathways of methane oxidation (see below) can also, in principle, contribute to establishing the isotopic composition of methane in a given environment.

Blair (1998) found a strong relationship between the isotopic composition of methane and the deposition flux of metabolizable organic carbon for a range of different marine sediment environments, with higher deposition fluxes leading to less ^{13}C enrichment in the methane. The reasons for this were not clear, but Blair (1998) speculated that it could relate to a possible control of methane production rate on fractionation or to a relationship between methanogenic pathway and carbon flux.

Surveying the results from culture experiments, there are clear differences in the extents of fractionation between the different methane formation pathways. Thus, organisms using the CO_2/H_2 pathway fractionate ($\varepsilon_{CO2\text{-methane}}$) between 25 and 79‰ (Botz *et al.*, 1996; Chidthaisong *et al.*, 2002; and references in Whiticar *et al.*, 1986; Games *et al.*, 1978). Some of this variability seems to relate to between-strain differences. Some variability may also be due to temperature (Whiticar *et al.*, 1986), with temperature presumably influencing the physiology of the organisms. Consistent with this, Botz *et al.*, (1996) observed a modest 3‰ decrease in fractionation per 10 °C increase in temperature during methanogenesis by three different pure cultures from the genus *Methanococcus,* where a relationship between temperature and fractionation was found both between cultures, and for a given culture grown at different temperatures. For example, *M. thermolithotrophicus* displayed an average fractionation of 62‰ at 45 °C, 61‰ at 55 °C, and 56‰ at 65 °C, when grown in the same fermenter system.

During acetoclastic methanogenesis, the methanogen *Methanosarcina barkeri* displayed low fractionations of only around 21‰ (Krzycki *et al.*, 1987; Gelwicks *et al.*, 1989). Similar low fractionations of around 19‰ were also found for natural populations of methanogens growing on acetate from Wintergreen Lake, Michigan (Gelwicks *et al.*, 1994). Incubation experiments on natural sediments, where methane production pathways have been constrained, tend to yield higher fractionations for acetoclastic methanogenesis in the range of around 40 to 45‰ (Avery *et al.*, 1999). The database is small, but it seems to confirm that acetoclastic methanogenesis produces lower fractionations than methane produced from CO_2/H_2.

In an interesting series of experiments, Kryzcki *et al.* (1987) explored the fractionations produced by the same organism, *M. barkeri*, for a variety of different substrates, including methanol, acetate, and CO_2/H_2. The highest fractionations were associated with methanol reduction ($\varepsilon_{\text{methanol-methane}}$ = 75‰), the next highest with CO_2/H_2 ($\varepsilon_{CO_2\text{-methane}}$ = 46‰), and the lowest with acetate ($\varepsilon_{\text{acetate-methane}}$ = 21‰). Furthermore, Summons *et al.* (1998) found *M. barkeri* to fractionate by 50‰ during trimethylamine disproportionation ($\varepsilon_{\text{trimethylamine-methane}}$), with even higher fractionations of 80‰ produced from trimethylamine by *Methanococcoides burtonii.*

Thus, the high end of fractionations associated with CO_2/H_2, methanol, and trimethylamine as substrates approach the values found in nature (Figure 10.13). Of these, however, methanol and trimethylamine probably do not contribute greatly to the methane produced (Lovley and Klug, 1986). Overall, the CO_2/H_2 pathway and even mixtures of the CO_2/H_2 and acetate pathways can account for many occurrences of highly ^{13}C-enriched methane, as long as the upper range of fractionation is assumed. Future work will undoubtedly concentrate on quantifying methane production pathways, the isotopic fractionations associated with each pathway, and the isotopic

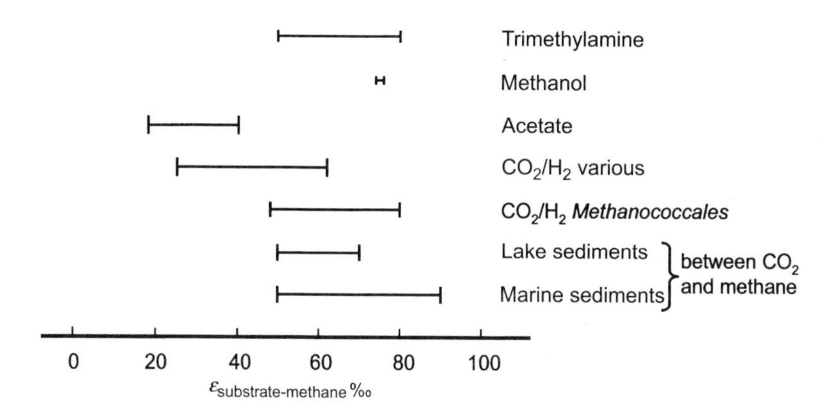

Figure 10.13 Summary of isotope fractionations by different groups of methanogens compared to the difference between the isotopic compositions of methane and DIC ($\Delta_{DIC\text{-}methane}$) in lake and marine sediments. References are in the text.

composition of methane to explore for closure in the methane isotope budget.

5.2. Methanotrophy

Isotope effects are also associated with the aerobic and anaerobic oxidation of methane. In virtually all cases, the isotopically lighter $^{12}CH_4$ is preferentially oxidized, leaving the residual methane ^{13}C enriched. In pure culture experiments, aerobic methanotrophy produces fractionations ($\varepsilon_{CH4\text{-}DIC}$) in the range of 5 to 31‰ (Barker and Fritz, 1981; Whiticar, 1999). The experiments of Coleman *et al.* (1981) demonstrated a temperature dependence on fractionation in which organisms grown at 24 °C produced a fractionation of 25‰, and a smaller fractionation of about 13‰ was associated with methane oxidation at 11 °C. The direction and magnitude of a temperature effect, however, may depend on the physiology of individual organisms. Field experiments produce a similar range in fractionation with evidence also for a temperature effect. Thus, for experiments on methane oxidation in tundra soils King *et al.* (1989) measured fractionations of 27‰ at 14 °C and a reduced fractionation of 16‰ at 4 °C. By contrast, Tyler *et al.* (1994) found a small negative relationship between fractionation and temperature during methane uptake in forest soils. In these experiments fractionations were around 25‰ at air temperatures of 7 to 10 °C, decreasing to 16 to 20‰ at higher air temperatures of 20 to 25 °C.

Tyler *et al.* (1994) attributed these different types of fractionation results to different fractionation mechanisms. The forest soils are water unsaturated, and Tyler *et al.* (1994) argued that fractionations arise from the diffusional transport of methane into the soils, where a higher diffusion coefficient for $^{12}CH_4$ should produce fractionations in the range observed. There should be little temperature dependence with this mechanism of fractionation. By contrast, in the tundra soils and in the pure culture experiments, fractionations are biologically mediated, and the observed temperature dependency has a biological cause.

Fractionations also accompany the production of microbial biomass during methanotrophy. For methanotrophs utilizing the RuMP pathway of carbon assimilation, fractionations depended on whether a soluble (sMMO) or particulate (pMMO) form of the methane monooxygenase enzyme is used (Jahnke *et al.*, 1999). With pMMO, fractions of $\Delta_{biomass} = \delta^{13}C_{methane} - \delta^{13}C_{biomass}$ of around 24‰ were found, and with the sMMO fractionations were smaller, at around 13‰. Fractionation into the membrane lipids was of a similar magnitude, and fractionations increased with decreasing temperature. With the serine pathway of carbon assimilation, fractionations were more variable, generally smaller than with the RuMP pathway, and even negative (−7 to −8‰) when CH_4 was limiting. With the serine pathway, there were generally larger fractionations into the lipids than into the total biomass.

The direct determination of isotope fractionation during AOM has not yet been reported, but estimates have been made from the diagenetic modeling of concentration and isotope profiles. Generally, the isotopic composition of methane shifts to heavier values through the zone of AOM. In two modeling studies, fractionations of around 9‰ (Alperin and Reeburgh, 1988) and 12‰ (Martens *et al.*, 1999) have been obtained. Overall, these fractionations are somewhat lower than those obtained during aerobic methanotrophy. Differences in fractionation should not be surprising, however, given the completely different microbiological control of each process. We anticipate that culture-based fractionation results will soon be available from *in vitro*-type experiments on AOM as conducted by Nauhaus *et al.* (2002).

Chapter 11

The Phosphorus Cycle

1. INTRODUCTION

Phosphorus is a vital functional and structural component of all living organisms. It occurs universally in living cells as phosphate in essential biomolecules such as nucleic acids (DNA and RNA), in energy transfer systems (NAD(P)H and ATP), and in cell membranes (phospholipids). Thus, phosphorus is an essential nutrient for biomass production. Phosphorus has four possible oxidation states (−3, 0, +3, and +5), but only the +5 state is stable under natural conditions in aqueous solutions. Phosphorus is not involved in redox cycling, except for some rare exceptions. Our main interest here, therefore, is the factors influencing the availability of phosphorus to organisms in nature. We consider the biochemical processes by which phosphorus is incorporated into cells and focus on those processes responsible for both phosphorus release and phosphorus sequestration in nature. Many of these are microbially mediated, and we highlight the role of microbes in the phosphorus cycle.

0-12-026147-2
© 2005 Elsevier Inc.
All rights reserved

2. THE GLOBAL PHOSPHORUS CYCLE

We begin by taking a broad look at the global phosphorus cycle. Many of the points raised here are considered in more detail in the following Sections. Phosphorus is the tenth most abundant element on Earth, with an average crustal abundance of 0.1% by weight. Within the lithosphere, phosphorus is found as phosphate in approximately 300 naturally occurring minerals. Among these, apatite [$Ca_5(PO_4)_3(F,Cl,OH)$] is by far the most abundant (Ruttenberg, 1993; Fölmi, 1996). In the Earth's crust, most phosphorus is contained in sediments (including soils) (Schenk and Jackson, 2002) and sedimentary rocks, of which a small fraction (0.00025%) (Table 11.1, Figure 11.1) is stored in economic deposits, primarily mined for use by the fertilizer industry (Jahnke, 1994). The phosphorus reservoirs containing living and dead organics, as well as dissolved inorganic phosphorus, collectively comprising the biologically active pools, account for only about 0.01% of the phosphorus deposited in sediments and sedimentary rocks. Of the biologically active phosphorus pools, only about 1% is found in living biomass (Table 11.1).

The main source of phosphorus to the oceans is riverine runoff, where phosphorus comes from the weathering of sedimentary deposits and the decomposition of terrestrial organics (Benitez-Nelson, 2000). Much of the riverine phosphorus is either adsorbed in inorganic form onto particles or bound in organic particles, which are rapidly sedimented in near-shore deposits (Figure 11.1). Some of this phosphorus is released to the marine water column as sediment organics are oxidized (Compton *et al.*, 2000). However, a major portion of the phosphorus enters the open oceans in dissolved inorganic and organic forms.

The global average concentration of dissolved inorganic phosphorus in the oceans is 2.3 μM, and this pool generally accounts for 50–100% of the total dissolved phosphorus (depending on location), with the remaining bound in organic form (Benitez-Nelson, 2000; Karl and Björkman, 2002). There is also a small net atmospheric transport of phosphorus to the ocean from land via

Table 11.1 Major phosphorus reservoirs on Earth

Type of pool	Location	Pool size (g P)
Aerosols	Atmosphere	2.8×10^9
Living biomass (LPOP)	Aquatic and terrestrial	2.8×10^{15}
Dead organics (DPOP + DOP)	Aquatic and terrestrial	3.0×10^{17}
Dissolved inorganic (PO_4^{3-})	Aquatic and terrestrial	
Inorganic (mostly apatite)	Sediments and sedimentary rock	4.0×10^{21}
Inorganic (apatite)	Earths core and mantle	6.0×10^{24}

From Jahnke (1994), Chameides and Perdue (1997), and Compton *et al.* (2000).

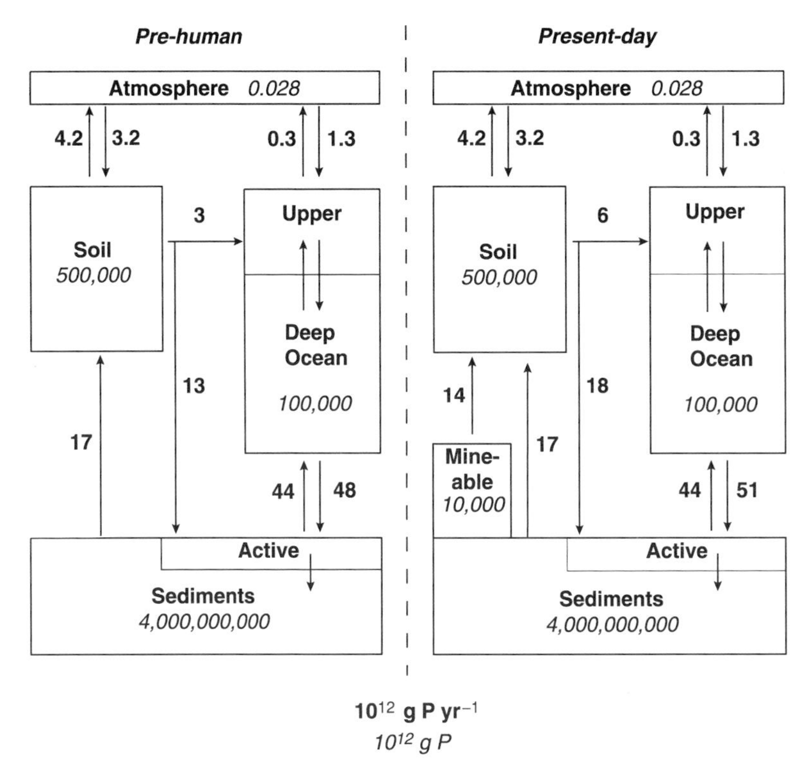

Figure 11.1 The current best estimates of the pre-human and present-day global phosphorus cycle. Fluxes are given as 10^{12} g yr^{-1} and reservoirs as 10^{12} g. All fluxes and pools represent the sum of living and dead, as well as particulate and dissolved phosphorus. Modified from Jahnke (1994), Howarth *et al.* (1995), Chameides and Perdue (1997), Benitez-Nelson (2000), and Compton *et al.* (2000).

dust and aerosols, but only about one-third of this material is soluble in sea water (Graham and Duce, 1982). Upwelling, vertical advection and eddy diffusion deliver phosphorus from the deep ocean to fuel primary production in the surface layer of the ocean. In total, marine plankton contain 50–70 × 10^{12} g P, about 2% as much phosphorus as in the terrestrial living biomass, which is dominated by long-lived forests (Compton *et al.*, 2000). If we consider the problem dynamically, marine primary production incorporates about 1200 × 10^{12} g P yr^{-1} on a global scale, which is three to four times higher than the terrestrial incorporation rate (Colman and Holland, 2000). Thus, the turnover time of the living oceanic biomass is short, ranging from a few days for prokaryotes to about a week for phytoplankton and a few months for zooplankton (Benitez-Nelson and Buesseler, 1999). By comparison, the average turnover time for terrestrial living biomass is at least 10 years.

Dissolved phosphorus is generally delivered upward from the deep ocean to the surface in a ratio with nitrogen similar to the average Redfield ratio of plankton (N:P of 16:1). A slight nitrogen deficit is frequently observed, however, supporting the suggestion that nitrogen, and not phosphorus, limits primary production in the open ocean (Falkowski, 2000). By contrast, from a geochemical perspective over long time scales, phosphorus is generally considered the principal limiting nutrient (e.g., Broecker and Peng, 1982; Van Cappellen and Ingall, 1996; Bjerrum and Canfield, 2002). This is because variability in factors such as ocean anoxia and weathering on land can change the inventory of phosphorus in the oceans (Van Cappellen and Ingall, 1996). As shown in Chapter 7, the inventory of nitrogen is controlled mainly by internal ocean processes, including denitrification and nitrogen fixation. It is believed that the balance of these processes will adjust the nitrogen inventory to match the phosphorus inventory to near-Redfield stoichiometry. Thus, if the phosphorus inventory increases, nitrogen fixation will boost the size of the nitrogen pool. If the phosphorus inventory drops relative to nitrogen, then nitrogen fixation rates will be suppressed until denitrification reduces the nitrogen inventory to near-Redfield stoichiometry. In this view, phosphorus availability is the driver, and internal ocean processes will adjust the nitrogen inventory to balance that of phosphorus. As the inventories of both these principal nutrients are dynamic, they need not be in exact Redfield balance.

A large amount of phosphorus is transferred to the deep ocean when biogenic particles sink from the surface. Although a substantial part of the phosphorus is regenerated by decomposition within the water column, some of the organic debris settles onto the sediment surface. A fraction of this organic material is buried in the sediment, ultimately removing phosphorus from the biological cycle (Delaney, 1998; Benitez-Nelson, 2000). In total, several important mechanisms remove phosphorus from the marine water column, including organic phosphorus burial, burial into special inorganic phosphorus-rich deposits known as phosphorites, the formation of finely dispersed inorganic precipitates in sediments, and adsorption onto clays and iron oxyhydroxide particles (Föllmi, 1996). Our understanding of the magnitudes of these different phosphorus sinks is uncertain, giving rise to a range in calculated ocean phosphorus residence times from 10,000 to 80,000 years (Froelich *et al.*, 1982; Ruttenberg, 1993; Delaney, 1998; Benitez-Nelson, 2000).

3. MICROBIAL PHOSPHORUS METABOLISM

Phosphorus is taken up as inorganic phosphate by almost all microorganisms for use in the synthesis of nucleic acids, nucleotides (ATP), phospholipids, phosphosugars, polyphosphates, several cofactors, some proteins, and other cell components. Microorganisms have evolved active membrane-transport systems for the regulated acquisition of phosphate from the environment, such as the Pts phosphate transport system of *Escherichia coli* (Silver and Walderhaug, 1992). Not all of this phosphorus, however, is used immediately by microorganisms. Some prokaryotes, for example, can store excess phosphate in the form of condensed polyphosphate strands (volutin) when they are temporarily limited by other nutrients. As conditions allow growth to resume, the polyphosphates are degraded and phosphate is mobilized for synthesis of ATP and other cellular constituents (Figure 11.2).

Phosphorus bound in organic form is usually not directly available to living organisms, although small organic molecules can sometimes be assimilated (Cotner and Wetzel, 1992). Most organic phosphorus must first be released from the organic moiety by mineralization before it is available for

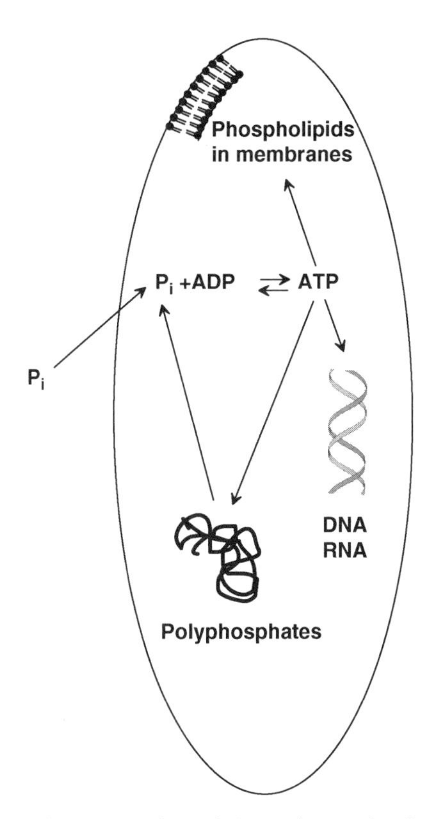

Figure 11.2 Schematic presentation of the major pools of phosphorus and phosphorus transformations in a prokaryote cell. P_i; for inorganic phosphorus.

assimilation. This is accomplished through hydrolytic action by a variety of special enzymes (Jansson *et al.*, 1988; Shan *et al.*, 1994). For example, phosphate is liberated from nucleic acids by nucleases, which yield nucleotides, followed by the action of nucleotidases, which yield nucleosides and inorganic phosphate. In another example, phosphate liberation from compounds such as phosphoproteins, phospholipids, and glycerol phosphates requires the enzymes phosphomonoesterases and phosphodiesterases (Nelson and Cox, 2000).

Assimilation of inorganic phosphorus occurs through energy-demanding ATP formation. The processes of ATP formation have been outlined in previous Chapters (see Chapters 3 and 4) and are not reviewed here. The further synthesis of phosphate-containing biomolecules typically proceeds through an initial reaction between a carbinol group and ATP, forming monomeric phosphate esters in the presence of an appropriate kinase. The process shown in Equation 11.1 is catalyzed by glucokinase.

$$\text{glucose} + \text{ATP} \rightarrow \text{glucose 6-phosphate} + \text{ADP} \tag{11.1}$$

As mentioned previously, phosphorus compounds are rarely used in microbial redox metabolism, as phosphorus is most stable only in the +5 oxidation state in aquatic environments. Nevertheless, an interesting finding has revealed that a sulfate reducer related to *Desulfospira joergensenii*, isolated from marine sediments, has a metabolism based on the oxidation of phosphite (+3) to phosphate, coupled to the reduction of sulfate to sulfide (Schink and Friedrich, 2000) (see also Chapter 9):

$$4HPO_3^{2-} + SO_4^{2-} + H^+ \rightarrow 4HPO_4^{2-} + HS^- \tag{11.2}$$

The ecological significance of this pathway is unclear, but it is probably small due to the low availability of phosphite in the environment.

4. PHOSPHORUS GEOCHEMISTRY AND ADSORPTION REACTIONS

Inorganic phosphorus occurs in both soluble and insoluble forms in aquatic environments. Soluble inorganic phosphorus exists as one of the dissociated forms of the triprotic phosphoric acid (H_3PO_4). The most common form in nature is, from Equation 11.3, either $H_2PO_4^-$ or HPO_4^{2-}, depending on the pH (Stumm and Morgan, 1996).

$$H_2PO_4^- \rightleftharpoons HPO_4^{2-} + H^+, K_a = 10^{-7.2} \tag{11.3}$$

Thus, at a pH of 8, typical of marine surface water, 16% of the inorganic phosphorus is in the form of $H_2PO_4^-$ and the remainder is HPO_4^{2-} (the fully protonated and fully deprotonated forms are insignificant at near-neutral pH), while at a pH of 7, $H_2PO_4^-$ accounts for 37% of the total phosphorus. Inorganic phosphorus, however, is not infinitely soluble; it forms insoluble precipitates with many divalent cations of which Ca^{2+} and Fe^{2+} are the most important in natural environments. The most common authigenic phosphate minerals are vivianite, $Fe_3(PO_4)_2 \cdot 8H_2O$, and carbonate fluorapatite (CFA), $Ca_5(PO_4)_3(X)$, where X consists of variable proportions of CO_3^{2-}, F^-, and OH^-.

In addition, orthophosphate adsorbs onto the oxides of Fe^{3+} and Al^{3+} and some positively charged clay mineral surfaces. The adsorption is reversible and can generally be described by simple adsorption isotherms. For example, the adsorption of phosphorus onto iron oxides formed during the oxidation of Fe^{2+} from mid-ocean hydrothermal systems can be described by the following linear isotherm:

$$P_{ads} = K_{ads}(P_{sol}/Fe_{ox}) \qquad (11.4)$$

where P_{ads} (μM) is the amount of adsorbed phosphate, K_{ads} (μM^{-1}) is the adsorption coefficient, P_{sol} (μM) is the concentration of orthophosphate in sea water, and Fe_{ox} (μM) is the concentration of Fe oxides. For iron oxide particles formed from hydrothermal solutions throughout the ocean, K_{ads} is remarkably constant, with an average value of $0.07 \pm 0.01\ \mu M^{-1}$ (Feely *et al.*, 1998). Absorption is also pH sensitive. For example, maximum adsorption occurs between phosphate and goethite at pH values of 5 to 6, dropping to about 60% of this amount at a pH of 9 (Figure 11.3) (Hawke *et al.*, 1989). The adsorption of phosphate onto aluminum oxide surfaces is also pH sensitive. For example, Chen *et al.* (1973) found the maximum adsorption between phosphate and kaolinite at pH values of 4 to 5, with adsorption dropping substantially above a pH of 6. McLaughlin *et al.* (1981) studied the adsorption of phosphate onto a range of poorly crystalline to well-crystallized Al and Fe oxides and oxyhydroxides and found that the adsorption was highly surface area dependent. They also found that the adsorption capacities for various Fe and Al oxides (and oxyhydroxides) were similar at constant pH and surface area (McLaughlin *et al.*, 1981).

The adsorption of phosphate, particularly onto iron oxides, has the potential to greatly influence the global cycling of phosphorus. Recent estimates reveal that adsorption onto iron oxides accounts for about 9% of the phosphorus removed from the marine water column (Bjerrum and Canfield, 2002). Thus, the biogeochemistries of iron and phosphorus are tightly coupled, as shown in Section 5.

Fractionation of phosphate in sediment by sequential extraction

To understand the burial and diagenesis of phosphorus in aquatic sediments, it is necessary to separate, identify, and quantify the solid-phase reservoirs of sedimentary phosphorus. Various sequential extraction procedures have been suggested (Boström *et al.*, 1982; Petterson *et al.*, 1988; Ruttenberg, 1992; Jensen and Thamdrup, 1993). A sequential extraction always starts with the weakest extractant and, after a number of steps, ends up with the strongest in the last step. The procedure presented here (Figure 11.4), is based on Lucotte and d'Anglejan (1985), as adapted by Jensen and Thamdrup (1993). This is similar to the technique presented by Ruttenberg (1992), and all of them are widely used in both freshwater and marine sediments.

In step 1 (exch-RP), subsamples of sediment are shaken with anoxic 1 M $MgCl_2$ (freshwater sediment) or anoxic 0.46 M NaCl (marine sediment) to extract loosely adsorbed reactive phosphate and reactive as well as non-reactive porewater phosphate. In step 2 (Fe-RP), the sediment is shaken in an anoxic bicarbonate–dithionite solution (BD: 0.11 M $NaHCO_3$ + 0.11 M $Na_2S_2O_4$, pH 7) to extract RP adsorbed to reducible Fe and Mn species. In step 3 (ads-RP), the sediment is treated with 0.1 M NaOH to extract RP adsorbed to clay minerals, aluminum oxides, and humic acids. Some phosphorus associated with organisms, such as polyphosphates, will also be extracted in this step. In step 4 (Ca-P), the sediment is shaken with 0.5 M HCl to extract calcium-bound phosphorus (e.g., apatite). In step 5 (org-P), the final step, the remaining sediment is combusted at 520 °C and subsequently is boiled in 1 M HCl to extract the residual phosphorus. This pool represents the most refractory of the organic phosphorus components. An example of results from phosphate fractionation in lake and marine sediments is shown in Figure 11.5. Phosphorus is mostly found as org-P and Fe-RP in lake sediment, whereas marine sediment is dominated by Ca-P and ads-RP.

We emphasize that sequential extraction methods are operationally defined, and the fractions obtained in the various extraction steps, such as steps 1 and 3, typically contain more than one well-defined phosphorus pool. In other cases, one specific pool, such as organic-bound phosphorus, is extracted by different steps based on the age and reactivity of the organic matter and is therefore split into different fractions. As a consequence, careful execution of the extraction procedures according to the standardized guidelines is necessary to obtain comparable results within and between studies.

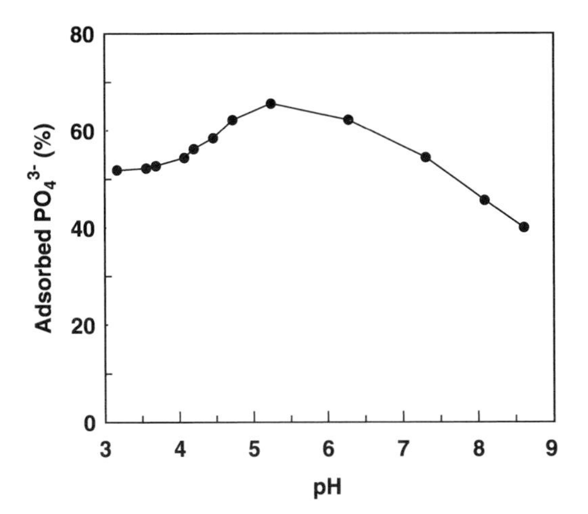

Figure 11.3 Percent adsorption of phosphate (30 μM) by goethite (0.25 g liter^{-1}) as a function of pH in water containing major seawater ions. Modified from Hawke *et al.* (1989).

5. THE ROLE OF MICROBES IN PHOSPHORUS CYCLING

Microbial processes largely control the mobilization and immobilization of phosphorus in aquatic environments. During organic matter mineralization, organic-bound phosphorus is partly assimilated and partly released as dissolved inorganic phosphorus (DIP) by the heterotrophic microbial community. The partitioning between release and uptake depends primarily on the C:P ratio of the organic substrates. Tezuka (1990) studied the fate of phosphorus in natural assemblages of freshwater microbes grown aerobically in mineral media supplemented with simple organic C, N, and P compounds (glucose, asparagine, and sodium glycerophosphate) to give a wide range of C:P ratios. No net phosphate mineralization occurred when the C:P ratio of the organic matter was above 80; all phosphorus in this situation was assimilated and incorporated into microbial biomass (Figure 11.6). However, the C:P ratio of the microbial biomass was strongly dependent on the C:P ratio of the available substrate, ranging from 31:1 to 515:1 when substrate C:P ranged from 60:1 to 1000:1. The low C:P ratios are lower than the Redfield ratio for plankton (106:1). Therefore, some of the variability in the microbial C:P ratios is probably due to the capability of prokaryotes to internally store excess phosphorus in the form of polyphosphates (Gächter and Meyer, 1993). When the C:P ratio of the organic substrate was below around 60, an increasing proportion of phosphate

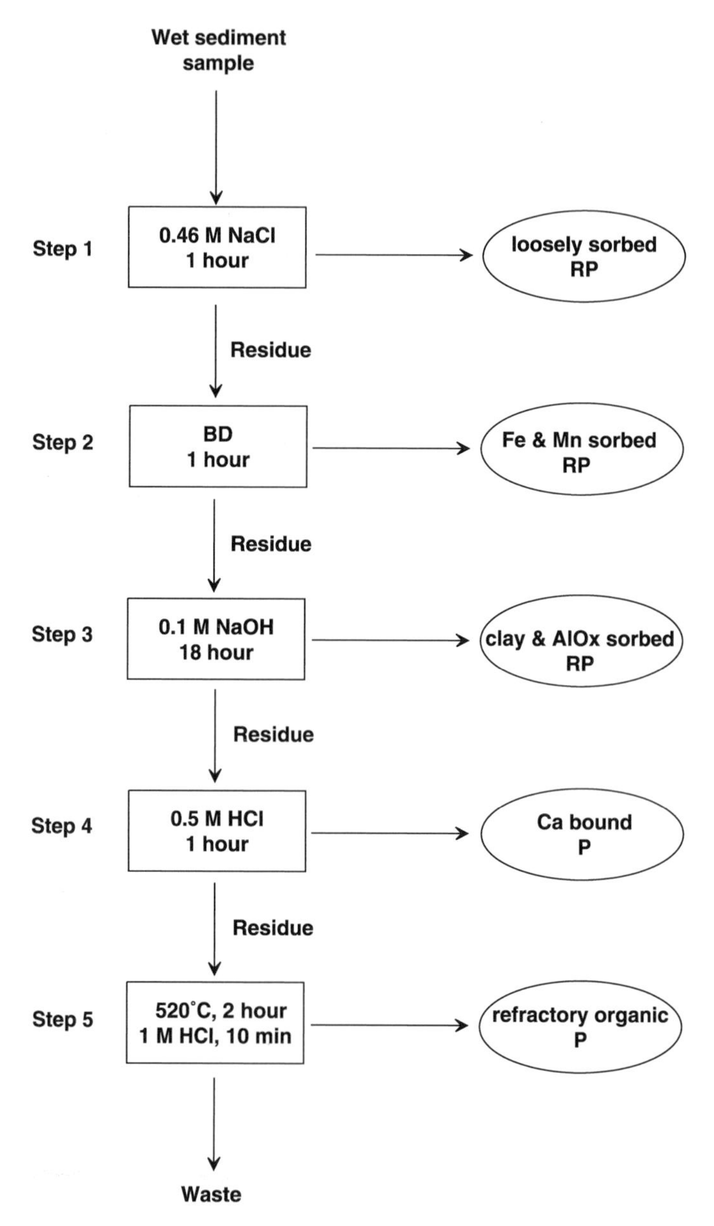

Figure 11.4 Simplified presentation of the sequential extraction scheme used to determine phosphorus pools in marine sediments. Steps 1 and 2 are performed under anoxic conditions. Modified from Jensen and Thamdrup (1993).

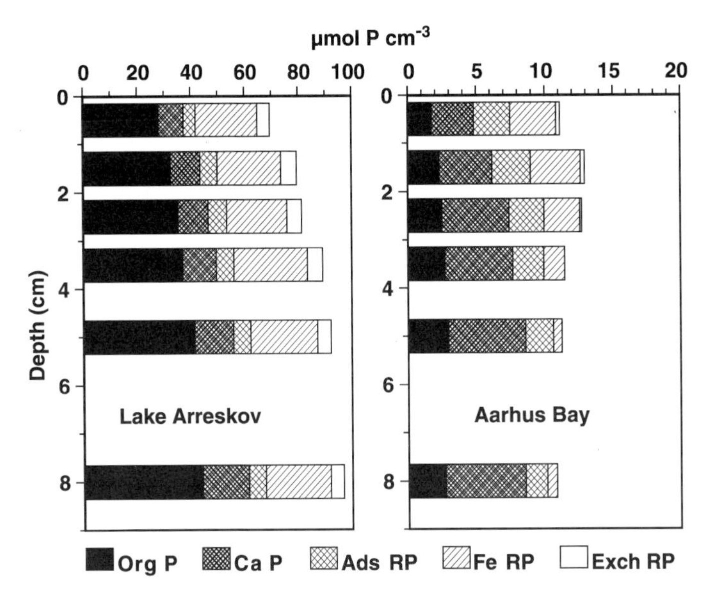

Figure 11.5 Particulate phosphorus fractions in profundal sediments of Lake Arreskov (3 m water depth; Andersen and Ring, 1999) and in sediments of the central Aarhus Bay (16 m water depth; Jensen *et al.*, 1995).

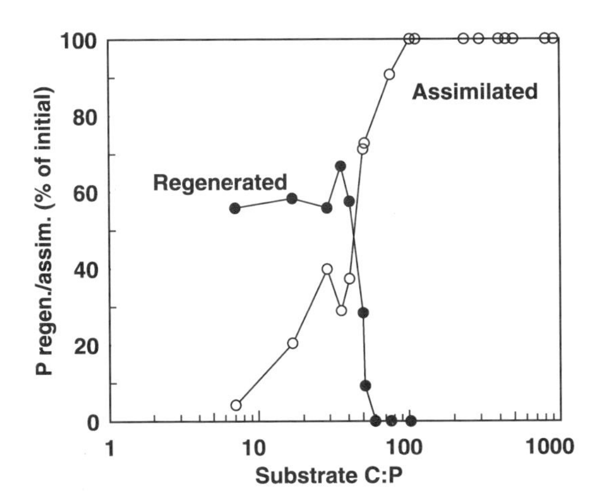

Figure 11.6 Percentage of regenerated (liberated to solution) and assimilated (incorporated) phosphate as a function of the C:P ratio of organic substrates in cultures of aerobic bacteria. Modified from Tezuka (1990).

was released in inorganic form (Figure 11.6), although little was liberated (regenerated) at a higher C:P ratio.

The boundary in C:P ratio between mobilization and immobilization in the aerobic cultures studied by Tezuka (1990), however, is lower than that found in many natural environments. Thus, the microbial degradation of phytoplankton having a C:P ratio of 106 (Redfield ratio) is known to result in significant phosphorus liberation (Garber, 1984). Net liberation of phosphate has also been shown to occur in sediments with an organic C:P ratio approaching 400 (Andersen and Jensen, 1992). There are two likely explanations for these discrepancies. First, the C:P ratio of bulk organic matter present in nature does not reflect the C:P ratio of the fraction actually undergoing degradation. A large fraction of the organic matter in most aquatic environments is composed of materials resistant to mineralization, such as structural polymers and old, previously degraded organic matter (Canfield, 1994). Therefore, the C:P ratio of the organic matter degrading is often lower than the bulk organic matter (Luckge *et al.*, 1999). Second, microbial growth yield is relatively high under oxic conditions, resulting in significant immobilization of phosphorus into microbial biomass. Under anoxic conditions, such as in sediments, anaerobic prokaryotes (e.g., sulfate reducers and associated fermenters) must degrade some four times the organic carbon or more to gain the same amount of energy for growth. As a consequence, microbes are expected to immobilize much less phosphorus under anoxic conditions given the same organic substrate (Howarth *et al.*, 1995).

Insoluble forms of inorganic phosphorus (Ca-, Al-, Mn-, and Fe-bound phosphate) may be solubilized directly or indirectly by the action of prokaryotes. Reducing conditions resulting from heterotrophic microbial activity promote the dissimilatory reduction of Fe(III). Consequently, a large fraction of phosphate immobilized onto ferric oxyhydroxides is released in dissolved form (Figure 11.7) (Anschutz *et al.*, 1998). Reduction of Fe(III) can also be mediated indirectly by microorganisms when the sulfide produced by sulfate reduction reacts with Fe oxides (see Chapters 8 and 9). The resulting sulfide-mediated Fe(III) reduction, and the subsequent precipitation of Fe sulfides, has the potential to release virtually all iron-bound phosphorus in deeper sediment layers (Jensen *et al.*, 1995). Much of this phosphorus may, however, be re-trapped onto Fe oxides at the sediment surface. Indeed, mobilized phosphate can be recycled several times across the redox boundary before escaping the sediment (Sundby *et al.*, 1992).

The active production of organic acids and CO_2 by fermentation and the chemolithoautotrophic oxidation of sulfide and ammonium may generate a thin pH minimum layer in near-surface sediments. In this zone the pH may approach 7 (see Canfield and Raiswell, 1991), and in certain sediments, such as those found in salt marshes and mangrove forests, pH may be as low as 5 (Cai *et al.*, 2000). As the adsorption of phosphate onto iron oxyhydroxides is

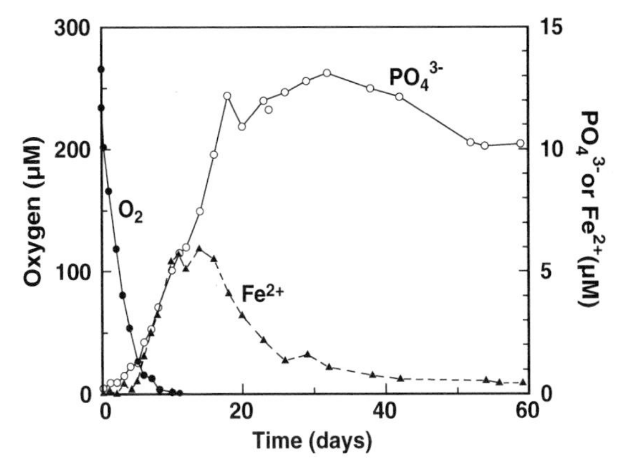

Figure 11.7 Temporal changes in overlying water concentrations of oxygen, DIP, and dissolved iron (Fe^{2+}) inside benthic chambers placed at 6 m depth in Gullmarsfjorden, Sweden. Sediment respiration was allowed to consume oxygen, which reached zero after about 10 days. When oxygen approached zero, DIP and dissolved iron was released in comparable amounts, indicating that adsorbed phosphate was liberated when iron oxyhydroxides were reduced. The subsequent disappearance of dissolved iron after 15 days was caused by precipitation with sulfide generated by sulfate reduction in the sediment. Modified after Sundby *et al.* (1986).

generally favored by low pH (Figure 11.3) (Carpenter and Smith, 1984), the oxic–anoxic interface may, under these conditions, form a very effective phosphorus trap when sufficient reactive Fe(III) is present. Surface sediments, therefore, can form the so-called "iron cap." The trap is so effective in some environments that the DIP flux is reversed, with water column phosphate actually diffusing into the sediment (McManus *et al.*, 1997). By contrast, high near-surface pH values resulting from phototrophic activity in microbial mat environments, for example, reduce the adsorption capacity of Fe oxides, potentially displacing and releasing adsorbed phosphate (Andersen, 1975).

6. PHOSPHORITE FORMATION

Sediments composed primarily of carbonate fluorapatite (phosphorite) are formed in highly productive continental margins such as the upwelling regions of the eastern Pacific and the south eastern Atlantic Oceans (Cook *et al.*, 1990). Accordingly, it is speculated that phosphorite could represent a significant sink for seawater phosphate in certain periods of Earth's history (e.g., Föllmi, 1996). Phosphorites occur as slowly growing (1 to 10 mm per thousand years) nodules, pebbles, slabs, and conglomerates. Where modern phosphorites form, the sediment porewaters are enriched in DIP (Jahnke *et al.*, 1983; Froelich *et al.*, 1988; Fölmi, 1996). Some of the DIP enrichment comes from rapid organic matter mineralization, and some probably also comes from iron-phosphate redox cycling near the sediment water interface, as well as fish bone dissolution (Van Cappellen and Berner, 1988; Heggie *et al.*, 1990; Van Cappellen and Berner, 1991).

Much of the carbonate fluorapatite forming into phosphorites is precipitated from DIP-enriched sediment porewaters by the reaction shown in Equation 11.5 (e.g. Froelich *et al.*, 1988; Van Cappellen and Berner, 1991).

$$10Ca^{2+} + (6-x)PO_4^{3-} + xCO_3^{2-} + (2+x)F^- \rightarrow Ca_{10}(PO_4)_{6-x}(CO_3)_xF_{2+x} \quad (11.5)$$

The formation pathway may be complex, involving metastable intermediate compounds. Indeed, the chemical profiles of F^- and PO_4^{3-} in modern sediments supporting carbonate fluorapatite precipitation do not show simple equilibrium with respect to a single solid phase (Froelich *et al.*, 1988). Off the coast of Peru, porewater profiles suggest that first a metastable phosphate-rich phase is precipitated, followed by slow incorporation of F^- (and carbonate), forming more stable carbonate fluorapatite (Froelich *et al.*, 1988).

While prevailing opinion holds that phosphorite is formed *in situ* by inorganic processes with only an indirect role of microbes, there may also be a microbial connection, at least in some cases (Nathan *et al.*, 1993). Bacterial genera, such as *Pseudomonas* and *Acinetobacter*, sequester phosphate as polyphosphate under oxic conditions. The polyphosphate is hydrolyzed and phosphate is released under anoxic conditions, possibly contributing to DIP enrichment and authigenic phosphorite generation (Nathan *et al.*, 1993). Slow sediment deposition rates may also be necessary to concentrate slow-growing phosphorite layers and nodules (e.g., Föllmi *et al.*, 1996). Altogether, taken together, the processes of phosphorite formation are not that well understood.

7. PHOSPHORUS CYCLING IN AQUATIC ENVIRONMENTS

We have outlined some of the basic considerations governing phosphorus cycling in the previous Sections. However, the specifics of phosphorus cycling may vary strongly depending on local conditions (Jensen *et al.*, 1995; Reynolds and Davies, 2001). For this reason, we next compare and contrast phosphorus cycling in a number of representative freshwater and marine systems.

7.1. Contrasting phosphorus dynamics in freshwater and marine systems

While similarities exist, there are also some fundamental differences in the phosphorus cycle between freshwater and marine environments. Phosphorus is generally the major nutrient limiting phytoplankton growth in freshwater lakes, whereas nitrogen limits production in the coastal ocean (Hecky and Kilham, 1988; Vitousek and Howarth, 1991) and in many areas of the open ocean. Caraco *et al.* (1989) examined 23 different coastal marine and freshwater environments and found that the switch from N to P limitation is linked to the control of sulfate concentration on the release of DIP from sediments. They found that DIP release increased when high sulfate availability stimulated sulfate reduction and induced the dissolution of Fe oxides and the subsequent precipitation of Fe sulfides (Figure 11.8A).

Gunnars and Blomqvist (1997) observed that when shifting from anoxic to oxic conditions, near-bottom DIP remained in solution above brackish water sediments. By contrast, active DIP removal was observed in freshwater systems subjected to the same switch. This contrasting behavior was explained by the different availability of dissolved Fe in the anoxic bottom waters. From their study, (Figure 11.8B) and from the literature, Gunnars and Blomqvist (1997) found that the dissolved Fe:P ratio in anoxic bottom water was generally much lower (Fe:P ratio of 0.008-1.1) in marine systems than in freshwater systems (Fe:P ratio of 1-280). The deficiency of reactive iron (low Fe:P ratios) in marine systems is related to the presence of sulfate and the scavenging and immobilization of Fe as iron sulfides (Caraco *et al.*, 1989). In such a situation, DIP will not be scavenged by Fe oxides at the onset of oxic conditions.

7.2. Water column phosphorus cycling

Open ocean surface waters (0–100 m depth) typically have low DIP ($<0.1\,\mu$M) and elevated dissolved organic phosphorus (DOP) ($>0.2\,\mu$M) concentrations (Figure 11.9). However, when moving down the water column, DIP and DOP exhibit an inverse relationship (Loh and Bauer, 2000), with gradually increasing DIP and decreasing DOP levels down to about 1000 m ($>3\,\mu$M DIP and $<0.1\,\mu$M DOP). This general pattern is caused by DIP uptake, net biomass production, and net release of DOP in surface waters followed by the mineralization of both dissolved and particulate organic phosphorus with depth.

Most of the DIP assimilated by phytoplankton in the euphotic zone of lakes, and especially in the open ocean, is locally regenerated back to DIP,

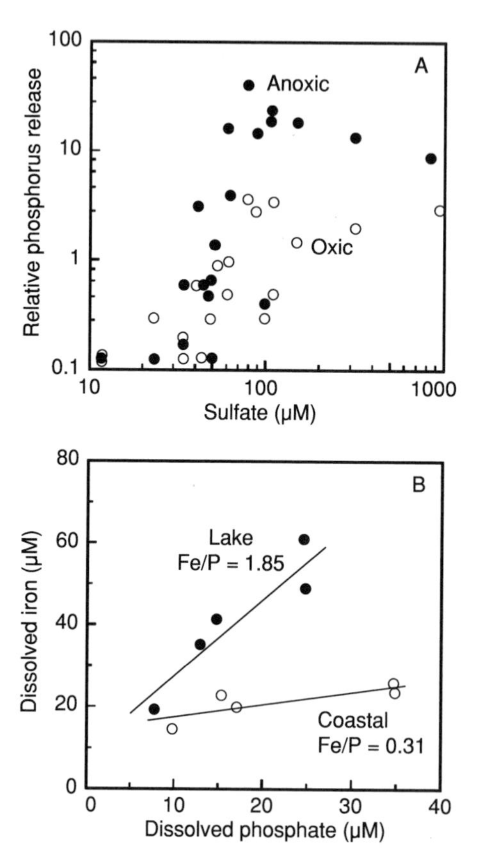

Figure 11.8 (A) The ratio of DIP to DIC release from sediments as a function of sulfate concentration in surface waters of 23 different marine and freshwater systems. Release under both oxic (open symbols) and temporary anoxic conditions (solid symbols) is shown. Modified from Caraco *et al.* (1989). (B) Concentration of dissolved iron plotted against the concentration of DIP in a 25 cm overlying water column after 12 days of anoxic sediment incubation. Five lake sediment cores were obtained from Lake Mälaren, Sweden (filled symbols), and five coastal marine sediment cores were obtained from Yttre Hållsfjärden in the Swedish archipelago (Baltic Sea) (open symbols). Lines are drawn according to least squares linear regression, and the slope (Fe/P) is given. Modified from Gunnars and Blomqvist (1997).

sustaining high primary production. Thus, typically less than 10% of the assimilated phosphorus is exported from the euphotic zone of the open ocean (Karl and Björkman, 2002). The uptake of DIP by phytoplankton and subsequent DOP release are usually coupled with the uptake and mineralization of DOP back to DIP by microorganisms. Björkman *et al.*

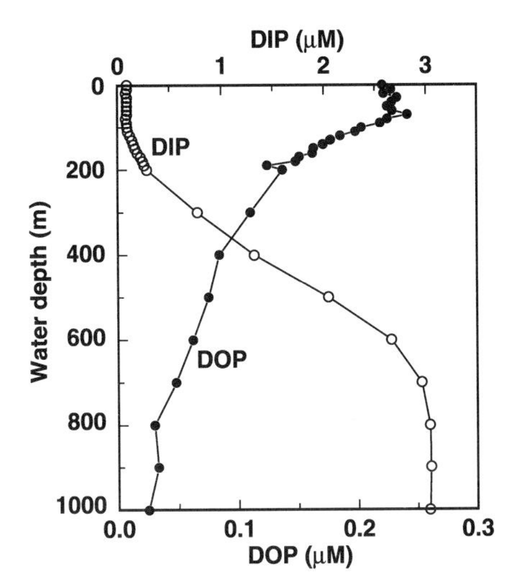

Figure 11.9 Vertical distribution of DIP and DOP in the North Pacific off Hawaii. Modified from Karl and Björkman (2002).

(2000) found that net DOP production in the North Pacific subtropical gyre is 10 to 40% of the net DIP uptake and that the turnover time of the DOP pool is 60–300 days.

Low-molecular-weight and labile DOP can directly serve as a phosphorus source for the growth of many planktonic algae and prokaryotes without the need for prior hydrolytic alteration (Wanner, 1993). Larger DOP molecules, however, must be enzymatically hydrolyzed at the cell surface or in the surrounding water before the generated DIP can be assimilated. Directly assimilated DOP can be used as biosynthetic precursors and thus enhance growth efficiency (Rittenberg and Hespell, 1975). Microalgae and heterotrophic prokaryotes may therefore compete for both DIP and DOP in phosphorus-limited open waters (Rhee, 1972).

In coastal marine environments, there is generally a tight metabolic coupling between microalgae that generate organic phosphorus and heterotrophic prokaryotes that regenerate DIP from organic sources. This algal–prokaryote coupling is, however, less pronounced in most nitrogen-poor aquatic environments. Here prokaryotes are frequently both the dominant producers (cyanobacteria) and consumers (heterotrophs) of organic phosphorus (Sellner, 1997; Karl *et al.*, 2002; Karl and Björkman, 2002).

Polymeric DOP is recycled rather slowly compared with the small but rapidly assimilated nucleotide pool, and DOP recycling is probably the "bottleneck" controlling phosphorus availability in DIP-limited environments. This is supported by observations indicating that microbial regeneration of DNA, and other high-molecular-weight DOP compounds, is an important source of phosphorus for phytoplankton growth in a variety of DIP-limited marine and freshwater systems (Berdalet *et al.*, 1996; Clark *et al.*, 1999; Matsui *et al.*, 2001). Thus, in the phosphorus-limited Sandfjord, western Norway, Thingstad *et al.* (1993) observed DOP:DIP concentration ratios of 10:1 to 100:1 with a strong dominance of polymeric DOP compounds such as DNA.

7.3. Phosphorus cycling in marine sediments

One of the most detailed studies of sediment phosphorus cycling in coastal marine environments is from Aarhus Bay, Denmark (Jensen *et al.*, 1995). Sedimentation of organic phosphorus in the form of phytoplankton detritus is greatest in this area during the spring bloom period in March (Figure 11.10A). Phytoplankton growth becomes limited after the spring bloom period as the organic products of the spring bloom settle to the sediment surface. Most of the phosphorus depositing onto the sediment during the spring bloom is mineralized and temporarily retained in the oxidized surface sediment by adsorption to Fe oxyhydroxides. Little DIP is released from the sediment until bottom water oxygen is depleted during summer as a result of water column stratification and high sediment oxygen demand induced by elevated temperatures (Figure 11.10B). The high DIP flux from the sediment in late summer and fall is associated with the reduction of Fe oxyhydroxides and the release of adsorbed DIP. At this time, the near-surface zone of the sediment containing oxidized iron narrows to a thin surface film.

The annual release of DIP from Aarhus Bay sediments (about 33 mmol m^{-2} yr^{-1}) contributes significantly to the phosphorus requirement for phytoplankton growth (about 51 mmol P m^{-2} yr^{-1}). A substantial amount of phosphorus is buried in Fe-bound form in the surface sediments (18 mmol m^{-2} yr^{-1}), and this phosphorus is gradually converted to apatite during early diagenesis. Accordingly, about 35% of the phosphorus sedimenting onto the sediment surface is permanently retained. This compares with a retention efficiency for organic carbon of only 21%. Such preferential burial of phosphorus, compared to carbon, is common in coastal marine sediments underlying an oxic water column, and it may be an important mechanism by which reactive phosphorus is removed from the marine ecosystem. In addition, sediments underlying oxic waters typically have a low organic C:P ratio. Two possible explanations for this have been offered (Ingall *et al.*, 1993). First, there may be an increased ability of microorganisms to store phosphorus in well-oxygenated environments, leading to the production of low C:P ratio bacterial biomass. Second, extensive oxidation of sedimentary

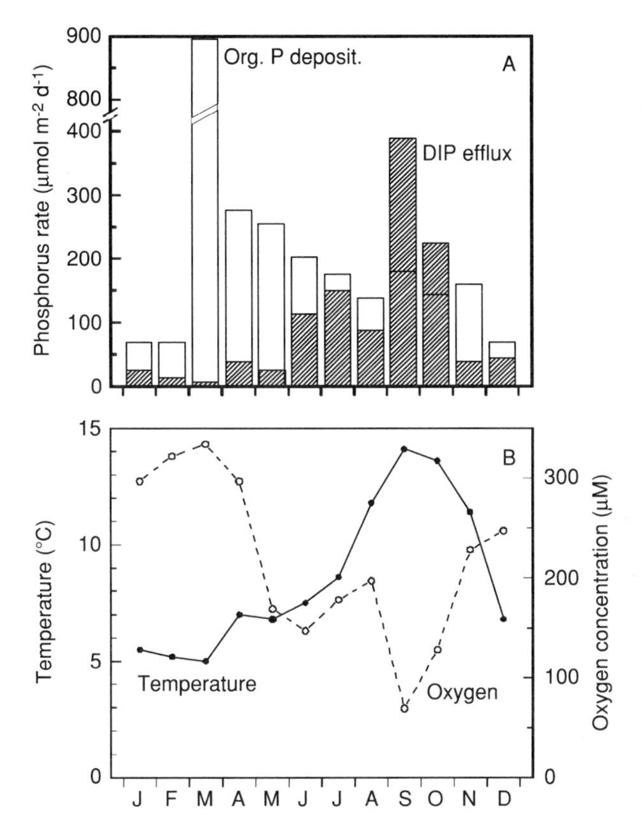

Figure 11.10 (A) Seasonal variation in the sedimentation of particulate phosphorus (Org. P deposit) and the release of dissolved inorganic phosphorus (DIP efflux) from sediment in Aarhus Bay, Denmark. (B) Seasonal variation in bottom water temperature and bottom water oxygen concentration in Aarhus Bay. Modified from Jensen *et al.* (1995).

organic matter in oxic sediments leads to the formation of residual organic phases with low C:P ratios.

At sites on the continental shelf off California and North Carolina as well as off the coast of Peru, oxygen-depleted bottom water significantly enhances phosphorus regeneration from sediments receiving high loads of organic carbon (Ingall and Jahnke, 1994). Generally, it appears that while sediments underlying oxygen-deficient bottom water experience enhanced organic carbon preservation (e.g., Canfield, 1994), the preservation of total phosphorus is reduced (Ingall and Jahnke, 1997). The reduced retention of phosphorus under anoxic conditions is also evident from ancient sediments as Ingall *et al.* (1993) observed average atomic C_{org}:P_{org} ratios of about 150 for bioturbated (oxic bottom water) shales and 3900 for laminated (anoxic bottom water) shales. Thus, the organic matter preserved in the bioturbated sediments is much enriched in phosphorus relative to the laminated shales (Murphy *et al.*, 2000). The high organic C:P ratios in anoxic sediments can be explained by a combination of three mechanisms (Ingall *et al.*, 1993): (1) limited ability of prokaryotes to store phosphorus under anoxic conditions, (2) preferential regeneration of phosphorus during anoxic degradation of sedimentary organic matter, and (3) slow microbial oxidation and thus enhanced preservation of organic carbon in anoxic sediments.

Taken together, the modern and ancient sediment results suggest that during times of ocean anoxia, phosphorus retention in sediments is reduced. This would increase the supply of phosphorus for primary production and ultimately increase the burial rates of organic carbon in sediments (Ingall *et al.*, 1993). Organic carbon burial provides a source of atmospheric oxygen, so there is a positive feedback between ocean anoxia and oxygen production through organic carbon burial (Van Cappellen and Ingall, 1996). This means that the system may self-regulate such that low oxygen levels promote ocean anoxia, while the specifics of phosphorus geochemistry under these conditions act to increase carbon burial and thus atmospheric oxygen concentrations.

The enhanced release of phosphorus under anoxic conditions can also impact the state of eutrophication of coastal areas. Release of phosphorus from the sediments may be significantly enhanced when eutrophication causes oxygen depletion in bottom waters, thereby providing additional phosphorus for a further increase in biological productivity.

7.4. Combating high phosphorus loading to eutrophic lakes

Phytoplankton production is believed to be phosphorus limited in most temperate lakes. However, decades of high nutrient discharges from agriculture and sewage have led to severe phosphorus loading in many lakes, increasing primary production and resulting in large inventories of both inorganic and organic phosphorus in lake sediments. The first and most obvious measure to combat human-induced eutrophication is to decrease the external nutrient load. However, reductions of the external phosphorus load do not always lead to a satisfactory reduction of DIP levels in lake water (Søndergaard *et al.*, 1993). The internal loading of phosphorus from sediments may slow down the recovery of lakes, maintaining an unacceptable degree of eutrophication for many years.

The most important factors controlling the release of DIP from sediments are bottom water temperature, oxygen and nitrate concentrations, and sediment Fe content (Boström *et al.*, 1988; Jensen and Andersen, 1992;

Smolders *et al.*, 2001). The interdependence of these factors normally causes enhanced internal phosphorus loading during warm summer months. Thus, enhanced organic matter supply and elevated microbial activity follow high summer temperatures. High temperatures and increased rates of microbial metabolism can reduce bottom water oxygen and nitrate levels, encouraging Fe reduction and the release of Fe-bound phosphorus (Figure 11.11).

A number of techniques have been used to reduce the internal phosphorus loading from lake sediments. Dredging of sediment has been highly successful, but is very cumbersome and costly, and is not a realistic solution for larger lakes (Björk, 1988). Alternatively, nitrate can be added in surplus to inhibit sediment Fe reduction in an attempt to prevent DIP release (Figure 11.12A) (Ripl, 1976). The precipitation and immobilization of phosphorus through the addition of iron chlorides ($FeCl_2$ or $FeCl_3$) to the sediment may be a useful technique in iron-poor lakes (Figure 11.12B) (Smolders *et al.*, 2001). However, as the beneficial effects of these latter two techniques are only temporary, adverse impacts of a continued artificial introduction of large amounts of nitrate or iron chlorides will make the use of these techniques undesirable in most cases.

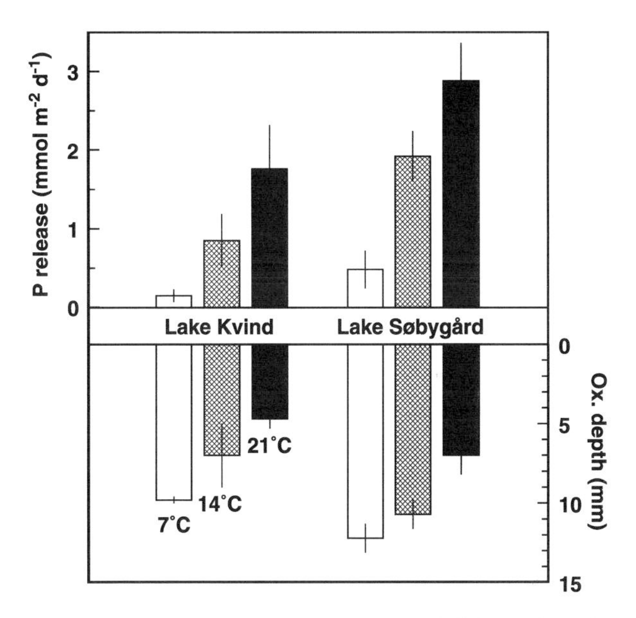

Figure 11.11 DIP release and thickness of the oxidized surface layer in undisturbed sediment from Lake Kvind and Lake Søbygård, Denmark, when incubated at three different temperatures. Modified from Jensen and Andersen (1992).

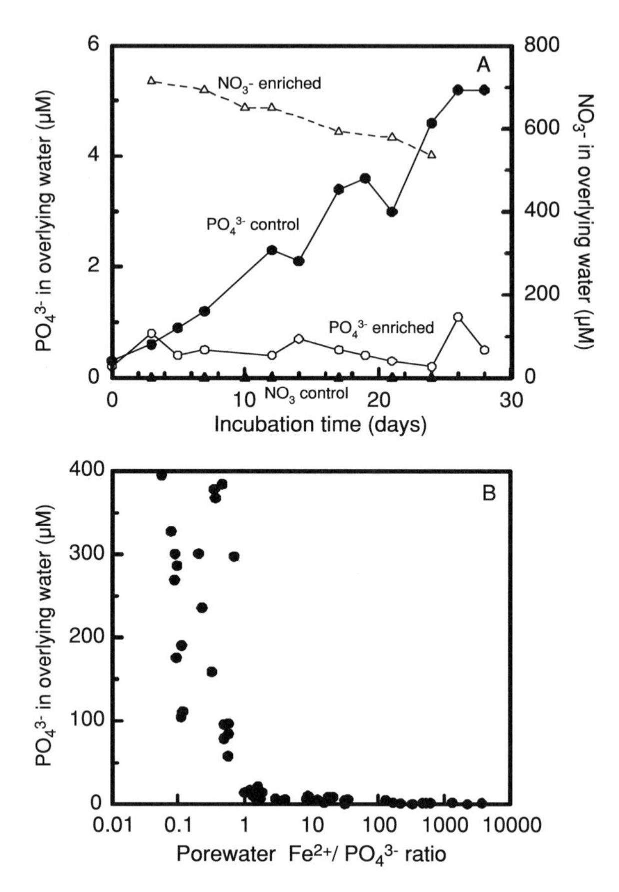

Figure 11.12 (A) Temporal pattern of DIP and nitrate concentration in water overlying sediment from the Harvey Estuary, Australia. The overlying water was either unamended (solid symbols) or amended initially with nitrate to a concentration of about 700 μM. Modified from McAuliffe *et al.* (1998). (B) Concentration of DIP in water overlying sediment from De Bruuk, Netherlands, as a function of the molar iron:DIP ratio in the sediment porewater. Modified from Smolders *et al.* (2001).

Chapter 12

The Silicon Cycle

1. INTRODUCTION

Silicon is the second most abundant element in the Earth's crust, 27% by weight (Faure, 1986), exceeded only by oxygen. The position of silicon in the periodic table of elements shows a clear relationship with carbon, sharing the same possible valence states (+4, +2 and −4). In natural environments, however, Si is most stable in the +4 form, in which it is found in various forms of solid silicon dioxide (silica, SiO_2), as mineral silicates and as the dissociated anions of silicic acid (H_4SiO_4). Silica forms the skeleton structures of a variety of aquatic plankton, including many diatoms, radiolarians, and silicoflagellates, as well as the spicules of sponges. Therefore, silica is an important nutrient controlling primary production in aquatic systems, and its cycling is controlled by the interplay between biological and physiochemical processes. In this Chapter, we outline some of the basics surrounding the chemistry of dissolved silica and the mineralogy of biological phases formed from silica. We consider the processes by which silica is formed into biogenic structures and also the processes controlling the dissolution and preservation of biogenic silica in aquatic environments, including the role of microorganisms.

ADVANCES IN MARINE BIOLOGY VOL 48
0-12-026147-2
© 2005 Elsevier Inc.
All rights reserved

Finally, we present some case studies of silica cycling in aquatic systems and review the evolution of the silica cycle through geologic time.

2. OVERVIEW OF SILICA CHEMISTRY AND MINERALOGY

Silica is a constituent of hundreds of important rock-forming minerals and is universally distributed throughout the Earth's crust. Of most interest here, however, is the involvement of silica in biological transformations, and therefore dissolved silica is a logical starting point. In solution, silica exists as silicic acid, H_4SiO_4 (sometimes written as $Si(OH)_4$), and its dissociation products. The dissociation constants (k_A) for silicic acid are as follows (Stumm and Morgan, 1996):

$$H_4SiO_4 \rightleftharpoons H^+ + H_3SiO_4^-,\ k_A = 10^{-9.5} \tag{12.1}$$

$$H_3SiO_4^- \rightleftharpoons H^+ + H_2SiO_4^{2-},\ k_A = 10^{-12.6} \tag{12.2}$$

It is obvious that up to a pH of 9, undissociated silicic acid is the predominant form. At around neutral pH and at concentrations of 1 to 2 mM, silicic acid readily polymerizes, forming oligomers of polysilicic acids. The polymerization process can be viewed as the removal of water between adjacent silicates to form siloxane linkages, as shown in the following example:

$$3\left(\begin{array}{c} OH \\ | \\ HO-Si-OH \\ | \\ OH \end{array}\right) \rightarrow \begin{array}{c} OH \quad\quad OH \quad\quad OH \\ | \quad\quad\quad | \quad\quad\quad | \\ HO-Si-O-Si-O-Si-OH \\ | \quad\quad\quad | \quad\quad\quad | \\ OH \quad\quad OH \quad\quad OH \end{array} + 2H_2O \tag{12.3}$$

Once the solubility of solid silicate phases is exceeded, the siloxane-bonded amorphous silica polymers form a gelatinous precipitate. Through time, the precipitate expels water and begins to crystallize, forming opal ($SiO_2{\cdot}nH_2O$), which is also the product of biological precipitation. During early sediment diagenesis, opal may be reprecipitated as the crystalline phase opal-CT, which may further recrystallize to form quartz (Kastner *et al.*, 1977; Kastner, 1979).

3. OVERVIEW OF THE GLOBAL SILICA CYCLE

The primary source of dissolved silica in aquatic environments is the weathering of rock-forming silicate minerals by acidic dissolution. As an example, the weathering of sodium feldspar (albite, $NaAlSi_3O_8$), forming the clay kaolinite ($Al_2Si_5O_5(OH)_4$), is shown in Equation 12.4 (see Section 12.5 for more on the role of microbes in weathering):

$$2NaAlSi_3O_8 + 9H_2O + 2H^+ \rightarrow Al_2Si_2O_5(OH)_4 + 2Na^+ + 4H_4SiO_4 \quad (12.4)$$

Both soluble secondary silicates (such as kaolinite) and dissolved silica (H_4SiO_4) are formed. Overall, weathering potentially provides 5.6×10^{12} mol yr^{-1} of dissolved silica (DSi) to the ocean via river discharge, with a further 0.5×10^{12} moles yr^{-1} introduced via atmospheric deposition (Tréguer *et al.*, 1995) (Figure 12.1).

There are many significant, recent advances in our understanding of the global biogeochemical Si cycle (Conley, 2002). A sizable pool of biogenic Si in terrestrial systems is contained as phytoliths in living tissues of growing plants, which remain in the soil after decomposition of the organic tisssues. Although studies of the terrestrial biogeochemical Si cycle are limited (Bartoli, 1983; Meunier *et al.*, 1999), the annual fixation of phytolith silica ranges from 60×10^{12} to 200×10^{12} mol yr^{-1}. This amount rivals the amount fixed annually by diatoms in the oceans (see below). Internal recycling of the phytolith pool is intense, and this pool buffers the dissolved silicate flux by rivers to the oceans.

A substantial part of the riverine load of DSi (0.6×10^{12} mol yr^{-1}), however, does not directly reach the ocean but rather is removed by biological uptake in estuaries. Hydrothermal venting and basalt dissolution introduces a further 0.6×10^{12} mol yr^{-1} to the deep ocean (Tréguer *et al.*, 1995). The net input of Si to the global ocean (6.1×10^{12} mol yr^{-1}) is almost in balance with the output via net deposition on the sea floor. The oceanic reservoir has an average DSi concentration of about 70 μM, which translates into a total reservoir size of about 10^{17} moles. The overall steady-state residence time for silica in the ocean ($10^{17}/6.1 \times 10^{12}$) is, therefore, approximately 16,000 years.

There is a great deal of internal oceanic cycling of silica, which is not represented in the residence time calculation generated above. Thus, diatoms, radiolarians, and silicoflagellates take up about 240×10^{12} mol DSi yr^{-1} from sea water to build their biogenic silica (BSi) skeletons (Nelson *et al.*, 1995). The residence time for silica relative to biological uptake, therefore, averages only ($10^{17}/240 \times 10^{12}$), about 400 years. This implies that silica delivered to the ocean passes through the biological uptake and dissolution cycle an average of about 40 times before permanent burial in sediments. Biological uptake depletes the surface waters with respect to DSi and produces undersaturation with respect to the solid BSi formed by planktonic

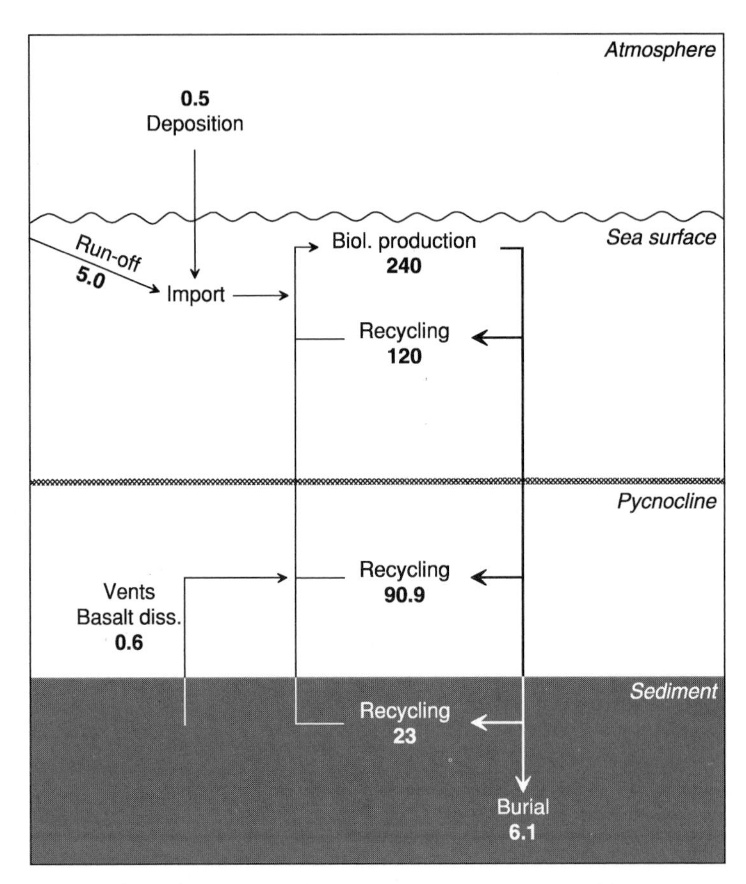

Figure 12.1 Flow diagram showing the silicon cycle in the world oceans (10^{12} moles yr^{-1}). Heavy arrows represent solid phase biogenic silica (BSi) and light arrows represent dissolved silica (DSi). Processes from three compartments are depicted: (1) the upper mixed water masses (including the photic zone), (2) water masses below the pycnocline, and (3) underlying sediments. "Deposition" represents dry and wet atmospheric DSi inputs; "Run-off" represents net river and diffuse DSi discharges from land to the open ocean (import = run-off + deposition); "Biol. production" represents the incorporation of DSi into BSi by primarily diatoms in the photic zone; "Recycling" represents dissolution of BSi to DSi in the three oceanic compartments; "Vents and basalt diss." represents the DSi input via hydrothermal vents and dissolution of basalt and other silicon minerals in the seabed, and "Burial" represents BSi, which is permanently buried and subsequently converted to lithogenic silica forms in sediments. Adapted from Tréguer *et al.* (1995).

organisms. When these organisms die, their skeletons sink and at the same time begin to dissolve. Indeed, about 50% of these skeletal structures dissolve within the upper 100 m of the water column, while about 40% dissolves when the remaining skeletons sink through deeper water (Van Cappellen *et al.*, 2002).

The portion of the BSi production escaping dissolution in the water column (~10%) ultimately reaches the sediment, where further dissolution and transformation occurs. Thus, most of the BSi depositing onto sediments (about 80%) is dissolved, liberating 23×10^{12} moles yr^{-1} of dissolved Si back into the water, representing about 8% of the total biological production (Koning *et al.*, 1997). The remaining BSi either forms into aluminosilicate minerals through a process known as reverse weathering (e.g., Mackenzie and Kump, 1995) or is crystallized. Thus, only about 2% of the biological silica production is permanently preserved and enters the geological cycle, while the remaining 98% is recycled (Tréguer *et al.*, 1995; Sayles *et al.*, 2001; Nelson *et al.*, 2002).

4. SILICA FORMATION BY DIATOMS

Although a number of different organisms, including diatoms, radiolarians, and silicoflagellates, have the potential to assimilate DSi and precipitate biogenic silica, the frustules formed by aquatic diatoms dominate the global production of BSi. Diatoms are found in all aquatic environments, including oceans, lakes, rivers, and photic sediments. They contribute 20–25% of the global primary production and 90% of the biogenic silica production. The remaining BSi is almost exclusively of radiolarian origin (Krumbein and Werner, 1983).

Diatom cells are enclosed in a wall of silica consisting of two perforated valves, which may have thickened ribs and spines. The shape of the valves may be pennate (elongated) or centric (circular). Diatoms reproduce their ornate frustules from generation to generation under strict genetic control, resulting in unique frustule morphology for each of the more than 3000 known species (Pickett-Heaps *et al.*, 1990). During asexual cell division, each offspring receives one of the two valves and forms the missing valve itself. Because the newly formed valve is usually smaller than the existing one (Figure 12.2), the average cell size of the population becomes progressively smaller. To combat this march into the infinitesimal, diatoms possess a sexual reproduction, with auxospore formation, when cells of a certain

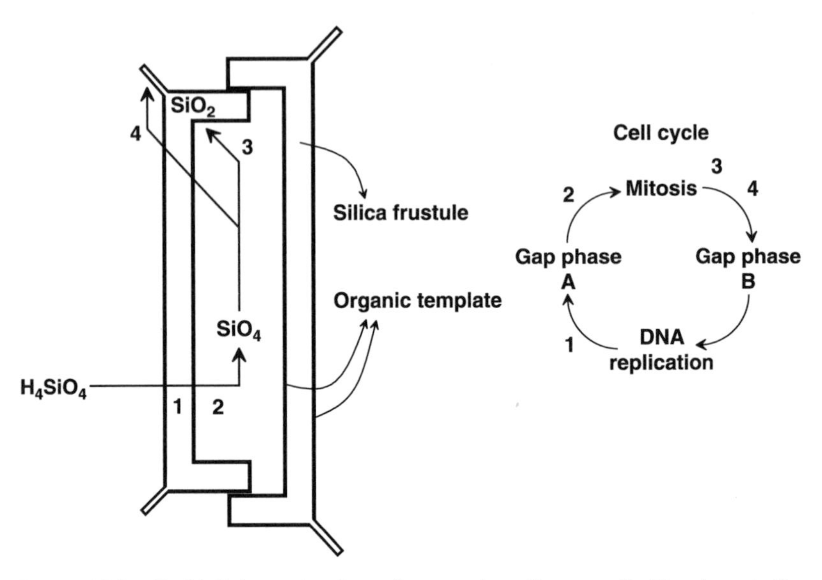

Figure 12.2 (Left) Schematic view of a growing diatom cell. Numbers indicate general events in the silicification process of the growing frustule: (1) active uptake of silicic acid across the cell membrane, (2) assimilation of silicic acid into a silica precursor and transport to the sites of active valve formation, (3) regulation of silicon polycondensation under the formation of the immature frustule by the genetically controlled organic template, and (4) formation of the mature frustule by silicon deposition in girdles, bands, and setae. (Right) Timing of the four silicification events during the diatom cell cycle. Modified from Ragueneau *et al.* (2000).

smallness appear. Auxospores then give rise to new vegetative cells of the maximum size for the species (Davidovich, 1994).

The frustules of marine diatoms may contain up to 96% silica oxide, with only small amounts of Al, Fe, and H_2O (Ehrlich, 1996). The cell-specific silica content, on the other hand, varies up to an order of magnitude depending on the species and frustule structure. Thus, silica comprises, for example, 3% dry weight in *Bellerochea yucatanensis* and 40% dry weight in *Coscinodiscus asteromphalus*. The availability of nutrients in the environment and the size of the cell may, in addition, cause considerable variation within species (Krumbein and Werner, 1983; Taylor, 1985). There is, on average, one order of magnitude greater silica content (per biovolume) in freshwater diatoms compared to marine species (Conley *et al.*, 1989). Salinity is partly, but not completely, responsible for this difference. Other factors such as different buoyancy strategies and the generally lower DSi content of marine waters could also contribute.

Measurement of biogenic silica production

Over the last 20 years biogenic silica production rates have been measured in a variety of aquatic environments via tracer incubations using either the stable isotope ^{30}Si or the radioisotope ^{32}Si. The first methods were based on the uptake of the stable isotope ^{30}Si followed by relatively time-consuming and tedious mass spectrometric determination (Nelson and Goering, 1977). More recently, Tréguer *et al.* (1991) introduced a method for determination of silica production using ^{32}Si (a weak β-emitter, $E_{max} = 0.227$ MeV, with a half-life of 134 yr). The activity of ^{32}Si can be measured either by Cerenkov counting of the daughter isotope ^{32}P after secular equilibrium is approached several months after the uptake experiment has been performed (Tréguer *et al.*, 1991), or by direct liquid scintillation counting (Brzezinski and Phillips, 1997).

Determination of pelagic silica production by the use of isotopes can be made with about the same spatial and temporal resolution achievable for ^{14}C measurements of primary production. Measurements are usually carried out on water samples collected from different depths in the euphotic zone. The samples are subsequently drawn into clear polycarbonate bottles (never glass), injected with the selected isotope and mixed well. The incubation bottles are then either hung on a free-floating array at the depths from which the water was sampled or incubated in the laboratory under *in situ* light and temperature. After 10 to 24 h, incubations are terminated by filtration through 0.6 μm polycarbonate filters under gentle vacuum (Brzezinski and Phillips, 1997). Filters holding the particulate material (diatoms) from ^{30}Si experiments are dried before the silica is converted to $BaSiF_6$ and analyzed by solid-phase mass spectrometry. Filters holding particulate material from ^{32}Si experiments are covered with HF to dissolve biogenic silica and to prevent ^{32}P adhesion before addition of scintillation cocktail and counting on a scintillation counter. Since the maximum energy of β emission from ^{32}P (1.709 MeV) is sufficiently different from that of ^{32}Si, the two isotopes can be separated and quantified using dual-label scintillation counting techniques. This latter procedure then allows the isotope to be used for simultaneous determination of biogenic silica and organic phosphorus production (Figure 12.3).

Although the ^{32}Si and ^{30}Si techniques generally provide similar results for silica production (Figure 12.4), the less elaborate ^{32}Si is preferable since a larger number of incubations can be conducted during a given study, allowing greater spatial and temporal resolution. It should be noted, however, that ^{32}Si is very expensive and not always available for purchase.

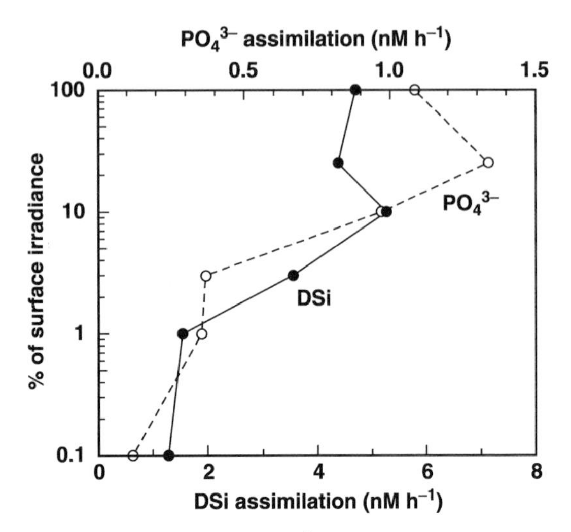

Figure 12.3 Assimilation of DSi and PO_4^{3-} determined simultaneously by the ^{32}Si technique as a function of irradiance in the euphotic zone of the Weddell Sea, Southern Ocean. Modified from Tréguer *et al.* (1991).

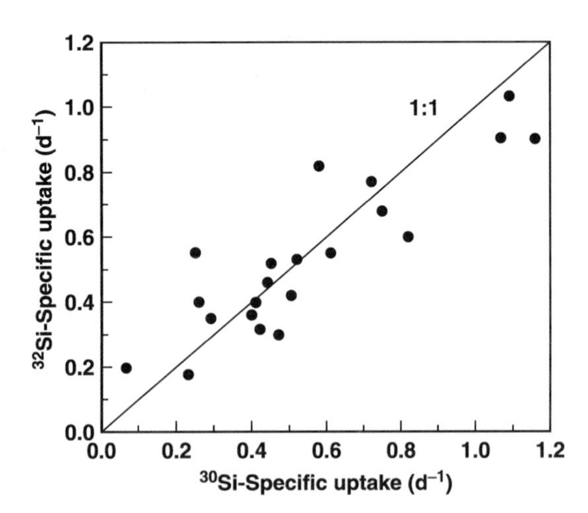

Figure 12.4 Specific uptake of Si (normalized to the standing stock of biogenic silica) from parallel incubations of Monterey Bay, California, surface water using either ^{32}Si or ^{30}Si as a tracer. The solid line represents the 1:1 relationship. Modified from Brzezinski and Phillips (1997).

The biochemistry and molecular biology of valve formation is currently under study (Kröger and Sumper, 2000; Ragueneau *et al.*, 2000), and we offer the following general description of the pathway. Silicon in the form of silicic acid is taken up by cells via active membrane transport involving a Na^{+}-dependent transporter with the expenditure of energy (ATP) (Figure 12.2). After assimilation, silicon is transported to special silica-deposition vesicles where active valve formation occurs. The nucleation and growth of diatom valves requires little or no input of energy. Protein or polyamine templates, which may directly interact with silicon during the polymerization process, mediate the intricate architecture of the frustules (Sumper, 2002). Once completed, the mature valve is extruded to the exterior of the cell, and it is coated with polysaccharides. Specific phases of frustule formation occur during certain periods of the diatom cell cycle. While silicic acid uptake and silicon assimilation immediately precede cell division (mitosis), new valves are constructed during cell division. The exact timing of the subsequent girdle band deposition and formation of structures such as setae, on the other hand, appears to be species specific (Pickett-Heaps *et al.*, 1990).

The rate of DSi uptake by diatoms, and the rate of cell division (growth) both increase with increasing extracellular DSi concentration according to Michaelis-Menten (silica uptake) or Monod (growth) kinetics (Goering *et al.*, 1973). Many diatoms have a lower half-saturation constant for growth (K_s) than for DSi uptake (K_m). For temperate and tropical species, K_s typically varies between 0.04 and 1 μM DSi, whereas K_m varies between 0.5 and 5 μM DSi (Nelson and Dortch, 1996). As a consequence, diatoms can maintain growth at DSi concentrations that limit uptake. Diatoms manage this situation by producing thinner frustules and thus lowering their Si:C ratio.

The rate of BSi production has been measured in many marine systems (Rageneau *et al.*, 2000; DeMaster, 2002). The grand mean oceanic BSi production rate is 1.5 to 2.1 mmol m^{-2} d^{-1}. This average rate covers a large range of spatial and temporal variability, from low production rates in oligotrophic open oceanic regions of the Atlantic and Pacific Oceans to high rates in coastal upwelling regions along west coasts of America (both North and South) and Africa (Table 12.1). Diatoms grow rapidly and dominate phytoplankton populations whenever conditions are optimal for their growth, such as in upwelling plumes, zones of equatorial divergence, river plumes, and tidal coastal areas where the DSi supply is great (Ragueneau *et al.*, 2000). Diatoms also grow rapidly during spring blooms and in ice-edge blooms. Temperate seas are typically characterized by a high abundance of diatoms during the spring bloom, progressing to dominance by non-diatom species (e.g., dinoflagellates) in summer and back again to diatoms in autumn (Jones and Gowen, 1990; Reid *et al.*, 1990). When nutrient supply is curtailed through stratification, DSi limitation is the

Table 12.1 Rates of biogenic silica (BSi) production in oceans. Data in each region are given as the low and high of mean values (mmol Si m^{-2} d^{-1}) from locations around the world

Region	Low	High
Coastal upwelling areas	22.0	222.0
Other coastal areas	6.0	24.0
Open oceanic regions	0.5	6.5
Southern Ocean	2.0	37.0

Modified from Ragueneau *et al.* (2000).

primary control on phytoplankton community structure, causing diatom populations to diminish during the summer (Kelly Gerreyn *et al.*, 2004).

5. SILICON BIOGEOCHEMISTRY

5.1. Silica dissolution

The dissolution of biogenic silica in aquatic environments occurs when biosiliceous debris (e.g., diatom skeletons) is exposed to undersaturated water. BSi dissolution in water is a simple physicochemical process, controlled by key variables such as the concentration of DSi, temperature, pH, and frustule surface area. A variety of different expressions have been proposed to describe the dissolution of biogenic silica (e.g., Greenwood *et al.*, 2001). Many of these expressions approximate a first order-type rate equation, in which the dissolution rate, V_{diss} (h^{-1}), in its simplest form, is represented by Equation 12.5 (Hurd and Birdwhistell, 1983; Greenwood *et al.*, 2001):

$$V_{diss} = kS(C_{sat} - C) \tag{12.5}$$

where k is the temperature- and pressure-dependent first order rate constant (cm h^{-1}), S represents the specific surface area of the solid silica source (e.g., frustules) (cm^2 mol^{-1}), C_{sat} is the concentration of DSi at saturation (mol cm^{-3}), and C is the ambient DSi concentration. Dissolution rate decreases rapidly with increasing DSi concentrations, and no dissolution occurs when DSi concentrations are equal to or higher than C_{sat} (Figure 12.5). In addition, dissolution rate increases by nearly two orders of magnitude with increasing temperature from 0 and 25 °C (assuming constant C). Most of this increase is caused by increasing k, although some is also related to increasing C_{sat} (Figure 12.6). There is also a strong dependency of V_{diss} on pH, with a pronounced minimum V_{diss} between pH 4 and 6 (Figure 12.7).

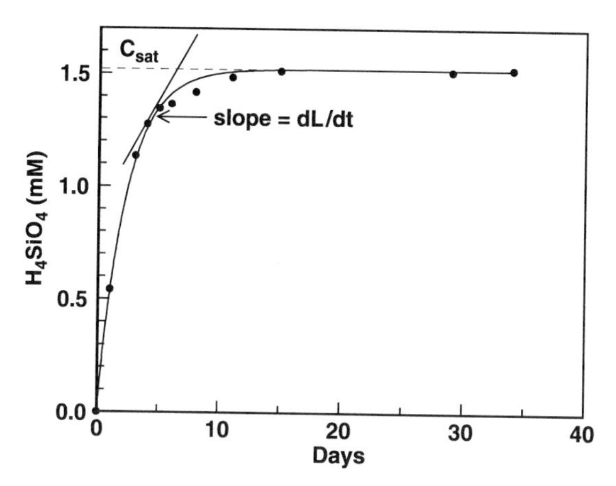

Figure 12.5 Development of DSi during BSi dissolution in a closed water system. The concentration of DSi asymptotically approaches the saturation level (C_{sat}) after about 15 days. The slope of a tangent to the curve at any point provides the dissolution rate (dL/dt with units of mMd^{-1}). Modified from Greenwood *et al.* (2001).

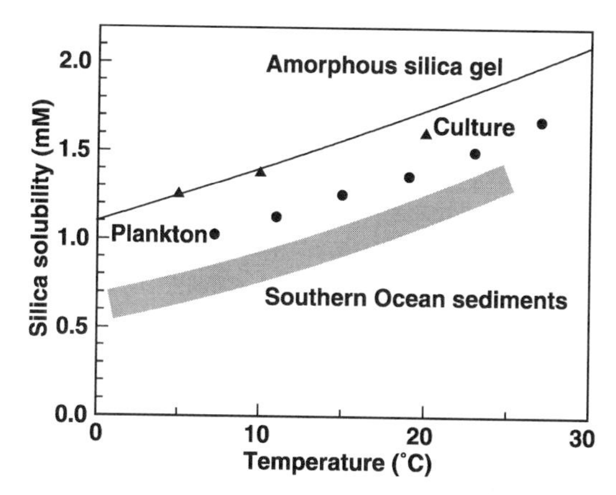

Figure 12.6. Temperature dependence of biogenic silica (BSi) solubility in (1) diatom cultures (triangles), (2) phytoplankton from surface waters off Hawaii (circles), and (3) surficial Southern Ocean sediments (grey zone). For comparison, the solubility of amorphous silica gel is indicated by the full line. Modified from Dixit *et al.* (2001).

However, Greenwood *et al.* (2001) warn that kinetic parameters derived from one set of biogenic silicates may not explain the dissolution behavior of another set due to differences in trace element content, degree of organic coating, and possibly opal crystallinity.

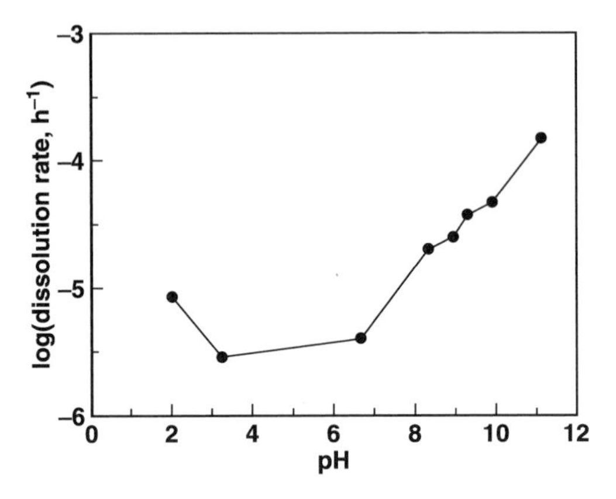

Figure 12.7 The variation of the initial dissolution rate of silica from the diatom *Cyclotella cryptica* with pH in a 0.7 M NaCl solution. Modified from Greenwood *et al.* (2001).

The specific surface area of biogenic silica is highly dependent on the size and morphology of the hard parts and mostly falls within the range of 12×10^4 to $250 \times 10^4\,cm^2\,mol^{-1}$ (Dixit *et al.*, 2001). Surface area strongly affects solubility, and this dependence is described by the Gibbs-Kelvin equation:

$$\ln C_{eq}(s) = \ln C_{eq}(s=0) + \left(\frac{2d}{3RT}\right)S \qquad (12.6)$$

Where R is the gas constant (8.314 J K^{-1} mol^{-1}), T is the absolute temperature (°K), d is the solid-solution interfacial free energy (J m^{-2}), $C_{eq}(s)$ is the solubility of biogenic silica having a specific molar surface area of S, and C_{eq} (s = 0) is the bulk solubility, that is, the limiting value when the interfacial free energy term is neglibible. Overall, biogenic silica has a 16% higher solubility with a specific surface area of 200×10^4 cm^2 mol^{-1}, compared to a surface area of 25×10^4 cm^2 mol^{-1} (Dixit *et al.*, 2001). Thus, as hard parts are transferred through the water column to the sediment surface there is a preferential disappearance of diatom species with high specific surface areas and delicate frustule structures (Barker *et al.*, 1994). For this reason, the assemblage of hard parts in ocean surface waters displays higher average dissolution rates than silica tests in underlying sediments (at similar temperature and pH).

Numerous studies have measured dissolution rates of biogenic silica in open oceanic waters using radiolabeled Si. The lowest measured values for V_{diss} in marine surface waters of about $3 \times 10^{-4}\,h^{-1}$ have been obtained in

Antarctic areas. The highest rates of 5×10^{-3} to 8×10^{-3} h^{-1} have been found in coastal waters off northwest Africa and in a Gulf Stream warm-core ring (Ragueneau *et al.*, 2000). This range in dissolution rates reflects, at least partially, differences in water temperature, as surface water DSi concentrations in all cases are considerably lower than C_{sat}. Sediment dissolution rates have been derived mostly from diagenetic diffusion reaction models of porewater DSi profiles (Boudreau, 1990; Koning *et al.*, 1997; Nelson and Brzezinski, 1997) or, more recently, from flow-through reactors (Van Cappellen and Qiu, 1997; Gallinari *et al.*, 2002). Rates of biogenic silica dissolution in surface sediments are quite variable, with rates at least an order of magnitude lower than in the water column and decreasing to virtually zero at sediment depth (Koning *et al.*, 1997). The biogenic silica preserved in sediments thus provides an important record of past sea surface productivity and temperature (Ragueneau *et al.*, 2000).

Diatom frustules supplied from the water column initially dissolve during sediment burial. As a result, DSi accumulates in the porewaters until an asymptotic level corresponding to the apparent silica saturation concentration is reached (Figure 12.8). Below this depth, no further silica dissolution

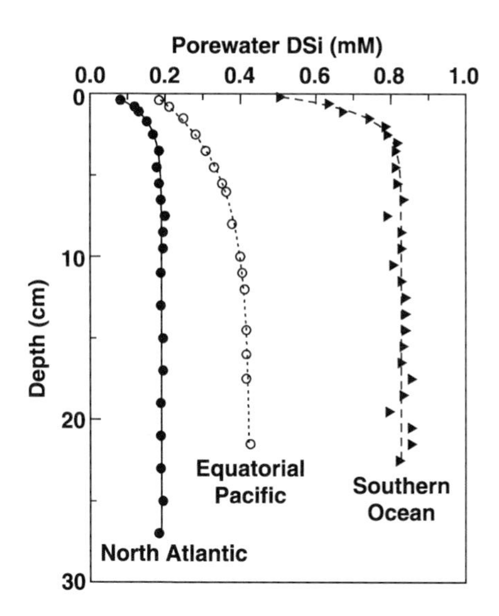

Figure 12.8 Depth dependence of dissolved silica (DSi) in porewater of sediment from the North Atlantic, the Equatorial Pacific, and the Southern oceans. The lines represent the best fit by a diagenetic diffusion-reaction model. DSi approaches asymptotically the "apparent porewater C_{sat}" level within the upper 4, 10, and 3 cm of the three sediments. The detrital:biogenic silica ratios of the three sediments are 26, 2.3, and 0.1 (see text for more details). Modified from Van Cappellen and Qiu (1997) and Gallinari *et al.* (2002).

or DSi accumulation occurs. These apparent saturation concentrations of DSi exhibit large spatial variability, and, in most cases, DSi is greatly undersaturated with respect to the theoretical *in situ* solubility of BSi (Dixit *et al.*, 2001; Gallinari *et al.*, 2002). This inconsistency may reflect an aging process of the silica surfaces rendering the biosiliceous, fragments less reactive. Since the dissolution rate of BSi is proportional to surface charge, the aging of biogenic silica may be associated with a decrease in the abundance of ionizable surface silanols (Dixit and van Cappellen, 2002). Hydroxyls of silanol groups ($>SiOH^0$) formed by hydration of silica surfaces can adsorb cations and anions by protonation and deprotonation. Blockage of these reactive surface sites with the build-up of organic or inorganic coatings and adsorption of inhibitors, as well as dissolution and reprecipitation (Van Cappellen and Qiu, 1997), renders biogenic silica particles less soluble. In addition, the structural incorporation of aluminum and other metal ions from porewaters is also known to reduce the solubility of BSi (van Beusekom *et al.*, 1997).

The departure of apparent C_{sat} from the theoretical equilibrium may also be linked to concurrent precipitation of authigenic (detrital) mineral phases during the dissolution of biogenic silica. This is supported by observations of a reduction in apparent BSi solubility with increasing detrital:biogenic silica ratio in sediments (Figure 12.8) (Gallinari *et al.*, 2002). Thus, when the detrital:biogenic ratio is high, reprecipitation reactions dominate, lowering the apparent C_{sat} to considerably below the level obtained when detrital:biogenic silica ratios are low.

While various mechanisms, as outlined above, act to enhance biogenic silica preservation, others can act to increase the transport of DSi from sediments, drawing DSi to levels below apparent porewater saturation. Most burrow-dwelling animals, such as polychaetes, bivalves and crustaceans, dramatically increase the exchange of porewater solutes between the sediment and the overlying water through a combination of excavation, burrowing, feeding, and ventilation (Aller, 2001). The presence of benthic infauna, therefore, stimulates the release of DSi to the overlying water and enhances sedimentary BSi dissolution several-fold (Aller and Yingst, 1985; Marinelli, 1992). Furthermore, DSi fluxes attributable to macrofauna are positively correlated with the density of burrow or tube structures (Aller, 1982) (Figure 12.9) and with the activity level of the burrow inhabitants (Marinelli, 1992). While the preservation of BSi is highly efficient in sediments underlying anoxic bottom waters in meromictic lakes and euxinic marine basins and fjords, preservation is reduced in oxic environments supporting active infaunal populations ventilating DSi from the sediment.

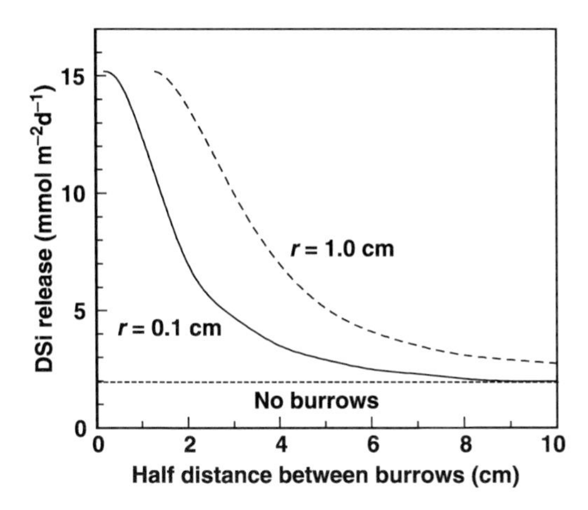

Figure 12.9 Theoretical rates of dissolved silicate release from marine sediment as a function of half-distance between irrigated infaunal burrows. Results from two different burrow sizes (radius 0.1 and 1.0 cm, respectively) and sediment with no burrows (dotted line) are shown for comparison. Modified from Aller (1982).

5.1.1. Microbes and biogenic silica dissolution

Microbes can significantly enhance the rate of biogenic silica dissolution. The dissolution rates of silica frustules vary considerably from one diatom species to the other (Tréguer *et al.*, 1989), and these differences depend to a large extent on the thickness and composition of organic and inorganic surface coatings (Kamatani *et al.*, 1988). Any process removing these coatings from the frustule surface exposes silica directly to sea water and will increase the dissolution rate significantly. Microbes accelerate biogenic silica dissolution by colonizing and enzymatically degrading the organic coating of diatom frustules (Figure 12.10). The organic coatings covering diatom frustules contain glycoproteins as a significant component. These are readily hydrolyzed by microbial protease enzymes, and the dissolution of the biogenic silica is hastened. Indeed, the proteases excreted by microbes colonizing frustule surfaces are normally the dominant ectohydrolase present, and protease activity strongly correlates with the silica dissolution rate (Bidle and Azam, 2001).

Species composition, abundance, and metabolic state are important variables in microbe-mediated Si regeneration and cause considerable variations in the dissolution rate (Smith *et al.*, 1992; Bidle and Azam, 1999). Thus, diatom aggregation into marine snow may enhance silica solubilization due to the high prokaryote abundance in the aggregates. Specific bacterial phylotypes associated with dissolving diatom aggregates include representatives from the Cytophaga/Flavobacteria/Bacteriodes as well as members of the

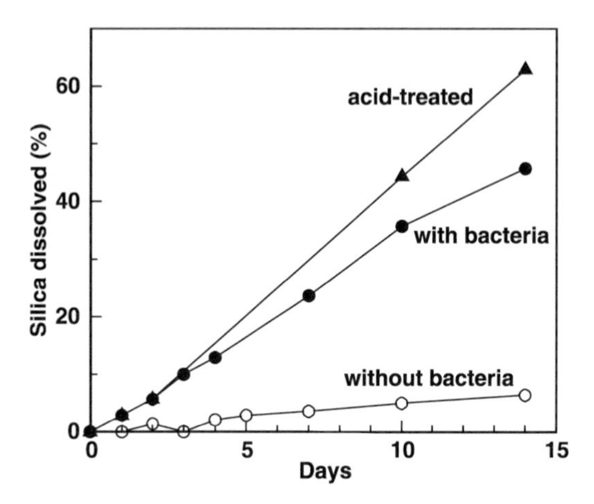

Figure 12.10 Time course of biogenic silica (BSi) dissolution when diatom detritus is incubated under various conditions: (1) untreated diatoms incubated in sterile sea water (without bacteria), (2) untreated diatoms incubated with natural bacterial assemblages, (with bacteria) and (3) diatoms pre-treated by acid to remove organic coatings from frustules and incubated in sterile sea water (acid-treated). Modified from Bidle and Azam (2001).

α- and γ-proteobacteria (Bidle *et al.*, 2003). Particularly, phylotypes within the γ-proteobacteria are known to be closely linked to biogenic silica dissolution (Bidle and Azam, 2001).

Also, trophic interactions, such as protozoan grazing and viral infection, which control microbial abundance and activity, may regulate biogenic silica dissolution. Protozoan grazers can also enhance Si regeneration by extracting virtually all of the organic matter from ingested diatoms, leaving exposed and fragmented frustules for rapid dissolution (Jacobsen and Anderson, 1986).

5.2. Microbes and silicate weathering

The destruction, degradation, and solubilization of mineral silicates may also be directly or indirectly controlled by microbial activity. Biofilms of free-living prokaryotes, fungi, and algae, as well as lichens, living on and within siliceous rocks, modify rates and mechanisms of chemical and physical weathering. Thus, these organisms contribute to soil and sediment formation and release dissolved silica to surface waters (Banfield *et al.*, 1999; Bennett *et al.*, 2001). The involvement of microorganisms in accelerated weathering of minerals may be the coincidental result of microbial metabolism, but it also represents a potential competitive advantage by providing the organisms involved with essential nutrients such as phosphorus. For example, fresh nutrients are exposed and released when microorganisms accelerate the dissolution of feldspars. It has been hypothesized that feldspars containing trace nutrients may weather most quickly, leaving behind a clay residuum and feldspars without nutritional value (Bennett *et al.*, 2001).

The mode of microbial attack on mineral silicates, such as feldspars, is quite diverse and includes the action of microbially produced organic and inorganic compounds. A general model of dissolution involves three major steps: (1) initial rapid exchange of charge-balancing cations (K^+, Na^+, and Ca^{2+}) for protons at the mineral surface, (2) hydrolysis and formation of an activated Si-O-Si or Al-O complex, leading to collapse of the lattice structure, and (3) detachment of silica and aluminum from the remaining framework and exposure of fresh surface for reaction (Wollast, 1967; Helgeson *et al.*, 1984).

The dissolution rate of siliceous rocks is a complex function of pH. The lowest rates of dissolution occur at neutral pH, while the rate increases with increasing proton concentrations (low pH) and also with increasing hydroxyl ion (high pH) concentrations (Wollast and Chou, 1992). The acid-base reactions on mineral surfaces stimulate cation exchange by increasing Fe and Al solubility and mobility. Thus, pH values of 3–4 have been detected in proximity to prokaryote cells within cleavages in biotite when the pH of the bulk solution was 7. This lowered pH corresponds to a 10- to 1000-fold increase in dissolution rate at the mineral surface (Barker *et al.*, 1998).

Microbes can also catalyze mineral weathering by producing organic ligands. These include metabolic by-products, extracellular enzymes, chelators, and organic acids. These substances can influence silicate mineral dissolution by decreasing pH, by forming destabilizing surface complexes, which weaken metal–oxygen bonds, and/or by complexing metals in solution, thereby decreasing solution saturation state (Banfield *et al.*, 1999; Bennett *et al.*, 2001). Relatively dilute solutions of compounds such as oxalic acid, citric acid, and acetic acid can enhance silicate dissolution by up to one order of magnitude (Bennett and Casey, 1994; Blake and Walter, 1996; Stillings *et al.*, 1996). It appears that rates of dissolution in solutions containing organic acids may be up to 10 times greater than the rates in solutions containing inorganic acids at the same pH (Welch and Ullman, 1993).

Microbes also affect mineral weathering by producing high-molecular-weight polymers. Mucoid polymers contain much water and increase the contact time between water and the mineral surface. This allows acids and ligands more intimate contact with the mineral surface and a diffusive outlet for ions diffusing away from the surfaces (Barker and Banfield, 1996). High-molecular-weight polysaccharides (e.g., alginates) may, like low-molecular-weight organic acids, increase the extent of silica dissolution by complexing with ions in solution and thus lowering solution saturation state (Welch *et al.*, 1999).

6. SILICON CYCLING IN AQUATIC ENVIRONMENTS

6.1. Silica depletion

Nitrogen and phosphorus are generally considered the two most important nutrients controlling overall algal production in aquatic environments (Ryther and Dunstan, 1971), but dissolved Si availability can also regulate the species composition and abundance of planktonic diatoms (Figure 12.11). The presence of diatoms depends on access to DSi, whereas other types of phytoplankton, such as dinoflagellates, coccolithophores, and cyanobacteria, do not require DSi. Thus, where DSi is limiting, non-siliceous species usually dominate phytoplankton communities. Some ocean areas, such as the eastern equatorial Pacific, are described as productivity paradoxes, because phytoplankton production and abundance are lower than expected from the NO_3^- availability. In these areas, unused NO_3^- is present and low surface water concentrations of DSi appear to limit diatom growth. Dugdale and Wilkerson (2001) have argued that excess NO_3^- arises from rapid nitrogen regeneration in the euphotic zone, compared to Si, where diatoms are actively grazed by zooplankton and rapidly exported to depth as fecal debris. Why other phytoplankton species do not take over is unclear, except for possible limitation by other micronutrients such as iron (see Chapter 8).

There has been a general trend for decreasing levels of DSi in rivers, lakes, and coastal marine areas during the last half of the 20th century. Examples

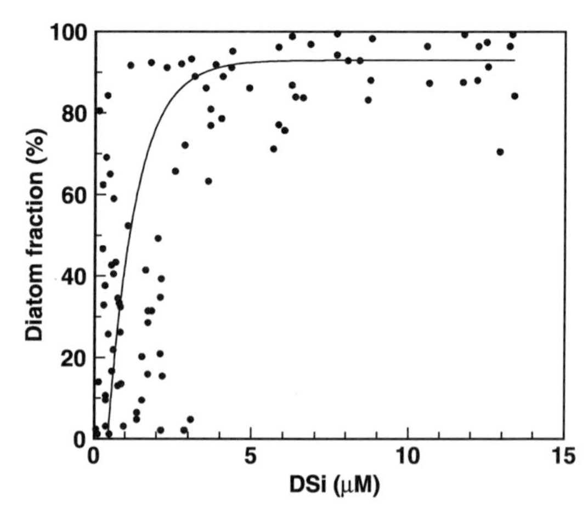

Figure 12.11 Diatom fraction of total phytoplankton cells from experimental mesocosms as a function of DSi concentration. Modified from Egge and Aksnes (1992).

include the Mississippi River and Lake Michigan (Figure 12.12), where DSi concentrations have decreased by about 50 and 80%, respectively, from the mid-1950s to the early 1970s, while a 40% decrease has been observed in the Baltic Sea from the mid-1960s to the mid-1980s (Jochem and Babenerd, 1989; Conley *et al.*, 1993).

The depletion in DSi is related to high discharges of nitrogen and phosphorus and to the regulation of rivers by damming. Schelske and Stoermer (1971) formulated a model for Si depletion, the "silica depletion syndrome." This model rests on the assumptions that increased nutrient (nitrogen and phosphorus) loading initially causes an increase in diatom production and that a large fraction of the produced diatoms is permanently lost from the water column before silica dissolution occurs. As a result, DSi in surface waters is lowered. Numerous recent lines of evidence from rivers, lakes, and coastal marine areas reinforce the relationship between nutrient loading and DSi depletion (Conley *et al.*, 1993). Lake Michigan, for example, has received increasing amounts of anthropogenically derived nitrogen and phosphorus during the last decades, while DSi originates from naturally occurring weathering processes that are largely unaffected by human activity (Schelske and Stoermer, 1971). The increased loading of nutrients, relative to DSi, has allowed increased diatom production and subsequent removal of biogenic silica via sedimentation. This has depleted the reservoir of DSi in the lake (Figure 12.12). At the same time, excess nutrients have allowed other non-Si dependent phytoplankton to flourish.

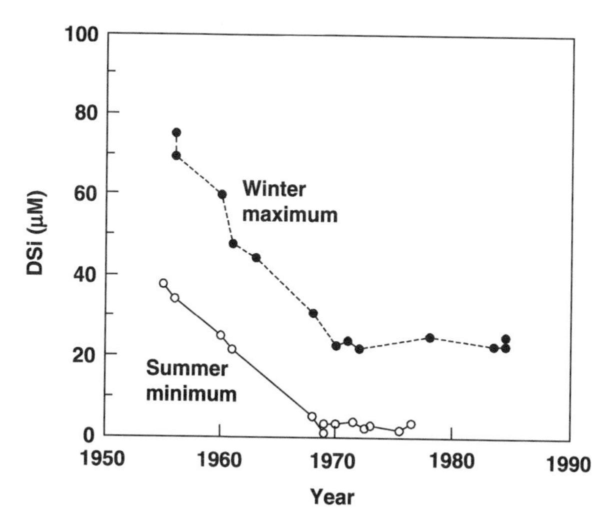

Figure 12.12 Long-term temporal changes in concentrations of DSi in Lake Michigan from 1950 to 1976 (summer minimum) and 1984 (winter maximum). Modified from Schelske (1988).

DSi limitation has inhibited diatom growth in many rivers, lakes, and coastal marine areas, with novel and often toxic phytoplankton species taking their place. These changes in phytoplankton community structure have altered food web dynamics with sometimes devastating effects on fisheries. The classic food chains, in which diatoms are grazed by zooplankton (copepods), contribute directly to large fishable populations. Other non-diatom-based food chains are undesirable because these phytoplankton species either remain largely ungrazed or fuel fish populations that are ecologically and economically inferior (Humborg *et al.*, 2000). Furthermore, blooms of toxic flagellates have become more frequent in many coastal areas, with serious implications for marine life (Richardson, 1997). The toxic haptophyte *Chrysochromulina polylepis*, for example, caused massive fish kills in the Skagerrak, Denmark/Norway, in 1988 (Maestrini and Graneli, 1991). There is no question that the depletion of DSi over the last few decades is of anthropogenic origin, as the most affected areas are highly influenced by human activities. For example, species of the dinoflagellate *Pfiesteria* have caused numerous fish kills along the rapidly developing South Carolina (USA) coastline during the last two decades (Lewitus and Holland, 2003). The *Pfiesteria* toxin has even been linked to human neurological illnesses (Burkholder and Glasgow, 1997). Consequently, a South Carolina Task Group on Harmful Algae was formed in 1997 to develop a coordinated state strategy to respond to possible *Pfiesteria*-related events.

DSi depletion is further exaggerated by the impact of large-scale damming of rivers. Because lake sediments retain some proportion of diatoms produced in the water column, the creation of reservoirs by damming results in declining DSi in rivers discharging to coastal areas. This "artificial-lake effect" (van Bennekom and Salomons, 1981) has been observed in many river systems. For example, DSi decreased 200 μM in the River Nile, Egypt, after completion of the Aswan Dam (Wahby and Bishara, 1980), and the construction of the Glen Canyon Dam on the Colorado River resulted in a 120 μM decrease in DSi (Mayer and Gloss, 1980). A more spectacular example is the Danube River and its impact on the Black Sea (Humborg *et al.*, 2000).

The Danube, which supplies about 70% of the freshwater discharge to the Black Sea, was dammed in 1970–1972 about 1000 km upstream by the "Iron Gate" Dam on the Yugoslavian-Romanian border. Available data from the pre- and post-dam periods suggests significant changes in DSi inputs to the Black Sea as a result of water and sediment storage in reservoirs. Retention in the upstream reservoir led to an overall decrease in DSi from 140 μM above the dam to 58 μM below the dam. As a consequence, DSi concentrations in Black Sea coastal waters have decreased from average winter values of 55 μM before the dam was constructed to 20 μM in the post-dam period (Figure 12.13). As dissolved inorganic nitrogen concentrations during the

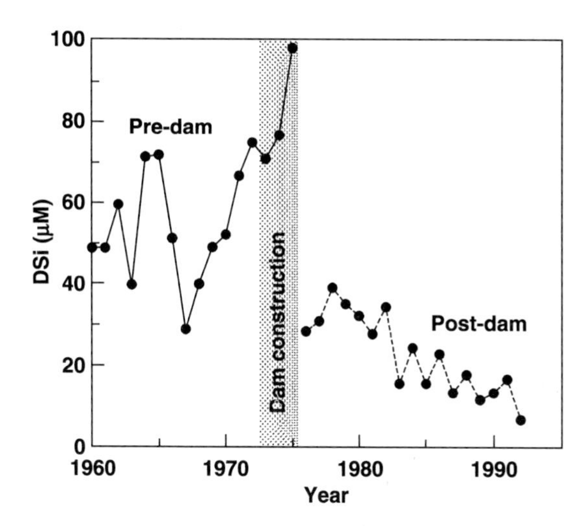

Figure 12.13 Time series (1960–1992) of average winter DSi concentrations in the Black Sea at a coastal station 60 nautical miles south of the Danube Delta. Modified from Humborg *et al.* (2000).

same time span have increased from 1.3 to 7.9 μM due to increased sewage discharges below the dam, the Si:N ratio of available nutrients has decreased from 42 to 2.8. The changes correlate well with increases in phytoplankton bloom frequency, cell densities, and the number of bloom-forming phytoplankton species other than diatoms in the Black Sea. Increasing riverine discharge of nitrogen and phosphorus, combined with reductions in river DSi concentrations as a consequence of damming, could lead to diatom depletion in coastal waters around the world.

6.2. Silica and the biological "carbon pump"

Variations in the marine silica cycle may influence atmospheric CO_2 levels and thus global climate. Atmospheric CO_2 is dissolved in the surface ocean, where it is taken up by phytoplankton via photosynthesis. This process requires nutrients such as dissolved silicate that are available in surface waters only in limited amounts. In a balanced situation, the same amount of CO_2 will be released back to the atmosphere via respiration. However, because of the biological "carbon pump" some of the organic carbon fixed by photosynthesis is exported from the surface waters to the deep ocean and the underlying sediments as POC and DOC (Tréguer and Pondaven, 2000). Consequently, together with the high solubility of CO_2 in sea water, the "carbon pump" is believed capable of influencing concentrations of atmospheric CO_2 (Tréguer and Pondavan, 2000).

The two major phytoplankton players in the biological "carbon pump" are diatoms and coccolithophores (Tréguer and Pondaven, 2000). While diatoms need DSi to build their siliceous cell walls, coccolithophores have shells of calcite. An efficient carbon pump means net withdrawal of CO_2 from the atmosphere during the formation of organic matter by reducing the partial pressure of dissolved CO_2 in surface water. However, carbonate shell formation by coccolithophores has the opposite effect. Although carbonate production reduces the total dissolved inorganic carbon in the water, the reaction increases the partial pressure of CO_2 at the ocean surface: $Ca^{2+} + 2\ HCO_3^- \rightarrow CaCO_3 + CO_2 + H_2O$, thus driving CO_2 from the ocean to the atmosphere. Dominance of diatoms over coccolithophores means decreased calcite production and less release of CO_2 from the ocean.

The loss of biogenic silica by sediment burial is replaced mostly by DSi inputs from rivers. If the "silica depletion syndrome" in coastal water continues (as mentioned above), a general change in marine phytoplankton composition from diatoms to non-siliceous species such as coccolithophores may very well occur. Such change can affect the global biogeochemical element cycling and have consequences for the role of the ocean as a sink for CO_2.

7. EVOLUTION OF THE SILICA CYCLE OVER GEOLOGICAL TIME

The global cycle of silicon has changed significantly over geological time from an early control by purely geochemical processes to control by biological processes, as occurs now (Siever, 1991). The concentration of DSi in the modern ocean is very low, ranging from less than 1 μM in open ocean surface waters to hundreds of μM in bottom waters. Although the highest concentrations in deep waters are close to saturation with quartz, the majority of deep and surface waters is at least one order of magnitude undersaturated with respect to quartz and close to two orders of magnitude undersaturated with respect to BSi. The patterns of DSi in the modern ocean cannot be explained from mineral equilibria, but are governed by efficient removal by diatoms in surface waters and regeneration by dissolution in deep waters.

Up to around 550 million years ago, late in the Precambrian, organisms had not yet evolved with the ability to secrete silica. Up to this time, silicification was probably governed by mineral equilibria with clay minerals, zeolites, and opaline silica and interactions between sea water and basalt

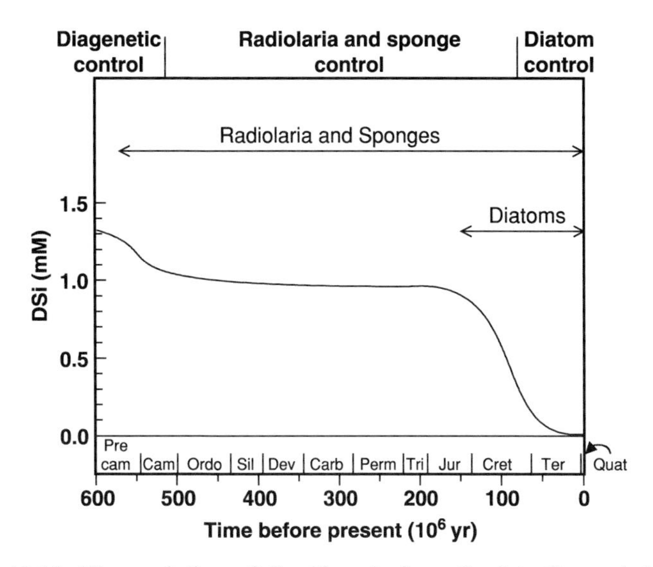

Figure 12.14 The evolution of the Si cycle from the late Precambrian to the present. Temporal ranges of the most important silica dependent organisms (radiolaria, sponges, and diatoms) and the hypothesized dissolved silicate concentration of the ocean are shown. Modified from Siever (1991).

during hydrothermal circulation (Siever, 1991). The DSi concentration of the oceans was likely in the range of 1.3–1.5 mM (Figure 12.14). With the evolution during the Cambrian of radiolarians in the pelagic ocean and siliceous sponges in marine shelf areas, DSi was actively removed by biological process. Together, these two groups reduced the DSi in surface waters to a level below that in the Precambrian, probably about 1 mM. The radiolarians and sponges subsequently controlled the Si cycle without significant changes for about 300 million years.

Diatoms and silicoflagellates evolved in the late Jurassic, but were initially unimportant contributors to the Si cycle. Following an intense radiation of species through the Cretaceous, diatoms in particular became globally important for the oceanic Si cycle near the beginning of the Tertiary. The gradually increasing dominance of diatoms in the surface ocean during the Tertiary resulted in drastically reduced DSi concentrations (Siever, 1991). As a consequence, the abundance and silicification of sponges and radiolarians was reduced (Harper and Knoll, 1975; Maldonado *et al.*, 1999). The extraordinary efficiency of diatoms in extracting DSi from sea water reduced oceanic DSi concentrations to the present very low levels.

Chapter 13

Microbial Ecosystems

1. INTRODUCTION

We have explored in this volume a wide range of topics concerning the relationship between microorganisms and their role in elemental cycling in nature. In individual Chapters we have focused on the cycling of specific elements. We believe this approach is logical and that it provides focus to the discussion. We have tried to highlight the areas where element cycles interconnect, but we have yet to consider this interconnectedness at the level of whole ecosystems (with the exception of the autotrophic ecosystems discussed in Chapter 4). This is what we do here. We highlight three different ecosystems in which microorganisms play a prominent role in elemental cycling. In what follows we discuss microbial mats, stratified water bodies, and mangrove forests. In microbial mats and stratified water bodies, microorganisms are the main drivers of elemental cycling. These examples allow us to explore how chemical energy and light are efficiently utilized in nature and to discuss how specific microbial populations adapt to maximize this

ADVANCES IN MARINE BIOLOGY VOL 48
0-12-026147-2
© 2005 Elsevier Inc.
All rights reserved

efficiency. Mangrove forests are somewhat different. These ecosystems are driven by the interaction between macroorganisms (such as mangrove trees and crabs) and the microbes involved in the terminal phases of carbon mineralization. These interactions, occurring across great divides in the Tree of Life, are highlighted here.

2. MICROBIAL MATS

Microbial mats appear deceptively simple. They are typically thin blue-green films of organic matter in out-of-the-way places such as ephemeral pools and ponds, some intertidal sands, hydrothermal hot springs, and the bottoms of salt evaporating ponds. They are easily overlooked, yet, as explored in more detail below, photosynthetic mats can photosynthesize at rates approaching those of hardwood forests. Furthermore, steep chemical gradients of metabolic products can develop over distances of only hundreds of micrometers. All of this activity would be barely discernible if not for the development of a variety of microsensors used for fine-scale measurement of light, pH, and chemical constituents such as O_2 and H_2S (Revsbech and Jørgensen, 1986; Jørgensen and Des Marais, 1988).

Microbial mats are not all phototrophic, and active chemotrophic communities can be found in remote settings such as deep-sea hydrothermal vent areas and sediments underlying oxygen minimum zones as found, for example, off the coasts of Chile and Peru (Fossing *et al.*, 1995). In another example, huge microbial towers are found in the anoxic portions of the Black Sea where locally abundant supplies of methane are oxidized in conjunction with sulfate reduction in the poorly understood process of anaerobic methane oxidation (Michaelis *et al.*, 2002) (see Chapter 10). Chemotrophic-type microbial mat communities are discussed elsewhere in the text (Chapters 4 and 9) and are not considered further here. Our immediate concern is microbial communities driven by photosynthesis.

Both oxygenic and anoxygenic phototrophic communities can be responsible for primary carbon production when light is available. Anoxygenic communities are most important when there is a ready supply of sulfide, such as found in some terrestrial hydrothermal settings (e.g., Jørgensen and Nelson, 1988; Castenholz and Pierson, 1995). Generally, however, microbial mats are somewhat rare, and their rather limited distribution can be ascribed to grazing and sediment churning by animals. Thus, mats are most common where animal activity is restricted, for example, by high salinities, high temperatures, and environmental instability such as that caused by frequent wetting and drying. In what follows we consider aspects of the structure, ecology, and biogeochemistry of microbial mat systems.

2.1. Structure of microbial mat ecosystems

Photosynthetic microbial mats offer an unparalleled view into the intimate associations between prokaryote populations in nature. These populations may be observed microscopically or probed with a variety of different chemical and light microsensors. A typical microbial mat is vertically stratified, with the dominant populations placed relative to their requirements for light and chemical interfaces, such as for example, between oxygen and sulfide. A composite drawing of the vertical structure of a microbial mat from high-salinity evaporating ponds of Exportadora de Sal, Guerrero Negro, Baja California, Mexico, demonstrates these points (Figure 13.1). Here, we see a surface population of diatoms, which is apparently well adapted to the high light incident onto the mat surface during the day. Just below is an assemblage of different cyanobacteria, in this case mostly filamentous types, which are responsible for most of the primary production within the mat. A peak in oxygen concentration is found associated with the highest density of cyanobacteria.

The cyanobacterial population density decreases dramatically with depth. In this region oxygen concentrations also decrease due to heterotrophic respiration within the mat and the oxidation of chemically reduced metabolic products such as sulfide and ammonia, which diffuse from below. At the oxygen–sulfide interface we find, in this case, a population of filamentous green non-sulfur bacteria. Other anoxygenic photosynthetic populations might also be found depending on the spectral characteristics of light in this region (see Chapter 9). Colorless sulfur bacteria might also be found, and they will be particularly prominent when oxygen reaches below the zone of light penetration (Jørgensen and Des Marais, 1986a). Below the region where oxygen penetrates, an assemblage of largely heterotrophic prokaryotes will be engaged in organic carbon mineralization, and when sulfate is available, the most important terminal mineralization process will be sulfate reduction. This, in turn, supplies the sulfide fueling the sulfide-oxidizing populations just above. In all, these zones of dominant populations, ranging from primary-producing cyanobacteria to sulfate reducers, are frequently condensed over a depth scale of 2 to 5 mm (Figure 13.1) (see also Minz *et al.*, 1999).

The above discussion has emphasized the vertical distribution of the dominant microbial populations. The true population structure, however, is far more complex than this. For example, aerobic heterotrophs and even

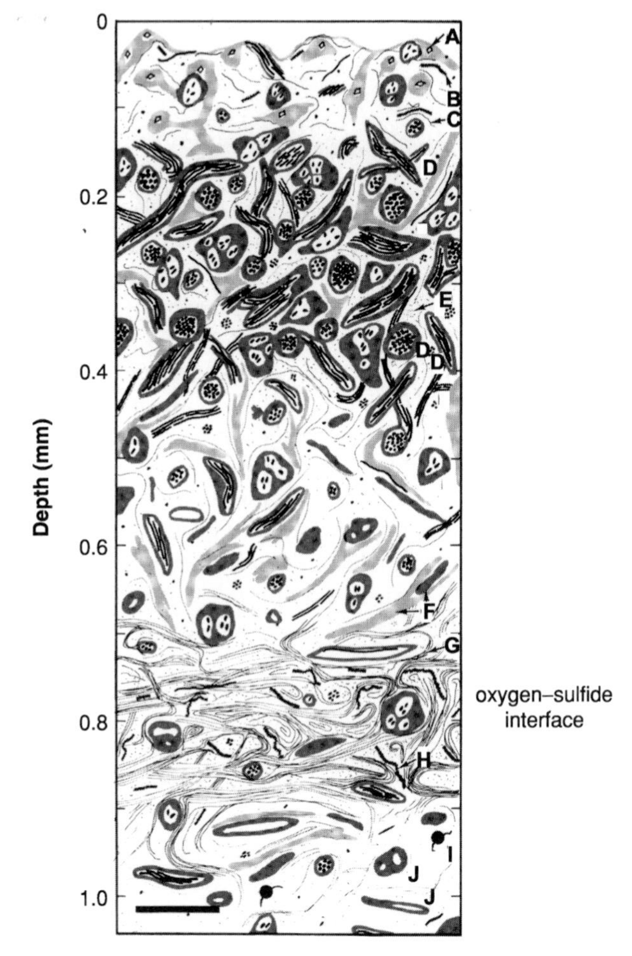

Figure 13.1 Composite drawing of the microbial members of a typical photosynthetic microbial mat from Guerrero Negro, Baja California, Mexico. The drawing was assembled from numerous TEM observations. The oxygen–sulfide interface is located during the day at about 0.8 mm depth. The letters refer to the following: A, diatoms; B, *Spirulina* sp. (cyanobacteria); C, *Oscillatoria* spp. (cyanobacteria); D, *Microcoleus chthonoplastes* (cyanobacteria); E, non-photosynthetic bacteria; F, fragments of bacterial mucilage; G, *Chloroflexus* spp. (green non-sulfur bacteria, capable of anoxygenic photosynthesis); H, *Beggiatoa* spp. (nonphotosynthetic sulfide-oxidizing bacteria); I, unidentified grazer; J, abandoned cyanobacterial sheaths. Adapted from Canfield and Des Marais (2001).

sulfate reducers (see Chapter 9) (Canfield and Des Marais, 1991) are found in the upper oxygen-containing zone. Indeed, Minz *et al.* (1999) found the largest amount of *Desulfovibrio*-like 16S rRNA in the upper oxygen-rich part of a microbial mat from Solar Lake, Sinai. Other sulfate-reducer populations, including those with *Desulfonema*-like and *Desulfobacteriaceae*-like rRNA, were concentrated above and below the oxycline. Some of these sulfate reducers may be engaged in aerobic sulfate reduction, as has been reported for these mats (Fründ and Cohen, 1992), or they may possibly conduct aerobic metabolism during the day, switching to anaerobic metabolism at night as the oxycline moves to the mat surface.

Further evidence for population heterogeneity comes from direct microscopic observations. Using epifluorescence microscopy on DAPI-stained cells, Sørensen (2002) found that small, non-cyanobacterial cells dominated the two prominent zones of oxygenic photosynthesis in a stratified microbial ecosystem established within gypsum crust growing in a production saltern from Eilat, Israel. Likewise, in the lower bright red band of the crust, housing purple-sulfur bacteria, non-phototrophic cells dominated by numbers. Still, in each of these phototrophic horizons, the phototrophs dominated by biovolume. Transmission electron microscopy (TEM) and scanning electron microscopy (SEM) observations of the phototrophic portions of other microbial mats reveal a similarly complex population structure (e.g., Jørgensen *et al.*, 1983; Fenchel and Kühl,1999).

Observations from molecular techniques further emphasize the extreme diversity among microbial mat populations. In recent work, Spear *et al.* (2003) extracted and amplified 16S rDNA from greenhouse-maintained mats from Guerrero Negro. They found a high degree of species and functional diversity within the upper cyanobacterial layer of the mat. Indeed, as with the gypsum crust study described above, cyanobacteria did not apparently dominate the microbial population by numbers. Furthermore, in the classic study by Ward *et al.* (1990), the diversity of molecular isolates (16S rRNA) was far greater than anticipated based on standard culturing techniques and microscopy. This was one of the first demonstrations of the great diversity of microbial populations in nature.

In another study, Nübel *et al.* (1999) focused specifically on cyanobacteria (discussed also in Chapter 2) and assessed diversity in eight different microbial mats. Diversity was gauged from morphological observations, pigment analysis, and DNA extraction followed by 16S rDNA amplification and separation. A similar picture of diversity was obtained with each approach, and generally, while a single population dominated a given phototrophic community, up to nine different species of cyanobacteria could be detected in a single mat. Some of them, however, were quite rare.

The main members of the mat community define the physical characteristics of a mat. Thus, long, filamentous, ensheathed cyanobacteria such as

Lyngbya sp. form tough mats that are extremely resistant to shear, breakage, and desiccation, and *Lyngbya* mats are typically found in intertidal areas where physical conditions are rough. Likewise, mats dominated by *Microcoleus chthonoplastes*, whose filaments bundle into common sheaths (see Figure 13.1), also form tough, resilient, leathery mats. Unicellular cyanobacteria, such as the halophilic *Aphanothece halophytica* and related species, form mats with thick photic zones (e.g., Jørgensen *et al.*, 1983; Jørgensen and Des Marais, 1986), in which phototrophic members are embedded in copious amounts of thick extracellular polymeric substances (EPS) (mostly polysaccharides) (e.g., Oren, 2000). The EPS stabilizes the mat. Furthermore, it holds water, helping mat organisms resist drought, and it may also help scavenge important trace metals (see Stal, 2000). However, EPS may form as a by-product of unbalanced growth under nitrogen limitation (e.g., Stal, 2000); therefore, these useful properties could be as much a matter of chance as a matter of design. Mats formed by unicellular cyanobacteria tend not to form in high-energy environments, as they are less resistant to physical stress than mats formed by filamentous organisms (e.g., Stal, 1994).

To conclude, dominant microbial populations, varying on depth scales of hundreds of micrometers, both shape and respond to the chemistry of the mat environment. Within individual prominent layers, dominant populations share with other populations in the efficient cycling of carbon and nutrients. The physical nature of the mat, be it soft or leathery, is dictated by the morphological characteristics of the most important cyanobacterial members. Not surprisingly, most of the organisms involved in microbial mat biogeochemical cycling are unknown in culture collections.

2.2. Microbial mat ecology

Microbial mats are dynamic and must be understood in terms of cyclic patterns of activity and interaction as driven by diel variations in light and temperature. We explore below these forcing factors and their influence on mat ecology.

The microbial mat community is exposed to changes in light intensity during the course of the day. Commonly, phototroph activity, measured as total (depth integrated) gross rates of photosynthesis, increases with increasing light intensity up until the maximum intensity is reached at midday. This was found, for example, for mats from Guerrero Negro, Baja California, Mexico, where light intensities incident onto the mat surface up to 800 μmol photons m^{-2} s^{-1} were encountered (Canfield and Des Marais, 1993). Similarly, Revsbech *et al.* (1983) found a linear dependence of gross photosynthesis, with incident light levels up to 1650 μmol photons m^{-2} s^{-1} for Solar Lake mats. This is not always the case, however, as photosaturation is

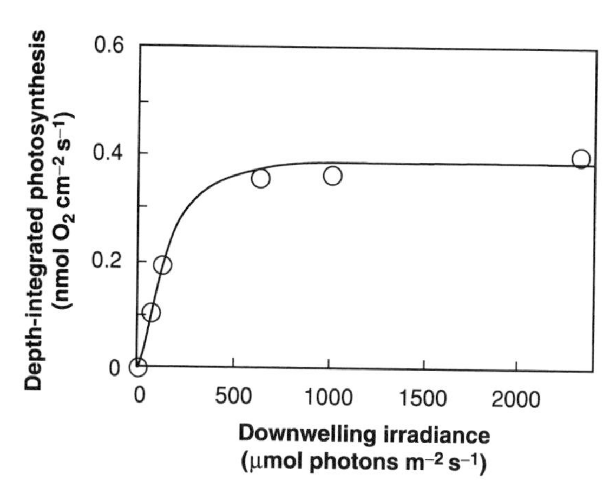

Figure 13.2 Relationship between depth-integrated rates of photosynthesis and downwelling irradiance for Solar Lake mats incubated at 25 °C. Photosaturation occurs at around 500 to 600 μmol photons m^{-2} s^{-1}. Adapted from Wieland and Kühl (2000a).

sometimes also encountered. In this case the phototrophic community reaches a limit of activity at which further increases in light level produce no increase in photosynthetic rate. Thus, in another work on Solar Lake mats, photosaturation of gross primary production was clearly seen at incident light levels of around 650 μmol photons m^{-2} s^{-1} (Figure 13.2) (Wieland and Kühl, 2000a). However, even in mats experiencing photosaturation, the phototrophic community responds to light at intensities far beyond the saturation level for individual cyanobacterial species. For cyanobacterial cultures, light saturation of photosynthesis typically occurs at light levels of only around 50 μmol photons $m^{-2}s^{-1}$, and photo-inhibition of photosynthesis may occur at higher light levels (e.g., Putt and Prézelin, 1985).

Therefore, the light environment in microbial mats, even over shallow depths of 1 mm or less, is considerably different from the light incident onto the mat surface. The development of micro-scale fiber optic light probes has allowed the light environment within a mat to be explored over depth scales of less than 100 μm (e.g., Jørgensen and Des Marais, 1988). Such probes reveal that 90% of the incident solar radiation in Solar Lake and Guerrero Negro mats is absorbed by 0.2 to 0.5 mm depth, with clear absorption maxima correlating with the absorption band for Chl *a* at 670 nm (Jørgensen and Des Marais, 1988; Kühl *et al.*, 1994) (see also Chapter 9). Strong absorption also occurs in the regions between 430 and 530 nm, corresponding to adsorption by Chl *a* and accessory pigments. Therefore, organisms located at depths of a few hundreds of micrometers within the mat may never experience photosaturation, even though saturating light levels are incident upon the mat surface. Also, as the light intensity increases, so does the depth of the photic zone (e.g., Revsbech *et al.*,1983; Canfield and Des Marais, 1993; Wieland and Kühl, 2000a). Thus, even if the upper layers of the phototrophic community experience photosaturation, a thickening photic zone at higher light levels could generate a positive relationship between total rates of photosynthesis and light.

Visible light is quickly attenuated with depth in photosynthetic mat communities due to the absorption by Chl *a* and the accessory pigments used by oxygenic photosynthetic organisms. Less absorbed in the upper photic zone is light in the near-infrared range. This is used by many different anoxygenic photosynthetic bacteria, which frequently become prominent at depths where visible light can no longer sustain oxygenic phototrophs (e.g., Jørgensen and Des Marais, 1988; Kühl and Fenchel, 2000). This aspect of anoxygenic phototrophic adaptation to light in microbial communities is explored in more detail in Chapter 9.

The matrix of microbial mat organisms is also in flux during the day, as some mat organisms actively shift position in response to changing light intensity. For example, *Oscillatoria* sp. and *Spirulina subsalsa* actively migrate to the mat surface at low light levels and in the dark, and migrate into the mat in the light (Garcia-Pichel *et al.*, 1994; Nadeau *et al.*, 1999). The depth of migration is correlated with light intensity incident upon the mat surface, and it appears that these cyanobacterial species seek to optimize their light environment within the mat, both to maximize their photosynthetic output and to avoid excessive UV irradiation. Migratory behavior in response to light, however, is not a general behavioral feature of cyanobacteria (Garcia-Pichel *et al.*, 1994). Indeed, if it was, a microbial mat would be a constantly moving entity with individual population members in constant flux, trying to gain advantage over other members only to see their position erode as other members moved to enhance their own standing. This constant motion would serve no real benefit to the total community and would probably sap the community of energy. It appears that only a select few members of the oxygenic phototrophic community have adapted migration as a method of enhancing their light environment. The bulk of the phototrophic community is apparently adapted to constantly shifting levels of light during the day.

Other members of the mat community also migrate day and night. For example, sulfide oxidizers of the genus *Beggiatoa*, which are often found in mats, can follow the oxygen–sulfide interface as it moves up and down in a diel cycle (Garcia-Pichel *et al.*, 1994) (see below). Motility has also been observed with the purple sulfur bacterium *Marichromatium gracile* (Jørgensen, 1982a; Thar and Kühl, 2001), an organism that exhibits complex migratory behavior. In the dark, *M. gracile* concentrates within the oxygen–sulfide interface, exhibiting a strong phobic response to oxygen at concentrations above about 10 μM (Thar and Kühl, 2001) and a lesser phobic response to high sulfide concentrations. In the light, *M. gracile* still avoids oxygen and spreads into the anoxic zone; it also reverses its direction when swimming into regions of lower light intensity. These combined responses keep the organism situated in the oxygen–sulfide interface at night where it oxidizes the sulfide with oxygen. In the day, *M. gracile* seeks the highest light intensity it can encounter while still remaining in the sulfide it oxidizes phototrophically.

Temperature is another deciding factor regulating mat ecology. In the broadest sense, temperature delineates the boundaries of the region in which phototrophic mat communities can be found. Thus, the upper temperature limit for both cyanobacteria and anoxygenic phototrophic bacteria is in the range of 70 to 73 °C (Brock, 1994). Phototrophic mats cannot be established at temperatures above this, and indeed, in thermal areas in Yellowstone National Park and Iceland, mats develop only when outflowing hydrothermal water cools to below the upper temperature limit for phototrophic growth (Jørgensen and Nelson, 1988; Brock, 1994). Microbial mats growing at these high temperatures are functionally similar to those growing in the mesophilic temperature range; however, some key microbial metabolisms might be lacking. For example, thermophilic sulfur-compound-disproportionating organisms have not been yet been described. However, these organisms may also be present but have simply not yet been identified.

Temperature also influences rates of microbial mat activity. Temperature response has mostly been studied with mats growing in the mesophilic temperature range, where Q_{10} responses are typically 2 to 3 (e.g., Canfield and Des Marais, 1993; Wieland and Kühl, 2000a,b). For example, in the study of Canfield and Des Marais (1993), both gross photosynthesis and sulfate reduction rates in *Microcoleous chthonoplastes*-dominated mats from Guerrero Negro were determined at temperatures ranging from 17 to 30 °C. Both autotrophic and heterotrophic processes scaled with temperature by similar amounts and with a Q_{10} response of about 3. Therefore, total mat activity changed with temperature, but the relative magnitudes of autotrophic and heterotrophic processes remained relatively constant. For Solar Lake mats (also dominated by *M. chthonoplastes*) incubated at temperatures from 25 to 40 °C, the gross rate of photosynthesis increased from 25 to 30 °C, but decreased from 30 to 40 °C. Also, some uncoupling between phototrophic and heterotrophic processes was observed. Thus, sulfate reduction rates continued to increase as temperature was increased above 30 °C, and at higher temperatures, less oxygen was used to oxidize organic substrate and more was used to oxidize sulfide. Possibly these imbalances represent transient responses to temperature variations outside of the range to which the mat was adapted. Prolonged exposure to high temperatures could produce a better adapted species composition.

In another study, Epping and Kühl (2000) explored the temperature response from 15 to 30°C in *M. chthonoplastes*-dominated mats from the salterns of the Alfacs Peninsula in Spain. In these mats, a temperature optimum in gross photosynthesis occurred at 20°C, and as in the Solar Lake mats, heterotrophic activity continued to increase with temperature up to 30°C, the maximum explored. At the higher temperatures the surface of the mat was populated by sulfide-oxidizing *Beggiatoa* sp., and the photic zone was sulfidic until relatively high incident irradiances of about 300 μmol photons $m^{-2} s^{-1}$ were reached. At lower light intensities it is possible that the phototrophic community was engaged in anoxygenic photosynthesis encouraged by the enhanced sulfide availability from sulfate reduction at high temperatures (see Chapter 4) (Epping and Kühl, 2000). Under these conditions, net oxygenic photosynthesis was suppressed (it is unknown whether gross oxygenic photosynthesis was completely inhibited). It appears that relatively high light intensities were required to activate oxygenic photosynthesis to rates high enough to overwhelm the sulfide flux and to allow for net oxygen production. The significance of this situation in nature is unknown.

2.3. Mat biogeochemistry

Rates of primary production in microbial mats can be enormous. In the most active mats, millimeter thin layers of cyanobacteria have gross rates of photosynthesis comparable to rates in hardwood forests and tropical rain forests (Table 13.1). As noted above, these high rates of oxygen production influence the mat chemical environment in a profound way. Oxygen concentrations can build to levels several times greater than air saturation during the day (Figure 13.3) (e.g., Revsbech *et al.*, 1983; Canfield and Des Marais, 1993). Remarkably, oxygen can accumulate to these high concentrations and fall again to nothing in just 2 mm of mat depth (e.g., Revsbech *et al.*, 1983; Canfield and Des Marais, 1993). Associated with high oxygen concentrations are high pH values, which can approach or even exceed 10 (Revsbech *et al.*, 1983; Revsbech *et al.*, 1988). These high pHs are a result of active CO_2 uptake by the photosynthesizing cyanobacteria. Indeed, the pHs are so high that CO_2 is virtually absent in the photic zone and the cyanobacteria must obtain most of their carbon by pumping HCO_3^- into the cell (Badger and Price, 1992) (for isotope fractionation consequences see Chapter 4). Typically, sulfide accumulates just below the region where oxygen disappears (Figure 13.3). This creates a well-defined oxygen–sulfide interface used by non-phototrophic sulfide oxidizers, while the sulfide is utilized by anoxygenic phototrophs.

The examples discussed above are not uncommon, but are not representative of all mats found in nature. In some mats, such as those growing at very high salinities, for example, rates of primary production may be quite low (Table 13.1) (Canfield *et al.*, 2004; Sørensen *et al.*, 2004) and oxygen concentrations may increase only marginally above air saturation. In our own work on phototrophic communities established in gypsum crusts, primary production rates were only a fraction of those found in high productivity mats such as those found in Solar Lake or Guerrero Negro (Table 13.1) (Canfield *et al.*, 2004). Also, instead of a few millimeters, oxygen penetrated to 3 to 4 cm depth (Canfield *et al.*, 2004; Sørensen *et al.*, 2004).

In the dark, without oxygenic photosynthesis, oxygen is consumed within mats, and the sulfide–oxygen interface migrates to near the mat surface (Figure 13.3). As mentioned above, colorless sulfur-oxidizing bacteria are often conspicuous members of the surface mat community at night. The pH values also return to values more typical of sediments, ranging, usually, from 7 to 8. These generalizations, however, do not account for all mats. In our own work on submerged *Lyngbya* mats, oxygen persisted overnight in the photic zone in large bubbles trapped by the network of interlocking *Lyngbya* filaments (Canfield and Des Marais, unpublished). Also, in the high salinity gypsum crusts mentioned above, oxygen persisted overnight in regions of the photic zone where respiration was not active (Canfield *et al.*, 2004).

With rapid metabolic rates and small spatial dimensions, there is a great deal of physical and biological interconnectedness between cycles of different elements within mats. In marine mats, the most important elements cycled are C, S, and O, and we consider below some of the significant relationships between these elements. We often think of mats as flat surfaces, like sediments, whose activities are described by element fluxes across the mat–water interface. This is part of the story, but there is also a great deal of internal cycling within the mat, and we can illustrate this by considering the biogeochemistry of carbon (Figure 13.4). During the day inorganic, carbon is fixed within the mat by photosynthesis (and chemolithoautotrophic processes that are probably minor), and there are a variety of sources for the carbon. There is, naturally, the dissolved inorganic carbon (DIC) diffusing into the mat from the overlying water, but there are also internal carbon sources, including mineralization by aerobic and anaerobic heterotrophs. Thus, DIC sources include sulfate reduction both within the oxic zone (see Chapter 9) and below, and O_2 respiration, including aerobic heterotrophy, photorespiration, and dark respiration by the cyanobacteria.

Each of these DIC sources was assessed in Guerrero Negro mats by measuring exchange fluxes into and out of the mat, by measuring rates of sulfate reduction within the mat, and by considering the O_2 budget (see below) which was used to constrain rates of O_2 respiration. The calculation details are outside of the present discussion and can be found in Canfield and Des Marais (1993, 1994). However, the conclusions are pertinent, and surprisingly, in two of the three cases explored, a majority of the DIC used to

Table 13.1 Primary production in microbial mats compared to terrestrial ecosystems

Mats	Temp. (°C)	[a]O_2 prod rate maximum (nmol $cm^{-2}h^{-1}$)	[a]O_2 prod rate daily average (nmol cm^{-2} h^{-1})	Notes	References
Artificial mat	20 to 25	3.5	—	lab experiment	1
Tagus Estuary, Portugal	27 to 32	2.5	1.8	*in situ* study-diatom mat	2
Um-El-Yums Sabka, Egypt	38	3.2	—	planar optrodes	3
Guerrero Negro					
Nov. 89	20	0.7	0.5	natural light	4
	30	1.8	1.0	natural light	4
April 90	17	1.3	0.9	natural light	4
April 90	30	5.9	3.0	natural light	4
Nov. 90	20	2.3	1.2	natural light	4
Artificial mat	35	0.8	—	lab experiment	5
Hot Spring, Iceland, Sta A	60	0.8	—	lab experiment	6
Sta B	52	3.0	—	lab experiment	6
La Salada de Chiprana, Spain	—	1.6	—	*in situ* analysis	7
Guerrero Negro					
P2	21 to 25	1.6	—	lab experiment, *in situ* salinity	8
P4	21 to 25	3.2	—	lab experiment, *in situ* salinity	8
P6	21 to 25	1.0	—	lab experiment, *in situ* salinity	8
NC2	21 to 25	2.8	—	lab experiment, *in situ* salinity	8
NC3	21 to 25	2.2	—	lab experiment, *in situ* salinity	8

Mushroom Spring, Yellowstone NP	68	2.5	—	natural light	9
Solar Lake	25	2.4	1.4	natural light	10
Solar Lake	27.5	1.3	—	lab experiment	11
Gypsum crust, Israel					
P200	21	0.3	0.18	200 ppt salinity	12
P201	21	0.05	0.03	230 ppt salinity	12
Terrestrial Ecosystems					
Alfalfa field	—	—	4.6[b]	—	13
Pine plantation	—	—	2.3[b]	—	13
Oak-pine forest	—	—	2.2[b]	—	13
Rain forest, Puerto Rico	—	—	8.5[b]	—	13

[a]Gross rates of primary oxygen production.

[b]Calculated from yearly averages assuming 12 hour day and a 1:1 stoichiometry between carbon fixation and O_2 production.

References: 1, Fenchel (1998); 2, Brotas *et al.* (2003); 3, Glud *et al.* (1999); 4, Canfield and Des Marais (1993); 5, Pringault and Garcia-Pichel (2000); 6, Jørgensen and Nelson (1988); 7, Jonkers *et al.* (2003); 8, Garcia-Pichel (1999); 9, Ferris *et al.* (2003); 10, Revsbech *et al.* (1983); 11, Wieland and Kühl (2000a); 12, Canfield and Sørsensen (unpublished); 13, Odum (1971).

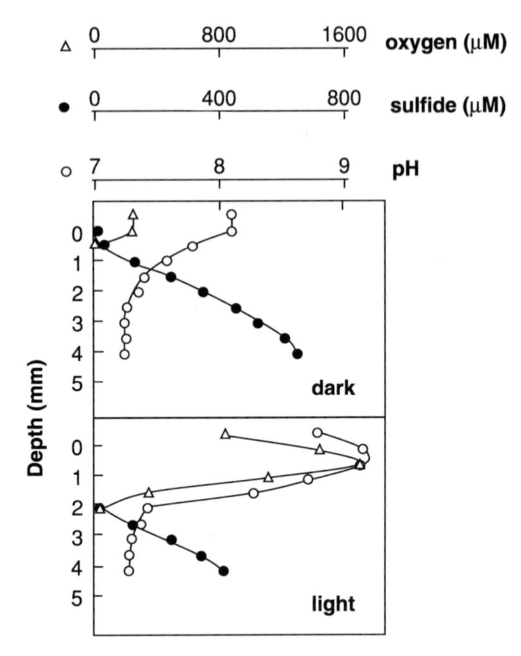

Figure 13.3 Distributions of oxygen, sulfide, and pH for pieces of Solar Lake mats incubated in the dark and in the light. Redrafted from Revsbech *et al.* (1983).

fuel primary production came from organic carbon mineralization within the mat (Table 13.2). Both sulfate reduction and oxic respiration proved to be important internal sources of DIC, with variable relative magnitudes.

The daytime geochemical cycling of oxygen parallels that of DIC. The only oxygen source is primary production; oxygen sinks include various forms of oxygen respiration in the mat, sulfide oxidation, and the efflux of oxygen from the mat (Figure 13.4). As with DIC, exchange across the mat–water interface tells only part of the story, and processes within the mat were either the most important sinks for oxygen or sinks of equal magnitude to the oxygen efflux. This again demonstrates the great deal of internal elemental cycling within the mat. Further evidence for intensive internal cycling comes from the nighttime behavior of ammonia. At night, with no active phototrophic community to sequester nutrients, and net mineralization, nutrient loss from the mat might be expected. To test this, ammonia concentrations were monitored in nighttime flux incubations (Canfield and Des Marais, 1994), and no ammonia loss was observed. This could, however, have resulted from active nitrification of the ammonia to nitrate. The incubation was allowed to go anoxic, precluding nitrification, and still no ammonia loss occurred. Therefore, the nighttime mat community was actively

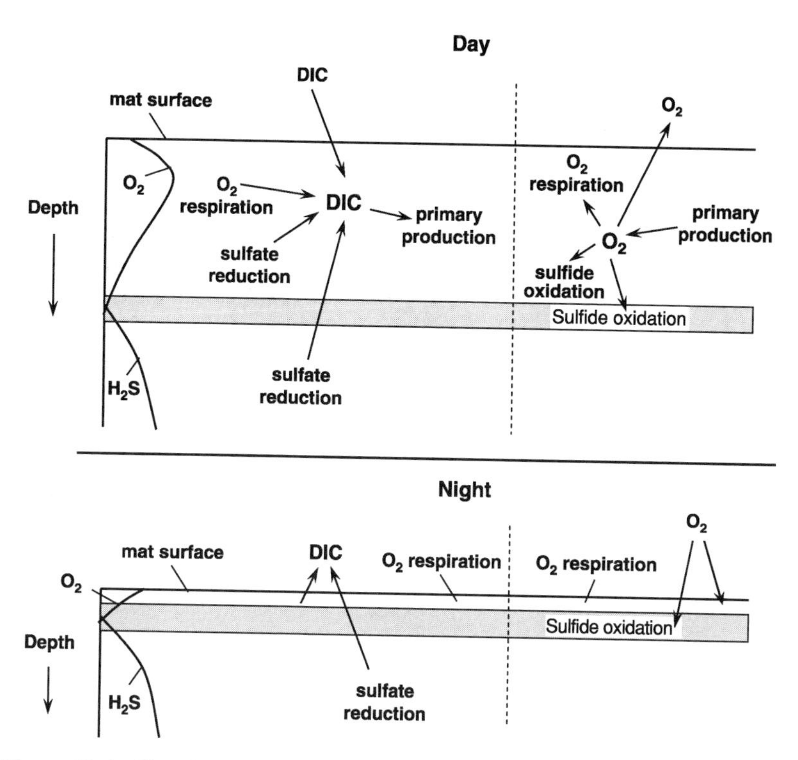

Figure 13.4 The day and night cycling of dissolved inorganic carbon (DIC) and oxygen in a photosynthetic microbial mat. Arrows indicate sources and sinks, and general aspects of microbial mat chemistry are also shown. Redrawn from Canfield and Des Marais (1993).

Table 13.2 Daytime DIC budget in Guerrero Negro mats

	April 1990, 17 °C	April 1990, 30 °C	Nov 1990, 20 °C
DIC sinks			
Oxygenic photosynthesis	75.7 (8.3)[a]	236 (26)	84.2 (9.8)
Anoxygenic photosynthesis	5.6 (2.4)	23 (5.8)	0
Total	81.3 (8.4)	259 (27)	84.2 (9.8)
DIC sources			
Diffusion from overlying water	34.5 (1.8)	98 (4.9)	69 (3.5)
Sulfate reduction	10.9 (1.3)	45 (5.4)	41.6 (13)
Oxic respiration	53.5 (8.4)	158 (27)	−3.2 (16)
Total	98.9 (8.7)	301 (28)	107.4 (21)

[a]One standard deviation.
From Canfield and Des Marais (1994).

sequestering ammonia, a valuable strategy for maintaining high activity in a nutrient-poor environment.

At night, oxygen diffuses into the mat and is used to oxidize sulfide and organic matter. At the same time, the DIC liberated from oxic respiration and sulfate reduction diffuses from the mat (Figure 13.3). These Guerrero Negro mats display very little net growth, as the fluxes of O_2 from the mat during the day nearly match the fluxes of O_2 into the mat at night, and vice versa for DIC (Canfield and Des Marais, 1993). Therefore, organic matter is very efficiently recycled in these mats. Curiously, the flux of DIC out of the mat at night was always much greater than the flux of O_2 in, and during the day, the flux of DIC into the mat was much greater than the flux of O_2 out (the opposite is expected during the breakdown of organic matter containing reduced nitrogen compounds). Canfield and Des Marais (1993) concluded that this imbalance likely occurred due to the storage of oxidized photosynthetic intermediates such as glycolate during the day and their oxidation at night. Fenchel (1998) observed a similar imbalance in O_2 and DIC fluxes in artificial mats and attributed this to the accumulation of an oxygen debt as reduced S and Fe compounds during the night and their oxidation during the day. This is a plausible explanation for the artificial mats but probably does not explain the imbalance in the Guerrero Negro mats, where reduced sulfur species were monitored on a diel basis and changes in their concentration were insufficient to account for the O_2–DIC imbalance.

3. STRATIFIED WATER BODIES

Physically stratified water bodies are frequently chemically stratified too, and as in microbial mats, layers of dominant microbial populations often develop. Also as in mats, these populations will position themselves to take advantage of available light and chemical interfaces between redox-active elements. Stratification is typically induced by gradients in temperature and/or salinity. In what follows, we consider the structure of microbial populations in chemically stratified water columns, as well as aspects of microbial ecology and biogeochemical cycling. We focus our attention on water bodies in which the stratification is more or less permanent.

3.1. Microbial stratification and interaction

There is a great deal of variability in the intensity of carbon cycling and in the chemical nature of stratified water columns. On one extreme are water columns such as those found in Darwin Bay, Galapagos (Richards and

Broenkow, 1971), and Golfo Dulce, Costa Rica (Dalsgaard *et al.*, 2003), where oxygen disappears at depth and nitrate concentrations are reduced due to denitrification, but sulfide either does not accumulate or accumulates only weakly in the very bottom of the basin. These water columns resemble oxygen-depleted oxygen minimum zones like those found in the eastern Pacific or Indian Oceans. At the other extreme are hyper-sulfidic water columns as seen, for example, in Lake Mahoney, British Columbia, where sulfide accumulates up to concentrations of 35 mM in the lake bottom waters (Overmann *et al.*, 1996b). The nature of microbial ecosystems in these different stratified water columns varies depending on the chemistry of the water column and the proximity of prominent redox interfaces to light.

3.1.1. Golfo Dulce

We were introduced Golfo Dulce in the last Section, and the major features of its water chemistry are shown in Figure 13.5. Unlike many stratified marine and lake water columns, the decrease in oxygen concentration is very gradual. Anoxygenic photosynthesis is unlikely here, as electron donors such as reduced sulfur species and Fe^{2+} are not found in the light-containing region of the water column (which constitutes the upper 60 m).

The main feature of this water column is the broad oxygen-free region between about 80 and 160 m, where denitrification is the dominant heterotrophic process of carbon mineralization (Figure 13.5). Curiously, ammonium, a normal product of anaerobic carbon mineralization, does not accumulate. Instead, ammonium is actively oxidized, anaerobically, with nitrite by the newly discovered process known as anammox (*an*aerobic *amm*onium *ox*idation; see Chapter 7) (Dalsgaard *et al.*, 2003). Therefore, there is a tight coupling between denitrification and anammox, in which denitrification supplies the ammonium used by the anammox bacteria. In such a commensalism, anammox should contribute 29% of the N_2 production. In Golfo Dulce, anammox produces at least this much, and in some cases even more, of the total N_2 production as determined by direct simultaneous measurements of denitrification and anammox activity with ^{15}N-labeled compounds (Dalsgaard *et al.*, 2003). As the Golfo Dulce water column shares a similar chemistry with oxygen-depleted oxygen-minimum zones around the world, anammox could be a globally important pathway of N_2 production (Dalsgaard *et al.*, 2003).

Sulfide accumulates, albeit to only low concentrations of around 2 μM, in the lower 20 m of the water column. Therefore, a reaction front is established between nitrate (and nitrite) and sulfide, and sulfide oxidation through nitrate reduction by colorless-sulfur bacteria is a logical removal pathway

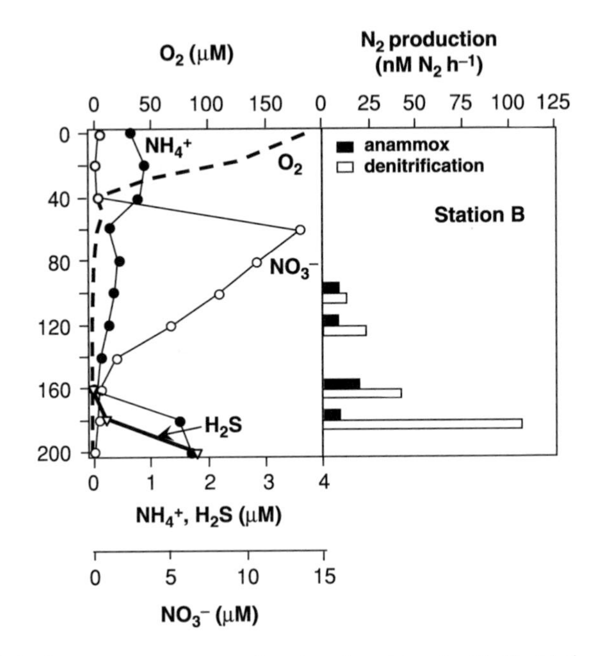

Figure 13.5 Important aspects of water chemistry in Golfo Dulce, a stratified marine basin in Costa Rica. Data are from Dalsgaard *et al.* (2003) and are taken from Station B, the inner station closest to the river outlet. In addition to the water chemistry data shown, there is a nitrite maximum of around 2 μM at 160 m depth. Rates of denitrification and anammox activity are also shown. The high rates of denitrification at 180 m depth may be the result of nitrate reduction with sulfide.

for sulfide (see Chapter 9). Indeed, vacuolated colorless sulfur bacteria were observed in bottom sediments from the deep part of the basin (Kuever, personal communication), which may well have conducted nitrate-mediated sulfide oxidation.

3.1.2. The Black Sea

The Black Sea is the world's largest anoxic basin, and for this reason it has received a great deal of attention. In the Black Sea, fresh water from surface runoff mixes with and overlays Mediterranean water that spills over the Bosporus and settles to depth in the basin. Therefore, salinity is the main factor controlling density stratification. Associated with the pycnocline is a strong chemocline, where, historically, an interface between oxygen and sulfide was established in the central part of the basin at around 120 m depth. Over the last few decades the depth of oxygen penetration has shoaled to around 70 to 100 m (Murray *et al.*, 1989; Kuypers *et al.*, 2003), although the depth at which sulfide accumulates has not significantly changed (Murray *et al.*, 1995). Therefore, sulfide and oxygen now rarely meet, at least in the central part of the basin (Oguz *et al.*, 2001). Instead, O_2 and sulfide are separated by a pronounced nitrate maximum (see Figure 13.6). Within this relatively narrow region of 20 to 40 m depth, elemental cycling by microbes is intense, diverse, and quite complex. Critical questions include how elemental cycles are linked and, in particular, how sulfide is oxidized and whether unique microbial physiologies can be identified.

We start with sulfide oxidation. Both phototrophic and nonphototrophic sulfide-oxidizing organisms have been identified within the Black Sea chemocline. A peak in BChl *e* concentrations, as well as carotenoids associated with green sulfur bacteria, is concentrated at the region where sulfide first begins to accumulate into the water column (Repeta *et al.*, 1989). Indeed, depth-integrated BChl *e* concentrations exceed those for Chl *a*, implying a potentially significant role for anoxygenic photosynthetic sulfide oxidation by *Chlorobium* spp. As discussed in Chapter 9, the very low light-adapted species *Chlorobium phaeobacteroides* was isolated from this region of the water column (Overmann, 1992). In addition, Jannasch *et al.* (1991) isolated from the chemocline a number of colorless sulfide-oxidizing bacterial strains of the genus *Thiomicrospira*. These strains were obligate aerobes and autotrophs and could oxidize sulfide, thiosulfate, tetrathionate, and elemental sulfur to sulfate. These organisms could not use nitrate, Fe oxides or Mn oxides as alternate electron acceptors. Therefore, their presence in the water column where sulfide and oxygen do not meet is somewhat enigmatic unless they are transported laterally from areas in the water column where O_2 and sulfide do meet, such as near the inlet of the Bosphorus (Oguz *et al.*, 2001). Alternatively, they may engage in an unexpected metabolism.

Direct determinations of sulfide oxidation rates in the dark demonstrated a peak in sulfide oxidation rate well within the sulfidic waters and considerably below the penetration depths of oxygen or nitrate (Jørgensen *et al.*, 1991). This result is also enigmatic, as obvious electron acceptors for the oxidation of sulfide are apparently lacking. However, on closer inspection, manganese may provide an important electron shuttle between oxygen and sulfide, thus promoting at least part of the sulfide oxidation (Murray *et al.*, 1995). To visualize how this might work, dissolved Mn^{2+} diffuses up into the zone containing oxygen (see Figure 13.6), where it is oxidized to particulate MnO_2, which settles into the sulfide-containing waters, oxidizing sulfide and re-reducing the MnO_2 to Mn^{2+} (Murray *et al.*, 1995). The details of this process and the possible role of microbes are unknown. Oguz *et al.* (2001) has recently proposed a similar model in which NO_3^- oxidizes the Mn^{2+} instead of oxygen. As discussed in Chapter 7, however, there is no direct evidence for this reaction in nature.

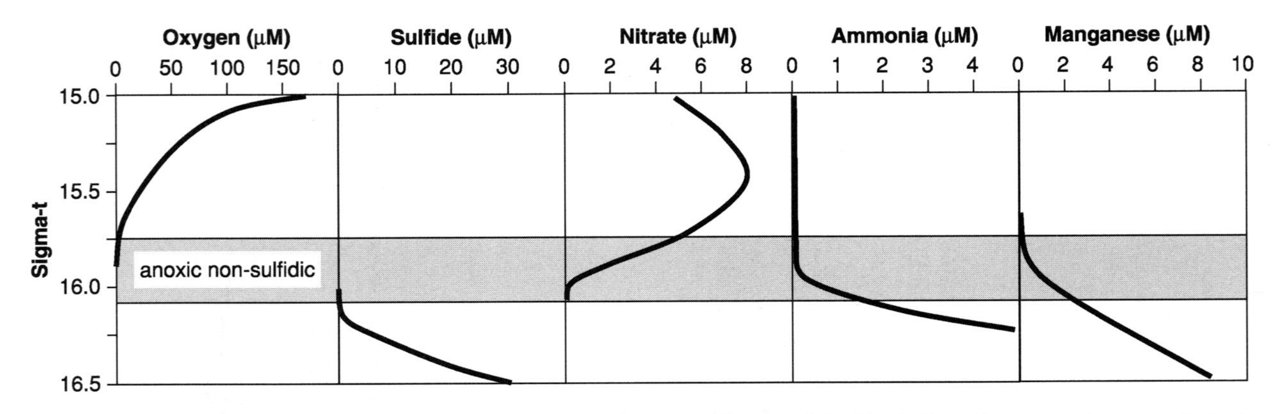

Figure 13.6 Important aspects of the water chemistry from the central basin of the Black Sea. Concentrations are plotted against sigma-t, which is measure of water density, rather than depth. This is because concentrations are often more stable along density surfaces than they are with water depth. Average profiles are shown, and these have been redrafted from Murray *et al.* (1995).

Whether Mn^{2+} is re-oxidized by oxygen or nitrate, the Mn^{2+} flux is insufficient to promote the complete oxidation to sulfate of the sulfide diffusing from below (see Murray *et al.*, 1995). Thus, other processes must be involved in sulfide oxidation, and anoxygenic phototrophic oxidation is a strong possibility. Rates of this process, unfortunately, have yet to be determined.

There is a sharp interface between nitrate and ammonium in the anoxic waters (Figure 13.6), suggesting a possible reaction front, and recently Kuypers *et al.* (2003) measured anammox activity in the vicinity of this interface. Also found in the same vicinity was an abundance of lipids from *Planctomycete* organisms, which are the only known organisms to conduct the anammox reaction (see Chapter 7). In the Black Sea, anammox is apparently supported by the diffusion of nitrate and ammonium into a common reaction front rather than from the ammonium liberated by denitrification as in Golfo Dulce (see above). Anammox is apparently a significant pathway of both ammonium oxidation and nitrate reduction in the Black Sea. Anammox activity is probably not, however, distributed equally in the chemocline throughout the basin and would not likely be prominent in the western part of the basin, where oxygen and ammonium form a reaction front. The nitrate formed in this part of the basin may be advected into the interior of the basin, where a substantial local nitrate source is unlikely, as ammonium does not contact oxygen (see Figure 13.6). Some nitrate, however, may come from ammonium liberated during aerobic organic matter mineralization in the chemocline, but the strength of this nitrate source needs to be evaluated.

Rates of sulfate reduction are low in the Black Sea and are extremely difficult to measure with standard radiotracer techniques. Sulfate reduction, however, does occur in the sulfidic waters, with maximum rates between 100 and 300 m depth (e.g., Jørgensen *et al.*, 1991; Lein *et al.*, 1991; Albert *et al.*, 1995). There is very little convergence between rate measurements from different studies in the Black Sea water column (see Albert *et al.*, 1995), with both volume-specific and depth-integrated rates varying by over one order of magnitude. Some of this variability might be real, but some might also be due to methodological difficulties. Therefore, we say little more about sulfate reduction rates in the Black Sea water column here, although this process is of obvious importance in carbon cycling. Unfortunately, there has been little work on the microbiology of sulfate reducers in the Black Sea.

Methane accumulates in the Black Sea water column to concentrations up to 11 μM by 500 m depth, and it remains at this concentration to the bottom of the basin (Reeburgh *et al.*, 1991). Active anaerobic methane oxidation (AOM) also occurs within the anoxic basin, where the oxidation is presumably coupled to sulfate reduction, and comprises the most important sink for methane within the Black Sea. The source of methane to the anoxic water column is shelf and slope sediments. The residence time for methane within the anoxic Black Sea water column is short, at only 5 to 20 yr. The organisms responsible for methane oxidation within the Black Sea water column are, unfortunately, unknown.

3.1.3. Mahoney Lake

Mahoney Lake is one of the best-studied stratified lakes. There is a pronounced salinity gradient in the lake: the upper mixolimnion has salinities ranging from around 8 to 16, increasing sharply to 39 in the lower monimolimnion (Overmann, 1997). Strong chemical and microbial stratification accompanies this density stratification, and the distribution of light, BChl *a* and important aspects of water chemistry are shown in Figure 13.7. Gradients are clearly sharp within the lake, and within less than 20 cm (approximately 6.7 to 6.9 m depth), light is completely absorbed and a strong peak in BChl *a* is found. All of this occurs within the vicinity of the O_2–sulfide interface at 6.75 m. The lake waters within this region are dark purple in color due to an intense "bacterial plate" of the purple sulfur bacterium *Amoebobacter purpureus* (up to 4×10^8 cells ml^{-1}), which is also responsible for the BChl *a* maximum. Other anoxygenic phototroph species have been identified by microscopy and culturing techniques, but *A. purpureus* accounts for 97.9% of the anoxygenic phototrophic biomass. Clearly, *A. purpureus* is thriving on sulfide oxidation with the Mahoney Lake chemocline (Overmann, 1997). Light within the spectral region of 515 to 625 nm reaches the depth of the bacterial plate (Overmann *et al.*, 1991); *A. purpureus*,

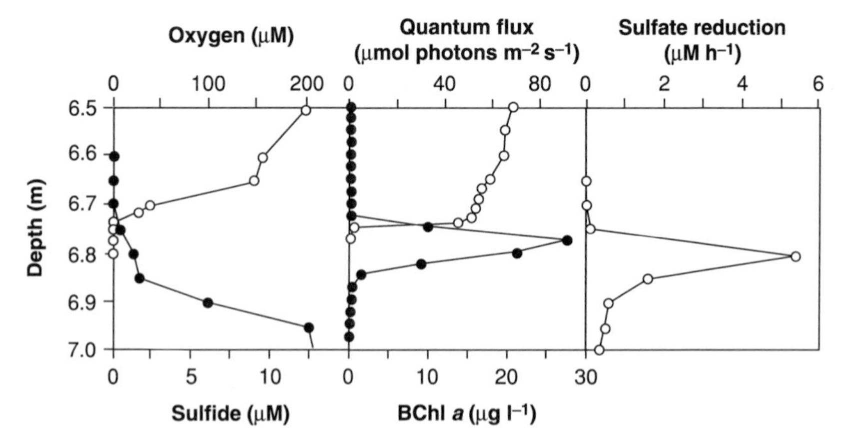

Figure 13.7 Concentrations of oxygen, sulfide, and BChl *a*, together with light intensity and sulfate reduction rate, during the summer in Mahoney Lake, British Columbia. Redrawn from Overmann (1997).

containing the carotenoid okeneone, is particularly well adapted to these wavelengths.

The biomass of *A. purpureus* is so thick that light does not penetrate the whole of the bacterial plate, and at any given time only about 10% of the *A. purpureus* cells obtains enough light for anoxygenic photosynthesis (Overmann *et al.*, 1991). This means that alternative metabolisms, such as possibly fermentation, are performed by *A. purpureus* when in the dark. Cells of *A. purpureus* are gas vacuolated, giving the cells positive buoyancy and helping them to maintain position within the bacterial plate (Overmann, 1992). In addition, cells are probably not stationary, and all viable cells probably visit the photic zone, at least periodically.

There is also a substantial population of chemoorganoheterotrophic and chemolithoautotrophic organisms within the bacterial plate, reaching cell numbers up to 2×10^8 cells ml^{-1}, or about half of the cell density of *A. purpureus* (Overmann, 1997). Some of these cells may be involved in non-phototrophic sulfide oxidation, which seems to be important during some times of the year. Overmann *et al.* (1994) estimated that up to 50% of the sulfide oxidized during the summer months was accomplished by non-phototrophic pathways (Figure 13.8).

Considering the sulfur cycle more broadly, sulfide diffusing from the sulfidic waters fuels, during the summer, only about half of the total sulfide oxidized within the bacterial plate (Figure 13.8). Significant rates of sulfate reduction were measured within the bacterial plate (Figure 13.7), and this process delivers the rest of the sulfide, which is subsequently oxidized (Overmann *et al.*, 1994). Sulfate reduction rates are correlated with the population size of *A. purpureus*, and the oxidation of dead microbial cells probably fuels most of the sulfate reduction in the bacterial plate. There is, therefore, an important coupling between anoxygenic photosynthesis and sulfate reduction in the lake. There is also a direct, unexpected link between the anoxygenic phototrophic population and the microbial food web in the upper mixolimnion of the lake. As mentioned above, cells of *A. purpureus* are gas vacuolated, and they occasionally ascend in aggregates into the upper water column (Overmann *et al.*, 1999), where they provide an important food source for zooplankton. Indeed, in some time intervals, the transfer of carbon by *A. purpureus* into the mixolimnion exceeds rates of oxygenic photosynthesis (Overmann *et al.*, 1999).

3.2. Competition among anoxygenic phototrophic populations

As discussed above, populations of anoxygenic phototrophs are an important feature of many stratified water bodies. The specific types of phototrophs present (e.g., green sulfur bacteria vs. purple sulfur bacteria)

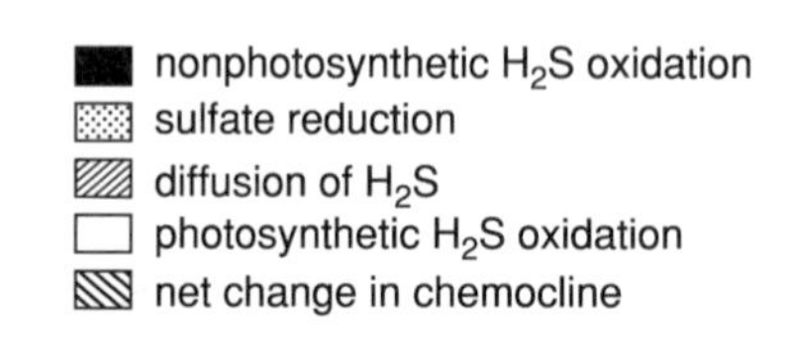

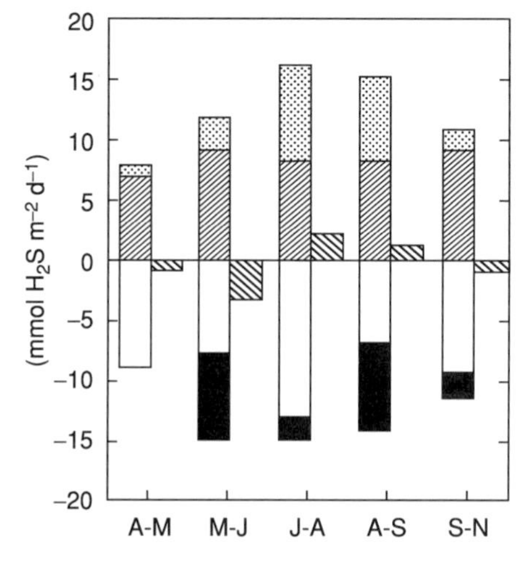

Figure 13.8 The processing of sulfide within the chemocline of Mahoney Lake during the summer. Shown are both sulfide sources and sulfide sinks. Redrawn from Overmann *et al.* (1996b).

and their stratification in the water column depend on several factors, including the spectral quality and intensity of light incident upon the phototrophic community, the adaptation of the population to the available light and specific physiological adaptations. Some of these considerations are explored in more detail below.

As discussed in Chapter 9, the spectral quality of light varies considerably with depth in a water column (see Fig 9.17), becoming increasingly enriched in the yellow-green part (450 to 600 nm) of the spectrum as more of the light is absorbed by water. In this spectral range, carotenoids are the main light-absorbing pigments, and the depth distribution of anoxygenic phototrophic species depends, to some extent, on the carotenoid pigments they contain. In a broad survey of the depth distributions of anoxygenic phototrophs in a variety of stratified lakes, van Gemerden and Mas (1995) found lycopene- and rhodopin-containing purple sulfur bacteria to be most dominant in the shallowest waters, while okenone- and spirilloxanthin-containing species are found deeper. The carotenoids lycopene and rhodopin have absorption maxima at somewhat shorter wavelengths, 448 to 506 nm, compared to okenone and spirilloxanthin, with absorption maxima from 465 to 530 nm (see Castenholz *et al.*, 1992). As the shorter wavelengths are preferentially absorbed by water (see Figure 9.17) and lost with depth, one might expect that organisms containing carotenoids absorbing at the shorter wavelengths will be found higher in the water column. This is apparently the case.

Sometimes a layer of green sulfur bacteria is found directly below a prominent layer of purple sulfur bacteria, which is usually of the family *Chromatiaceae*. One finds such stratification, for example, in Lake Vilar in Spain (Guerrero *et al.*, 1985). The reasons for this stratification are not completely clear, but they may be related to the lower light adaptation of *Chlorobium* sp., particularly the brown species, compared to *Chromatium* sp. (see Chapter 9). Also, members of the genus *Chlorobium* are obligate anaerobes with a limited metabolic capability, compared to many purple sulfur bacteria that can tolerate, and in some cases even utilize, oxygen (see Chapter 9). Thus, *Chlorobium* may also seek some distance from the oxic–anoxic interface, particularly if the interface is unstable.

Populations of anoxygenic phototrophic populations in nature are not, however, necessarily found exclusive of one another. Multiple species are sometimes found living in the same space, and interesting strategies are often used to allow for this cohabitation. We look, for example, at the co-occurrence of two size classes of purple sulfur bacteria, with large-celled organisms such as *Chromatium okenii* and *Chromatium weissei*, coexisting with small-celled organisms such as *Chromatium vinosum* and *Chromatium minutissimum* (Van Gemerden and Mas, 1995). These different size classes of organisms represent different physiological types. Thus, large *Chromatium* species have low u_{max} (they are slow growers) and grow best in dim light and when they are exposed to dark–light cycles (Pfennig and Trüper, 1992). By contrast, small-celled species grow faster and do best under constant illumination. If, for example, large-celled *Chromatium weissei* and small-celled *Chromatium vinosum* are cultured together in a chemostat with constant illumination, the higher growth rate of *Chromatium visosum* will ensure that it always outgrows its larger-celled relative (Van Gemerden, 1974; Van Gemerden and Mas, 1995). However, these two organisms coexist when constant illumination is replaced by light–dark cycles, and the reasons behind this provide a fascinating example of ecophysiological adaptation.

The secret lies in subtle differences in how the two organisms store and utilize sulfur over a diel cycle. Thus, the small-celled *Chromatium vinosum* has a higher sulfide affinity but tends to oxidize sulfide completely to sulfate, whereas *Chromatium weissei* tends to oxidize sulfide incompletely to elemental sulfur, which is stored in its cell. On top of this, diel fluctuations in sulfide concentration in the chemostat are observed, in which sulfide increases over the dark period when there is no sulfide demand, and it decreases to low concentrations in the light period when sulfide demand is high. This mimics natural conditions. During the initial light phase, when sulfide concentrations are high, large-celled *Chromatium weissei* actively oxidizes sulfide, storing a large reservoir of elemental sulfur. As sulfide levels drop, *Chromatium vinosum* dominates sulfide oxidation due its lower sulfide affinity. All is not lost for *Chromatium weissei*, however, as it now oxidizes its accumulated store of elemental sulfur, maintaining growth even under sulfide-limiting conditions.

In another interesting example, stable co-cultures of the green sulfur bacterium *Chlorobium limicola* f.*thiosulfatophilum* (DSM 249) could be maintained in chemostat with the purple sulfur bacterium *Chromatium vinosum* (DSM 185), thus explaining the frequent co-occurrence of purple and green sulfur bacteria during blooms in lake water columns (van Gemerden and Beeftink, 1981). In this case the *Chlorobium* has the highest sulfide affinity. Furthermore, *Chlorobium* produces elemental sulfur and deposits it externally during its sulfide oxidation, and polysulfides are formed as dissolved sulfide reacts with the elemental sulfur (Visscher and van Gemerden, 1988). The *Chlorobium* species and the *Chromatium* species oxidize polysulfides equally well, and for *Chlorobium*, but not *Chromatium*, there is some inhibition of polysulfide utilization with increasing sulfide levels. Therefore, even though *Chlorobium* has a higher affinity for sulfide, *Chromatium* has a chance because it uses polysulfides, a by-product of *Chlorobium* metabolism, and it does this as well or better than the *Chlorobium*.

3.3. Significance of anoxygenic photosynthesis

The significance of anoxygenic photosynthesis in total carbon production in stratified basins varies widely. For example, in Green Lake, New York, anoxygenic photosynthesis accounts for over 80% of the total carbon production (Culver and Brunskill, 1969), in Lake Magil'noe (Kil'din Island, Barents Sea; Ivanov *et al.*, 2001) it accounts for 67%, while in other lakes, the contribution may be as small as a few percent (Overmann, 1997) (see Table 13.3). Anoxygenic photosynthesis is usually viewed as secondary carbon production. The rationale is that the carbon fueling sulfate reduction in the anoxic part of the water column, and the nutrients liberated there, are ultimately derived from oxygenic photosynthesis in the upper water column (e.g., Overmann, 1997). On close inspection, however, it can be seen that the situation is not so straightforward. For example, in Green Lake, Lake Mahoney, and Lake Mogil'noe (see Table 13.3), a majority of the sulfide fueling anoxygenic photosynthesis is derived from the oxidation of organic carbon produced by anoxygenic photosynthesis. Hence, anoxygenic photosynthesis is driven by the vast reservoir of sulfide and nutrients accumulated from the oxidation of anoxygenic photosynthetic biomass and not directly by the carbon made available by oxygenic photosynthesis.

Table 13.3 Primary productivity by anoxygenic phototrophs

Water body	Productivity (g C m^{-2} y^{-1})	% of total primary production
Fayetteville Green Lake	239	83
Mahoney Lake	34	73
Lake Mogil'noe	219	67
Smith Hole	35	50
Waldsea	32	46
Lake Cisó	56	25
Lake Cadagno	272	23
Deadmoose Lake	14	17
Big Soda Lake	50	10
Paul	17	6
Rose	33	6
Knaack Lake	17	5
Mirror	60	4
Mittlerer Buchensee	—	4
Lake Vechten	6	4
Peter	11	3
Valle de San Juan	14	3
Fish	13	1
Mary	2	0.3

Data from van Gemerden and Mas (1995); Overmann (1997); Camacho and Vincente (1998); Camacho *et al.* (2001); Ivanov *et al.* (2001).

If we make an analogy with the oceans, nutrients from the deep ocean basins are advected and diffused (by eddy diffusion) into the upper water column, fueling a large percentage of what is known as new production (see Chapters 5 and 12). By some measures, new production is the most important indicator of the primary production rate, providing an estimate of the export rate of carbon out of the upper ocean water column. Yet, in anoxic basins this analogous situation leads, by some definitions, to secondary production. Facing this dichotomous situation, we feel it is most parsimonious to view anoxygenic photosynthesis as a true measure of primary production of equivalent importance to oxygenic photosynthesis. Both, therefore, should be considered in describing the carbon dynamics and productivity of a stratified water column.

In what follows we try to gain insight into some of the factors regulating the relative significance of oxygenic photosynthesis vs. anoxygenic photosynthesis in stratified water columns. We begin with the following equalities:

$$\text{AnoxP} = 2a\text{SR} \tag{13.1}$$

$$\text{AnoxP} + \text{OxP} = \text{TCprod} \tag{13.2}$$

$$2\text{SR} = b(\text{TCprod} - \text{OxR}) \tag{13.3}$$

The first equation states that rates of carbon fixation by anoxygenic photosynthesis (AnoxP) are equivalent to rates of sulfate reduction (SR) as modified by the proportion *a* of the sulfide produced that becomes available to the phototrophic populations. The remaining proportion is fixed as mineral sulfides in the bottom sediments. The factor 2 represents the stoichiometric relationship between the moles organic carbon fixed per mole sulfide oxidized (see Equation 18 in Chapter 9). Equation 13.2 states simply that total rates of carbon production (TCprod) are equivalent to rates of carbon fixation by anoxygenic photosynthesis and oxygenic photosynthesis (OxP). In the final equation (Equation 13.3) sulfate reduction is driven by the carbon production that escapes oxic respiration (OxR). The factor *b* represents the fraction of this carbon preserved in sediments, and the numeral 2 is the stoichiometric relationship between the moles of carbon oxidized per mole sulfate reduced during sulfate reduction (reverse of Equation 18 in Chapter 9). Substituting Equation 13.3 into Equation 13.1 and dividing by TCprod, we get the following:

$$\frac{\mathrm{AnoxP}}{\mathrm{TCprod}} = ab\left(1 - \frac{\mathrm{OxR}}{\mathrm{TCprod}}\right) \tag{13.4}$$

We see immediately from Equation 13.4 that the significance of anoxygenic photosynthesis is reduced when there is substantial sulfide and carbon preservation in the sediments. Usually sulfide and carbon preservation is relatively small compared to rates of carbon and sulfur turnover (Canfield, 1989b), but this has not been evaluated in stratified lakes. Otherwise, the relationship quite reasonably states that the proportion of total carbon production by anoxygenic photosynthesis is inversely proportional to the proportion of total carbon production oxidized by oxic respiration. In other words, the more of the total carbon production oxidized by oxic respiration, the less is oxidized by sulfate reduction and the less sulfide is available to fuel anoxygenic photosynthesis. In this view, anoxygenic photosynthesis is most important when there is an efficient transfer of photosynthetic products into the anoxic zone, minimizing the extent of oxic respiration. This occurs, in turn, when carbon production is truly concentrated at the oxic zone, as occurs in many stratified water bodies such as Green Lake and Lake Mogil'noe. In cases like this, the upper water body is probably nutrient starved, with the anoxic zone providing the main source of nutrients to the photic zone.

Parkin and Brock (1980) also reported an impressive positive relationship between light intensity in the zone of anoxygenic photosynthesis and the proportion of total carbon production channeled through anoxygenic photosynthesis. This observation led to the logical conclusion that light intensity was the principal factor controlling the relative significance of anoxygenic photosynthesis in total carbon production. It is uncertain, however, whether light intensity is a primary or secondary factor influencing the significance of anoxygenic photosynthesis. For example, in turbid lakes much of the turbidity is due to high concentrations of oxygenic phototrophs fueled probably by large external sources of nutrients. Hence, in these cases it may be nutrient availability rather than light *per se* that controls the significance of anoxygenic photosynthesis. Furthermore, clear lakes have small upper-lake populations of oxygenic phototrophs, due likely to a limited external supply of nutrients, so it is no surprise that oxygenic phototrophs are relatively less significant in total carbon production in these systems. A full understanding of the factors controlling the relative significance of anoxygenic photosynthesis in stratified waters will require a complete accounting of nutrient dynamics, but to our knowledge studies of this nature have yet to be performed.

4. MANGROVE FORESTS

4.1. What is a mangrove forest?

Mangrove forests are a dominant ecosystem along nearly 75% of the world's tropical and subtropical coastlines (Por, 1984; Kathiresan and Bingham, 2001). They are characterized by trees growing in sea water at the interface between land and sea, particularly in sheltered areas with large tidal excursions (Figure 13.9). Lugo and Snedaker (1974) have classified mangrove forests into five types:

1. Riverine forests: floodplains along river drainages, which are inundated by most high tides and flooded during the wet season.
2. Basin forests: partially impounded depressions, which are inundated by few high tides during the dry season, with most high tides during the wet season.
3. Fringe forests: shorelines with steep elevation gradients, which are inundated and flushed by all high tides.
4. Overwash forests: low islands and small peninsulas, which are completely overwashed on all high tides.
5. Dwarf forests: topographic flats above mean high water, which are tidally inundated only during wet season and are dry and salty for most of the year.

Mangrove forests are highly productive, particularly in tropical areas (net primary production of 100–500 mmol C m^{-2} d^{-1}) and support abundant planktonic and benthic communities, as well as high rates of microbial decomposition (Christensen *et al.*, 1978; Alongi *et al.*, 1999, 2000b). They

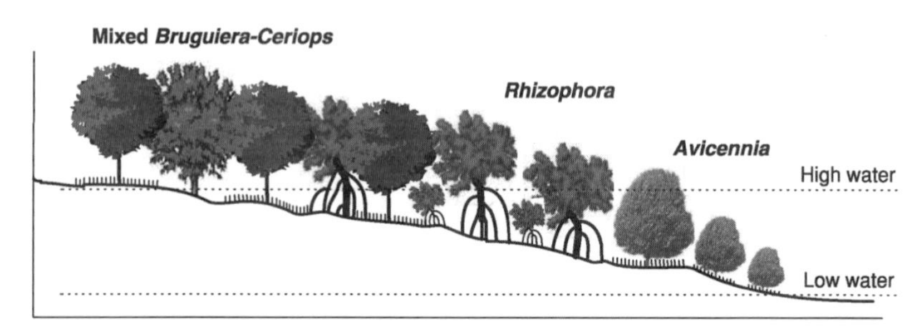

Figure 13.9 Characteristic zonation of mangrove trees on open shorelines in southeast Asia and northern Australia.

also serve as important habitats and provide a nutrient source for juvenile fish and crustacean species (Robertson and Duke, 1987; Qasim and Wafar, 1990; Kathiresan and Bingham, 2001). Mangrove forests also exchange dissolved and particulate nutrients with adjacent coastal waters. The extent and direction of this exchange is highly dependent on a number of factors, including geomorphology, tidal regime, climate, and freshwater inputs (Hemminga *et al.*, 1994; Rivera-Monroy *et al.*, 1995; Robertson and Alongi, 1995; Alongi, 1996).

The special feature of mangrove forests is trees adapted to growth in anoxic intertidal sediment saturated with sea water (Kathiresan and Bingham, 2001). Specific adaptations of these trees include unique developments of the root system, including pneumatophores (e.g., *Avicennia* spp.) and prop roots (e.g., *Rhizophora* spp; Kitaya *et al.*, 2002), which supply the heavily aerenchymated (air-storing tissue) roots with oxygen from the air and provide physical support for growth in soft mud. Some species may reach heights of more than 30 m. Salt from the sea water is either excluded at the roots (e.g., *Rhizophora* spp.) or excreted by glands on the leaves (e.g., *Avicennia* spp.) (Passioura *et al.*, 1992; Clarke and Allaway, 1993). The benthic fauna associated with mangrove trees is dominated by various burrowing decapods, such as leaf-eating crabs (*Grapsidae*) and fiddler crabs (*Ocypodidae*), which occur in densities of 15–70 m^{-2} in Southeast Asian mangrove forests (Macintosh *et al.*, 2002). They actively dig and maintain burrows in the sediment as a refuge from predation and environmental extremes and as a place to store food (Giddins *et al.*, 1986; Warren, 1990; Dittmann, 1996). The burrows affect sediment topography and biogeochemistry by modifying particle size distribution, drainage, redox conditions, and organic matter availability (Mouton and Felder, 1996; Botto and Iribarne, 2000).

Mangrove sediments usually consist of fine-grained mud with an organic carbon content ranging from 3 to 6 mmol g^{-1} (Table 13.4). Mangrove trees

Table 13.4 Examples of sediment composition from various mangrove locations around the world

	Org. C[a] (mmol g^{-1})	Org. N[b] (μmol g^{-1})	C:N[c]	Median grain size (μm)	Particles <63 μm (%)	References
Andaman Island	5.6–10.7	0.22–0.34	25–31	—	—	1
Mekong Delta, Vietnam	3.3–5.2	0.18–0.27	18–19	—	—	2
Bangrong, Thailand	2.5–3.6	0.10–0.17	21–25	50–70	50–70	3
Indus Delta, Pakistan	0.7–0.9	0.03–0.04	22–23	—	—	4
Hinchinbrook Island, Australia	5.2–10.5	0.16–0.28	33–38	—	—	5
Western Australia, Australia	1.0–5.9	0.05–0.24	20–25	—	30–90	6
Everglades, USA	3.3–4.3	0.15–0.19	22–23	—	—	7
Kumarakam, India	0.3–4.1	—	—	98–205	21–82	8
Cross River, Nigeria	4.3	—	—	—	67	9
Umngazana Estuary, S. Africa	1.2–4.1	—	—	74–136	10–82	10

[a]Organic carbon.
[b]Organic nitrogen.
[c]Molar ratio of organic carbon to organic nitrogen.
References: 1, Mongia and Ganeshamurthy (1989); 2, Alongi *et al.* (2000a); 3, Kristensen *et al.* (2000); 4, Kristensen *et al.* (1992); 5, Alongi (1996); 6, Alongi *et al.* (2000b); 7, Chen and Twilley (1999); 8, Badarudeen *et al.* (1996); 9, Ukpong (1994); 10, Dye (1983).

actively colonize areas of mud accumulation, and their establishment leads to faster sediment accretion (Zimmermann and Thom, 1982). Although mangrove forests export organic matter and nutrients (see later), they are also efficient sediment sinks (Woodroffe, 1992). The physical processes involved in the exchange of both organic matter and sediment are essentially the same, principally freshwater and tidal currents. Incoming water receives an abundant supply of fine-grained sediment originating at tidal flats along the coast. This particle-charged water enters the mangrove forest during flood tides via a network of creeks. When the turbid tidal water enters the forest, the dense mesh of prop roots and pneumatophores dampens water currents and traps sediment effectively, while the fine roots act as important sediment binders (Wolanski *et al.*, 1992). Rates of sediment accumulation in various mangrove forests typically range between 3 and 9 mm yr^{-1} (Spenceley, 1982; Lynch *et al.*, 1989; Woodroffe, 1990).

4.2. Food webs and detritus handling

The main source of carbon and nutrients for decomposer food webs in mangrove environments is litter from trees (leaves, propagules, and twigs) deposited at the forest floor and subsurface root growth (Alongi, 1998). In a variety of mangrove forests, litterfall deposits 50–100 mmol C m^{-2} d^{-1} and 0.4–1.3 mmol N m^{-2} d^{-1} (Poovachiranon and Tantichodok, 1991; Alongi, 1998; Clough *et al.*, 2000). The living below-ground root biomass constitutes 9–35% of the total living tree biomass (Alongi and Dixon, 2000) and is expected to generate detritus within the sediment of a magnitude comparable to litterfall. The contribution of detrital carbon from macroalgae, benthic microalgae, phytoplankton, and epiphytes is low (these sources contribute less than 14% of the total forest primary production; Alongi, 1998) in most mangrove forests due to severe light limitation under the tree canopy. However, algae has a high nitrogen content, with C:N ratio of ~7–10, compared with much higher C:N ratios in mangrove litter of about 100. Therefore, algal primary producers contribute more than 50% of the organic nitrogen to mangrove detrital food webs.

Stable carbon isotopes reveal that mangrove-derived detritus (tree litter) is important for many, but not all, macroinvertebrates, as some also consume algae as their main food source (Rodelli *et al.*, 1984; Newell *et al.*, 1995; Bouillon *et al.*, 2002). For example, while fiddler crabs (*Ocypodidae*) and various gastropods are important consumers of benthic microalgae, sesarmid crabs (*Grapsidae*) are remarkable consumers of mangrove litter (Dye and Lasiak, 1987; Robertson *et al.*, 1992). Indeed, in a variety of mangrove environments, litter consumption and burial by sesarmid crabs can remove 30–90% of the annual litterfall (Robertson, 1986; Micheli, 1993; Slim *et al.*, 1997; Thongtham *et al.*, in press). In general, crabs consume about half of the litter immediately, while the remainder is pulled into burrows, promoting microbial colonization and the subsequent leaching of tannins. The litter stored in burrows is later consumed by the crabs or is left for further microbial degradation. Such effective retention of litter by crabs conserves nutrients within mangrove forests by reducing the amount of litter available for export (Figure 13.10).

Some of the fallen litter is, nevertheless, exported from the mangrove forest to the adjacent coastal areas by tidal currents. Estimates of litter export vary from 1 to 40% of the total litterfall in forests from different geographical regions (Robertson *et al.*, 1992; Suraswadi *et al.*, unpublished). The quantity

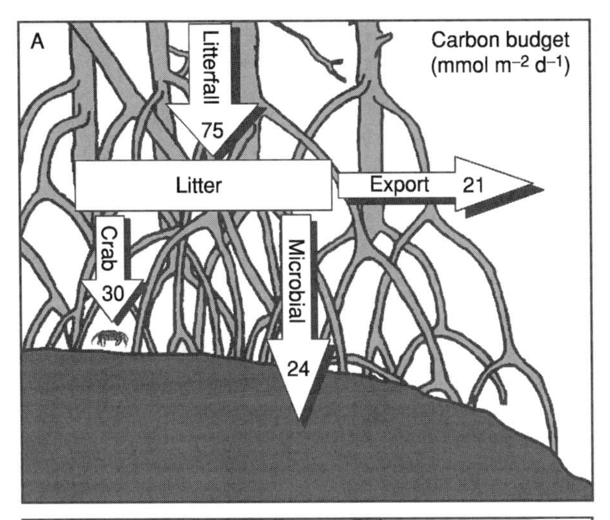

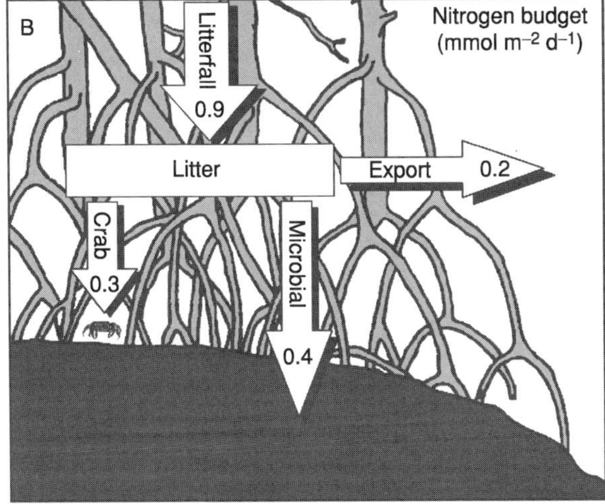

Figure 13.10 Average carbon and nitrogen budgets for litter in Southeast Asian and northern Australian mangrove forests dominated by *Rhizophora* spp. Data obtained from Robertson and Daniel (1989); Poovachiranon and Tantichodok (1991); Alongi (1998); Clough *et al.* (2000); and Thongtham *et al.* (in press).

of litter available for export is dependent not only on the forest type and hydrodynamics, but also on the degree of litter retention and processing by crabs. Litter retained within mangrove forests, either passively trapped by roots or handled by crabs, will eventually enter the microbial food chain, either in the form of uneaten remains buried in the sediment or as fecal material and crab carcasses (Giddins *et al.*, 1986; Robertson, 1986; Lee, 1997).

Microbial decomposition of detritus is slow in mangrove sediments compared with environments fed by marine detritus of algal origin (Alongi, 1998). Furthermore, marine mycelial decomposers belonging to eumycotes (fungi) and oomycotes (protoctista) are important in mangrove ecosystems (Newell, 1996). This is by contrast to other marine ecosystems where prokaryotes dominate degradation processes. Thus, the biomass of fungi associated with decaying leaves of *Rhizophora* sp. is about 1–5 mg g^{-1} dry mass (Blum *et al.*, 1988), greatly exceeding the biomass of prokaryotes associated with leaves at around 0.7 mg g^{-1} dry mass (Newell and Fell, 1992). Detritus derived from mangrove litter is generally of poor nutritional value for virtually all microbial decomposers, as it contains high levels of tannins, lignin, and structural carbohydrates (cellulose and lignocellulose, which account for about 50% of the carbon in living mangrove leaves). All of these inhibit microbial degradation (Benner *et al.*, 1990; Lee *et al.*, 1990). The low nutrient content of mangrove litter further constrains the activity of microbial decomposers. Thus, microbes must acquire nitrogen from outside sources such as the assimilation of dissolved nitrogen from sediment porewaters to maintain growth and effectively degrade carbon-rich mangrove detritus.

The decay of mangrove litter in sediments begins with significant leaching of soluble organic substances (Figure 13.11). Newly fallen mangrove litter loses 20–40% of the organic carbon by leaching when submerged in

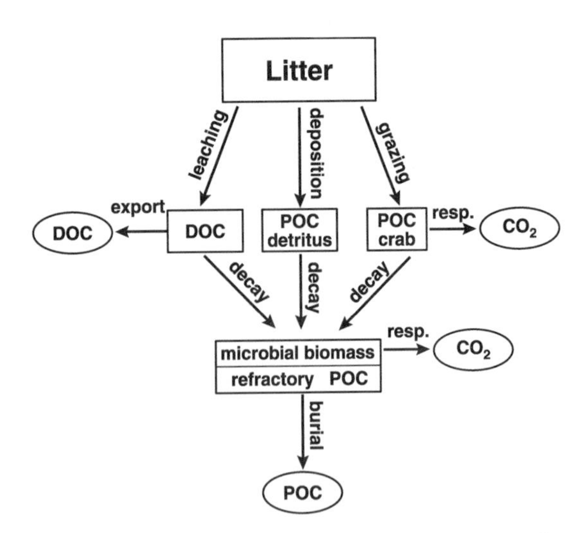

Figure 13.11 Schematic presentation of the fate of mangrove litter deposited at the sediment surface in typical Southeast Asian and northern Australian mangrove forests.

sea water for 10–14 days (Camilleri and Ribi, 1986; Twilley *et al.*, 1997). Leached carbon consists of non-lignocellulose components (Neilson and Richards, 1989) as well as tannins and other phenolic compounds (up to 18% of the dissolved organic matter, DOM, in mangrove leachate) (Benner *et al.*, 1986). Much of the leached DOM is labile and is rapidly consumed by the benthic microbial community (Kristensen and Pilgaard, 2001), while some is exported to the mangrove waterways. A fraction of the assimilated carbon from DOM is converted into microbial biomass, while the remainder is lost by respiration. Fresh mangrove leachates are degraded efficiently under oxic and nutrient-replete conditions, with conversion efficiencies of up to 90% into microbial biomass (Benner *et al.*, 1986). However, the microbial community in nutrient-limited mangrove sediments, including aerobic as well as anaerobic organisms, incorporates DOM with an average efficiency of only 35% (Boto *et al.*, 1989).

Further decomposition of the remaining particulate material occurs by extracellular enzymatic hydrolysis mediated by bacterial and mycelial decomposers followed by assimilation of the solubilized compounds (Figure 13.11). The polysaccharide (i.e., cellulosic) components of lignocellulose are generally degraded about twice as fast as the lignin, enriching mangrove detritus in lignin-derived carbon with time (Benner and Hodson, 1985). While cellulose and lignin are readily degradable in oxic environments, these compounds are degraded slowly under anoxic conditions. Lignin, for example, has a half-life of more than 150 yr in anoxic mangrove sediment (Dittmar and Lara, 2001). The formation of old and refractory material in mangrove sediments can be observed visually as lignified and humified (spongy) litter fragments. Accordingly, Dittmar and Lara (2001) estimated that the average age of organic carbon in the upper 1.5 m of the sediment in the Foru do Meio mangrove forest, Brazil is between 400 and 770 yr.

4.3. Fate of carbon and nitrogen in sediment

Sedimentary organic matter in mangrove environments is enriched in nitrogen compared with fresh litter, as indicated by a molar C:N ratio of 20–30 (Table 13.4) compared with about 100 for leaf litter (Kristensen *et al.*, 1995). A rapid initial release of carbon, and retention or enrichment of nitrogen, must occur during the early degradation of mangrove litter before it is buried into the sediment (Twilley *et al.*, 1997; Wafar *et al.*, 1997). Thus, a loss of about 75% of carbon from the litter is necessary to obtain the low sedimentary C:N ratio if other sources and sinks of carbon and nitrogen are excluded (Dittmar and Lara, 2001). Indeed, fast initial carbon removal is encouraged by the action of leaf-eating crabs (Kristensen and Pilgaard, 2001). As labile components are removed, the reactivity of the remaining carbon and nitrogen in the detritus decreases rapidly. After burial into the sediment, there is an increase in humic compounds and other refractory geopolymers. First, we will consider processes controlling the fate of carbon, while nitrogen will be dealt with in detail below.

Most of the organic carbon oxidation in mangrove sediments is mediated by anaerobic microbial processes. Aerobic respiration occurs near the sediment surface, around crab burrows and along oxic root surfaces. However, the consumption of O_2 at these interfaces is usually so rapid that O_2 rarely penetrates more than 1 mm into the sediment (Kristensen *et al.*, 1994). A wide variety of anaerobic microorganisms is responsible for carbon oxidation below the oxic zone of mangrove sediments. While aerobic respirers consume litter and algal detritus deposited at or near the sediment surface, anaerobic respirers are fueled by litter buried by leaf-eating crabs and sedimentation and by below-ground root production in the form of dead biomass and DOC excretion (Alongi, 1998).

Rates of total carbon oxidation in mangrove sediments typically range from 20 to 60 mmol m^{-2} d^{-1} (Kristensen *et al.*, 2000; Alongi *et al.*, 2001) and are in the low range of rates from intertidal environments in colder climates (Mackin and Swider, 1989; Magenheimer *et al.*, 1996). The relatively low rates are partly a product of high degrees of carbon preservation (up to 50% of the deposited carbon; Kristensen *et al.*, 1995), which result from the refractory nature of mangrove detritus. Aerobic respiration and anaerobic sulfate reduction are traditionally viewed as the most important respiration processes in mangrove sediments (Alongi, 1998), with a share of 40–50% each (13.2). Because of high rates of sulfate reduction, most mangrove sediments contain high levels of reduced inorganic sulfur. Sulfide generated by sulfate reduction in mangrove sediments is rapidly re-oxidized or precipitated with iron to form pyrite (FeS_2), and negligible amounts of iron monosulfides (FeS) are formed (Holmer *et al.*, 1994). Elemental sulfur (S^0) can also be an important sulfur phase in some mangrove sediments.

Denitrification, manganese respiration, and iron respiration have usually been considered unimportant for carbon oxidation in mangrove sediments, although denitrification may be important in areas impacted by sewage (Corredor and Morell, 1994). Recent evidence suggests, however, that the role of iron respiration for carbon oxidation may be comparable to, or higher, than sulfate reduction in many iron-rich mangrove environments (Table 13.5). As sulfate reduction usually is inhibited in the presence of more potent electron acceptors (e.g., O_2 and Fe oxides; Canfield *et al.*, 1993), this process becomes inferior to iron respiration when oxidizing roots and infaunal burrows increase the Fe oxide content deep in vegetated and bioturbated mangrove sediments (Figure 13.12) (Nielsen *et al.*, 2003). However, the relative roles of the major carbon mineralization vary considerably within and between mangrove environments. The deciding factors include sediment

Table 13.5 Contribution of electron acceptors used for microbial carbon oxidation in selected mangrove forests. Units are mmol C m^{-2} d^{-1}[a]

	Bangrong, Thailand[d]	Mekong Delta, Vietnam[e]	Dampier, Western Australia[f]
Aerobic respiration	30.2 (57)[b]	23.7 (44)	23.1 (46)
Denitrification	<0.1[g] (<1)	2.8 (5)	n.m.[c]
Manganese reduction	n.m.	1.2 (2)	0.02 (~0)
Iron reduction	16.6 (31)	~0 (0)	~0 (0)
Sulfate reduction	5.9 (11)	26.0 (51)	27.6 (54)
Methanogenesis	n.m.	n.m.	n.m.
Total	52.7	53.7	50.7

[a]The anaerobic microbial respiration processes are measured independently by various techniques, while aerobic respiration is determined as the difference between total benthic CO_2 release and the sum of all anaerobic processes.
[b]Values in parenthesis indicate the percentage contribution of each process.
[c]n.m. – not measured.
[d]Annual average in mid-intertidal zone of *Rhizophora* spp. forest (Kristensen *et al.*, 2000).
[e]November data from 35-year-old *Rhizophora apiculata* forest near shrimp farm (Alongi *et al.*, 2000a).
[f]Annual average in mid-intertidal zone of *Rhizophora stylosa* forest (Alongi *et al.*, 2000b).
[g]Obtained from the nearby Ao Nam Bor mangrove forest (Kristensen *et al.*, 1998).

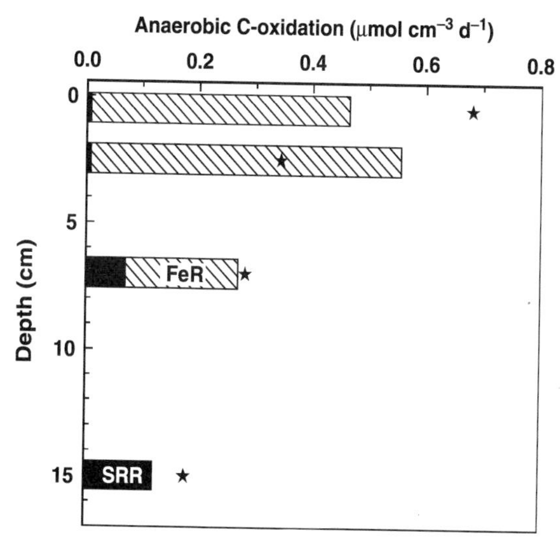

Figure 13.12 Anaerobic carbon oxidation with depth in the iron-rich sediment of a mid-intertidal mangrove forest (Bangrong) in southern Thailand. Only the most important anaerobic processes, iron respiration (FeR) and sulfate reduction (SRR), are included. Stars indicate the total anaerobic carbon oxidation rate. Modified from Kristensen *et al.* (2000).

composition, intensity and style of bioturbation, tidal inundation frequency, sediment composition, and sulfate, iron and organic carbon availability.

The role of anaerobic decomposition pathways, particularly sulfate reduction, generally decreases with increasing tidal elevation in mangrove forests (Figure 13.13). Less frequently inundated high intertidal sediments have a more extensive oxic–anoxic interface than those in the low intertidal regions. In high intertidal sediments there is generally a high abundance of burrowing crabs, and crab burrows, and the rapid drainage of water during low tide exposes oxygen to sediment that may otherwise be anoxic (Kristensen *et al.*, 1992; Holmer *et al.*, 1999). These air-filled interstices provide sites of rapid oxygen consumption and thus increase the overall oxygen uptake by the entire sediment system. In these sediments, then, sulfate reduction is hampered and oxygen uptake is stimulated, particularly during air exposure. Some of the oxygen is not consumed by heterotrophic bacteria, but is rather used chemically or by chemoautotrophic organisms to reoxidize reduced metabolites, such as sulfide and iron in the sediment.

Mangrove forests are efficient in retaining and recycling nitrogen. Several mechanisms are at play, including the re-absorption or re-translocation of nitrogen by mangrove trees prior to leaf fall, the burial of fallen leaves by crabs and the efficient uptake of dissolved nitrogen by microorganisms (Kristensen *et al.*, 1995). Accordingly, mangrove environments appear poor in dissolved inorganic nitrogen (DIN) compared to other intertidal systems, and the sediments generally act as sinks of DIN (Rivera-Monroy *et al.*, 1995; Alongi, 1996; Rivera-Monroy and Twilley, 1996). The uptake of DIN by vegetated mangrove sediments and the low DIN concentrations in porewaters (Figure 13.14) may, furthermore, imply an uptake by tree roots (Alongi, 1996; Kristensen *et al.*, 1998). However, a substantial part of the DIN demand must be assigned to microbial assimilation during the degradation of nitrogen-poor mangrove detritus (Alongi, 1996; Rivera-Monroy and Twilley, 1996). As microorganisms generally need substrates with an elemental C:N ratio below 10 for maintenance and growth (see Chapter 7), mangrove litter and other tree materials with C:N ratios of 100 or more are insufficient substrates. Accordingly, the cycling of inorganic nitrogen must be rapid and efficient to support microbial growth in mangrove sediments. The turnover times for NH_4^+ are therefore short; estimates range from <1 day in near-surface sediment to about 1 week at 10 cm depth (Nedwell *et al.*, 1994; Kristensen *et al.*, 2000). The microbial demand for nitrogen is particularly high near the sediment–water interface, as indicated by the influx of NH_4^+ during inundation (Figure 13.15), while porewater profiles (Figure 13.14) indicate a considerable release from the sediment (Rivera-Monroy *et al.*, 1995; Holmer *et al.*, 2001).

In accordance with the low porewater NO_3^- levels (<3 to $5\,\mu M$), sediment nitrification and denitrification rates in pristine mangrove forests are very

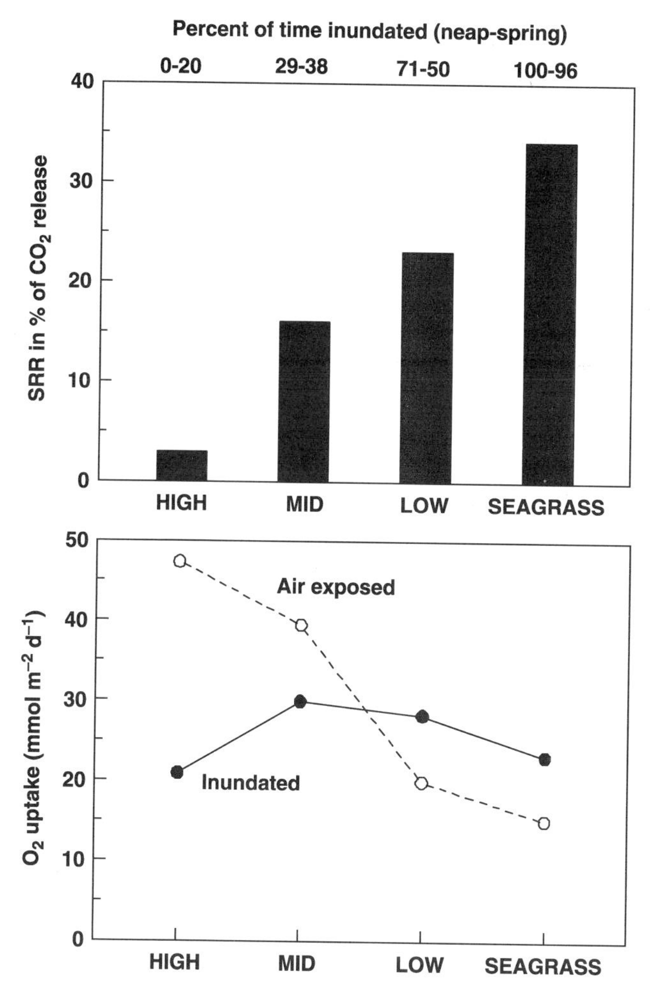

Figure 13.13 (Upper panel) The contribution of carbon oxidation by sulfate reduction (SRR) to the total sedimentary CO_2 released along a transect from the upper intertidal through the low intertidal forest and into the low intertidal seagrass beds outside the Bangrong mangrove forest, Thailand. The percentage of time inundated is indicated for neap and spring tidal periods. (Lower panel) Sediment O_2 uptake measured during inundation and air exposure along the above-mentioned transect. Modified from Holmer *et al.* (1999).

low (Figure 13.15) (Rivera-Monroy and Twilley, 1996; Kristensen *et al.*, 1998; Alongi *et al.*, 2000a). However, mangrove areas receiving sewage discharges show relatively high denitrification rates one to two orders of magnitude higher than in pristine areas (Corredor and Morell, 1994).

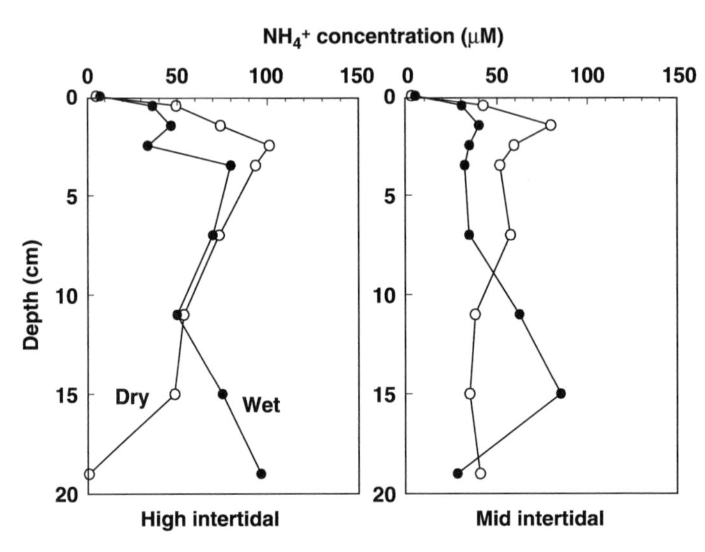

Figure 13.14 NH_4^+ in sediment porewaters of high intertidal and mid-intertidal mangrove areas in southern Thailand. "Dry" indicates the dry season (November to April) and "Wet" indicates the rainy season (May to October). Modified from Holmer *et al.* (2001).

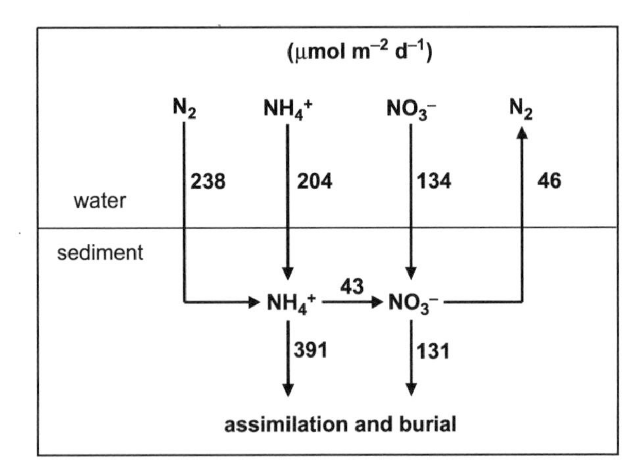

Figure 13.15 Fluxes and transformations of dissolved inorganic nitrogen in the sediment of a mid-intertidal mangrove forest (Ao Nam Bor) from southern Thailand. Modified from Kristensen *et al.* (1998).

Mangrove sediments support relatively low rates of N fixation compared with most intertidal sediments (Howarth *et al.*, 1988a,b; Welsh *et al.*, 1996; Kristensen *et al.*, 1998; Alongi *et al.*, 2000a). Most of the N fixation within the sediment is mediated by sulfate reducers, although a substantial fraction of the activity is also linked to microorganisms associated with mangrove tree roots (Kristensen *et al.*, 1998). In any case, N fixation rates are up to 10 times higher than rates of denitrification, implying that nitrogen is gained, rather than lost, by microbial processes in mangrove sediments.

4.4. Creeks: dynamic conduits to the ocean

The concentration of nutrients in the creek water of pristine tropical mangrove swamps is generally low but varies both in time and space as a result of differences in hydrodynamics, freshwater input, degree of solar insolation, and plankton productivity (Ovalle *et al.*, 1990; Alongi *et al.*, 1992; Bava and Seralathan, 1998; Trott and Alongi, 1999; Ayukai *et al.*, 2000). Indeed, mangrove waterways exhibit variable rates of primary production, depending on turbidity and nutrient availability (Table 13.6). A large autochthonous input of materials, together with particulate and dissolved materials from other sources, supports an active heterotrophic pelagic community (Alongi, 1998; Kristensen and Suraswadi, 2002). These ample supplies of organic substrates may cause creeks to be net heterotrophic, even during clear and sunny days (Ayukai and Alongi, 2000; Kristensen and Suraswadi, 2002).

Organic matter and nutrients in both particulate and dissolved forms are generally exported from mangrove forests via creeks (Wolanski *et al.*, 1992; Hemminga *et al.*, 1994; Rivera-Monroy *et al.*, 1995), although in some cases inorganic nitrogen and phosphorus may also be imported (Simpson *et al.*, 1997; Alongi, 1998; Ayukai *et al.*, 1998). The export of dissolved inorganic nutrients by tidal water frequently occurs even when the forest floor and creek bottoms appear to be significant sinks of these compounds (Kristensen *et al.*, 1998; Rivera-Monroy *et al.*, 1999; Holmer *et al.*, 2001). In such cases,

Table 13.6 Mean depth integrated water column gross primary production (GPP), respiration (RSP), and daily net primary production (DNPP) in mangrove creeks (mmol C m^{-2} d^{-1})

	GPP	RSP	DNPP	K^a	References
Laet Creek, Thailand	28	−57	−29	2.6–5.1	1
Bangrong, Thailand	53	−83	−30	1.0–2.5	2
Estero Pargo, Mexico	98	−22	76	0.5–2.4	3

[a]The light attenuation coefficient (K) with water depth is presented according to $E_d(z) = E_d(0)$ exp $(-Kz)$, where $E_d(0)$ and $E_d(z)$ are the downward irradiance at 0 and z meters depth.
References: 1, Ayukai and Alongi (2000); 2, Kristensen and Suraswadi (2002); 3, Rivera-Monroy *et al.* (1998).

nutrients originate from the microbial mineralization of organic substrates within the creek itself. This is particularly true in turbid mangrove creeks (Ayukai and Alongi, 2000; Kristensen and Suraswadi, 2002). Although more than 50% of the carbon fixed by phytoplankton is excreted back into the water as DOC and DON, it can support only a fraction of the heterotrophic demand. Up to 90% of the microbial activity in creek water may instead be fueled by organic substrates derived from vascular plants (Moran *et al.*, 1991; Bano *et al.*, 1997).

When biological activity within mangrove forests are limited by nitrogen and phosphorus availability, these elements are imported by tidal currents via the creeks. Nutrient deficiency has been observed in a number of mangrove environments (Mohammed and Johnstone, 1995; Wolanski *et al.*, 2000), but the nutrient in question (N versus P limitation) depends on local conditions. It has been inferred from observations of low PO_4^{3-} levels that many southeast Asian and Australian mangrove environments are limited by the availability of P (Alongi, 1996; Trott and Alongi, 1999; Holmer *et al.*, 2001). The exceptionally high productivity forests in Papua New Guinea, on the other hand, have high concentrations of sediment-extractable phosphorus (Alongi, 1998). Thus, phosphorus limitation appears to be a regional phenomenon in the tropics coinciding with highly weathered and phosphorus-poor soils.

Appendices

APPENDIX 1: USEFUL CONSTANTS AND CONVERSIONS

$R = 8.314\,kJ\,mol^{-1}\,K^{-1} = 1.987 \times 10^{-3}\,kcal\,mol^{-1}\,K^{-1}$
$0°\,K = -273.15°C$
$100\,kPa = 0.9869\,atm$
$1\,cal = 4.1840\,joule$
$F = 96.53\,kJ/volt = 23.07\,kcal/volt$

0-12-026147-2
© 2005 Elsevier Inc.
All rights reserved

APPENDIX 2

ΔG^0 ΔH^0 and S^0 values for inorganic chemical species of biological and environmental interest in aquatic systems. Valid at 25°C, 1 atm pressure (adapted from Stumm and Morgan, 1996, with modifications)

Species	ΔG_f^0 (kJ mol^{-1})	ΔH_f^0 (kJ mol^{-1})	S^0 (kJ mol^{-1})	Reference
Al	0	0	28.3	1
Al^{3+} (aq)	−489.4	−531.0	−308	1
$AlOH^{2+}$ (aq)	−698			2
$Al(OH)_2^+$ (aq)	−911			2
$Al(OH)_3$ (aq)	−1115			2
$Al(OH)_4^-$ (aq)	−1325			2
$Al(OH)_3$ (amorph)	−1139			1
AlOOH (boehmite)	−922	−1000	17.8	1
$Al(OH)_3$ (gibbsite)	−1155	−1293	68.4	1
$Al_2Si_2O_5(OH)_4$ (kaolinite)	−3799	−4120	203	1
$KAl_3Si_3O_{10}(OH)_2$ (muscovite)	−1341			1
$Mg_5Al_2Si_3O_{10}(OH)_8$ (chlorite)	−1962			1
As (α-metal)	0	0	35.1	3
H_3AsO_4 (aq)	−766.0	−898.7	206	3
$H_2AsO_4^-$ (aq)	−748.5	−904.5	117	3
$HAsO_4^{2-}$ (aq)	−707.1	−898.7	3.8	3
AsO_4^{3-} (aq)	−636.0	−870.3	−145	3
$H_2AsO_3^-$ (aq)	−587.4			3
Ba^{2+} (aq)	−560.7	−537.6	9.6	1
$BaSO_4$ (barite)	−1362	−1473	132	1
$BaCO_3$ (witherite)	−1132	−1211	112	1
H_3BO_3 (aq)	−968.7	−1072	162	3
$B(OH)_4^-$ (aq)	−1153.3	−1344	102	3
Br_2 (aq)	3.93	−259	130.5	3
Br^- (aq)	−104.0	−121.5	82.4	3
C (graphite)	0	0	152	3
C (diamond)	3.93	−2.59	130.5	3
CO_2 (g)	−394.37	−393.5	213.6	3
H_2CO_3 (aq)	−623.2	−699.6	187.0	1[a]
H_2CO_3 ("true")	−607.1			2
HCO_3^- (aq)	−586.8	−692.0	91.2	2
CO_3^{2-} (aq)	−527.9	−677.1	−56.9	3
Ca^{2+} (aq)	−553.54	−542.83	−53	1
$CaOH^+$ (aq)	−718.4			3

(continued)

Appendix 2 *continued*

Species	ΔG_f^0 (kJ mol^{-1})	ΔH_f^0 (kJ mol^{-1})	S^0 (kJ mol^{-1})	Reference
$CaCO_3$ (calcite)	−1128.8	−1207.4	91.7	1
$CaCO_3$ (aragonite)	−1127.8	−1207.4	88.0	1
$CaMg(CO_3)_2$ (dolomite)	−2161.7	−2324.5	155.2	1
$CaSiO_3$ (wollastonite)	−1549.9	−1635.2	82	1
$CaSO_4$ (anhydrite)	−1321.7	−1434.1	106.7	1
$CaSO_4 \cdot 2\ H_2O$ (gypsum)	−1797.2	−2022.6	194.1	1
$Ca_5(PO_4)_3OH$ (hydroxyapatite)	−6338.4	−6721.6	390.4	1
Cd^{2+} (aq)	−77.28	−75.90	−73.2	1
Cl^- (aq)	−131.3	−167.2	56.5	3
Cl_2 (g)	0	0	223.0	3
Cl_2 (aq)	6.90	−23.4	121	3
Co^{2+} (aq)	−54.4	−58.2	−113	1
Co^{3+} (aq)	−134	−92	−305	1
Cr^{2+} (aq)		−143.5		3
Cr^{3+} (aq)	−215.5	−256.0	308	3
$HCrO_4^-$ (aq)	−764.8	−878.2	184	1
CrO_4^{2-} (aq)	−727.9	−881.1	50	1
$Cr_2O_7^{2-}$ (aq)	−1301	−1490	262	1
Cu^+ (aq)	50.0	71.7	40.6	3
Cu^{2+} (aq)	65.5	64.8	−99.6	3
CuS (covellite)	−53.6	−53.1	66.5	3
Cu_2S (α)	−86.2	−79.5	121	3
F_2 (g)	0	0	202	3
F^- (aq)	−278.8	−332.6	−13.8	3
HF(aq)	−296.8	320.0	88.7	3
HF_2^- (aq)	−578.1	−650	92.5	3
Fe (metal)	0	0	27.3	3
Fe^{2+} (aq)	−78.87	−89.10	138	3
$FeOH^+$ (aq)	−277.4	324.7	29	3
$Fe(OH)_2$ (aq)	−441.0	–	–	3
Fe^{3+} (aq)	−4.60	−48.5	−316	3
$FeOH^{2+}$ (aq)	−229.4	−324.7	−29.2	3
$Fe(OH)_2^+$ (aq)	−438	250.8	142	3
$Fe(OH)_3$ (aq)	−659.4	–	–	3
$Fe(OH)_4^-$ (aq)	−842.2	–	34.5	3
$Fe_2(OH)_2^{4+}$ (aq)	−467.27	612.1	356.0	3
FeS_2 (pyrite)	−160.2	−171.5	52.9	1
FeS_2 (marcasite)	−158.4	−169.4	53.9	1
FeS (pyrrhotite)	−100.4	−100	60.29	3
FeS (mackinawite)	−93.0			4
Fe_3S_4 (greigite)	−290			4

(continued)

Appendix 2 *continued*

Species	ΔG_f^0 (kJ mol^{-1})	ΔH_f^0 (kJ mol^{-1})	S^0 (kJ mol^{-1})	Reference
FeO (s)	−251.1	−272.0	59.8	1
$Fe(OH)_2$ (precip.)	−486.6	−569	87.9	3
α-Fe_2O_3 (hematite)	−742.7	−824.6	87.4	1
Fe_3O_4 (magnetite)	−1012.6	−1115.7	146	1
α-FeOOH (goethite)	−488.6	−559.3	60.5	1
FeOOH (amorph)	−462			2
$Fe(OH)_3$ (amorph)	−699(−712)			2
$FeCO_3$ (siderite)	−699.7	−737.0	105	1
Fe_2SiO_4 (fayalite)	−1379.4	−1479.3	148	1
H_2 (g)	0	0	130.6	3
Hg_2^{2+} (aq)	153.6	172.4	84.5	3
Hg^{2+} (aq)	164.4	171.0	−32.2	3
Hg_2Cl_2 (calomel)	−210.8	−265.2	−192.4	3
HgS (metacinnabar)	−43.3	−46.7	96.2	3
I_2 (crystal)	0	0	116	3
I_2 (aq)	16.4	22.6	137	3
I^- (aq)	−51.59	−55.19	111	3
I_3^- (aq)	−51.5	−51.5	239	3
HIO (aq)	−99.2	−138	95.4	3
IO^- (aq)	−38.5	−107.5	−5.4	3
HIO_3 (aq)	−132.6	−211.3	167	3
IO_3^- (aq)	−128.0	−221.3	118	3
Mg^{2+} (aq)	−454.8	−466.8	−138	1
$MgOH^+$ (aq)	−626.8			2
$MgOH_2$ (aq)	−769.4	−926.8	−149	3
$MgOH_2$ (brucite)	−833.5	−924.5	63.2	1
$MgCO_3$ (magnesite)	−1012.1	−1095.8	65.7	3
Mn^{2+} (aq)	−228.0	−220.7	−73.6	1
$Mn(OH)_2$ (precip.)	−616			5
Mn_3O_4 (hausmannite)	−1228.2	−1387.8	155.6	3
MnOOH (α-manganite)	−557.7			5
MnO_2 (manganate) (IV)				
$MnO_{1.7}$–MnO_2	−465.14	−520.3	53.06	3
MnO_2 (pyrolusite)	−465.1	−520.0	53	3
$MnCO_3$ (rhodochrosite)	−816.7	−894.1	85.8	3
MnS (albandite)	−218.1	−213.8	782	3
$MnSiO_3$ (rhodonite)	−1243	−1319	131	3
N_2 (g)	0	0	191.5	3
NO (g)	86.57	90.25	210.6	2
NO_2 (g)	51.3	33.2	240.0	2
N_2O (g)	104.2	82.0	220	3

(*continued*)

Appendix 2 *continued*

Species	ΔG_f^0 (kJ mol^{-1})	ΔH_f^0 (kJ mol^{-1})	S^0 (kJ mol^{-1})	Reference
NH_3 (g)	−16.48	−46.1	192	3
NH_3 (aq)	−26.57	−80.29	111	3
NH_4^+ (aq)	−79.37	−132.5	113.4	3
HNO_2 (aq)	−42.97	−119.2	153	3
NO_2^- (aq)	−37.2	−104.6	140	3
HNO_3 (aq)	−111.3	−207.3	146	3
NO_3^- (aq)	−111.3	−207.3	146.4	3
O_2 (g)	0	0	205	3
O_2 (aq)	16.32	−11.17	111	3
O_3 (g)	163.2	142.7	239	3
O_3 (aq)		125.9		3
$O_2^{\cdot -}$	31.84			3
$HO_2^{\cdot}$ (aq)	4.44			3
H_2O_2 (g)	−105.6	136.31	232.6	3
H_2O_2 (aq)	−134.1	−191.17	143.9	3
HO_2^- (aq)	−67.4	−160.33	23.8	3
$OH^{\cdot}$ (g)	34.22	38.95	183.64	3
$OH^{\cdot}$ (aq)	7.74			3
OH^- (aq)	−157.29	−230.0	−10.75	3
PO_4^{3-} (aq)	−1018.8	−1277.4	−222	3
HPO_4^{2-} (aq)	−1089.3	−1292.1	−33.4	3
$H_2PO_4^-$ (aq)	−1130.4	−1296.3	90.4	3
H_3PO_4 (aq)	−1142.6	−1288.3	158	3
Pb^{2+} (aq)	−24.39	−1.67	10.5	3
PbS	−98.7	−100.4	91.2	3
S (rhombic)	0	0	31.8	3
SO_2 (g)	−300.2	−296.8	248	3
SO_3 (g)	−371.1	−395.7	257	3
H_2S (g)	−33.56	−20.63	205.7	3
H_2S (aq)	−27.87	−39.75	121.3	3
S^{2-} (aq)	85.8	33.0	−14.6	3
HS^- (aq)	12.05	−17.6	62.8	3
SO_3^{2-} (aq)	−486.6	−635.5	−29	3
HSO_3^- (aq)	−527.8	−626.2	140	3
H_2SO_3 (aq)	−537.9	−608.8	232	3[b]
H_2SO_3 (aq) ("true")	~ 534.5			2
SO_4^{2-} (aq)	−744.6	−909.2	20.1	3
HSO_4^- (aq)	−756.0	−887.3	132	3
SeO_3^{2-} (aq)	−369.9	−509.2	12.6	3
$HSeO_3^-$ (aq)	−431.5	−514.5	135	3
H_2SeO_3 (aq)	−426.2	−507.5	208	3
SeO_4^{2-} (aq)	−441.4	−599.1	54.0	3
$HSeO_4^-$ (aq)	−452.3	−581.6	149	3

(continued)

Appendix 2 *continued*

Species	ΔG_f^0 (kJ mol^{-1})	ΔH_f^0 (kJ mol^{-1})	S^0 (kJ mol^{-1})	Reference
SiO_2 (α, quartz)	−856.67	−910.94	41.8	3
SiO_2 (α, cristobalite)	−855.88	−909.48	42.7	3
SiO_2 (α, tridymite)	−855.29	−909.06	43.5	3
SiO_2 (amorph)	−850.73	−903.49	46.9	3
H_4SiO_4 (aq)	−1308.0	−1468.6	180	3
Sr^{2+} (aq)	−559.4	−545.8	−33	2
Zn^{2+} (aq)	−147.0	−153.9	112	3
$ZnOH^+$ (aq)	−330.1			3
$Zn(OH)_2$ (aq)	−522.3			3
$ZnCO_3$ (smithsonite)	−731.6	−812.8	82.4	3

[a]Includes CO_2 (aq) + [H_2CO_3] "true."
[b]Includes SO_2 (aq) + [H_2SO_3] "true."

1, Wagman *et al.* (1982)

2, Robie *et al.* (1978)

3, Stumm and Morgan (1996)

4, Drever (1988)

5, Bricker (1965).

APPENDIX 3

Thermodynamic properties of selected organic compounds at 25° C, 1 atm pressure

	ΔG^0 (kJ mol^{-1})	ΔH^0 (kJ mol^{-1})	S^0 (kJ mol^{-1})	Reference
Hydrocarbons				
CH_4 (g) (methane)	−50.72	−74.81	186.3	1
CH_4 (aq)	−34.33	−89.04	83.7	1
C_2H_6 (g) (ethane)	−32.82	−84.68	229.6	1
C_2H_6 (aq)	−17.01	−102.09	118.4	1
C_2H_4 (g) (ethene)	68.15	10.57	219.56	1
C_2H_6 (aq)	81.36	36.36	122.2	1
Alcohols				
CH_3OH (aq) (methanol)	−175.31	−245.9	133.1	1
C_2H_5OH (aq) (ethanol)	−181.64	−288.3	148.5	1
C_3H_7OH (aq) (n-propanol)	−175.81	–	–	2
C_3H_7Oh (aq) (iso-propanol)	−185.94	–	–	2
C_4H_9OH (aq) (n-butanol)	−171.8	–	–	2
$C_3H_5(OH)_3$ (aq) (glycerol)	−488.5	–	–	2
$C_3H_5(OH)_3$ (l)	−477.1	–	–	2
$C_3H_5(OH)_3$ (l) (mannitol)	−942.61			2
$C_3H_5(OH)_3$ (l) (sorbitol)	−942.7			2
Aldehydes				
CH_2O (aq) (formaldehyde)	−102.53	108.57	218.77	1
CH_3CHO (g) (acetaldehyde)	−128.9	−166.2	250.3	1
CH_3CHO (aq)	−139.9	–	–	2
Ketones				
$C(CH_3)_2O$ (acetone)	−161.17	–	–	2
Monocarboxylic acids				
HCO_2^- (aq) (formate)	−351.0	−425.6	92	1
HCO_2H (aq) (formic acid)	−372.3	−425.4	163	1
CH_3COO^- (aq) (acetate)	−369.3	−486.0	86.6	1
CH_3COOH (aq) (acetic acid)	−396.5	−485.8	178.7	1
$CH_3CH_2COO^-$ (aq) (propionate)	−361.1	–	–	2
$CH_3(CH_2)_2COO^-$ (aq) (butyrate)	−352.6	–	–	2
$CH_3(CH_2)_3COO^-$ (aq) (valerate)	−344.3	–	–	2
$CH_3(CH_2)_4COO^-$ (aq) (caproate)	−336	–	–	2
$CH_3(CH_2)_{14}COOH_{(s)}$ (palmitic acid)	−305	–	–	2
$HOCH_2COO^-$ (aq) (glycolate)	−531			2
$HCCH_3OHCOO^-$ (lactate)	−517.8	–	–	2
$(CH_2)_2OHCOO^-$ (aq) (β-hydroxypropionate)	−518.4	–	–	2
CH_3COCOO^- (aq) (pyruvate)	−474.63	–	–	2
Dicarboxylic acids				
$(COO)_2^{2-}$ (aq) (oxalate)	−673.9	−825.1	45.6	1
$(CH_2)_2(COO)_2^{2-}$ (aq) (succinate)	−690.2	–	–	2

(continued)

Appendix 3 *continued*

	ΔG^0 (kJ mol^{-1})	ΔH^0 (kJ mol^{-1})	S^0 (kJ mol^{-1})	Reference
$(CH)_2(COO)_2^{2-}$ (aq) (fumarate)	−604.2	–	–	2
$C_2H_3OH(COO)_2^{2-}$ (aq) (malate)	−845.1	–	–	2
$CH_2CO(COO)_2^{2-}$ (aq) (oxaloacetate)	−797.2	–	–	2
$(CH_2)_2CO(COO)_2^{2-}$ (aq) (α-ketoglutarate)	−797.6	–	–	2
Tricarboxylic acids				
$(CH_2)_2COHCOOH(COO)_2^{2-}$ (aq) (citrate)	−1.168	–	–	2
$HCOHCH_2C_2HOOH(COO)_2^{2-}$ (aq) (isocitrate)	−1.162	–	–	2
Carbohydrates				
$C_5H_{10}O_5$ (aq) (D-ribose)	−757.3	–	–	2
$C_6H_{12}O_6$ (aq) (α-D-glucose)	−917.22	–	–	2
$C_6H_{12}O_6$ (aq) (D-fructose)	−915.4	–	–	2
$C_{11}H_{22}O_{11}$ (aq) (α-lactose)	−1.515	–	–	2
$C_{11}H_{22}O_{11}$ (aq) (β-lactose)	−1.497	–	–	2
$C_{12}H_{22}O_{11}$ (aq) (sucrose)	−1.552	–	–	2
Amino acids				
L-alanine (aq)	−371.5	–	–	2
L-arginine (s)	−240.2	–	–	2
L-asparagine (aq) × H_2O	−764	–	–	2
L-aspartic acid (aq)	−721.3	–	–	2
L-aspartate (aq)	−700.4	–	–	2
L-cysteine (aq)	−339.8	–	–	2
L-cystine (aq)	−666.9	–	–	2
L-glutamic acid (aq)	−723.8	–	–	2
L-glutamate (aq)	−699.6	–	–	2
L-glutamine (aq)	−529.7	–	–	2
glucine (aq)	−370.8	–	–	2
glucine$^+$ (aq)	−384.2	–	–	2
glucine$^-$ (aq)	−315	–	–	2
L-leucine (aq)	−343.1	–	–	2
L-isoleucine (aq)	−343.9	–	–	2
L-methionine (aq)	−502.9	–	–	2
L-phenylalanine (aq)	−207.1	–	–	2
L-serine (aq)	−510.9	–	–	2
L-threonine (aq)	−514.6	–	–	2
L-threonine (c)	−550.2	–	–	2
L-tryptophane (aq)	−112.6	–	–	2
L-tyrosine (aq)	−370.7	–	–	2
L-valine (aq)	−356.9	–	–	2
Aromatic compounds				
C_6H_6 (g) (benzene)	129.7	82.9	269.2	1
C_6H_6 (l)	124.7	49	172	1

(*continued*)

Appendix 3 *continued*

	ΔG^0 (kJ mol^{-1})	ΔH^0 (kJ mol^{-1})	S^0 (kJ mol^{-1})	Reference
C_6H_6OH (s) (phenol)	−47.6	–	–	2
$C_6H_4O_2$ (s) (quinone)	−87.3	–	–	2
$C_6H_6O_2$ (s) (hydroquinone)	−207	–	–	2
$C_6H_4CH_3OH$ (g) (o-cresol)	37.1	–	–	2
$C_6H_4CH_3OH$ (g) (m-cresol)	−40.54	–	–	2
$C_6H_4CH_3OH$ (g) (p-cresol)	−32.1	–	–	2
$C_6H_5CH_3$ (l) (toluene)	114.22	–	–	2
C_6H_5COOH (s) (benzoic acid)	−246	–	–	2
$C_6H_5CH_2OH$ (l) (benzyl alcohol)	−31	–	–	2
Other compounds				
CH_3Cl (aq)	−51.4	−101.7	144.8	1
CH_2Cl_2 (l)	−67.3	−121.5	177.8	1
$CHCl_3$ (l)	−73.7	−134.5	201.7	1
CCl_4 (l)	−65.2	−135.4	216.4	1
CCl_4 (g)	−60.6	−102.9	309.9	1
CH_3NH_2 (aq) (methylamine)	20.8	−70.2	123.4	1
$CO(NH_2)_2$ (s) (urea)	−197.3	−333.5	104.6	1
$CO(NH_2)_2$ (aq)	−203.8	–	–	2
SCN^- (aq) (thiocyanate)	92.7	76.4	144.3	1
$(CH_3)_2O$ (g) (dimethylether)	−112.6	−184.1	266.4	1
CH_3CN (l) (acetonitrile)	77.2	31.4	149.6	1
$(CH_3)_3NH^+$ (aq) hydrated (trimethylamine)	37.4	−112.9	196.6	1
$(C_2H_5)_2S$ (g) (diethyl sulfide)	18.29	−83.1	368.13	1

1, Wagman *et al.* (1982)

2, Thauer *et al.* (1977).

References

Achouak, W., Normand, P., and Heulin, T. (1999). Comparative phylogeny of *rrs* and *nifH* genes in the *Bacillaceae*. *International Journal of Systematic Bacteriology* **49**, 961–967.

Achtnich, C., Bak, F., and Conrad, R. (1995). Competition for electron donors among nitrate reducers, ferric iron reducers, sulfate reducers, and methanogens in anoxic paddy soil. *Biology and Fertility of Soils* **19**, 65–72.

Adams, M. W. W. (1998). The influence of environment and metabolic capacity on the size of a microorganism."Size Limits of Very Small Microorganisms (Panel 2".) Space Studies Board, National Academy of Sciences, Washington, DC.

Aeckersberg, F., Bak, F., and Widdel, F. (1991). Anaerobic oxidation of saturated hydrocarbons to CO_2 by a new type of sulfate-reducing bacterium. *Archives of Microbiology* **156**, 5–14.

Akagi, J. M. (1995). Respiratory sulfate reduction. *In* "Sulfate-Reducing Bacteria" (L. L. Barton, ed.), pp. 89–109. Plenum Press, New York.

Albert, D. B., Taylor, C., and Martens, C. S. (1995). Sulfate reduction rates and low molecular weight fatty acid concentrations in the water column and surficial sediments of the Black Sea. *Deep-Sea Research* **42**, 1239–1260.

Albrecht-Gary, A. M. and Crumbliss, A. L. (1998). Coordination chemistry of siderophores: Thermodynamics and kinetics of iron chelation and release. *In* "Bacterial Iron Transport: Mechanisms, Genetics, and Regulation:" Metal Ions in Biological Systems" (A. Sigel and H. Sigel, eds), pp. 239–327. Marcel Dekker, New York.

Aller, R. C. (1982). The effects of macrobenthos on chemical properties of marine sediment and overlying water. *In* "Animal-Sediment Relations" (P. L. McCall and J. S. Tevesz, eds), pp. 53–102. Plenum Press, New York.

Aller, R. C. (1990). Bioturbation and manganese cycling in hemipelagic sediments. *Philosophical Transactions of the Royal Society of London A* **331**, 51–68.

Aller, R. C. (1994a). Bioturbation and remineralization of sedimentary organic matter: Effects of redox oscillation. *Chemical Geology* **114**, 331–345.

Aller, R. C. (1994b). The sedimentary Mn cycle in Long Island Sound: Its role as intermediate oxidant and the influence of bioturbation, O_2 and C_{org} flux on diagenetic reaction balance. *Journal of Marine Research* **52**, 259–295.

Aller, R. C. (2001). Transport and reactions in the bioirrigated zone. *In* "The Benthic Boundary Layer: Transport Processes and Biogeochemistry" (B. P. Boudreau and B. B. Jørgensen, eds), pp. 269–301. Oxford University Press, Oxford.

Aller, R. C. and Rude, P. D. (1988). Complete oxidation of solid phase sulfides by manganese and bacteria in anoxic marine sediments. *Geochimica et Cosmochimica Acta* **52**, 751–765.

Aller, R. C. and Yingst, J. Y. (1985). Effects of the marine deposit-feeders *Heteromastus filiformus* (poluchaeta), *Macoma balthica* (bivalvia), and *Tellina texana* (bivalvia) on averaged sedimentary solute transport, reaction rates, and microbial distributions. *Journal of Marine Research* **43**, 615–645.

Aller, R. C., Blair, N. E., and Rude, P. D. (1996). Remineralization rates, recycling, and storage of carbon in Amazon shelf sediments. *Continental Shelf Research* **16**, 753–786.

Aller, R. C., Hall, P. O. J., Rude, P. D., and Aller, J. Y. (1998). Biogeochemical heterogeneity and suboxic diagenesis in hemipelagic sediments of the Panama Basin. *Deep-Sea Research Part* I-*Oceanographic Research Papers* **45**, 133–165.

Alongi, D. M. (1996). The dynamics of benthic nutrient pools and fluxes in tropical mangrove forests. *Journal of Marine Research* **54**, 123–148.

Alongi, D. M. (1998). Coastal Ecosystem Processes. CRC Press, Boca Raton.

Alongi, D. M. and Dixon, P. (2000). Mangrove primary production and above- and below-ground biomass in Sawi Bay, southern Thailand. *Phuket Marine Biological Center Special Publications* **22**, 31–38.

Alongi, D. M., Boto, K. G., and Robertson, A. I. (1992). Nitrogen and phosphorus cycles. *In* "Tropical Mangrove Ecosystems" (A. I. Robertson and D. M. Alongi, eds), pp. 251–292. American Geophysical Union, Washington, DC.

Alongi, D. M., Tirendi, F., Dixon, P., Trott, L. A., and Brunskill, G. J. (1999). Mineralization of organic matter in intertidal sediments of a tropical semi-enclosed delta. *Estuarine Coastal and Shelf Science* **48**, 451–467.

Alongi, D. M., Tirendi, E., and Clough, B. F. (2000a). Below-ground decomposition of organic matter in forests of the mangroves *Rhizophora stylosa* and *Avicennia marina* along the arid coast of Western Australia. *Aquatic Botany* **68**, 97–122.

Alongi, D. M., Tirendi, F., Trott, L. A., and Xuan, T. T. (2000b). Benthic decomposition rates and pathways in plantations of the mangrove *Rhizophora apiculata* in the Mekong delta, Vietnam. *Marine Ecology-Progress Series* **194**, 87–101.

Alongi, D. M., Wattayakorn, G., Pfitzner, J., Tirendi, F., Zagorskis, I., Brunskill, G. J., Davidson, A., and Clough, B. F. (2001). Organic carbon accumulation and metabolic pathways in sediments of mangrove forests in southern Thailand. *Marine Geology* **179**, 85–103.

Alperin, M. J. and Reeburgh, W. S. (1984). Geochemical observations supporting anaerobic methane oxidation. *In* "4th International Symposium on Microbial Growth on One Carbon Compounds" (R. L. Crawford and R. S. Hanson, eds). American Society of Microbiology, University of Minnesota, Minneapolis, Minnesota.

Alperin, M. J. and Reeburgh, W. S. (1985). Inhibition experiments on anaerobic methane oxidation marine sediment diagenesis. *Applied and Environmental Microbiology* **50**, 940–945.

Alperin, M. J. and Reeburgh, W. S. (1988). Carbon and hydrogen isotope fractionation resulting from anaerobic methane oxidation. *Global Biogeochemical Cycles* **2**, 279–288.

Alt, J. C. and Shanks, IIIW. C. (1998). Sulfur in serpentinized oceanic peridotites: Serpentinization processes and microbial sulfate reduction. *Journal of Geophysical Research* **103**, 9917–9929.

Altabet, M. A. and Francois, R. (1994). Sedimentary nitrogen isotopic ratio as a recorder for surface ocean nitrate utilization. *Global Biogeochemical Cycles* **8**, 103–116.

Altekar, W. and Rajagopalan, R. (1990). Ribulose bisphosphate caroboxylase in halophilic *Archaebacteria*. *Archives of Microbiology* **153**, 169–174.

Amann, R., Ludwig, W., and Schleifer, K. H. (1995). Phylogenetic identification and *in-situ* detection of individual microbial-cells without cultivation. *Science* **59**, 143–169.

Amaral, J. A., Archambault, C., Richards, S. R., and Knowles, R. (1995). Denitrification associated with Groups I and II methanotrophs in a gradient enrichment system. *FEMS Microbiology Ecology* **18**, 289–298.

Amon, R. M. W., Fitznar, H. P., and Benner, R. (2001). Linkages among the bioreactivity, chemical composition, and diagenetic state of marine dissolved organic matter. *Limnology and Oceanography* **46**, 287–297.

Anbar, A. D. and Knoll, A. H. (2002). Proterozoic ocean chemistry and evolution: A bioinorganic bridge? *Science* **297**, 1137–1142.

Anbar, A. D., Roe, J. E., Barling, J., and Nealson, K. H. (2000). Nonbiological fractionation of iron isotopes. *Science* **288**, 126–128.

Andersen, F. Ø. and Jensen, H. S. (1992). Regeneration of inorganic phosphorus and nitrogen from decomposition of seston in a freshwater sediment. *Hydrobiologia* **228**, 71–81.

Andersen, F. Ø. and Ring, P. (1999). Comparison of phosphorus release from littoral and profundal sediment in a shallow, eutrophic lake. *Hydrobiologia* **408/409**, 175–183.

Andersen, J. M. (1975). Influence of pH on release of phosphorus from lake sediments. *Archiv für Hydrobiologie* **76**, 411–419.

Andresen, M. and Kristensen, E. (2002). The importance of bacteria and microalgae in the diet of the deposit-feeding polychaete *Arenicola marina*. *Ophelia* **56**, 179–196.

Angert, E. R., Clements, K. D., and Pace, N. R. (1993). The largest bacterium. *Nature* **362**, 239–241.

Anschutz, P., Zhong, S., Sundby, B., Mucci, A., and Gobeil, C. (1998). Burial efficiency of phosphate and the geochemistry of iron in continental margin sediments. *Limnology and Oceanography* **43**, 53–64.

Anthonisen, A. C., Loehr, R. C., Prakasam, T. B. S., and Srinath, E. G. (1976). Inhibition of nitrification by ammonia and nitrous-acid. *Journal Water Pollution Control Federation* **48**, 835–852.

Archibald, F. (1983). *Lactobacillus plantarum*, an organism not requiring iron. *FEMS Microbiology Letters* **19**, 29–32.

Archibald, F. and Roy, B. (1992). Production of manganic chelates by laccase from the lignin-degrading fungus *Trametes (Coriolus) versicolor*. *Applied and Environmental Microbiology* **58**, 1496–1499.

Arnold, R. G., Hoffmann, M. R., Dichristina, T. J., and Picardal, F. W. (1990). Regulation of dissimilatory Fe(III) reduction activity in *Shewanella putrefaciens*. *Applied and Environmental Microbiology* **56**, 2811–2817.

Arnosti, C., Repeta, D. J., and Blough, N. V. (1994). Rapid bacterial-degradation of polysaccharides in anoxic marine systems. *Geochimica et Cosmochimica Acta* **58**, 2639–2652.

Asada, K. (1999). The water-water cycle in chloroplasts: Scavenging of active oxygens and dissipation of excess photons. *Annual Review of Plant Physiology and Plant Molecular Biology* **50**, 601–639.

Atlas, R. M., and Bartha, R. (1998). Microbial Ecology. Fundamentals and Applications p. 694. Benjamin/Cummings Science Publishing, Menlo Park, California.

Avery, Jr.G. B. and Martens, C. S. (1999). Controls on the stable isotopic composition of methane produced in a tidal freshwater estuarine sediment. *Geochimica et Cosmochimica Acta* **63**, 1075–1082.

Avery, Jr.G. B., Shannon, R. D., White, D. C., Martens, C. S., and Alperin, M. J. (1999). Effect of seasonal changes in the pathways of methanogenesis on the $d^{13}C$ values of pore water methane. *Global Biogeochemical Cycles* **13**, 475–484.

Avery, Jr.G. B., Shannon, R. D., White, J. R., Martens, C. S., and Alperin, M. J. (2002). Controls on methane production in a tidal freshwater estuary and a peatland: Methane production via acetate fermentation and CO_2 reduction. *Biogeochemistry* **62**, 19–37.

Ayukai, T. and Alongi, D. M. (2000). Pelagic carbon fixation and heterotrophy in shallow coastal water of Sawi Bay, southern Thailand. *Phuket Marine Biological Center Special Publications* **22**, 39–50.

Ayukai, T., Miller, D., Wolanski, E., and Spagnol, S. (1998). Fluxes of nutrients and dissolved and particulate organic carbon in two mangrove creeks in northeastern Australia. *Mangroves and Salt Marshes* **2**, 223–230.

Ayukai, T., Wolanski, E., Wattayakorn, G., and Alongi, D. M. (2000). Organic carbon and nutrient dynamics in mangrove creeks and adjacent coastal waters of Sawi bay, southern Thailand. *Phuket Marine Biological Center Special Publications* **22**, 51–62.

Azam, F., Fenchel, T., Field, J. G., Gray, J. S., Meyerreil, L. A., and Thingstad, F. (1983). The ecological role of water-column microbes in the sea. *Marine Ecology-Progress Series* **10**, 257–263.

Badarudeen, A., Damodaran, K. T., Sajan, K., and Padmalal, D. (1996). Texture and geochemistry of the sediments of a tropical mangrove ecosystem, southwest coast of India. *Environmental Geology* **27**, 164–169.

Badger, M. R. and Andrews, T. J. (1987). Co-evolution of Rubisco and CO_2 concentrating mechanisms. *In* "Progress in Photosynthesis Research" (J. Biggens, ed.), pp. 600–609. Martinus Nijhoff Publishers, Dordrecht.

Badger, M. R. and Price, G. D. (1992). The CO_2 concentrating mechanism in cyanobacteria and microalgae. *Plant Physiology* **84**, 606–615.

Badger, M. R., Andrews, T. J., Whitney, S. M., Ludwig, M., Yellowlees, D. C., Leggat, W., and Price, G. D. (1998). The diversity and coevolution of Rubisco, plastids, pyrenoids, and chloroplast-based CO_2-concentrating mechanisms in algae. *Canadian Journal of Botany* **76**, 1052–1071.

Bak, F. and Pfennig, N. (1987). Chemolithotrophic growth of *Desulfovibrio sulfodismutans* sp. nov. by disproportionation of inorganic sulfur compounds. *Archives of Microbiology* **147**, 184–189.

Bak, F. and Pfennig, N. (1991). Microbial sulfate reduction in littoral sediment of Lake Constance. *FEMS Microbiology Ecology* **85**, 31–42.

Baldy, V., Gessner, M. O., and Chauvet, E. (1995). Bacteria, fungi and the breakdown of leaf-litter in a large river. *Oikos* **74**, 93–102.

Banfield, J. F., Barker, W. W., Welch, S. A., and Taunton, A. (1999). Biological impact on mineral dissolution: Application of the lichen model to understanding mineral weathering in the rhizosphere. *Proceedings of the National Academy of Sciences USA* **96**, 3404–3411.

Bano, N., Nisa, M. U., Khan, N., Saleem, M., Harrison, P. J., Ahmed, S. I., and Azam, F. (1997). Significance of bacteria in the flux of organic matter in the tidal creeks of the mangrove ecosystem of the Indus River delta, Pakistan. *Marine Ecology-Progress Series* **157**, 1–12.

Barford, C. C., Montoya, J. P., Altabet, M. A., and Mitchell, R. (1999). Steady-state nitrogen isotope effects of N_2 and N_2O production in *Paracoccus denitrificans*. *Applied and Environmental Microbiology* **65**, 989–994.

Barker, J. F. and Fritz, P. (1981). Carbon isotope fractionation during microbial methane oxidation. *Nature* **293**, 289–291.

Barker, P., Fontes, J. C., Gasse, F., and Druart, J. C. (1994). Experimental dissolution of diatom silica in concentrated salt-solutions and implications for paleoenvironmental reconstruction. *Limnology and Oceanography* **39**, 99–110.

Barker, W. W. and Banfield, J. F. (1996). Biologically versus inorganically mediated weathering reactions: Relationships between minerals and extracellular microbial polymers in lithobiontic communities. *Chemical Geology* **132**, 55–69.

Barker, W. W., Welch, S. A., Chu, S., and Banfield, J. F. (1998). Experimental observations of the effects of bacteria on aluminosilicate weathering. *American Mineralogist* **83**, 1551–1563.

Barns, S. M., Delwiche, C. F., Palmer, J. D., and Pace, N. R. (1996). Perspectives on archaeal diversity, thermophily and monophyly from environmental rRNA sequences. *Proceedings of the National Academy of Sciences USA* **93**, 9188–9193.

Bartoli, F. (1983). The biogeochemical cycle of silicon in two temperate forest ecosystems. *Ecological Bulletins* **35**, 469–476.

Bartoli, M., Nizzoli, D., Welsh, D. T., and Viaroli, P. (2000). Short-term influence of recolonisation by the polycheate worm *Nereis succinea* on oxygen and nitrogen fluxes and denitrification: A microcosm simulation. *Hydrobiologia* **431**, 165–174.

Bassler, B. L. and Miller, M. B. (2001). Quorum sensing. *In* "The Prokaryotes: An Evolving Electronic Resource for the Microbiological Community" (A. Balows, H. G. Trüper, M. Dworkin, W. Harder, and K.-H. Schleifer, eds), 3rd Ed., release 3.5, November 2. Springer, New York.

Bauer, J. E., Druffel, E. R. M., Wolgast, D. M., and Griffin, S. (2002). Temporal and regional variability in sources and cycling of DOC and POC in the northwest Atlantic continental shelf and slope. *Deep-Sea Research Part* II-*Topical Studies in Oceanography* **49**, 4387–4419.

Baughn, A. D. and Malamy, M. H. (2004). The strict anaerobe *Bacteroides fragilis* grows in and benefits from nanomolar concentrations of oxygen. *Nature* **427**, 441–444.

Bava, K. A. and Seralathan, P. (1999). Interstitial water and hydrochemistry of a mangrove forest and adjoining water system, south west coast of India. *Environmental Geology* **38**, 47–52.

Baxter, N. J., Hirt, R. P., Bodrossy, L., Kovacs, K. L., Embley, T. M., Prosser, J. I., and Murrell, J. C. (2002). The ribulose-1.5-bisphosphate carboxylase/oxygenase gene cluster of *Methylococcus capsulatus* (Bath). *Archives of Microbiology* **177**, 279–289.

Beale, S. I. (1999). Enzymes of chlorophyll biosynthesis. *Photosynthesis Research* **60**, 43–73.

Beard, B. L., Johnson, C. M., Cox, L., Sun, H., Nealson, K. H., and Aguilar, C. (1999). Iron isotope biosignatures. *Science* **285**, 1889–1892.

Beard, B. L., Johnson, C. M., Skulan, J. L., Nealson, K. H., Cox, L., and Sun, H. (2003). Application of Fe isotopes to tracing the geochemical and biological cycling of Fe. *Chemical Geology* **14137**, 1–31.

Beardsley, C., Pernthaler, J., Wosniok, W., and Amann, R. (2003). Are readily culturable bacteria in coastal North Sea waters suppressed by selective grazing mortality? *Applied and Environmental Microbiology* **69**, 2624–2630.

Beaumont, V. and Robert, F. (1999). Nitrogen isotope ratios of kerogens in Precambrian cherts: A record of the evolution of atmosphere chemistry? *Precambrian Research* **96**, 63–82.

Bédard, C. and Knowles, R. (1989). Physiology, biochemistry, and specific inhibitors of CH_4, NH_4^+, and CO oxidation by methanotrophs and nitrifiers. *Microbiological Reviews* **53**, 68–84.

Béjà, O., Spudich, E. N., Spudich, J. L., Leclerc, M., and DeLong, E. F. (2001). Proteorhodopsin phototrophy in the ocean. *Nature* **411**, 786–789.

Bell, R. T. and Riemann, B. (1989). Adenine incorporation into DNA as a measure of microbial production in freshwaters. *Limnology and Oceanography* **34**, 435–444.

Bender, M. and Conrad, R. (1992). Kinetics of CH_4 oxidation in oxic soils exposed to ambient air or high CH_4 mixing ratios. *FEMS Microbiology Ecology* **101**, 261–270.

Bender, M. and Conrad, R. (1993). Kinetics of methane oxidation in oxic soils. *Chemosphere* **26**, 687–696.

Bender, M. L. and Grande, K. D. (1987). Production, respiration, and the isotope geochemistry of O_2 in the upper water column. *Global Biogeochemical Cycles* **1**, 49–59.

Bender, M. L. and Heggie, D. T. (1984). Fate of organic carbon reaching the deep-sea floor: A status report. *Geochimica et Cosmochimica Acta* **48**, 977–986.

Bender, M., Sowers, T., and Labeyrie, L. (1994). The Dole effect and its variations during the last 130,000 years as measured in the Vostok ice core. *Global Biogeochemical Cycles* **8**, 363–376.

Benitez-Nelson, C. R. (2000). The biogeochemical cycling of phosphorus in marine systems. *Earth Science Reviews* **51**, 109–135.

Benitez-Nelson, C. R. and Buesseler, K. O. (1999). Variability of inorganic and organic phosphorus turnover rates in the coastal ocean. *Nature* **398**, 502–505.

Benner, R. and Hodson, R. E. (1985). Microbial degradation of the leachable and lignocellulosic components of leaves and wood from *Rhizophora mangle* in a tropical mangrove swamp. *Marine Ecology Progress Series* **23**, 221–230.

Benner, R., Peele, E. R., and Hodson, R. E. (1986). Microbial utilization of dissolved organic-matter from leaves of the red mangrove, *Rhizophora mangle*, in the Fresh Greek estuary, Bahamas. *Estuarine Coastal and Shelf Science* **23**, 607–619.

Benner, R., Hatcher, P. G., and Hedges, J. I. (1990). Early diagenesis of mangrove leaves in a tropical estuary: Bulk chemical characterization using solid-state ^{13}C NMR and elemental analyses. *Geochimica et Cosmochimica Acta* **54**, 2003–2013.

Bennett, P. C. and Casey, W. H. (1994). Organic acids and the dissolution of silicates. *In* "The Role of Organic Acids in Geological Processes" (E. D. Pittman and M. Lewan, eds), pp. 162–201. Springer, Berlin.

Bennett, B. A., Smith, C. R., Glaser, B., and Maybaum, H. L. (1994). Faunal community structure of a chemoautotrophic assemblage on whale bones in the deep northeast Pacific Ocean. *Marine Ecology Progress Series* **108**, 205–223.

Bennett, P. C., Rogers, J. R., and Choi, W. J. (2001). Silicates, silicate weathering, and microbial ecology. *Geomicrobiology Journal* **18**, 3–19.

Benschop, J. J., Badger, M. R., and Price, G. D. (2003). Characterisation of CO_2 and HCO_3^- uptake in the cyanobacterium *Synechocystis* sp PCC6803. *Photosynthesis Research* **77**, 117–126.

Benz, M., Brune, A., and Schink, B. (1998). Anaerobic and aerobic oxidation of ferrous iron at neutral pH by chemoheterotrophic nitrate-reducing bacteria. *Archives of Microbiology* **169**, 159–165.

Berdalet, E., Marrase, C., Estrada, M., Arin, L., and MacLean, M. L. (1996). Microbial community responses to nitrogen- and phosphorous-deficient nutrient inputs: Microplankton dynamics and biochemical characterization. *Journal of Plankton Research* **18**, 1627–1641.

Berelson, W., McManus, J., Coale, K., Johnson, K., Burdige, D., Kilgore, T., Colodner, D., Chavez, F., Kudela, R., and Boucher, J. (2003). A time series of benthic flux measurements from Monterey Bay, CA. *Continental Shelf Research* **23**, 457–481.

Berg, P., Rysgaard, S., and Thamdrup, B. (2003). Dynamic modeling of early diagenesis and nutrient cycling. A case study an Arctic marine sediment. *American Journal of Science* **303**, 905–955.

Berman-Frank, I., Cullen, J. T., Shaked, Y., Sherrell, R. M., and Falkowski, P. G. (2001a). Iron availability, cellular iron quotas, and nitrogen fixation in *Trichodesmium*. *Limnology and Oceanography* **46**, 1249–1260.

Berman-Frank, I., Lundgren, P., Chen, Y.-B., Küpper, H., Kolber, Z., Bergman, B., and Falkowski, P. (2001b). Segregation of nitrogen fixation and oxygenic photosynthesis in the marine cyanobacterium *Trichodesmium*. *Nature* **294**, 1534–1537.

Berman-Frank, I., Lundgren, P., and Falkowski, P. (2003). Nitrogen fixation and photosynthetic oxygen evolution in cyanobacteria. *Research in Microbiology* **154**, 157–164.

Berner, R. (1964). An idealized model of dissolved sulfate distribution in recent sediments. *Geochimica et Cosmochimica Acta* **28**, 1497–1503.

Berner, R. A. (1980). Early Diagenesis: A Theoretical Approach, p. 241. Princeton University Press, Princeton, New Jersey.

Berner, R. A. (1981). Authigenic mineral formation resulting from organic matter decomposition in modern sediments. *Fortschritte der Mineralogie* **59**, 117–135.

Berner, R. A. (1982). Burial of organic carbon and pyrite sulfur in the modern ocean: Its geochemical and environmental significance. *American Journal of Science* **282**, 451–473.

Berner, R. A. (1984). Sedimentary pyrite formation: An update. *Geochimica et Cosmochimica Acta* **48**, 605–615.

Berner, R. A. (1989). Biogeochemical cycles of carbon and sulfur and their effect on atmospheric oxygen over phanerozoic time. *Global and Planetary Change* **75**, 97–122.

Berner, R. A. and Canfield, D. E. (1989). A new model for atmospheric oxygen over Phanerozoic time. *American Journal of Science* **289**, 333–361.

Berner, R. A., Petsch, S. T., Lake, J. A., Beerling, D. J., Popp, B. N., Lane, R. S., Laws, E. A., Westley, M. B., Cassar, N., Woodward, F. I., and Quick, W. P. (2000). Isotope fractionation and atmospheric oxygen: Implications for phanerozoic O_2 evolution. *Science* **287**, 1630–1633.

Beveridge, T. J. (1989). Role of cellular design in bacterial metal accumulation and mineralization. *Annual Review of Microbiology* **43**, 147–171.

Bidle, K. D. and Azam, F. (1999). Accelerated dissolution of diatom silica by natural marine bacterial assemblages. *Nature* **397**, 508–512.

Bidle, K. D. and Azam, F. (2001). Bacterial control of silicon regeneration from diatom detritus: Significance of bacterial ectohydrolases and species identity. *Limnology and Oceanography* **46**, 1606–1623.

Bidle, K. D., Brzezinski, M. A., Long, R. A., Jones, J. L., and Azam, F. (2003). Diminished efficiency in the oceanic silica pump caused by bacteria-mediated silica dissolution. *Limnology and Oceanography* **48**, 1855–1868.

Bjerrum, C. J. and Canfield, D. E. (2002). Ocean productivity before about 1.9 Gyr ago limited by phosphorus adsorption onto iron oxides. *Nature* **417**, 159–162.

Björk, S. (1988). Redevelopment of lake ecosystems. A case study approach. *Ambio* **17**, 90–98.

Björkman, K., Thomson-Bulldis, A. L., and Karl, D. (2000). Phosphorus dynamics in the North Pacific subtropical gyre. *Aquatic Microbial Ecology* **22**, 185–198.

Blackburn, T. H. (1979a). Method for measuring rates of NH_4^+ turnover in anoxic marine sediments, using a ^{15}N-NH_4^+ dilution technique. *Applied and Environmental Microbiology* **37**, 760–765.

Blackburn, T. H. (1979b). Nitrogen/carbon ratios and rates of ammonia turnover in anoxic sediments. *In* "Microbial Degradation of Pollutants in Marine Environments" (A. W. Bourquin and P. H. Pritchard, eds), pp. 148–153. EPA, Gulf Breeze.

Blackburn, T. H. (1983). The microbial nitrogen cycle. *In* "Microbial Geochemistry" (W. E. Krumbein, ed.), pp. 63–89. Blackwell Scientific Publications, Oxford.

Blackburn, T. H. and Henriksen, K. (1983). Nitrogen cycling in different types of sediments from Danish waters. *Limnology and Oceanography* **28**, 477–493.

Blackburn, T. H., Hall, P. O. J., Hulth, S., and Landen, A. (1996). Organic-N loss by efflux and burial associated with a low efflux of inorganic N and with nitrate assimilation in Arctic sediments (Svalbard, Norway). *Marine Ecology-Progress Series* **141**, 283–293.

Blair, N. (1998). The $d^{13}C$ of biogenic methane in marine sediments: The influence of C_{org} deposition rate. *Chemical Geology* **152**, 139–150.

Blake, R. E. and Walter, L. M. (1996). Effects of organic acids on the dissolution of orthoclase at 80°C and pH 6. *In* "Chemical and Biological Control on Mineral Growth and Dissolution Kinetics" (L. L. Stillings, ed.), pp. 91–102. Elsevier, Amsterdam.

Blankenship, R. E. (1992). Origin and early evolution of photosynthesis. *Photosynthesis Research* **33**, 91–111.

Blankenship, R. E. (2002). Molecular Mechanisms of Photosynthesis, p. 321. Blackwell Science Ltd, Oxford.

Blankenship, R. E. and Hartman, H. (1998). The origin and evolution of oxygenic photosynthesis. *Trends in Biochemical Sciences* **23**, 94–97.

Blum, J. S., Burns, A. B., Buzzelli, J., Stolz, J. F., and Oremland, R. S. (1998). *Bacillus arsenicoselenatis*, sp. nov., and *Bacillus selenitireducens*, sp. nov.: Two haloalkalophiles from Mono Lake, California that respire oxyanions of selenium and arsenic. *Archives of Microbiology* **171**, 19–30.

Blum, L. K., Mills, A. L., Zieman, J. C., and Zieman, R. T. (1988). Abundance of bacteria and fungi in seagrass and mangrove detritus. *Marine Ecology-Progress Series* **42**, 73–78.

Blöchl, E., Rachel, R., Burggraf, S., Hafenbradl, D., Jannasch, H. W., and Stetter, K. O. (1997). *Pyrolobus fumarii*, gen. and sp. nov., represents a novel group of archaea, extending the upper temperature limit for life to 113°C. *Extremophiles* **1**, 14–21.

Boetius, A., Ravenschlag, K., Schubert, C. J., Rickert, D., Widdel, F., Gieseke, A., Amann, R., Jørgensen, B. B., Witte, U., and Pfannkuche, O. (2000). A marine microbial consortium apparently mediating anaerobic oxidation of methane. *Nature* **407**, 623–626.

Boga, H. I. and Brune, A. (2003). Hydrogen-dependent oxygen reduction by homoacetogenic bacteria isolated from termite guts. *Applied and Environmental Microbiology* **69**, 779–786.

Bond, D. R. and Lovley, D. R. (2002). Reduction of Fe(III) oxide by methanogens in the presence and absence of extracellular quinones. *Environmental Microbiology* **4**, 115–124.

Bond, D. R., Holmes, D. E., Tender, L. M., and Lovley, D. R. (2002). Electrode-reducing microorganisms that harvest energy from marine sediments. *Science* **295**, 483–485.

Bonin, P. (1996). Anaerobic nitrate reduction to ammonium in two strains isolated from costal marine sediment: A dissimilatory pathway. *FEMS Microbiology Ecology* **19**, 27–38.

Bonin, P., Omnes, P., and Chalamet, A. (1998). Simultaneous occurrence of denitrification and nitrate ammonification in sediments of the French Mediterranean Coast. *Hydrobiologia* **389**, 169–182.

Boone, D. R., Whitman, W. B., and Rouvière, P. (1993). Diversity and taxonomy of methanogens. *In* "Methanogenesis" (J. G. Ferry, ed.). Chapman & Hall, Inc., London.

Booth, I. R. (1999). Adaptation to extreme environments. *In* "Biology of the Prokaryotes" (J. W. Lengeler, G. Drews, and H. G. Schlegel, eds), pp. 652–671. Blackwell Science, Thieme, Stuttgart, New York.

Boström, B., Jansson, M., and Forsberg, C. (1982). Phosphorus release from lake sediments. *Archiv für Hydrobiologie* **18**, 5–59.

Boström, B., Andersen, J. M., Fleischer, S., and Jansson, M. (1988). Exchange of phosphorus across the sediment-water interface. *Hydrobiologia* **170**, 229–244.

Boto, K. G., Alongi, D. M., and Nott, A. L. J. (1989). Dissolved organic carbon-bacteria interactions at sediment-water interface in a tropical mangrove system. *Marine Ecology-Progress Series* **51**, 243–251.

Botto, F. and Iribarne, O. (2000). Contrasting effects of two burrowing crabs (*Chasmagnathus granulata* and *Uca uruguayensis*) on sediment composition and transport in estuarine environments. *Estuarine Coastal and Shelf Science* **51**, 141–151.

Botz, R., Pokojski, H.-D., Schmitt, M., and Thomm, M. (1996). Carbon isotope fractionation during bacterial methanogenesis by CO_2 reduction. *Organic Geochemistry* **25**, 255–262.

Boudreau, B. P. (1990). Modelling early diagenesis of silica in non-bioturbated sediments. *Deep-Sea Research* **37**, 1543–1567.

Boudreau, B. P. (1994). Is burial velocity a master parameter for bioturbation? *Geochimica et Cosmochimica Acta* **58**, 1243–1249.

Boudreau, B. P. (1997). Diagenetic Models and Their Implementation, p. 414. Springer, New York.

Boudreau, B. P. and Ruddick, B. R. (1991). On a reactive continuum representation of organic matter diagenesis. *American Journal of Science* **291**, 507–538.

Boudreau, B. P., Mucci, A., Sundby, B., Luther, G. W., and Silverberg, N. (1998). Comparative diagenesis at three sites on the Canadian continental margin. *Journal of Marine Research* **56**, 1259–1284.

Bouillon, S., Koedam, N., Raman, A. V., and Dehairs, F. (2002). Primary producers sustaining macro-invertebrate communities in intertidal mangrove forests. *Oecologia* **130**, 441–448.

Bowen, H. J. M. (1979). Environmental Chemistry of the Elements. Academic Press, London.

Bowman, J. (2000). The Methanotrophs—The families Methylococcaceae and Methylocystaceae. *In* "The Prokaryotes: An Evolving Electronic Resource for the Microbiological Community" (M. Dworkin, A. Balows, H. G. Trüper, W. Harder, and K.-H. Schleifer, eds), 3rd Ed., release 3.1, January 20, 2000. Springer, New York.

Braker, G., Fesefeldt, A., and Witzel, K. P. (1998). Development of PCR primer systems for amplification of nitrite reductase genes (*nirK* and *nirS*) to detect denitrifying bacteria in environmental samples. *Applied and Environmental Microbiology* **64**, 3769–3775.

Braker, G., Zhou, J. Z., Wu, L. Y., Devol, A. H., and Tiedje, J. M. (2000). Nitrite reductase genes (*nirK* and *nirS*) as functional markers to investigate diversity of denitrifying bacteria in Pacific Northwest marine sediment communities. *Applied and Environmental Microbiology* **66**, 2096–2104.

Braker, G., Ayala-del-Rio, H. L., Devol, A. H., Fesefeldt, A., and Tiedje, J. M. (2001). Community structure of denitrifiers, bacteria, and Archaea along redox gradients in Pacific Northwest marine sediments by terminal restriction fragment length polymorphism analysis of amplified nitrite reductase (*nirS)* and 16S rRNA genes. *Applied and Environmental Microbiology* **67**, 1893–1901.

Brand, L. E., Sunda, W. G., and Guillard, R. R. L. (1983). Limitation of marine phytoplankton reproductive rates by zinc, manganese, and iron. *Limnology and Oceanography* **28**, 1182–1198.

Brandes, J. A. and Devol, A. H. (1995). Simultaneous nitrate and oxygen respiration in coastal sediments: Evidence for discrete diagenesis. *Journal of Marine Research* **53**, 771–797.

Brandes, J. A. and Devol, A. H. (1997). Isotopic fractionation of oxygen and nitrogen in coastal marine sediment. *Geochimica et Cosmochimica Acta* **61**, 1793–1801.

Brandes, J. A. and Devol, A. H. (2002). A global marine-fixed nitrogen isotopic budget: Implications for Holocene nitrogen cycling. *Global Biogeochemical Cycles* **16**, 67 1–67 14.

Brandes, J. A., Devol, A. H., Yoshinari, T., Jayakumar, D. A., and Naqvi, S. W. A. (1998). Isotopic composition of nitrate in the central Arabian Sea and eastern tropical North Pacific: A tracer for mixing and nitrogen cycles. *Limnology and Oceanography* **43**, 1680–1689.

Brandt, K. K., Vester, F., Jensen, A. N., and Ingvorsen, K. (2001). Sulfate reduction dynamics and enumeration of sulfate-reducing bacteria in hypersaline sediments of the Great Salt Lake (Utah, USA). *Microbial Ecology* **41**, 1–11.

Brasier, M. D., Green, O. R., Jephcoat, A. P., Kleppe, A. K., van Krankendonk, M. J., Lindsay, J. F., Steele, A., and Grassineau, N. V. (2002). Questioning the evidence for Earth's oldest fossils. *Nature* **416**, 76–81.

Braun, S. T., Proctor, L. M., Zani, S., Mellon, M. T., and Zehr, J. P. (1999). Molecular evidence for zooplankton-associated nitrogen-fixing anaerobes based on amplification of the nifH gene. *FEMS Microbiology Ecology* **28**, 273–279.

Braun, V. and Killmann, H. (1999). Bacterial solutions to the iron supply problem. *Trends in Biochemical Sciences* **24**, 104–109.

Braun, V., Hantke, K., and Koster, W. (1998). Bacterial iron transport: Mechanisms, genetics, and regulation. *In* "Iron Transport and Storage in Microorganisms, Plants, and Animals. Metal Ions in Biological Systems" (A. Sigel and H. Sigel, eds), pp. 67–145. Marcel Dekker, New York.

Breed, R. S., Murray, E. G. D., and Parker Hitchens, A. (eds.) (1948). Bergey's Manual of Determinative Bacteriology. William and Wilkins, New York.

Brettar, I., Moore, E. R. B., and Hofle, M. G. (2001). Phylogeny and abundance of novel denitrifying bacteria isolated from the water column of the central Baltic Sea. *Microbial Ecology* **42**, 295–305.

Bricker, O. (1965). Some stability relations in the system Mn–O_2–H_2O at 25° and one atmosphere total pressure. *The American Mineralogist* **50**, 1296–1354.

Brierley, E. D. R. and Wood, M. (2001). Heterotrophic nitrification in an acid forest soil: Isolation and characterisation of a nitrifying bacterium. *Soil Biology and Biochemistry* **33**, 1403–1409.

Brioukhanov, A. L., Thauer, R. K., and Netrusov, A. I. (2002). Catalase and superoxide dismutase in the cells of strictly anaerobic microorganisms. *Microbiology* **71**, 281–285.

Brock, T. D. (1994). Life at High Temperatures, p. 31. Yellowstone Association for Natural Science, History & Education, Inc., Yellowstone National Park, WY.

Brocks, J. J., Logan, G. A., Buick, R., and Summons, R. E. (1999). Archean molecular fossils and the early rise of eukaryotes. *Science* **285**, 1033–1036.

Brocks, J. J., Buick, R., Summons, R. E., and Logan, G. A. (2003). A reconstruction of Archean biological diversity based on molecular fossils from the 2.78 to 2.45 billion-year-old Mount Bruce Supergroup, Hamersley Basin, Western Australia. *Geochimica et Cosmochimica Acta* **67**, 4321–4335.

Broda, E. (1977). Two kinds of lithotrophs missing in nature. *Zeitschrift Fur Allgemeine Mikrobiologie* **17**, 491–493.

Broecker, W. S. and Peng, T.-H. (1982). Tracers in the Sea. Eldigio, Palisades, N.Y.

Brotas, V., Risgaard-Petersen, N., Serôdio, J., Ottosen, L., Dalsgaard, T., and Ribeiro, J. (2003). *In situ* measurements of photosynthetic activity and respiration of intertidal benthic microalgal communities undergoing vertical migration. *Ophelia* **57**, 13–26.

Brouwers, G. J., Vijgenboom, E., Corstjens, P., De Vrind, J. P. M., and de Vrind-de Jong, E. W. (2000). Bacterial Mn^{2+} oxidizing systems and multicopper oxidases: An overview of mechanisms and functions. *Geomicrobiology Journal* **17**, 1–24.

Brown, J. R. and Doolittle, W. F. (1997). *Archaea* and the prokaryote-to-eukaryote transition. *Microbiology and Molecular Biology Review* **61**, 456–502.

Bruland, K. W., Donat, J. R., and Hutchins, D. A. (1991). Interactive influences of bioactive trace metals on biological production in oceanic waters. *Limnology and Oceanography* **36**, 1555–1577.

Brunet, R. C. and Garcia-Gil, L. J. (1996). Sulfide-induced dissimilatory nitrate reduction to ammonia in anaerobic freshwater sediments. *FEMS Microbiology Ecology* **21**, 131–138.

Bryan, B. A., Shearer, G., Skeeters, J. L., and Kohl, D. H. (1983). Variable expression of the nitrogen isotope effect associated with denitrification of nitrate. *The Journal of Biological Chemistry* **258**, 8613–8617.

Bryantseva, I. A., Gorlenko, V. M., Tourova, T. P., Kuznetsov, B. B., Lysenko, A. M., Bykova, S. A., Gal'chenko, V. F., Mityushina, L. L., and Osipov, G. A. (2000). *Heliobacterium sulfidophilum* sp. nov. and *Heliobacterium undosum* sp. nov.: Sulfide-oxidizing Heliobacteria from thermal sulfidic springs. *Microbiology* **69**, 325–334.

Brylinsky, M. (1977). Release of dissolved organic–matter by some marine macrophytes. *Marine Biology* **39**, 213–220.

Brzezinski, M. A. and Phillips, D. R. (1997). Evaluation of ^{32}Si as a tracer for measuring silica production rates in marine waters. *Limnology and Oceanography* **42**, 856–865.

Buchanan, B. B. (1992). Carbon dioxide assimilation in oxygenic and anoxygenic photosynthesis. *Photosynthesis Research* **33**, 147–162.

Buchholz-Cleven, B. E. E., Rattunde, B., and Straub, K. L. (1997). Screening for genetic diversity of isolates of anaerobic Fe(II)-oxidizing bacteria using DGGE and whole-cell hybridization. *Systematic and Applied Microbiology* **20**, 301–309.

Buckel, W. (1999). Anaerobic energy metabolism. *In* "Biology of the Prokaryotes" (J. W. Lengeler, G. Drews, and H. G. Schlegel, eds). Blackwell Science, New York.

Buesseler, K. O., Andrews, J. E., Pike, S. M., and Charette, M. A. (2004). The effects of iron fertilization on carbon sequestration in the Southern Ocean. *Science* **304**, 414–417.

Buffett, B. A. (2000). Clathrate hydrates. *Annual Review of Earth and Planetary Science* **28**, 477–507.

Buffle, J. (1990). The analytical challenge posed by fulvic and humic compounds. *Analytica Chimica Acta* **232**, 1–2.

Buick, R. (1992). The antiquity of oxygenic photosynthesis: Evidence from stromatolites in sulphate-deficient Archean lakes. *Nature* **255**, 74–77.

Bullen, T. D., White, A. F., Childs, C. W., Vivit, D. V., and Schulz, M. S. (2001). Demonstration of significant abiotic iron isotope fractionation in nature. *Geology* **29**, 699–702.

Burdige, D. J. (2002). Sediment pore waters. *In* "Biogeochemistry of Marine Dissolved Organic Matter" (D. Hansell and C. Carlson, eds), pp. 611–663. Academic Press, New York.

Burdige, D. J. and Gardner, K. G. (1998). Molecular weight distribution of dissolved organic carbon in marine sediment pore waters. *Marine Chemistry* **62**, 45–64.

Burdige, D. J. and Nealson, K. H. (1986). Chemical and microbiologcal studies of sulfide-mediated manganese reduction. *Geomicrobiological Journal* **4**, 361–387.

Burdige, D. J., Dhakar, S. P., and Nealson, K. H. (1992). Effects of manganese oxide mineralogy on microbial and chemical manganese reduction. *Geomicrobiology Journal* **10**, 27–48.

Burkholder, J. M. and Glasgow, H. B. (1997). The ichthyotoxic dinoflagellate, *Pfiesteria piscicidae*: Behavior, impacts, and environmental controls. *Limnology and Oceanography* **42**, 1052–1075.

Burns, R. G. and Burns, V. M. (1979). Manganese oxides. *In* "Marine Minerals. Reviews in Mineralogy" (R. G. Burns, ed.), pp. 1–46. Mineralogical Society of America, Washington, D.C.

Button, D. K. (1985). Kinetics of nutrient-limited transport and microbial growth. *Microbiological Reviews* **49**, 270–297.

Button, D. K. (1986). Affinity of organisms for substrate. *Limnology and Oceanography* **31**, 453–456.

Cai, W.-J., Zhao, P., and Wang, Y. (2000). pH and pCO_2 microelectrode measurements and the diffusive behavior of carbon dioxide species in coastal marine sediments. *Marine Chemistry* **70**, 133–148.

Cajal-Medrano, R. and Maske, H. (1999). Growth efficiency, growth rate and the remineralization of organic substrate by bacterioplankton—Revisiting the Pirt model. *Aquatic Microbial Ecology* **19**, 119–128.

Call, H. P. and Mücke, I. (1997). History, overview and applications of mediated lignolytic systems, especially laccase-mediator-systems (Lignozym(R)-process). *Journal of Biotechnology* **53**, 163–202.

Calvert, S. E. and Pedersen, T. F. (1992). Organic carbon accumulation and preservation in marine sediments: How important is anoxia? *In* "Organic Matter: Productivity, Accumulation, and Preservation in Recent and Ancient Sediments" (J. K. Whelan and J. W. Farrington, eds), pp. 231–263. Columbia University Press, New York.

Camacho, A. and Vicente, E. (1998). Carbon photoassimilation by sharply stratified phototrophic communities at the chemocline of Lake Arcaas (Spain). *FEMS Microbiology Ecology* **25**, 11–22.

Camacho, A., Erez, J., Chicote, A., Flórin, M., Squires, M. M., Lehmann, C., and Bachofen, R. (2001). Microbial microstratification, inorganic carbon

photoassimilation and dark carbon fixation at the chemocline of the meromictic Lake Cadagno (Switzerland) and its relevance to the food web. *Aquatic Sciences* **63**, 91–106.
Camilleri, J. C. and Ribi, G. (1986). Leaching of dissolved organic-carbon (DOC) from dead leaves, formation of flakes from DOC, and feeding on flakes by crustaceans in mangroves. *Marine Biology* **91**, 337–344.
Canfield, D. E. (1989a). Reactive iron in marine sediments. *Geochimica et Cosmochimica Acta* **53**, 619–632.
Canfield, D. E. (1989b). Sulfate reduction and oxic respiration in marine sediments: Implications for organic carbon preservation in euxinic environments. *Deep-Sea Research* **36**, 121–138.
Canfield, D. E. (1991). Sulfate reduction in deep-sea sediments. *American Journal of Science* **291**, 177–188.
Canfield, D. E. (1993). Organic matter oxidation in marine sediments. *In* "Interactions of C, N, P and S Biogeochemical Cycles and Global Change" (R. Wollast, F. T. Mackenzie, and L. Chou, eds), pp. 333–363. Springer, Berlin.
Canfield, D. E. (1994). Factors influencing organic carbon preservation in marine sediments. *Chemical Geology* **114**, 315–329.
Canfield, D. E. (1997). The geochemistry of river particulates from the continental United States: Major elements. *Geochimica et Cosmochimica Acta* **61**, 3349–3365.
Canfield, D. E. (1998). A new model for Proterozoic ocean chemistry. *Nature* **396**, 450–453.
Canfield, D. E. (2001). Biogeochemistry of sulfur isotopes. *In* "Reviews in Mineralogy and Geochemistry" (J. W. Valley and D. R. Cole, eds), pp. 607–636. Mineralogical Society of America, Blacksburg, VA.
Canfield, D. E. and Des Marais, D. J. (1991). Aerobic sulfate reduction in microbial mats. *Science* **251**, 1471–1473.
Canfield, D. E. and Des Marais, D. J. (1993). Biogeochemical cycles of carbon, sulfur, and free oxygen in a microbial mat. *Geochimica et Cosmochimica Acta* **57**, 3971–3984.
Canfield, D. E. and Des Marais, D. J. (1994). Cycling of carbon, sulfur, oxygen and nutrients in a microbial mat. *In* "Microbial Mats: Structure, Development and Environmental Significance" (L. Stal and P. Caumette, eds), pp. 255–263. Springer, Berlin.
Canfield, D. E. (2004). The evolution of the Earth surface sulfur reservoir. *American Journal of Science* **304**, 839–861.
Canfield, D. E. and Raiswell, R. (1991). Carbonate precipitation and dissolution: Its relevance to fossil preservation. *In* "Topics in Geobiology" (P. A. Allison and D. E. G. Briggs, eds), pp. 411–453. Plenum Press, New York.
Canfield, D. E. and Raiswell, R. (1999). The evolution of the sulfur cycle. *American Journal of Science* **299**, 697–723.
Canfield, D. E. and Teske, A. (1996). Late Proterozoic rise in atmospheric oxygen concentration inferred from phylogenetic and sulphur-isotope studies. *Nature* **382**, 127–132.
Canfield, D. E. and Thamdrup, B. (1994). The production of ^{34}S-depleted sulfide during bacterial disproportionation of elemental sulfur. *Science* **266**, 1973–1975.
Canfield, D. E., Raiswell, R., Westrich, J. T., Reaves, C. M., and Berner, R. A. (1986). The use of chromium reduction in the analysis of reduced inorganic sulfur in sediments and shales. *Chemical Geology* **54**, 149–155.
Canfield, D. E., Raiswell, R., and Bottrell, S. (1992). The reactivity of sedimentary iron minerals toward sulfide. *American Journal of Science* **292**, 659–683.

Canfield, D. E., Jørgensen, B. B., Fossing, H., Glud, R., Gundersen, J., Ramsing, N. B., Thamdrup, B., Hansen, J. W., Nielsen, L. P., and Hall, P. O. J. (1993a). Pathways of organic carbon oxidation in three continental margin sediments. *Marine Geology* **113**, 27–40.

Canfield, D. E., Thamdrup, B., and Hansen, J. W. (1993b). The anaerobic degradation of organic matter in Danish coastal sediments: Fe reduction, Mn reduction and sulfate reduction. *Geochimica et Cosmochimica Acta* **57**, 3867–3883.

Canfield, D. E., Thamdrup, B., and Fleischer, S. (1998). Isotope fractionation and sulfur metabolism by pure and enrichment cultures of elemental sulfur disproportionating bacteria. *Limnology and Oceanography* **43**, 253–264.

Canfield, D. E., Sørensen, K. B., and Oren, A. (2004). Biogeochemistry of a gypsum-encrusted microbial ecosystem. *Geobiology* **2**, 133–150.

Capone, D. G. (1988). Benthic nitrogen fixation. *In* "Nitrogen Cycling in Coastal Marine Environments" (T. H. Blackburn and J. Sørensen, eds), pp. 85–123. John Wiley and Sons, Chichester.

Capone, D. G. and Carpenter, E. J. (1982). Nitrogen fixation in the marine environment. *Science* **217**, 1140–1142.

Capone, D. G. and Kiene, R. P. (1988). Comparison of microbial dynamics in marine and freshwater sediments: Contrasts in anaerobic carbon catabolism. *Limnology and Oceanography* **33**, 725–749.

Capone, D. G., Zehr, J. P., Paerl, H. W., Bergman, B., and Carpenter, E. J. (1997). *Trichodesmium*, a globally significant marine cyanobacterium. *Science* **276**, 1221–1229.

Caraco, N. F., Cole, J. J., and Likens, G. E. (1989). Evidence for sulfate-controlled phosphorus from sediments of aquatic systems. *Nature* **341**, 316–318.

Carlsen, H. N., Jørgensen, L., and Degn, H. (1991). Inhibition by ammonia of methane utilization in *Methylococcus capsulatus* (Bath). *Applied Microbiology and Biotechnology* **35**, 124–127.

Carpenter, E. J., Harvey, H. R., Fry, B., and Capone, D. G. (1997). Biogeochemical tracers of the marine cyanobacterium *Trichodesmium*. *Deep-Sea Research I* **44**, 27–38.

Carpenter, P. D. and Smith, J. D. (1984). Effect of pH, iron and humic acid on the estuarine behaviour of phosphate. *Environmental Technology Letters* **6**, 65–72.

Castenholz, R. W. and Garcia-Pichel, F. (2000). Cyanobacterial responses to UV-radiation. *In* "The Ecology of Cyanobacteria" (B. A. Whitton and M. Potts, eds), pp. 591–611. Klüwer Academic Publishers, Dordrecht.

Castenholz, R. W. and Pierson, B. K. (1995). Ecology of thermophilic anoxygenic phototrophs. *In* "Anoxygenic Photosynthetic Bacteria" (R. E. Blankenship, M. T. Madigan, and C. E. Bauer, eds), pp. 87–103. Kluwer Academic Publishers, Dordrecht.

Castenholz, R., Bauld, J., and Pierson, B. (1992). Photosynthetic activity in modern microbial mat-building communities. *In* "The Proterozoic Biosphere. A Multidisciplinary Study" (J. W. Schopf and C. Klein, eds), pp. 279–285. The Press Syndicate of the University of Cambridge, Cambridge.

Castresana, J. and Saraste, M. (1995). Evolution of energetic metabolism: The respiration-early hypothesis. *Trends in Biochemical Sciences* **20**, 443–448.

Castro, H. F., Williams, N. H., and Ogram, A. (2000). Phylogeny of sulfate-reducing bacteria. *FEMS Microbiology Ecology* **31**, 1–9.

Cavanaugh, C. M. (1983). Symbiotic chemotrophic bacteria in marine invertebrates from sulfide rich habitats. *Nature* **302**, 58–61.

Cavanaugh, C. M. (1993). Methanotroph-invertebrate symbioses in the marine environment: Ultrastructural, biochemical and molecular studies. *In* "Microbial

Growth on C_1 Compounds" (J. C. Murrell and D. P. Kelly, eds), pp. 315–328. Intercept Press Ltd., Andover, United Kingdom.
Cavanaugh, C. M., Gardiner, S. L., Jones, M. L., Jannasch, H. W., and Waterbury, J. B. (1981). Prokaryotic cells in the hydrothermal vent tube worm Riftia-Pachyptila Jones – possible chemoautotrophic symbionts. *Science* **213**, 340–342.
Cavanaugh, C. M., Levering, P. R., Maki, J. S., Michell, R., and Lidstrom, M. E. (1987). Symbiosis of methylotrophic bacteria and deep-sea mussels. *Nature* **325**, 346–348.
Chambers, L. A. and Trudinger, P. A. (1975). Are thiosulfate and tetrathionate intermediates in disimilatory sulfate reduction? *Journal of Bacteriology* **123**, 36–40.
Chambers, R. M. (1997). Porewater chemistry associated with *Phragmites* and *Spartina* in a Connecticut tidal marsh. *Wetlands* **17**, 360–367.
Chameides, W. L. and Perdue, E. M. (1997). Biogeochemical Cycles. A Computer-Interactive Study of Earth System Science and Global Change. Oxford University Press, New York.
Chanton, J. P. (1985). Sulfur mass balance and isotopic fractionation in an anoxic marine sediment. Ph.D. Thesis, University of North Carolina, Chapel Hill.
Chaudhuri, S. K., Lack, J. G., and Coates, J. D. (2001). Biogenic magnetite formation through anaerobic biooxidation of Fe(II). *Applied and Environmental Microbiology* **67**, 2844–2848.
Chen, R. H. and Twilley, R. R. (1999). A simulation model of organic matter and nutrient accumulation in mangrove wetland soils. *Biogeochemistry* **44**, 93–118.
Chen, Y.-S. R., Butler, J. N., and Stumm, W. (1973). Adsorption of phosphate on alumina and kaolinite from dilute aqueous solutions. *Journal of Colloid and Interface Science* **43**, 421–436.
Cherrier, J., Bauer, J. E., Druffel, E. R. M., Coffin, R. B., and Chanton, J. P. (1999). Radiocarbon in marine bacteria: Evidence for the ages of assimilated carbon. *Limnology and Oceanography* **44**, 730–736.
Chidthaisong, A., Chin, K.-J., Valentine, D. L., and Tyler, S. C. (2002). A comparison of isotope fractionation of carbon and hydrogen from paddy field rice roots and soil bacterial enrichments during CO_2/H_2 methanogenesis. *Geochimica et Cosmochimica Acta* **66**, 983–995.
Chien, Y. T. and Zinder, S. H. (1994). Cloning, DNA-sequencing, and characterization of a *nifD*-homologous gene from the archaeon *Methanosarcina-Barkeri* 227 which resembles *nifD1* from the eubacterium *Clostridium pasteurianum*. *Journal of Bacteriology* **176**, 6590–6598.
Chien, Y. T. and Zinder, S. H. (1996). Cloning, functional organization, transcript studies, and phylogenetic analysis of the complete nitrogenase structural genes (*nifHDK2*) and associated genes in the archaeon *Methanosarcina barkeri* 227. *Journal of Bacteriology* **178**, 143–148.
Childers, S. E. and Lovley, D. R. (2001). Differences in Fe(III) reduction in the hyperthermophilic archaeon, *Pyrobaculum islandicum*, versus mesophilic Fe(III)-reducing bacteria. *FEMS Microbiology Letters* **195**, 253–258.
Childers, S. E., Ciufo, S., and Lovley, D. R. (2002). Geobacter metallireducens accesses insoluble Fe(III) oxide by chemotaxis. *Nature* **416**, 767–769.
Childress, J. J., Fischer, C. R., Brooks, J. M., Kennicut, M. C., Bidigare, R., and Anderson, A. E. (1986). A methonotrophic marine molluscan (Bivalvia *Mytilidae*) symbiosis: Mussels fueled by gas. *Science* **233**, 1306–1308.
Chisholm, S. W., Olson, R. J., Zettler, E. R., Goericke, R., Waterbury, J. B., and Welschmeyer, N. A. (1988). A novel free-living prochlorophyte abundant in the oceanic euphotic zone. *Nature* **334**, 340–343.

Christensen, B. (1978). Biomass and primary production of *Rhizophora-Apiculata* Bl. in a mangrove in southern Thailand. *Aquatic Botany* **4**, 43–52.
Christensen, J. P. and Rowe, G. T. (1984). Nitrification and oxygen consumption in northwest Atlantic deep-sea sediments. *Journal of Marine Research* **42**, 1099–1116.
Christensen, J. P., Smethie, W. M., and Devol, A. H. (1987). Benthic nutrient regeneration and denitrification on the Washington continental shelf. *Deep-Sea Research* **34**, 1027–1047.
Christensen, P. B., Revsbech, N. P., and Sand-Jensen, K. (1994). Microsensor analysis of oxygen in the rhizosphere of the aqutic macrophyte *Litorella uniflora* (L.) Ascherson. *Plant Physiology* **105**, 847–852.
Christensen, P. B., Rysgaard, S., Sloth, N. P., Dalsgaard, T., and Schwaerter, S. (2000). Sediment mineralization, nutrient fluxes, denitrification and dissimilatory nitrate reduction to ammonium in an estuarine fjord with sea cage trout farms. *Aquatic Microbial Ecology* **21**, 73–84.
Christianson, D. W. (1997). Structural chemistry and biology of manganese metalloenzymes. *Progress in Biophysics and Molecular Biology* **67**, 217–252.
Chrost, R. H. and Faust, M. A. (1983). Organic-carbon release by phytoplankton: Its composition and utilization by bacterioplankton. *Journal of Plankton Research* **5**, 477–493.
Clark, L. L., Ingall, E. D., and Benner, R. (1999). Marine organic phosphorus cycling: Novel insights from nuclear magnetic resonance. *American Journal of Science* **299**, 724–737.
Clarke, G. D. P., Beiko, R. G., Ragan, M. A., and Charlebois, R. L. (2002). Inferring genome trees by using a filter to eliminate phylogenetically discordant sequences and a distance matrix based on mean normalized BLASTP scores. *Journal of Bacteriology* **184**, 2072–2080.
Clarke, P. J. and Allaway, W. G. (1993). The regeneration niche of the gray mangrove (*Avicennia marina*): Effects of salinity, light and sediment factors on establishment, growth and survival in the field. *Oecologia* **93**, 548–556.
Cline, J. D. and Kaplan, I. R. (1975). Isotopic fractionation of dissolved nitrate during denitrification in the eastern tropical North Pacific Ocean. *Marine Chemistry* **3**, 271–299.
Clough, B. F., Tan, D. T., Phuong, D. X., and Buu, D. C. (2000). Canopy leaf area index and litter fall in stands of mangrove Rhizophora apiculata of different age in the Mekong Delta, Vietnam. *Aquatic Botany* **66**, 311–320.
Coale, K. H., Johnson, K. S., Fitzwater, S. E., Gordon, R. M., Tanner, S., Chavez, F. P., Ferioli, L., Sakamoto, C., Rogers, P., Millero, F., Steinberg, P., Nightingale, P., Cooper, D., Cochlan, W. P., Landry, M. R., Constantinou, J., Rollwagen, G., Trasvina, A., and Kudela, R. (1996). A massive phytoplankton bloom induced by an ecosystem-scale iron fertilization experiment in the equatorial Pacific Ocean. *Nature* **383**, 495–501.
Coale, K. H., Johnson, K. S., Chavez, F. P., Buesseler, K. O., Barber, R. T., Brzezinski, M. A., Cochlan, W. P., Millero, F. J., Falkowski, P. G., Bauer, J. E., Wanninkhof, R. H., Kudela, R. M., Altabet, M. A., Bidigare, R. R., Wang, X., Chase, Z., Strutton, P. G., Friederich, G. E., Gorbunov, M. Y., Lance, V. P., Hilting, A. K., Hiscock, M. R., Demarest, M., Hiscock, W. T., Sullivan, K. F., Tanner, S. J., Gordon, M., Hunter, C. N., Elrod, V. A., Fitzwater, S. E., Jones, J. L., Tozzi, S., Koblizek, M., Roberts, A. E., Herndon, J., Brewster, J., Ladizinsky, N., Smith, G., Cooper, D., Timothy, D., Brown, S. L., Selph, K. E., Sheridan, C. C., Twining, B. S., and Johnson, Z. I. (2004). Southern ocean iron

enrichment experiment: Carbon cycling in high- and low-Si waters. *Science* **304**, 408–414.

Coates, J. D., Ellis, D. J., Blunt-Harris, E. L., Gaw, C. V., Roden, E. E., and Lovley, D. R. (1998). Recovery of humic-reducing bacteria from a diversity of environments. *Applied and Environmental Microbiology* **64**, 1504–1509.

Codispoti, L. A. and Christensen, J. P. (1985). Nitrification, denitrification and nitrous-oxide cycling in the eastern tropical South-Pacific Ocean. *Marine Chemistry* **16**, 277–300.

Codispoti, L. A., Brandes, J. A., Christensen, J. P., Devol, A. H., Naqvi, S. W. A., Paerl, H. W., and Yoshinary, T. (2001). The oceanic fixed nitrogen and nitrous oxide budgets: Moving targets as we enter the anthropocene? *Scientia Marina* **65**(Suppl. 2), 85–105.

Cohen, Y., Jørgensen, B. B., Padan, E., and Shilo, M. (1975). Sulphide-dependent anoxygenic photosynthesis in the cyanobacterium *Oscillatoria limnetica*. *Nature* **257**, 489–492.

Cohen, Y., Jørgensen, B. B., Revsbech, N. P., and Poplawski, R. (1986). Adaptation to hydrogen sulfide of oxygenic and anoxygenic photosynthesis among cyanobacteria. *Applied and Environmental Microbiology* **51**, 398–407.

Coleman, D. C., Risatti, J. B., and Schoell, M. (1981). Fractionation of carbon and hydrogen isotopes by methane-oxidizing bacteria. *Geochimica et Cosmochimica Acta* **45**, 1033–1037.

Colman, A. S. and Holland, H. D. (2000). The global diagenetic flux of phosphorus from marine sediments to the oceans: Redox sensitivity and the control of atmospheric oxygen levels. *SEMP (Society of Sedimentary Geology) Special Publication* **66**, 53–75.

Compton, J., Mallinson, D., Glenn, C. R., Filippelli, G., Föllmi, K., Shields, G., and Zanin, Y. (2000). Variations in the global phosphorus cycle. *SEMP (Society of Sedimentary Geology) Special Publication* **66**, 21–33.

Conley, D. J. (2002). Terrestrial ecosystems and the global biogeochemical silica cycle. *Global Biogeochemical Cycles* **16**, Art. No. 1121.

Conley, D. J., Kilham, S. S., and Theriot, E. (1989). Differences in silica content between marine and freshwater diatoms. *Limnology and Oceanography* **34**, 205–213.

Conley, D. J., Schelske, C. L., and Stoermer, E. F. (1993). Modification of the biogeochemical cycle of silica with eutrophication. *Marine Ecology-Progress Series* **101**, 179–192.

Conley, D. J., Stalnacke, P., Pitkanen, H., and Wilander, A. (2000). The transport and retention of dissolved silicate by rivers in Sweden and Finland. *Limnology and Oceanography* **45**, 1850–1853.

Conrad, R. (1999). Contribution of hydrogen to methane production and control of hydrogen concentrations in methanogenic soils and sediments. *FEMS Microbiology Ecology* **28**, 193–202.

Conrad, R. and Rothfuss, F. (1991). Methane oxidation in the soil surface-layer of a flooded rice field and the effect of ammonium. *Biology and Fertility of Soils* **12**, 28–32.

Conrad, R., Schink, B., and Phelps, T. J. (1986). Thermodynamics of H_2-consuming and H_2-producing metabolic reactions in diverse methanogenic environments under *in situ* conditions. *FEMS Microbiology Letters* **38**, 353–360.

Contreras, M. L. and Escamilla, E. (1999). An unusual cytochrome o'-type cytochrome c oxidase in a *Bacillus cereus* cytochrome a_3 mutant has a very high affinity for oxygen. *Microbiology* **145**, 1563–1573.

Cook, P. J., Shergold, J. H., Burnett, W. C., and Riggs, S. R. (1990). Phosphorite research: A historical overview. *Geological Society of London, Special Publication* **52**, 1–22.

Corliss, J. B., Dymond, J., Gordon, L. I., Edmond, J. M., von Herzen, R. P., Ballard, R. D., Green, K., Williams, D., Bainbridge, A., Crane, K., and van Andel, T. H. (1979). Submarine thermal springs on the Galapagos Rift. *Science* **203**, 1073–1082.

Cornell, R. M. and Schwertmann, U. (1996). The Iron Oxides, p. 573. VCH, Weinheim.

Corredor, J. E. and Morell, J. M. (1994). Nitrate depuration of secondary sewage effluents in mangrove sediments. *Estuaries* **17**, 295–300.

Cotner, J. B. and Wetzel, R. G. (1992). Uptake of dissolved inorganic and organic phosphorus compounds by phytoplankton and bacterioplankton. *Limnology and Oceanography* **37**, 232–243.

Cottrell, M. T. and Kirchman, D. L. (2000). Community composition of marine bacterioplankton determined by 16S rRNA gene clone libraries and fluorescence in situ hybridization. *Applied and Environmental Microbiology* **66**, 5116–5122.

Cottrell, M. T. and Kirchman, D. L. (2003). Contribution of major bacterial groups to bacterial biomass production (thymidine and leucine incorporation) in the Delaware estuary. *Limnology and Oceanography* **48**, 168–178.

Coyne, M. S., Arunakumari, A., Averill, B. A., and Tiedje, J. M. (1989). Immunological identification and distribution of dissimilatory heme *cd1* and non-heme copper nitrite reductases in denitrifying bacteria. *Applied and Environmental Microbiology* **55**, 2924–2931.

Craig, H. (1961). Isotopic variations in meteoric waters. *Science* **133**, 1702–1703.

Crespi, B. J. (2001). The evolution of social behavior in microorganisms. *Trends in Ecology and Evolution* **16**, 178–183.

Crill, P. M. and Martens, C. S. (1983). Spatial and temporal fluctuations of methane production in anoxic coastal marine sediments. *Limnology and Oceanography* **28**, 1117–1130.

Crill, P. M. and Martens, C. S. (1986). Methane production from bicarbonate and acetate in an anoxic marine sediment. *Geochimica et Cosmochimica Acta* **50**, 2089–2097.

Croal, L. R., Johnson, C. M., Beard, B. L., and Newman, D. K. (2004). Iron isotope fractionation by Fe(II)-oxidizing photoautotrophic bacteria. *Geochimica et Cosmochimica Acta* **68**, 1227–1242.

Culver, D. A. and Brunskill, G. J. (1969). Fayetteville Green Lake, New York. V. Studies of primary production and zooplankton in a meromictic marl lake. *Limnology and Oceanography* **14**, 862–873.

Cypionka, H. (1994). Novel metabolic capacities of sulfate-reducing bacteria, and their activities in microbial mats. *In* "Microbial Mats: Structure, Development, and Environmental Significance" (L. J. Stal and P. Caumette, eds). Springer, Berlin.

Cypionka, H. (1995). Solute transport and cell energetics. *In* "Sulfate-Reducing Bacteria" (L. L. Barton, ed.), pp. 151–184. Plenum Press, New York.

Cypionka, H. (2000). Oxygen respiration by *Desulfovibrio* species. *Annual Review of Microbiology* **54**, 827–848.

Cypionka, H., Smock, A. M., and Bottcher, M. E. (1998). A combined pathway of sulfur compound disproportionation in *Desulfovibrio desulfuricans*. *FEMS Microbiology Letters* **166**, 181–186.

Czepiel, P. M., Crill, P. M., and Harriss, R. C. (1995). Environmental-factors influencing the variability of methane oxidation in temperate zone soils. *Journal of Geophysical Research-Atmospheres* **100**, 9359–9364.

Dacey, J. W. H. and Klug, M. J. (1979). Methane efflux from lake-sediments through water lilies. *Science* **203**, 1253–1255.

Dahl, C. and Trüper, H. G. (1994). Enzymes of dissimilatory sulfide oxidation in phototrophic sulfur bacteria. *In* "Methods in Enzymology. Inorganic Microbial Sulfur Metabolism" (H. D. Peck, Jr and J. LeGall, eds), pp. 400–421. Academic Press, San Diego.

Dalsgaard, T. and Thamdrup, B. (2002). Factors controlling anaerobic ammonium oxidation with nitrite in marine sediments. *Applied and Environmental Microbiology* **68**, 3802–3808.

Dalsgaard, T., Canfield, D. E., Petersen, J., Thamdrup, B., and Acuña-Gonzalez, J. (2003). N_2 production by the anammox reaction in the anoxic water column of Golfo Dulce, Costa Rica. *Nature* **422**, 606–608.

Damgaard, L. R., Revsbech, N. P., and Reichardt, W. (1998). Use of an oxygen-insensitive microscale biosensor for methane to measure methane concentration profiles in a rice paddy. *Applied and Environmental Microbiology* **64**, 864–870.

Das, A., Mishra, A. K., and Roy, P. (1992). Anaerobic growth on elemental sulfur using dissimilar iron reduction by autotrophic *Thiobacillus ferrooxidans*. *FEMS Microbiology Letters* **97**, 167–172.

Davidovich, N. A. (1994). Factors controlling the size of initial cells in diatoms. *Russian Journal of Plant Physiology* **41**, 220–224.

Davies, S. H. R. and Morgan, J. J. (1989). Manganese(II) oxidation kinetics on metal oxide surfaces. *Journal of Colloid and Interface Science* **129**, 63–77.

de Bie, M. J. M., Speksnijder, A., Kowalchuk, G. A., Schuurman, T., Zwart, G., Stephen, J. R., Diekmann, O. E., and Laanbroek, H. J. (2001). Shifts in the dominant populations of ammonia-oxidizing beta-subclass Proteobacteria along the eutrophic Schelde estuary. *Aquatic Microbial Ecology* **23**, 225–236.

de Boer, W. and Laanbroek, H. J. (1989). Ureolytic nitrification at low pH by *Nitrosospira* species. *Archives of Microbiology* **152**, 178–181.

Dedysh, S. N., Liesack, W., Khmelenina, V. N., Suzina, N. E., Trotsenko, Y. A., Semrau, J. D., Bares, A. M., Panikov, N. S., and Tiedje, J. M. (2000). *Methylocella palustris* gen. nov., sp nov., a new methane-oxidizing acidophilic bacterium from peat bags, representing a novel subtype of serine-pathway methanotrophs. *International Journal of Systematic and Evolutionary Microbiology* **50**, 955–969.

Dedysh, S. N., Khmelenina, V. N., Suzina, N. E., Trotsenko, Y. A., Semrau, J. D., Liesack, W., and Tiedje, J. M. (2002). *Methylocapsa acidiphila* gen. nov., sp. nov., a novel methane-oxidizing and dinitrogen-fixing acidophilic bacterium from *Sphagnum* bog. *International Journal of Systematic and Evolutionary Microbiology* **52**, 251–261.

Deer, W. A., Howie, R. A., and Zussman, J. (1966). An Introduction to the Rock-Forming Minerals, p. 528. Longmans, London.

del Giorgio, P. A. and Cole, J. J. (1998). Bacterial growth efficiency in natural aquatic systems. *Annual Review of Ecology and Systematics* **29**, 503–541.

del Giorgio, P. A. and Duarte, C. M. (2002). Respiration in the open ocean. *Nature* **420**, 379–384.

Delaney, M. L. (1998). Phosphorus accumulation in marine sediments and the oceanic phosphorus cycle. *Global Biogeochemical Cycles* **12**, 563–572.

DeLiong, E. F. (1992). Archaea in coastal marine environments. *Proceedings of the National Academy of Sciences USA* **89**, 5685–5689.

Delong, E. F., Wu, K. Y., Prezelin, B. B., and Jovine, R. V. (1994). High abundance of Archaea in antarctic marine picoplankton. *Nature* **371**, 695–697.

Delwiche, C. C. (1970). The nitrogen cycle. *Scientific American* **223**, 136–146.

Delwiche, C. C. and Steyn, P. L. (1970). Nitrogen isotope fractionation in soils and microbial reactions. *Environmental Science & Technology* **4**, 929–935.

DeMaster, D. J. (2002). The accumulation and cycling of biogenic silica in the Southern Ocean: Revisiting the marine silica budget. *Deep-Sea Research Part II-Topical Studies in Oceanography* **49**, 3155–3167.

DeNiro, M. J. and Epstein, S. (1981). Influence of diet on the distribution of nitrogen isotopes in animals. *Geochimica et Cosmochemica Acta* **45**, 341–351.

Des Marais, D. J. and Canfield, D. E. (1994). The carbon isotope biogeochemistry of microbial mats. *In* "Structure, Development and Environmental Significance of Microbial Mats" (L. J. Stal and P. Caumette, eds), pp. 289–298. Springer, Berlin.

Detmers, J., Brüchert, V., Habicht, K. S., and Kuever, J. (2001). Diversity of sulfur isotope fractionations by sulfate-reducing prokaryotes. *Applied and Environmental Microbiology* **67**, 888–894.

Devereux, R. and Mundform, G. W. (1994). A phylogenetic tree of 16S rRNA sequences from sulfate reducing bacteria in a sandy marine sediment. *Applied and Environmental Microbiology* **60**, 3437–3439.

Devereux, R. and Stahl, D. A. (1993). Phylogeny of sulfate-reducing bacteria and a perspective for analyzing their natural communities. *In* "The Sulfate-Reducing Bacteria: Contemporary Perspectives" (J. M. Odom and R. Singleton, Jr., eds), pp. 131–160. Springer, New York.

Devol, A. H. (1983). Methane oxidation rates in the anaerobic sediments of Saanich Inlet. *Limnology and Oceanography* **28**, 738–742.

Devol, A. H., Anderson, J. J., Kuivila, K., and Murray, J. W. (1984). A model for coupled sulfate reduction and methane oxidation in the sediments of Saanich Inlet. *Geochimica et Cosmochimica Acta* **48**, 993–1004.

de Vries, S. and Schröder, I. (2002). Comparison between the nitric oxide reductase family and its aerobic relatives, the cytochrome oxidases. *Biochemical Society Transactions* **30**, 662–667.

de Vrind, J. P. M., Boogerd, F. C., and de Vrind-de Jong, E. W. (1986). Manganese reduction by a marine *Bacillus* species. *Journal of Bacteriology* **167**, 30–34.

D'Hondt, S., Rutherford, S., and Spivack, A. J. (2002). Metabolic activity of subsurface life in deep-sea sediments. *Science* **295**, 2067–2070.

DiChristina, T. J. and DeLong, E. F. (1993). Design and application of rRNA-targeted oligonucleotide probes for the dissimilatory iron- and manganese-reducing bacterium *Shewanella putrefaciens*. *Applied and Environmental Microbiology* **59**, 4152–4160.

DiChristina, T. J., Moore, C. M., and Haller, C. A. (2002). Dissimilatory Fe(III) and Mn(IV) reduction by *Shewanella putrefaciens* requires ferE, a homolog of the pulE (gspE) type II protein secretion gene. *Journal of Bacteriology* **184**, 142–151.

Dickens, G. R., Castillo, M. M., and Walker, J. C. G. (1997). A blast of gas in the latest Paleocene: Simulating first-order effects of massive dissociation of oceanic methane hydrate. *Geology* **25**, 259–262.

Dilling, W. and Cypionka, H. (1990). Aerobic respiration in sulfate-reducing bacteria. *FEMS Microbiology Letters* **71**, 123–128.

Dittmann, S. (1996). Effects of macrobenthic burrows on infaunal communities in tropical tidal flats. *Marine Ecology-Progress Series* **134**, 119–130.

Dittmar, T. and Lara, R. J. (2001). Molecular evidence for lignin degradation in sulfate-reducing mangrove sediments (Amazonia, Brazil). *Geochimica et Cosmochimica Acta* **65**, 1417–1428.

Dixit, S. and Van Cappellen, P. (2002). Surface chemistry and reactivity of biogenic silica. *Geochimica et Cosmochimica Acta* **66**, 2559–2568.

Dixit, S., Van Cappellen, P., and van Bennekom, A. J. (2001). Processes controlling solubility of biogenic silica and pore water build-up of silicic acid in marine sediments. *Marine Chemistry* **73**, 333–352.

D'mello, R., Hill, S., and Poole, R. K. (1996). The cytochrome *bd* quinol oxidase in *Escherichia coli* has an extremely high oxygen affinity and two oxygen-binding haems: Implications for regulation of activity *in vivo* by oxygen inhibition. *Microbiology* **142**, 755–763.

Dole, M., Lane, G. A., Rudd, D. P., and Zaukelies, D. A. (1954). Isotopic composition of atmospheric oxygen and nitrogen. *Geochimica et Cosmochimica Acta* **6**, 65–78.

Doolittle, R. F., Feng, D. F., Tsang, S., Cho, G., and Little, E. (1996). Determining divergence times of major kingdoms of living organisms with a protein clock. *Science* **271**, 470–477.

Doolittle, W. F. (1999). Phylogenetic classification and the universal tree. *Science* **284**, 2124–2128.

Dow, C. S. and Swoboda, U. K. (2000). Cyanotoxins. *In* "The Ecology of Cyanobacteria" (B. A. Whitton and M. Potts, eds), pp. 613–632. Klüwer Academic Publishers, Dordrecht.

Dubilier, N., Mülders, C., Ferdelman, T., de Beer, D., Pernthaler, A., Thiermann, F., Krieger, J., Giere, O., and Amann, R. (2001). Endosymbiotic sulphate-reducing and sulphide-oxidizing bacteria in an oligochaete worm. *Nature* **411**, 298–302.

Duce, R. A. and Tindale, N. W. (1991). Atmospheric transport of iron and its deposition in the ocean. *Limnology and Oceanography* **36**, 1715–1726.

Ducklow, H. W. and Carlson, C. A. (1992). Oceanic bacterial production. *In* "Advances in Microbial Ecology" (K. C. Marshall, ed.), pp. 113–181. Plenum Press, New York.

Duffy, P. B., Eby, M., and Weaver, A. J. (2001). Climate model simulations of effects of increased atmospheric CO_2 and loss of sea ice on ocean salinity and tracer uptake. *Journal of Climate* **14**, 520–532.

Dugdale, R. C. and Goering, J. J. (1967). Uptake of new and regenerated forms of nitrogen in primary productivity. *Limnology and Oceanography* **12**, 196–206.

Dugdale, R. C. and Wilkerson, F. P. (2001). Sources and fates of silicon in the ocean: The role of diatoms in the climate and glacial cycles. *Scientia Marina* **65**, 141–152.

Dworkin, M. (1992). Prokaryotic life cycles. *In* "The Prokaryotes" (A. Balows, H. G. Trüper, W. Harder, M. Dworkin, and K.-H. Schleifer, eds), pp. 209–240. Springer, New York.

Dye, A. H. (1983). Oxygen consumption by sediments in a southern African mangrove swamp. *Estuarine, Coastal and Shelf Science* **17**, 473–478.

Dye, A. H. and Lasiak, T. A. (1987). Assimilation efficiencies of fiddler-crabs and deposit-feeding gastropods from tropical mangrove sediments. *Comparative Biochemistry and Physiology* **87A**, 341–344.

Eady, R. R. (1996). Structure-function relationships of alternative nitrogenases. *Chemical Reviews* **96**, 3013–3030.

Edwards, K. J., Bond, P. L., Gihring, T. M., and Banfield, J. F. (2000). An archaeal iron-oxidizing extreme acidophile important in acid mine drainage. *Science* **287**, 1796–1799.

Edwards, K. J., Rogers, D. R., Wirsen, C. O., and McCollom, T. M. (2003). Isolation and characterization of novel psychrophilic, neutrophilic, Fe-oxidizing, chemolithoautotrophic a- and g-*proteobacteria* from the deep sea. *Applied and Environmental Microbiology* **69**, 2906–2913.

Egge, J. K. and Aksnes, D. L. (1992). Silicate as regulating nutrient in phytoplankton competition. *Marine Ecology-Progress Series* **83**, 281–289.

Ehrenberg, C. G. (1836). Vorläufige Mitteilungen über das Wirklige Vorkommen fossiler Infusorien und ihre große Verbreitung. *Poggendorfs Annalen der Physik und Chemie* **38**, 213–227.

Ehrenreich, A. and Widdel, F. (1994). Anaerobic oxidation of ferrous iron by purple bacteria, a new type of phototrophic metabolism. *Applied and Environmental Microbiology* **60**, 4517–4526.

Ehrlich, H. L. (1996). Geomicrobiology, 3rd Ed. Marcel Dekker, New York.

Eilers, H., Pernthaler, J., Glöckner, F. O., and Amann, R. (2000). Culturability and *in situ* abundance of pelagic bacteria from the North Sea. *Applied and Environmental Microbiology* **66**, 3044–3051.

Eisenmann, E., Beuerle, J., Sulger, K., Kroneck, P. M. H., and Schumacher, W. (1995). Lithotrophic growth of sulfurospirillum-deleyianum with sulfide, as electron-donor coupled to respiratory reduction of nitrate to ammonia. *Archives of Microbiology* **164**, 180–185.

Ellis, J. E., Williams, A. G., and Lloyd, D. (1989). Oxygen consumption by ruminal microorganisms: Protozoal and bacterial conditions. *Applied and Environmental Microbiology* **55**, 2583–2587.

Emerson, D. (2000). Microbial oxidation of Fe(II) and Mn(II) at circumneutral pH. *In* "Environmental Metal-Microbe Interactions" (D. R. Lovley, ed.), pp. 31–52. ASM Press, Washington, D.C.

Emerson, D. and Moyer, C. (1997). Isolation and characterization of novel iron-oxidizing bacteria that grow at circumneutral pH. *Applied and Environmental Microbiology* **63**, 4784–4792.

Emerson, D. and Moyer, C. L. (2002). Neutrophilic Fe-oxidizing bacteria are abundant at the Loihi Seamount hydrothermal vents and play a major role in Fe oxide deposition. *Applied and Environmental Microbiology* **68**, 3085–3093.

Emerson, D. and Revsbech, N. P. (1994a). Investigation of an iron-oxidizing microbial mat community located near Aarhus, Denmark: Field studies. *Applied and Environmental Microbiology* **60**, 4022–4031.

Emerson, D. and Revsbech, N. P. (1994b). Investigation of an iron-oxidizing microbial mat community located near Aarhus, Denmark: Laboratory studies. *Applied and Environmental Microbiology* **60**, 4032–4038.

Emerson, S., Kalhorn, S., Jacobs, S., Tebo, B. M., Nealson, K., and Rosson, R. A. (1982). Environmental oxidation rate of manganese(II): Bacterial catalysis. *Geochimica et Cosmochimica Acta* **46**, 1073–1079.

Epping, E. and Kühl, M. (2000). The response of photosynthesis and oxygen consumption to short-term changes in temperature and irradiance in a cyanobacterial mat (Ebro Delta, Spain). *Environmental Microbiology* **2**, 465–474.

Ernstsen, V., Binnerup, S. J., and Sørensen, J. (1998). Reduction of nitrate in clayey subsoils controlled by geochemical and microbial barriers. *Geomicrobiology Journal* **15**, 195–207.

Evans, M. C. W., Buchanan, B. B., and Arnon, D. I. (1966). A new ferrodoxin-dependent carbon reduction cycle in a photosynthetic bacterium. *Proceedings of the National Academy of Sciences USA* **55**, 928–934.

Falkowski, P. G. (1997). Evolution of the nitrogen cycle and its influence on the biological sequestration of CO_2 in the ocean. *Nature* **387**, 272–275.

Falkowski, P. G. and Raven, J. A. (1997). Aquatic Photosynthesis, p. 375. Blackwell Science, Malden, MA.

Falkowski, P., Scholes, R. J., Boyle, E., Candell, J., Canfield, D. E., Elser, J., Gruber, N., Hibbard, K., Hogsberg, P., Linder, S., Mackenzie, F. T., Moore, B., Pedersen, T., Rosenthal, Y., Seitzinger, S., Smetacek, V., and Steffen, W. (2000). Global carbon cycle: A test of our knowledge of earth as a system. *Science* **290**, 291–296.

Fani, R., Gallo, R., and Lio, P. (2000). Molecular evolution of nitrogen fixation: The evolutionary history of the *nifD, nifK, nifE*, and *nifN* genes. *Journal of Molecular Evolution* **51**, 1–11.

Farquhar, G. D., Ehleringer, J. R., and Hubick, K. T. (1989). Carbon isotope discrimination and photosynthesis. *Annual Review of Plant Physiology and Plant Molecular Biology* **40**, 503–507.

Farquhar, G. D., Lloyd, J., Taylor, J. A., Flanagan, L. B., Syvertsen, J. P., Hubick, K. T., Wong, S. C., and Ehleringer, J. R. (1993). Vegetation effects on the isotope composition of oxygen in atmospheric CO_2. *Nature* **363**, 439–443.

Farquhar, J., Bao, H. M., and Thiemens, M. (2000). Atmospheric influence of Earth's earliest sulfur cycle. *Science* **289**, 756–758.

Farquhar, J., Wing, B. A., McKeegan, K. D., Harris, J. W., Cartigny, P., and Thiemens, M. H. (2002). Mass-independent sulfur of inclusions in diamond and sulfur recycling on early Earth. *Science* **298**, 2369–2372.

Farrell, J. W., Pedersen, T. F., Calvert, S. E., and Nielsen, B. (1995). Glacial-interglacial changes in nutrient utilization in the equatorial Pacific Ocean. *Nature* **377**, 514–517.

Fasham, M. J. R., Balino, B. M., Bowles, M. C., Anderson, R., Archer, D., Bathmann, U., Boyd, P., Buesseler, K., Burkill, P., Bychkov, A., Carlson, C., Chen, C. T. A., Doney, S., Ducklow, H., Emerson, S., Feely, R., Feldman, G., Garcon, V., Hansell, D., Hanson, R., Harrison, P., Honjo, S., Jeandel, C., Karl, D., Le Borgne, R., Liu, K. K., Lochte, K., Louanchi, F., Lowry, R., Michaels, A., Monfray, P., Murray, J., Oschlies, A., Platt, T., Priddle, J., Quinones, R., Ruiz-Pino, D., Saino, T., Sakshaug, E., Shimmield, G., Smith, S., Smith, W., Takahashi, T., Treguer, P., Wallace, D., Wanninkhof, R., Watson, A., Willebrand, J., and Wong, C. S. (2001). A new vision of ocean biogeochemistry after a decade of the Joint Global Ocean Flux Study (JGOFS). *Ambio Special Issue* **10**, 4–31.

Faure, G. (1986). Principles in Istotope Geology, p. 608. John Wiley & Sons, New York.

Feely, R. A., Trefry, J. H., Lebon, G. T., and German, C. R. (1998). The relationship between P/Fe and V/Fe ratios in hydrothermal precipitates and dissolved phosphate in seawater. *Geophysical Research Letters* **25**, 2253–2256.

Fenchel, T. (1987). Ecology of Protozoa. The Biology of Free-Living Phagotrophic Protists, p. 197. Science Tech Publishers, Madison, Wisconsin.

Fenchel, T. (1994). Motility and chemosensory behaviour of sulphur bacterium *Thiovolum majus*. *Microbiology* **140**, 3109–3116.

Fenchel, T. (1998). Artificial cyanobacterial mats: Cycling of C, O, and S. *Aquatic Microbial Ecology* **14**, 253–259.

Fenchel, T. (2002). Microbial behaviour in the heterogeneous world. *Science* **296**, 1068–1071.

Fenchel, T. and Bernard, C. (1993). A purple protist. *Nature* **362**, 300.

Fenchel, T. and Blackburn, T. H. (1979). Bacteria and Mineral Cycling. Academic Press, London.

Fenchel, T. and Finlay, B. F. (1995). Ecology and Evolution in Anoxic Worlds. Oxford University Press, Oxford.

Fenchel, T. and Glud, R. N. (1998). Veil architecture in a sulphide-oxidizing bacterium enhances countercurrent flux. *Nature* **394**, 367–369.

Fenchel, T. and Kühl, M. (1999). Livet før det højere liv. *Naturens Verden* **10**, 20–35.

Fenchel, T., King, G. M., and Blackburn, T. H. (1998). Bacterial Biogeochemistry: The Ecophysiology of Mineral Cycling, p. 307. Academic Press, San Diego, CA.

Feng, D.-F., Cho, G., and Doolittle, R. F. (1997). Determining divergence times with a protein clock: Update and reevaluation. *Proceedings of the National Academy of Sciences USA* **94**, 13028–13033.

Ferdelman, T. G., Fossing, H., Neuman, K., and Schulz, H. D. (1999). Sulfate reduction in surface sediments of the southeast Atlantic continental margin between 15°38′S and 27° 57′S (Angola and Namibia). *Limnology and Oceanography* **44**, 650–661.

Ferris, M. J., Kühl, M., Wieland, A., and Ward, D. M. (2003). Cyanobacterial ecotypes in different optical microenvironments of a 68°C hot spring mat community revealed by 16S–23S rRNA internal transcribed spacer region variation. *Applied and Environmental Microbiology* **69**, 2893–2898.

Field, C. B., Behrenfeld, M. J., Randerson, J. T., and Falkowski, P. (1998). Primary production of the biosphere: Integrating terrestrial and oceanic components. *Science* **281**, 237–240.

Field, C. D. (1999). Mangrove rehabilitation: Choice and necessity. *Hydrobiologia* **413**, 47–52.

Findlay, S. (1993). Thymidine incorporation into DNA as an estimate of sediment bacterial production. *In* "Aquatic Microbial Ecology" (P. F. Kemp, B. F. Sherr, E. B. Sherr, and J. J. Cole, eds). Lewis Publishers, Boca Raton.

Findlay, S., Pace, M. L., Lints, D., Cole, J. J., Caraco, N. F., and Peierls, B. (1991). Weak coupling of bacterial and algal production in a heterotrophic ecosystem: The Hudson River estuary. *Limnology and Oceanography* **36**, 268–278.

Findlay, S. E. G., Meyer, J. L., and Edwards, R. T. (1984). Measuring bacterial production via rate of incorporation of [^{3}H]thymidine into DNA. *Journal of Microbiological Methods* **2**, 57–72.

Finneran, K. T., Forbush, H. M., VanPraagh, C. V. G., and Lovley, D. R. (2002a). *Desulfitobacterium metallireducens* sp. nov., an anaerobic bacterium that couples growth to the reduction of metals and humic acids as well as chlorinated compounds. *International Journal of Systematic and Evolutionary Microbiology* **52**, 1929–1935.

Finneran, K. T., Housewright, M. E., and Lovley, D. R. (2002b). Multiple influences of nitrate on uranium solubility during bioremediation of uranium-contaminated subsurface sediments. *Environmental Microbiology* **4**, 510–516.

Finster, K., Bak, F., and Pfennig, N. (1994). *Desulfuromonas acetexigens* sp. nov., a dissimilatory sulfur-reducing eubacterium from anoxic freshwater sediments. *Archives of Microbiology* **161**, 328–332.

Finster, K., Liesack, W., and Tindall, B. J. (1997a). *Desulfospira joergensenii*, gen. nov., sp. nov., a new sulfate-reducing bacterium isolated from marine surface sediment. *Systematic Applied Microbiology* **20**, 201–208.

Finster, K., Liesack, W., and Tindall, B. J. (1997b). *Sulforospirillum arcachonense* sp. nov., a new microaerophilic sulfur-reducing bacterium. *International Journal of Systematic Bacteriology* **47**, 1212–1217.

Finster, K., Liesack, W., and Thamdrup, B. (1998). Elemental sulfur and thiosulfate disproportionation by *desulfocapsa sulfoexigens* sp. nov., a new anaerobic

bacterium isolated from marine surface sediment. *Applied and Environmental Microbiology* **64**, 119–125.
Fitz-Gibbon, S. T. and House, C. H. (1999). Whole genome-based phylogenetic analysis of free-living microorganisms. *Nucleic Acids Research* **27**, 4218–4222.
Föllmi, K. B. (1996). The phosphorus cycle, phosphogenesis and marine phosphate-rich deposits. *Earth Science Reviews* **40**, 55–124.
Focht, D. D. and Verstraete, W. (1977). Biochemical ecology of nitrification and denitrification. *Advances in Microbial Ecology* **1**, 135–214.
Fossing, H. and Jørgensen, B. B. (1989). Measurement of bacterial sulfate reduction in sediments: Evaluation of a single-step chromium reduction method. *Biogeochemistry* **8**, 205–222.
Fossing, H., Gallardo, V. A., Jørgensen, B. B., Hüttel, M., Nielsen, L. P., Schulz, H., Canfield, D. E., Forster, S., Glud, R. N., Gundersen, J. K., Küver, J., Ramsing, N. B., Teske, A., Thamdrup, B., and Ulloa, O. (1995). Concentration and transport of nitrate by the mat-forming sulphur bacterium *Thioploca*. *Nature* **374**, 713–715.
Francis, C. A. and Tebo, B. M. (2001). cumA multicopper oxidase genes from diverse Mn(II)-oxidizing and non-Mn(II)-oxidizing Pseudomonas strains. *Applied and Environmental Microbiology* **67**, 4272–4278.
Francis, C. A. and Tebo, B. M. (2002). Enzymatic manganese(II) oxidation by metabolically dormant spores of diverse *Bacillus* species. *Applied and Environmental Microbiology* **68**, 874–880.
Francis, C. A., Co, E. M., and Tebo, B. M. (2001). Enzymatic manganese(II) oxidation by a marine a-proteobacterium. *Applied and Environmental Microbiology* **67**, 4024–4029.
Francois, R., Altabet, M. A., and Burkle, L. H. (1992). Glacial to interglacial changes in surface nitrate utilization in the Indian sector of the Southern Ocean as recorded by sediment d^{15}N. *Paleoceanography* **7**, 589–606.
Frederiksen, T.-M. and Finster, K. (2003). Sulfite-oxidoreductase is involved in the oxidation of sulfite in Desulfocapsa sulfoexigens during disproportionation of Thiosulfate and elemental sulfur. *Biodegradation* **14**, 189–198.
Friedl, G., Wehrli, B., and Manceau, A. (1997). Solid phases in the cycling of manganese in eutrophic lakes: New insights from EXAFS spectroscopy. *Geochimica et Cosmochimica Acta* **61**, 275–290.
Froelich, P. N., Klinkhammer, G. P., Bender, M. L., Luedtke, N. A., Heath, G. R., Cullen, D., Dauphin, P., Hammond, D., Hartman, B., and Maynard, V. (1979). Early oxidation of organic matter in pelagic sediments of the eastern equatorial Atlantic: Suboxic diagenesis. *Geochimica et Cosmochimica* **43**, 1075–1090.
Froelich, P. N., Bender, M. L., Luedtke, N. A., Heath, G. R., and DeVries, T. (1982). The marine phosphorus cycle. *American Journal of Science* **282**, 474–511.
Froelich, P. N., Arthur, M. A., Burnett, W. C., Deakin, M., Hensley, V., Jahnke, R., Kaul, L., Kim, K.-H., Roe, K., Soutar, A., and Vathakanon, C. (1988). Early diagenesis of organic matter in Peru continental margin sediments: Phosporite precipitation. *Marine Geology* **80**, 309–343.
Fröstl, J. M. and Overmann, J. (2000). Phylogenetic affiliation of the bacteria that constitute phototrophic consortia. *Archives of Microbiology* **174**, 50–58.
Fründ, C. and Cohen, Y. (1992). Diurnal cycles of sulfate reduction under oxic conditions in cyanobacterial mats. *Applied and Environmental Microbiology* **58**, 70–77.
Fuchs, G. (1986). CO_2 fixation in acetogenic bacteria: Variations on a theme. *FEMS Microbiology Reviews* **39**, 181–213.

Fuchs, G. (1989). Alternative pathways of autotrophic CO_2 fixation. *In* "Autotrophic Bacteria" (H. G. Schlegel and B. Bowien, eds), pp. 365–382. Science Tech Publishers, Madison, WI.

Fuchs, G. (1999). Oxidation of organic compounds. *In* "Biology of the Prokaryotes" (J. W. Lengeler, G. Drews, and H. G. Schlegel, eds), pp. 187–233. Blackwell Science, New York.

Fuchs, G. and Kröger, A. (1999). Growth and Nutrition. *In* "Biology of the Prokaryotes" (J. W. Lengeler, G. Drews, and H. G. Schlegel, eds), pp. 88–108. Blackwell Science, New York.

Fuchs, G. and Stupperich, E. (1985). Evolution of autotrophic CO_2 fixation. *In* "Evolution of Prokaryotes" (K. H. Schleifer and E. Stackebrandt, eds), pp. 235–251. Academic Press, London.

Fuchs, G., Thauer, R., Ziegler, H., and Stichler, W. (1979). Carbon isotope fractionation by *Methanobacterium thermoautotrophicum*. *Archives of Microbiology* **120**, 135–139.

Fuhrman, J. (2003). Genome sequences from the sea. *Nature* **424**, 1001–1002.

Fuhrman, J. A., McCallum, K., and Davis, A. A. (1992). Novel major archaebacterial group from marine plankton. *Nature* **356**, 148–149.

Furnes, H. and Staudigel, H. (1999). Biological mediation in ocean crust alternation: How deep is the deep biosphere? *Earth and Planetary Science Letters* **166**, 97–103.

Furnes, H., Muehlenbachs, K., Torsvik, T., Thorseth, I. H., and Tumyr, O. (2001a). Microbial fractionation of carbon isotopes in altered basaltic glass from the Atlantic Ocean, Lau Basin and Costa Rica Rift. *Chemical Geology* **173**, 313–330.

Furnes, H., Staudigel, H., Thorseth, I. H., Torsvik, T., Muehlenbachs, K., and Tumyr, O. (2001b). Bioalteration of basaltic glass in a oceanic crust. *Geochemistry, Geophysics, Geosystems* **2**, paper number (2000G)C000150.

Fuseler, K., Krekeler, D., Sydow, U., and Cypionka, H. (1996). A common pathway of sulfide oxidation by sulfate-reducing bacteria. *FEMS Microbiology Letters* **144**, 129–134.

Gächter, R. and Meyer, J. S. (1993). The role of microorganisms in sediment phosphorus dynamics in relation to mobilization and fixation of phosphorus. *Hydrobiologia* **253**, 103–121.

Gallinari, M., Ragueneau, O., Corrin, L., DeMaster, D. J., and Treguer, P. (2002). The importance of water column processes on the dissolution properties of biogenic silica in deep-sea sediments I. Solubility. *Geochimica et Cosmochimica Acta* **66**, 2701–2717.

Galtier, N., Tourasse, N., and Gouy, M. (1999). A nonhyperthermophilic common ancestor to extant life forms. *Science* **283**, 220–221.

Gamble, T. N., Betlach, M. R., and Tiedje, J. M. (1977). Numerically dominant denitrifying bacteria from world soils. *Applied and Environmental Microbiology* **33**, 926–939.

Games, L. M., Hayes, J. M., and Gunsalus, P. (1978). Methane-producing bacteria: Natural fractionations of the stable carbon isotopes. *Geochimica et Cosmochimica Acta* **42**, 1295–1297.

Garber, J. H. (1984). Laboratory study of nitrogen and phosphorus remineralization during the decomposition of coastal plankton and seston. *Estuarine, Coastal and Shelf Science* **18**, 685–702.

García, H. E. and Gordon, L. I. (1992). Oxygen solubility in seawater—Better fitting equations. *Limnology and Oceanography* **37**, 1307–1312.

Garcia, J.-L., Patel, B. K. C., and Ollivier, B. (2000). Taxonomic, phylogenetic, and ecological diversity of methanogenic Archaea. *Anaerobe* **6**, 206–226.

García-Horsman, J. A., Barquera, B., Rumbley, J., Ma, J. X., and Gennis, R. B. (1994). The superfamily of heme-copper respiratory oxidases. *Journal of Bacteriology* **176**, 5587–5600.
Garcia-Pichel, F. and Castenholz, R. W. (1991). Characterization and biological implications of scytonemin, a cyanobacterial sheath pigment. *Journal of Phycology* **27**, 395–409.
Garcia-Pichel, F., Mechling, M., and Castenholz, R. W. (1994). Diel migrations of microorganisms within a benthic, hypersaline mat community. *Applied and Environmental Microbiology* **60**, 1500–1511.
Garcia-Pichel, F., Kühl, M., Nübel, U., and Muyzer, G. (1999). Salinity-dependent limitation of photosynthesis and oxygen exchange in microbial mats. *Journal of Phycology* **35**, 227–238.
Garcia-Ruiz, R., Pattinson, S. N., and Whitton, B. A. (1998). Kinetic parameters of denitrification in a river continuum. *Applied and Environmental Microbiology* **64**, 2533–2538.
Garrels, R. M. and Christ, C. L. (1965). Solutions, Minerals and Equilibrium, p. 450. Harper & Row, New York.
Geertz-Hansen, O., Sand-Jensen, K., Hansen, D. F., and Christiansen, A. (1993). Growth and grazing control of abundance of the marine macroalgae, *Ulva lactuca* L. in a eutrophic Danish estuary. *Aquatic Botany* **46**, 101–109.
Gelwicks, J. T., Risatti, J. B., and Hayes, J. M. (1989). Carbon isotope effects associated with autotrophic acetogenesis. *Organic Geochemistry* **14**, 441–446.
Ghiorse, W. C. (1984). Biology of iron-depositing and manganese-depositing bacteria. *Annual Review of Microbiology* **38**, 515–550.
Gibson, J. and Harwood, C. S. (1995). Degradation of aromatic compounds by nonsulfur purple bacteria. *In* "Anoxygenic Photosynthetic Bacteria" (R. E. Blankenship, M. T. Madigan, and C. E. Bauer, eds), pp. 991–1003. Klüwer Academic Publishers, The Netherlands.
Giddins, R. L., Lucas, J. S., Neilson, M. J., and Richards, G. N. (1986). Feeding ecology of the mangrove crab *Neosarmatium-Smithi* (Crustacea, Decapoda, Sesarmidae). *Marine Ecology-Progress Series* **33**, 147–155.
Gilbert, B. and Frenzel, P. (1998). Rice roots and CH_4 oxidation: The activity of bacteria, their distribution and the microenvironment. *Soil Biology & Biochemistry* **30**, 1903–1916.
Giovannoni, S. and Rappé, M. S. (2000). Evolution, diversity, and molecular ecology of marine prokaryotes. *In* "Microbial Ecology of the Oceans" (D. L. Kirchman, ed.), pp. 47–84. Wiley-Liss Inc., New York.
Giovannoni, S. J., Britschgi, T. B., Moyer, C. L., and Field, K. G. (1990). Genetic diversity in Sargasso Sea bacterioplankton. *Nature* **345**, 60–63.
Giudici-Orticoni, M.-T., Leroy, G., Nitschke, W., and Bruschi, M. (2000). Characterization of a new dihemic c_4-type cytochrome isolated from *Thiobacillus ferrooxidans*. *Biochemistry* **39**, 7205–7211.
Glasby, G. P. (2000). Manganese: Predominant role in nodules and crusts. *In* "Marine Geochemistry" (H. D. Schultz and M. Zabel, eds), pp. 335–372. Springer, Berlin.
Glöckner, F. O., Fuchs, G., and Amann, R. (1999). Bacterioplankton compositions of lakes and oceans: A first comparison based on fluorescence *in situ* hybridization. *Applied and Environmental Microbiology* **65**, 3721–3726.
Glud, N. R., Kühl, M., Kohls, O., and Ramsing, N. B. (1999). Heterogeneity of oxygen production and compsumption in a photosynthetic microbial mat as studied by planar optodes. *Journal of Phycology* **35**, 270–279.

Glud, R. N., Gundersen, J. K., Jørgensen, B. B., Revsbech, N. P., and Schulz, H. D. (1994). Diffusive and total oxygen uptake of deep-sea sediments in the eastern South Atlantic Ocean: *In situ* and laboratory measurements. *Deep-Sea Research* **41**, 1767–1788.

Glud, R. N., Risgaard-Petersen, N., Thamdrup, B., Fossing, H., and Rysgaard, S. (2000). Benthic carbon mineralization in a high-Arctic sound (Young Sound, NE Greenland). *Marine Ecology-Progress Series* **206**, 59–71.

Glud, R. N., Tengberg, A., Kühl, M., Hall, P. O. J., and Klimant, I. (2001). An *in situ* instrument for planar O_2 optode measurements at benthic interfaces. *Limnology and Oceanography* **46**, 2073–2080.

Glud, R. N., Gundersen, J. K., Roy, H., and Jørgensen, B. B. (2003). Seasonal dynamics of benthic O_2 uptake in a semienclosed bay: Importance of diffusion and faunal activity. *Limnology and Oceanography* **48**, 1265–1276.

Goering, J. J., Nelson, D. M., and Carter, J. A. (1973). Silicic acid uptake by natural populations of marine phytoplankton. *Deep-Sea Research* **20**, 777–789.

Gogarten, J. P., Kibak, H., and Dittrich, P. (1989). Evolution of the vacuolar H^+-ATPase: Implications for the origin of eukaryotes. *Proceedings of the National Academy of Sciences USA* **86**, 6661–6665.

Goldman, J. C., Caron, D. A., and Dennett, M. R. (1987). Regulation of gross growth efficiency and ammonium regeneration in bacteria by substrate C:N ratio. *Limnology and Oceanography* **32**, 1239–1252.

Goreau, T. J., Kaplan, W. A., Wofsy, S. C., McElroy, M. B., Valois, F. W., and Watson, S. W. (1980). Production of NO_2^- and N_2O by nitrifying bacteria at reduced concentrations of oxygen. *Applied and Environmental Microbiology* **40**, 526–532.

Gottschalk, G. (1986). Bacterial Metabolism, 2nd Ed. Springer, New York.

Graham, W. F. and Duce, R. A. (1982). The atmospheric transport of phosphorus to the Western North Atlantic. *Atmospheric Environment* **16**, 1089–1097.

Granick, S. (1965). Evolution of heme and chlorophyll. *In* "Evolving Genes and Proteins" (V. Bryson and H. J. Vogel, eds), pp. 67–88. Academic Press, New York.

Grassle, J. F. (1986). The ecology of deep-sea hydrothermal vent communities. *Advances in Marine Biology* **23**, 301–363.

Greenwood, J. E., Truesdale, V. W., and Rendell, A. R. (2001). Biogenic silica dissolution in seawater—*in vitro* chemical kinetics. *Progress in Oceanography* **48**, 1–23.

Gribsholt, B., Kostka, J. E., and Kristensen, E. (2003). Impact of fiddler crabs and plant roots on sediment biogeochemistry in a Georgia saltmarsh. *Marine Ecology-Progress Series* **259**, 237–251.

Grinenko, V. A. and Thode, H. G. (1970). Sulfur isotope effects in volcanic gas mixture. *Canadian Journal of Earth Science* **7**, 1402–1409.

Gros, O., Frenkiel, L., and Felbeck, H. (2000). Sulfur-oxidizing endosymbiosis in *Divaricella quadrisulcata (Bivalvia: Lucinidae)*: Morphological, ultrastructural, and phylogenetic analysis. *Symbiosis* **29**, 293–317.

Gruber, N. and Sarmiento, J. L. (1997). Global patterns of marine nitrogen fixation and denitrification. *Global Biogeochemical Cycles* **11**, 235–266.

Guerrero, M. A. and Jones, R. D. (1996). Photoinhibition of marine nitrifying bacteria. 1. Wavelength-dependent response. *Marine Ecology-Progress Series* **141**, 183–192.

Guerrero, R., Montesinos, E., Pedrós-Alió, C., Esteve, I., Mas, J., van Gemerden, H., Hofman, P. A. G., and Bakker, J. F. (1985). Photosynthetic sulfur bacteria

in two Spanish lakes: Vertical distribution and limiting factors. *Limnology and Oceanography* **30**, 919–931.
Gunnars, A. and Blomqvist, S. (1997). Phosphate exchange across the sediment-water interface when shifting from anoxic to oxic conditions—An experimental comparison of freshwater and brackish-marine systems. *Biogeochemistry* **37**, 203–226.
Gunnarsson, T., Sundin, P., and Tunlid, A. (1988). Importance of leaf litter fragmentation for bacterial-growth. *Oikos* **52**, 303–308.
Guo, L., Santschi, P. H., and Warnken, K. W. (1995). Dynamics of dissolved organic carbon (DOC) in oceanic environments. *Limnology and Oceanography* **40**, 1392–1403.
Gupta, R. S., Mukhtar, T., and Signh, B. (1999). Evolutionary relationship among photosynthetic prokaryotes (*Heliobacterium chlorum, Chloroflexus aurantiacus, cyanobacteria, Chlorobium tepidum* and proteobacteria): Implications regarding the origin of photosynthesis. *Molecular Microbiology* **32**, 893–906.
Guy, R. D., Fogel, M. L., and Berry, J. A. (1993). Photosynthetic fractionation of the stable isotopes of oxygen and carbon. *Plant Physiology* **101**, 37–47.
Haaker, H. and Klugkist, J. (1987). The bioenergetics of electron-transport to nitrogenase. *FEMS Microbiology Reviews* **46**, 57–71.
Habicht, K. S. and Canfield, D. E. (1997). Sulfur isotope fractionation during bacterial sulfate reduction in organic-rich sediments. *Geochimica et Cosmochimica Acta* **61**, 5351–5361.
Habicht, K. S. and Canfield, D. E. (2001). Isotope fractionation by sulfate-reducing natural populations and the isotopic composition of sulfide in marine sediments. *Geology* **29**, 555–558.
Habicht, K. S., Canfield, D. E., and Rethmeier, J. (1998). Sulfur isotope fractionation during bacterial reduction and disproportionation of thiosulfate and sulfite. *Geochimica et Cosmochimica Acta* **62**, 2585–2595.
Habicht, K. S., Gade, M., Thamdrup, B., Berg, P., and Canfield, D. E. (2002). A calibration of sulfate levels in the Archean Ocean. *Science* **298**, 2372–2374.
Hafenbradl, D., Keller, M., Dirmeier, R., Rachel, R., Rossnagel, P., Burggraf, S., Huber, H., and Stetter, K. O. (1996). *Ferroglobus placidus* gen. nov., sp. nov, a novel hyperthermophilic archaeum that oxidizes Fe^{2+} at neutral pH under anoxic conditions. *Archives of Microbiology* **166**, 308–314.
Hallbeck, L. and Pedersen, K. (1991). Autotrophic and mixotrophic growth of *Gallionella ferruginea*. *Journal of General Microbiology* **137**, 2657–2661.
Hallberg, R. and Ferris, F. G. (2004). Biomineralization by *Gallionella*. *Geomicrobiology Journal* **21**, 325–330.
Hallin, S. and Lindgren, P. E. (1999). PCR detection of genes encoding nitrile reductase in denitrifying bacteria. *Applied and Environmental Microbiology* **65**, 1652–1657.
Halverson, G. P., Hoffman, P. F., Schrag, D. P., and Kaufman, A. J. (2002). A major perturbation of the carbon cycle before the Ghaub glaciation (Neoproterozoic) in Namibia: Prelude to the snowball Earth? *Geochemistry, Geophysics, Geosystems* **3**, art. no. 1035.
Hansen, H. C. B., Koch, C. B., NanckeKrogh, H., Borggaard, O. K., and Sørensen, J. (1996). Abiotic nitrate reduction to ammonium: Key role of green rust. *Environmental Science & Technology* **30**, 2053–2056.
Hansen, I. S., Ærtebjerg, G., Richardson, K., Heilmann, J. P., Olsen, O. V., and Pedersen, F. B. (1994). Effects of Reduced Nitrogen Input on Oxygen Conditions

in the Inner Danish Waters (in Danish). Havforskning fra Miljøstyrelsen 29. Danish Environmental Protection Agency, Copenhagen.

Hansen, J. I., Henriksen, K., and Blackburn, T. H. (1981). Seasonal distribution of nitrifying bacteria and rates of nitrification in coastal marine-sediments. *Microbial Ecology* **7**, 297–304.

Hansen, K. and Kristensen, E. (1997). Impact of macrofaunal recolonization on benthic metabolism and nutrient fluxes in a shallow marine sediment previously overgrown with macroalgal mats. *Estuarine, Coastal and Shelf Science* **45**, 613–628.

Hansen, K., Mouridsen, S., and Kristensen, E. (1998). The impact of *Chironomus plumosus* larvae on organic matter decay and nutrient (N,P) exchange in a shallow eutrophic lake sediment following a phytoplankton sedimentation. *Hydrobiologia* **364**, 65–74.

Hansen, L. K., Jakobsen, R., and Postma, D. (2001). Methanogenesis in a shallow sandy aquifer, Rømø, Denmark. *Geochimica et Cosmochimica Acta* **65**, 2925–2935.

Hansen, T. A. (1993). Carbon metabolism of sulfate-reducing bacteria. *In* "The Sulfate-Reducing Bacteria: Contemporary Perspectives" (J. M. Odom and R. Singleton, eds), pp. 21–41. Springer, New York.

Hanson, R. S. and Hanson, T. E. (1996). Methanotrophic bacteria. *Microbiological Reviews* **60**, 439–471.

Hanson, T. E. and Tabita, F. R. (2001). A ribulose-1,5-bisphosphate carboxylase/oxygenase (RubisCO)-like protein from *Chlorobium tepidum* that is involved with sulfur metabolisme and the response to oxidative stress. *Proceedings of the National Academy of Sciences USA* **98**, 4397–4402.

Hao, O. J., Richard, M. G., Jenkins, D., and Blanch, H. W. (1983). The half-saturation coefficient for dissolved oxygen—A dynamic method for its determination and its effect on dual species competition. *Biotechnology and Bioengineering* **25**, 403–416.

Harder, W., Kuenen, J. G., and Matin, A. (1977). A review: Microbial selection in continuous culture. *Journal of Applied Bacteriology* **43**, 1–24.

Hardy, R. W. F. and Burns, R. C. (1968). Biological nitrogen fixation. *Annual Review of Biochemistry* **37**, 331–358.

Harper, H. E. and Knoll, A. H. (1975). Silica, diatoms, and cenozoic radiolarian evolution. *Geology* **3**, 175–177.

Hauck, S., Benz, M., Brune, A., and Schink, B. (2001). Ferrous iron oxidation by denitrifying bacteria in profundal sediments of a deep lake (Lake Constance). *FEMS Microbiology Ecology* **1269**, 1–8.

Hawke, D., Carpenter, P. D., and Hunter, K. A. (1989). Competitive adsorption of phosphate on goethite in marine electrolytes. *Environmental Science & Technology* **23**, 187–191.

Head, I. M., Hiorns, W. D., Embley, T. M., McCarthy, A. J., and Saunders, J. R. (1993). The phylogeny of autotrophic ammonia-oxidizing bacteria as determined by analysis of 16S ribosomal-RNA gene-sequences. *Journal of General Microbiology* **139**, 1147–1153.

Hecky, R. E. and Kilham, P. (1988). Nutrient limitation of phytoplankton in fresh water and marine environments—A review of recent evidence on the effects of enrichment. *Limnology and Oceanography* **33**, 796–822.

Hedges, J. I. (1992). Global biogeochemical cycles: Progress and problems. *Marine Chemistry* **39**, 67–93.

Hedges, J. I. and Keil, R. G. (1995). Sedimentary organic matter preservation: An assessment and speculative synthesis. *Marine Chemistry* **49**, 81–115.

Hedges, J. I., Clark, W. A., and Cowie, G. L. (1988). Influence of oxygen exposure time on organic carbon preservation in continental margin sediments. *Nature* **391**, 572–574.

Hedges, J. I., Hu, F. S., Devol, A. H., Hartnett, H. E., Tsamakis, E., and Keil, R. G. (1999). Sedimentary organic matter preservation: A test for selective degradation under oxic conditions. *American Journal of Science* **299**, 529–555.

Hedges, J. I., Eglinton, G., Hatcher, P. G., Kirchman, D. L., Arnosti, C., Derenne, S., Evershed, R. P., Kogel-Knabner, I., de Leeuw, J. W., Littke, R., Michaelis, W., and Rullkotter, J. (2000). The molecularly uncharacterized component of nonliving organic matter in natural environments. *Organic Geochemistry* **31**, 945–958.

Hedges, J. I., Baldock, J. A., Gélinas, Y., Lee, C.-T., Peterson, M. L., and Wakeham, S. G. (2002). The biochemical and elemental compositions of marine plankton: A NMR perspective. *Marine Chemistry* **78**, 47–63.

Heggie, D. T., Skyring, G. W., O'Brien, G. W., Reimers, C. E., Herczeg, A., Moriarty, D. J. W., Burnett, W. C., and Milnes, A. R. (1990). Organic carbon cycling and modern phosphorite formation on the East Australian continental margin: An overview. *Geological Society of London, Special Publication* **52**, 87–117.

Heising, S. and Schink, B. (1998). Phototrophic oxidation of ferrous iron by a *Rhodomicrobium* vannielii strain. *Microbiology–UK* **144**, 2263–2269.

Heising, S., Richter, L., Ludwig, W., and Schink, B. (1999). *Chlorobium ferrooxidans* sp nov., a phototrophic green sulfur bacterium that oxidizes ferrous iron in coculture with a "*Geospirillum*" sp strain. *Archives of Microbiology* **172**, 116–124.

Helgeson, H. C., Murphy, W. M., and Aagaard, P. (1984). Thermodynamic and kinetic constraints on reaction-rates among minerals and aqueous-solutions. II. Rate constants, effective surface-area, and the hydrolysis of feldspar. *Geochimica et Cosmochimica Acta* **48**, 2405–2432.

Hemminga, M. A., Slim, F. J., Kazungu, J., Ganssen, G. M., Nieuwenhuize, J., and Kruyt, N. M. (1994). Carbon outwelling from a mangrove forest with adjacent seagrass beds and coral reefs (Gazi Bay, Kenya). *Marine Ecology-Progress Series* **106**, 291–301.

Henrichs, S. M. (1992). Early diagenesis of organic matter in marine sediments: Progress and perplexity. *Marine Chemistry* **39**, 119–149.

Henrichs, S. M. and Reeburgh, W. S. (1987). Anaerobic mineralization of marine sediment organic matter: Rates and the role of anaerobic processes in the oceanic carbon economy. *Geomicrobiology Journal* **5**, 191–237.

Henriksen, K. and Kemp, W. M. (1988). Nitrification in estuarine and coastal marine sediments. *In* "Nitrogen Cycling in Coastal Marine Environments" (T. H. Blackburn and J. Sørensen, eds), pp. 207–249. John Wiley and Sons, Chichester.

Henriksen, K., Hansen, J. I., and Blackburn, T. H. (1981). Rates of nitrification, distribution of nitrifying bacteria, and nitrate fluxes in different types of sediment from Danish waters. *Marine Biology* **61**, 299–304.

Henriksen, K., Blackburn, T. H., Lomstein, B. A., and McRoy, C. P. (1993). Rates of nitrification, distribution of nitrifying bacteria and inorganic N fluxes in the northern Bering-Chukchi shelf sediments. *Continental Shelf Research* **13**, 629–651.

Henry, E. A., Devereaux, R., Maki, J. S., Gilmour, C. C., Woese, C. R., Mandelco, L., Schauder, R., Remsen, C. C., and Mitchell, R. (1994). Characterization of a new thermophilic sulfate-reducing bacterium. *Archives of Microbiology* **161**, 62–69.

Hentschel, U., Berger, E. C., Bright, M., Felbeck, H., and Ott, J. A. (1999). Metabolism of nitrogen and sulfur in ectosymbiotic bacteria of marine nematodes (Nematoda, Stilbonematinae). *Marine Ecology Progress Series* **183**, 149–158.

Herbert, R. A. (1999). Nitrogen cycling in coastal marine ecosystems. *FEMS Microbiology Reviews* **23**, 563–590.

Hilario, E. and Gogarten, J. P. (1993). Horizontal transfer of ATPase genes—The tree of life becomes a net of life. *BioSystems* **31**, 111–119.

Hillis, D. M., Allard, M. W., and Miyamoto, M. M. (1993). Analysis of DNA sequence data: Phylogenetic inference. *Methods of Enzymology* **224**, 456–487.

Hinrichs, K.-U., Hayes, J. M., Sylva, S. P., Brewer, P. G., and DeLong, E. F. (1999). Methane-consuming archaebacteria in marine sediments. *Nature* **398**, 802–805.

Hodson, R. E., Christian, R. R., and Maccubbin, A. E. (1984). Lignocellulose and lignin in the salt-marsh grass spartina-alterniflora: Initial concentrations and short-term, post-depositional changes in detrital matter. *Marine Biology* **81**, 1–7.

Hoehler, T. M., Alperin, M. J., Albert, D. B., and Martens, C. S. (1994). Field and laboratory studies of methane oxidation in an anoxic marine sediment: Evidence for a methanogen-sulfate reducer consortium. *Global Biogeochemical Cycles* **8**, 451–463.

Hoehler, T. M., Alperin, M. J., Albert, D. B., and Martens, C. S. (1998). Thermodynamic control on hydrogen concentration in anoxic sediments. *Geochimica et Cosmochimica Acta* **62**, 1745–1756.

Hoehler, T. M., Bebout, B. M., and Des Marais, D. J. (2001). The role of microbial mats in the production of reduced gases on the early Earth. *Nature* **412**, 324–327.

Hoffman, P. F., Kaufman, A. J., Halverson, G. P., and Schrag, D. P. (1998). A neoproterozoic snowball Earth. *Science* **281**, 1342–1346.

Hofmann, H. J. (1976). Precambrian microflora, Belcher Islands, Canada: Significance and systematics. *Journal of Paleontology* **50**, 1040–1073.

Hoh, C. Y. and Cord-Ruwisch, R. (1996). A practical kinetic model that considers endproduct inhibition in anaerobic digestion processes by including the equilibrium constant. *Biotechnology and Bioengineering* **51**, 597–604.

Holland, H. D. (1984). The Chemical Evolution of the Atmosphere and Oceans, p. 582. Princeton University Press, Princeton.

Holloway, J. M., Dahlgren, R. A., and Casey, W. H. (2001). Nitrogen release from rock and soil under simulated field conditions. *Chemical Geology* **174**, 403–414.

Holmer, M. and Storkholm, P. (2001). Sulphate reduction and sulphur cycling in lake sediments: A review. *Freshwater Biology* **46**, 431–451.

Holmer, M., Kristensen, E., Banta, G., Hansen, K., Jensen, M. H., and Bussawarit, N. (1994). Biogeochemical cycling of sulfur and iron in sediments of a south-east Asian mangrove, Phuket Island, Thailand. *Biogeochemistry* **26**, 145–161.

Holmer, M., Andersen, F. Ø., Holmboe, N., Kristensen, E., and Thongtham, N. (1999). Transformation and exchange processes in the Bangrong forest-seagrass bed system, Thailand. Seasonal and spatial variations in benthic matabolism and sulfur biogeochemistry. *Aquatic and Microbial Ecology* **20**, 203–212.

Holmer, M., Andersen, F. Ø., Holmboe, N., Kristensen, E., and Thongtham, N. (2001). Spatial and temporal variability in benthic processes along a mangrove-seagrass transect near Bangrong Mangrove, Thailand. *Wetlands Ecology and Management* **9**, 141–158.

Holmes, D. E., Finneran, K. T., O'Neil, R. A., and Lovley, D. R. (2002). Enrichment of members of the family Geobacteraceae associated with stimulation of dissimilatory metal reduction in uranium-contaminated aquifer sediments. *Applied and Environmental Microbiology* **68**, 2300–2306.

Holo, H. (1989). *Chloroflexus aurantiacus* secretes 3-hydroxypropionate, a possible intermediate in the assimilation of CO_2 and acetate. *Archives of Microbiology* **151**, 252–256.

Holo, H. and Sirevåg, R. (1986). Autotrophic growth and CO_2 fixation of *Chloroflexus aurantiacus*. *Archives of Microbiology* **145**, 173–180.

Horikoshi, K. (1999). Alkaliphiles: Some applications of their products for biotechnology. *Microbiology and Molecular Biology Reviews* **63**, 735–750.

Horne, A. J. and Goldman, C. R. (1994). Limnology. 2nd Ed. McGraw-Hill, New York.

Horrigan, S. G., Montoya, J. P., Nevins, J. L., and McCarthy, J. J. (1990). Natural isotopic composition of dissolved inorganic nitrogen in the Chesapeake Bay. *Estuarine, Coastal and Shelf Science* **30**, 393–410.

House, C. H. (1999). Carbon Isotope Fractionation by Diverse Extant and Fossil Prokaryotes and Microbial Phylogenetic Diversity Revealed through Genonics. Ph.D. Thesis, UCLA, Los Angeles.

House, C. H., Runnegar, B., and Fitz-Gibbon, S. T. (2003). Geobiological analysis using whole genome-based tree building applied to the Bacteria, Archaea, and Eukarya. *Geobiology* **1**, 15–26.

House, C. H., Schopf, J. W., and Stetter, K. O. (2003). Carbon isotopic fractionation by Archaeans and other thermophilic prokaryotes. *Organic Geochemistry* **34**, 345–356.

Howarth, R. W. (1979). Pyrite: Its rapid formation in a salt marsh and its importance in ecosystem metabolism. *Science* **203**, 49–51.

Howarth, R. W., Marino, R., and Cole, J. J. (1988a). Nitrogen fixation in freshwater, estuarine, and marine ecosystems. 2. Biogeochemical controls. *Limnology and Oceanography* **33**, 669–687.

Howarth, R. W., Marino, R., Lane, J., and Cole, J. J. (1988b). Nitrogen fixation in fresh-water, estuarine, and marine ecosystems. 1. Rates and importance. *Limnology and Oceanography* **33**, 688–701.

Howarth, R. W., Jensen, H. S., Marino, R., and Postma, D. (1995). Transport to and processing of P in near-shore and oceanic waters. *In* "Phosphorus in the Global Environment" (H. Thiessen, ed.). John Wiley and Sons, Chichester.

Huber, H., Hohn, M. J., Rachel, R., Fuchs, T., Wimmer, V. C., and Stetter, K. O. (2002). A new phylum of Archaea represented by a nanosized hyperthermophilic symbiont. *Nature* **417**, 63–67.

Huber, R., Langworthy, T. A., König, H., Thomm, M., Woese, C. R., Sleytr, U. B., and Stetter, K.-O. (1986). *Thermotoga maritima* sp. nov. represents a new genus of unique extremely thermophilic eubacteria growing up to 90°C. *Archives of Microbiology* **144**, 324–333.

Huelsenbeck, J. P. and Rannala, B. (1997). Phylogenetic methods come of age: Testing hypotheses in an evolutionary context. *Science* **276**, 227–232.

Huettel, M. and Gust, G. (1992). Impact of bioroughness on interfacial solute exchange in permeable sediments. *Marine Ecology-Progress Series* **89**, 253–267.

Huettel, M., Forster, S., Klöser, S., and Fossing, H. (1996). Vertical migration in the sediment-dwelling bacteria *Thioploca spp.* in overcoming diffusion limitations. *Applied and Environmental Microbiology* **62**, 1863–1872.

Hughes, J. B., Hellmann, J. J., Ricketts, T. H., and Bohannan, B. J. M. (2001). Counting the uncountable: Statistical approaches to estimating microbial diversity. *Applied and Environmental Microbiology* **67**, 4399–4406.

Hulston, J. R. and Thode, H. G. (1965). Variations in the S^{33}, S^{34}, S^{36} contents of meteorites and their relation to chemical and nuclear effects. *Journal of Geophysical Research* **70**, 3475–3484.

Hulth, S., Aller, R. C., and Gilbert, F. (1999). Coupled anoxic nitrification/manganese reduction in marine sediments. *Geochimica et Cosmochimica Acta* **63**, 49–66.
Hulthe, G., Hulth, S., and Hall, P. O. J. (1998). Effect of oxygen on degradation rate of refractory and labile organic matter in continental margin sediments. *Geochimica et Cosmochimica Acta* **62**, 1319–1328.
Humborg, C., Conley, D. J., Rahm, L., Wulff, F., Cociasu, A., and Ittekkot, V. (2000). Silicon retention in river basins: Far-reaching effects on biogeochemistry and aquatic food webs in coastal marine environments. *Ambio* **29**, 45–50.
Hurd, D. C. and Birdwhistell, S. (1983). On producing a more general model for biogenic silica dissolution. *American Journal of Science* **283**, 1–28.
Hurlbut, C. S. and Klein, C. (1977). Manual of Mineralogy, 19th Ed., p. 532. John Wiley and Sons, New York.
Hutchins, D. A., Witter, A. E., Butler, A., and Luther, G. W. (1999). Competition among marine phytoplankton for different chelated iron species. *Nature* **400**, 858–861.
Imhoff, J. F. (1992). Taxonomy, phylogeny, and general ecology of anoxygenic phototrophic bacteria. *In* "Photosynthetic Prokaryotes" (N. H. Mann and N. G. Carr, eds), pp. 53–92. Plenum Press, New York.
Imhoff, J. F. (1995). Taxonomy and physiology of phototrophic purple bacteria and green sulfur bacteria. *In* "Anoxygenic Photosynthetic Bacteria" (R. E. Blankenship, M. T. Madigan, and C. E. Bauer, eds), pp. 1–15. Kluwer, Netherlands.
Imhoff, J. F. and Trüper, H. G. (1992). The genus Rhodospirillum and related genera. *In* "The Prokaryotes" (A. Balows, H. G. Trüper, M. Dworkin, W. Harder, and K.-H. Schleifer, eds), 2nd Ed., pp. 2141–2155. Springer, New York.
Imlay, J. A. (2002). What biological purpose is served by superoxide reductase? *Journal of Biological Inorganic Chemistry* **7**, 659–663.
Imlay, J. A. and Fridovich, I. (1991). Assay of metabolic superoxide production in *Escherichia coli*. *Journal of Biological Chemistry* **266**, 6957–6965.
Indrebø, G., Pengerud, B., and Dundas, I. (1979). Microbial activities in a permanently stratified estuary. I. Primary production and sulfate reduction. *Marine Biology* **51**, 295–304.
Ingall, E. and Jahnke, R. (1994). Evidence for enhanced phosphorus regeneration from marine sediments overlain by oxygen depleted waters. *Geochimica et Cosmochimica Acta* **58**, 2571–2575.
Ingall, E. D. and Jahnke, R. A. (1997). Influence of water column anoxia on the elemental fractionation of carbon and phosphorus during sediment diagenesis. *Marine Geology* **139**, 219–229.
Ingall, E. D., Bustin, R. M., and Van Cappellen, P. (1993). Influence of water column anoxia on the burial and preservation of carbon and phosphorus in marine shales. *Geochimica et Cosmochimica Acta* **57**, 303–316.
Ingledew, W. J. (1982). *Thiobacillus ferrooxidans*: The bioenergetics of an acidophilic chemolithotroph. *Biochimica et Biophysica Acta* **683**, 89–117.
Ingvorsen, K. and Jørgensen, B. B. (1984). Kinetics of sulfate uptake by freshwater and marine species of *Desulfovibrio*. *Archives of Microbiology* **139**, 61–66.
Ingvorsen, K., Zehnder, A. J. B., and Jørgensen, B. B. (1984). Kinetics of sulfate and acetate uptake by *Desulfobacter postgatei*. *Applied and Environmental Microbiology* **47**, 403–408.
Iriarte, A., de la Sota, A., and Orive, E. (1998). Seasonal variation of nitrification along a salinity gradient in an urban estuary. *Hydrobiologia* **362**, 115–126.
Isaksen, M. F. and Jørgensen, B. B. (1996). Adaptation of psychrophilic and psychrotrophic sulfate-reducing bacteria to permanently cold marine environments. *Applied and Environmental Microbiology* **62**, 408–414.

Ishii, M., Miyake, T., Satoh, T., Sugiyams, H., Oshima, Y., Kodama, T., and Igarashi, Y. (1996). Autotrophic carbon dioxide fixation in *Acidianus brierleyi*. *Archives of Microbiology* **166**, 368–371.

Ivanov, M. V., Rusanov, I. I., Pimenov, N. V., Bairamov, I. T., Yusupov, S. K., Savvichev, A. S., Lein, A. Y., and Sapozhnikov, V. V. (2001). Microbial processes of the carbon and sulfur cycles in Lake Mogil'noe. *Microbiology* **70**, 583–593.

Iversen, N. and Jørgensen, B. B. (1985). Anaerobic methane oxidation rates at the sulfate-methane transition in marine sediments from Kattegat and Skagerrak (Denmark). *Limnology and Oceanography* **30**, 944–955.

Iwabe, N., Kuma, K., Hasegawa, M., Osawa, S., and Miyata, T. (1989). Evolutionary relationship of archaebacteria, eubacteria and eukaryotes inferred from phylogenetic trees of duplicated genes. *Proceedings of the National Academy of Sciences USA* **86**, 9355–9359.

Jacobson, D. M. and Anderson, D. M. (1986). Thecate heterotrophic dinoflagellates: Feeding behaviour and mechanisms. *Journal of Phycology* **22**, 249–258.

Jaenicke, R. and Sterner, R. (2002). Life at high temperatures. *In* "The Prokaryotes: An Evolving Electronic Resource for the Microbiological Community" (A. Balows, H. G. Trüper, M. Dworkin, W. Harder, and K. H. Schleifer, eds), 3rd Ed., release 3.9. April 1, 2002. Springer, New York.

Jaffe, D. A. (2000). The nitrogen cycle. *In* "Earth System Science" (M. C. Jacobsen, R. J. Charlson, H. Rohde, and G. H. Orians, eds), pp. 322–342. Academic Press, San Diego.

Jahnke, L. L., Summons, R. E., Hope, J. M., and Des Marais, D. J. (1999). Carbon isotopic fractionation in lipids from methanotrophic bacteria. II: The effects of physiology and environmental parameters on the biosynthesis and isotopic signatures of biomarkers. *Geochimica et Cosmochimica Acta* **63**, 79–93.

Jahnke, R. A. (1994). The phosphorus cycle. *In* "Global Biogeochemical Cycles" (S. S. Butcher, R. J. Charlson, G. H. Orians, and G. V. Wolfe, eds), pp. 301–315. Academic Press, London.

Jahnke, R. A., Emerson, S. R., Roe, K. K., and Burnett, W. C. (1983). The present-day formation of apatite in Mexican continental-margin sediments. *Geochimica et Cosmochimica Acta* **47**, 259–266.

Jain, R., Rivera, M. C., and Lake, J. A. (1999). Horizontal gene transfer among genomes: The complexity hypothesis. *Proceedings of the National Academy of Sciences USA* **96**, 3801–3806.

Jannasch, H. W. (1967). Enrichments of aquatic bacteria in continuous culture. *Archiv für Mikrobiologie* **59**, 165–173.

Jannasch, H. W. (1979). Microbial turnover of organic-matter in the deep-sea. *Bioscience* **29**, 228–232.

Jannasch, H. W. and Mottl, M. J. (1985). Geomicrobiology of deep-sea hydrothermal vents. *Science* **229**, 717–725.

Jannasch, H. W., Nelson, D. C., and Wirsen, C. O. (1989). Massive natural occurence of unusually large bacteria (*Beggiatoa* sp.) at a hydrothermal deep-sea vent site. *Nature* **342**, 834–836.

Jannasch, H. W., Wirsen, C. O., and Molyneaux, S. J. (1991). Chemoautotrophic sulfur-oxidizing bacteria from the Black Sea. *Deep-Sea Research* **38**, S1105–S1120.

Janssen, P. H., Schuhmann, A., Bak, F., and Liesack, W. (1996). Fermentation of inorganic sulfur compounds by the sulfate-reducing bacterium *Desulfocapsa thiozymogenes* gen. nov., sp. nov. *Archives of Microbiology* **166**, 184–192.

Jansson, M., Olsson, H., and Pettersson, K. (1988). Phosphatases—Origin, characteristics and function in lakes. *Hydrobiologia* **170**, 157–175.

Jenkins, M. C. and Kemp, W. M. (1984). The coupling of nitrification and denitrification in two estuarine sediments. *Limnology and Oceanography* **29**, 609–619.

Jenney, F. E., Verhagen, M., Cui, X. Y., and Adams, M. W. W. (1999). Anaerobic microbes: Oxygen detoxification without superoxide dismutase. *Science* **286**, 306–309.

Jensen, H. S. and Andersen, F. Ø. (1992). Importance of temperature, nitrate, and pH for phosphate release from aerobic sediments of 4 shallow, eutrophic lakes. *Limnology and Oceanography* **37**, 577–589.

Jensen, H. S. and Thamdrup, B. (1993). Iron-bound phosphorus in marine-sediments as measured by bicarbonate-dithionite extraction. *Hydrobiologia* **253**, 47–59.

Jensen, H. S., Mortensen, P. B., Andersen, F. O., Rasmussen, E., and Jensen, A. (1995). Phosphorus cycling in a coastal marine sediment, Aarhus Bay, Denmark. *Limnology and Oceanography* **40**, 908–917.

Jensen, K., Revsbech, N. P., and Nielsen, L. P. (1993). Microscale distribution of nitrification activity in sediment determined with a shielded microsensor for nitrate. *Applied and Environmental Microbiology* **59**, 3287–3296.

Jensen, K., Sloth, N. P., Risgaard-Petersen, N., Rysgaard, S., and Revsbech, N. P. (1994). Estimation of nitrification and denitrification from microprofiles of oxygen and nitrate in model sediment systems. *Applied and Environmental Microbiology* **60**, 2094–2100.

Jensen, M. M., Thamdrup, B., Ryesgaard, S., Holmer, M., and Fossing, H. (2003). Rates and regulation of microbial iron reduction in sediments of the Baltic-North Sea transition. *Biogeochemistry* **65**, 295–317.

Jerlov, N. G. (1976). Marine Optics, p. 231. Elsevier Scientific Publications.

Jetten, M. S. M. (2001). New pathways for ammonia conversion in soil and aquatic systems. *Plant and Soil* **230**, 9–19.

Jetten, M. S. M., Logemann, S., Muyzer, G., Robertson, L. A., de Vries, S., van Loosdrecht, M. C. M., and Kuenen, J. G. (1997). Novel principles in the microbial conversion of nitrogen compounds. *Antonie Van Leeuwenhoek International Journal of General and Molecular Microbiology* **71**, 75–93.

Jetten, M. S. M., Strous, M., van de Pas-Schoonen, K. T., Schalk, J., van Dongen, U., van de Graaf, A. A., Logemann, S., Muyzer, G., van Loosdrecht, M. C. M., and Kuenen, J. G. (1998). The anaerobic oxidation of ammonium. *FEMS Microbiology Reviews* **22**, 421–437.

Jetten, M. S. M., Wagner, M., Fuerst, J., van de Pas-Schoonen, K. T., Kuenen, G., and Strous, M. (2001). Microbiology and application of the anaerobic ammonium oxidation. ('anammox') processe. *Current Opinion in Biotechnology* **12**, 283–288.

Jin, Q. and Bethke, C. M. (2003). A new rate law describing microbial respiration. *Applied and Environmental Microbiology* **69**, 2340–2348.

Jochem, F. and Babenerd, B. (1989). Naked *Dictyocha Speculum*—A new type of phytoplankton bloom in the Western Baltic. *Marine Biology* **103**, 373–379.

Johnson, K. S., Gordon, R. M., and Coale, K. H. (1997). What controls dissolved iron concentrations in the world ocean? *Marine Chemistry* **57**, 137–161.

Johnson, P. W. and Sieburth, J. M. (1979). Chroococcoid cyanobacteria in the sea—Ubiquitous and diverse phototropic biomass. *Limnology and Oceanography* **24**, 928–935.

Jones, J. G., Davison, W., and Gardener, S. (1984). Iron reduction by bacteria—Range of organisms involved and metals reduced. *FEMS Microbiology Letters* **21**, 133–136.

Jones, K. J. and Gowen, R. J. (1990). Influence of stratification and irradiance regime on summer phytoplankton composition in coastal and shelf seas of the British Isles. *Estuarine Coastal and Shelf Science* **30**, 557–567.

Jones, M. L. (1981). Riftia pachyptila Jones: Observations on the vestimentiferan worm from the Galapagos rift. *Science* **213**, 333–336.

Jones, R. D. and Morita, R. Y. (1985). Low-temperature growth and whole-cell kinetics of a marine ammonium oxidizer. *Marine Ecology-Progress Series* **21**, 239–243.

Jonkers, H. M., Ludwig, R., De Wit, R., Pringault, O., Muyzer, G., Niemann, H., Finke, N., and de Beer, D. (2003). Structural and functional analysis of a microbial mat ecosystem from a unique permanent hypersaline inland lake: "La Salada de Chiprana" (NE Spain). *FEMS Microbiology Ecology* **44**, 175–189.

Joye, S. B. and Hollibaugh, J. T. (1995). Influence of sulfide inhibition of nitrification on nitrogen regeneration in sediments. *Science* **270**, 623–625.

Joye, S. B., Smith, S. V., Hollibaugh, J. T., and Paerl, H. W. (1996). Estimating denitrification rates in estuarine sediments: A comparison of stoichiometric and acetylene based methods. *Biogeochemistry* **33**, 197–215.

Jørgensen, B. B. (1977a). Bacterial sulfate reduction within reduced microniches of oxidized marine sediments. *Marine Biology* **41**, 7–17.

Jørgensen, B. B. (1977b). Distribution of colorless sulfur bacteria (*Beggiatoa* spp.) in a coastal marine sediment. *Marine Biology* **41**, 19–28.

Jørgensen, B. B. (1977c). The sulfur cycle of a coastal marine sediment (Limfjorden, Denmark). *Limnology and Oceanography* **22**, 814–832.

Jørgensen, B. B. (1978a). A comparison of methods for the quantification of bacterial sulfate reduction in coastal marine sediments I. Measurement with radiotracer techniques. *Geomicrobiology Journal* **1**, 11–27.

Jørgensen, B. B. (1978b). A comparison of methods for the quantification of bacterial sulfate reduction in coastal marine sediments. II. Calculation from mathematical models. *Geomicrobiology Journal* **1**, 29–47.

Jørgensen, B. B. (1982a). Ecology of the bacteria of the sulphur cycle with special reference to anoxic-oxic interface environments. *Philosophical Transactions of The Royal Society of London Series B* **298**, 543–561.

Jørgensen, B. B. (1982b). Mineralization of organic matter in the sea bed—The role of sulfate reduction. *Nature* **296**, 643–645.

Jørgensen, B. B. (1983). Processes at the sediment-water interface. *In* "The Major Biogeochemical Cycles and Their Interactions" (B. Bolin and R. B. Cook, eds), pp. 477–515. John Wiley, Chichester.

Jørgensen, B. B. (1989). Biogeochemistry of chemoautotrophic bacteria. *In* "Autotrophic Bacteria" (H. G. Schlegel and B. Bowien, eds), pp. 117–146. Science Tech Publishers and Springer, Madison, WI.

Jørgensen, B. B. (1990a). A thiosulfate shunt in the sulfur cycle of marine sediments. *Science* **249**, 152–154.

Jørgensen, B. B. (1990b). The sulfur cycle of freshwater sediments: Role of thiosulfate. *Limnology and Oceanography* **35**, 1329–1342.

Jørgensen, B. B. (1996). Case study—Aahus Bay. *In* "Eutrophication in Coastal Marine Ecosystems" (B. B. Jørgensen and K. Richardson, eds), pp. 137–154. American Geophysical Union, Washington DC.

Jørgensen, B. B. (2000). Bacteria and marine biogeochemistry. *In* "Marine Geochemistry" (H. D. Schulz and M. Zabel, eds), pp. 173–207. Springer, Berlin.

Jørgensen, B. B. and Bak, F. (1991). Pathways and microbiology of thiosulfate transformations and sulfate reduction in a marine sediment (Kattegat, Denmark). *Applied and Environmental Microbiology* **57**, 847–856.

Jørgensen, B. B. and Cohen, Y. (1977). Solar Lake (Sinai). 5. The sulfur cycle of the benthic cyanobacterial mats. *Limnology and Oceanography* **22**, 657–666.
Jørgensen, B. B. and Des Marais, D. J. (1986a). Competition for sulfide among colorless and purple sulfur bacteria in cyanobacterial mats. *FEMS Microbiology Ecology* **38**, 179–186.
Jørgensen, B. B. and Des Marais, D. J. (1986b). A simple fiber-optic microprobe for high resolution light measurements: Application in marine sediment. *Limnology and Oceanography* **31**, 1376–1383.
Jørgensen, B. B. and Des Marais, D. J. (1988). Optical properties of benthic photosynthetic communities: Fiber-optic studies of cyanobacterial mats. *Limnology and Oceanography* **33**, 99–113.
Jørgensen, B. B. and Gallardo, V. A. (1999). *Thioploca* spp.: Filamentous sulfur bacteria with nitrate vacuoles. *FEMS Microbiology Ecology* **28**, 301–313.
Jørgensen, B. B. and Nelson, D. C. (1988). Bacterial zonation, photosynthesis, and spectral light distribution in hot spring microbial mats of Iceland. *Microbial Ecology* **16**, 133–147.
Jørgensen, B. B. and Revsbech, N. P. (1983). Colorless sulfur bacteria, *Beggiatoa* spp. and *Thiovulum* spp., in O_2 and H_2S microgradients. *Applied and Environmental Microbiology* **45**, 1261–1270.
Jørgensen, B. B. and Revsbech, N. P. (1985). Diffusive boundary layers and the oxygen uptake of sediments and detritus. *Limnology and Oceanography* **30**, 111–122.
Jørgensen, B. B. and Sørensen, J. (1985). Seasonal cycles of O_2, NO_3^- and SO_4^{2-} reduction in estuarine sediments—the significance of an NO_3^- reduction maximum in spring. *Marine Ecology-Progress Series* **24**, 65–74.
Jørgensen, B. B., Kuenen, J. G., and Cohen, Y. (1979). Microbial transformations of sulfur compounds in a stratified lake (Solar Lake, Sinai). *Limnology and Oceanography* **24**, 799–822.
Jørgensen, B. B., Revsbech, N. P., and Cohen, Y. (1983). Photosynthesis and structure of benthic microbial mats: Microelectrode and SEM studies of four cyanobacterial communities. *Limnology and Oceanography* **28**, 1075–1093.
Jørgensen, B. B., Fossing, H., Wirsen, C. O., and Jannasch, H. W. (1991). Sulfide oxidation in the anoxic Black Sea chemocline. *Deep-Sea Research* **38**, S1083–S1103.
Jørgensen, B. B., Isaksen, M. F., and Jannasch, H. W. (1992). Bacterial sulfate reduction above 100°C in deep sea hydrothermal vent sediments. *Science* **258**, 1756–1757.
Kaebernick, M. and Neilan, B. A. (2001). Ecological and molecular investigations of cyanotoxin production. *FEMS Microbiology Ecology* **35**, 1–9.
Kajander, E. O. and Çiftçioglu, N. (1998). Nanobacteria: An alternative mechanism for pathogenic intra- and extracellular calcification and stone formation. *Proceedings of the National Academy of Sciences USA* **95**, 8274–8279.
Kajander, E. O., Kuronen, I., Akerman, K. K., Pelttari, A., and Ciftcioglu, N. (1997). Nanobacteria from blood: The smallest culturable autonomously replicating agent on Earth. *Proceedings of the International Society for Optical Engineering* **3111**, 420–428.
Kalhorn, S. and Emerson, S. (1984). The oxidation state of manganese in surface sediments of the deep sea. *Geochimica et Cosmochimica Acta* **48**, 897–902.
Kálmán, L., LoBrutto, R., Allen, J. P., and Williams, J. C. (2003). Manganese oxidation by modified reaction centers from Rhodobacter sphaeroides. *Biochemistry* **42**, 11016–11022.
Kamatani, A., Ejiri, N., and Treguer, P. (1988). The dissolution kinetics of diatom ooze from the Antarctic Area. *Deep-Sea Research* **35**, 1195–1203.

Kana, T. M. (1993). Rapid oxygen cycling in *Trichodesmium thiebautii*. *Limnology and Oceanography* **38**, 18–24.
Kandler, O. (1994). Cell wall biochemistry and three-domain concept of life. *Systematic and Applied Microbiology* **16**, 501–509.
Kaplan, A. and Reinhold, L. (1999). CO_2 concentrating mechanisms in photosynthetic microorganisms. *Annual Review of Plant Physiology and Plant Molecular Biology* **50**, 539–570.
Kaplan, W. A. (1983). Nitrification. *In* "Nitrogen in the Marine Environment" (E. J. Carpenter and D. G. Capone, eds), pp. 139–190. Academic Press, New York.
Kappler, A. and Newman, D. K. (2004). Formation of Fe(III)-minerals by Fe(II)-oxidizing photoautotrophic bacteria. *Geochimica et Cosmochimica Acta* **68**, 1217–1226.
Karl, D., Letelier, R., Tupas, L., Dore, J., Christian, J., and Hebel, D. (1997). The role of nitrogen fixation in biogeochemical cycling in the subtropical North Pacific Ocean. *Nature* **388**, 533–538.
Karl, D., Michaels, A., Bergman, B., Capone, D., Carpenter, E., Letelier, R., Lipschultz, F., Paerl, H., Sigman, D., and Stal, L. (2002). Dinitrogen fixation in the world's oceans. *Biogeochemistry* **57**, 47–98.
Karl, D. M. (1993). Microbial RNA and DNA synthesis derived from the assimilation of [2. 3H]-adenine. *In* "Methods in Aquatic Microbial Ecology" (P. F. Kemp, B. F. Sherr, E. B. Sherr, and J. J. Cole, eds), pp. 471–480. Lewis Publishers, Boca Raton, Florida.
Karl, D. M. and Björkman, K. M. (2002). Dynamics of DOP. *In* "Biogeochemistry of Marine Dissolved Organic Matter" (D. A. Hansell and C. A. Carlson, eds). Academic Press, Amsterdam.
Karl, D. M. and Novitsky, J. A. (1988). Dynamics of microbial growth in surface layers of a coastal marine sediment ecosystem. *Marine Ecology-Progress Series* **50**, 169–176.
Karl, D. M. and Winn, C. D. (1984). Adenine metabolism and nucleic acid synthesis: Applications to microbiological oceanography. *In* "Heterotrophic Activity in the Sea" (J. E. Hobbie and P. J. L. Williams, eds), pp. 197–215. Plenum Publishing Corp., New York.
Karner, M. B., DeLong, E. F., and Karl, D. M. (2001). Archaeal dominance in the mesopelagic zone of the Pacific Ocean. *Nature* **409**, 507–510.
Karnholz, A., Kusel, K., Gossner, A., Schramm, A., and Drake, H. L. (2002). Tolerance and metabolic response of acetogenic bacteria toward oxygen. *Applied and Environmental Microbiology* **68**, 1005–1009.
Kashefi, K. and Lovley, D. (2003). Extending the upper temperature limit for life. *Science* **310**, 934.
Kashefi, K., Holmes, D. E., Reysenbach, A. L., and Lovley, D. R. (2002). Use of Fe(III) as an electron acceptor to recover previously uncultured hyperthermophiles: Isolation and characterization of *Geothermobacterium ferrireducens* gen. nov., sp nov. *Applied and Environmental Microbiology* **68**, 1735–1742.
Kaspar, H. F., Asher, R. A., and Boyer, I. C. (1985). Microbial nitrogen transformations in sediments and inorganic nitrogen fluxes across the sediment water interface on the South Island West Coast, New Zealand. *Estuarine Coastal and Shelf Science* **21**, 245–255.
Kasting, J. F. and Walker, J. C. G. (1981). Limits on oxygen concentration in the prebiotic atmosphere and the rate of abiotic fixation of nitrogen. *Journal of Geophysical Research-Oceans and Atmospheres* **86**, 1147–1158.

Kasting, J. F., Pollack, J. B., and Crisp, D. (1984). Effects of high CO_2 levels on surface-temperature and atmospheric oxidation-state of the early Earth. *Journal of Atmospheric Chemistry* **1**, 403–428.

Kastner, M. (1979). Silica polymorphs. *Reviews in Mineralogy* **6**, 99–109.

Kastner, M., Keene, J. B., and Gieskes, J. M. (1977). Diagenesis of siliceous oozes. I. Chemical controls on rate of opal-A to opal-CT transformation—experimental study. *Geochimica et Cosmochimica Acta* **41**, 1041.

Kathiresan, K. and Bingham, B. L. (2001). Biology of mangroves and mangrove ecosystems. *Advances in Marine Biology* **40**, 81–251.

Kehres, D. G. and Maguire, M. E. (2003). Emerging themes in manganese transport, biochemistry and pathogenesis in bacteria. *FEMS Microbiology Reviews* **27**, 263–290.

Kelley, D. S., Baross, J. A., and Delaney, J. R. (2002). Volcanoes, fluids, and life at mid-ocean ridge spreading centers. *Annual Review of Earth and Planetary Sciences* **30**, 385–491.

Kelly, D. P. and Wood, A. P. (2000). Reclassification of some species of *Thiobacillus* to the newly designated genera *Acidithiobacillus* gen. nov., Halothiobacillus gen. nov. and Thermithiobacillus gen.nov. *International Journal of Systematic and Evolutionary Microbiology* **50**, 511–516.

Kelly, D. P., McDonald, I. R., and Wood, A. P. (2000). Proposal for the reclassification of *Thiobacillus novellus* as *Starkeya novella* gen. nov., comb. nov., in the alpha-subclass of the Proteobacteria. *International Journal of Systematic and Evolutionary Microbiology* **50**, 1797–1802.

Kelly-Gerreyn, B. A., Anderson, T. R., Holt, J. T., Gowen, R. J., and Proctor, R. (2004). Phytoplankton community structure at contrasting sites in the Irish Sea: A modeling investigation. *Estuarine Coastal and Shelf Science* **59**, 363–383.

Kemp, P. F., Lee, S., and LaRoche, J. (1993). Estimating the growth rate of slowly growing marine bacteria from RNA content. *Applied and Environmental Microbiology* **59**, 2594–2601.

Kennett, J. P., Cannariato, K. G., Hendy, I. L., and Behl, R. J. (2003). Methane Hydrates in Quaternary Climate Change: The Clathrate Gun Hypothesis, p. 216. American Geophysical Union, Washington, DC.

Kiddon, J., Bender, M. L., Orchardo, J., Caron, D. A., Goldman, J. C., and Dennett, M. (1993). Isotopic fractionation of oxygen by respiring marine organisms. *Global Biogeochemical Cycles* **7**, 679–694.

King, D. W., Lounsbury, H. A., and Millero, F. J. (1995). Rates and mechanism of Fe(II) oxidation at nanomolar total iron concentrations. *Environmental Science & Technology* **29**, 818–824.

King, G. M., Roslev, P., and Skovgaard, H. (1990). Distribution and rate of methane oxidation in sediments of the Florida Everglades. *Applied and Environmental Microbiology* **56**, 2902–2911.

King, S. L., Quay, P. D., and Lansdown, J. M. (1989). The $^{13}C/^{12}C$ kinetic isotope effect for soil oxidation of methane at ambient atmospheric concentrations. *Journal of Geophysical Research* **94**, 18273–18277.

Kirchman, D. L. (1992). Incorporation of thymidine and leucine in the subarctic Pacific: Application to estimating bacterial production. *Marine Ecology-Progress Series* **82**, 301–309.

Kirchman, D. L. (1993). Leucine incorporation as a measure of biomass production by heterotrophic bacteria. *In* "Aquatic Microbial Ecology" (P. F. Kemp, B. F. Sherr, E. B. Sherr, and J. J. Cole, eds), pp. 509–512. Lewis Publishers, Boca Raton.

Kirchman, D. L. (2000). Uptake and regeneration of inorganic nutrients by marine heterotrophic bacteria. *In* "Microbial Ecology of the Oceans" (D. L. Kirchman, ed.). Wiley-Liss, Inc., New York.

Kirchman, D. L. (2002). The ecology of Cytophaga-Flavobacteria in aquatic environments. *FEMS Microbiology Ecology* **39**, 91–100.

Kirchman, D. L., Suzuki, Y., Garside, C., and Ducklow, H. W. (1991). High turnover rates of dissolved organic carbon during a spring phytoplankton bloom. *Nature* **352**, 612–614.

Kitaya, Y., Yabuki, K., Kiyota, M., Tani, A., Hirano, T., and Aiga, I. (2002). Gas exchange and oxygen concentration in pneumatophores and prop roots of four mangrove species. *Trees-Structure and Function* **16**, 155–158.

Klein, M., Friedrich, M., Roger, A. J., Hugenholtz, P., Fishbain, S., Abicht, H., Blackall, L. L., Stahl, D. A., and Wagner, M. (2001). Multiple lateral transfers of dissimilatory sulfite reductase genes between major lineages of sulfate-reducing prokaryotes. *Journal of Bacteriology* **183**, 6028–6035.

Klinkhammer, G. P., McManus, J., Colbert, D., and Rudnicki, M. D. (2000). Behavior of terrestrial dissolved organic matter at the continent-ocean boundary from high-resolution distributions. *Geochimica et Cosmochimica Acta* **64**, 2765–2774.

Klumpp, D. W., Salitaespinosa, J. S., and Fortes, M. D. (1992). The role of epiphytic periphyton and macroinvertebrate grazers in the trophic flux of a tropical seagrass community. *Aquatic Botany* **43**, 327–349.

Knoblauch, C., Jørgensen, B. B., and Harder, J. (1999). Community size and metabolic rates of psychrophilic sulfate-reducing bacteria in arctic marine sediments. *Applied and Environmental Microbiology* **65**, 4230–4233.

Knoll, A. H. (2003). Life on a Young Planet. The First Three Billion Years of Evolution on Earth, p. 277. Princeton University Press, Princeton and Oxford.

Knoll, A. H. and Bauld, J. (1989). The evolution of ecological tolerance in Prokaryotes. *Transactions of the Royal Society of Edinburgh: Earth Sciences* **80**, 209–223.

Koike, I. and Sørensen, J. (1988). Nitrate reduction and denitrification in marine sediments. *In* "Nitrogen Cycling in Coastal Marine Environments" (T. H. Blackburn and J. Sørensen, eds), pp. 251–273. John Wiley and Sons, Chichester.

Kolber, Z. S., Plumley, F. G., Lang, A. S., Beatty, J. T., Blankenship, R. E., VanDover, C. L., Vetriani, C., Koblizek, M., Rathgeber, C., and Falkowski, P. G. (2001). Contribution of aerobic photoheterotrophic bacteria to the carbon cycle in the ocean. *Science* **292**, 2492–2495.

Kondratieva, E. N., Pfennig, N., and Trüper, H. G. (1992). The photosynthetic prokaryotes. *In* "The Prokaryotes" (A. Balows, H. G. Trüper, M. Dworkin, W. Harder, and K. H. Schleifer, eds), 2nd Ed. Springer, Berlin.

Koning, E., Brummer, G. J., Van Raaphorst, W., Van Bennekom, J., Helder, W., and Van Iperen, J. (1997). Settling, dissolution and burial of biogenic silica in the sediments off Somalia (northwestern Indian Ocean). *Deep-Sea Research* II-*Topical Studies in Oceanography* **44**, 1341–1360.

Koops, H. P. and Pommerening-Roser, A. (2001). Distribution and ecophysiology of the nitrifying bacteria emphasizing cultured species. *FEMS Microbiology Ecology* **37**, 1–9.

Kostka, J. E. and Nealson, K. H. (1995). Dissolution and reduction of magentite by bacteria. *Environmental Science and Technology* **29**, 2535.

Kostka, J. E., Luther, G. W., and Nealson, K. H. (1995). Chemical and biological reduction of Mn(III)-pyrophosphate complexes: Potential importance of dissolved Mn(III) as an environmental oxidant. *Geochimica et Cosmochimica Acta* **59**, 885–894.

Kostka, J. E., Stucki, J. W., Nealson, K. H., and Wu, J. (1996). Reduction of structural Fe(III) in smectite by a pure culture of *Shewanella putrefaciens* strains MR-1. *Clays and Clay Minerals* **44**, 522–529.

Kostka, J. E., Thamdrup, B., Glud, R. N., and Canfield, D. E. (1999). Rates and pathways of carbon oxidation in permanently cold Arctic sediments. *Marine Ecology-Progress Series* **180**, 7–21.

Kostka, J. E., Dalton, D. D., Skelton, H., Dollhopf, S., and Stucki, J. W. (2002a). Growth of iron(III)-reducing bacteria on clay minerals as the sole electron acceptor and comparison of growth yields on a variety of oxidized iron forms. *Applied and Environmental Microbiology* **68**, 6256–6262.

Kostka, J. E., Gribsholt, B., Petrie, E., Dalton, D., Skelton, H., and Kristensen, E. (2002b). The rates and pathways of carbon oxidation in bioturbated saltmarsh sediments. *Limnology and Oceanography* **47**, 230–240.

Kowalchuk, G. A., Stephen, J. R., DeBoer, W., Prosser, J. I., Embley, T. M., and Woldendorp, J. W. (1997). Analysis of ammonia-oxidizing bacteria of the b subdivision of the class *Proteobacteria* in coastal sand dunes by denaturing gradient gel electrophoresis and sequencing of PCR-amplified 16S ribosomal DNA fragments. *Applied and Environmental Microbiology* **63**, 1489–1497.

Krämer, M. and Cypionka, H. (1989). Sulfate formation via ATP sulfurylase in thiosulfate- and sulfite-disproportionating bacteria. *Archives of Microbiology* **151**, 232–237.

Krekeler, D. and Cypionka, H. (1995). The preferred electron acceptor of *Desulfovibrio desulfiricans* CSN. *FEMS Microbiology Ecology* **17**, 271–278.

Kristensen, E. (1990). Characterization of biogenic organic-matter by stepwise thermogravimetry (STG). *Biogeochemistry* **9**, 135–159.

Kristensen, E. (1993). Seasonal variations in Benthic community metabolism and nitrogen dynamics in a shallow, organic-poor Danish lagoon. *Estuarine, Coastal and Shelf Science* **36**, 565–586.

Kristensen, E. and Hansen, K. (1995). Decay of plant detritus in organic-poor marine sediment: Production rates and stoichiometry of dissolved C and N compounds. *Journal of Marine Research* **53**, 675–702.

Kristensen, E. and Hansen, K. (1999). Transport of carbon dioxide and ammonium in bioturbated (Nereis diversicolor) coastal, marine sediments. *Biogeochemistry* **45**, 147–168.

Kristensen, E. and Holmer, M. (2001). Decomposition of plant materials in marine sediments exposed to different electron acceptors (O_2, NO_3^-, SO_4^{2-}), with emphasis on substrate origin, degradation kinetics, and the role of bioturbation. *Geochimica et Cosmochimica Acta* **65**, 419–433.

Kristensen, E. and Pilgaard, R. (2001). The role of fecal pellet deposition by leaf-eating sesarmid crabs on litter decomposition in a mangrove sediment (Phuket, Thailand). *In* "Organism-Sediment Interactions" (J. Y. Aller, S. A. Woodin, and R. C. Aller, eds), pp. 369–384. University of South Carolina Press, Columbia.

Kristensen, E. and Suraswadi, P. (2002). Carbon, nitrogen and phosphorus dynamics in creek water of a southeast Asian mangrove forest. *Hydrobiologia* **474**, 197–211.

Kristensen, E., Holmer, M., and Bussarawit, N. (1991a). Benthic metabolism and sulfate reduction in a Southeast-Asian mangrove swamp. *Marine Ecology-Progress Series* **73**, 93–103.

Kristensen, E., Jensen, M. H., and Aller, R. C. (1991b). Direct measurement of dissolved inorganic nitrogen exchange and denitrification in individual polychaete (*Nereis virens*) burrows. *Journal of Marine Research* **49**, 355–377.

Kristensen, E., Devol, A. H., Ahmed, S. I., and Saleem, M. (1992). Preliminary study of benthic metabolism and sulfate reduction in a mangrove swamp of the Indus delta, Pakistan. *Marine Ecology-Progress Series* **90**, 287–297.

Kristensen, E., King, G. M., Holmer, M., Banta, G. T., Jensen, M. H., Hansen, K., and Bussarawit, N. (1994). Sulfate reduction, acetate turnover and carbon metabolism in sediments of the Ao-Nam-Bor mangrove, Phuket, Thailand. *Marine Ecology-Progress Series* **109**, 245–255.

Kristensen, E., Holmer, M., Banta, G. T., Jensen, M. H., and Hansen, K. (1995). Carbon, nitrogen and sulfur cycling in sediments of the Ao Nam Bor mangrove forest, Phuket, Thailand: A review. *Phuket Marine Biological Center Research Bulletin* **60**, 37–64.

Kristensen, E., Jensen, M. H., Banta, G. T., Hansen, K., Holmer, M., and King, G. M. (1998). Transformation and transport of inorganic nitrogen in sediments of a southeast Asian mangrove forest. *Aquatic Microbial Ecology* **15**, 165–175.

Kristensen, E., Devol, A. H., and Hartnett, H. E. (1999). Organic matter diagenesis in sediment on the continental shelf and slope of the Eastern Tropical and temperate North Pacific. *Continental Shelf Research* **19**, 1331–1351.

Kristensen, E., Andersen, F. Ø., Holmboe, N., Holmer, M., and Thongtham, N. (2000). Carbon and nitrogen mineralization in sediments of Bangrong area, Phuket, Thailand. *Aquatic Microbial Ecology* **22**, 199–213.

Kröger, N. and Sumper, M. (2000). The biochemistry of silica formation in diatoms. *In* "Biomineralization" (E. Bäuerlein, ed.), pp. 151–170. Wiley-VCH, Weinheim.

Krohn, M. D., Evans, J., and Robinson, G. R., Jr. (1988). Mineral-bound ammonium in black shales of the Triassic Cumnock Formation, Deep River Basin, North Carolina. *U.S. Geological Survey Bulletin* **1776**, 86–98.

Krull, E. S. and Retallack, G. J. (2000). Delta C-13 depth profiles from paleosols across the Permian-Triassic boundary: Evidence for methane release. *Geological Society of America Bulletin* **112**, 1459–1472.

Krulwich, T. A. (2000). Alkaliphilic Prokaryotes. *In* "The Prokaryotes: An Evolving Electronic Resource for the Microbiological Community" (M. Dworkin, A. Balows, H. G. Trüper, W. Harder, and K.-H. Schleifer, eds), 3rd Ed. release 3,7, November 2, 2001. Springer, New York.

Krumbein, W. E. and Werner, D. (1983). The microbial silica cycle. *In* "Microbial Geochemistry" (W. E. Krumbein, ed.), pp. 125–157. Blackwell Scientific Publishers, Oxford.

Krumholz, L. R., McKinley, J. P., Ulrich, G. A., and Suflita, J. M. (1997). Confined subsurface microbial communities in Cretaceous rock. *Nature* **386**, 64–66.

Krzycki, J. A., Kenealy, W. R., DeNiro, M. J., and Zeikus, J. G. (1987). Stable carbon isotopes fractionation by *Methanosarcina barkeri* during methanogenesis from acetate, methanol, or carbon dioxide-hydrogen. *Applied and Environmental Microbiology* **53**, 2597–2599.

Kuenen, G. (1999). Oxidation of inorganic compounds by chemolithotrophs. *In* "Biology of the Prokaryotes" (J. W. Lengeler, G. Drews, and H. G. Schlegel, eds), pp. 234–260. Blackwell Science, New York.

Kühl, M. and Fenchel, T. (2000). Bio-optical characteristics and the vertical distribution of photosynthetic pigments and photosynthesis in an artificial cyanobacterial mat. *Microbial Ecology* **40**, 94–103.

Kühl, M. and Jørgensen, B. B. (1994). The light field of microbenthic communities: Radience distribution and microscale optics of sandy coastal sediments. *Limnology and Oceanography* **39**, 1368–1398.

Kühl, M., Lassen, C., and Jørgensen, B. B. (1994). Optical properties of microbial mats: Light measurements with fiber-optic microprobes. *In* "Microbial Mats: Structure, Development and Environmental Significance" (L. Stal and P. Caumette, eds), pp. 305–311. Springer, Berlin.

Kuivila, K. M., Murray, J. W., and Devol, A. H. (1988). Methane production, sulfate reduction and competition for substrates in the sediments of Lake Washington. *Geochemica et Cosmochimica Acta* **53**, 409–416.

Kuivila, K. M., Murray, J. W., and Devol, A. H. (1990). Methane production in the sulfate-depleted sediments of two marine basins. *Geochimica et Cosmochimica Acta* **54**, 403–411.

Küsel, K., Dorsch, T., Acker, G., and Stackebrandt, E. (1999). Microbial reduction of Fe(III) in acidic sediments: Isolation of *Acidiphilium cryptum* JF-5 capable of coupling the reduction of Fe(III) to the oxidation of glucose. *Applied and Environmental Microbiology* **65**, 3633–3640.

Kustka, A., Carpenter, E. J., and Sanudo-Wilhelmy, S. A. (2002). Iron and marine nitrogen fixation: Progress and future directions. *Research in Microbiology* **153**, 255–262.

Kuypers, M. M. M., Sliekers, A. O., Lavik, G., Schmid, M., Jorgensen, B. B., Kuenen, J. G., Damste, J. S. S., Strous, M., and Jetten, M. S. M. (2003). Anaerobic ammonium oxidation by anammox bacteria in the Black Sea. *Nature* **422**, 608–611.

Kvenvolden, K. A. (1988a). Methane hydrate—A major reservoir of carbon in a shallow geosphere? *Chemical Geology* **71**, 41–51.

Kvenvolden, K. A. (1988b). Methane hydrates and global climate. *Global Biogeochemical Cycles* **2**, 221–229.

Kvenvolden, K. A. (1993). Gas hydrates—Geological perspective and global change. *Reviews of Geophysics* **31**, 173–187.

Landing, W. M. and Bruland, K. W. (1987). The contrasting biogeochemistry of iron and manganese in the Pacific Ocean. *Geochimica et Cosmochimica Acta* **51**, 29–43.

Lansdown, J. M., Quay, P. D., and King, S. L. (1992). CH_4 production via CO_2 reduction in a temperate bog: A source of ^{13}C-depleted CH_4. *Geochimica et Cosmochimica Acta* **56**, 3493–3503.

LaRoche, J., van der Staay, G. W. M., Partensky, F., Ducret, A., Aebersold, R., Li, R., Golden, S. S., Hiller, R. G., Wrench, P. M., Larkum, A. W. D., and Green, B. R. (1996). Independent evolution of the prochlorophyte and green plant chlorophyll a/b light-harvesting proteins. *Proceedings of the National Academy of Sciences USA* **93**, 15244–15248.

Lee, C. and Henrichs, S. M. (1993). How the nature of dissolved organic-matter might affect the analysis of dissolved organic-carbon. *Marine Chemistry* **41**, 105–120.

Lee, K. H., Moran, M. A., Benner, R., and Hodson, R. E. (1990). Influence of soluble components of red mangrove (*Rhizophora mangle*) leaves on microbial decomposition of structural (lignocellulosic) leaf components in seawater. *Bulletin of Marine Science* **46**, 374–386.

Lee, S. Y. (1997). Potential trophic importance of the faecal material of the mangrove sesarmine crab Sesarma messa. *Marine Ecology-Progress Series* **159**, 275–284.

Le Faou, A., Rajagopal, B. S., Daniels, L., and Fauque, G. (1990). Thiosulfate, polythionates and elemental sulfur assimilation and reduction in the bacterial world. *FEMS Microbiology Reviews* **75**, 351–382.

Lefèvre, D., Denis, M., Lambert, C. E., and Miquel, J.-C. (1996). Is DOC the main source of organic matter remineralization in the ocean water column? *Journal of Marine Systems* **7**, 281–291.

Lein, A. Y., Ivanov, M. V., and Vainshtein, M. B. (1990). Hydrogen sulfide balance in the deep water zone of the Black Sea. *Microbiology* **59**, 453–460.
Le Mer, J. and Roger, P. (2001). Production, oxidation, emission and consumption of methane by soils: A review. *European Journal of Soil Biology* **37**, 25–50.
Lemos, R. S., Gomes, C. M., Santana, M., LeGall, J., Xavier, A. V., and Teixeira, M. (2001). The 'strict' anaerobe *Desulfovibrio gigas* contains a membrane-bound oxygen-reducing respiratory chain. *FEBS Letters* **496**, 40–43.
Lewis, B. L. and Landing, W. M. (1991). The biogeochemistry of manganese and iron in the Black Sea. *Deep-Sea Research Part A-Oceanographic Research Papers* **38**, S773–S803.
Lewitus, A. J. and Holland, A. F. (2003). Initial results from a multi-institutional collaboration to monitor harmful algal blooms in South Carolina. *Environmental Monitoring and Assessment* **81**, 361–371.
Li, J.-H., Purdy, K. J., Takii, S., and Hayashi, H. (1999a). Seasonal changes in ribosomal RNA of sulfate-reducing bacteria and sulfate reducing activity in a freshwater lake sediment. *FEMS Microbiology Ecology* **28**, 31–39.
Li, L. N., Kato, C., and Horikoshi, K. (1999b). Bacterial diversity in deep-sea sediments from different depths. *Biodiversity and Conservation* **8**, 659–677.
Lidstrom, M. E. (2001). Aerobic methylotrophic prokaryotes. *In* "The Prokaryotes" (M. Dworkin, A. Balows, H. G. Trüper, W. Harder, and K.-H. Schleifer, eds), 3rd Ed., release 3.7, November 11. Springer, New York.
Lilley, M. D., Butterfield, D. A., Olson, J. E., Lupton, J. E., Macko, S. A., and McDuff, R. E. (1993). Anomalous CH_4 and NH_4 concentrations at an insedimented mid-ocean-ridge hydrothermal system. *Nature* **364**, 45–47.
Lin, J. and Gerstein, M. (2002). Whole-genome trees based on the occurence of folds and orthologs: Implications for comparing genomes on different levels. *Genome Research* **10**, 808–818.
Liu, K.-K. and Kaplan, I. R. (1984). Denitrification rates and availability of organic matter in marine environments. *Earth and Planetary Science Letters* **68**, 88–100.
Liu, M. C. and Peck, H. D. (1981). The isolation of a hexaheme cytochrome from *Desulfovibrio desulfuricans* and its identification as a new type of nitrite reductase. *Journal of Biological Chemistry* **256**, 3159–3164.
Llobet-Brossa, E., Rosselló-Móra, R., and Amann, R. (1998). Microbial community composition of Wadden Sea sediments as revealed by fluorescence in situ hybridization. *Applied and Environmental Microbiology* **64**, 2691–2696.
Lloyd, J. R. (2003). Microbial reduction of metals and radionuclides. *FEMS Microbiology Reviews* **27**, 411–425.
Loh, A. N. and Bauer, J. E. (2000). Distribution, partitioning and fluxes of dissolved and particulate organic C, N and P in the eastern North Pacific and Southern Oceans. *Deep-Sea Research Part I-Oceanographic Research Papers* **47**, 2287–2316.
Lohse, L., Helder, W., Epping, E. H. G., and Balzer, W. (1998). Recycling of organic matter along shelf-slope transect across the N. W. European Continental Marg (Goban Spur). *Progress in Oceanography* **42**, 77–110.
Lombard, M., Fontecave, M., Touati, D., and Niviere, V. (2000). Reaction of the desulfoferrodoxin from *Desulfoarculus baarsii* with superoxide anion: Evidence for a superoxide reductase activity. *Journal of Biological Chemistry* **275**, 115–121.
Lomstein, B. A., Jensen, A.-G. U., Hansen, J. W., Andreasen, J. B., Hansen, L. S., Berntsen, J., and Kunzendorf, H. (1998). Budgets of sediment nitrogen and carbon cycling in the shallow water of Knebel Vig, Denmark. *Aquatic Microbial Ecology* **14**, 69–80.

Lorenzen, J., Larsen, L. H., Kjaer, T., and Revsbech, N. P. (1998). Biosensor determination of the microscale distribution of nitrate, nitrate assimilation, nitrification, and denitrification in a diatom-inhabited freshwater sediment. *Applied and Environmental Microbiology* **64**, 3264–3269.

Lovley, D. R. (1985). Minimum threshhold for hydrogen metabolism in methanogenic bacteria. *Applied and Environmental Microbiology* **49**, 1530–1531.

Lovley, D. R. (1991). Dissimilatory Fe(III) and Mn(IV) reduction. *Microbiological Reviews* **55**, 259–287.

Lovley, D. R. (2000a). Anaerobic benzene degradation. *Biodegradation* **11**, 107–116.

Lovley, D. R. (2000b). Dissimilatory Fe(III)- and Mn(IV)-reducing prokaryotes. *In* "The Prokaryotes" (M. Dwork, ed.). Springer, New York.

Lovley, D. R. (2000c). Fe(III) and Mn(IV) reduction. *In* "Environmental Metal-Microbe Interactions" (D. R. Lovely, ed.), pp. 3–30. ASM Press, Washington, D. C.

Lovley, D. R. (2000d). Fe(III)- and Mn(IV)-reducing prokaryotes. *In* "The Prokaryotes" (E. Stackebrandt, ed.). Springer, New York.

Lovley, D. R. (2002). Dissimilatory metal reduction: From early life to bioremediation. *ASM News* **68**, 231–237.

Lovley, D. R. and Goodwin, S. (1988). Hydrogen concentrations as an indicator of the predominant terminal electron-accepting reactions in aquatic sediments. *Geochimica et Cosmochimica Acta* **52**, 2993–3003.

Lovley, D. R. and Klug, M. J. (1986). Model for the distribution of sulfate reduction and methanogensis in freshwater sediments. *Geochimica et Cosmochimica Acta* **50**, 11–18.

Lovley, D. R. and Phillips, E. J. P. (1987a). Competitive mechanisms for inhibition of sulfate reduction and methane prodution in the zone of ferric iron reduction in sediments. *Applied and Environmental Microbiology* **53**, 2636–2641.

Lovley, D. R. and Phillips, E. J. P. (1987b). Rapid assay for microbially reducible ferric iron in aquatic sediments. *Applied and Environmental Microbiology* **53**, 1536–1540.

Lovley, D. R. and Phillips, E. J. P. (1988). Novel mode of microbial energy metabolism: Organic carbon oxidation coupled to dissimilatory reduction of iron or manganese. *Applied and Environmental Microbiology* **54**, 1472–1480.

Lovley, D. R. and Phillips, E. J. P. (1994). Novel processes for anaerobic sulfate production from elemental sulfur by sulfate-reducing bacteria. *Applied and Environmental Microbiology* **60**, 2394–2399.

Lovley, D. R., Dwyer, D. F., and Klug, M. J. (1982). Kinetic analysis of competition between sulfate reducers and methanogens for hydrogen in sediments. *Applied and Environmental Microbiology* **43**, 1373–1379.

Lovley, D. R., Coates, J. D., Blunt-Harris, E. L., Phillips, E. J. P., and Woodward, J. C. (1996). Humic substances as electron acceptors for microbial respiration. *Nature* **382**, 445–448.

Lovley, E. D., Roden, E. E., Phillips, E. J. P., and Woodward, J. C. (1993). Enzymatic iron and uranium reduction by sulfate-reducing bacteria. *Marine Geology* **113**, 41–53.

Luckge, A., Ercegovac, M., Strauss, H., and Littke, R. (1999). Early diagenetic alteration of organic matter by sulfate reduction in Quaternary sediments from the northeastern Arabian Sea. *Marine Geology* **158**, 1–13.

Lucotte, M. and d'Anglejan, B. (1985). A comparison of several methods for the determination of iron hydroxides and associated ortho-phosphates in estuarine particulate matter. *Chemical Geology* **48**, 257–264.

Lugo, A. E. and Snedaker, S. C. (1974). The ecology of mangroves. *Annual Review of Ecology and Systematics* **5**, 39–65.
Luther, G. W. and Popp, J. I. (2002). Kinetics of the abiotic reduction of polymeric manganese dioxide by nitrite: An anaerobic nitrification reaction. *Aquatic Geochemistry* **8**, 15–36.
Luther, G. W., III, Church, T., and Powell, D. (1991). Sulfur speciation and sulfide oxidation in the water column of the Black Sea. *Deep-Sea Research* **38**, S1121–S1137.
Luther, G. W., III, Sundby, B., Lewis, B. L., and Brendel, P. J. (1997). Interactions of manganese with the nitrogen cycle: Alternative pathways to dinitrogen. *Geochimica et Cosmochimica Acta* **61**, 4043–4052.
Lynch, J. C., Meriwether, J. R., McKee, B. A., Veraherrera, F., and Twilley, R. R. (1989). Recent accretion in mangrove ecosystems based on ^{137}Cs and ^{210}Pb. *Estuaries* **12**, 284–299.
Macfarlane, G. T. and Herbert, R. A. (1984). Dissimilatory nitrate reduction and nitrification in estuarine sediments. *Journal of General Microbiology* **130**, 2301–2308.
Macintosh, D. J., Ashton, E. C., and Havanon, S. (2002). Mangrove rehabilitation and intertidal biodiversity: A study in the Ranong mangrove ecosystem, Thailand. *Estuarine Coastal and Shelf Science* **55**, 331–345.
MacIntyre, H. L., Kana, T. M., Anning, T., and Geider, R. J. (2002). Photoacclimation of photosynthesis irradiance response curves and photosynthetic pigments in microalgae and cyanobacteria. *Journal of Phycology* **38**, 17–38.
Mackenzie, F. T. (1995). Our Changing Planet. Prentice Hall, Upper Saddle River, NJ.
Mackenzie, F. T. and Kump, L. R. (1995). Reverse weathering, clay mineral formation, and oceanic element cycles. *Science* **270**, 586–587.
Mackin, J. E. and Aller, R. C. (1984). Ammonium adsorption in marine sediments. *Limnology and Oceanography* **29**, 250–257.
Mackin, J. E. and Swider, K. T. (1989). Organic matter decomposition pathways and oxygen consumption in coastal marine sediments. *Journal of Marine Research* **47**, 681–716.
Madigan, M. T. (1992). The family heliobacteriaceae. *In* "The Prokaryotes" (A. Balows, H. G. Trüper, M. Dworkin, W. Harder, and K.-H. Schleifer, eds), 2nd Ed., pp. 1981–1992. Springer, New York.
Madigan, M. T., Takigiku, R., Lee, R. G., Gest, H., and Hayes, J. M. (1989). Carbon isotope fractionation by thermophilic phototrophic sulfur bacteria: Evidence for autotrophic growth in natural populations. *Applied and Environmental Microbiology* **55**, 639–644.
Madigan, M. T., Martinko, J. M., and Parker, J. (2000). Brock's Biology of Microorganisms. 9th Ed. Prentice Hall, NJ.
Madigan, M. T., Martinko, J. M., and Parker, J. (2003). Brock's Biology of Microorganisms. 10th Ed. Prentice Hall, NJ.
Madrid, V. M., Aller, J. Y., Aller, R. C., and Chistoserdov, A. Y. (2001). High prokaryote diversity and analysis of community structure in mobile mud deposits off French Guiana: Identification of two new bacterial candidate divisions. *FEMS Microbiology Ecology* **37**, 197–209.
Maestrini, S. Y. and Graneli, E. (1991). Environmental conditions and ecophysiological mechanisms which led to the 1988 *Chrysochromulina polylepis* bloom: An hypothesis. *Oceanologica Acta* **14**, 397–413.

Magenheimer, J. F., Moore, T. R., Chmura, G. L., and Daoust, R. J. (1996). Methane and carbon dioxide flux from a macrotidal salt marsh, Bay of Fundy, New Brunswick. *Estuaries* **19**, 139–145.

Maidak, B. L., Cole, J. R., Parker, C. T., Jr., Garrity, G. M., Larsen, L. N., Li, B., Lilburn, T. G., McCaughey, M. J., Olsen, G. J., Overbeek, R., Pramanik, S., Schmidt, T. M., Tiedje, J. M., and Woese, C. R. (1999). A new version of the RDP (Ribosomal Database Project). *Nucleic Acids Research* **27**, 171–173.

Maldonado, M., Carmona, M. G., Uriz, M. J., and Cruzado, A. (1999). Decline in Mesozoic reef-building sponges explained by silicon limitation. *Nature* **401**, 785–788.

Mandernack, K. W. and Tebo, B. M. (1993). Manganese scavanging and oxidation at hydrothermal vents and in vent plumes. *Geochimica et Cosmochimica Acta* **57**, 3907–3923.

Mandernack, K. W., Post, J., and Tebo, B. M. (1995). Manganese mineral formation by bacterial spores of the marine *Bacillus*, strain SG-1—Evidence for the direct oxidation of Mn(II) to Mn(IV). *Geochimica et Cosmochimica Acta* **59**, 4393–4408.

Mannino, A. and Harvey, H. R. (2000). Biochemical composition of particles and dissolved organic matter along an estuarine gradient: Sources and implications for DOM reactivity. *Limnology and Oceanography* **45**, 775–788.

Mansfield, S. D. and Barlocher, F. (1993). Seasonal-variation of fungal biomass in the sediment of a salt-marsh in New Brunswick. *Microbial Ecology* **26**, 37–45.

Margulis, K. and Schwartz, K. V. (1998). Five Kingdoms. An Illustrated Guide to the Phyla of Life on Earth, 3rd Ed., p. 520. W. H. Freeman and Company, New York.

Margulis, L. (1970). Origin of Eukaryotic Cells. Yale University Press, New Haven, CT.

Marinelli, R. L. (1992). Effects of polychaetes on silicate dynamics and fluxes in sediments: Importance of species, animal activity and polychaete effects on benthic diatoms. *Journal of Marine Research* **50**, 745–779.

Mariotti, A., Germon, J. C., Hubert, P., Kaiser, P., Letolle, R., Tardieux, A., and Tardieux, P. (1981). Experimental determination of nitrogen kinetic isotope fractionation: Some principles; illustration for the denitrification and nitrification processes. *Plant and Soil* **62**, 413–430.

Martens, C. S. and Berner, R. A. (1974). Methane production in the interstitial waters of sulfate-depleted marine sediments. *Science* **185**, 1167–1169.

Martens, C. S. and Berner, R. A. (1977). Interstitial water chemistry of anoxic Long Island Sound sediments. 1. Dissolved gases. *Limnology and Oceanography* **22**, 10–25.

Martens, C. S. and Valklump, J. (1980). Biogeochemical cycling in an organic-rich coastal marine basin. 1. Methane sediment-water exchange processes. *Geochimica et Cosmochimica Acta* **44**, 471–490.

Martens, C. S., Albert, D. B., and Alperin, M. J. (1999). Stable isotope tracing of anaerobic methane oxidation in the gassy sediments of Eckernförde Bay, German Baltic Sea. *American Journal of Science* **299**, 589–610.

Martin, J. H. and Knauer, G. A. (1973). Elemental composition of plankton. *Geochimica et Cosmochimica Acta* **37**, 1639–1653.

Martin, J. H., Gordon, R. M., and Fitzwater, S. E. (1990). Iron in Antarctic waters. *Nature* **345**, 156–158.

Martin, J. H., Coale, K. H., Johnson, K. S., Fitzwater, S. E., Gordon, R. M., Tanner, S. J., Hunter, C. N., Elrod, V. A., Nowicki, J. L., Coley, T. L., Barber, R. T., Lindley, S., Watson, A. J., Vanscoy, K., Law, C. S., Liddicoat, M. I., Ling, R.,

Stanton, T., Stockel, J., Collins, C., Anderson, A., Bidigare, R., Ondrusek, M., Latasa, M., Millero, F. J., Lee, K., Yao, W., Zhang, J. Z., Friederich, G., Sakamoto, C., Chavez, F., Buck, K., Kolber, Z., Greene, R., Falkowski, P., Chisholm, S. W., Hoge, F., Swift, R., Yungel, J., Turner, S., Nightingale, P., Hatton, A., Liss, P., and Tindale, N. W. (1994). Testing the iron hypothesis in ecosystems of the equatorial Pacific ocean. *Nature* **371**, 123–129.

Matsui, K., Honjo, M., and Kawabata, Z. (2001). Estimation of the fate of dissolved DNA in thermally stratified lake water from the stability of exogenous plasmid DNA. *Aquatic Microbial Ecology* **26**, 95–102.

Matthijs, H. C. P. and Lubberding, H. J. (1988). Dark respiration in cyanobacteria. *In* "Biochemistry of the Algai and Cyanobacteria" (L. J. Rogers and J. R. Gallon, eds), pp. 131–145. Clarendon Press, Oxford.

Mayer, L. M. (1994). Surface-area control of organic-carbon accumulation in continental-shelf sediments. *Geochimica et Cosmochimica Acta* **58**, 1271–1284.

Mayer, L. M. and Gloss, S. P. (1980). Buffering of silica and phosphate in a turbid river. *Limnology and Oceanography* **25**, 12–22.

Maynard Smith, J. (1998). Evolutionary Genetics, 2nd Ed., p. 330. Oxford University Press, Oxford.

McAuliffe, T. F., Lukatelich, R. J., McComb, A. J., and Qiu, S. (1998). Nitrate applications to control phosphorus release from sediments of a shallow eutrophic estuary: An experimental evaluation. *Marine and Freshwater Research* **49**, 463–473.

McCaig, A. E., Glover, L. A., and Prosser, J. I. (1999). Molecular analysis of bacterial community structure and diversity in unimproved and improved upland grass pastures. *Applied and Environmental Microbiology* **65**, 1721–1730.

McDonald, I. R., Kelly, D. P., Murrell, J. C., and Wood, A. P. (1996). Taxonomic relationships of *Thiobacillus halophilus, T. aquaesulis*, and other species of *Thiobacillus*, as determined using 16S rDNA sequencing. *Archives of Microbiology* **166**, 394–398.

McGlathery, K. J., Risgaard-Petersen, N., and Christensen, P. B. (1998). Temporal and spatial variation in nitrogen fixation activity in the eelgrass *Zostera marina* rhizosphere. *Marine Ecology-Progress Series* **168**, 245–258.

McHatton, S. C., Barry, J. P., Jannasch, H. W., and Nelson, D. C. (1996). High nitrate concentrations in vacuolate, autotrophic marine Beggiatoa spp. *Applied and Environmental Microbiology* **62**, 954–958.

McKay, D. S., Gibson, E. K., Thomas-Keprta, K. L., Vali, H., Romanek, C. S., Clemett, S. J., Chillier, X. D. F., Maechling, C. R., and Zare, R. N. (1996). Search for past life on Mars—Possible relic biogenic activity in Martian meteorite ALH84001. *Science* **273**, 924–930.

McLaughlin, J. R., Ryden, J. C., and Syers, J. K. (1981). Sorption of inorganic-phosphate by iron-containing and aluminum-containing components. *Journal of Soil Science* **32**, 365–377.

McManus, J., Berelson, W. M., Coale, K. H., Johnson, K. S., and Kilgore, T. E. (1997). Phosphorus regeneration in continental margin sediments. *Geochimica et Cosmochimica Acta* **61**, 2891–2907.

Megonigal, J. P. and Schlesinger, W. H. (2002). Methane-limited methanotrophy in tidal freshwater swamps. *Global Biogeochemical Cycles* **16**, Art. No. 1088.

Megraw, S. R. and Knowles, R. (1987). Methane production and consumption in a cultivated humisol. *Biology and Fertility of Soils* **5**, 56–60.

Menendez, C., Bauer, Z., Huber, H., Gad'on, N., Stetter, K.-O., and Fuchs, G. (1999). Presence of acetyl coenzyme A (CoA) carboxylase and propionyl-CoA carboxylase in autotrophic *Crenarchaeota* and indication for operation of a

3-hydroxypropionate cycle in autotrophic carbon fixation. *Journal of Bacteriology* **181**, 1088–1098.
Merrett, M. J. and Armitage, T. L. (1982). The effect of oxygen concentration on photosynthetic biomass production by algae. *Planta* **155**, 95–96.
Meunier, J. D., Colin, F., and Alarcon, C. (1999). Biogenic silica storage in soils. *Geology* **27**, 835–838.
Mevel, G. and Prieur, D. (1998). Thermophilic heterotrophic nitrifiers isolated from Mid-Atlantic Ridge deep-sea hydrothermal vents. *Canadian Journal of Microbiology* **44**, 723–733.
Michaelis, N. and Menten, M. I. (1913). Die Kinetik der Invertinwirkung. *Biochemische Zeitschrift* **49**, 333–369.
Michaelis, W., Seifert, R., Nauhaus, K., Treude, T., Thiel, V., Blumenberg, M., Knittel, K., Gieseke, A., Peterknecht, K., Pape, T., Boetius, A., Amann, R., Jørgensen, B. B., Widdel, F., Peckmann, J., Pimenov, N. V., and Gulin, M. B. (2002). Microbial reefs in the Black Sea fueled by anaerobic oxidation of methane. *Science* **297**, 1013–1015.
Micheli, F. (1993). Feeding ecology of mangrove crabs in North Eastern Australia—Mangrove litter consumption by *Sesarma messa* and *Sesarma smithii*. *Journal of Experimental Marine Biology and Ecology* **171**, 165–186.
Michelson, A. R., Jacobson, M. E., Scranton, M. I., and Mackin, J. E. (1989). Modeling the distribution of acetate in anoxic estuarine sediments. *Limnology and Oceanography* **34**, 747–757.
Middelboe, M., Nielsen, T. G., and Bjørnsen, P. K. (2002). Viral and bacterial production in the North Water: *In situ* measurements, batch culture experiments and characterization and distribution of a virus host system. *Deep-Sea Research II* **49**, 5063–5079.
Middelburg, J. J. (1989). A simple rate model for organic matter decomposition in marine sediments. *Geochimica et Cosmochimica Acta* **53**, 1577–1581.
Miller, M. B. and Bassler, B. L. (2001). Quorum sensing in bacteria. *Annual Review of Microbiology* **55**, 165–199.
Millero, F. J. (1991). The oxidation of H_2S in Black Sea waters. *Deep-Sea Research* **38**, S1139–S1150.
Millero, F. J. (1996). Chemical Oceanography, 2nd Ed. CRC Press, Boca Raton.
Millero, F. J., Izaguirre, M., and Sharma, V. K. (1987). The effect of ionic interaction on the rates of oxidation in natural waters. *Marine Chemistry* **22**, 179–191.
Minz, D., Fishbain, S., Green, S. J., Muyzer, G., Yehuda, C., Rittmann, B. E., and Stahl, D. A. (1999). Unexpected population distribution in a microbial mat community: Sulfate-reducing bacteria localized to the highly oxic chemocline in constrast to a Eukaryotic preference for anoxia. *Applied and Environmental Microbiology* **65**, 4659–4665.
Miziorko, H. M. and Lorimar, G. H. (1983). Ribulose-1,5-bisphosphate carboxylase-oxygenase. *Annual Review of Biochemistry* **52**, 507–535.
Mohammed, S. M. and Johnstone, R. W. (1995). Spatial and temporal variations in water column nutrient concentrations in a tidally dominated mangrove creek: Chwaka Bay, Zanzibar. *Ambio* **24**, 482–486.
Mongia, A. D. and Ganeshamurthy, A. N. (1989). Typical differences between the chemical characteristics of *rhizophora* and *avicennia* mangrove forest soils in South Andamans. *Agrochimica* **33**, 464–470.
Monod, J. (1942). Recherches sur la Croissance des Cultures Bacteriennes. Hermann, Paris.

Montoya, J. P. and McCarthy, J. J. (1995). Isotopic fractionation during nitrate uptake by phytoplankton grown in continuous culture. *Journal of Plankton Research* **17**, 439–464.

Montoya, J. P., Carpenter, E. J., and Capone, D. G. (2002). Nitrogen fixation and nitrogen isotope abundances in zooplankton of the oligotrophic North Atlantic. *Limnology and Oceanography* **47**, 1617–1628.

Moore, L. R. and Chisholm, S. W. (1999). Photophysiology of the marine cyanobacterium *Prochlorococcus*: Ecotypic differences among cultured isolates. *Limnology and Oceanography* **44**, 628–638.

Moore, L. R., Goericke, R., and Chisholm, S. W. (1995). Comparative physiology of *Synechococcus* and *Prochlorococcus*—Influence of light and temperature on growth, pigments, fluorescence and absorptive properties. *Marine Ecology-Progress Series* **116**, 259–275.

Moore, L. R., Post, A. F., Rocap, G., and Chisholm, S. W. (2002). Utilization of different nitrogen sources by the marine cyanobacteria *Prochlorococcus* and *Synechococcus*. *Limnology and Oceanography* **47**, 989–996.

Moran, M. A., Wicks, R. J., and Hodson, R. E. (1991). Export of dissolved organic matter from a mangrove swamp ecosystem: Evidence from natural fluorescence, dissolved lignin phenols, and bacterial secondary production. *Marine Ecology-Progress Series* **76**, 175–184.

Morell, M. K., Paul, K., Kane, H. J., and Andrews, T. J. (1992). Rubisco: Maladapted or misunderstood? *Australian Journal of Botany* **40**, 431–441.

Mori, T. and Johnson, C. H. (2001). Circadian programming in cyanobacteria. *Seminars in Cell & Developmental Biology* **12**, 271–278.

Moriarty, D. J. W. (1986). Measurement of bacterial growth rates in aquatic systems from rates of nucleic acid synthesis. *In* "Advances in Microbial Ecology" (K. C. Marshall, ed.), pp. 245–292. Plenum Press, New York.

Morita, R. Y. (1997). Bacteria in Oligotrophic Environments: Starvation-Surviving Lifestyle, p. 529. Chapman & Hall Microbiology Series. Chapman and Hall, New York.

Morris, R. M., Rappé, M. S., Connon, S. A., Vergin, K. L., Siebold, W. A., Carlson, C. A., and Giovannoni, S. J. (2002). SAR11 clade dominates ocean surface bacterioplankton communities. *Nature* **420**, 806–810.

Moses, C. O. and Herman, J. S. (1991). Pyrite oxidation at circumneutral pH. *Geochimica et Cosmochimica Acta* **55**, 471–482.

Mountfort, D. O., Kaspar, H. F., Downes, M., and Asher, R. A. (1999). Partitioning effects during terminal carbon and electron flow in sediments of a low-salinity meltwater pond near Bratina Island, McMurdo Ice Shelf, Antarctica. *Applied and Environmental Microbiology* **65**, 5493–5499.

Mountfort, D. O., Kaspar, H. F., Asher, R. A., and Sutherland, D. (2003). Influences of pond geochemistry, temperature, and freeze-thaw on terminal anaerobic processes occurring in sediments of six ponds of the McMurdo Ice Shelf, near Bratina Island, Antarctica. *Applied and Environmental Microbiology* **69**, 583–592.

Moussard, H., L'Haridon, S., Tindall, B. J., Banta, A., Schumann, P., Stackebrandt, E., Reysenbach, A. L., and Jeanthon, C. (2004). *Thermodesulfatator indicus* gen. nov., sp. nov., a novel thermophilic chemolithoautotrophic sulfate-reducing bacterium isolated from the central Indian Ridge. *International Journal of Systematic and Evolutionary Microbiology* **54**, 227–233.

Mouton, E. C. and Felder, D. L. (1996). Burrow distributions and population estimates for the fiddler crabs *Uca spinicarpa* and *Uca longisignalis* in a Gulf of Mexico salt marsh. *Estuaries* **19**, 51–61.

Murphy, A. E., Sageman, B. B., Hollander, D. J., Lyons, T. W., and Brett, C. E. (2000). Black shale deposition and faunal overturn in the Devonian Appalachian basin: Clastic starvation, seasonal water-column mixing, and efficient biolimiting nutrient recycling. *PaleOceanography* **15**, 280–291.

Murray, J. W., Balistieri, L. S., and Paul, B. (1984). The oxidation state of manganese in marine sediments and ferromanganese nodules. *Geochimica et Cosmochimica Acta* **48**, 1237–1247.

Murray, J. W., Jannasch, H. W., Honjo, S., Anderson, R. F., Reeburgh, W. S., Top, Z., Friederich, G. E., Codispoti, L. A., and Izdar, E. (1989). Unexpected changes in the oxic/anoxic interface in the Black Sea. *Nature* **338**, 411–413.

Murray, J. W., Codispoti, L. A., and Friederich, G. E. (1995). Oxidation-reduction environments—The suboxic zone in the Black Sea. *Advances in Chemistry Series* **244**, 157–176.

Muyzer, G., de Wall, E. C., and Uitterlinden, A. G. (1993). Profiling of complex microbial populations by denaturing gradient gel electrophoresis analysis of polymerase chain-amplified genes coding for 16S rRNA. *Applied and Environmental Microbiology* **59**, 695–700.

Myers, C. R. and Myers, J. M. (1992). Localization of cytochromes to the outer membrane of anaerobically grown *Shewanella putrefaciens* MR-1. *Journal of Bacteriology* **174**, 3429–3438.

Myers, C. R. and Myers, J. M. (1997). Outer membrane cytochromes of *Shewanella putrefaciens* MR-1, Spectral analysis, and purification of the 83-kDa c-type cytochrome. *Biochimica et Biophysica Acta-Biomembranes* **1326**, 307–318.

Myers, C. R. and Nealson, K. H. (1988). Bacterial manganese reduction and growth with manganese oxide as the sole electron acceptor. *Science* **240**, 1319–1321.

Myers, J. M. and Myers, C. R. (2001). Role for outer membrane cytochromes OmcA and OmcB of *Shewanella putrefaciens* MR-1 in reduction of manganese dioxide. *Applied and Environmental Microbiology* **67**, 260–269.

Nadeau, T.-L., Howard-Williams, C., and Castenholz, R. W. (1999). Effects of solar UV and visible irradiance on photosynthesis and vertical migration of *Oscillatoria* sp. (Cyanobacteria) in an Antarctic microbial mat. *Aquatic Microbial Ecology* **20**, 231–243.

Nagata, T. and Kirchman, D. L. (1992). Release of macromolecular organic complexes by heterotrophic marine flagellates. *Marine Ecology-Progress Series* **83**, 233–240.

Namsaraev, Z. B., Gorlenko, V. M., Namsaraev, B. B., Buryukhaev, S. P., and Yurkov, V. V. (2003). The structure and biogeochemical activity of the phototrophic communities from the Bol'sherechenskii alkaline hot spring. *Russian Journal of Microbiology* **72**, 193–202.

Nathan, Y., Bremner, J. M., Lowenthal, R. E., and Monteiro, P. (1993). Role of bacteria in phosphorite genesis. *Geomicrobiology Journal* **11**, 69–76.

Nauhaus, K., Boetius, A., Krüger, M., and Widdel, F. (2002). *In vitro* demonstration of anaerobic oxidation of methane coupled to sulphate reduction in sediment from a marine gas hydrate area. *Environmental Microbiology* **4**, 296–305.

Nealson, K. H. and Hastings, J. W. (1979). Bacterial bioluminescence: Its control and ecological significance. *Microbiological Reviews* **43**, 496–518.

Nealson, K. H., Myers, C. R., and Wimpee, B. B. (1991). Isolation and identification of manganese-reducing bacteria and estimates of microbial Mn(IV)-reducing potential in the Black Sea. *Deep-Sea Research* **38**, S907–S920.

Nedwell, D. B. (1999). Effect of low temperature on microbial growth: Lowered affinity for substrates limits growth at low temperature. *FEMS Microbiology Ecology* **30**, 101–111.

Nedwell, D. B., Blackburn, T. H., and Wiebe, W. J. (1994). Dynamic Nature of the Turnover of Organic-Carbon, Nitrogen and Sulfur in the Sediments of a Jamaican Mangrove Forest. *Marine Ecology-Progress Series* **110**, 223–231.

Neilson, M. J. and Richards, G. N. (1989). Chemical composition of degrading mangrove leaf litter and changes produced after consumption by mangrove crab *Neosarmatium smithi* (Crustacea, Decapoda, Sesarmidae). *Journal of Chemical Ecology* **15**, 1267–1283.

Nelson, D. C., Jørgensen, B. B., and Revsbech, N. P. (1986). Growth pattern and yield of a chemoautotrophic *Beggiatoa* sp. in oxygen-sulfide microgradients. *Applied and Environmental Microbiology* **52**, 225–233.

Nelson, D. L. and Cox, M. M. (2000). Lehninger Principles of Biochemistry. Worth Publ., New York.

Nelson, D. M. and Brzezinski, M. A. (1997). Diatom growth and productivity in an oligotrophic midocean gyre: A 3-yr record from the Sargasso Sea near Bermuda. *Limnology and Oceanography* **42**, 473–486.

Nelson, D. M. and Dortch, Q. (1996). Silicic acid depletion and silicon limitation in the plume of the Mississippi River: Evidence from kinetic studies in spring and summer. *Marine Ecology-Progress Series* **136**, 163–178.

Nelson, D. M. and Goering, J. J. (1977). Stable isotope tracer method to measure silicic-acid uptake by marine-phytoplankton. *Analytical Biochemistry* **78**, 139–147.

Nelson, D. M., Treguer, P., Brzezinski, M. A., Leynaert, A., and Queguiner, B. (1995). Production and dissolution of biogenic silica in the ocean: Revised global estimates, comparison with regional data and relationship to biogenic sedimentation. *Global Biogeochemical Cycles* **9**, 359–372.

Nelson, D. M., Anderson, R. F., Barber, R. T., Brzezinski, M. A., Buesseler, K. O., Chase, Z., Collier, R. W., Dickson, M. L., Francois, R., Hiscock, M. R., Honjo, S., Marra, J., Martin, W. R., Sambrotto, R. N., Sayles, F. L., and Sigmon, D. E. (2002). Vertical budgets for organic carbon and biogenic silica in the Pacific sector of the Southern Ocean, 1996–1998. *Deep-Sea Research Part II-Topical Studies in Oceanography* **49**, 1645–1674.

Nelson, K. E., Clayton, R. A., Gill, S. R., Gwinn, M. L., Dodson, R. J., Haft, D. H., Hickey, E. K., Peterson, L. D., Nelson, W. C., Ketchum, K. A., McDonald, L., Utterback, T. R., Malek, J. A., Linher, K. D., Garrett, M. M., Stewart, A. M., Cotton, M. D., Pratt, M. S., Phillips, C. A., Richardson, D., Heidelberg, J., Sutton, G. G., Fleischmann, R. D., Eisen, J. A., White, O., Salzberg, S. L., Smith, H. O., Venter, J. C., and Fraser, C. M. (1999). Evidence for lateral gene transfer between Archaea and bacteria from genome sequence of *Thermotogo maritima*. *Nature* **399**, 323–329.

Newell, R. I. E., Marshall, N., Sasekumar, A., and Chong, V. C. (1995). Relative importance of benthic microalgae, phytoplankton, and mangroves as sources of nutrition for penaeid prawns and other coastal invertebrates from Malaysia. *Marine Biology* **123**, 595–606.

Newell, S. Y. (1996). Established and potential impacts of eukaryotic mycelial decomposers in marine/terrestrial ecotones. *Journal of Experimental Marine Biology and Ecology* **200**, 187–206.

Newell, S. Y. and Fell, J. W. (1992). Ergosterol content of living and submerged, decaying leaves and twigs of red mangrove. *Canadian Journal of Microbiology* **38**, 979–982.

Nicholson, J. A. M., Stolz, J. F., and Pierson, B. K. (1987). Structure of a microbial mat at Great Sippowissett Marsh, Cape Cod, Massachusetts. *FEMS Microbiology Ecology* **45**, 343–364.

Nielsen, J. L. and Nielsen, P. H. (1998). Microbial nitrate-dependent oxidation of ferrous iron in activated sludge. *Environmental Science & Technology* **32**, 3556–3561.

Nielsen, L. B., Finster, K., Welsh, D. T., Donelly, A., Herbert, R. A., de Wit, R., and Lomstein, B. A. (2001). Sulphate reduction and nitrogen fixation rates associated with roots, rhizomes and sediments from *Zostera noltii* and *Spartina maritima* meadows. *Environmental Microbiology* **3**, 63–71.

Nielsen, L. P. (1992). Denitrification in sediment determined from nitrogen isotope pairing. *FEMS Microbiology Ecology* **86**, 357–362.

Nielsen, L. P., Christensen, P. B., Revsbech, N. P., and Sørensen, J. (1990). Denitrification and photosynthesis in stream sediment studied with microsensor and whole-core techniques. *Limnology and Oceanography* **35**, 1135–1144.

Nielsen, O. I., Kristensen, E., and Macintosh, D. J. (2003). Impact of fiddler crabs (*Uca* spp.) on rates and pathways of benthic mineralization in deposited mangrove shrimp pond waste. *Journal of Experimental Marine Biology and Ecology* **289**, 59–81.

Nienhuis, P. H. and Groenendijk, A. M. (1986). Consumption of eelgrass (*Zostera marina*) by birds and invertebrates—An annual budget. *Marine Ecology Progress Series* **29**, 29–35.

Nold, S. C. and Zwart, G. (1998). Patterns and governing forces in aquatic microbial communities. *Aquatic Ecology* **32**, 17–35.

Nordstrom, D. K. and Southam, G. (1997). Geomicrobiology of sulfide mineral oxidation. *In* "Reviews in Mineralogy" (J. F. Banfield and K. H. Nealson, eds), pp. 361–390. The Mineralogical Society of America, Washington D.C.

Nouchi, I., Mariko, S., and Aoki, K. (1990). Mechanism of methane transport from the rhizosphere to the atmosphere through rice plants. *Plant Physiology* **94**, 59–66.

Novitsky, J. A. and Morita, R. Y. (1976). Morphological characterization of small cells resulting from nutrient starvation of a psychrophilic marine vibrio. *Applied and Environmental Microbiology* **32**, 617–622.

Nübel, U., Garcia-Pichel, F., Kühl, M., and Muyzer, G. (1999). Quantifying microbial diversity: Morphotypes, 16S rRNA genes, and carotenoids of oxygenic phototrophs in microbial mats. *Applied and Environmental Microbiology* **65**, 422–430.

Nüsslein, B., Chin, K.-J., Eckert, W., and Conrad, R. (2001). Evidence for anaerobic syntrophic acetate oxidation during methane production in the profundal sediment of subtropical Lake Kinneret (Israel). *Environmental Microbiology* **3**, 460–470.

Ochman, H. and Wilson, A. C. (1987). Evolution in bacteria: Evidence for a universal substitution rate in cellular genomes. *Journal of Molecular Evolution* **26**, 74–86.

Odum, E. P. (1971). Fundamentals of Ecology, 3rd Ed. W. B. Saunders Company, Philadelphia.

Oguz, T., Murray, J. W., and Callahan, A. E. (2001). Modeling redox cycling across the suboxic-anoxic interface zone in the Black Sea. *Deep-Sea Research I* **48**, 761–787.

Ollivier, B., Hatchilcan, C. E., Prensier, G., Guezennec, J., and Garcia, J. L. (1991). Desulfohalobium-retbaense gen. nov. sp. nov., a halophilic sulfate-reducing bacterium from sediments of a hypersaline lake in Senegal. *International Journal of Systematic Bacteriology* **41**, 74–81.

Olsen, G. J. and Woese, C. R. (1993). Ribosomal RNA: A key to phylogeny. *FASEB Journal* **7**, 113–123.

Olson, R. J. (1981a). ^{15}N tracer studies of the primary nitrite maximum. *Journal of Marine Research* **39**, 203–226.

Olson, R. J. (1981b). Differential photoinhibition of marine nitrifying bacteria—A possible mechanism for the formation of the primary nitrite maximum. *Journal of Marine Research* **39**, 227–238.
Omnes, P., Slawyk, G., Garcia, N., and Bonin, P. (1996). Evidence of denitrification and nitrate ammonification in the River Rhone plume (northwestern Mediterranean Sea). *Marine Ecology-Progress Series* **141**, 275–281.
Oren, A. (1999). Bioenergetic aspects of halophilism. *Microbiology and Molecular Biology Reviews* **63**, 334–348.
Oren, A. (2000). Salts and brines. *In* "The Ecology of Cyanobacteria. Their Diversity in Time and Space" (B. A. Whitton and M. Potts, eds), pp. 281–306. Kluwer Academic Publishers, Boston.
Oren, A. (2001). The bioenergetic basis for the decrease in metabolic diversity at increasing salt concentrations: Implications for the functioning of salt lake ecosystems. *Hydrobiologia* **466**, 61–72.
Orphan, V. J., Hinrichs, K.-U., Ussler, W., III, Paull, C. K., Taylor, L. T., Sylva, S. P., Hayes, J. M., and DeLong, E. F. (2001a). Comparative analysis of methane-oxidizing Archaea and sulfate-reducing bacteria in anoxic marine sediments. *Applied and Environmental Microbiology* **67**, 1922–1934.
Orphan, V. J., House, C. H., Hinrichs, K.-U., McKeegan, K. D., and DeLong, E. F. (2001b). Methane-consuming Archaea revealed by directly coupled isotopic and phylogenetic analysis. *Science* **293**, 484–487.
Orphan, V. J., House, C. H., Hinrichs, K.-U., McKeegan, K. D., and Delong, E. F. (2002). Multiple archaeal groups mediae methane oxidation in anoxic cold seep sediments. *Proceedings of the National Academy of Sciences USA* **99**, 7663–7668.
Orso, S., Gouy, M., Navarro, E., and Normand, P. (1994). Molecular phylogenetic analysis of *nitrobacter* spp. *International Journal of Systematic Bacteriology* **44**, 83–86.
Ort, D. R. and Baker, N. R. (2002). A photoprotective role for O_2 as an alternative electron sink in photosynthesis? *Current Opinion in Plant Biology* **5**, 193–198.
Osborne, J. P. and Gennis, R. B. (1999). Sequence analysis of cytochrome *bd* oxidase suggests a revised topology for subunit I. *Biochimica et Biophysica Acta* **1410**, 32–50.
Otte, S., Kuenen, J. G., Nielsen, L. P., Paerl, H. W., Zopfi, J., Schultz, H. N., Teske, A., Strotman, B., Gallardo, V. A., and Jørgensen, B. B. (1999). Nitrogen, carbon, and sulfur metabolism in natural *Thioploca* samples. *Applied and Environmental Microbiology* **65**, 3148–3157.
Ottley, C. J., Davison, W., and Edmunds, W. M. (1997). Chemical catalysis of nitrate reduction by iron(II). *Geochimica et Cosmochimica Acta* **61**, 1819–1828.
Ottosen, L. D. M., Risgaard-Petersen, N., and Nielsen, L. P. (1999). Direct and indirect measurements of nitrification and denitrification in the rhizosphere of aquatic macrophytes. *Aquatic Microbial Ecology* **19**, 81–91.
Ouverney, C. C. and Fuhrman, J. A. (2000). Marine planktonic Archaea take up amino acids. *Applied and Environmental Microbiology* **66**, 4829–4833.
Ovalle, A. R. C., Rezende, C. E., Lacerda, L. D., and Silva, C. A. R. (1990). Factors affecting the hydrochemistry of a mangrove tidal creek, Sepetiba Bay, Brazil. *Estuarine Coastal and Shelf Science* **31**, 639–650.
Overmann, J. (1992). Phylum BXI. Chlorobi phy. nov. Family I. "Chlorobiaceae" Green sulfur bacteria. *In* "Bergey's Manual of Systematic Bacteriology," (D. R. Boone, R. W. Castenholz, and G. M. Garrity, eds), 2nd Ed., pp. 601–604. Springer, New York.

Overmann, J. (1997). Mahoney lake: A case study of the ecological significance of phototrophic sulfur bacteria. *In* "Advances in Microbial Ecology" (J. G. Jones, ed.), pp. 251–288. Plenum Press, New York.

Overmann, J. and Schubert, K. (2002). Phototrophic consortia: Model systems for symbiotic interrelations between prokaryotes. *Archives of Microbiology* **177**, 201–208.

Overmann, J., Beatty, J. T., Hall, K. J., Pfennig, N., and Northcote, T. G. (1991). Characterization of a dense, purple sulfur bacterial layer in a meromictic salt lake. *Limnology and Oceanography* **36**, 846–859.

Overmann, J., Cypionka, H., and Pfennig, N. (1992). An extremely low-light-adapted phototrophic sulfur bacterium from the Black Sea. *Limnology and Oceanography* **37**, 150–155.

Overmann, J., Beatty, J. T., and Hall, K. J. (1994). Photosynthetic activity and population dynamics of *Amoebacter purpureus* in a mermictic saline lake. *FEMS Microbiology Ecology* **15**, 309–320.

Overmann, J., Beatty, J. T., and Hall, K. J. (1996a). Purple sulfur bacteria control the growth of aerobic heterptrophic bacterioplankton in a meromictic salt lake. *Applied and Environmental Microbiology* **62**, 3251–3258.

Overmann, J., Beatty, J. T., Krouse, H. R., and Hall, K. J. (1996b). The sulfur cycle in the chemocline of a meromictic salt lake. *Limnology and Oceanography* **41**, 147–156.

Overmann, J., Tuschak, C., Frostl, J. M., and Sass, H. (1998). The ecological niche of the consortium "pelochromatium roseum." *Archives of Microbiology* **169**, 120–128.

Overmann, J., Hall, K. J., Northcote, T. G., Ebenhoh, W., Chapman, M. A., and Beatty, T. (1999). Structure of the aerobic food chain in a meromictic lake dominated by purple sulfur bacteria. *Archiv für Hydrobiologie* **144**, 127–156.

Owens, N. J. P. (1987). Natural variations in N–15 in the marine environment. *Advances in Marine Biology* **24**, 389–451.

Pace, N. R. (1997). A molecular view of microbial diversity and the biosphere. *Science* **276**, 734–740.

Pace, N. R., Olsen, G. J., and Woese, C. R. (1986). Ribosomal RNA phylogeny and the primary lines of evolutionary descent. *Cell* **45**, 325–326.

Paerl, H. W. (2000). Marine plankton. *In* "The Ecology of Cyanobacteria. Their Diversity in Time and Space" (B. A. Whitton and M. Potts, eds). Klüwer Academic Publishers, Dordrecht.

Paerl, H. W. and Zehr, J. P. (2000). Marine nitrogen fixation. *In* "Microbial Ecology of the Oceans" (D. L. Kirchman, ed.), pp. 387–426. Wiley-Liss Inc., New York.

Pakulski, J. D., Coffin, R. B., Kelley, C. A., Holder, S. L., Downer, R., Aas, P., Lyons, M. M., and Jeffrey, W. H. (1996). Iron stimulation of Antarctic bacteria. *Nature* **383**, 133–134.

Palenik, B. and Haselkorn, R. (1992). Multiple evolutionary origins of prochlorophytes, the chlorophyll b-containing prokaryotes. *Nature* **355**, 265–267.

Palmer, J. D. and Delwiche, C. F. (1996). Second-hand chloroplasts and the case of the disappearing nucleus. *Proceedings of the National Academy of Sciences USA* **93**, 7432–7435.

Panikov, N. S. (1995). Microbial Growth Kinetics, p. 378. Chapman and Hall, London.

Papen, H., Vonberg, R., Hinkel, I., Thoene, B., and Rennenberg, H. (1989). Heterotrophic Nitrification by Alcaligenes-Faecalis-NO_2^-, NO_3^-, N_2O, and NO production in exponentially growing cultures. *Applied and Environmental Microbiology* **55**, 2068–2072.

Park, R. and Epstein, S. (1960). Carbon isotope fractionation during photosynthesis. *Geochimica et Cosmochimica Acta* **21**, 110–126.
Parkes, R. J., Gibson, G. R., Mueller-Harvey, I., Buckingham, W. J., and Herbert, R. A. (1989). Determination of the substrates for sulphate-reducing bacteria within marine and estuarine sediments with different rates of sulphate reduction. *Journal of General Microbiology* **135**, 175–187.
Parkes, R. J., Cragg, B. A., Fry, J. C., Herbert, R. A., and Wimpenny, J. W. T. (1990). Bacterial biomass and activity in deep sediment layers from the Peru margin. *Philosophical Transactions of the Royal Society of London Series A* **331**, 139–153.
Parkes, R. J., Cragg, B. A., Bale, S. J., Getliff, J. M., Goodman, K., Rochelle, P. A., Fry, J. C., Weightman, A. J., and Harvey, S. M. (1994). Deep bacterial biosphere in Pacific Ocean sediments. *Nature* **371**, 410–413.
Parkes, R. J., Cragg, B. A., and Wellsbury, P. (2000). Recent studies on bacterial populations and processes in marine sediments: A review. *Hydrogeology Journal* **8**, 11–28.
Parkin, T. B. and Brock, T. D. (1980). Photosynthetic bacterial production in lakes: The effects of light intensity. *Limnology and Oceanography* **25**, 711–718.
Parsek, M. R. and Greenberg, E. P. (1999). Quorum sensing signals in development of *Pseudomonas aeruginosa* biofilms. *Methods in Enzymology* **310**, 43–55.
Partensky, F., Blanchot, J., and Vaulot, D. (1999a). Differential distribution and ecology of *Prochlorococcus* and *Synechococcus* in oceanic waters: A review. *Bulletin de l'Institut Océanographique, Monaco* **19**, 457–475.
Partensky, F., Hess, W. R., and Vaulot, D. (1999b). *Prochlorococcus*, a marine photosynthetic prokaryote of global significance. *Microbiology and Molecular Biology Reviews* **63**, 106–127.
Paschinger, H., Paschinger, J., and Gaffron, H. (1974). Photochemical disproportionation of sulfur into sulfide and sulfate by *Chlorbium limnola* forma *thiosulfatophilum*. *Archives of Microbiology* **96**, 341–351.
Passioura, J. B., Ball, M. C., and Knight, J. H. (1992). Mangroves may salinize the soil and in so doing limit their transpiration rate. *Functional Ecology* **6**, 476–481.
Pattison, S. N., Garcia-Ruiz, R., and Whitton, B. A. (1998). Spatial and seasonal variation in denitrification in the Swale-Ouse system, a river continuum. *Science of the Total Environment* **210/211**, 289–305.
Pauer, J. J. and Auer, M. T. (2000). Nitrification in the water column and sediment of a hypereutrophic lake and adjoining river system. *Water Research* **34**, 1247–1254.
Pavlov, A. A., Kasting, J. F., Brown, L. L., Rages, K. A., and Freedman, R. (2000). Greenhouse warming by CH_4 in the atmosphere of early earth. *Journal of Geophysical Research-Planets* **105**, 11981–11990.
Pedersen, A. G. U., Berntsen, J., and Lomstein, B. A. (1999). The effect of eelgrass decomposition on sediment carbon and nitrogen cycling: A controlled laboratory experiment. *Limnology and Oceanography* **44**, 1978–1992.
Pedersen, K. (1997). Microbial life in deep granitic rock. *FEMS Microbiology Ecology* **20**, 399–414.
Pedersen, K. (2000). Exploration of deep intraterrestrial microbial life: Current perspectives. *FEMS Microbiology Ecology* **185**, 9–16.
Pel, R., Oldenhuis, R., Brand, W., Vos, A., Gottschal, J. C., and Zwart, K. B. (1997). Stable-isotope analysis of a combined nitrification-denitrification sustained by thermophilic methanotrophs under low-oxygen conditions. *Applied and Environmental Microbiology* **63**, 474–481.

Pelegri, S. P., Nielsen, L. P., and Blackburn, T. H. (1994). Denitrification in estuarine sediment stimulated by the irrigation activity of the amphipod *Corophium volutator*. *Marine Ecology-Progress Series* **105**, 285–290.

Pennock, J. R., Velinsky, D. J., Ludlam, J. M., Sharp, J. H., and Fogel, M. L. (1996). Isotopic fractionation of ammonium and nitrate during uptake by *Skeletonema costatum*: Implications for $d^{15}N$ dynamics under bloom conditions. *Limnology and Oceanography* **41**, 451–459.

Pereira, M. M., Santana, M., and Teixeira, M. (2001). A novel scenario for the evolution of haem-copper oxygen reductases. *Biochimica et Biophysica Acta-Bioenergetics* **1505**, 185–208.

Perna, N. T., Plunkett, G., III, Burland, V., Mau, B., Glasner, J. D., Rose, D. J., Mayhew, G. F., Evans, P. S., Gregor, J., Kirkpatrick, H. A., Pósfal, G., Hackett, J., Klink, S., Boutin, A., Shao, Y., Miller, L., Grotbeck, E. J., Davis, N. W., Lim, A., Dimalanta, E. T., Potamousis, K. D., Apodaca, J., Anantharaman, T. S., Lin, J., Yen, G., Schwartz, D. C., Welch, R. A., and Blattner, F. R. (2001). Genome sequence of enterohaemorrhagic *Escherichia coli* 0157.H7. *Nature* **409**, 529–533.

Petit, J. R., Jouzel, J., Raynaud, D., Barkov, N. I., Barnola, J.-M., Basile, I., Benders, M., Chappellaz, J., Davis, M., Delaygue, G., Delmotte, M., Kotlyakov, V. M., Legrand, M., Lipenkov, V. Y., Lorius, C., Pépin, L., Ritz, C., Saltzman, E., and Stievenard, M. (1999). Climate and atmospheric history of the past 420,000 years from the Vostok ice core, Antarctica. *Nature* **399**, 429–436.

Petterson, K., Boström, B., and Jacobsen, O.-S. (1988). Phosphorus in sediments—Speciation and analysis. *In* "Phosphorus in Freshwater Ecosystems. Developments in Hydrobiology 48" (G. Persson and M. Jannsson, eds), pp. 91–101. Klüwer Academic Publishers, Dordrecht.

Pfeiffer, W. J. and Wiegert, R. G. (1991). Grazers on *Spartina* and their predators. *In* "The Ecology of a Salt Marsh" (L. R. Pomeroy and R. G. Wiegert, eds), pp. 87–109. Springer, New York.

Pfennig, N. (1975). The phototrophic bacteria and their role in the sulfur cycle. *Plant Soil* **43**, 1–16.

Pfennig, N. (1977). Phototrophic green and purple bacteria: A comparative systematic survey. *Annual Reviews of Microbiology* **31**, 275–290.

Pfennig, N. (1989). Ecology of phototrophic purple and green sulfur bacteria. *In* "Autotrophic Bacteria" (H. G. Schlegel and B. Bowien, eds), pp. 97–116. Springer, Madison WI.

Pfennig, N. and Trüper, H. G. (1992). The family Chromatiaceae. *In* "The Prokaryotes" (A. Balows, H. G. Trüper, M. Dworkin, W. Harder, and K.-H. Schleifer, eds). Springer, New York.

Philippe, H. and Laurent, J. (1998). How good are deep phylogenetic trees? *Current Opinion in Genetics & Development* **8**, 616–623.

Pickett-Heaps, J., Schmid, A. M. M., and Edgar, L. A. (1990). The cell biology of diatom valve formation. *In* "Progress in Phycological Research" (F. E. Round and D. J. Chapman, eds), pp. 1–169. Biopress, Bristol, UK.

Pielou, E. C. (1969). An Introduction to Mathematical Ecology. Wiley-Interscience, a Division of John Wiley & Sons, New York.

Piepenburg, D., Blackburn, T. H., von Dorrien, C. F., Gutt, J., Hall, P. O. J., Hulth, S., Kendall, M. A., Opalinski, K. W., Rachor, E., and Schmid, M. K. (1995). Partitioning of benthic community respiration in the Arctic (northwestern Barents Sea). *Marine Ecology-Progress Series* **118**, 199–213.

Pierson, B. K. (2001). Phylum BVI. Chloroflexi phy. nov. Family I. "Chloroflexaceae". *In* "Bergey's Manual of Systematic Bacteriology" (D. R. Boone, R. W. Castenholz, and G. M. Garrity, eds), 2nd Ed., pp. 427–429. Springer, New York.

Pierson, B. K. and Castenholz, R. W. (1992). The family *Chloroflexaceae*. *In* "The Prokaryotes" (A. Balows, H. G. Trüper, M. Dworkin, W. Harder, and K.-H. Schleifer, eds), pp. 3754–3774. Springer, New York.

Pierson, B. K. and Castenholz, R. W. (1995). Taxonomy and physiology of filamentous anoxygenic phototrophs. *In* "Anoxygenic Photosynthetic Bacteria" (R. E. Blankenship, M. T. Madigan, and C. E. Bauer, eds), pp. 31–47. Kluwer Academic Publishers, The Netherlands.

Pierson, B. K., Parenteau, M. N., and Griffin, B. M. (1999). Phototrophs in high-iron-concentration microbial mats: Physiological ecology of phototrophs in an iron-depositing hot spring. *Applied and Environmental Microbiology* **65**, 5474–5483.

Pirt, S. J. (1975). Principles of microbe and cell cultivation. Blackwell Publishers, Oxford.

Ploug, H., Zimmermann-Timm, H., and Schweitzer, B. (2002). Microbial communities and respiration on aggregates in the Elbe Estuary, Germany. *Aquatic Microbial Ecology* **27**, 241–248.

Pomeroy, L. E. and Deibel, D. (1986). Temperature regulation of bacterial activity during the spring bloom in Newfoundland coastal waters. *Science* **233**, 359–361.

Pomeroy, L. R. and Wiebe, W. J. (1988). Energetics of microbial food webs. *Hydrobiologia* **159**, 7–18.

Pomeroy, L. R. and Wiebe, W. J. (2001). Temperature and substrates as interactive limiting factors for marine heterotrophic bacteria. *Aquatic Microbial Ecology* **23**, 187–204.

Pond, D. W., Bell, M. V., Dixon, D. R., Fallick, A. E., Segonzac, M., and Sargent, J. R. (1998). Stable-carbon-isotope composition of fatty acids in hydrothermal vent mussels containing methanotrophic and thiotrophic bacterial endosymbionts. *Applied and Environmental Microbiology* **64**, 370–375.

Poole, R. K. (1994). Oxygen reactions with bacterial oxidases and globins: Binding, reduction and regulation. *Antonie van Leeuwenhoek International Journal of General and Molecular Microbiology* **65**, 289–310.

Poole, R. K. and Hill, S. (1997). Respiratory protection of nitrogenase activity in *Azotobacter vinelandii*—Roles of the terminal oxidases. *Bioscience Reports* **17**, 303–317.

Poovachiranon, S. and Tantichodok, P. (1991). The role of sesarmid crabs in the mineralization of leaf litter of *Rhizophora apiculata* in a mangrove, southern Thailand. *Phuket Marine Biological Center Research Bulletin* **56**, 63–74.

Popp, B. N., Laws, E. A., Bidigare, R. R., Dore, J. E., Hanson, K. L., and Wakeham, S. G. (1998). Effect of phytoplankton cell geometry on carbon isotopic fractionation. *Geochimica et Cosmochimica Acta* **62**, 69–77.

Por, F. D. (1984). The ecosystem of the mangal: General considerations. *In* "Hydrobiology of the Mangal" (F. D. Por and I. Dor, eds), pp. 1–14. W. Junk Publishers, The Hague.

Posey, J. E. and Gherardini, F. C. (2000). Lack of a role for iron in the Lyme disease pathogen. *Science* **288**, 1651–1653.

Postgate, J. (1998). Nitrogen Fixation, 3rd Ed. Cambridge University Press, Cambridge.

Postgate, J. R. (1984). The Sulfate-Reducing Bacteria, p. 151. Cambridge University Press, London.

Postgate, J. R. and Hunter, J. R. (1962). The survival of starved bacteria. *Journal of General Microbiology* **29**, 233–262.

Postma, D. (1985). Concentrations of Mn and separation from Fe in sediments. 1. Kinetics and stoichiometry of the reaction between birnessite and dissolved Fe(II) at 10 °C. *Geochimica et Cosmochimica Acta* **49**, 1023–1033.

Postma, D. (1990). Kinetics of nitrate reduction by detrital Fe(II)-silicates. *Geochimica et Cosmochimica Acta* **54**, 903–908.

Postma, D. (1993). The reactivity of iron oxides in sediments—A kinetic approach. *Geochimica et Cosmochimica Acta* **57**, 5027–5034.

Postma, D. and Jakobsen, R. (1996). Redox zonation: Equilibrium constraints on the Fe(III)/SO_4-reduction interface. *Geochimica et Cosmochimica Acta* **60**, 3169–3175.

Potts, M. (1999). Mechanisms of desiccation tolerance in cyanobacteria. *European Journal of Phycology* **34**, 319–328.

Poulton, S. W. and Raiswell, R. (2002). The low-temperature geochemical cycle of iron: From continental fluxes to marine sediment deposition. *American Journal of Science* **302**, 774–805.

Prather, M. J., Derwent, R., Ehhalt, D., Fraser, P., Sanhueza, E., and Zhou, X. (1995). Other trace gases and atmospheric chemistry. *In* "Climate Change (1994)" (J. T. Houghton, L. G. Meira-Filho, J. P. Bruce, H. Lee, B. A. Callander, E. F. Haites, N. Harris, and K. Mashell, eds), pp. 73–126. Cambridge University Press, Cambridge.

Preisig, O., Zufferey, R., ThonyMeyer, L., Appleby, C. A., and Hennecke, H. (1996). A high-affinity *cbb*$_3$-type cytochrome oxidase terminates the symbiosis-specific respiratory chain of *Bradyrhizobium japonicum*. *Journal of Bacteriology* **178**, 1532–1538.

Preston, C. M., Wu, K. Y., Molinski, T. F., and DeLong, E. F. (1996). A psychrophilic crenarchaeon inhabits a marine sponge: *Cenarchaeum symbiosum* gen. nov., sp. nov. *Proceedings of the National Academy of Sciences USA* **93**, 6241–6246.

Preuss, A., Schauder, R., Fuchs, G., and Stichler, W. (1989). Carbon isotope fractionation by autotrophic bacteria with three different CO_2 fixation pathways. *Zeitschrift für Naturforschung* **44c**, 397–402.

Price, N. M. and Morel, F. M. M. (1998). Biological cycling of iron in the ocean. *Metal Ions in Biological Systems* **35**, 1–36.

Pringault, O. and Garcia-Pichel, F. (2000). Monitoring of oxygenic and anoxygenic photosynthesis in a unicyanobacterial biofilm, grown in benthic gradient chamber. *FEMS Microbiology Ecology* **33**, 251–258.

Pronk, J. T., Debruyn, J. C., Bos, P., and Kuenen, J. G. (1992). Anaerobic growth of *Thiobacillus ferrooxidans*. *Applied and Environmental Microbiology* **58**, 2227–2230.

Putt, M. and Prezelin, B. B. (1985). Observations of diel patterns of photosynthesis in cyanobacteria and nanoplankton in the Santa Barbara Channel during El Nino. *Journal of Plankton Research* **7**, 779–790.

Pyzik, A. J. and Sommer, S. S. (1981). Sedimentary iron monosulfides: Kinetics and mechanim of formation. *Geochimica et Cosmochimica Acta* **45**, 687–698.

Qasim, S. Z. and Wafar, V. M. (1990). Marine resources in the trophics. *In* "Trophical Resources: Ecology and Development" (J. I. Furtado, W. B. Morgan, J. R. Pfafflin, and K. Ruddle, eds), pp. 141–169. Harwood Academic Publishers, London.

Quandt, L., Gottschalk, G., Ziegler, H., and Stichler, W. (1977). Isotope discrimination by photosynthetic bacteria. *FEMS Microbiology Letters* **1**, 125–128.

Rabus, R., Nordhaus, R., Ludwig, W., and Widdel, R. (1993). Complete oxidation of toluene under strictly anoxic conditions by a new sulfate-reducing bacterium. *Applied and Environmental Microbiology* **59**, 1444–1451.

Ragueneau, O., Treguer, P., Leynaert, A., Anderson, R. F., Brzezinski, M. A., DeMaster, D. J., Dugdale, R. C., Dymond, J., Fischer, G., Francois, R., Heinze, C., Maier-Reimer, E., Martin-Jezequel, V., Nelson, D. M., and Queguiner, B. (2000). A review of the Si cycle in the modem ocean: Recent progress and missing gaps in the application of biogenic opal as a paleoproductivity proxy. *Global and Planetary Change* **26**, 317–365.

Rai, A. N., Söderback, E., and Bergman, B. (2000). Cyanobacterium-plant symbioses. *New Phytologist* **147**, 449–481.

Rappé, M. S., Connon, S. A., Vergin, K. L., and Giovannoni, S. J. (2002). Cultivation of the ubiquitous SAR11 marine bacterioplankton clade. *Nature* **418**, 630–633.

Ratering, S. and Schnell, S. (2000). Localization of iron-reducing activity in paddy soil by profile studies. *Biogeochemistry* **48**, 341–365.

Raven, J. A. (1988). The iron and molybdenum use efficiencies of plant growth with different energy, carbon and nitrogen sources. *New Phytologist* **109**, 279–287.

Raven, J. A. (1991). Physiology of inorganic C acquisition and implications for resource use efficiency by marine phytoplankton: Relation to increased CO_2 and temperature. *Plant, Cell and Environment* **14**, 779–794.

Raven, J. A. (1994). Carbon fixation and carbon availability in marine phytoplankton. *Photosynthesis Research* **39**, 259–273.

Raven, J. A., Johnston, A. M., Parsons, R., and Kübler, J. (1994). The influence of natural and experimental high O_2 concentrations on O_2-evolving phototrophs. *Biological Reviews* **69**, 61–94.

Ravenschlag, K., Sahm, K., Pernthaler, J., and Amann, R. (1999). High bacterial diversity in permanently cold marine sediments. *Applied and Environmental Microbiology* **65**, 3982–3989.

Rawlings, D. E., Tributsch, H., and Hansford, G. S. (1999). Reasons why '*Leptospirillum*'-like species rather than *Thiobacillus ferroxidans* are the dominant iron-oxidizing bacteria in many commercial processes for the biooxidation of pyrite and related ores. *Microbiology-UK* **145**, 5–13.

Raymond, J., Zhaxybayeva, O., Gogarten, J. P., Gerdes, S. Y., and Blankenship, R. E. (2002). Whole-genome analysis of photosynthetic prokaryotes. *Science* **298**, 1616–1620.

Redfield, A. C. (1958). The biological control of chemical factors in the environment. *American Scientist* **46**, 205–222.

Reeburgh, W. S. (1969). Observations of gases in Chesapeake Bay sediments. *Limnology and Oceanography* **14**, 368–375.

Reeburgh, W. S. (1976). Methane consumption in cariaco trench waters and sediments. *Earth and Planetary Science Letters* **28**, 337–344.

Reeburgh, W. S. (1980). Anaerobic methane oxidation: Rate depth distributions in Skan Bay sediments. *Earth and Planetary Science Letters* **47**, 345–352.

Reeburgh, W. S., Ward, B. B., Whalen, S. C., Sandbeck, K. A., Kilpatrick, K. A., and Kerkhof, L. J. (1991). Black-sea methane geochemistry. *Deep-Sea Research Part A-Oceanographic Research Papers* **38**, S1189–S1210.

Rees, G. N., Grassia, G. S., Sheehy, A. J., Dwivedi, P. P., and Patel, B. K. C. (1995). *Desulfacinum infernum* gen. nov., sp. nov., a thermophilic sulfate-reducing bacterium from a petroleum reservoir. *International Journal of Systematic Bacteriology* **45**, 85–89.

Reichenbach, H. and Dworkin, M. (1981). The order Cytophagales (with addenda on the genera *Herpetisiphon, Saprospiran*, and *Flexithrix*). *In* "The Prokaryotes" (M. P. Starr, H. Stolp, H. G. Trüper, A. Belows, and H. G. Schlegel, eds). Springer, Berlin.

Reichenbach, H. and Dworkin, M. (1992). The Myxobacteria. *In* "The Prokaryotes" (A. Balows, H. G. Trüper, M. Dworkin, W. Harder, and K.-H. Schleifer, eds), pp. 3416–3487. Springer, New York.

Reid, P. C., Lancelot, C., Gieskes, W. W. C., Hagmeier, E., and Weichert, G. (1990). Phytoplankton of the North Sea and its dynamics—A review. *Netherlands Journal of Sea Research* **26**, 295–331.

Reimers, C. E., Tender, L. M., Fertig, S., and Wang, W. (2001). Harvesting energy from the marine sediment-water interface. *Environmental Science & Technology* **35**, 192–195.

Reinfelder, J. R., Kraepiel, A. M. L., and Morel, F. M. M. (2000). Unicellular C4 photosynthesis in a marine diatom. *Nature* **407**, 996–999.

Ren, T., Amaral, J. A., and Knowles, R. (1997). The response of methane consumption by pure cultures of methanotrophic bacteria to oxygen. *Canadian Journal of Microbiology* **43**, 925–928.

Repeta, D. J., Simpson, D. J., Jørgensen, B. B., and Jannasch, H. W. (1989). Evidence for anoxygenic photosynthesis from the distribution of bacteriochlorophylls in the Black Sea. *Nature* **342**, 69–72.

Revsbech, N. J., Jørgensen, B. B., Blackburn, T. H., and Cohen, Y. (1983). Microelectrode studies of the photosynthesis and O_2, H_2S, and pH profiles of a microbial mat. *Limnology and Oceanography* **28**, 1062–1074.

Revsbech, N. P. (1988). Benthic primary production and oxygen profiles. *In* "Nitrogen Cycling in Coastal Marine Environments" (T. H. Blackburn and J. Sørensen, eds), pp. 69–83. John Wiley & Sons Ltd.

Revsbech, N. P. and Jørgensen, B. B. (1986). Microelectrodes: Their use in microbial ecology. *In* "Advances in Microbial Ecology" (K. C. Marshall, ed.), pp. 293–352. Plenum Publishing Corporation, New York.

Reynolds, C. S. and Davies, P. S. (2001). Sources and bioavailability of phosphorus fractions in freshwaters: A British perspective. *Biological Reviews* **76**, 27–64.

Rhee, G. Y. (1972). Competition between an Alga and an Aquatic Bacterium for Phosphate. *Limnology and Oceanography* **17**, 505–514.

Rice, C. W. and Hempfling, W. P. (1978). Oxygen-limited continuous culture and respiratory energy-conservation in *Escherichia coli*. *Journal of Bacteriology* **134**, 115–124.

Richards, F. A. (1965). Chemical observations in some anoxic, sulfide-bearing basins and fjords. *In* "Advances in Water Pollution Research" (E. A. Pearson, ed.), pp. 215–233. Pergamon Press, London, Edinburgh, New York, Frankfurt-am-Main.

Richards, F. A. and Broenkow, W. W. (1971). Chemical changes, including nitrate reduction, in Darwin Bay, Galapagos archipelago, over a 2-month period, 1969. *Limnology and Oceanography* **16**, 758–765.

Richardson, K. (1996). Carbon flow in the water column case study: The southern Kattegat. *In* "In Eutrophication in Coastal Marine Ecosystems" (B. B. Jørgensen and K. Richardson, eds), pp. 95–114. American Geophysical Union, Washington, DC.

Richardson, K. (1997). Harmful or exceptional phytoplankton blooms in the marine ecosystem. *Advances in Marine Biology* **31**, 301–385.

Riemann, L. and Middelboe, M. (2002). Viral lysis of marine bacterioplanckton: Implications for organic matter cycling and bacterial clonal composition. *Ophelia* **56**, 57–68.

Ripl, W. (1976). Biochemical oxidation of polluted lake sediment with nitrate—A new lake restoration method. *AMBIO* **3**, 132–135.

Risatti, J. B., Capman, W. C., and Stahl, D. A. (1994). Community structure of a microbial mat: The phylogenetic dimension. *Proceedings of the National Academy of Sciences USA* **91**, 10173–10177.

Risgaard-Petersen, N. and Jensen, N. K. (1997). Nitrification and denitrification in the rhizosphere of the aquatic macrophyte *Lobelia dortmanna* L. *Limnology and Oceanography* **42**, 529–537.

Rittenberg, S. C. and Hespell, R. B. (1975). Energy efficiency of intraperiplasmic growth of bdellovibrio bacteriovorus. *Journal of Bacteriology* **121**, 1158–1165.

Rivera-Monroy, V. H. and Twilley, R. R. (1996). The relative role of denitrification and immobilization in the fate of inorganic nitrogen in mangrove sediments (Terminos Lagoon, Mexico). *Limnology and Oceanography* **41**, 284–296.

Rivera-Monroy, V. H., Day, J. W., Twilley, R. R., Veraherrera, F., and Coronadomolina, C. (1995). Flux of nitrogen and sediment in a fringe mangrove forest in Terminos Lagoon, Mexico. *Estuarine Coastal and Shelf Science* **40**, 139–160.

Rivera-Monroy, V. H., Torres, L. A., Bahamon, N., Newmark, F., and Twilley, R. R. (1999). The potential use of mangrove forests as nitrogen sinks of shrimp aquaculture pond effluents: The role of denitrification. *Journal of the World Aquaculture Society* **30**, 12–25.

Robertson, A. I. (1986). Leaf-burying crabs: Their influence on energy flow and export from mixed mangrove forests (*Rhizophora* spp.) in north eastern Australia. *Journal of Experimental Marine Biology and Ecology* **102**, 237–248.

Robertson, A. I. (1991). Plant animal interactions and the structure and function of mangrove forest ecosystems. *Australian Journal of Ecology* **16**, 433–443.

Robertson, A. I. and Alongi, D. M. (1995). Role of riverine mangrove forests in organic carbon export to the tropical coastal ocean: A preliminary mass balance for the Fly Delta (Papua New Guinea). *Geo-Marine Letters* **15**, 134–139.

Robertson, A. I. and Daniel, P. A. (1989). The influence of crabs on litter processing in high intertidal mangrove forests in tropical Australia. *Oecologia* **78**, 191–198.

Robertson, A. I. and Duke, N. C. (1987). Insect herbivory on mangrove leaves in North Queensland. *Australian Journal of Ecology* **12**, 1–7.

Robertson, A. I., Alongi, D. M., and Boto, K. G. (1992). Food chains and carbon fluxes. *In* "Tropical Mangrove Ecosystems. Coastal and Estuarine Studies 41" (A. I. Robertson and D. M. Alongi, eds), pp. 293–326. American Geophysical Union, Washington, DC.

Robertson, L. A. and Kuenen, J. G. (1984). Aerobic denitrification: A controversy revived. *Archives of Microbiology* **139**, 351–354.

Robertson, L. A. and Kuenen, J. G. (1992). The colorless sulfur bacteria. *In* "The Prokaryotes" (A. Balows, H. G. Trüper, M. Dworkin, W. Harder, and K.-H. Schleifer, eds), 2nd Ed., pp. 385–413. Springer, New York.

Robertson, L. A., Dalsgaard, T., Revsbech, N. P., and Kuenen, J. G. (1995). Confirmation of 'aerobic denitrification' in batch cultures, using gas chromatography and N-15 mass spectrometry. *FEMS Microbiology Ecology* **18**, 113–119.

Robie, R. A., Hemingway, B. S., and Fisher, J. R. (1978). "Thermodynamic Properties of Minerals and Related Substances at 298.15 K and I Bar (10^5 Pascals) Pressure and at Higher Temperatures." U.S. Geological Survey Bulletin No. 1452, pp. 456.

Robinson, J. J., Scott, K. M., Swanson, S. T., O'Leary, M. H., Horken, K., Tabita, F. R., and Cavanaugh, C. M. (2003). Kinetic isotope effect and characterization of form II RubisCO from the chemoautotrophic endosymbionts of the hydrothermal vent tubeworm *Riftia pachyptila*. *Limnology and Oceanography* **48**, 48–54.

Rocap, G., Larimer, F. W., Lamerdin, J., Malfatti, S., Chain, P., Ahlgren, N. A., Arellano, A., Coleman, M., Hauser, L., Hess, W. R., Johnson, Z. I., Land, M., Lindell, D., Post, A. F., Regala, W., Shah, M., Shaw, S. L., Steglich, C., Sullivan, M. B., Ting, C. S., Tolonen, A., Webb, E. A., Zinser, E. R., and Chisholm, S. W. (2003). Genome divergence in two *Prochlorococcus* ecotypes reflects oceanic niche differentiation. *Nature* **424**, 1042–1047.

Rodelli, M. R., Gearing, J. N., Gearing, P. J., Marshall, N., and Sasekumar, A. (1984). Stable isotope ratio as a tracer of mangrove carbon in Malaysian ecosystems. *Oecologia* **61**, 326–333.

Roden, E. E. (2003). Diversion of electron flow from metahnogenesis to crystalline Fe(III) oxide reduction in carbon-limited cultures of wetland sediment microorganisms. *Applied and Environmental Microbiology* **69**, 5702–5706.

Roden, E. E. and Urrutia, M. M. (2002). Influence of biogenic Fe(II) on bacterial crystalline Fe(III) oxide reduction. *Geomicrobiology Journal* **19**, 209–251.

Roden, E. E. and Wetzel, R. G. (1996). Organic carbon oxidation and suppression of methane production by microbial Fe(III) oxide reduction in vegetated and unvegetated freshwater wetland sediments. *Limnology and Oceanography* **41**, 1733–1748.

Roden, E. E. and Wetzel, R. G. (2002). Kinetics of microbial Fe(III) oxide redction in freshwater wetland sediments. *Limnology and Oceanography* **47**, 198–211.

Roden, E. E. and Wetzel, R. G. (2003). Competition between Fe(III)-reducing and methanogenic bacteria for acetate in iron-rich freshwater sediments. *Microbial Ecology* **45**, 252–258.

Roden, E. E. and Zachara, J. M. (1996). Microbial reduction of crystalline iron(III) oxides: Influence of oxide surface area and potential for cell growth. *Environmental Science & Technology* **30**, 1618–1628.

Roden, E. E., Urratia, M. M., and Mann, C. (2000). Bacterial reductive dissolution of crystalline Fe(III) oxide in continuous-flow column reactors. *Applied and Environmental Microbiology* **66**, 1062–1065.

Rodrigo, M. A., Miracle, M. R., and Vicente, E. (2001). The meromictic Lake La Cruz (Central Spain). Patterns of stratification. *Aquatic Sciences* **63**, 406–416.

Roe, J. E., Anbar, A. D., and Barling, J. (2003). Nonbiological fractionation of Fe isotopes: Evidence of an equilibrium isotope effect. *Chemical Geology* **195**, 69–85.

Roenneberg, T. and Carpenter, E. J. (1993). Daily rhythm of O_2-evolution in the cyanobacterium *Trichodesmium thiebautii* under natural and constant conditions. *Marine Biology* **117**, 693–697.

Roger, P. A. and Ladha, J. K. (1992). Biological N_2 fixation in wetland rice fields: Estimation and contribution to nitrogen balance. *Plant and Soil* **141**, 41–55.

Röling, W. F. M., van Breukelen, B. M., Braster, M., Lin, B., and van Verseveld, H. W. (2001). Relationships between microbial community structure and hydrochemistry in a landfill leachate-polluted aquifer. *Applied and Environmental Microbiology* **67**, 4619–4629.

Romanenko, W. I. (1965). The relation between the consumption of oxygen and CO_2 by heterotrophic bacteria during the growth in the presence of peptone. *Microbiologia* **33**, 134–139.

Rooney-Varga, J. N., Anderson, R. T., Fraga, J. L., Ringelberg, D., and Lovley, D. R. (1999). Microbial communities associated with anaerobic benzene degradation in a petroleum-contaminated aquifer. *Applied and Environmental Microbiology* **65**, 3056–3063.

Rose, A. L. and Waite, T. D. (2002). Kinetic model for Fe(II) oxidation in seawater in the absence and presence of natural organic matter. *Environmental Science & Technology* **36**, 433–444.
Rosenfeld, J. K. (1979). Ammonium adsorption in nearshore anoxic sediments. *Limnology and Oceanography* **24**, 356–364.
Rosson, R. A. and Nealson, K. H. (1982). Manganese binding and oxidation by spores of a marine *Bacillus*. *Journal of Bacteriology* **151**, 1027–1034.
Roswall, T. (1983). The nitrogen cycle. *In* "The Major Biogeochemical Cycles and Their Interactions" (B. Bolin and R. B. Cooke, eds), pp. 46–50. John Wiley and Sons, Chichester.
Rothfuss, F. and Conrad, R. (1993). Vertical profiles of CH_4 concentrations, dissolved substrates and processes involved in CH_4 production in a flooded Italian rice field. *Biogeochemistry* **18**, 137–152.
Roychoudhury, A. N., Viollier, E., and Van Cappellen, P. (1998). A plug flow-through reactor for studying biogeochemical reactions in undisturbed aquatic sediments. *Applied Geochemistry* **13**, 269–280.
Ruby, E. G. (1992). The genus *Bdellovibrio*. *In* "The Prokaryotes" (A. Balows, H. G. Trüper, M. Dworkin, W. Harder, and K.-H. Schleifer, eds), pp. 3400–3415. Springer, New York.
Ruby, E. G. (1996). Lessons from a cooperative, bacterial-animal association: The *Vibrio fischeri-Euprymna scolopes* light organ symbiosis. *Annual Review of Microbiology* **50**, 591–624.
Rue, E. L. and Bruland, K. W. (1995). Complexation of iron(III) by natural organic ligands in the central North Pacific as determined by a new competitive ligand equilibration adsorptive cathodic stripping voltammetric method. *Marine Chemistry* **50**, 117–138.
Rueter, J. G. (1988). Iron stimulation of photosynthesis and nitrogen-fixation in *Anabaena*-7120 and *Trichodesmium* (Cyanophyceae). *Journal of Phycology* **24**, 249–254.
Runnegar, B. (1986). Molecular palaeontology. *Palaeontology* **29**, 1–24.
Russell, J. B. and Cook, G. M. (1995). Energetics of bacterial growth: Balance of anabolic and catabolic reactions. *Microbioological Reviews* **59**, 48–62.
Russell, M. J. and Hall, A. J. (1997). The emergence of life from iron monosulphide bubbles at a submarine hydrothermal redox and pH front. *Journal of the Geological Society* **154**, 377–402.
Russell-Hunter, W. D. (1970). Aquatic Productivity: An Introduction to Some Basic Aspects of Biological Oceanography and Limnology. Macmillan, London.
Ruttenberg, K. C. (1992). Development of a sequential extraction method for different forms of phosphorus in marine-sediments. *Limnology and Oceanography* **37**, 1460–1482.
Ruttenberg, K. C. (1993). Reassessment of the oceanic residence time of phosphorus. *Chemical Geology* **107**, 405–409.
Rysgaard, S., Risgaard-Petersen, N., Nielsen, L. P., and Revsbech, N. P. (1993). Nitrification and denitrification in lake and estuarine sediments measured by the ^{15}N dilution technique and isotope pairing. *Applied and Environmental Microbiology* **59**, 2093–2098.
Rysgaard, S., Risgaard-Petersen, N., Sloth, N. P., Jensen, K., and Nielsen, L. P. (1994). Oxygen regulation of nitrification and denitrification in sediments. *Limnology and Oceanography* **39**, 1643–1652.
Rysgaard, S., Christensen, P. B., and Nielsen, L. P. (1995). Seasonal variation in nitrification and denitrification in estuarine sediment colonized by benthic

microalgae and bioturbating infauna. *Marine Ecology-Progress Series* **126**, 111–121.

Rysgaard, S., Thamdrup, B., Risgaard-Petersen, N., Fossing, H., Berg, P., Christensen, P. B., and Dalsgaard, T. (1998). Seasonal carbon and nutritient mineralization in a high-Arctic coastal marine sediment, Young Sound, Northeast Greenland. *Marine Ecology-Progress Series* **175**, 261–276.

Rysgaard, S., Fossing, H., and Jensen, M. M. (2001). Organic matter degradation through oxygenic respiration, denitrification, and manganese, iron, and sulfate reduction in marine sediments (the Kattegat and the Skagerrak). *Ophelia* **55**, 77–91.

Ryther, J. H. and Dunstan, W. M. (1971). Nitrogen, phosphorus and eutrophication in the coastal marine sediment. *Science* **171**, 1008–1013.

Sagemann, J., Jørgensen, B. B., and Greeff, O. (1998). Temperature dependence and rates of sulfate reduction in cold sediments of Svalbard, Arctic Ocean. *Geomicrobiology Journal* **15**, 85–100.

Sand-Jensen, K., Prahl, C., and Stokholm, H. (1982). Oxygen release from roots of submerged aquatic macrophytes. *Oikos* **38**, 349–354.

Santschi, P. H., Guo, L. D., Baskaran, M., Trumbore, S., Southon, J., Bianchi, T. S., Honeyman, B., and Cifuentes, L. (1995). Isotopic evidence for the contemporary origin of high-molecular-weight organic-matter in oceanic environments. *Geochimica et Cosmochimica Acta* **59**, 625–631.

Saraste, M. and Castresana, J. (1994). Cytochrome-oxidase evolved by tinkering with denitrification enzymes. *FEBS Letters* **341**, 1–4.

Sarbu, S. M., Kane, T. C., and Kinkle, B. K. (1996). A chemoautotrophic based cave ecosystem. *Science* **272**, 1953–1955.

Sass, H., Cypionka, H., and Babenzien, H.-D. (1997). Vertical distribution of sulfate-reducing bacteria at the oxic-anoxic interface in sediments of the oligotrophic Lake Stechlin. *FEMS Microbiology Ecology* **22**, 245–255.

Sass, A. M., Eschemann, A., Kühl, M., Thar, R., Sass, H., and Cypionka, H. (2002). Growth and chemosensory behavior of sulfate-reducing bacteria in oxygen-sulfide gradients. *FEMS Microbiology Ecology* **40**, 47–54.

Sayles, F. L., Martin, W. R., Chase, Z., and Anderson, R. F. (2001). Benthic remineralization and burial of biogenic SiO_2, $CaCO_3$, organic carbon, and detrital material in the Southern Ocean along a transect at 170 degrees West. *Deep-Sea Research Part* II-*Topical Studies in Oceanography* **48**, 4323–4383.

Scala, D. J. and Kerkhof, L. J. (1999). Diversity of nitrous oxide reductase (nosZ) genes in continental shelf sediments. *Applied and Environmental Microbiology* **65**, 1681–1687.

Schauder, S. and Bassler, B. L. (2001). The languages of bacteria. *Genes and Development* **15**, 1468–1480.

Schelske, C. L. (1988). Historic trends in Lake-Michigan silica concentrations. *Internationale Revue Der Gesamten Hydrobiologie* **73**, 559–591.

Schelske, C. L. and Stoermer, E. F. (1971). Eutrophication, silica depletion, and predicted changes in algal quality in Lake Michigan. *Science* **173**, 423–424.

Schenk, H. J. and Jackson, R. B. (2002). Rooting depths, lateral root spreads and below-ground/above-ground allometries of plants in water-limited ecosystems. *Journal of Ecology* **90**, 480–494.

Scherer, S. and Neuhaus, K. (2002). Life at low temperatures. *In* "The Prokaryotes: An Evolving Electronic Resource for the Microbiological Community" (A. Balows, H. G. Trüper, M. Dworkin, W. Harder, and K. H. Schleifer, eds), 3rd Ed., release 3.9, April 1, 2002. Springer, New York.

Schiff, J. A. and Fankhauser, H. (1981). Assimilatory sulfate reduction. *In* "Biology of Inorganic Nitrogen and Sulfur" (H. Bothe and A. Trebst, eds), pp. 153–168. Springer, New York.

Schink, B. (1992). The genus *Pelobacter*. *In* "The Prokaryotes" (A. Balows, H. Trüper, M. Dworkin, W. Harder, and K.-H. Schleifer, eds), 2nd Ed., pp. 583–624. Springer, Berlin.

Schink, B. (1997). Energetics of syntrophic cooperation in methanogenic degradation. *Microbiology and Molecular Biology Reviews* **61**, 262–280.

Schink, B. and Friedrich, M. (2000). Bacterial metabolism—Phosphite oxidation by sulphate reduction. *Nature* **406**, 37.

Schink, B., Thiemann, V., Laue, H., and Friedrich, M. W. (2002). *Desulfotignum phosphitoxidans* sp. nov., a new marine sulfate reducer that oxidizes phosphite to phosphate. *Archives of Microbiology* **177**, 381–391.

Schippers, A. and Jorgensen, B. B. (2001). Oxidation of pyrite and iron sulfide by manganese dioxide in marine sediments. *Geochimica et Cosmochimica Acta* **65**, 915–922.

Schippers, A. and Jørgensen, B. B. (2002). Biogeochemistry of pyrite and iron sulfide oxidation in marine sediments. *Geochimica et Cosmochimica Acta* **66**, 85–92.

Schlegel, H. G. (1976). "Allgemeine Microbiologie." 4th Ed. Georg Thieme Verlag, Stuttgart.

Schleper, C., Puehler, G., Holz, I., Gambacorta, A., Janekovic, D., Santarius, U., Klenk, H.-P., and Zillig, W. (1995). *Picrophilus* gen-nov, fam-nov—A novel aerobic, heterotrophic, thermoacidophilic genus and family comprising archaea capable of growth around pH-0. *Journal of Bacteriology* **177**, 7050–7059.

Schlesinger, W. H. (1997). "Biogeochemistry." 2nd Ed. Academic Press, San Diego.

Schlosser, D. and Höfer, C. (2002). Laccase-catalyzed oxidation of Mn^{2+} in the presence of natural Mn^{3+} chelators as a novel source of extracellular H_2O_2 production and its impact on manganese peroxidase. *Applied and Environmental Microbiology* **68**, 3514–3521.

Schmaljohann, R. (1991). Oxidation of various potential-energy sources by the methanotrophic endosymbionts of siboglinum-poseidoni (pogonophora). *Marine Ecology Progress Series* **76**, 143–148.

Schmidt, T. M., Arieli, B., Cohen, Y., Padan, E., and Strohl, W. R. (1987). Sulfur metabolism in *Beggiatoa alba*. *Journal of Bacteriology* **169**, 5466–5472.

Scholten, J. C. M. and Conrad, R. (2000). Energetics of syntrophic propionate oxidation in defined batch and chemostat cocultures. *Applied and Environmental Microbiology* **66**, 2934–2942.

Schopf, J. W. (1992). Paleobiology of the Archean. *In* "The Proterozoic Biosphere. A Multidisciplinary Study" (J. W. Schopf and C. Klein, eds). Cambridge University Press, Cambridge.

Schopf, J. W., Kudryavtsev, A. B., Agresti, D. G., Wdowiak, T. J., and Czaja, A. D. (2002). Laser-Raman imagery of Earth's earliest fossils. *Nature* **416**, 73–76.

Schrag, D. P., Berner, R. A., Hoffman, P. F., and Halverson, G. P. (2002). On the initiation of a snowball Earth. *Geochemistry Geophysics Geosystems G3* **3**, art. no. 1036, doi:10.1029/2001GC000219.

Schramm, A., Larsen, L. H., Revsbech, N. P., Ramsing, N. B., Amann, R., and Schleifer, K.-H. (1996). Structure and function of a nitrifying biofilm as determined by *in situ* hybridization and the use of microelectrodes. *Applied and Environmental Microbiology* **62**, 4641–4647.

Schulz, H. N. and Jørgensen, B. B. (2001). Big bacteria. *Annual Review of Microbiology* **55**, 105–137.

Schulz, H. N., Jørgensen, B. B., Fossing, H., and Ramsing, N. B. (1996). Community structure of filamentous, sheath-building sulfur bacteria, *Thioploca* spp., off the coast of Chile. *Applied and Environmental Microbiology* **62**, 1855–1862.

Schulz, H. N., Brinkhoff, T., Ferdelman, T. G., Hernández Mariné, M., Teske, A., and Jørgensen, B. B. (1999). Dense populations of a giant sulfur bacterium in Namibian Shelf sediments. *Science* **284**, 493–495.

Schulz, S. and Conrad, R. (1996). Influence of temperature on pathways to methane production in the permanently cold profundal sediment of Lake Constance. *FEMS Microbiology Ecology* **20**, 1–14.

Schönheit, P. and Schäfer, T. (1995). Metabolism of hyperthermophiles. *World Journal of Microbiology & Biotechnology* **11**, 26–57.

Seaver, L. C. and Imlay, J. A. (1999). Hydrogen peroxide fluxes and compartmentalization inside growing *Escherichia coli. Journal of Bacteriology* **183**, 7182–7189.

Seaver, L. C. and Imlay, J. A. (2001). Alkyl hydroperoxide reductase is the primary scavenger of endogenous hydrogen peroxide in *Escherichia coli. Journal of Bacteriology* **183**, 7173–7181.

Sebacher, D. I., Harriss, R. C., and Bartlett, K. B. (1985). Methane emissions to the atmosphere through aquatic plants. *Journal of Environmental Quality* **14**, 40–46.

Segal, M. G. and Sellers, R. M. (1984). Redox reactions at solid-liquid interfaces. *Advances in Inorganic and Bioinorganic Interfaces* **3**, 97–129.

Segerer, A. H. and Stetter, K. O. (1992). The order sulfolobales. *In* "The Prokaryotes" (A. Balows, H. G. Trüper, M. Dworkin, W. Harder, and K.-H. Schleifer, eds), 2nd Ed., pp. 684–701. Springer, New York.

Seitzinger, S. P. (1988). Dentrification in freshwater and coastal marine ecosystems: Ecological and geochemical significance. *Limnology and Oceanography* **33**, 702–724.

Seitzinger, S. P. (1990). Denitrification in aquatic sediments. *In* "Denitrification in Soil and Sediment" (N. P. Revsbech and J. Sørensen, eds), pp. 301–322. Plenum Press, New York.

Seitzinger, S. P. (1993). Denitrification and nitrification rates in aquatic sediments. *In* "Handbook of Methods in Aquatic Microbial Ecology" (P. F. Kemp, B. F. Sherr, E. B. Sherr, and J. J. Cole, eds), pp. 633–641. Lewis Publishers, Boca Raton.

Seitzinger, S. P. and Garber, J. H. (1987). Nitrogen fixation and $^{15}N_2$ calibration of the acetylene reduction assay in coastal marine sediments. *Marine Ecology-Progress Series* **37**, 65–73.

Seitzinger, S. P. and Giblin, A. E. (1996). Estimating denitrification in North Atlantic continental shelf sediments. *Biogeochemistry* **35**, 235–260.

Seitzinger, S. P. and Kroeze, C. (1998). Global distribution of nitrous oxide and N inputs in freshwater and coastal marine ecosystems. *Global Biogeochemical Cycles* **12**, 93–113.

Seitzinger, S. P., Nixon, S. W., and Pilson, M. E. Q. (1984). Denitrification and nitrous oxide production in a coastal marine ecosystem. *Limnology and Oceanography* **29**, 73–83.

Selje, N., Simon, M., and Brinkhoff, T. (2004). A newly discovered *Roseobacter* cluster in temperate and polar oceans. *Nature* **427**, 445–448.

Sellner, K. G. (1997). Physiology, ecology, and toxic properties of marine cyanobacteria blooms. *Limnology and Oceanography* **42**, 1089–1104.

Serger, A. H. and Stetter, K. O. (1992). The order Sulfolobales. *In* "The Prokaryotes" (A. Balows, H. G. Trüper, M. Dworkin, W. Harder, and K. H. Schleifer, eds), 2nd Ed. Springer, Berlin.

Shan, Y., McKelvie, I. D., and Hart, B. T. (1994). Determination of alkaline phosphatase-hydrolyzable phosphorus in natural-water systems by enzymatic flow-injection. *Limnology and Oceanography* **39**, 1993–2000.
Shapiro, J. A. (1998). Thinking about bacterial populations as multicellular organisms. *Annual Review of Microbiology* **52**, 81–104.
Shock, E. L. (1996). Hydrothermal systems as environments for the emergence of life. *In* "Evolution of Hydrothermal Ecosystems on Earth (and Mars?)" (G. R. Bock and J. A. Goode, eds), pp. 40–52. John Wiley & Sons.
Siever, R. (1991). Silica in the oceans: Biological-geochemical interplay. *In* "Scientists on Gaia" (S. H. Schneider and P. J. Boston, eds), pp. 287–295. MIT Press, Cambridge.
Sigalevich, P., Baev, M. V., Teske, A., and Cohen, Y. (2000). Sulfate reduction and possible aerobic metabolism of the sulfate-reducing bacterium *Desulfovibrio oxyclinae* in a chemostat coculture with *Marinobacter* sp. strain MB under exposure to increasing oxygen concentrations. *Applied and Environmental Microbiology* **66**, 5013–5018.
Sigman, D. M., Altabet, M. A., McCorkle, D. C., Francois, R., and Fischer, G. (2000). The $d^{15}N$ of nitrate in the Southern Ocean: Nitrogen cycling and circulation in the ocean interior. *Journal of Geophysical Research* **105**, 19599–19614.
Silberman, J. D., Simpson, A. G. B., Kulda, J., Cepicka, I., Hampl, V., Johnson, P. J., and Roger, A. J. (2002). Retortamonad flagellates are closely related to diplomonads—Implications for the history of mitochondrial function in eukaryote evolution. *Molecular Biology and Evolution* **19**, 777–786.
Silver, S. and Walderhaug, M. (1992). Gene-regulation of plasmid-determined and chromosome-determined inorganic-ion transport in bacteria. *Microbiological Reviews* **56**, 195–228.
Simpson, J. H., Gong, W. K., and Ong, J. E. (1997). The determination of the net fluxes from a mangrove estuary system. *Estuaries* **20**, 103–109.
Sinnighe Damsté, J. S., Strous, M., Rijpstra, W. I. C., Hopmans, E. C., Geenevasen, J. A. J., van Duin, A. C. T., van Niftrik, L. A., and Jetten, M. S. M. (2002). Linearly concantenated cyclobitane lipids form a dense bacterial membrane. *Nature* **419**, 708–712.
Sirevåg, R., Buchanan, B. B., Berry, J. A., and Troughton, J. H. (1977). Mechanisms of CO_2 fixation in bacterial photosynthesis studied by the carbon isotope fractionation technique. *Archives of Microbiology* **112**, 35–38.
Sjodin, A. L., Lewis, W. M., and Saunders, J. F. (1997). Denitrification as a component of the nitrogen budget for a large plains river. *Biogeochemistry* **39**, 327–342.
Skerker, J. M. and Berg, H. C. (2001). Direct observation of extension and retraction of type IV pili. *Proceedings of the National Academy of Sciences USA* **98**, 6901–6904.
Skov, M. W. and Hartnoll, R. G. (2002). Paradoxical selective feeding on a low-nutrient diet: Why do mangrove crabs eat leaves? *Oecologia* **131**, 1–7.
Slim, F. J., Hemminga, M. A., Ochieng, C., Jannink, N. T., de la Moriniere, E. C., and vanderVelde, G. (1997). Leaf litter removal by the snail *Terebralia palustris* (Linnaeus) and sesarmid crabs in an East African mangrove forest (Gazi Bay, Kenya). *Journal of Experimental Marine Biology and Ecology* **215**, 35–48.
Slomp, C. P., Malschaert, J. F. P., Lohse, L., and Van Raaphorst, W. (1997). Iron and manganese cycling in different sedimentary environments on the North Sea continental margin. *Continental Shelf Research* **17**, 1083–1117.

Smith, A. J. (1982). Modes of cyanobacterial carbon metabolism. *In* "The Biology of Cyanobacteria" (N. G. Carr and B. A. Whitton, eds). Blackwell Science, Oxford.

Smith, D. C., Simon, M., Alldredge, A. L., and Azam, F. (1992). Intense hydrolytic enzyme activity on marine aggregates and implications for rapid particle dissolution. *Nature* **359**, 139–142.

Smith, R. L. and Oremland, R. S. (1987). Big Soda Lake (Nevada). 2. Pelagic sulfate reduction. *Limnology and Oceanography* **32**, 793–803.

Smolders, A. J. P., Lamers, L. P. M., Moonen, M., Zwaga, K., and Roelofs, J. G. M. (2001). Controlling phosphate release from phosphate-enriched sediments by adding various iron compounds. *Biogeochemistry* **54**, 219–228.

Snoeyenbos-West, O. L., Nevin, K. P., Anderson, R. T., and Lovley, D. R. (2000). Enrichment of *Geobacter* species in response to stimulation of Fe(III) reduction in sandy aquifer sediments. *Microbial Ecology* **39**, 153–167.

Sobolev, D. and Roden, E. E. (2001). Suboxic deposition of ferric iron by bacteria in opposing gradients of Fe(II) and oxygen at circumneutral pH. *Applied and Environmental Microbiology* **2001**, 1328–1334.

Sogin, M. L. (1989). Evolution of Eukaryotic microorganisms and their small subunit ribosomal RNA's. *American Zoologist* **29**, 487–499.

Sogin, M. L. (1991). Early evolution and the origin of eukaryotes. *Current Opinion in Genetics and Development* **1**, 457–463.

Sogin, M. L. (1997). History assignment: When was the mitochondrion founded? *Current Opinion in Genetics & Development* **7**, 792–799.

Søndergaard, M., Kristensen, P., and Jeppesen, E. (1993). 8 years of internal phosphorus loading and changes in the sediment phosphorus profile of Lake Sobygaard, Denmark. *Hydrobiologia* **253**, 345–356.

Søndergaard, M., Williams, P. J. L., Cauwet, G., Riemann, B., Robinson, C., Terzic, S., Woodward, E. M. S., and Worm, J. (2000). Net accumulation and flux of dissolved organic carbon and dissolved organic nitrogen in marine plankton communities. *Limnology and Oceanography* **45**, 1097–1111.

Sonne-Hansen, J. and Ahring, B. (1999). *Thermodesulfobacterium hveragerdense* sp. nov., and the *Thermodesulfovibrium islandicus* sp. nov., two thermophillic sulfate reducing bacteria isolated from Icelandic hot springs. *Systematic Applied Microbiology* **22**, 559–564.

Soon, W., Baliunas, S., Idso, S. B., Kondratyev, K. Y., and Posmentier, E. S. (2001). Modeling climatic effects of anthropogenic carbon dioxide emissions: Unknowns and uncertainties. *Climate Research* **18**, 259–275.

Sørensen, J. (1978). Denitrification rates in a marine sediment as measures by acetylene inhibition technique. *Applied and Environmental Microbiology* **36**, 139–143.

Sørensen, J. and Thorling, L. (1991). Stimulation by Lepidocrocite (g-FeOOH) of Fe(II)-dependent nitrite reduction. *Geochimica et Cosmochimica Acta* **55**, 1289–1294.

Sørensen, J., Christensen, D., and Jørgensen, B. B. (1981). Volatile fatty acids and hydrogen as substrates for sulfate-reducing bacteria in anaerobic marine sediment. *Applied Environmental Microbiology* **42**, 5–11.

Sørensen, K. B. (2002). Microbial Ecology in Gradient Systems. Ph.D. Thesis, University of Southern Denmark, Odense.

Sørensen, K. B. and Canfield, D. E. (2004). Annual fluctuations in stable sulfur isotope fractionation in Mariager Fjord are caused by sulfur disproportionation. *Geochimica et Cosmochimica Acta* **68**, 503–515.

Sørensen, K. B., Canfield, D. E., and Oren, A. (2004). Salinity responses of benthic microbial communities in a solar saltern (Eilat, Israel). *Applied and Environmental Science* **70**, 1608–1616.

Sorokin, D. Y., Muyzer, G., Brinkhoff, T., Kuenen, G., and Jetten, M. S. M. (1998). Isolation and characterisation of a novel facultatively alkaliphilic *Nitrobacter* species, *N. alkalicus* sp. nov. *Archives of Microbiology* **170**, 345.

Spear, J. R., Ley, R. E., Berger, A. B., and Pace, N. P. (2003). Complexity in natural microbial ecosystems: The Guerrero Negro Experience. *Biological Bulletin* **204**, 168–173.

Spenceley, A. P. (1982). Sedimentation patterns in a mangal on Magnetic Island near Townsville, North Queensland, Australia. *Singapore Journal of Tropical Geography* **3**, 100–107.

Sprent, J. I. and Sprent, P. (1990). Nitrogen Fixing Organisms: Pure and Applied Aspects. Chapman and Hall, London.

Stackebrandt, E., Stahl, D. A., and Devereux, R. (1995). Taxonomic relationships. *In* "Sulfate-Reducing Bacteria" (L. L. Barton, ed.), pp. 49–87. Plenum, New York.

Stal, L. J. (1994). Microbial mats in coastal environments. *In* "Microbial Mats. Structure. Development and Environmental Significance" (L. J. Stal and P. Caumette, eds), pp. 21–32. Springer, Heidelberg.

Stal, L. J. (2000). Cyanobacterial mats and stromatolites. *In* "The Ecology of Cyanobacteria. Their Diversity in Time and Space" (B. A. Whitton and M. Potts, eds), pp. 61–120. Klüwer Academic Publishers, Boston.

Stal, L. J. and Moezelaar, R. (1997). Fermentation in cyanobacteria. *FEMS Microbiology Ecology* **21**, 179–211.

Steele, L. P., Krummel, P. B., and Langenfelds, R. L. (eds.) (2002). Atmospheric CH_4 concentrations from sites in the CSIRO Atmospheric Research GASLAB air sampling network (October 2002 version). *In* "Trends: A Compendium of Data on Global Change." Carbon Dioxide Information Analysis Center, Oak Ridge National Laboratory, U.S. Department of Energy, Oak Ridge, TN.

Stehr, G., Böttcher, B., Dittberner, P., Rath, G., and Koops, H.-P. (1995). The ammonia-oxidizing nitrifying population of the river elbe estuary. *FEMS Microbiology Ecology* **17**, 177–186.

Stephen, J. R., Kowalchuk, G. A., Bruns, M.-A. V., McCaig, A. E., Phillips, C. J., Embley, T. M., and Prosser, J. I. (1998). Analysis of b-subgroup proteobacterial ammonia oxidizer populations in soil by denaturing gradient gel electrophoresis analysis and hierarchical phylogenetic probing. *Applied and Environmental Microbiology* **64**, 2958–2965.

Stetter, K. O. (1992). The genus *Archaeoglobus*. *In* "The Prokaryotes" (A. Balows, H. G. Trüper, M. Dworkin, W. Harder, and K.-H. Schleifer, eds), 2nd Ed., pp. 707–711. Springer, Berlin.

Stetter, K. O. (1996). Hyperthermophiles in the history of life. *In* "Evolution of Hydrothermal Ecosystems on Earth (and Mars?)" (G. R. Bock and J. A. Goode, eds), pp. 1–10. John Wiley and Sons, New York.

Stetter, K. O. (1998). Smallest cell sizes within hyperthermophilic archaea ("Archaebacteria"). *In* "Size Limits of Very Small Microorganisms: Proceedings of a workshop" (A. H. Knoll, ed.). Space Studies Board, National Academy Press, Washington, DC.

Stetter, K. O. and Gaag, G. (1983). Reduction of molecular sulphur by methanogenic bacteria. *Nature* **305**, 309–311.

Stetter, K.-O., Segerer, A., Zillig, W., Huber, G., Fiala, G., Huber, R., and König, H. (1986). Extremely thermophilic sulfur-metabolizing archaebacteria. *Systematic Applied Microbiology* **7**, 393–397.

Stevens, T. (1997). Lithoautotrophy in the subsurface. *FEMS Microbiology Reviews* **20**, 327–337.

Stevens, T. O. and McKinley, J. P. (1995). Lithoautotrophic microbial ecosystems in deep basalt aquifers. *Science* **270**, 450–454.

Stevenson, A. K., Kimble, L. K., Woese, C. R., and Madigan, M. T. (1997). Characterization of new phototrophic heliobacteria and their habitats. *Photosynthetic Research* **53**, 1–12.

Stewart, W. D. P., Fitzgerald, G. P., and Burris, R. H. (1967). *In situ* studies on nitrogen fixation with acetylene reduction technique. *Science* **158**, 1426–1432.

Stillings, L. L., Drever, J. I., Brantley, S. L., Sun, Y., and Oxburgh, R. (1996). Rates of feldspar dissolution at pH 3–7 with 0–8 mM oxalic acid. *Chemical Geology* **132**, 79–90.

Stone, A. T., Godtfredsen, K. L., and Deng, B. (1994). Sources and reactivity of reductants encountered in aquatic environments. *In* "Chemistry of Aquatic Systems" (G. Bidoglio and W. Stumm, eds), pp. 337–374. ECSC, EEC, EAEC, Brussels.

Stouthamer, A. H. (1988). Dissimilatory reduction of oxidized nitrogen compounds. *In* "Biology of Anaerobic Microorganisms" (A. J. B. Zehnder, ed.), pp. 245–303. John Wiley and Sons, New York.

Straub, K. L., Benz, M., Schink, B., and Widdel, F. (1996). Anaerobic, nitrate-dependent microbial oxidation of ferrous iron. *Applied and Environmnetal Microbiology* **62**, 1458–1460.

Straub, K. L. and Buchholz-Cleven, B. E. E. (1998). Enumeration and detection of anaerobic ferrous iron-oxidizing, nitrate-reducing bacteria from diverse European sediments. *Applied and Environmental Microbiology* **64**, 4846–4856.

Straub, K. L., Rainey, F. A., and Widdel, F. (1999). *Rhodovulum iodosum* sp. nov, and *Rhodovulum robiginosum* sp. nov., two new marine phototrophic ferrous-iron-oxidizing purple bacteria. *International Journal of Systematic Bacteriology* **49**, 729–735.

Straub, K. L., Benz, M., and Schink, B. (2001). Iron metabolism in anoxic environments at near neutral pH. *FEMS Microbiology Ecology* **34**, 181–186.

Strauss, E. A., Mitchell, N. L., and Lamberti, G. A. (2002). Factors regulating nitrification in aquatic sediments: Effects of organic carbon, nitrogen availability, and pH. *Canadian Journal of Fisheries and Aquatic Sciences* **59**, 554–563.

Strauss, G. and Fuchs, G. (1993). Enzymes of a novel autotrophic CO_2 fixation pathway in the phototrophic bacterium *Chloroflexus aurantiacus*, the 3-hydroxypropionate cycle. *European Journal of Biochemistry* **215**, 633–643.

Striegl, R. G., McConnaughey, T. A., Thorstenson, D. C., Weeks, E. P., and Woodward, J. C. (1992). Consumption of atmospheric methane by desert soils. *Nature* **357**, 145–147.

Strom, S. L., Benner, R., Ziegler, S., and Dagg, M. J. (1997). Planktonic grazers are a potentially important source of marine dissolved organic carbon. *Limnology and Oceanography* **42**, 1364–1374.

Stroo, H. F., Klein, T. M., and Alexander, M. (1986). Heterotrophic nitrification in an acid forest soil and by an acid-tolerant fungus. *Applied and Environmental Microbiology* **52**, 1107–1111.

Strous, M., vanGerven, E., Kuenen, J. G., and Jetten, M. (1997). Effects of aerobic and microaerobic conditions on anaerobic ammonium-oxidizing (Anammox) sludge. *Applied and Environmental Microbiology* **63**, 2446–2448.

Strous, M., Fuerst, J. A., Kramer, E. H. M., Logemann, S., Muyzer, G., van de Pas-Schoonen, K. T., Webb, R., Kuenen, J. G., and Jetten, M. S. M. (1999a). Missing lithotroph identified as new planctomycete. *Nature* **400**, 446–449.

Strous, M., Kuenen, J. G., and Jetten, M. S. M. (1999b). Key physiology of anaerobic ammonium oxidation. *Applied and Environmental Microbiology* **65**, 3248–3250.

Stumm, W. and Morgan, J. J. (1981). Aquatic Chemistry. John Wiley & Sons, New York.

Stumm, W. and Morgan, J. J. (1996). "Aquatic Chemistry: Chemical Equilibria and Rates in Natural Waters"3rd Ed. John Wiley and Sons, Inc., New York.

Suess, E. (1980). Particulate organic-carbon flux in the oceans—Surface productivity and oxygen utilization. *Nature* **288**, 260–263.

Sulzberger, B., Suter, D., Siffert, C., Banwart, S., and Stumm, W. (1989). Dissolution of Fe(III)(hydr)oxides in natural waters: Laboratory assessment on the kinetics controlled by surface coordination. *Marine Chemistry* **28**, 127–144.

Summit, M. and Baross, J. A. (2001). A novel microbial habitat in the mid-ocean ridge subseafloor. *Proceedings of the National Academy of Sciences USA* **98**, 2158–2163.

Summons, R. E., Franzmann, P. D., and Nichols, P. D. (1998). Carbon isotopic fractionation associated with methylotrophic methanogenesis. *Organic Geochemistry* **28**, 465–475.

Summons, R. E., Jahnke, L. L., Hope, J. M., and Logan, G. A. (1999). 2-Methylhopanoids as biomarkers for cyanobacterial oxygenic photosynthesis. *Nature* **400**, 554–557.

Sumper, M. (2002). A phase separation model for the nanopattering of diatom biosilica. *Science* **295**, 2430–2433.

Sunda, W. G. and Kieber, J. (1994). Oxidation of humic substances by manganese oxides yields low-molecular weight organic substrates. *Nature* **367**, 62–64.

Sundby, B. and Silverberg, N. (1981). Pathways of manganese in an open estuarine system. *Geochimica et Cosmochimica Acta* **45**, 293–307.

Sundby, B., Anderson, L. G., Hall, P. O. J., Iverfeldt, A., Vanderloeff, M. M. R., and Westerlund, S. F. G. (1986). The effect of oxygen on release and uptake of cobalt, manganese, iron and phosphate at the sediment-water interface. *Geochimica et Cosmochimica Acta* **50**, 1281–1288.

Sundby, B., Gobeil, C., Silverberg, N., and Mucci, A. (1992). The phosphorus cycle in coastal marine-sediments. *Limnology and Oceanography* **37**, 1129–1145.

Suter, D., Siffert, C., Sulzberger, B., and Stumm, W. (1988). Catalytic dissolution of iron(III)(hydr)oxides by oxalic acid in the presence of Fe(II). *Naturwissenschaften* **75**, 571–573.

Suzuki, J. Y., Bollivar, D. W., and Bauer, C. E. (1997). Genetic analysis of chlorophyll biosynthesis. *Annual Review of Genetics* **31**, 61–89.

Svensson, J. M. (1997). Influence of *Chironomus plumosus* larvae on ammonium flux and denitrification (measured by the acetylene blockage- and the isotope pairing-technique) in eutrophic lake sediment. *Hydrobiologia* **346**, 157–168.

Tabita, F. R. (1995). The biochemistry and metabolic regulation of carbon metabolism and CO_2 fixation in purple bacteria. *In* "Anoxygenic Photosynthetic Bacteria" (R. E. Blankenship, M. T. Madigan, and C. E. Bauer, eds), pp. 885–914. Kluwer, Netherlands.

Tabita, F. R. (1999). Microbial ribulose 1,5-bisphosphate carboxylase/oxygenase: A different perspective. *Photosynthesis Research* **60**, 1–28.

Takahashi, T., Feely, R. A., Weiss, R. F., Wanninkhof, R. H., Chipman, D. W., Sutherland, S. C., and Takahashi, T. T. (1997). Global air-sea flux of CO_2, an estimate based on measurements of sea-air pCO_2 difference. *Proceedings of the National Academy of Sciences USA* **94**, 8292–8299.

Taylor, N. J. (1985). Silica incorporation in the diatom *Coscinodiscus granii* as affected by light intensity. *British Phycological Journal* **20**, 365–374.

Taylor, S. C., Dalton, H., and Dow, C. S. (1981). Ribulose-1,5-bisphosphate carboxylase/oxygenase and carbon assimilation in *Methylococcus capsulatus* (Bath). *Journal of General Microbiology* **122**, 89–94.

Tebo, B. M., Ghiorse, W. C., van Waasbergen, L. G., Siering, P. L., and Caspi, R. (1997). Bacterially mediated mineral formation: Insights into manganese(II) oxidation from molecular genetic and biochemical studies. *In* "Geomicrobiology: Interactions between Microbes and Minerals. Reviews in Mineralogy" (J. F. Banfield and K. H. Nealson, eds), pp. 225–266. Mineralogical Society of America, Washington, D.C.

Tempest, D. W. (1978). Biochemical significance of microbial-growth yields: A reassessment. *Trends in Biochemical Sciences* **3**, 180–184.

Teske, A., Alm, E., Regan, J. M., Toze, S., Rittmann, B. E., and Stahl, D. A. (1994). Evolutionary relationships among ammonia and nitrite oxidizing bacteria. *Journal of Bacteriology* **176**, 6623–6630.

Teske, A., Ramsing, N. B., Kuver, J., and Fossing, H. (1996a). Phylogeny of Thioploca and related filamentous sulfide-oxidizing bacteria. *Systematic and Applied Microbiology* **18**, 517–526.

Teske, A., Wawer, C., Muyzer, G., and Ramsing, N. B. (1996b). Distribution of sulfate-reducing bacteria in a stratified fjord (Mariager Fjord, Denmark) as evaluated by most-probable-number counts and denaturing gradient gel eletrophoresis of PCR-amplified ribosomal DNA fragments. *Applied and Environmental Microbiology* **62**, 1405–1415.

Tezuka, Y. (1990). Bacterial regeneration of ammonium and phosphate as affected by the carbon – nitrogen – phosphorus ratio of organic substrates. *Microbial Ecology* **19**, 227–238.

Thamdrup, B. (2000). Microbial manganese and iron reduction in aquatic sediments. *Advances in Microbial Ecology* **16**, 41–84.

Thamdrup, B. and Canfield, D. E. (1996). Pathways of carbon oxidation in continental margin sediments off central Chile. *Limnology and Oceanography* **41**, 1629–1650.

Thamdrup, B. and Canfield, D. E. (2000). Benthic respiration in aquatic sediments. *In* "Methods in Ecosystem Science" (O. Sala, H. Mooney, R. Jackson, and R. Howarth, eds), pp. 86–103. Springer, New York.

Thamdrup, B. and Dalsgaard, T. (2000). The fate of ammonium in anoxic manganese oxide-rich marine sediment. *Geochimica et Cosmochimica Acta* **64**, 4157–4164.

Thamdrup, B. and Dalsgaard, T. (2002). Production of N_2 through anaerobic ammonium oxidation coupled to nitrate reduction in marine sediments. *Applied and Environmental Microbiology* **68**, 1312–1318.

Thamdrup, B. and Fleischer, S. (1998). Temperature dependence of oxygen respiration, nitrogen mineralization, and nitrification in Arctic sediments. *Aquatic Microbial Ecology* **15**, 191–199.

Thamdrup, B., Finster, K., Hansen, J. W., and Bak, F. (1993). Bacterial disproportionation of elemental sulfur coupled to chemical reduction of iron or manganese. *Applied and Environmental Microbiology* **59**, 101–108.

Thamdrup, B., Finster, K., Fossing, H., Hansen, J. W., and Jørgensen, B. B. (1994a). Thiosulfate and sulfite distribution in porewater marine sediments related to manganese, iron, and sulfur geochemistry. *Geochimica et Cosmochimica Acta* **58**, 67–73.

Thamdrup, B., Fossing, H., and Jørgensen, B. B. (1994b). Manganese, iron, and sulfur cycling in a coastal marine sediment, Aarhus Bay, Denmark. *Geochimica et Cosmochimica Acta* **58**, 5115–5129.

Thamdrup, B., Glud, R. N., and Hansen, J. W. (1994c). Manganese oxidation and *in situ* manganese fluxes from a coastal sediment. *Geochimica et Cosmochimica Acta* **58**, 2563–2570.

Thamdrup, B., Hansen, J. W., and Jørgensen, B. B. (1998). Temperature dependence of aerobic respiration in a coastal sediment. *FEMS Microbiology Ecology* **25**, 189–200.

Thamdrup, B., Roselló-Mora, R., and Amann, R. (2000). Microbial manganese and sulfate reduction in Black Sea shelf sediment. *Applied and Environmental Microbiology* **66**, 2888–2897.

Thar, R. and Kühl, M. (2001). Motility of Marichromatium gracile in response to light, oxygen, and sulfide. *Applied and Environmental Microbiology* **67**, 5410–5419.

Thar, R. and Kühl, M. (2003). Bacteria are not too small for spatial sensing of chemical gradients: An experimental evidence. *Proceeding of the National Academy of Science USA* **100**, 5748–5753.

Thauer, R. K. (1998). Biochemistry of methanogenesis: A tribute to Marjory Stephenson. *Microbiology-UK* **144**, 2377–2406.

Thauer, R. K. and Kunow, J. (1995). Sulfate-reducing Archaea. *In* "Sulfate-Reducing Bacteria" (L. L. Barton, ed.), pp. 33–48. Plenum Press, New York.

Thauer, R. K., Jungermann, K., and Decker, K. (1977). Energy conservation in chemotrophic anaerobic bacteria. *Bacteriological Reviews* **41**, 100–180.

Therkildsen, M. S., King, G. M., and Lomstein, B. A. (1996). Urea production and turnover following the addition of AMP, CMP, RNA and a protein mixture to a marine sediment. *Aquatic Microbial Ecology* **10**, 173–179.

Thingstad, T. F. (2000). Control of bacterial growth in idealized food webs. *In* "Microbial Ecology of the Oceans" (D. L. Kirchman, ed.), pp. 229–260. Wiley-Liss, Inc., New York.

Thingstad, T. F., Skjoldal, E. F., and Bohne, R. A. (1993). Phosphorus Cycling and Algal-Bacterial Competition in Sandsfjord, Western Norway. Marine Ecology-Progress Series **99**, 239–259.

Thomsen, U. (2001). "Microbial Iron and Sulfate Reduction in Aquatic Sediments." Ph.D. Thesis, University of Southern Denmark, Odense.

Thomsen, U., Thamdrup, B., Stahl, D. A., and Canfield, D. E. (2004). Pathways of organic carbon oxidation in a deep lacustrine sediment, Lake Michigan. *Limnology and Oceanography* **49**, 2046–2057.

Thongtham, N., Kristensen, E., and Phungprasan, S.-Y. (in press). Leaf removal by leaf-eating sesarmid crabs in the Bangrong mangrove forest, Phuket, Thailand; with emphasis on the feeding ecology of *Neoepisesarma versicolor*. *Asian Marine Biology*.

Thurston, C. F. (1994). The structure and function of fungal laccases. *Microbiology-UK* **140**, 19–26.

Tijhuis, L., van Loosdrecht, M. C. M., and Heijnen, J. J. (1993). A thermodynamically based correlation for maintainance Gibbs energy requirements in aerobic and anaerobic chemotrophic growth. *Biotechnology and Bioengineering* **42**, 509–519.

Tolbert, N. E. (1994). Role of photosynthesis and photorespiration in regulating atmospheric CO_2 and O_2. *In* "Regulation of Atmospheric CO_2 and O_2 by Photosynthetic Carbon Metabolism" (N. E. Tolbert and J. Preiss, eds), pp. 8–33. Oxford University Press, New York, Oxford.

Torn, M. S. and Chapin III, F. S. (1993). Environmental and biotic controls over methane flux from arctic tundra. *Chemosphere* **26**, 357–368.

Tortell, P. D., Maldonado, M. T., Granger, J., and Price, N. M. (1999). Marine bacteria and biogeochemical cycling of iron in the oceans. *FEMS Microbiology Ecology* **29**, 1–11.

Treguer, P. (2002). Silica and the cycle of carbon in the ocean. *Comptes Rendus GeoScience* **334**, 3–11.

Treguer, P. and Pondaven, P. (2000). Silica control of carbon dioxide. *Nature* **406**, 358–359.

Treguer, P., Kamatani, A., Gueneley, S., and Queguiner, B. (1989). Kinetics of dissolution of antarctic diatom frustules and the biogeochemical cycle of silicon in the Southern Ocean. *Polar Biology* **9**, 397–403.

Treguer, P., Lindner, L., van Bennekom, A. J., Leynaert, A., Panouse, M., and Jacques, G. (1991). Production of biogenic silica in the Weddell-Scotia Seas measured with ^{32}Si. *Limnology and Oceanography* **36**, 1217–1227.

Treguer, P., Nelson, D. M., van Bennekom, A. J. V., DeMaster, D. J., Leynaert, A., and Queguiner, B. (1995). The silica balance in the world ocean: A reestimate. *Science* **268**, 375–379.

Troelsen, H. and Jørgensen, B. B. (1982). Seasonal dynamics of elemental sulfur in two coastal sediments. *Estuarine, Coastal and Shelf Science* **15**, 255–266.

Trott, L. A. and Alongi, D. M. (1999). Variability in surface water chemistry and phytoplankton biomass in two tropical, tidally dominated mangrove creeks. *Marine and Freshwater Research* **50**, 451–457.

Trüper, H. G. and Fischer, U. (1982). Anaerobic oxidation of sulphur compounds as electron donors for bacterial photosynthesis. *Philosophical Transaction of the Royal Society of London B* **298**, 529–542.

Trüper, H. G. and Pfennig, N. (1992). The family Chlorobiaceae. *In* "The Prokaryotes" (A. Balows, H. G. Trüper, M. Dworkin, W. Harder, and K.-H. Schleifer, eds), pp. 3583–3592. Springer, New York.

Turick, C. E., Tisa, L. S., and Caccavo, F. (2002). Melanin production and use as a soluble electron shuttle for Fe(III) oxide reduction and as a terminal electron acceptor by Shewanella algae BrY. *Applied and Environmental Microbiology* **68**, 2436–2444.

Turick, C. E., Caccavo, F., and Tisa, L. S. (2003). Electron transfer from *Shewanella algae* BrY to hydrous ferric oxide is mediated by cell-associated melanin. *FEMS Microbiology Letters* **220**, 99–104.

Turley, C. M. and Mackie, P. J. (1994). Biogeochemical significance of attached and free-living bacteria and the flux of particles in the NE Atlantic Ocean. *Marine Ecology-Progress Series* **115**, 191–203.

Twilley, R. R., Pozo, M., Garcia, V. H., RiveraMonroy, V. H., and Bodero, R. Z. A. (1997). Litter dynamics in riverine mangrove forests in the Guayas River estuary, Ecuador. *Oecologia* **111**, 109–122.

Tyler, S. C., Crill, P. M., and Brailsford, G. W. (1994). $^{13}C/^{12}C$ fractionation of methane during oxidation in a temperate forested soil. *Geochimica et Cosmochimica Acta* **58**, 1625–1633.

Tyson, G. W., Chapman, J., Hugenholtz, P., Allen, E. A., Ram, R. J., Richardson, P. M., Solovyev, V. V., Rubin, E. M., Rokhsar, D. S., and Banfield, J. F. (2004).

Community structure and metabolism through reconstruction of microbial genomes from the environment. *Nature* **428**, 37–43.
Ukpong, I. E. (1994). Soil-vegetation interrelationships of mangrove swamps as revealed by multivariate analyses. *Geoderma* **64**, 167–181.
Unden, G. (1999). Aerobic respiration and regulation of aerobic/anaerobic metabolism. *In* "Biology of the Prokaryotes" (J. W. Lengeler, G. Drews, and H. G. Schlegel, eds), pp. 261–277. Blackwell Science, New York.
Unden, G. and Bongaerts, J. (1997). Alternative respiratory pathways of *Escherichia coli*: Energetics and transcriptional regulation in response to electron acceptors. *Biochimica et Biophysica Acta* **1320**, 217–234.
Urbach, E., Robertson, D. L., and Chisholm, S. W. (1992). Multiple evolutionary origins of prochlorophytes within the cyanobacterial radiation. *Nature* **355**, 267–270.
Vadstein, O. (2000). Heterotrophic, planktonic bacteria and cycling of phosphorus—Phosphorus requirements, competitive ability, and food web interactions. *Advances in Microbial Ecology* **16**, 115–167.
Valentine, J. S., Wertz, D. L., Lyons, T. J., Liou, L. L., Goto, J. J., and Gralla, E. B. (1998). The dark side of dioxygen biochemistry. *Current Opinion in Chemical Biology* **2**, 253–262.
van Bennekom, A. J. and Salomons, W. (1981). Pathways of nutrients and organic matter from land to ocean through rivers. *In* "River Inputs to Ocean Systems" (J.-M. Martin, J. D. Burton, and D. Eisma, eds), pp. 33–51. UNEP/UNESCO, Rome.
Van Beusekom, J. E. E., Van Bennekom, A. J., Treguer, P., and Morvan, J. (1997). Aluminium and silicic acid in water and sediments of the Enderby and Crozet Basins. *Deep-Sea Research Part II-Topical Studies in Oceanography* **44**, 987–1003.
Van Cappellen, P. and Berner, R. A. (1988). A mathematical model for the early diagenesis of phosphorus and fluorine in marine sediments: Apatite precipitation. *American Journal of Science* **288**, 289–333.
Van Cappellen, P. and Berner, R. A. (1991). Fluorapatite crystal-growth from modified seawater solutions. *Geochimica et Cosmochimica Acta* **55**, 1219–1234.
Van Cappellen, P. and Ingall, E. D. (1996). Redox stabilization of the atmosphere and oceans by phosphorus-limited marine productivity. *Science* **271**, 493–496.
Van Cappellen, P. and Qiu, L. (1997). Biogenic silica dissolution in sediments of the Southern Ocean. I. Solubility. *Deep-Sea Research II* **44**, 1109–1128.
Van Cappellen, P., Viollier, E., Roychoudhury, A., Clark, L., Ingall, E., Lowe, K., and Dichristina, T. (1998). Biogeochemical cycles of manganese and iron at the oxic-anoxic transition of a stratified marine bas (Orca Basin, Gulf of Mexico). *Environmental Science & Technology* **32**, 2931–2939.
Van Cappellen, P., Dixit, S., and van Beusekom, J. (2002). Biogenic silica dissolution in the oceans: Reconciling experimental and field-based dissolution rates. *Global Biogeochemical Cycles* **16**, 1075.
van Cleemput, O. and Baert, L. (1983). Nitrite stability influenced by iron compounds. *Soil Biology & Biochemistry* **15**, 137–140.
van de Graaf, A. A., Mulder, A., de Bruijn, P., Jetten, M. S. M., Robertson, L. A., and Kuenen, J. G. (1995). Anaerobic oxidation of ammonia is a biologically mediated process. *Applied and Environmental Microbiology* **61**, 1246–1251.
van de Graaf, A. A., de Bruijn, P., Robertson, L. A., Jetten, M. S. M., and Kuenen, J. G. (1997). Metabolic pathway of anaerobic ammonium oxidation on the basis of N-15 studies in a fluidized bed reactor. *Microbiology-UK* **143**, 2415–2421.

van den Ende, F. P. (1997). "Microbial Ecology of Oxygen-Sulfide Interfaces in Marine Benthic Ecosystems." Doctoraat in de Wiskunde en Natuurwetenschappen Thesis, Rijksuniversiteit Groningen, Groningen.

van de Peer, Y., Neefs, J.-M., De Rijk, P., De Vos, P., and De Wachert, R. (1994). About the order of divergence of the major bacterial taxa during evolution. *Systematic and Applied Microbiology* **17**, 32–38.

Van Dover, C. L. (2000). "The Ecology of Deep-Sea Hydrothermal Vents." Princeton University Press, Princeton, NJ.

Van Gemerden, H. (1974). Coexistence of organisms competing for the same substrate: An example among the purple sulfur bacteria. *Microbial Ecology* **1**, 104–119.

Van Gemerden, H. and Beeftink, H. H. (1981). Coexistence of chlorobium and chromatium in a sulfide-limited continuous culture. *Archives of Microbiology* **129**, 32–34.

Van Gemerden, H. and Mas, J. (1995). Ecology of phototrophic sulfur bacteria. *In* "Anoxygenic Photosynthetic Bacteria" (R. E. Blankenship, M. T. Madigan, and C. E. Bauer, eds), pp. 49–85. Klüwer Academic Publishers, Dordrecht.

van Raaphorst, W. and Malschaert, J. F. P. (1996). Ammonium adsorption in superficial North Sea sediments. *Continental Shelf Research* **16**, 1415–1435.

Vanzella, A., Guerrero, M. A., and Jones, R. D. (1989). Effect of CO and light on ammonium and nitrite oxidation by chemolithotrophic bacteria. *Marine Ecology-Progress Series* **57**, 69–76.

Velinsky, D. J., Fogel, M. L., Todd, J. F., and Tebo, B. M. (1991). Isotopic fractionation of dissolved ammonium at the oxygen-hydrogen sulfide interface in anoxic waters. *Geophysical Research Letters* **18**, 649–652.

Venter, J. C., Remington, K., Heidelberg, J. F., Halpern, A. L., Rusch, D., Eisen, J. A., Wu, D. Y, Paulsen, I., Nealson, K., White, O., Peterson, J., Hoffman, J., Parsons, R., Baden-Tillson, H., Pfannkoch, C., Rogers, Y. H., and Smith, H. O. (2004). Environmental genome shotgun sequencing of the Sargasso Sea. *Science* **304**, 66–74.

Visscher, P. T., Reid, R. P., and Bebout, B. M. (2000). Microscale observations of sulfate reduction: Correlation of microbial activity with lithified micritic laminae in modern marine stromatolites. *Geology* **28**, 919–922.

Vitousek, P. M. and Howarth, R. W. (1991). Nitrogen limitation on land and in the sea—How can it occur. *Biogeochemistry* **13**, 87–115.

Vitousek, P. M., Aber, J. D., Howarth, R. W., Likens, G. E., Matson, P. A., Schindler, D. W., Schlesinger, W. H., and Tilman, D. G. (1997). Human alteration of the global nitrogen cycle: Sources and consequences. *Ecological Applications* **7**, 737–750.

Vogels, G. D. and van der Drift, C. (1976). Degradation of purines and pyrimidines by microorganisms. *Bacteriological Reviews* **40**, 403–468.

Von Damm, K. L. (1990). Seafloor hydrothermal activity: Black smoker chemistry and chimneys. *Annual Review of Earth and Planetary Science* **18**, 173–204.

Voss, M., Dippner, J. W., and Montoya, J. P. (2001). Nitrogen isotope patterns in the oxygen-deficient waters of the Eastern Tropical North Pacific Ocean. *Deep-Sea Research I* **48**, 1905–1921.

Voytek, M. A. and Ward, B. B. (1995). Detection of ammonium-oxidizing bacteria of the beta-subclass of the class proteobacteria in aquatic samples with the PCR. *Applied and Environmental Microbiology* **61**, 2811.

Wächtershäuser, G. (1988). Pyrite formation, the first energy source for life: A hypothesis. *Systematic and Applied Microbiology* **10**, 207–210.

Wada, E. and Hattori, A. (1976). Natural abundance of ^{15}N in particulate organic matter in the North Pacific Ocean. *Geochimica et Cosmochemica Acta* **40**, 249–251.
Wafar, S., Untawale, A. G., and Wafar, M. (1997). Litter fall and energy flux in a mangrove ecosystem. *Estuarine Coastal and Shelf Science* **44**, 111–124.
Wagman, D., Evans, W. H., Parker, V. B., Schumm, R. H., Harlow, I., Bailey, S. M., Churney, K. L., and Nuttall, R. L. (1982). The NBS tables of chemical thermodynamic properties: Selected values for inorganic and C1 and C2 organic substances in SI units. *Journal of Physical and Chemical Reference Data* **11**, 1–392.
Wagner, M., Roger, A. J., Flax, J. L., Brusseau, G. A., and Stahl, D. A. (1998). Phylogeny of dissimilatory sulfite reductases supports an early origin of sulfate respiration. *Journal of Bacteriology* **180**, 2975–2982.
Wahby, S. D. and Bishara, N. F. (1980). The effect of the River Nile on Mediterranean water, before and after the construction of the High Dam at Aswan. *In* "Proceedings of a SCOR Workshop on River Inputs to Ocean Systems" (J.-M. Martin, J. D. Burton, and D. Eisma, eds), pp. 311–318. UNESCO, Paris.
Wakeham, S. G., Lee, C., Hedges, J. I., Hernes, P. J., and Peterson, M. L. (1997). Molecular indicators of diagenetic status in marine organic matter. *Geochimica et Cosmochimica Acta* **61**, 5363–5369.
Walsby, A. E. (1985). The permeability of heterocysts to the gases nitrogen and oxygen. *Proceedings of the Royal Society of London Series B-Biological Sciences* **226**, 345–366.
Wang, X. C., Druffel, E. R. M., Griffin, S., Lee, C., and Kashgarian, M. (1998). Radiocarbon studies of organic compound classes in plankton and sediment of the northeastern Pacific Ocean. *Geochimica et Cosmochimica Acta* **62**, 1365–1378.
Wanner, B. L. (1993). Gene-regulation by phosphate in enteric bacteria. *Journal of Cellular Biochemistry* **51**, 47–54.
Ward, B. B. (1996). Nitrification and denitrification: Probing the nitrogen cycle in aquatic environments. *Microbial Ecology* **32**, 247–261.
Ward, B. B. (2000). Nitrification and the marine nitrogen cycle. *In* "Microbial Ecology of the Ocean" (D. L. Kirchman, ed.), pp. 427–453. Wiley-Liss, New York.
Ward, D. M. and Castenholz, R. W. (2000). Cyanobacteria in geothermal habitats. *In* "The Ecology of Cyanobacteria. Their Diversity in Time and Space" (B. A. Whitton and M. Potts, eds). Klüwer Academic Publishers, Dordrecht.
Ward, D. M., Weller, R., and Bateson, M. M. (1990). 16S ribosomal-RNA sequences reveal numerous uncultured microorganisms in a natural community. *Nature* **345**, 63–65.
Warford, A. L., Kosiur, D. R., and Doose, P. R. (1979). Methane production in Santa Barbara basin sediments. *Geomicrobiology Journal* **1**, 117–137.
Warren, J. H. (1990). The use of open burrows to estimate abundances of intertidal estuarine crabs. *Australian Journal of Ecology* **15**, 277–280.
Waser, N. A. D., Harrison, P. J., Nielsen, B., Calvert, S. E., and Turpin, D. H. (1998). Nitrogen isotope fractionation during the uptake and assimilation of nitrate, nitrite, ammonium, and urea by a marine diatom. *Limnology and Oceanography* **43**, 215–224.
Waterbury, J. B., Watson, S. W., Guillard, R. R. L., and Brand, L. E. (1979). Widespread occurrence of a unicellular, marine, planktonic, cyanobacterium. *Nature* **277**, 293–294.
Watson, A., Stephen, K. D., Nedwell, D. B., and Arah, J. R. M. (1997). Oxidation of methane in peat: Kinetics of CH_4 and O_2 removal and the role of plant roots. *Soil Biology and Biochemistry* **29**, 1257–1267.

Watson, G. M. F., Yu, J.-P., and Tabita, F. R. (1999). Unusual ribulose 1,5-bisphosphate carboxylase/oxygenase of anoxic *Archaea*. *Journal of Bacteriology* **181**, 1569–1575.

Watson, S. W., Bock, E., Valois, F. W., Waterbury, J. B., and Schlosser, U. (1986). *Nitrospira-Marina* Gen-Nov Sp-Nov—A chemolithotrophic nitrite-oxidizing bacterium. *Archives of Microbiology* **144**, 1–7.

Weber, A. and Jørgensen, B. B. (2002). Bacterial sulfate reduction in hydrothermal sediments of the Guaymas Basin, Gulf of California, Mexico. *Deep-Sea Research I* **49**, 827–841.

Weber, K. A., Picardal, F. W., and Roden, E. E. (2001). Microbially catalyzed nitrate-dependent oxidation of biogenic solid-phase Fe(II) compounds. *Environmental Science & Technology* **35**, 1644–1650.

Wedepohl, K. H. (1995). The composition of the continental crust. *Geochimica et Cosmochimica Acta* **59**, 1217–1232.

Weiss, M. S., Abele, U., Weckesser, J., Welte, W., Schiltz, E., and Schulz, G. E. (1991). Molecular architecture and electrostatic properties of a bacterial porin. *Science* **254**, 1627–1630.

Welch, S. A. and Ullman, W. J. (1993). The effect of organic-acids on plagioclase dissolution rates and stoichiometry. *Geochimica et Cosmochimica Acta* **57**, 2725–2736.

Welch, S. A., Barker, W. W., and Banfield, J. F. (1999). Microbial extracellular polysaccharides and plagioclase dissolution. *Geochimica et Cosmochimica Acta* **63**, 1405–1419.

Wellsbury, P., Goodman, K., Cragg, B. A., and Parkes, R. J. (2000). The geomicrobiology of deep marine sediments from Blake Ridge containing methane hydrate (Sites 994, 995, and 997). *In* "Ocean Drilling Program, Scientific Results" (C. K. Paull, R. Matsumoto, P. J. Wallace, and W. P. Dillon, eds), pp. 379–391. College Station, TX.

Wellsbury, P., Herbert, R. A., and Parkes, R. J. (1994). Bacterial [*methyl*-^{3}H]thymidine incorporation in substrate-amended estuarine sediment slurries. *FEMS Microbiology Ecology* **15**, 237–248.

Welsh, D. T., Bourgues, S., deWit, R., and Herbert, R. A. (1996). Seasonal variations in nitrogen-fixation (acetylene reduction) and sulphate-reduction rates in the rhizosphere of *Zostera noltii*: Nitrogen fixation by sulphate reducing bacteria. *Marine Biology* **125**, 619–628.

Welsh, D. T., Castadelli, G., Bartoli, M., Poli, D., Careri, M., de Wit, R., and Viaroli, P. (2001). Denitrification in an intertidal seagrass meadow, a comparison of N^{15}-isotope and acetylene-block techniques: Dissimilatory nitrate reduction to ammonia as a source of N_2O? *Marine Biology* **139**, 1029–1036.

Wenzhöfer, F. and Glud, R. N. (2002). Benthic carbon mineralization in the Atlantic: A synthesis based on *in situ* data from the last decade. *Deep-Sea Research I* **49**, 1255–1279.

Westrich, J. T. (1983). "The Consequences and Controls of Bacterial Sulfate Reduction in Marine Sediments." Ph.D. Thesis, Yale University, New Haven, CT.

Westrich, J. T. and Berner, R. A. (1984). The role of sedimentary organic matter in bacterial sulfate reduction: The *G* model tested. *Limnology and Oceanography* **29**, 236–249.

Wetzel, R. G., Rich, P. H., Miller, M. C., and Allen, H. L. (1972). Metabolism of dissolved and particulate detrital carbon in a temperate hard-water lake. *Memorie dell'Istituto Italiano di Idrobiologia Supplement* **29**, 185–243.

Whalen, S. C. and Reeburgh, W. S. (1990). Consumption of atmospheric methane by tundra soils. *Nature* **346**, 160–162.

Whalen, S. C., Reeburgh, W. S., and Sandbeck, K. A. (1990). Rapid methane oxidation in a landfill cover soil. *Applied and Environmental Microbiology* **56**, 3405–3411.

Whitby, C. B., Saunders, J. R., Rodriguez, J., Pickup, R. W., and McCarthy, A. (1999). Phylogenetic differentiation of two closely related *Nitrosomonas* spp. that inhabit different sediment environments in an oligotrophic freshwater lake. *Applied and Environmental Microbiology* **65**, 4855–4862.

White, D. (1995). The Physiology and Biochemistry of Prokaryotes. Oxford University Press, New York.

Whitfield, M. (2001). Interaction between phytoplankton and trace metals in the ocean. *Advances in Marine Biology* **41**, 1–128.

Whiticar, M. J. (1999). Carbon and hydrogen isotope systematics of bacterial formation and oxidation of methane. *Chemical Geology* **161**, 291–314.

Whiticar, M. J., Faber, E., and Schoell, M. (1986). Biogenic methane formation in marine and freshwater environments: CO_2 reduction *vs* acetate fermentation—Isotope evidence. *Geochimica et Cosmochimica Acta* **50**, 693–709.

Whitman, W. B., Coleman, D. C., and Wiebe, W. J. (1998). Prokaryotes: The unseen majority. *Proceedings of the National Academy of Sciences USA* **95**, 6578–6583.

Whitman, W. B., Bowen, T. L., and Boone, D. R. (1999). The methanogenic bacteria. *In* "The Prokaryotes" (M. Dworkin, A. Balows, H. G. Trüper, W. Harder, and K.-H. Schleifer, eds), 3rd Ed., release 3.0, May 21. Springer, New York.

Whittaker, R. H. and Margulis, L. (1978). Protist classification and the kingdoms of organisms. *Biosystems* **10**, 3–18.

Whitton, B. A. and Potts, M. (2000). Introduction to the cyanobacteria. *In* "The Ecology of Cyanobacteria. Their Diversity in Time and Space" (B. A. Whitton and M. Potts, eds), pp. 1–11. Klüwer Academic Publishers, Dordrecht.

Widdel, F. (1988). Microbiology and ecology of sulfate- and sulfur-reducing bacteria. *In* "Biology of Anaerobic Organisms" (A. J. B. Zehnder, ed.), pp. 469–585. John Wiley, New York.

Widdel, F. and Hansen, T. A. (1992). The dissimilatory sulfate- and sulfur-reducing bacteria. *In* "The Prokaryotes: A Handbook on the Biology of Bacteria: Ecophysiology, Isolation, Identification, Applications" (A. Balows, ed.), 2nd Ed., pp. 583–624. Springer, New York.

Widdel, F. and Pfennig, N. (1977). New anaerobic, sporing, acetate-oxidizing sulfate-reducing bacterium, *desulfotomaculum (emend)acetoxidans*. *Archives of Microbiology* **112**, 119–122.

Widdel, F. and Pfennig, N. (1982). Studies on dissimilatory sulfate-reducing bacteria that decompose fatty acids II. Incomplete oxidation of propionate by *Desulfobulbus propionicus* gen. nov., sp. nov. *Archives of Microbiology* **131**, 360–365.

Widdel, F., Schnell, S., Heising, S., Ehrenreich, A., Assmus, B., and Schink, B. (1993). Ferrous iron oxidation by anoxygenic phototrophic bacteria. *Nature* **362**, 834–835.

Wieland, A. and Kühl, M. (2000a). Irradiance and temperature regulation of oxygenic photosynthesis and O_2 consumption in a hypersaline cyanobacterial mat (Solar Lake, Egypt). *Marine Biology* **137**, 71–85.

Wieland, A. and Kühl, M. (2000b). Short-term temperature effects on oxygen and sulfide cycling in a hypersaline cyanobacterial mat (Solar Lake; Egypt). *Marine Ecology-Progress Series* **196**, 87–102.

Willey, J. D. and Spivack, A. J. (1997). Dissolved silica concentrations and reactions in pore waters from continental slope sediments offshore from Cape Hatteras, North Carolina, USA. *Marine Chemistry* **56**, 227–238.

Winogradsky, S. (1888). Über Eisenbakterien. *Botanische Zeitung* **46**, 261–270.

Woese, C. R. (1987). Bacterial evolution. *Microbiological Reviews* **51**, 221–271.
Woese, C. R. (2000). Interpreting the universal phylogenetic tree. *Prooceedings of the National Academy of Science USA* **97**, 8392–8396.
Woese, C. R. and Fox, G. E. (1977). Phylogenetic structure of the prokaryotic domain: The primary kingdoms. *Proceedings of the National Academy of Science USA* **74**, 5088–5090.
Woese, C. R., Weisburg, W. G., Hahn, C. M., Paster, B. J., Zablen, L. B., Lewis, B. J., Macke, T. J., Ludwig, W., and Stackebrandt, E. (1985). The phylogeny of purple bacteria—The gamma-subdivision. *Systematic and Applied Microbiology* **6**, 25–33.
Woese, C. R., Kandler, O., and Wheelis, M. L. (1990). Towards a natural system of organisms: Proposal for the domains Archaea, Bacteria and Eucarya. *Proceedings of the National Academy of Sciences USA* **87**, 4576–4579.
Wolanski, E., Mazda, Y., and Ridd, P. (1992). Mangrove hydrodynamics. *In* "Trophical Mangrove Ecosystems. Coastal and Estuarine Studies 41" (A. I. Robertson and D. M. Alongi, eds), pp. 43–62. American Geophysical Union, Washington, D.C.
Wolanski, E., Spagnol, S., Thomas, S., Moore, K., Alongi, D. M., Trott, L., and Davidson, A. (2000). Modelling and visualizing the fate of shrimp pond effluent in a mangrove-fringed tidal creek. *Estuarine Coastal and Shelf Science* **50**, 85–97.
Wollast, R. (1967). Kinetics of alteration of k-feldspar in buffered solutions at low temperature. *Geochimica et Cosmochimica Acta* **31**, 635–648.
Wollast, R. and Chou, L. (1992). Surface reactions during the early stages of weathering of albite. *Geochimica et Cosmochimica Acta* **56**, 3113–3121.
Woodroffe, C. D. (1990). The impact of sea-level rise on mangrove shorelines. *Progress in Physical Geography* **14**, 483–520.
Woodroffe, C. D. (1992). Mangrove sediments and geomorphology. *In* "Trophical Mangrove Ecosystems. Coastal and Estuarine Studies 41" (A. I. Robertson and D. M. Alongi, eds), pp. 7–41. American Geophysical Union, Washington, D.C.
Wooldridge, K. G. and Williams, P. H. (1993). Iron uptake mechanisms of pathogenic bacteria. *FEMS Microbiology Reviews* **12**, 325–348.
Wuchter, C., Schouten, S., Boschker, H. T. S., and Damsté, J. H. S. (2003). Bicarbonate uptake by marine Crenarchaeota. *FEMS Microbiology Ecology* **219**, 203–207.
Xiong, J., Fischer, W. M., Inoue, K., Nakahara, M., and Bauer, C. E. (2000). Molecular evidence for the early evolution of photosynthesis. *Science* **289**, 1724–1730.
Xu, F., Kulys, J. J., Duke, K., Li, K. C., Krikstopaitis, K., Deussen, H. J. W., Abbate, E., Galinyte, V., and Schneider, P. (2000). Redox chemistry in laccase-catalyzed oxidation of N-hydroxy compounds. *Applied and Environmental Microbiology* **66**, 2052–2056.
Yao, W. S. and Millero, F. J. (1993). The rate of sulfide oxidation by $dMnO_2$ in seawater. *Geochimica et Cosmochimica Acta* **57**, 3359–3365.
Yao, W. S. and Millero, F. J. (1996). Oxidation of hydrogen sulfide by hydrous Fe(III) oxides in seawater. *Marine Chemistry* **52**, 1–16.
Young, J. P. W. (1992). Phylogenetic classification of nitrogen-fixing organisms. *In* "Biological Nitrogen Fixation" (G. Stacey, R. H. Burris, and H. J. Evans, eds), pp. 43–86. Chapman and Hall, New York.
Zehnder, A. J. B. and Brock, T. D. (1979). Methane formation and methane oxidation by methanogenic bacteria. *Journal of Bacteriology* **137**, 420–432.
Zehnder, A. J. B. and Zinder, S. H. (1980). The sulfur cycle. *In* "The Handbook of Environmental Chemistry" (O. Hutzinger, ed.), pp. 105–145. Springer, Heidelberg.

Zehr, J. P., Waterbury, J. B., Turner, P. J., Montoya, J. P., Omoregie, E., Steward, G. F., Hansen, A., and Karl, D. M. (2001). Unicellular cyanobacteria fix N_2 in the subtropical North Pacific Ocean. *Nature* **412**, 635–638.

Zelitch, I. (1975). Improving the efficiency of photosynthesis. *Science* **188**, 626–633.

Zhang, J.-Z. and Millero, F. J. (1993). The products from the oxidation of H_2S in seawater. *Geochimica et Cosmochimica Acta* **57**, 1705–1718.

Zillig, W., Palm, P., and Klenk, H. P. (1992). The nature of the common ancestor of the three domains of life and the origin of the Eucarya. *In* "Frontiers of Life" (J. K. T. T. Vân, J. C. Mounolow, J. Schneider, and C. McKay, eds), pp. 181–193. Proceedings of the 3rd Recontres de Blois, France.

Zimmermann, R. C. and Thom, B. G. (1982). Physiographic plant geography. *Progress in Physical Geography* **6**, 45–59.

Zinder, S. H. and Koch, M. (1984). Non-aceticlastic methanogenesis from acetate-acetate oxidation by a thermophilic syntrophic coculture. *Archives of Microbiology* **138**, 263–272.

Ziong, J., Fischer, W. M., Inoue, K., Nakahara, M., and Bauer, C. E. (2000). Molecular evidence for the early evolution of photosynthesis. *Science* **289**, 1724–1730.

Zopfi, J., Ferdelman, T. G., Jørgensen, B. B., Teske, A., and Thamdrup, B. (2001). Influence of water column dynamics on sulfide oxidation and other major biogeochemical processes in the chemocline of Mariager Fjord (Denmark). *Marine Chemistry* **74**, 29–51.

Zuckerkandl, E. and Pauling, L. (1965). Molecules as documents of evolutionary history. *Journal of Theoretical Biology* **8**, 357–366.

Zumft, W. G. (1997). Cell biology and molecular basis of denitrification. *Microbiology and Molecular Biology Reviews* **61**, 533–616.

TAXONOMIC INDEX

SUBJECT INDEX

Series Contents for Last Ten Years*

*The full list of contents for volumes 1–37 can be found in volume 38.